Aqueous-Phase Organometallic Catalysis

Edited by
B. Cornils and W. A. Herrmann

Further Titles of Interest:

B. Cornils, W. A. Herrmann (Eds.)
Applied Homogeneous Catalysis with Organometallic Compounds
A Comprehensive Handbook in Two Volumes

XXXVI, 1246 pages with over 1000 figures and 100 tables
1996. Hardcover, ISBN 3527-29286-1

M. Beller, C. Bolm (Eds.)
Transition Metals for Organic Synthesis
Building Blocks and Fine Chemicals

approx. 1200 pages with 1000 figures
1998. Hardcover, ISBN 3-527-29501-1

A. Togni, R. L. Halterman (Eds.)
Metallocenes

approx. 1000 pages in Two Volumes
1998. Hardcover, ISBN 3-527-29539-9

F. Diederich, P. J. Stang (Eds.)
Metal-catalyzed Cross-coupling Reactrom

XXII, 518 pages with 200 figures
1998. Hardcover, ISBN 3-527-29421-X

Aqueous-Phase Organometallic Catalysis

Concepts and Applications

Edited by
Boy Cornils and Wolfgang A. Herrmann

Weinheim · New York · Chichester
Brisbane · Singapore · Toronto

Prof. Dr. Boy Cornils
Hoechst AG
D-65926 Frankfurt/Main
Germany

Prof. Dr. Wolfgang A. Herrmann
Technische Universität München
Arcisstraße 21
D-80333 München
Germany

Cover picture: Homogeneous catalysis in aqueous phase: a view on the commercial oxo plant of Ruhrchemie at Oberhausen, Germany

Library of Congress Card No. applied for

A catalogue record for this book is available from the British Library

Die Deutsche Bibliothek – CIP-Einheitsaufnahme
Aqueous phase organometallic catalysis : concepts and applications / ed. by Boy Cornils und Wolfgang A. Herrmann. – Weinheim ; New York ; Chichester ; Brisbane ; Singapore ; Toronto : Wiley-VCH, 1998
ISBN 3-527-29478-3

Printed on acid-free and low chlorine paper

Composition, printing and binding: Konrad Triltsch, Druck- und Verlagsanstalt GmbH, D-97016 Würzburg
Printed in the Federal Republic of Germany

Preface

This book describes homogeneously catalyzed reactions under two major boundary conditions: the catalysts employed are *organometallic* complexes that are used in the *aqueous phase*. In this respect the book is restricted to one area of homogeneous catalysis and therefore – though substantially expanded and more detailed – to one special area of our previous book, *Applied Homogeneous Catalysis with Organometallic Complexes* (VCH, Weinheim, Germany, 1996).

The subject of the book is the use of water-soluble organometallic catalysts for chemical reactions. These catalysts are so far the sole successful means of implementing the idea of *heterogenization* of homogeneous catalysts by *immobilizing* them with the aid of liquid supports. They thus solve the cardinal problem of homogeneous catalysis, which lies in the expensive separation of catalyst as well as product that is inherent in the system: the catalyst used in the homogeneous phase is separated by simply decanting the aqueous catalyst phase from the organic phase of the substrates and reaction products. Since all attempts to heterogenize homogeneous catalysts by immobilizing them on solid supports ("anchoring") have to varying degrees been unsuccessful, only the use of homogeneous catalysts in aqueous solution and thus on liquid supports ("biphase operation") leads to a neat, inexpensive solution to the problem that conserves resources and is therefore environmentally friendly.

This book is restricted essentially to *aqueous*-phase catalyses and thus to one area of the more comprehensively defined two-phase catalyses. This restriction to the most recent and successful development of homogeneous catalysis takes account of the rapid technical advances in the process concept first described by Manassen et al. in 1973, which was followed in rapid succession in the 1970s by hesitant basic work and in 1984 by the first commercial implementation. This unusual sequence – industrial implementation in a 100 000 tonnes per year oxo plant for the hydroformylation of propylene **before** years of time-consuming basic research to determine mechanistic, kinetic and other data – demonstrates clearly the great leap forward that this process development represented in the field of homogeneous catalysis and in solving the central problem mentioned earlier. Since then other processes employing homogeneous catalysis have been converted to an aqueous two-phase procedure.

The development work intensified worldwide in various research groups in the years following the first commercial implementation at Ruhrchemie AG in Oberhausen. The obvious course of action was to let colleagues and specialists themselves report on their developments. This led to the formation of the international circle of contributors from the USA, France, the United Kingdom, China, Italy, Japan, India, Hungary and Germany which gives first-hand reports on its work.

One focus of the book is the hydroformylation process, the process involved in the first commercial implementation of aqueous-phase catalysis with its detailed descriptions of fundamental laws, special process features, and the present state of the art. Further focal points of the book are basic research on the complex catalysts (central atoms, ligands) and on the influence of the reaction conditions, solvents, and co-solvents, and a survey of other aqueous two-phase concepts and of proposed applications, with experimental examples and details. Environmental aspects are also considered.

We are sure that the outline chosen and the wide range of contributions from the authors give a multifaced and informative picture of the present state of developments in the field of aqueous two-phase catalysis, which presents not only the principles and accounts of the latest applications but also many aspects of spin-offs and alternative processes.

This description of ideas and process developments appears to us to be highly important for an appreciation of the potential of aqueous biphase catalysis. The familiar assessment of the most important aspects of heterogeneous and homogeneous catalysis demonstrates that only in a solution of the problem of continuous separation of catalyst and product, such as becomes possible with the processes involving aqueous immobilized catalysts, is the key to further progress found. Only *homogeneous* catalysts that can be handled without problems will give us scientists and developers confidence that the clear and sure mechanistic understanding of their mode of action and the possibility of easy variability of steric and/or electronic properties can be transferred to other immobilized, and thus easy-to-handle, catalysts. More optimistically, it is hoped that this will apply especially to those heterogenized catalysts that basically are derived from tailor-made homogeneous catalysts.

The sharp line of demarcation between homogeneous and heterogeneous catalysis would thus be blurred and the possibility opened up of combining in one species the advantages of homogeneous catalysts and none of the disadvantages of heterogeneous catalysts: heterogenized homogeneous catalysts would lead to equally advantageous results as homogenized heterogeneous catalysts – the long-awaited dream of catalysis research would be fulfilled!

We thank the team at WILEY-VCH, especially Mrs. Diana Boatman, Dr. Anette Eckerle, and Mrs. Claudia Grössl for their cooperation during preparation of this book and for helpful technical assistance.

Dipl.-Chem. Kolja Wieczorek is acknowledged for preparing all formulas, figures, and schemes; Dipl.-Chem. Thomas Weskamp for the total index.

Frankfurt–Höchst and München
Spring, 1998

Boy Cornils
Wolfgang A. Herrmann

Contents

Contributors

Prof. Dr. Jacques Augé
Université de Cergy-Pontoise
5, mail Gay-Lussac
Neuville-sur-Oise
F-95031 Cergy-Pontoise/France
Tel: +33/1 34 25 70 51
Fax: +33/1 34 25 70 51

Prof. Dr. Manfred Baerns
Institut für Angewandte Chemie
Berlin-Adlershof e. V.
Rudower Chaussee 5
D-12484 Berlin/Germany
Tel: +49/30 6392 4444
Fax: +49/30 6392 4454

Dr. Helmut Bahrmann
Celanese GmbH/Werk Ruhrchemie
Postfach 13 01 60
D-46128 Oberhausen/Germany
Tel: +49/208 693 2201
Fax: +49/208 693 2291

Prof. Dr. Arno Behr
Universität Dortmund
Fachbereich Chemietechnik
Lehrstuhl für Technische Chemie A
Emil-Figge-Straße 66
D-44227 Dortmund/Germany
Tel: +49/231 755 2310
Fax: +49/231 755 2311

Prof. Dr. Matthias Beller
Technische Universität München
Anorganisch-chemisches Institut
Lichtenbergstraße 4
D-85747 Garching/Germany
Tel: +49/89 2891 3096
Fax: +49/89 2891 3473

Dr. Bhachandra M. Bhanage
National Chemical Laboratory
Pune 411 008/India
Tel: +91/212 331 453
Fax: +91/212 330 233

Dr. Claudio Bianchini
Istituto per lo Studio della Stereochimica
ed Energetica dei Composti
CNR di Coordinazione
Via J. Nardi, 39
I-50132 Firenze/Italy
Tel: +39/55 243 990/24 5990
Fax: +39/55 247 8366

Dr. Sandra Bogdanovic
Celanese GmbH
C 660
D-65926 Frankfurt/Germany
Tel: +49/69 305 2240
Fax: +49/69 305 13684

Prof. Dr. Mario Bressan
Universita' G. D'Annunzio
Dipartimento di Scienze
Viale Pindaro 42
I-65127 Pescara/Italy
Tel: +39/85 453 7548
Fax: +39/85 453 7545

Dr. Henry R. Bryndza
DuPont Nylon
Experimental Station, Bldg. 302
P.O.Box 80328
Wilmington, Delaware 19880-0302/USA
E-mail: Bryndza@esvax.enet.duPont.com
Tel: +1/302 695 3761
Fax: +1/302 695 9084

Prof. Dr. Raghunath V. Chaudhari
National Chemical Laboratory
Pune 411 008/India
Tel: +91/212 346 135
Fax: +91/212 333 941

Priv.-Doz. Dr. Peter Claus
Institut für Angewandte Chemie
Berlin-Adlershof e.V.
Rudower Chaussee 5
D-12484 Berlin/Germany
Tel: +49/30 6392 4322
Fax: +49/30 6392 4350

Prof. Dr. Boy Cornils
Hoechst AG
F 821
D-65926 Frankfurt/Germany
Tel: +49/69 305 5683
Fax: +49/69 305 83128
privat:
Kirschgartenstraße 6
D-65719 Hofheim/Ts.
Tel/Fax: +49/6192 23502

Prof. Dr. Mark E. Davis
Chemical Engineering
California Institute of Technology
Pasadena, California 91125/USA
Tel: +1/626 395 4251
Fax: +1/626 568 8143
E-mail: mdavis@cheme.caltech.edu

Prof. Dr. Eckehard V. Dehmlow
Universität Bielefeld
Fakultät für Chemie
Universitätsstraße 25
D-33615 Bielefeld/Germany
Tel: +49/521 106 2051
Fax: +49/521 106 6146

Dr. Michel Dessoudeix
Ecole Nationale Supérieure de Chimie de Toulouse
Laboratoire de Catalyse et Chimie fine
118 Route de Narbonne
F-31077 Toulouse/France
Tel: +33/561 175 656
Fax: +33/561 175 600

Prof. Dr. Eckhard Dinjus
Forschungszentrum Karlsruhe
Abt. ITC-CPV
Postfach 3640
D-76021 Karlsruhe
Tel: +49/7247 822 400
Fax: +49/7247 822 244

Dipl.-Chem. Robert W. Eckl
Technische Universität München
Anorganisch-chemisches Institut
Lichtenbergstraße 4
D-85747 Garching/Germany
Tel: +49/89 289 13083
Fax: +49/89 289 13473

Dr. Carl D. Frohning
Celanese GmbH / Werk Ruhrchemie
Postfach 13 01 60
D-46128 Oberhausen/Germany
Tel: +49/208 693 2419
Fax: +49/208 693 2291

Dr. Franz Gaßner
Forschungszentrum Karlsruhe
Abt. ITC-CPV
Postfach 3640
D-76021 Karlsruhe
Tel: +49/7247 826 011
Fax: +49/7247 822 244

Prof. Dr. Robert H. Grubbs
Division of Chemistry and Chemical Engineering
California Institute of Technology
Pasadena, California 91125/USA
Tel: +1/626 395 6003
Fax: +1/626 564 9297

Dr. Steffen Haber
Clariant GmbH
Division Feinchemikalien
D 569
D-65926 Frankfurt/Germany
Tel: +49/69 305 5076
Fax: +49/69 305 17562

Prof. Dr. Brian E. Hanson
Virginia Polytechnic Institute
and State University
Department of Chemistry
College of Arts and Sciences
Blacksburg, Virginia 24061-0212/USA
Tel: +1/540 231 7206
Fax: +1/540 231 3255

Dr. John A. Harrelson, Jr.
DuPont Nylon
Experimental Station, Bldg. 302
P.O. Box 80328
Wilmington, Delaware 19880-0302/USA
Tel: +1/302 695 3761
Fax: +1/302 695 9084

Prof. Dr. Wolfgang A. Herrmann
Präsident der
Technischen Universität München
Arcisstraße 21
D-80333 München/Germany
Tel: +49/89 2892 2200
Fax: +49/89 2892 3399

Dr. István T. Horváth
Corporate Research Laboratories
Exxon Research and Engineering Company
Clinton Township, Route 22 East
Annandale, New Jersey 08801/USA
Tel: +1/908 730 2154
Fax: +1/908 730 3042

Prof. Dr. Zilin Jin
State Key Laboratory of Fine Chemicals
Dalian University of Technology
116012 Dalian/China
Tel: +86/411 467 1511
Fax: +86/411 363 3080

Prof. Dr. Ferenc Joó
Institute of Physical Chemistry
Kossuth Lajos University
P.O.Box 7
H-4010 Debrecen/Hungary
Tel: +36/52 316 666
Fax: +36/52 310 936

Prof. Dr. Philippe Kalck
Ecole Nationale Supérieure de Chimie
de Toulouse
Laboratoire de Catalyse et Chimie Fine
118 route de Narbonne
F-31077 Toulouse/France
Tel: +33/561 175 690
Fax: +33/561 175 600

Dr. Paul C.J. Kamer
University of Amsterdam
Institute for Molecular Chemistry
Inorganic Chemistry & Homogeneous
Catalysis
Nieuwe Achtergracht 166
NL-1018 WV Amsterdam/The Netherlands
Tel: +31/20 525 6495/6454
Fax: +31/20 525 6456

Dr. Agnes Kathó
Institute of Physical Chemistry
Kossuth Lajos University
P.O.Box 7
H-4010 Debrecen/Hungary
Tel: +36/52 316 666
Fax: +36/52 310 936

Prof. Dr. Shu Kobayashi
Department of Applied Chemistry
Faculty of Science
Science University of Tokyo (SUT)
Kagurazaka, Shinjuku-ku
Tokyo 162/Japan
Tel: +81/3 3260 4271 ext. 1101
Fax: +81/3 3260 4726

Dr. Christian W. Kohlpaintner
Celanese GmbH / Werk Ruhrchemie
Postfach 13 01 60
D-46128 Oberhausen/Germany
Tel: +49/208 693 2461
Fax: +49/208 693 2053

Dr. Jürgen G.E. Krauter
Technische Universität München
Anorganisch-chemisches Institut
Lichtenbergstraße 4
D-85747 Garching/Germany
Tel: +49/89 2891 3647
Fax: +49/89 2891 3473

Dr. Fritz E. Kühn
Technische Universität München
Anorganisch-chemisches Institut
Lichtenbergstraße 4
D-85747 Garching/Germany
Tel: +49/89 2891 3105
Fax: +49/89 2891 3473

Dr. Emile G. Kuntz
CPE-LYON – Laboratoire d'Electrochimie Analytique
43 Bd du 11 Novembre 1918
F-69616 Villeurbanne/France
Tel: +33/472 448 478
Fax: +33/472 448 479

Prof. Dr. Gábor Laurenczy
Institut de Chimie Minerale et Analytique
Batiment de Chimie (BCH)
CH-1015 Dorigny Lausanne/Suisse
Tel: +41/21 692 3858
Fax: +41/21 692 3865

Prof. Dr. Piet W. N. M. van Leeuwen
University of Amsterdam
Institute for Molecular Chemistry
Inorganic Chemistry & Homogeneous Catalysis
Nieuwe Achtergracht 166
NL-1018 WV Amsterdam/The Netherlands
Tel: +31/20 525 5419
Fax: +31/20 525 6456

Priv.-Doz. Dr. Walter Leitner
Max-Planck-Institut für Kohlenforschung
Kaiser-Wilhelm-Platz 1
D-45470 Mülheim/Germany
Tel: +49/208 306 2272
Fax: +49/208 306 2993

Dr. Gerhard M. Lobmaier
Grünau-Illertissen GmbH
Postfach 1063
D-39251 Illertissen/Germany
Tel: +49/730 31 3403
Fax: +49/730 31 3220

Prof. Dr. André Lubineau
Laboratoire de Chimie Organique Multifonctionnelle
Université de Paris-Sud
Bat. 420
F-91405 Orsay/France
Tel: +33/1 691 57233
Fax: +33/1 691 54715

Dr. David M. Lynn
Laboratories of Chemistry
California Institute of Technology
Pasadena, California 91125/USA
Tel: +1/626 395 6003
Fax: +1/626 564 9297

Dr. Andrea Meli
Istituto per lo Studio della Stereochimica ed Energetica dei Composti
CNR di Coordinazione
Via J. Nardi, 39
I-50132 Firenze/Italy
Tel: +39/55 243 990 245 990
Fax: +39/55 247 8366

Prof. Dr. Joseph S. Merola
Department of Chemistry
Virginia Polytechnic Institute and State University
Blacksburg, Virginia 24061-0212/USA
Tel: +1/540 231 4510
Fax: +1/540 231 3255

Prof. Dr. Eric Monflier
Université d'Artois
Faculté des Sciences J. Perrin/Laboratoire de Physicochimie des Interfaces – CRUAL
Rue Jean Souvraz
Sac postal 18
F-62307 Lens/France
Tel: +33/321 791 715
Fax: +33/321 791 717

Prof. Dr. André Mortreux
Université des Sciences et Technologies de Lille
Ecole Nationale Supérieure de Chimie de Lille
B.P. 108
F-59652 Villeneuve d'Ascq Cédex/France
Tel: +33/320 434 993
Fax: +33/320 436 585

Prof. Dr. Antonino Morvillo
Università di Padova
Dipartimento di Chimica Inorganica
Via Marzolo 1
I-35100 Padova/Italy
Tel: +39/49 827 5156
Fax: +39/49 827 5161

Prof. Dr. Günther Oehme
Institut für Organische Katalyseforschung
an der Universität Rostock e.V.
Buchbinderstraße 5–6
D-18055 Rostock/Germany
Tel: +49/381 466 930
Fax: +49/381 466 9324

Prof. Dr. Tamon Okano
Department of Materials Science
Faculty of Engineering
Tottori University
Tottori 680/Japan
Tel: +81/857 31 5260
Fax: +81/857 31 0881

Prof. Dr. Hélène Olivier
Institut Français du Pétrole
1–4 Avenue de bois Préau
F-92852 Rueil-Malmaison Cédex/France
Tel: +33/1 475 26779
Fax: +33/1 475 26055

Prof. Dr. Georgios Papadogianakis
University of Athens
Department of Chemistry
Industrial Chemistry Laboratory
Panepistimiopolis – Zografou
GR-15771 Athens/Greece
Tel: +30/1 728 4235
Fax: +30/1 724 9103

Prof. Dr. Peter J. Quinn
King's College London
Section of Biochemistry
Campden Hill
GB-London W8 7AH/Great Britain
Tel: +44/171 333 4408
Fax: +44/171 333 4500

Dr. Claus-Peter Reisinger
BAYER AG
Zentrale Forschung-SFV
Geb. R79
D-47812 Krefeld
Tel: +49/2151 888 667
Fax: +49/2151 888 658

Dr. Wolfgang C. Schattenmann
Anorganisch-chemisches Institut
der Technischen Universität München
Lichtenbergstraße 4
D-85747 Garching/Germany
Tel: +49/89 289 13092
Fax: +49/89 289 13473

Dr. Marcel Schreuder Goedheijt
University of Amsterdam
Institute for Molecular Chemistry
Inorganic Chemistry & Homogeneous
Catalysis
Nieuwe Achtergracht 166
NL-1018 WV Amsterdam/The Netherlands
Tel: +31/20 525 6960
Fax: +31/20 525 6456

Prof. Dr. Roger A. Sheldon
Delft University of Technology
Department of Organic Chemistry
and Catalysis
Julianalaan 136
NL-2628 BL Delft/The Netherlands
Tel: +31/15 278 2675
Fax: +31/15 278 1415

Prof. Dr. Denis Sinou
Université Claude-Bernard Lyon I
CPE Lyon
Laboratoire de Synthèse Asymétrique
U.M.R. U.C.B.L./C.N.R.S. 5622
43, Bd du 11 Novembre 1918
F-69622 Villeurbanne Cédex/France
Tel: +33/472 446263
Fax: +33/472 448160

Prof. Dr. Othmar Stelzer
Bergische Universität
Gauss-Straße 20
D-42097 Wuppertal
Tel: +49/202 439 2517
Fax: +49/202 439 2517

Dr. Dieter Vogt
Institut für Technische Chemie und Petrolchemie der Rheinisch-Westfälischen Technischen Hochschule Aachen (RWTH)
Templergraben 55
D-52056 Aachen/Germany
Tel: +49/241 806 480
Fax: +49/241 888 8177

Ernst Wiebus
Celanese GmbH / Werk Ruhrchemie
Postfach 13 01 60
D-46128 Oberhausen/Germany
Tel: +49/208 693 2459
Fax: +49/208 693 2041

Dr. Noriaki Yoshimura
Chemical Research Laboratory
Kuraray Co., Ltd.
2045-1 Sakazu, Kurashiki
710 Okayama/Japan
Tel: +81/86 423 2271
Fax: +81/86 422 4851

Dr. Xiaolai Zheng
State Key Laboratory of Fine Chemicals
Dalian University of Technology
Dalian 116012/China
Tel: +86/411 467 1511
Fax: +86/411 363 3080

1

Introduction

1 Introduction

Boy Cornils, Wolfgang A. Herrmann

"For it is one thing to invent a basically correct process, another to introduce it in industry".

„Denn [eines] ist es, ein prinzipiell richtiges Verfahren zu erfinden, ein anderes, es in die Industrie einzuführen."

Hermann Ost [1]

Well disposed critics think that heterogeneous catalysis [2] is still at a stage of blindly groping empiricism and therefore at the level of a "black art" [3]. This statement, which has not remained uncontradicted [4], is in complete unison with the result of a comparison of heterogeneous with homogeneous catalysis (cf. Table 1 [5]).

Echoing the above criticism, the comparison under "variability of steric and electronic properties" and "mechanistic understanding" shows an advanced understanding of elementary steps in homogeneous catalysis. Yet Table 1 also lists the major industrial disadvantage of homogeneous catalysis: the immense difficulty of catalyst recycling, which is reponsible for the fact that about 80 % of catalytic reactions still employ heterogeneous catalysts and only 20 % involve homogeneous catalysts. This is because it is inherently difficult to separate the molecularly dissolved homogeneous catalyst from the reaction products and

Table 1. Homogeneous versus heterogeneous catalysis [6].

	Homogeneous catalysis	Heterogeneous catalysis
Activity (relative to metal content)	High	Variable
Selectivity	High	Variable
Reaction conditions	Mild	Harsh
Service life of catalysts	Variable	Long
Sensitivity toward catalyst poisons	Low	High
Diffusion problems	None	May be important
Catalyst recycling	Expensive	Not necessary
Variability of steric and electronic properties of catalysts	Possible	Not possible
Mechanistic understanding	Plausible under random conditions	More or less impossible

any unconverted reactants in which the catalyst is likewise dissolved at a molecular level. Particularly, homogeneous organometallic catalysts while being recycled/worked-up, e.g., by using distillation or chemical techniques, suffer from thermal or chemical stress. Table 2 shows this and in detail in a comparison of homogeneous two versions of heterogeneous catalyses [6 b].

Table 2. Catalyst removal in homogeneous and heterogeneous catalysis [6 b].

	Homogeneous catalysis	Heterogeneous catalysis	
		Suspension	Fixed bed
Separation	Filtration after chemical decomposition Distillation Extraction	Filtration	No separation problems
Additional equipment required	Yes	Little	No
Catalyst recycling	Possible	Easy	Not necessary
Costs of catalyst losses	High	Minimal	Minimal
Catalyst concentration in product	Low	High	–

Tables 1 and 2 immediately suggest as a practical solution that a "heterogenization" of homogeneous catalysts, i.e., the immobilization or anchoring of dissolved catalysts on immobile, solid supports, may be a way of transferring many of the advantages of heterogeneous catalysis to homogeneous systems. In theory, a heterogenized (immobilized) homogeneous catalyst should behave like a heterogeneous catalyst and solve the problem of catalyst recycling, provided the attendant diffusion problems – the significant disadvantage of heterogeneous catalysis – prove tolerable.

Since solving this recycling problem is essential for the high-volume processes of homogeneous catalysis (for hydroformylation especially: annual output about 6.5 million tonnes [7]), legions of scientists have published innumerable papers demonstrating ways of achieving heterogenization by the anchoring of homogeneous catalysts. Many of the sometimes very ingenious methods tried to date (including poly- or copolymerization of catalytically active and polymerizable monomers, functionalization of suitable supports and introduction of catalytically active constituents, precipitation of metals, or impregnation of suitable supports with active catalyst precursors, etc. [8, 9]) did indeed lead to initially active, "heterogenized" catalysts. However, it was also found that, despite coordination-capable support groups and covalent bonds, all these catalysts have only a finite on-stream life due to the leaching which starts from the first minute of use. This leaching always affects not only the (usually costly) central atoms but also the (frequently more costly) ligands of homogeneous metal complex catalysts. The economic need to recover them more than offsets the saving due to simplified recycling. Whistling in the dark is not problem-solving: at present, despite sporadic news of success, there is no economical process for heterogenizing homogeneous catalysts of large industrial processes.

There is little mileage in looking for ostensibly more and more effective ligands and better and better optimization of support and catalyst precursor to ensure, on the one hand, adequate immobilization on the support (sufficient stability of the covalent bond between support matrix and central atom) and, on the other, adequate mobility for the catalytically active catalyst constituents (sufficient lability of the ligand sphere of the metal atom). All the results so far allow only the conclusion that the heterogenizing techniques used had to remain unsuccessful. The reason for this is that the various catalyst species undergo changes in spatial configuration as they pass through the catalytic cycle typical of a homogeneous process. The constant "mechanical" stress on the central atom ↔ ligand bonds and the constant change in the bond angles and lengths ultimately lead also to a weakening of the central atom ↔ support bond. This is conveniently demonstrated using the hydroformylation catalyzed by heterogenized cobalt carbonyls as an example (Fig. 1). The catalyst passes through the two forms of a trigonal-bipyramidal and of a tetrahedral cobalt carbonyl, which overstresses and weakens the heterogenizing bond Co ↔ support.

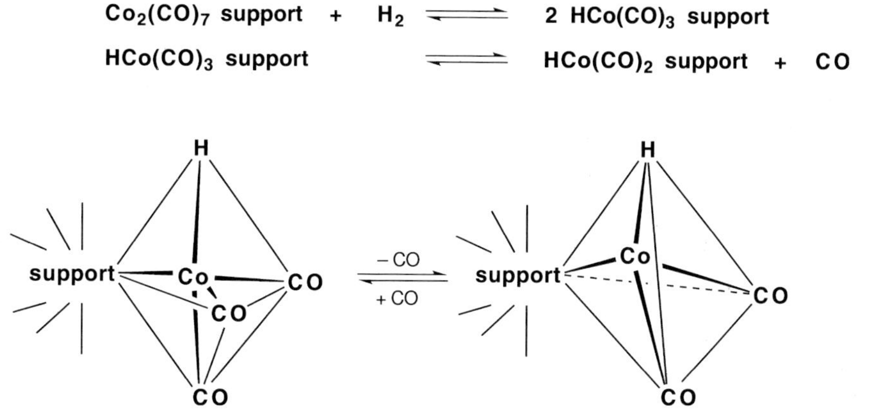

Figure 1. Extreme configuration changes of cobalt carbonyls in the course of the catalytic hydroformylation cycle.

A more elegant and ultimately more successful solution is the idea, probably first articulated and systematized by Manassen [10], although Papadogianakis and Sheldon [11] mistakenly credit Bailar [9], of an immobilization with the aid of a "liquid support". In 1972 Manassen suggested

"... the use of two immiscible liquid phases, one containing the catalyst and the other containing the substrate ..."

and hence the general form of biphase catalysis, which constitutes a logical development of the work in "molten salt media" (known today as "ionic liquids"; this term used to refer to high-melting, inorganic salts or salt mixtures [12]) described by Parshall [13]. Interestingly, the inventors of Shell's SHOP process, who had already worked on soluble, homogeneous complex catalysts in a biphase system some years earlier [14], cited the special method without particular emphasis, judging by the wording in the first patents. Shell seems not to have been aware immediately that it had laid its hands on the key to the novel technology of homogeneous catalysis by means of immobilized catalysts on liquid supports – albeit not water (cf. Section 7.1).

The basic principles of biphase catalysis is accordingly that the homogeneous catalyst is in solution in one of the phases and the reaction products are located in a second phase which is immiscible with the catalyst phase – "heterogeneous" – and are therefore easy to separate off (see below).

The specific form of aqueous biphase catalysis was very significantly stimulated by the work of Joó and Beck at Debrecen (in relation to hydrogenation especially [15]; cf. Section 6.2) and Kuntz at Rhône–Poulenc (hydroformylation, telomerization [16] cf. Section 6.1.1), following Manassen's work, which was then unfortunately merely theoretical. Remarkably, the fundamental papers of Joó and Kuntz created little interest and only found a wider echo in academic research once Ruhrchemie AG had managed to achieve industrial scale-up of aqueous biphase catalysis in an oxo process (cf. Section 6.1 [17]). In a drastic departure from the normal pattern, here basic research lagged considerably behind industrial research and application. Reviews, even recent ones, tend to concentrate more on the state of basic research than on that of the large-scale industrial processes [11, 18], and curiously there are reviews appearing even now which fail to cite the contributions made by industry (see [19]!). In addition, it has to be mentioned as typical of this very recent development of homogeneous catalysis that considerable areas of the art and of its advances are chronicled in patents. Anyone who knows of the reluctance of basic researchers to read patents knows what this means for the current awareness amongst workers in basic research on biphasic catalysis.

In a technology involving two liquid phases, one of which contains the metal complex catalyst in solution, the idea of using water as one of the phases is not necessarily obvious. Hydroformylation, in 1972 – the time of Manassen's idea – the most important application of homogeneous catalysis, utilized cobalt catalysts, whose handling sensitivity ruled out an aqueous phase. Or, as P. Cintas [20] wrote,

"At first, the idea of performing organometallic reactions in water might seem ridiculous, since it goes against the traditional belief that most organometallics are extremely sensitive to traces of air and moisture and rapidly decompose in water."

This is all the more surprising as the history of the oxo process actually prescribed the use of aqueous catalysts and catalyst precursors (aqueous cobalt salts as precursors of the earlier "Diaden" process [21], the at least partially aqueous cycle of the BASF and Kuhlmann process [22b], or the cleavage of solvent-soluble by-products and heavy ends of the oxo process with the aid of water-dissolved metal salts [22c, 23]).

The advantage of using water is that it is easy to separate from organic products, as indicated by Manassen in a continuation of the above quotation [10]:

"The two phases can be separated by conventional means and high degrees of dispersion can be obtained through emulsification. This ease of separation may be particularly advantageous in situations where frequent catalyst regeneration is required".

Figure 2 illustrates the enormous importance of the biphase technique for homogeneous catalysis: the aqueous catalyst solution is charged in the reactor with the reactants A and B, which react to form the solvent-dissolved reaction products C and D. C and D are less polar than the aqueous catalyst solution and are therefore simple to separate from the aqueous phase (which is recycled directly into the reactor) in the downstream phase separator (decanter).

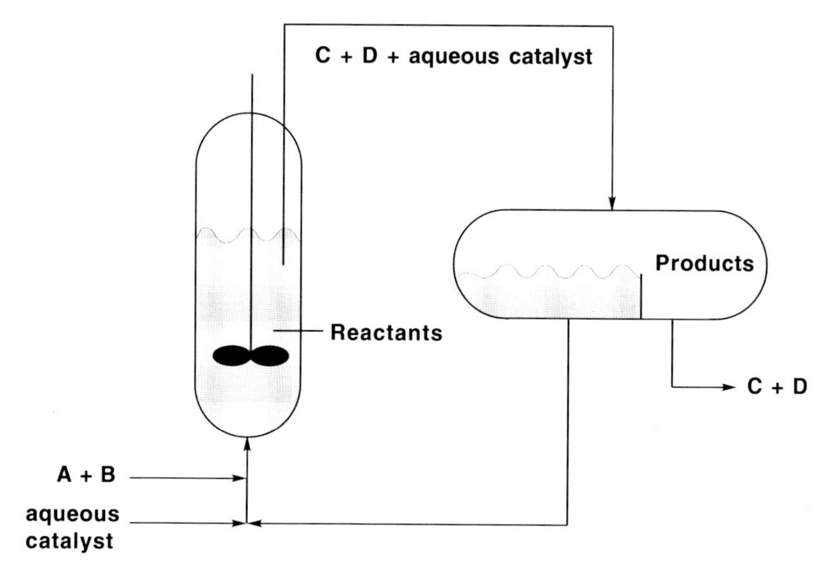

Figure 2. Principle of biphase catalysis illustrated for the reaction A + B → C + D.

The advantage of the "liquid support" water and of its high affinity for the metal complex catalyst is evident. The catalyst is heterogenized with respect to the organic reaction products C and D and therefore can not only be separated from the products in the "other phase" (possibly including unconverted reactants A and B), but also immediately thereafter starts a new cycle of the catalytic cycle process. Aqueous biphase catalysis is therefore – intentionally – located between heterogeneous and homogeneous catalysis, as illustrated in Figure 3. Special attention may be drawn to the somewhat confusing and ambiguous use of the terms "homogeneous" and "heterogeneous" in the context of homoge-

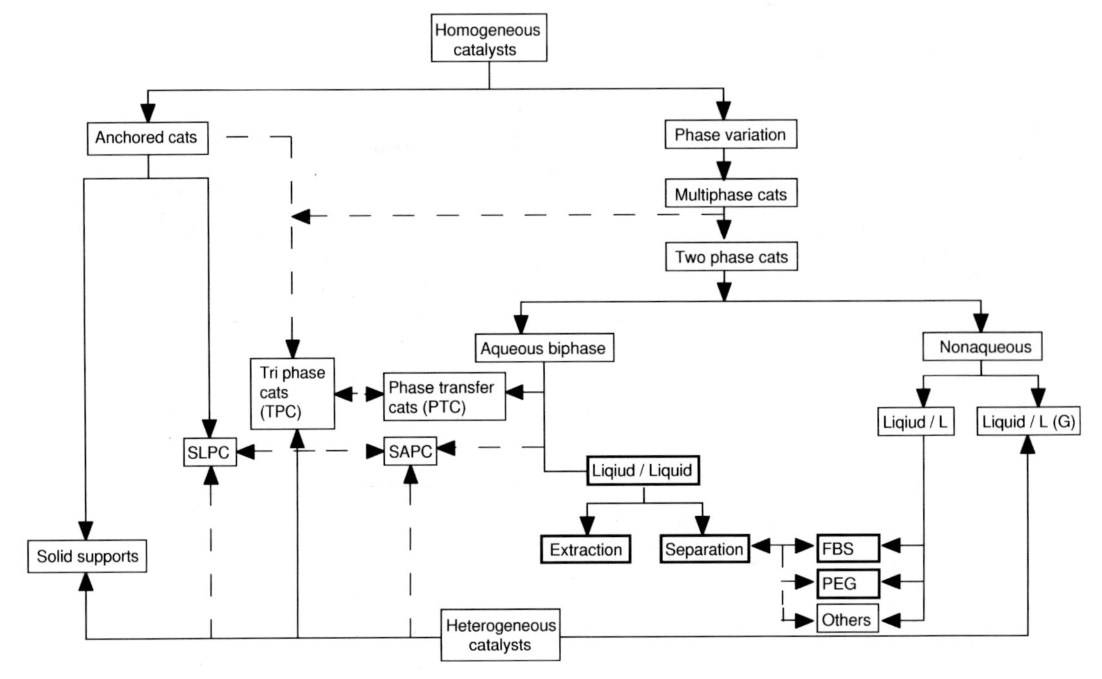

Figure 3. Positioning of aqueous biphase catalysis: different approaches according to the variation of the application phase of catalysts. FBS, fluorous biphase system (cf. Section 7.2 [24]; PEG, polyethylene glycol (cf. Sections 4.6.3 and 6.1.3.2); SAPC, and SLPC, cf. Section 4.7.

neous/heterogeneous catalysis and homogeneous/heterogeneous phase variation: the catalytic system works homogeneously despite heterogeneous phase variation.

The immobilization of the homogeneous catalyst with the aid of water as liquid support leads to appreciable technical simplifications, as illustrated in Figure 4 for the recycling of the catalyst of an industrial hydroformylation process (A = olefin, B = CO/H_2, C and D = butyraldehydes) [22a, 25].

For the overall process of hydroformylation, the lower expense of aqueous biphase catalysis compared with the "classic" process can easily be projected at about 50% of capital expenditure costs [17e]. Figure 5 illustrates the significant simplifications and savings in a comparison of the two process variants.

Table 2 predicts that, of the biphase processes, the aqueous version will attain particular importance because of the many advantages of water as the support. As a solvent, water has numerous anomalies (e.g., density anomaly, the only non-toxic and liquid "hydride" of the non-metals, pressure-dependence of the melting point, dielectric constant), and its two- or even three-dimensional structure is still not well understood (cf. Sections 2.1–2.3). Some of the known properties are listed below:

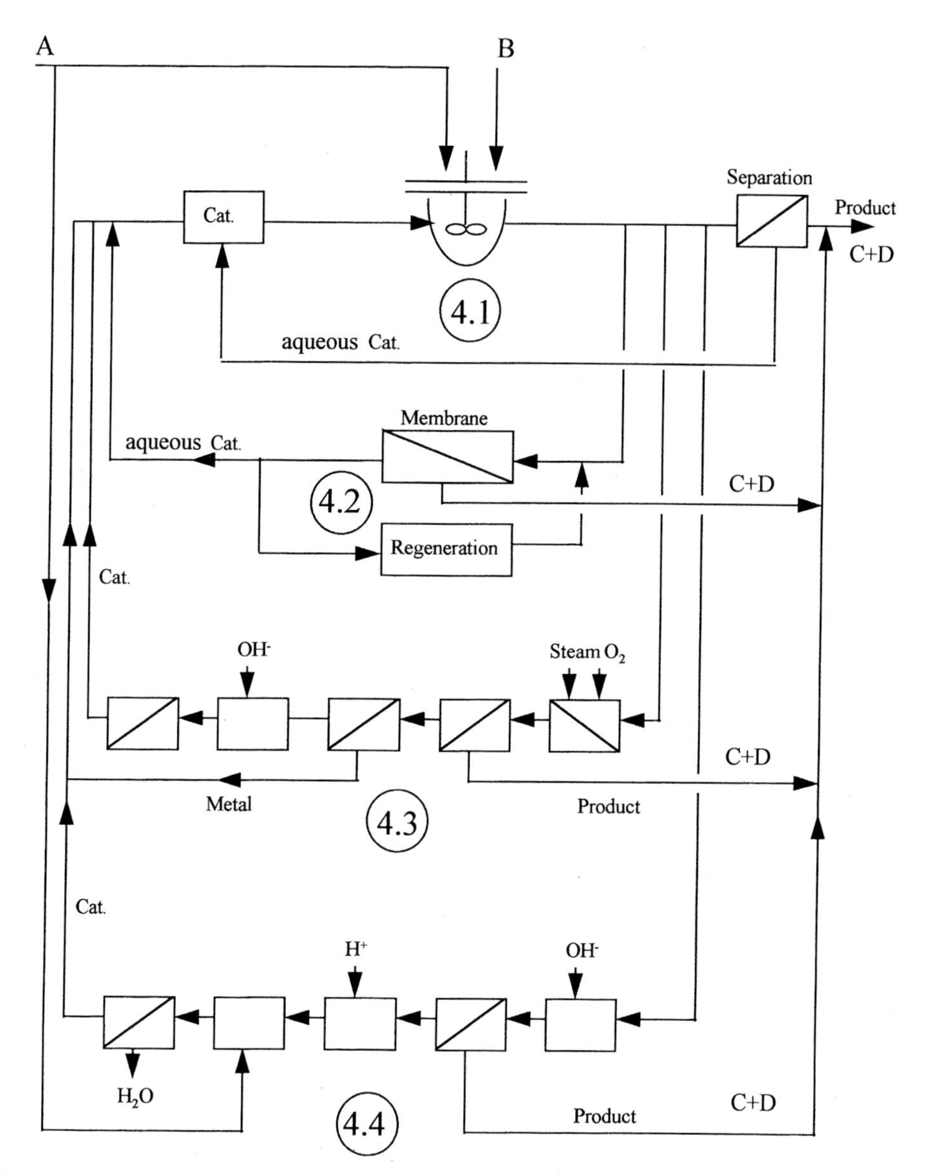

Figure 4. The different methods of separation and recycling of oxocatalysts for the reaction A + B → C + D. (4.1) aqueous biphase operation; (4.2) membrane technique; (4.3) thermal methods; (4.4) chemical treatment [22a].

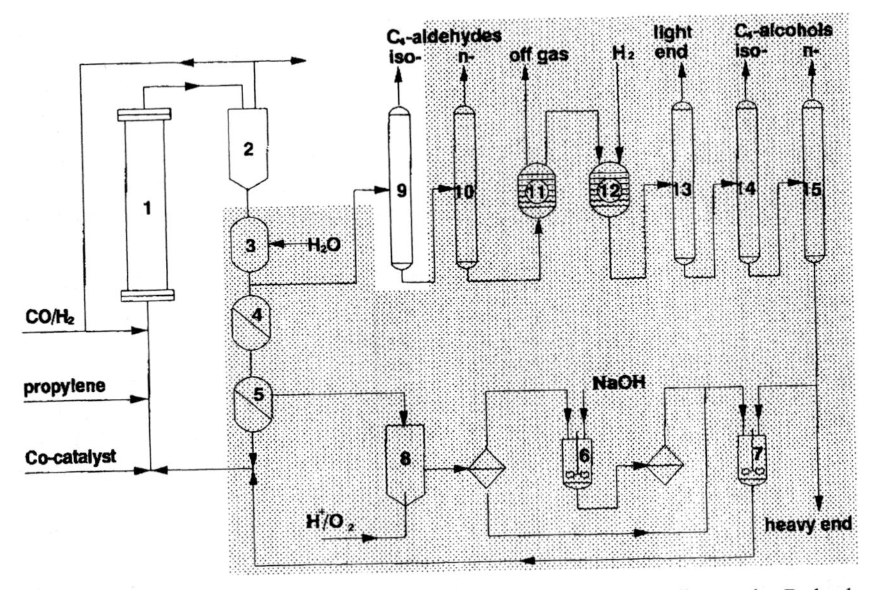

Figure 5. Schematic flow sheet of a "classic" oxo process (known earlier as the Ruhrchemie process [22a]). Those parts of the process and equipment which are dispensed with in the novel RCH/RP process are stippled [17e].

(1) Polar and easy to separate from apolar solvents or products; polarity may influence (improve) reactivity.
(2) Nonflammable, incombustible.
(3) Widely available in suitable quality.
(4) Odourless and colourless, making contamination easy to recognize.
(5) Formation of a hexagonal two-dimensional surface structure and a tetrahedron three-dimensional molecular network, which influence the mutual (in)solubility significantly; chaotropic compounds lower the order by H-bond breaking.
(6) High Hildebrand parameter, as unit of solubility of non-electrolytes in organic solvents.
(7) A density of 1 g/cm^3 provides a sufficient difference from most organic substances.
(8) Very high dielectric constant.
(9) High thermal conductivity, high specific heat capacity and high evaporation enthalpy.
(10) Low refractive index.
(11) High solubility for many gases, especially CO_2.
(12) Formation of hydrates and solvates.
(13) Highly dispersible and high tendency toward micelle formation, stabilization by additives.
(14) Amphoteric behavior in the Brønsted sense.
(15) Advantageous influence on chemical reactivity.

Besides its positive influence on reactivity (entry 15, [26]), the properties of water which are of direct significance for the aqueous two-phase processes are in particular the physiological (entries 2, 4), economic (1, 3, 6), ecological/safety (2, 4), technical (1, 6, 7, 9, 11, 12, 13) and physical properties (1, 5, 6, 8, 10, 12, 14). The various properties have multiple effects and are mutually reinforcing. For instance, water, whose high Hildebrand parameter [27] and high polarity have an advantageous effect on organic reactions (such as hydroformylation) [26], has a sufficiently high polarity and density difference from organic (reaction) products to allow separation of the phases following the homogeneously catalyzed reaction.

On the other hand, the high solubility for many compounds and gases – possibly augmented by solvate, hydrate, or hydrogen bond formation – facilitates reactions within the two-phase system. The chaotropic properties of many chemical compounds prevent the H_2O cage structures necessary for the formation of solvates, and facilitate the transfer of apolar molecules from both nonaqueous and aqueous phases.

Water does not ignite, does not burn, is odorless and colorless, and is ubiquitous: important prerequisites for the solvent of choice in catalytic processes. The dielectric constant or the refractive index can be important in specific reactions and their analytical monitoring. The favorable thermal properties make water doubly qualified as mobile support and as heat-transfer fluid, which is industrially exploited in the RCH/RP process (cf. Section 6.1 [17e–h]).

The decisive advance for aqueous biphase technology was to leave behind traditional ways of thinking and move from organically and hence homogeneously soluble to heterogeneous and water-soluble hydroformylation catalysts with the aid of water-soluble ligands [16a]. This water solubility meant more than the change in the ligand sphere (CO → L, e.g., phosphines) and hence the replacement of the "classic" oxo catalyst $HMe(CO)_4$ by those of the type $HMe(CO)_nL_m$ and also more than the switch of the central atoms (CO → Rh). Water not only dissolves the catalyst, but also acts as a moderator [22, 26]; this represents the decisive difference in relation to the chemical reaction of the hydroformylation process and the process control of the reaction. The "standard ligand" of aqueous biphase catalysis, triphenylphosphine trisulfonate (TPPTS, 3,3′,3″-phosphinidynetris(benzenesulfonic acid), trisodium salt; cf. Section 3.2.1), has particular significance here. To what extent the replacement of TPPTS by new ligands in the catalyst system has repercussions for the management of the aqueous biphase reaction will depend on their nature and possibly also the addition of additives and auxiliaries (cf. Sections 4.1–4.7; for acronyms cf. Section 3.2).

The prevalent view is that the feed olefins of the aqueous biphase technique require a certain minimum solubility in the aqueous catalyst phase for adequate conversion. The reactivity differences in hydroformylation between, for example, propene and octene are readily explained by the solubility differences between the two olefins (Figure 6 [28, 29]). This is also the basis of the many proposals for solubilizing solvents or cosolvents, which are meant to make possible the reaction of higher and hence less water-soluble olefins in the bulk

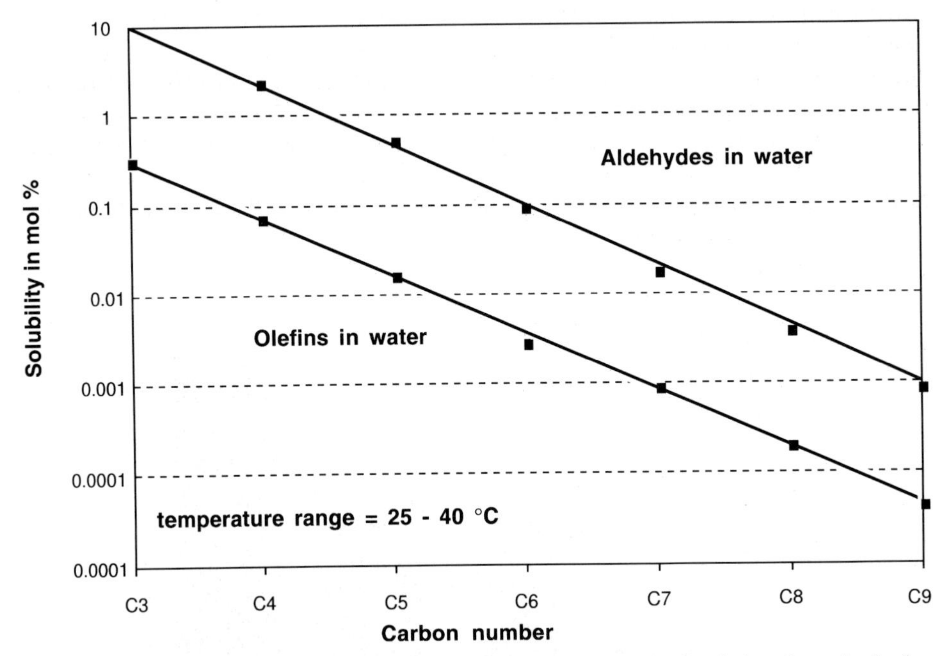

Figure 6. Solubility of olefins [30, 31 a] and of the aldehydes obtained therefrom by hydroformylation in water [31 b, c].

of the catalyst solution. On the other hand, there is much evidence arguing for the boundary layer between the two phases as the critical reaction site [30, 32], which makes the addition of surface-active ligands and/or measures which increase the width of the interfacial layer appear promising.

The evidence from kinetic studies (cf. Section 6.1.2) and from reaction measurements [32] indicates that this phase boundary layer is indeed to be considered the reaction site under normal circumstances. The dependence of the reactivity of aqueous systems on the solubility of the reactants in the aqueous catalyst solutions is of appreciable importance for the problem of universal applicability (cf., e.g., Sections 4.1, 4.2, 6.1.3.2).

Using water as the solvent has not only the advantage of having a "mobile support" and hence of a *de facto* "immobilization" of the catalyst while retaining a homogeneous form of reaction, but also has positive repercussions on the environmental aspects of hydroformylation (cf. Section 5.2).

Aqueous biphase processes will become more important in the future because of the immense advantages of this version of homogeneous catalysis. Dimerizations, telomerizations, hydrocyanations, aldolizations, Claisen condensations, and the great diversity of C–C coupling reactions are or will become targets, in addition to the syntheses already utilized now. Work in this area is described in Section 6.

Systematic studies on the oxo process in particular have shown the vast improvements which can be achieved through variation of the ligands [33]. The comparison of the standard TPPTS ligand of aqueous biphase technology with new ligands such as BISBIS, NORBOS, or BINAS shows up distinct differences. The hydroformylation results so far demonstrate that different requirements such as highest possible activity, highest *n*/iso ratio (ratio between linear [l or *n*-] and branched [b or *iso*-] compounds), lowest required Rh concentration, or lowest excess of ligands (in all cases BINAS ≫ TPPTS) may be achieved by different ligands, thus indicating bright prospects for future tailor-made oxo catalysts in biphasic operation (Figure 7 and Section 3.2).

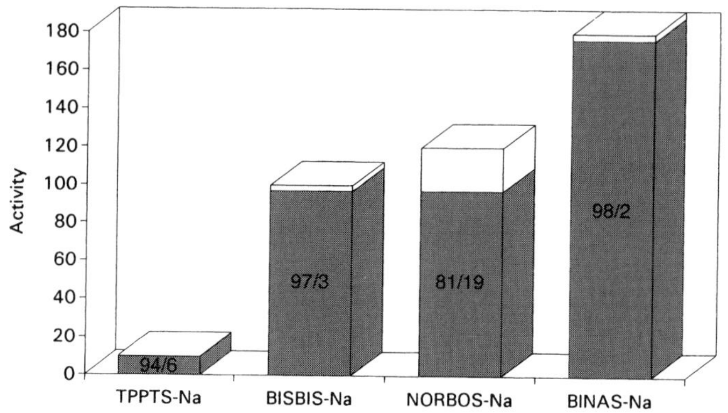

Figure 7. How do ligands of aqueous oxo catalysts compare with each other? Oxo catalyst $HRh(CO)L_3$, with L = TPPTS, BISBIS, NORBOS or BINAS [33].

To what extent asymmetric/enantiomeric reactions will play a role, only the future will decide; initial experiences are encouraging (cf. Section 6.9).

The use of biphase processes, starting from aqueous biphase catalysis, is only just beginning. As well as the versions which have been mentioned, other immiscible solvent mixtures will be used (depending on the requirements of the reactions or of the homogeneous catalysts), chosen on the one hand according to the principle of "like dissolves like" ("Similia similibus solvuntur," as the alchymists used to say) in respect of the catalyst solvent and, on the other, according to fundamental nonmiscibility studies, similarly to the miscibility diagram (Figure 8 [34]).

Section 7 reviews non-aqueous biphase processes and their variations. Sections 4.5 and 4.6 deal with micellar systems and various applications of phase transfer catalysis in relation to aqueous biphase catalysis. Interestingly, biphase techniques are also being utilized from the other side, that of heterogeneous catalysis [35].

It is likely that the spread of biphase processes [37] will increase the proportion of homogeneously catalyzed reactions and hence the importance of homogeneous catalysis in general. It will then also be possible to demonstrate in full the great advantages of homogeneous catalysis over the rather empirical meth-

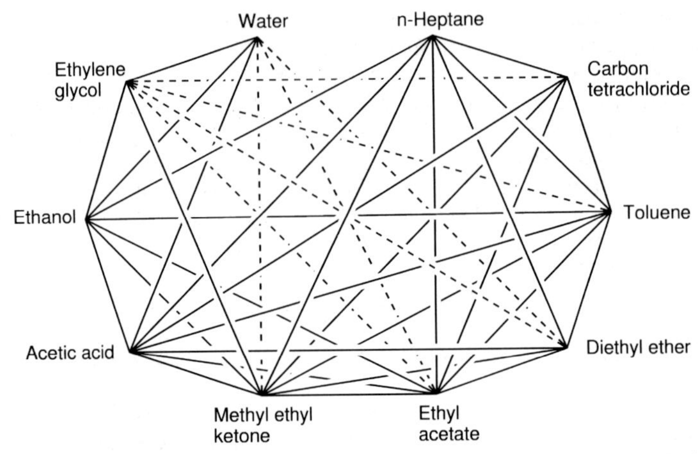

Figure 8. Miscibility of organic solvents. — miscible in all proportions; ---- limited miscibility; ..... little miscibility; no line, immiscible.

ods of heterogeneous catalysis and answer Heinemann's 1971 question, "Homogeneous and heterogeneous catalysis – common frontier or common territory?" [38], clearly in favor of the homogeneous version.

References

[1] H. Ost in a lecture about noxious industrial wastes, in *Z. Angew. Chem.* **1907**, *20*, 1689.

[2] G. Ertl, H. Knözinger, J. Weitkamp (Eds.) *Handbook of Heterogeneous Catalysis,* VCH, Weinheim, **1997**, 3 volumes.

[3] R. Schlögl, *Angew. Chem.* **1993**, *105*, 402.

[4] J. M. Thomas, K. I. Zamaraev, *Angew. Chem.* **1994**, *106*, 316.

[5] (a) B. Cornils, W. A. Herrmann (Eds.), *Applied Homogeneous Catalysis with Organometallic Compounds,* VCH, **1996**, 2 volumes; (b) cf. [5a], p. 5.

[6] (a) Anonymous, *Nachr. Chem. Tech. Lab.* **1979**, *27*, 257; (b) J. Falbe, H. Bahrmann, *Chem. Zeit* **1981**, *15*, 37; W. Keim in *Industrial Applications of Homogeneous Catalysis* (Eds.: A. Mortreux, F. Petit), p. 338, D. Reidel, Dordrecht, **1988**.

[7] W. Gick, Hoechst AG, Werk Ruhrchemie, private communication.

[8] (a) J. M. Basset, A. K. Smith, Catalysis by Supported Complexes, in *Homogeneous Catalysis* (Eds. M. Tsutsui, R. Ugo), Plenum Press, New York, **1977**, p. 69 ff.; (b) W. T. Ford, *CHEMTECH* **1984**, *14*, 436; (c) special issue of *J. Mol. Catal.* **1980**, *9*; (d) R. H. Grubbs, *CHEMTECH* **1977**, *7*, 512; (e) N. L. Holy, *CHEMTECH*, **1980**, *10*, 366; (f) D. E. Bergbreiter, *CHEMTECH* **1987**, *17*, 686; (g) E. Bayer, V. Schurig, *CHEMTECH* **1976**, *6*, 212; (h) J. Manassen, *Platin. Met. Rev.* **1971**, *15*, 142; (i) Z. M. Michalska, D. E. Webster, *CHEMTECH* **1975**, *5*, 117; (k) F. R. Hartley, *Supported Metal Complexes,* Reidel, Dordrecht, **1985**; (l) Y. L. Yermakov, B. N. Kusnetsov, V. A. Zakharov (Eds.), *Catalysis by Supported Complexes,* Elsevier, Amsterdam, **1981**.

[9] J. C. Bailar, *Catal. Rev.* **1974**, *10*, 17.

[10] J. Manassen in *Catalysis: Progress in Research* (Eds.: F. Bassolo, R. L. Burwell), Plenum Press, London, **1973**, p. 177, 183.

[11] G. Papadogianakis, R. A. Sheldon, *New J. Chem.* **1996**, *20*, 175.
[12] C. N. Kenney, *Catal. Rev. – Sci. Eng.* **1975**, *11*, 197.
[13] G. W. Parshall, *J. Am. Chem. Soc.* **1972**, *94*, 8716.
[14] (a) W. Keim, *Chem. Ing. Techn.* **1984**, *56*, 850; (b) W. Keim, *Stud. Surf. Sci. Catal.* **1989**, *44*, 321; (c) Shell (W. Keim, T. M. Shryne, R. S. Bauer, H. Chung, P. W. Glockner, H. van Zwet), DE 2054009 (**1969**).
[15] (a) F. Joó, L. Somsák, M. T. Beck, *Proc. Symp. Rhodium in Homogeneous Catalysis*, Sept. **1978**, Veszprém/Hungary, p. 5; (b) F. Joó, M. T. Beck, *React. Kin. Cat. Lett.* **1975**, *2*, 257; (c) F. Joó, Z. Tóth, M. T. Beck, *Inorg. Chim. Acta* **1977**, *25*, L61; (d) F. Joó, Z. Tóth, *J. Mol. Catal.* **1980**, *8*, 369.
[16] (a) Rhône–Poulenc (E. G. Kuntz), FR 2314910 (1975), FR 2349562 (1976), FR 2338253 (1976), FR 2338253 (1976), FR 2366237 (1976); (b) E. G. Kuntz, *CHEMTECH* **1987**, *17*(9), 570; (c) B. Cornils, E. G. Kuntz, *J. Organomet. Chem.* **1995**, *502*, 177.
[17] (a) B. Cornils, J. Falbe, *Proc. 4th Int. Symp. Homogeneous Catalysis,* Leningrad, Sept. **1984**, p. 487; (b) H. Bach, W. Gick, E. Wiebus, B. Cornils, *Prepr. Int. Symp. High-Pressure Chem. Eng.*, Erlangen/Germany, Sept. **1984**, p. 129; (c) H. Bach, W. Gick, E. Wiebus, B. Cornils, *Prepr. 8th ICC*, Berlin, **1984**, Vol. V, p. 417; [*C.A.* **1987**, *106*, 198051; cited also in A. Behr, M. Röper, *Erdgas Kohle* **1984**, *37*(11), 485]; (d) H. Bach, W. Gick, E. Wiebus, B. Cornils, *1st IUPAC Symp. Org. Chemistry,* Jerusalem, **1986**, Abstracts p. 295; (e) E. Wiebus, B. Cornils; *Chem. Ing. Techn.* **1994**, *66*, 916; (f) B. Cornils, E. Wiebus, *CHEMTECH* **1995**, *25*, 33; (g) E. Wiebus, B. Cornils, *Hydrocarb. Process.* **1996**, *75*(3), 63; (h) B. Cornils, E. Wiebus, *Recl. Trav. Chim. Pays-Bas* **1996**, *115*, 211.
[18] (a) W. A. Herrmann, C. W. Kohlpaintner, *Angew. Chem.* **1993**, *105*, 1588; *Angew. Chem., Int. Ed. Engl.* **1993**, *32*, 1524; (b) P. Kalck, F. Monteil, *Adv. Organomet. Chem.* **1992**, *34*, 219; (c) Abstracts of the NATO Advanced Research Workshop, *Aqueous Organometallic Chemistry and Catalysis,* Debrecen, Hungary, **1984**; I. T. Horváth, F. Joó, *Aqueous Organometallic Chemistry and Catalysis,* Kluwer, Dordrecht, **1995**; (d) special issue of *J. Mol. Catal.* **1997**, *116*; (e) P. W. N. M. van Leeuwen, *Cattech* **1998**, announced; (f) M. J. H. Russel, *Plat. Met. Rev.* **1988**, *32*, 179.
[19] (a) R. V. Chaudhari, A. Bhattacharya, B. M. Bhanage, *Catalysis Today* **1995**, *24*, 123; (b) D. J. Darensbourg, C. T. Bischoff, *Inorg. Chem. 1993* **1993**, *32*, 47 or *J. Organomet. Chem.* **1995**, *488*, 99.
[20] P. Cintas, *Chem. Eng. News* **1995**, March 20, 5.
[21] Survey (a) in: B. Cornils, W. A. Herrmann, M. Rasch, *Angew. Chem.* **1994**, *106*, 2219, *Angew. Chem. Int. Ed. Engl.* **1994**, *33*, 2144. Aqueous catalyst precursors in: (b) Chemische Verwertungsgesellschaft Oberhausen mbH (ChVW), GB 719.573 (1952); (c) ChVW (H. Nienburg), DE 948.150 (1953); (d) ChVW (H. Kolling, K. Büchner, H. Heger, E. Stiebling), DE 949.737 (1952); (e) ChVW, GB 711.696 (1950); (f) ChVW (W. Reppe, H. Eilbracht), DE 891.688 (1942); (g) ChVW (H. Fritzsche, O. Roelen), DE 888.097 (1943).
[22] (a) B. Cornils, Hydroformylation, in *New Syntheses with Carbon Monoxide* (Ed.: J. Falbe), Springer, Berlin, **1980**; (b) [21a] pp. 164, 165; (c) Ref. [21a] p. 157.
[23] Ruhrchemie AG (H. Tummes, J. Meis), US 3.462.500 (1969).
[24] B. Cornils, *Angew. Chem.* **1997**, *109*, 2147; *Angew. Chem. Int. Ed. Engl.* **1997**, *36*, 2057.
[25] (a) A. Behr, W. Keim, *Erdöl, Erdgas, Kohle* **1987**, *103*, 126; (b) A. Behr, *Henkel-Referate* **1995**, *31*, 31.
[26] (a) T. G. Southern, *Polyhedron* **1989**, *8*, 407; (b) C.-L. Li, *Chem. Rev.* **1993**, *93*, 2023; (c) A. Lubineau, *Chem. Ind.* **1996** (4), 123 and *Synthesis* **1994**, *8*, 741.
[27] J. H. Hildebrand, *J. Am. Chem. Soc.* **1935**, *57*, 866.

[28] B. Cornils, *Angew. Chem.* **1995**, *107*, 1709; *Angew. Chem., Int. Ed. Engl.* **1995**, *34*, 1575.
[29] D. M. Himmelblau, E. Arends, *Chem. Ing. Techn.* **1959**, *21*, 791.
[30] S. Partzsch, Hoechst AG, private communication.
[31] (a) C. McAuliffe, *J. Phys. Chem.* **1966**, *70*, 1267; (b) R. M. Stephenson, *J. Chem. Eng. Data* **1993**, *38*, 630; (c) P. L. Davis, *J. Gas Chromatogr.* **1968**, *6*(10), 518.
[32] K. Himmler, H. W. Bach, H. Bahrmann, C. D. Frohning, Th. Müller, O. Wachsen, E. Wiebus, B. Cornils, *Catal. Today,* in press.
[33] W. A. Herrmann, C. W. Kohlpaintner, H. Bahrmann, W. Konkol, *J. Mol. Catal.* **1992**, *73*, 191; W. A. Herrmann, C. W. Kohlpaintner, R. B. Manetsberger, H. Bahrmann, H. Kottmann, *J. Mol. Catal.* **1995**, *97*, 65; W. A. Herrmann, W. R. Thiel, J. G. Kuchler, *Chem. Ber.* **1990**, *123*, 1953; W. A. Herrmann, J. Kellner, H. Riepl, *J. Organomet. Chem.* **1990**, *389*, 103.
[34] G. Duve, O. Fuchs, H. Overbeck, *Lösemittel Hoechst*, 6th ed., Hoechst AG, Frankfurt/Main, **1976**, p. 49, also referred in C. Reichardt, *Solvents and Solvent Effects in Organic Chemistry,* VCH, Weinheim, Germany, **1990**, Fig. 2-2.
[35] Asahi Kasei Kogyo KK (K. Yamashita, H. Obana, I. Katsuta), EP 0.552.809 (1993).
[36] Cf. [5a], pp. 245, 258, 577, 601 and 644.
[37] H. Heinemann, *CHEMTECH* **1971**, *5*, 286.

2

Basic Aqueous Chemistry

2.1 Organic Chemistry in Water

André Lubineau, Jacques Augé

2.1.1 Introduction

Water is a natural solvent. Molecular interactions and biochemical transformations in living systems mostly occur in an aqueous environment. Nevertheless, the use of water as a solvent in modern synthetic chemistry had practically disappeared when the role of highly polar organic solvents endowed with solvation properties was recognized. The lack of water-solubility of organic compounds, along with the water-sensitivity of some reagents or reactive intermediates, prevented chemists from thinking about water as a solvent for organic synthesis. Its use was rediscovered in the 1980s when Breslow [1 a], and Kuntz for organometallic catalyzed reactions [1 b], showed that the rate of the cycloaddition of cyclopentadiene with methyl vinyl ketone in water was enhanced by a factor of more than 700 compared with the reaction in isooctane [1]. Since this seminal contribution, there has been an upsurge in interest in using water as the solvent, not only to enhance the reaction rates but also to perform organic reactions that would otherwise be impossible, or to elicit new selectivities. Several reviews have been devoted specially to such a use [2–8], which nevertheless does not exclude the possibility of further catalyzing the reactions with Lewis acids [9, 10].

Among the main advantages of using water as the solvent are the following:

- as the most abundant liquid that occurs on Earth, water is very cheap and more importantly it is not toxic, so it can be used in large amounts without any associated hazard;
- in water-promoted reactions, mild conditions can be sufficient and yields and selectivities can therefore be largely improved;
- water-soluble compounds such as carbohydrates can be used directly without the need for the tedious protection–deprotection process;
- water-soluble catalysts can be re-used after filtration, decantation or extraction of the water-insoluble products [1 b–f].

A remarkable feature of water-promoted reactions is that the reactants only need to be sparingly soluble in water, and most of the time, the effects of water occur under *biphasic* conditions. If the reactants are not soluble enough, co-solvents can be used as well as surfactants. Another possibility for inducing water-solubility lies in grafting a hydrophobic moiety (a sugar residue or carboxylate, for instance) onto the hydrophobic reactant.

This contribution encompasses the main concepts supporting the origin of the reactivity in water, along with some applications in organic synthesis with the exception of transition-metal-catalyzed reactions, which are fully described in Section 2.2.

2.1.2 Origin of the Reactivity in Water

The combination of a small size and a three-dimensional hydrogen-bonded network system is responsible for the complexity of the structure of water, which results in a large cohesive energy density (550 cal/mL or 22000 atm), a high surface tension and a high heat capacity. These three attributes give water its unique structure as a liquid, and give rise to the special properties known as hydrophobic effects, which play a critical role in the folding of biological macromolecules, in the formation and stabilization of membranes and micelles, or in the molecular recognition processes such as antibody–antigen, enzyme–substrate, and receptor–hormone binding. Since Breslow's discovery that the Diels–Alder reaction, which is known as insensitive to solvent effects, can be dramatically accelerated in aqueous solution, special attention was focused on the origin of the aqueous acceleration. Breslow suggested that hydrophobic packing of the reactants is likely to be responsible for the rate enhancement of Diels–Alder reactions [1 a, 2]. Of interest was the observation of good correlations between solubilities of the reactants and Diels–Alder rate constants [11]. The influence of hydrophobic effects on solubilities, reaction rates, and selectivities could be interpreted by the use of prohydrophobic and antihydrophobic additives [12]. In 1986, Lubineau assumed that a fundamental issue is the high cohesive energy of water; he postulated then that a kinetically controlled reaction between two non-polar molecules for which $\Delta V^{\neq}$ is negative must be accelerated in water [13]. The importance of the cohesive energy density to Diels–Alder reactions and Claisen rearrangements, both displaying a negative activation volume, was demonstrated by Gajewski [14, 15].

By measuring standard Gibbs energy of transfer from organic to aqueous solvents, Engberts and co-workers showed that enforced hydrophobic interaction due to a decrease in the overall hydrophobic surface area during the activation process plays an important role in the rate acceleration in water [16]. This effect was considered to be a consequence of the high cohesive energy density of water [6] and should be expressed in terms of pressure (cohesive pressure), but one must avoid any confusion with the internal pressure of water, which is small compared with other solvents [17].

Employing a self-consistent reaction field model, Cramer and Truhlar concluded that the hydrophobic effect is always accelerating in aqueous Claisen rearrangements, even if most of the activation stems from polarization contributions to the activation energy [18].

The importance of hydrophobic effects was recently emphasized in Monte-Carlo simulated Diels–Alder reactions, especially when both reactants are non-polar; when one of the reactant is a hydrogen-bond acceptor, enhanced hydrogen-bonding interaction and hydrophobic effect contribute equally to the rate enhancement [19]. With methyl vinyl ketone as dienophile model in Diels–Alder reactions, computed partial charges displayed greater polarization of the carbonyl bond in the transition state and consequently enhanced hydrogen bonding to the transition state; on the basis of Monte-Carlo simulations [20] and molecular orbital calculations [21], hydrogen bonding was proposed as the key factor controlling the variation of the acceleration for Diels–Alder reactions in water. Monte-Carlo simulations showed enhanced hydrogen bonding to the oxygen in the transition-state envelope of water molecules for Claisen rearrangements as well [22].

Such an enhanced hydrogen-bonding effect was invoked to explain the experimental differences of reactivity between dienophiles in some Diels–Alder reactions [23, 24] and to understand the acceleration in water of the retro Diels–Alder reaction, a reaction with a slightly negative activation volume [25].

In summary, the acceleration in water of reactions between neutral molecules arises from:

- an enforced hydrophobic effect, especially when apolar reactants are involved;
- a charge development in transition states, especially when one of the reactant is a hydrogen donor or acceptor.

In both cases a negative volume of activation is expected. Both contributions (hydrophobic effects and polarity) could be active in the same reaction (Figure 1), which means a greater destabilization of the hydrophobic reactants in the initial state than in the transition state, and a greater stabilization of a more polar transition state.

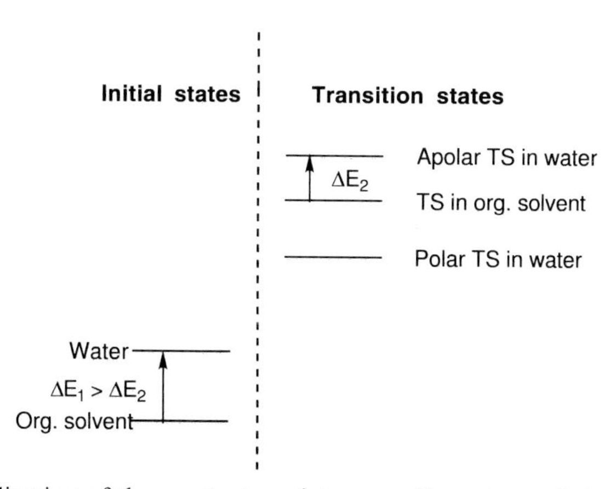

Figure 1. Destabilization of the reactants and, to a smaller extent, of the transition states in water versus organic solvents.

2.1.3 Pericyclic Reactions

2.1.3.1 Diels–Alder Reactions

In his pioneering work [1 a], Breslow studied the kinetics of the cycloaddition between cyclopentadiene and methyl vinyl ketone (Eq. (1) and Table 1). The implication of the hydrophobic effect in Diels–Alder reactions was extensively supported by the effect of cyclodextrins [26] and additives, such as lithium chloride (salting-out agent) or guanidinium chloride (salting-in agent), which respectively increases or decreases the rate of the reaction [27].

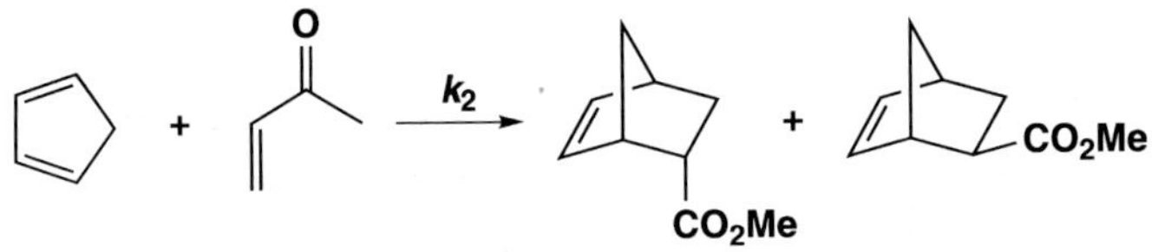

(1)

Table 1. Effect of different solvents according to Eq. (1).

Solvent	$10^5\ k_2$ [$M^{-1}\ s^{-1}$]	Solvent	*endo/exo*
Isooctane	5.94	Cyclopentadiene	3.85
Methanol	75.5	Ethanol	8.5
Water	4400	Water	22.5
4.86 M LiCl in H_2O	10800		
4.86 M $(NH_2)_3CCl$ in H_2O	4300		
β-Cyclodextrin [10 mM]	10900		
α-Cyclodextrin [10 mM]	2610		

By measuring the rate constant of the cycloaddition between cyclopentadiene and dimethylfumarate, Schneider [28] demonstrated the importance of solvophobic effects, quantified by solvophobic power (S_p) values [29], which originate from the standard free energy of transfer of alkanes from the gas phase to a given solvent. Such a sensitivity of Diels–Alder reactions to solvent hydrophobicity depends, however, on the nature of the reagents [30] and is more pronounced for the reactions with a more negative activation volume [31]. Evidence for presence of enforced hydrophobic effects was provided from the study of rate constants and activation parameters of Diels–Alder reactions in aqueous solutions [16].

By measuring activation parameters (Eq. (2) and Table 2), it has been shown that the acceleration arises from a favorable change of activation entropy, which is an indication of the implication of the hydrophobic effect [32]. Concentrated

$$\text{cyclopentadiene} + \text{methyl acrylate} \xrightarrow{k_2} \text{endo-}CO_2Me + \text{exo-}CO_2Me \qquad (2)$$

Table 2. Influence of solvents on Eq. (2).

	$10^5\ k_2$	ΔH_{act}	ΔS_{act}
Solvent	$[M^{-1}\ s^{-1}]$	$[kJ\ mol^{-1}]$	$[J\ mol^{-1}\ K^{-1}]$
Methanol	75	38.0 ± 1.0	-173.9 ± 3.4
Water	4400	38.0 ± 1.7	-140.9 ± 5.0

aqueous carbohydrate solutions (glucose and saccharose for instance) have been shown to accelerate the Diels–Alder reaction. The acceleration is even greater than that observed with saturated β-cyclodextrin solution (Eq. (3) and Table 3) [32].

$$\text{glycoside diene} + \text{methyl vinyl ketone} \longrightarrow \textbf{cycloadducts} \qquad (3)$$

Table 3. Acceleration of Eq. (3) by various carbohydrate solutions.

Additive	$10^5\ k_2$ $[M^{-1}\ s^{-1}]$	ΔH_{act} $[kJ\ mol^{-1}]$	ΔS_{act} $[J\ mol^{-1}\ K^{-1}]$
–	28.5	40.0 ± 0.6	-178.8 ± 2.1
Methanol	8.5	33.6 ± 0.8	-211.1 ± 2.6
LiCl (2.6 M)	57.8	39.3 ± 1.7	-175.1 ± 5.4
Glucose (2.6 M)	45	39.2 ± 0.3	-177.4 ± 1.1
Ribose (2.6 M)	35	36.7 ± 1.5	-188.3 ± 4.9
Glucose (3 M)	61.3		
Saccharose (2 M)	74.9		
Satd. β-cyclodextrin	40.2		

Another aspect of the influence of water as the solvent in Diels–Alder reactions is the higher *endo* selectivity observed by comparison with organic solvents (Table 1). The Diels–Alder reaction has a negative activation volume (ca. $-30\ cm^3\ mol^{-1}$) and of the two possible transition states, the *endo* one is the more compact. In the cycloaddition between ethyl maleate and cyclopentadiene, the *endo* selectivity was directly correlated with solvophobicity power [26], but in the cycloaddition between methyl acrylate and cyclopentadiene, the endoselectivity results were accounted for by means of a two-parameter S_p/E_T model [31]. Unlike Schneider's results [26], solvophobic (S_p) and polar (E_T) contributions, including hydrogen-bonding ability of the solvent, showed similar relative importance [33]. It is worth noting that the difference in compactness between the *endo* and the *exo* transition states of the ethyl maleate–cyclopentadiene cycloaddition ($\Delta\Delta V^{\ddagger} = 0.82\ \text{cm}^3\ \text{mol}^{-1}$) is greater than that for the methyl acrylate–cyclopentadiene cycloaddition ($\Delta\Delta V^{\ddagger} = 0.62\ \text{cm}^3\ \text{mol}^{-1}$), which could explain the greater hydrophobic influence of the former reaction [31].

A recent improvement in the rate of the aqueous Diels–Alder reaction came with the use of Lewis acid in aqueous media. The first study deals with the Diels–Alder reaction between cyclopentadiene and a bidentate dienophile. A large acceleration can be achieved by the combined use of copper(II) nitrate as a catalyst and water as a solvent [10, 34]. Lanthanide and scandium triflates [9, 35] as well as indium trichloride [36] were found to catalyze the Diels–Alder reaction in water.

The aqueous Diels–Alder reaction has been widely exploited at the preparative level since the first studies of Breslow. Two reviews [4, 6] were in part devoted to this aspect. In order to increase the water-solubility, dienes have been attached at the anomeric position of a carbohydrate. The sugar moiety, which induces chirality, can easily be removed after the cycloaddition either via acidic or enzymic hydrolysis [37].

2.1.3.2 Hetero Diels–Alder Reactions

Aqueous aza Diels–Alder reactions were first described in cycloadditions of iminium salts and dienes [38]. Likewise, iminium salts derived from amino acids react in excellent yields in aqueous medium [39]. Such an aqueous aza Diels–Alder reaction was found to be catalyzed by lanthanide(III) trifluoromethanesulfonates [40].

An aqueous solution of glyoxylic acid reacts with cyclopentadiene to provide α-hydroxy-γ-lactones (Eq. (4)); the more acidic the solution, the faster the reaction [41]. Such an aqueous hetero Diels–Alder reaction, which was extended to other dienes [42], was applied in various syntheses, including sesbanimides A and B [43], carbovir [44], mevinic acids [45], aristeromycin and carbodine [46], ketodeoxyoctulosonic acid (KDO) and analogs [47], and sialic acids [48]. Pyruv-

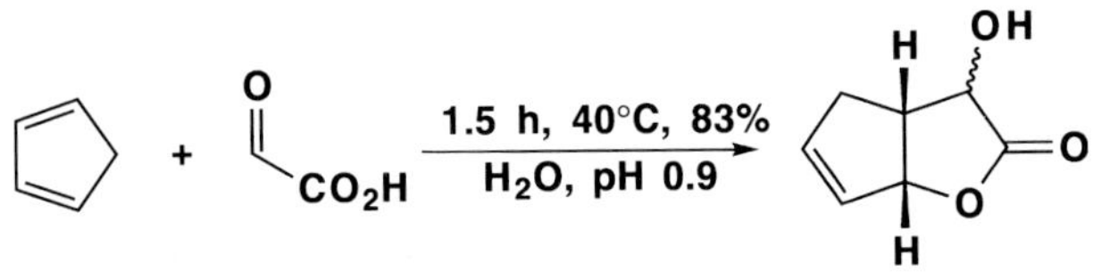

(4)

aldehyde, glyoxal, and even ketones like pyruvic acid also react with dienes in water [42].

The influence of a low pH on the rate of reaction was also observed in the hetero Diels–Alder of di(2-pyridyl)-1,2,4,5-tetrazine with substituted styrenes [49].

2.1.3.3 Other Cycloadditions

An unusual influence of water on the rate of 1,3-dipolar cycloadditions was first observed when 2,6-dichlorobenzonitrile *N*-oxide was allowed to react with 2,5-dimethyl-*p*-benzoquinone [50]. Likewise, bromonitrile oxide, generated in water at acidic pH, gave cycloadducts efficiently with water-soluble alkenes and alkynes [51]. In highly aqueous media remarkable accelerations for the cycloaddition of phenyl azide to norbornene were observed [52].

Whereas cycloaddition of azomethine ylids were usually conducted with careful exclusion of water, it was recently shown that the cycloaddition in water–tetrahydrofuran solution of stabilized ylids derived from ethyl sarcosinate with several dipolarophiles can occur in excellent yields [53].

The cycloaddition of α,α'-dibromo (or dichloro) ketones with furan (or cyclopentadiene) gave very good yields when the reaction was conducted in pure water with iron powder. Furthermore, in the presence of triethylamine as the base, monobromo (or chloro) ketones react to furan (or cyclopentadiene) in water to afford the corresponding cycloadducts in near-quantitative yields (Eq. (5)). In both cases, 2-oxyallyl cation, the formation of which is favored in water, was considered as the reactive intermediate [54].

promoter, H_2O (5)

X= Cl, Br ; Y= H, Br ; Z= O,C

isolated yields 66–88%

2.1.3.4 Claisen Rearrangements

The Claisen rearrangement, which displays a negative activation volume, is also accelerated in water. Thus, the non-enzymatic rearrangement of chorismate to prephenate occurs 100 times faster in water than in methanol [55]. The accelerating influence of water as a solvent on the rate of the Claisen rearrangement has been largely demonstrated on various substrates in organic synthesis [56]. Although otherwise impossible, the Claisen rearrangement of fenestrene was performed with success in mixed aqueous media [56].

The aqueous rate-enhanced Claisen rearrangement of glycoorganic compounds was shown to proceed in excellent yields [57]. The water solubility of the reactants was induced by grafting a free sugar onto the allyl vinyl ether moiety; moreover, the sugar functioned as a chiral template and gave highly crystalline diastereomers which were easily separable to yield pure enantiomers after enzymic hydrolysis. This method allows the preparation of either enantiomerically pure (*R*)- or (*S*)-1,3-diol (Scheme 1).

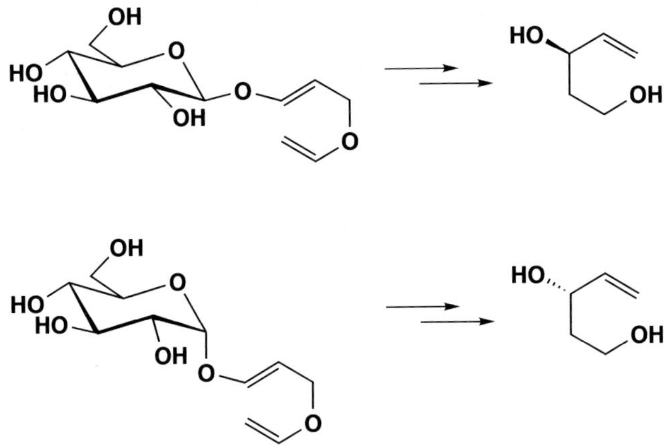

Scheme 1

2.1.4 Carbonyl Additions

2.1.4.1 Aldol-type Reactions

The Mukaiyama reaction is an aldol-type reaction between a silyl enol ether and an aldehyde in the presence of a stoichiometric amount of titanium chloride. The reaction, which displays a negative volume of activation, could be performed without acidic promoter under high pressure [58]. In this case, the major product is the *syn* hydroxy ketone, not as for the $TiCl_4$-promoted reactions which lead mostly to the *anti* addition product. Since the *syn* or *anti* selectivity is the result of two transition states with different activation volumes ($\Delta V^{\ddagger}_{syn} < \Delta^{\ddagger}_{anti}$), it was of great interest to investigate the aldol reaction in water. Indeed, the reaction of the silyl enol ether of cyclohexanone with benzaldehyde in aqueous medium was shown to proceed without any catalyst and under atmospheric pressure, with the same *syn* selectivity as under high pressure (Eq. (6) and Table 4). This is an indication of the implication of hydrophobic effects during the activation process [13]. Taking account of the competitive hydrolysis of the silyl enol ether, this aqueous reaction is remarkable. The method was extended to other carbonyl compounds, such as formaldehyde, substituted benzaldehydes and α,β-unsaturated aldehydes and ketones [59]. To improve the yields and therefore the scope of this aqueous aldolization, the use of lanthanide triflates as water-tolerant Lewis acids was recommended [9, 60]. After completion of the reaction nearly 100% of the catalyst is recovered from the aqueous layer and can be re-used quite easily. Other water-tolerant Lewis acids, including indium chloride [61] and tris(pentafluorophenyl) boron [62], were proposed as catalysts in the aqueous aldol reaction.

OSiMe3 —PhCHO→ (O, H, OH, Ph) *syn* + (O, H, OH, Ph) *anti* (6)

Table 4. Influence of solvents on Eq. (6).

Solvent	Temp. [°C]	Time	Conditions	Yield [%]	*syn*:*anti*
CH_2Cl_2	20	2 h	1 eq. $TiCl_4$	82	25:75
CH_2Cl_2	60	9 d	10000 atm.	90	75:25
H_2O	20	5 d	stirring	23	85:15
H_2O/THF (1:1)	20	5 d	stirring	45	74:26
H_2O/THF (1:1)	55	2 d	ultrasound	76	74:26

The Henry reaction is an aldol-type reaction between a nitroalkane and an aldehyde in the presence of a base. Since basic reagents are also catalysts for the aldol condensation, the nitroaldol reactions must be strictly controlled. An interesting alternative lies in the use of surfactants to perform the reaction in an aqueous medium [63]. The Reformatsky reaction, which involves α-haloketones and aldehydes, can be mediated by zinc, tin or indium in water; in the latter case the proportion of undesirable reduction products could be strongly reduced [64].

Considerable rate enhancements have been observed when water is used as solvent compared with alcoholic or hydrocarbon media for Mannich reactions, i.e. condensations of ketones with secondary amines in the presence of formaldehyde [65]. Allylsilanes [66] and allylstannanes [67] in aqueous media were used in organic synthesis under Mannich-like conditions. More recently, Kobayashi reported the catalysis of the reaction of vinyl ethers with iminium salts by ytterbium triflate in tetrahydrofuran–water mixtures [68].

2.1.4.2 Michael-type Reactions

The use of water as a solvent in the conjugate addition of 1,3-diketones was reported earlier [69] and applied more recently in organic synthesis [70]. Ytterbium triflate turned out to be an efficient catalyst for the Michael addition of various β-ketoesters to β-unsubstituted enones [71].

A huge acceleration of the Michael reaction of nitroalkanes with methyl vinyl ketone was mentioned when going from non-polar organic solvents to water. The hydrophobic effect could be at least to some extent involved, since additives, such as glucose or saccharose, accelerate the reaction even more [72]. Cetyltrimethylammonium chloride as an amphiphilic species which can influence the hydrophobic interactions was found to promote the Michael reaction of various nitroalkanes with conjugated enones in dilute aqueous solutions of sodium hydroxide [73].

There is a remarkable effect of water as the solvent on the rate of the conjugate addition of amines to α,β-unsaturated nitriles. The lack of apparent reactivity of α,β-unsaturated esters comes from the reverse reaction, which is particularly accelerated in water [74].

A related reaction known as Baylis–Hillman reaction, which also has a large negative activation volume, was found to be greatly accelerated in water, compared with usual organic solvents. The first step is the conjugate addition of a tertiary amine (1,4-diazabicyclo[2.2.2]octane; DABCO) to acrylonitrile (Scheme 2) which is fast in water. The second step, which is rate-determining, is accelerated via a process wherein the hydrophobic effect could be involved. Other structured solvents also enhance the Baylis–Hillman reaction, but to a small extent [75].

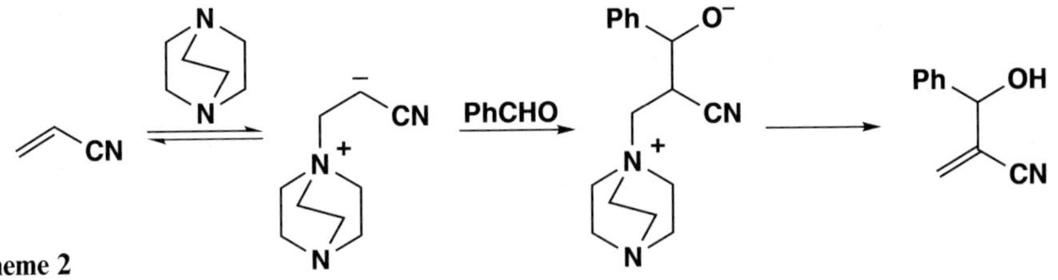

Scheme 2

2.1.4.3 Allylation Reactions

The allylation of aldehydes via organotin reagents displays a negative activation volume [76]. As a matter of fact, the allylation of benzaldehyde with diallyltin dibromide is accelerated by addition of water [77]. The reaction was extended to various aldehydes and ketones and to various allylic organotin dichlorides

[78] or tetraallyltin in acidic aqueous medium [79]. With scandium triflate as a catalyst, tetraallyltin [80] or tetraallylgermane [81] react smoothly in mixed aqueous solvents providing high yields of the corresponding homoallylic alcohols.

Since the observation that allylation of carbonyl compounds could be mediated by tin in aqueous medium [77], there has been an intensive development of the Barbier-type allylation reaction in water. Three metals were particularly investigated: zinc, tin, and indium. In the aqueous zinc-promoted allylation, allylzinc species are considered unlikely. The initiation of the reaction could be attributed to the formation of an allylic radical anion on the metal surface; this radical surface could then react with the carbonyl compound to give an alkoxide radical, which could add an electron and form the alcohols [82]. Allyl bromide or even chloride reacts with aldehydes and ketones in the presence of commercial zinc powder in a mixture of tetrahydrofuran and saturated ammonium chloride aqueous solution (Eq. (7)) [83].

$$\text{PhCHO} + \text{CH}_2{=}\text{CHCH}_2\text{Br} \xrightarrow[\text{r.t., 45 min., 95\%}]{\text{Zn, NH}_4\text{Cl aq. / THF}} \text{PhCH(OH)CH}_2\text{CH}{=}\text{CH}_2 \quad (7)$$

By contrast, tin- and indium-promoted allylation reactions might involve organometallic intermediates, since preformed organotin and organoindium are highly reactive toward aldehydes in aqueous medium [84]. The tin-mediated allylation reaction requires either acidic conditions, metallic aluminum as an additive [77], or ultrasonic waves [85]. When applied to carbohydrates, the sonoallylation proceeded with useful diastereoselectivity (*threo* selectivity) and made it possible to prepare higher-carbon sugars from water-soluble substrates directly in aqueous ethanol without protection [86].

A major improvement was realized with the use of indium, a metal with a very low first ionization potential (5.8 eV) which works without ultrasonic radiation even at room temperature [87]. As the zero-valent indium species is regenerated by either zinc, aluminum, or tin, a catalytic amount of indium trichloride together with zinc, aluminum [88], or tin [89] could be utilized in the allylation of carbonyl compounds in aqueous medium. The regeneration of indium after its use in an allylation process could be readily carried out by electrodeposition of the metal on an aluminum cathode [90]. Compared with tin-mediated allylation in ethanol–water mixtures, the indium procedure is superior in terms of reactivity and selectivity. Indium-mediated allylation of pentoses and hexoses, which were however facilitated in dilute hydrochloric acid, produced fewer by-products and were more diastereoselective. The reactivity and the diastereoselectivity are compatible with a chelation-controlled reaction [84, 91]. Indeed, the methodology was used to prepare 3-deoxy-D-glycero-D-galacto-nonulosonic acid (KDN) [92, 93], *N*-acetylneuraminic acid [93, 94], and analogs [95].

2.1.5 Oxido-reductions

2.1.5.1 Oxidations

Oxidation reactions have often been conducted in water using water-soluble oxidants such as potassium permanganate, sodium periodate, and sodium (or calcium) hyperchlorite. Moreover, many oxidations can be performed in aqueous conditions using peroxy acids such as *meta*-chloroperbenzoic acid (MCPBA) or *meta*-peroxyphthalic acid (MPPA), or with hydroperoxides in the presence of transition metals. Thus, peroxybenzoic acid and MCPBA quickly epoxidize olefins in good yields in aqueous hydrogen carbonate [96]. Although the epoxidation of polyolefinic alcohols in organic media is usually not regiospecific, good regio- and stereoselectivities are observed when using hydroperoxides in water in the presence of transition metals [97]. Likewise, polyolefinic alcohols are epoxized regioselectively by monoperoxyphthalic acid when controlling the pH of the medium [98].

2.1.5.2 Reductions

Apart from sodium borohydride, which is frequently used in water or water–alcohol mixtures to reduce ketones or aldehydes selectively, water is rarely used as the solvent in reductions, because of incompatibility with most reducing agents. However, samarium iodide reduction of ketones, as well as alkyl and aryl iodides is accelerated in water [99]. Likewise, the α-deoxygenation of unprotected aldonolactones is efficient when the SmI_2–tetrahydrofuran–water system is used [100].

A water-soluble tin hydride was used to reduce alkyl bromides in a phosphate buffer in the presence of a radical initiator or light (Eq. (8)) [101]. The more available tributyltin hydride, albeit insoluble in water, was successfully used as the reducing agent with or without a detergent as a solubilizing agent [102].

(8)

2.1.6 Outlook

The origin of specific reactivity in water lies both in the high cohesive energy density of water and in its ability to form hydrogen bonds. Radical or ionic reactions with a negative volume of activation can be facilitated in aqueous solutions. Water can be used with co-solvents in two-phase or one-phase systems, or with additives which enhance the hydrophobic effects (concentrated solutions of structure-making salts, sugars, etc.), the polarization of transition states (water-tolerant Lewis acids in catalytic amounts), or the solubility of reactants (surfactants). Smooth conditions are then possible, even for the more energy-demanding reactions. At high temperatures and pressures, dramatic changes occur in the physical properties of water, so that it can act as an acid–base bicatalyst, which could have ecological applications in recycling, regeneration, disposal and detoxification of chemicals [103].

The understanding of aqueous chemistry should favor the discovery of new selective transformations with benign environmental impacts.

References

[1] (a) D. C. Rideout, R. Breslow, *J. Am. Chem. Soc.* **1980**, *102*, 7816; (b) E. G. Kuntz, *CHEMTECH* **1987**, *17(a)*, 570; (c) B. Cornils, J. Falbe, *Proc 4th Int. Symp. on Homogeneous Catalysis,* Leningrad, Sept. **1984**, p. 487; (d) H. Bach, W. Gick, E. Wiebus, B. Cornils, *1st IUPAC Symp. Org. Chemistry*, Jerusalem, **1986**, Abstracts, p. 295; (e) E. Wiebus, B. Cornils, *Chem.-Ing.-Techn.* **1994**, *66*, 916; (f) B. Cornils, E. Wiebus, *CHEMTECH* **1995**, *25*, 33.
[2] R. Breslow, *Acc. Chem. Res.* **1991**, *24*, 159.
[3] P. A. Grieco, *Aldrichim. Acta* **1991**, *24*, 59.
[4] C.-J. Li, *Chem. Rev.* **1993**, *93*, 2023.
[5] W. A. Herrmann, C. W. Kohlpaintner, *Angew. Chem., Int. Ed. Engl.* **1993**, *32*, 1524.
[6] A. Lubineau, J. Augé, Y. Queneau, *Synthesis* **1994**, 741.
[7] C.-J. Li, *Tetrahedron* **1996**, *52*, 5643.
[8] M. Beller, B. Cornils, C. D. Frohning, C. W. Kohlpaintner, *J. Mol. Catal A* **1995**, *104*, 17.
[9] S. Kobayashi, *Snylett* **1994**, 689.
[10] J. B. F. N. Engberts, B. L. Feringa, E. Keller, S. Otto, *Recl. Trav. Chim. Pays-Bas* **1996**, *115*, 457.
[11] R. Breslow, Z. Zhu, *J. Am. Chem. Soc.* **1995**, *117*, 9923.
[12] R. Breslow, *Structure and Reactivity in Aqueous Solution* (Eds.: C. J. Cramer, D. G. Truhlar), ACS Symposium Series No. 568, **1994**, Chapter 20.
[13] A. Lubineau, *J. Org. Chem.* **1986**, *51*, 2142.
[14] J. J. Gajewski, *J. Org. Chem.* **1992**, *57*, 5500.
[15] J. J. Gajewski, N. L. Brichford, *Structure and Reactivity in Aqueous Solution* (Eds.: C. J. Cramer, D. G. Truhlar), ACS Symposium Series No. 568, **1994**, Chapter 16.
[16] W. Blokzijl, M. J. Blandamer, J. B. F. N. Engberts, *J. Am. Chem. Soc.* **1991**, *113*, 4241.
[17] M. R. J. Dack, *Chem. Soc. Rev.* **1975**, *4*, 211.

[18] C. J. Cramer, D. G. Truhlar, *J. Am. Chem. Soc.* **1992**, *114*, 8794.
[19] T. R. Furlani, J. Gao, *J. Org. Chem.* **1996**, *61*, 5492.
[20] J. F. Blake, W. L. Jorgensen, *J. Am. Chem. Soc.* **1991**, *113*, 7430.
[21] J. F. Blake, D. Lim, W. L. Jorgensen, *J. Org. Chem.* **1994**, *59*, 803.
[22] D. L. Severance, W. L. Jorgensen, *J. Am. Chem. Soc.* **1992**, *114*, 10966.
[23] S. Otto, W. Blokzijl, J. B. F. N. Engberts, *J. Org. Chem.* **1994**, *59*, 5372; G. K. van der Wel, J. W. Wijnen, J. B. F. N. Engberts, *J. Org. Chem.* **1996**, *61*, 9001.
[24] J. B. F. N. Engberts, *Pure Appl. Chem.* **1995**, *67*, 823.
[25] J. W. Wijnen, J. B. F. N. Engberts, *J. Org. Chem.* **1997**, *62*, 2039.
[26] H.-J. Schneider, N. K. Sangwan, *Angew. Chem., Int. Ed. Engl.* **1987**, *26*, 896.
[27] R. Breslow, C. J. Rizzo, *J. Am. Chem. Soc.* **1991**, *113*, 4340.
[28] H.-J. Schneider, N. K. Sangwan, *J. Chem. Soc., Chem. Commun.* **1986**, 1787.
[29] M. H. Abraham, *J. Am. Chem. Soc.* **1982**, *104*, 2085; M. H. Abraham, P. L. Grellier, R. A. McGill, *J. Chem. Soc., Perkin Trans 2* **1988**, 339.
[30] N. K. Sangwan, H.-J. Schneider, *J. Chem. Soc., Perkin Trans. 2* **1989**, 1223.
[31] C. Cativiela, J. I. Garcia, J. A. Mayoral, L. Salvatella, *Chem. Soc. Rev.* **1996**, 209 and references therein.
[32] A. Lubineau, H. Bienhaymé, Y. Queneau, M.-C. Scherrmann, *New J. Chem.* **1994**, *18*, 279.
[33] C. Cativiela, J. I. Garcia, J. A. Mayoral, L. Salvatella, *J. Chem. Soc., Perkin Trans. 2* **1994**, 847.
[34] S. Otto, F. Bertoncin, J. B. F. N. Engberts, *J. Am. Chem. Soc.* **1996**, *118*, 7702.
[35] S. Kobayashi, I. Hachiya, M. Araki, H. Ishitani, *Tetrahedron Lett.* **1993**, *34*, 3755.
[36] T.-P. Loh, J. Pei, M. Lin, *J. Chem. Soc., Chem. Commun.* **1996**, 2315.
[37] A. Lubineau, Y. Queneau, *J. Org. Chem.* **1987**, *52*, 1001; A. Lubineau, Y. Queneau, *Tetrahedron* **1989**, *45*, 6697.
[38] S. D. Larsen, P. A. Grieco, *J. Am. Chem. Soc.* **1985**, *107*, 1768; P. A. Grieco, S. D. Larsen, *J. Org. Chem.* **1986**, *51*, 3553.
[39] H. Waldmann, *Angew. Chem., Int. Ed. Engl.* **1988**, *27*, 274; H. Waldmann, M. Braun, *Liebigs Ann. Chem.* **1991**, 1045.
[40] L. Yu, D. Chen, P. G. Wang, *Tetrahedron Lett.* **1996**, *37*, 2169.
[41] A. Lubineau, J. Augé, N. Lubin, *Tetrahedron Lett.* **1991**, *32*, 7529.
[42] A. Lubineau, J. Augé, E. Grand, N. Lubin, *Tetrahedron* **1994**, *50*, 10265.
[43] P. A. Grieco, K. J. Henry, J. J. Nunes, J. E. Matt jr., *J. Chem. Soc., Chem. Commun.* **1992**, 368.
[44] R. A. MacKeith, R. McCague, H. F. Olivo, C. F. Palmer, S. M. Roberts, *J. Chem. Soc., Perkin Trans. 1* **1993**, 313.
[45] R. McCague, H. F. Olivo, S. M. Roberts, *Tetrahedron Lett.* **1993**, 3785.
[46] F. Burlina, A. Favre, J.-L. Fourrey, M. Thomas, *Bio Med. Chem. Lett.* **1997**, *7*, 247.
[47] A. Lubineau, J. Augé, N. Lubin, *Tetrahedron* **1993**, *49*, 4639.
[48] A. Lubineau, Y. Queneau, *J. Carbohydr. Chem.* **1995**, *14*, 1295.
[49] J. W. Wijnen, S. Zavarise, J. B. F. N. Engberts, *J. Org. Chem.* **1996**, *61*, 2001.
[50] Y. Inoue, K. Araki, S. Shiraishi, *Bull. Chem. Soc. Jpn.* **1991**, *64*, 3079.
[51] J. C. Rohloff, J. Robinson III, J. O. Gardner, *Tetrahedron Lett.* **1992**, *33*, 3113.
[52] J. W. Wijnen, R. A. Steiner, J. B. F. N. Engberts, *Tetrahedron Lett.* **1995**, *36*, 5389.
[53] A. Lubineau, G. Bouchain, Y. Queneau, *J. Chem. Soc., Perkin Trans. 1* **1995**, 2433.
[54] A. Lubineau, G. Bouchain, *Tetrahedron Lett.* **1997**, *38*, 8031.
[55] S. D. Copley, J. R. Knowles, *J. Am. Chem. Soc.* **1987**, *109*, 5008.
[56] E. Brandes, P. A. Grieco, J. J. Gajewski, *J. Org. Chem.* **1989**, *54*, 515; P. A. Grieco, E. B. Brandes, S. McCann, J. D. Clark, *J. Org. Chem.* **1989**, *54*, 5849.

[57] A. Lubineau, J. Augé, N. Bellanger, S. Caillebourdin, *Tetrahedron Lett.* **1990**, *31*, 4147; A. Lubineau, J. Augé, N. Bellanger, S. Caillebourdin, *J. Chem. Soc., Perkin Trans. 1* **1992**, 1631.
[58] Y. Yamamoto, K. Maruyama, K. Matsumoto, *J. Am. Chem. Soc.* **1983**, *105*, 6963.
[59] A. Lubineau, E. Meyer, *Tetrahedron* **1988**, *44*, 6065.
[60] S. Kobayashi, *Chem. Lett.* **1991**, 2187; S. Kobayashi, I. Hachiya, *Tetrahedron Lett.* **1992**, *33*, 1625; S. Kobayashi, I. Hachiya, *J. Org. Chem.* **1994**, *59*, 3590.
[61] T.-P. Loh, J. Pei, G.-Q. Cao, *J. Chem. Soc., Chem. Commun.* **1996**, 1819.
[62] K. Ishihara, N. Hananki, H. Yamamoto, *Synlett* **1993**, 577.
[63] R. Ballini, G. Bosica, *J. Org. Chem.* **1997**, *62*, 425.
[64] T. H. Chan, C.-J. Li, M. C. Lee, Z. Y. Wei, *Can. J. Chem.* **1994**, *72*, 1181.
[65] V. Tychopoulos, J. H. P. Tyman, *Snyth. Commun.* **1986**, *16*, 1401.
[66] S. D. Larsen, P. A. Grieco, W. F. Fobare, *J. Am. Chem. Soc.* **1986**, *108*, 3512.
[67] P. A. Grieco, A. Bahsas, *J. Org. Chem.* **1987**, *52*, 1378.
[68] S. Kobayashi, H. Ishitani, *J. Chem. Soc., Chem. Commun.* **1995**, 1379.
[69] U. Eder, G. Sauer, R. Wiechert, *Angew. Chem., Int. Ed. Engl.* **1971**, *10*, 496; Z. G. Hajos, D. R. Parrish, *J. Org. Chem.* **1974**, *39*, 1612.
[70] J.-F. Lavallée, P. Deslongchamps, *Tetrahedron Lett.* **1988**, *29*, 6033.
[71] E. Keller, B. L. Feringa, *Tetrahedron Lett.* **1996**, *37*, 1879.
[72] A. Lubineau, J. Augé, *Tetrahedron Lett.* **1992**, *33*, 8073.
[73] R. Ballini, G. Bosica, *Tetrahedron Lett.* **1996**, *37*, 8027.
[74] G. Jenner, *Tetrahedron* **1996**, *52*, 13557.
[75] J. Augé, N. Lubin, A. Lubineau, *Tetrahedron Lett.* **1994**, *35*, 7947.
[76] Y. Yamamoto, K. Maruyama, K. Matsumoto, *J. Chem. Soc., Chem. Commun.* **1983**, 489.
[77] J. Nokami, J. Otera, T. Sudo, R. Okawara, *Organometallics* **1983**, *2*, 191.
[78] A. Boaretto, D. Marton, G. Tagliavini, A. Gambaro, *J. Organomet. Chem.* **1985**, *286*, 9; D. Furlani, D. Marton, G. Tagliavini, M. Zordan, *J. Organomet. Chem.* **1988**, *341*, 345.
[79] A. Yanagisawa, H. Inoue, M. Morodome, H. Yamamoto, *J. Am. Chem. Soc.* **1993**, *115*, 10356.
[80] I. Hachiya, S. Kobayashi, *J. Org. Chem.* **1993**, *58*, 6958.
[81] T. Akiyama, J. Iwai, *Tetrahedron Lett.* **1997**, *38*, 853.
[82] C. Einhorn, J.-L. Luche, *J. Organomet. Chem.* **1987**, *322*, 177.
[83] C. Pétrier, J.-L. Luche, *J. Org. Chem.* **1985**, *50*, 910.
[84] E. Kim, D. M. Gordon, W. Schmid, G. M. Whitesides, *J. Org. Chem.* **1993**, *58*, 5500.
[85] C. Pétrier, J. Einhorn, J.-L. Luche, *Tetrahedron Lett.* **1985**, *26*, 1449.
[86] W. Schmid, G. M. Whitesides, *J. Am. Chem. Soc.* **1991**, *113*, 6674.
[87] C. J. Li, T. H. Chan, *Tetrahedron Lett.* **1991**, *32*, 7017.
[88] S. Araki, S.-J. Jin, Y. Idou, Y. Butsugan, *Bull. Chem. Soc. Jpn.* **1992**, *65*, 1736.
[89] T.-P. Loh, X. R. Li, *J. Chem. Soc., Chem. Commun.* **1996**, 1929.
[90] R. H. Prenner, W. H. Binder, W. Schmid, *Liebigs Ann. Chem.* **1994**, 73.
[91] L. A. Paquette, T. M. Mitzel, *Tetrahedron Lett.* **1995**, *36*, 6863; L. A. Paquette, T. M. Mitzel, *J. Am. Chem. Soc.* **1996**, *118*, 1931.
[92] T. H. Chan, C.-J. Li, *J. Chem. Soc., Chem. Commun.* **1992**, 747.
[93] T.-H. Chan, M.-C. Lee, *J. Org. Chem.* **1995**, *60*, 4228.
[94] D. M. Gordon, G. M. Whitesides, *J. Org. Chem.* **1993**, *58*, 7937.
[95] S.-K. Choi, S. Lee, G. M. Whitesides, *J. Org. Chem.* **1996**, *61*, 8739.
[96] F. Fringuelli, R. Germani, F. Pizzo, G. Savelli, *Tetrahedron Lett.* **1989**, *30*, 1427.

[97] Y. Gao, R. M. Hanson, J. M. Klunder, S. Y. Ko, H. Masamune, K. B. Sharpless, *J. Am. Chem. Soc.* **1987**, *109*, 5765.
[98] F. Fringuelli, R. Germani, F. Pizzo, F. Santinelli, G. Savelli, *J. Org. Chem.* **1992**, *57*, 1198.
[99] E. Hasegawa, D. P. Curran, *J. Org. Chem.* **1993**, *58*, 5008.
[100] S. Hanessian, C. Girard, *Synlett* **1994**, 861.
[101] J. Light, R. Breslow, *Tetrahedron Lett.* **1990**, *31*, 2957.
[102] U. Maitra, K. D. Sarma, *Tetrahedron Lett.* **1994**, *35*, 7861.
[103] B. Kuhlmann, E. M. Arnett, M. Siskin, *J. Org. Chem.* **1994**, *59*, 3098; A. R. Katritzky, S. M. Allin, M. Siskin, *Acc. Chem. Res.* **1996**, *29*, 399.

2.2 Organometallic Chemistry of Water

Wolfgang A. Herrmann

2.2.1 Introduction

Water plays a fundamental role in coordination chemistry. Not only do many, if not all, metals bind water molecules to fill up their coordination sphere; it has been through the kinetics of water-exchange processes at metal ions that the basics of the theory of coordination chemistry have been conveyed [1]. By way of contrast, relatively little is known of aqueous organometallic chemistry, because of the notorious lability of organometallics toward water. This section focuses on the key features of water in both aspects, i.e., coordination and organometallic chemistry. Short review articles have recently appeared [1e, f].

2.2.2 Water as a Solvent and Ligand

Water as a *solvent* offers opportunities as compared with organic solvents. It favors ionic reactions because of its high dielectric constant and the ability to solvate cations as well as anions. Beyond that, water is the ideal solvent for radical reactions since the strong O–H bonds (enthalpy 436 kJ/mol) are not easily attacked [1g].

Classified as a *ligand* for metal ions, water has decent crystal field splitting properties, standing between oxygen-bound anions and nitrogen donors such as pyridines in the spectrochemical series:

$$I^- < Br^- < Cl^- < F^- < OH^- < CH_3CO_2^- < \text{oxalate} < \mathbf{H_2O} < \text{pyridine} \approx NH_3 < NO_2^- < CN^-, CO, CNR$$

The water molecule is a good σ-donor ligand, while π-backbonding is negligible. For this reason, higher-valent transition metals form the more stable metal complexes, but the nature of the metal by itself is important, too (Table 1).

For first-row metals (+2), an extra destabilization due to electrons in e_g orbitals accounts for a ligand labilization. This effect is commonly referred to as "crystal field activation energy" (CFAE) [1a–c].

Table 1. Rates of water exchange of hexaquo metal complexes.

Metal	Rate [sec^{-1}]	Electron configuration
Cr^{2+}	7×10^9	d^4
Cr^{3+}	3×10^{-6}	d^3
Mn^{2+}	3×10^7	d^5
Fe^{2+}	3×10^6	d^6
Fe^{3+}	3×10^3	d^5
Co^{2+}	1×10^6	d^7
Ni^{2+}	3×10^4	d^8
Cu^{2+}	8×10^9	d^9
Rh^{3+}	4×10^{-8}	d^6

From these properties it is concluded that low-valent metals do not favor water in the ligand sphere. Note that $Cr(CO)_6$ is a stable compound because of the outstanding π-backbonding of carbon monoxide, while $\{Cr(H_2O)_6\}$ does not exist – quite contrary to the common $[Cr(H_2O)_6]^{3+}$. On the other hand, trivalent chromium does not form the (hypothetical) cationic carbonylchromium complex $\{[Cr(CO)_6]^{3+}\}$. This is, in short, one major reason why so little is known about typical organometallic water complexes.

Few organometallic aquo complexes have been isolated in substance. Examples are the carbonylrhenium(I) and (π-benzene)ruthenium(II) complexes **1a** and **1b**, respectively, and their congeners **1c–e** [2–5]. They are of course soluble in water and can be used as convenient starting materials for complexes exhibiting the respective organometallic backbones, e.g., the $[Re(CO)_3]^+$ cation from **1a** which is otherwise available only with difficulty [5]. The synthesis of aquo complexes from metal carbonyls proceeds via photolysis from the anhydrous parent compounds. The air-stable (!) rhenium complex **1a** is conveniently

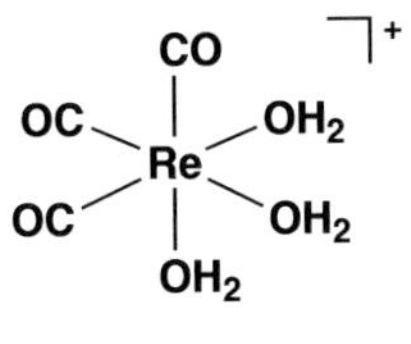

1a 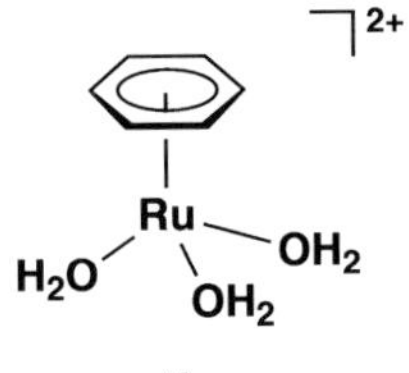 **1b**

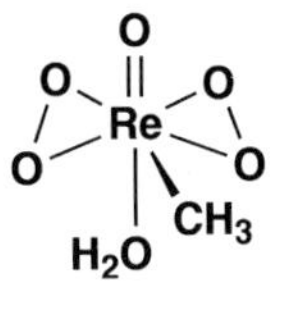

1c 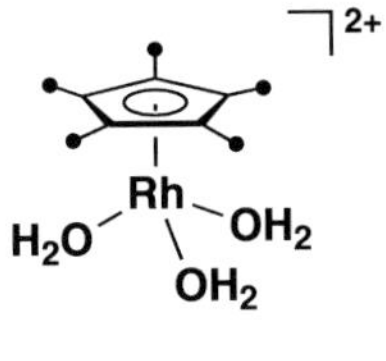 **1d** 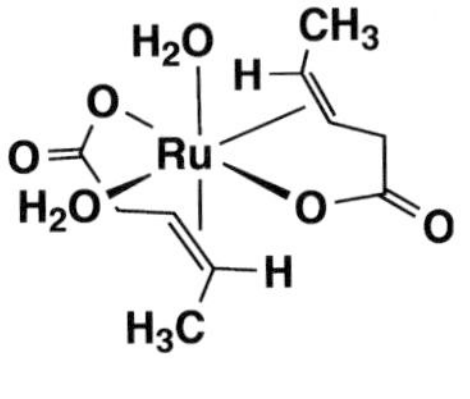 **1e**

generated from $[ReO_4]^-$ or $[ReOCl_4]^-$ and $[BH_3 \cdot \text{solvent}]$ in the presence of carbon monoxide [5a, b]. It is an outstanding precursor of products like $[L_3Re(CO)_3]^+$, with L = N- or P-donors.

2.2.3 Organometallic Reactions of Water

Metal–carbon (M–C) bonds are thermodynamically unstable with regard to their hydrolysis products. Water can attack M–C bonds either by proton transfer (H^+, electrophilic reaction) or via the oxygen (OH_2 or OH^-, nucleophilic reaction). Examples are shown in Scheme 1. Ligands such as carbon monoxide and ethylene are activated toward nucleophilic attack upon coordination to (low-valent) metals, e.g., Pd^{2+}. A number of C–C-bond forming reactions derive from this activation. Allyl ligands are generated by proton attack to the terminal 1,3-diene carbon groups (Scheme 1). In other cases, protonation of heteroatoms of metal-attached ligands is followed by elimination steps; for example, the allyl alcohol ligand $H_2C{=}C(CH_3)CH_2OH$ (η^2) is converted by $H[BF_4]$ into the allyl cation $[H_2CC(CH_3){=}CH_2]^+$, which is a standard route of making metal-allyl complexes [2].

High bond polarity yields increased reactivity with water. Thus, $Al(CH_3)_3$ and $In(CH_3)_3$ hydrolyze quickly. $Sb(CH_3)_3$ and $Sn(CH_3)_4$ are inert because of

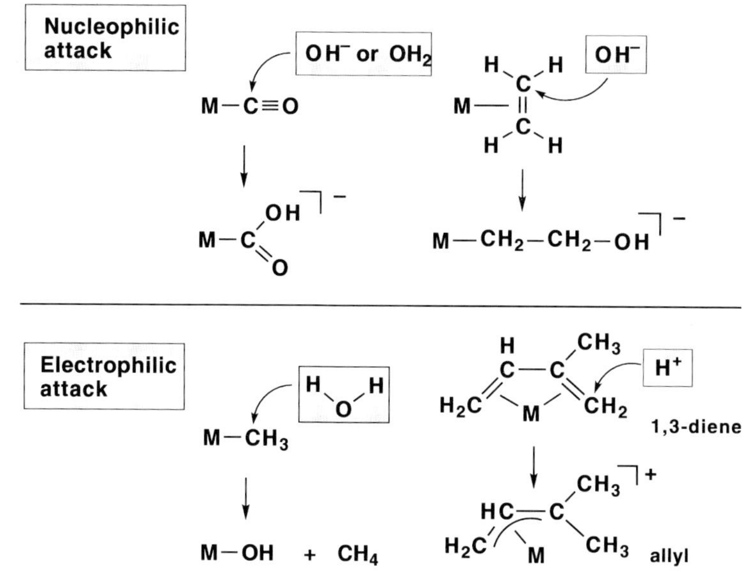

Scheme 1. Basic reactions of organometallic compounds in aqueous systems.

low bond polarity and efficient metal shielding (electron pair at Sb, coordination number 4 at Sn). $Si(CH_3)_4$ is water-stable (low bond polarity, good steric shielding), while SiH_4 hydrolyzes quickly due to inefficient shielding and nucleophilic attack, probably via 3d orbitals of the silicon.

However, metal hydrides hydrolyze only if they are ionic or coordinatively unsaturated. For example, NaH (ionic) instantaneously extrudes hydrogen upon contact with water whereas the covalent hydrides $Mn(CO)_5H$, $(\eta^5\text{-}C_5H_5)Fe(CO)_2H$, and $(\eta^5\text{-}C_5Me_5)ReH_6$ remain unchanged.

Since the M–R bond is normally polarized toward an anionic organyl group ($R^{\delta-}$), a metal hydroxide forms along with the respective hydrocarbon from metal alkyls [Eq. (1)].

$$\overset{\delta+}{M}-\overset{\delta-}{R} + H_2O \longrightarrow M-OH + R-H \tag{1}$$

$$2\ M-OH \longrightarrow M=O + H_2O-M \tag{2}$$

$$2\ M-OH \longrightarrow M-O-M + H_2O \tag{3}$$

$$2\ Re_2(CO)_{10} + 4\ H_2O \xrightarrow[-8\ CO]{h\nu} [(CO)_3ReOH]_4 + 2\ H_2 \tag{4}$$

Follow-up products may include metal oxides, be it in a mononuclear or in a di- and oligonuclear form [Eqs. (2)–(4)]. The tetranuclear rhenium(I) complex formed by photolysis [Eq. (4)] has a cubane-type structure [6]. Related complexes **2a–d** were made according to Scheme 2 [7].

Since many organometallics behave as Lewis bases due to electron-rich metals, protonation is a common reaction. For example, the tungsten hydride **3** undergoes reversible protonation at the metal, forming the water-soluble cationic hydride **4** [Eq. (5)]. In nickelocene **5**, a $20e^-$ complex, protonation occurs at the π-bonded cyclopentadienyl ligand; the intermediate **6** has a stable, isolable counterpart in the fully methylated derivative. Consecutive loss of cyclopentadiene forms the cation **7**, which adds to unchanged nickelocene forming the tripledecker sandwich **8** (Scheme 3).

$$\underset{\mathbf{3}}{Cp_2WH_2} \underset{OH^-}{\overset{H^+}{\rightleftharpoons}} \underset{\mathbf{4}}{[Cp_2WH_3]^+} \tag{5}$$

Suffice it to say that pronouncedly oxophilic metals such as the rare-earth elements are particularly sensitive to water. There are numerous cases where an oxo ligand has been introduced by accidental moisture present under the conditions of reaction [8]. There is good reason for chemistry of this type of metals to be performed under scrupulous glove-box conditions.

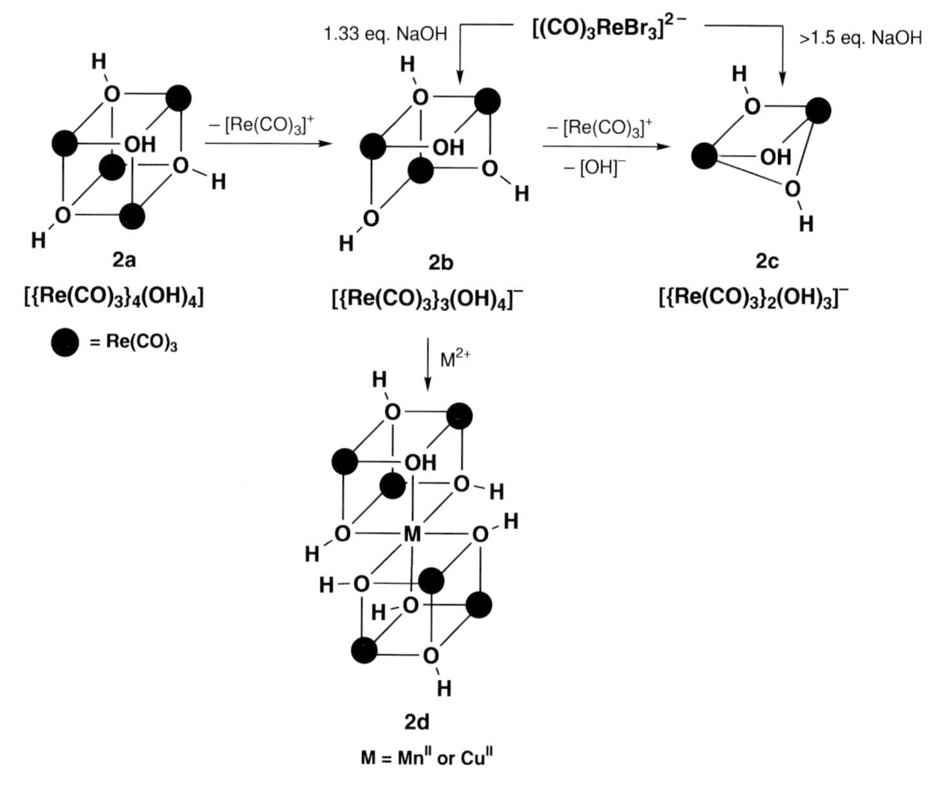

Scheme 2. Organometallic hydroxy complexes from hydrolysis and aggregation reactions.

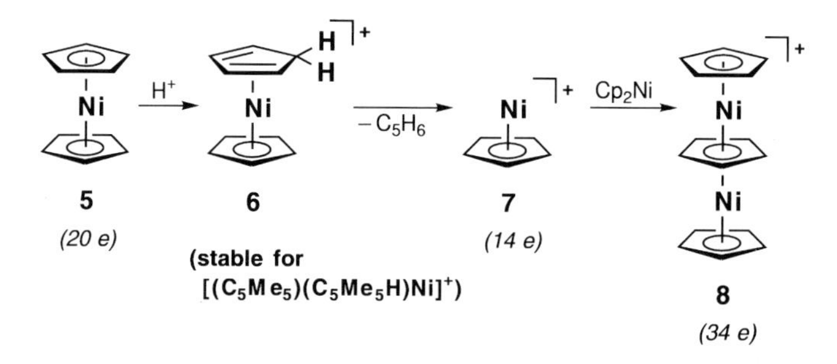

Scheme 3. Protonation of an electron-rich ligand of an organometallic complex.

There are also cases where the *ligand* reacts with water (cf. Section 2.2.6). Organic phosphanes, for example, may show up as phosphane oxides under certain circumstances [cf. Eq. (6)]. The oxidant is not always obvious in these cases, but it can be water.

$$L_nRh{-}P(C_6H_5)_3 + H_2O \longrightarrow O{=}P(C_6H_5)_3 + \ldots \quad (6)$$

Strong metalla-acids undergo protonation of water to form the corresponding anion; $HCo(CO)_4$ is such an example [Eq. (7)].

$$(CO)_4Co{-}H + H_2O \longrightarrow [(CO)_4Co]^- + H_3O^+ \qquad (7)$$

There are cases where the M–C bonds withstand cleavage. The organorhenium(VII) oxide **9a** rapidly exchanges the oxo ligands from water [Eq. (8)], and the water-soluble (ca. 30 g/L) methyltrioxorhenium(VII) (MTO) **10** forms adducts with water such as **11** [Eq. 9] [9a–f] before it undergoes aggregation to "polymeric MTO." Steric reasons are likely to account for this difference. MTO **10** is also an example of a water-stable metal organyl: even in boiling water, only 8% of the methyl groups are lost (as methane) after several hours [9a, e]!

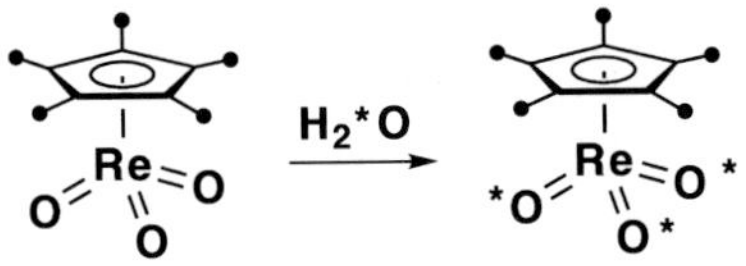

$H_2{}^*O$

9a → **9b** (8)

•— = CH_3; *O = ^{17}O

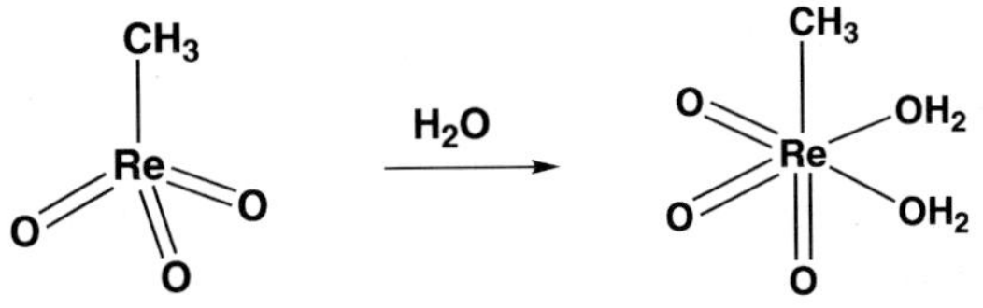

H_2O

10 → **11** (9)

Water can also *oxidize* organometallic complexes. For example, the platinum(IV) complex cation $[(C_6H_5)_2Pt(OH)]^+$ is generated from divalent platinum by water [9c].

2.2.4 Catalytic Reactions with Water

2.2.4.1 Water-gas Shift Reaction

A famous example is the water-gas shift reaction [10]. Efficient catalysts are late transition metals such as iron, e.g. $Fe(CO)_5$ [10].

Mechanistically, attack of water (hydroxide) occurs at the carbonyl ligands, with the catalytic cycle depending on the formation and lability of metallacarboxylic acids (Scheme 4). This reaction can interfere with typical CO reactions such as hydroformylation (C–C-bond formation) [11].

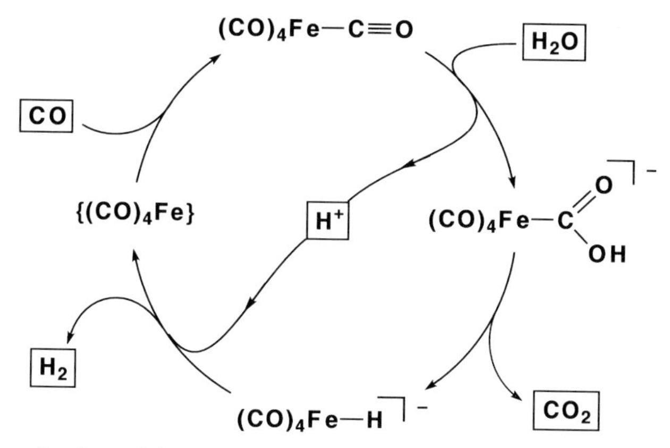

Scheme 4. The mechanism of the water-gas shift reaction (cf. [10] and references cited therein).

2.2.4.2 Wacker–Hoechst Acetaldehyde Process

Water is also involved as a substrate in the Wacker–Hoechst acetaldehyde process based on a partial, selective oxidation of ethylene [12]. According to Eq. (10), it is necessary to form the new C–O bond starting from ethylene (*trans*-stereochemistry), while the oxygen of Eq. (11) regenerates the catalyst ($Pd^0 \rightarrow Pd^{2+}$), but does not oxidize the ethylene as suggested by the net Eq. (12). Metal attachment of ethylene is the prerequisite to make it accessible to nucleophilic attack by water.

$$CH_2{=}CH_2 + H_2O + PdCl_2 \rightarrow CH_3{-}C({=}O)H + 2\,HCl + Pd \quad (10)$$

$$Pd + {}^1/_2\,O_2 + 2\,HCl \rightarrow PdCl_2 + H_2O \quad (11)$$

$$CH_2{=}CH_2 + {}^1/_2\,O_2 \rightarrow CH_3{-}C({=}O)H \quad (12)$$

2.2.4.3 Olefin Hydration

Otherwise, water has not been exploited as a cheap building block in catalytic reactions. Note, for example, that the highly desired *anti*-Markovnikov addition of water to α-olefins forming α-alcohols still awaits efficient catalysis [Eq. (13)] [11]. Very probably, an oxidative addition will activate the water for the nucleophilic attack at the (metal-bound) olefin [Eq. (14)] [2].

$$R{-}CH{=}CH_2 + H_2O \longrightarrow \underset{\textit{anti}\text{-Markownikoff addition}}{R{-}CH_2{-}CH_2{-}OH} \qquad (13)$$

$$L_nM^{(n+)} + H_2O \longrightarrow L_nM^{(n+)}{\leftarrow}O(H)H \longrightarrow L_nM^{(n+2)}(H)(OH) \qquad (14)$$

2.2.4.4 Hydrodimerization

Hydrodimerization is a special case of telomerization, where a (solvent) molecule A–B (the telogen, e.g., H_2O) reacts with n molecules of an unsaturated molecule M (the taxogen) to yield oligomers or polymers of relatively low molecular mass [Eqs. (15), (16)]. An important special case is the Kuraray 1-octanol process resulting from the products of Eq. (17) by subsequent hydrogenation. This industrially relevant reaction includes hydrodimerization of 1,3-butadiene [13]. Efficient catalysts are palladium–phosphine complexes, e.g., Pd^{2+}/TPPMS (TPPMS = $P(C_6H_4\text{-}m\text{-}SO_3^-Na^+)(C_6H_5)_2$). Little is as yet known on mechanisms.

$$A{-}B + n\,M \longrightarrow A{-}(M)_n{-}B \qquad (15)$$

$$H{-}OH + n\,M \longrightarrow H{-}(M)_n{-}OH \qquad (16)$$

$$H{-}OH + 2CH_2{=}CH{-}CH{=}CH_2 \longrightarrow CH_2{=}CH{-}(CH_2)_3{-}\underset{OH}{CH}{-}CH{=}CH_2 \qquad (17)$$

[+ 2,7-Octadienol(1)]

Telomerizations provide an opportunity to make functionalized organic products from simple, abundant precursor molecules, e.g., butadiene and water.

2.2.5 Water-soluble Metal Complexes

Following the pioneering work in the area of biphasic catalysis [14], a steady demand for water-soluble and, at the same time, water-compatible metal complexes has been recognized. Despite a broad array of solubilizing ligands (Section 3.2.1), sulfonated derivatives of ligands containing aryl groups have proven most successful, mostly because of the outstanding solubility in water. Notably, the standard tris(*m*-sulfonatophenyl)phosphine [$P(C_6H_4$-*m*-$SO_3^- Na^+)_3$; TPPTS] has a solubility of ca. 1.1 kg/L upon which the success of the catalyst system Rh/TPPTS depends in biphasic, aqueous hydroformylation [15] (cf. Section 6.1).

Numerous ionic organometallics, e.g., $Na[Re(CO)_5]$ or $[(\eta^5\text{-}C_5H_5)Fe(CO)_2\text{-}\{P(C_6H_5)_3\}]^+ I^-$, are soluble in water and can be precipitated by large counter-ions (e.g., $[P(C_6H_5)_4]^+$, $[B(C_6H_5)_4]^-$, $[(C_6H_5)_3P{=}N{=}P(C_6H_5)_3]^+$). Basic ions such as $[Re(CO)_5]^-$ give hydrido complexes upon protonation, e.g., $HRe(CO)_5$.

2.2.6 Perspectives

Organometallic entities display a variety of reactions with water, following their commonly observed thermodynamic instability. However, in many cases, kinetic barriers prevent these reactions from occurring, making organometallics seemingly stable toward water and, often, even toward protic aqueous acids and toward hydroxide. Even in cases where highly charged metal ions coordinate water, further degradation of the (organic) ligand sphere does not necessarily occur. Striking examples are the organorhenium(VII) oxide CH_3ReO_3, a series of alkylmercury compounds (e.g., $[(CH_3)_3Hg]^+$), and the well-known methylcobalamin. Also, hydrated alkylchromium and alkylcobalt complexes are known [1 d]. Therefore, water is much more compatible with organometallic compounds than has previously been assumed. We note that organic chemistry in water is just beginning to gain wider attention, too [16].

Specific reaction modes arising from the reactivity of water have to be taken into account. Consider, for example, the sulfonated hydroformylation catalyst **12**, which undergoes slow but significant P–C bond cleavage (Scheme 5). The

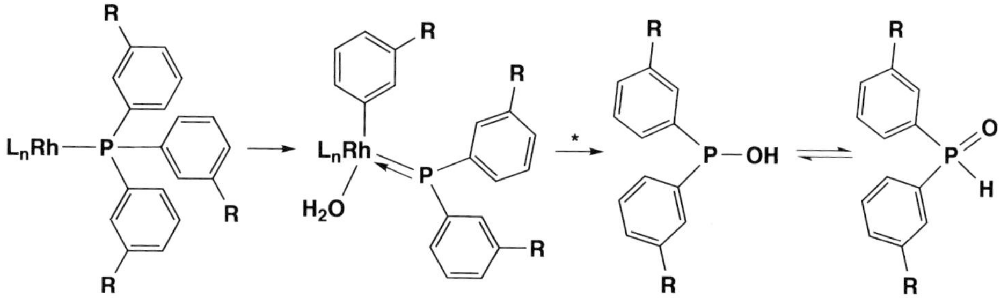

12 13 14 15

R = SO_3Na

Scheme 5. P–C bond cleavage at rhodium(I) hydroformylation catalysts in aqueous medium; see text. * Loss of the rhodium-containing entity.

resulting phosphinidene intermediate **13** can react with water to give the phosphinous acid **15** via tautomerization of the hydroxyphosphine **14** [14a]. Thus, one must be aware of oxygenated (side)products when working with aqueous organometallics. The field is far from being fully explored.

Supercritical water offers opportunities in selectivity for organometallic reactions. Thus, the cyclotrimerization of certain alkynes is very selective for benzene derivatives in supercritical water at 374 °C [17]. Organometallic, water-soluble anticancer reagents, normally in the context of amino acid ligands, are appearing more frequently in the literature now [18].

References

[1] (a) J. D. Atwood, *Inorganic and Organometallic Reaction Mechanisms*, Brooks/Cole, Monterey, Canada, **1985**; (b) F. Basolo, R. G. Pearson, *Mechanisms of Inorganic Reactions*, Wiley, New York, **1967**; (c) M. Gerloch, E. C. Constable, *Transition Metal Chemistry*, VCH, Weinheim, **1994**; (d) C. F. Baer, R. E. Mesmer, *The Hydrolysis of Cations*, Wiley, New York, **1976**; (e) D. M. Roundhill, *Advan. Organomet. Chem.* **1995**, *38*, 155; (f) F. Joó, Á. Kathó, *J. Mol. Catal.* **1997**, *116*, 3; (g) B. Cornils, C. W. Kohlpaintner, E. Wiebus, *Encyclopedia of Chemicals Processing and Design* (Eds.) J. McKetta, G. E. Weismantel), Dekker, New York, **1998**, in press.

[2] (a) J. P. Collman, L. S. Hegedus, J. R. Norton, R. G. Finke, *Principles and Applications of Organotransition Metal Chemistry*, University Science Books, Mill Valley, CA, **1987**; (b) Ch. Elschenbroich, A. Salzer, *Organometallics, A Concise Introduction*, 2nd ed., VCH, Weinheim, **1992**; (c) P. Powell, *Principles of Organometallic Chemistry*, 2nd ed., Chapman and Hall, London, **1988**; (d) R. H. Crabtree, *The Organometallic Chemistry of the Transition Metals*, John Wiley, New York, **1988**.

[3] M. Herberhold, G. Süß, J. Ellermann, H. Gäbelein, *Chem. Ber.* **1987**, *111*, 2931.

[4] (a) W. A. Herrmann, *J. Organomet. Chem.* **1995**, *500*, 149; (b) W. A. Herrmann, F. E. Kühn, *Acc. Chem. Res.* **1997**, *30*, 169; (c) C. Romão, F. E. Kühn, W. A. Herrmann, *Chem. Rev.*, in press.

[5] (a) R. Alberto, A. Egli, V. Abram, K. Hegetsweiler, V. Gramlich, P. A. Schubiger, *J. Chem. Soc., Dalton Trans.* **1994**, 2815; (b) R. Alberto, R. Schibli, A. Egli, P. A. Schubiger, W. A. Herrmann, G. Artus, V. Abram, T. A. Kade, *J. Organomet. Chem.* **1995**, *492*, 217; (c) W. A. Herrmann, R. Alberto, J. C. Bryan, A. P. Sattelberger, *Chem. Ber.* **1991**, *124*, 1107; (d) W. A. Howard, G. Parkin, *Polyhedron* **1993**, *12*, 1253; (e) W. A. Herrmann, R. W. Fischer, W. Scherer, M. U. Rauch, *Angew. Chem.* **1993**, *105*, 1209; *Angew. Chem., Int. Ed. Engl.* **1993**, *32*, 1157; (f) J. W. Steed, D. A. Tocker, *J. Chem. Soc., Chem. Commun.* **1991**, *22*, 1609; (g) G. Erker, C. Krüger, C. Starter, S. Werner, *J. Organomet. Chem.* **1989**, *377*, C55; (h) D. V. McGrath, R. H. Grubbs, J. W. Ziller, *J. Am. Chem. Soc.* **1991**, *113*, 3611; (i) D. M. Lynn, S. Kanaoka, R. H. Grubbs, *J. Am. Chem. Soc.* **1996**, *118*, 784; (j) S. Wache, *J. Organomet. Chem.* **1995**, *494*, 235; (k) M. Stehler-Röthlisberger, W. Hummel, P. Pittet, H. Bürgi, A. Ludi, M. E. Merbach, *Inorg. Chem.* **1988**, *27*, 1358; (l) M. S. Eisen, A. Haskel, H. Chen, M. M. Olmstead, D. P. Smith, M. F. Maestre, R. H. Fish, *Organometallics* **1995**, *14*, 2806; (m) S. Ogo, H. Chen, M. M. Olmstead, R. H. Fish, *Organometallics* **1996**, *15*, 2009.

[6] M. Herberhold, G. Süss-Fink, *Angew. Chem.* **1975**, *87*, 710; *Angew. Chem. Int. Ed. Engl.* **1975**, *14*, 700.

[7] W. A. Herrmann, A. Egli, E. Herdtweck, R. Alberto, F. Baumgärtner, *Angew. Chem.* **1996**, *108*, 486; *Angew. Chem., Int. Ed. Engl.* **1996**, *35*, 432.

[8] (a) *Topics in Current Chemistry* Vol. 179 (Ed.: W. A. Herrmann), Springer, Heidelberg, **1996**; b) W. A. Herrmann, N. W. Huber, O. Runte, *Angew. Chem.* **1995**, *106*, 2371; *Angew. Chem., Int. Ed. Engl.* **1995**, *34*, 2187.

[9] (a) W. A. Herrmann, R. W. Fischer, *J. Am. Chem. Soc.* **1994**, *117*, 3223; (b) W. A. Herrmann, W. Scherer, R. W. Fischer, J. Blümel, M. Kleine, W. Mertin, R. Gruehn, J. Mink, H. Boysen, Ch. C. Wilson, R. Ibberson, L. Bachmann, M. Mattner, *J. Am. Chem. Soc.* **1995**, *117*, 3231; (c) M. R. Mattner, W. A. Herrmann, R. Berger, Ch. Gerber, J. K. Gimzewski, *Adv. Mater* **1996**, *8*, 654; (d) H. S. Genin, K. A. Lawler, R. Hoffmann, W. A. Herrmann, R. W. Fischer, W. Scherer, *J. Am. Chem. Soc.* **1995**, *117*, 3244; (e) W. Scherer, Ph. D. Thesis, Technische Universität München, **1994**; (f) R. W. Fischer, Ph. D. Thesis, Technische Universität München, **1994**; (g) A. J. Canty, S. D. Fritsche, H. Jin, R. T. Honeyman, B. W. Skelton, A. H. White, *J. Organomet. Chem.* **1996**, *510*, 281.

[10] W. A. Herrmann, in: *Applied Homogeneous Catalysis by Organometallic Compounds* (Eds.: B. Cornils, W. A. Herrmann), VCH, Weinheim, **1996**, p. 957.

[11] (a) W. A. Herrmann, B. Cornils, in [10], p. 1167; (b) W. A. Herrmann, B. Cornils, *Angew. Chem.* **1997**, *109*, 1074; *Angew. Chem., Int. Ed. Engl.* **1997**, *36*, 1047.

[12] R. Jira, in [10], p. 374.

[13] N. Yoshimura, in [10], Vol. 1, p. 351.

[14] a) W. A. Herrmann, Ch. W. Kohlpaintner, *Angew. Chem.* **1993**, *105*, 1588; *Angew. Chem., Int. Ed. Engl.* **1993**, *32*, 1524; (b) C. D. Frohning, Ch. W. Kohlpaintner, in E. Wichers [10], p. 29.

[15] (a) B. Cornils, E. Wiebus, *Chem. Ing. Techn.* **1994**, *66*, 916; (b) B. Cornils, W. A. Herrmann, R. W. Eckl, *J. Mol. Catal.* **1997**, *116*, 27.

[16] A. Lubineau, J. Augé, Y. Queneau, *Synthesis* **1994**, *8*, 741.

[17] K. S. Jerome, E. J. Parsons, *Organometallics* **1993**, *12*, 2991.

[18] M. M. Harding, M. Prodigalidad, M. J. Lynch, *J. Med. Chem.* **1996**, *39*, 5012.

2.3 Characterization of Organometallic Compounds in Water

Gábor Laurenczy

2.3.1 Introduction

It is generally known that the majority of organometallic compounds are stable only in the absence of water, and many of them can be synthesized only under rigorously dry conditions. Although in most cases H_2O acts destructively on organometallics, in some cases the organometallic compounds resist water.

The growing importance of aqueous organometallic chemistry is well demonstrated by the increasing number of publications in this field: both catalytic applications in industrial processes and fundamental research requirements justify this explosion [1–5]. In this section a general overview is given to characterize compounds containing metal–carbon bonds in aqueous solutions. Nowadays pressure has become one of the common parameters used in several laboratories to help improve understanding of structures and processes: this is an additional variable besides temperature, concentration, pH, solvents, ionic strength, etc. Special attention will be given to the pressurization of organometallic systems in water. There are two completely different subsections: the first deals with the effect of the high hydrostatic pressure on organometallics and the second concerns characterization of reactions involving pressurized gases with aqueous organometallics.

2.3.2 General Survey

Nuclear magnetic resonance (NMR) spectroscopy has been proven the most versatile technique to study organometallic compounds both non-aqueous and aqueous solutions [6, 7]. To explore all the possibilities of ^1H NMR one has to either work in D_2O as solvent or use a water signal suppression technique.

Proton chemical shifts can give information about the structure. Generally, protons bound to carbons coordinated to a metal center show a low-field shift, about 1–4 ppm, compared with the metal-free environment. Metal hydrides usually have negative chemical shifts, sometimes down to −40 ppm. In a paramagnetic environment chemical shifts can be very large; the resonances are broadened and sometimes are not even detectable. Substituting protons by deuterium can have a dramatic effect on the rest of the ^{1}H NMR spectrum: since ^{2}H nuclei have a spin value of 1, so the multiplicity of the resonances increase. ^{13}C is naturally one of the most widely used nuclei to elucidate structure and dynamic behavior of organometallic compounds. ^{13}C NMR is applied both when the natural abundance and for compounds enriched in ^{13}C during the synthesis. ^{13}C nuclei have a spin of 1/2, $^{1}J(C,H)$ couplings giving direct information on the number of H atoms linked to the carbon: the CH signal is a doublet, the CH_2 is a triplet and the CH_3 gives a quartet in a ^{13}C spectrum. The $^{1}J(C,H)$ coupling constants are about 250 Hz, 160 Hz and 125 Hz in the alkynes, alkenes, and alkanes, respectively. Protons can be decoupled to simplify ^{13}C spectra. In aqueous solution ^{17}O NMR gives the possibility of studying water exchange and dynamic behavior of organometallic aquo complexes. Having a spin of 5/2, the ^{17}O signals are broad and the observed chemical shifts are between 1500 and −250 ppm referenced to the bulk water. Presence of other NMR-active nuclei in the compounds (^{31}P and ^{19}F having a spin of 1/2, ^{14}N and ^{2}H having a spin of 1) and sometimes the metal atoms (^{103}Rh, ^{59}Co, ^{195}Pt, etc.) extend the available information via NMR spectroscopy. To move to a higher magnetic field can simplify the strongly coupled spectra, because chemical shifts in hertz increase with the magnetic field but the coupling constants stay the same (in hertz). Measuring spin–lattice or longitudinal relaxation times, T_1, can help to distinguish between classical dihydrides and molecular hydrogen complexes, the latter having much shorter T_1 values. To resolve difficult structures two-dimensional NMR or special pulse sequences can sometimes be useful. Intramolecular exchange processes can broaden NMR signals, giving further possibilities for determining structures in solution. From variable-temperature NMR spectra the activation enthalpies and entropies of these dynamic processes can be calculated.

Infrared (IR) spectroscopy is particularly helpful to study organometallic compounds [8, 9] especially carbonyl complexes. This is also true in aqueous solutions, although one has to take into consideration the water absorbances in the IR spectral region. The CO stretching vibrations are intense and characteristic of the chemical environment. In the case of several carbonyls in the same molecule, the intensity ratios give structural information and can help to distinguish isomers. In water the useful spectral domain is limited by solvent absorbances and one must use a shorter optical pathlength and compensate for water absorbance.

In the UV-visible spectra of organometallic compounds the spectral bands generally are large and less specific absorbances, due either to the transition metal or to the ligands in the compounds. So optical spectroscopy is used to study kinetic processes, to follow fast reactions (stopped-flow method), or to determine equilibrium constants. UV–visible spectra of the organometallic compounds are often studied for photochemical purposes. Preparative-scale photolysis has proven to be a valuable synthetic technique in organometallic chemistry, and also in water [10]. Ligand substitution reactions are accelerated by UV light.

The acidity of the coordinated water in organometallic aquocomplexes varies over a large range. In aqueous solutions pH potentiometry is one of the most precise methods of determining the acid dissociation constants, K_a, for the organometal ions. Several stable oxidation states of the transition metal ions in organometallic compounds can exist in water. Cyclic voltammetry gives information on the reversibility of the oxidation/reduction processes.

Crystallography, chromatographic techniques, mass spectrometry, and elemental analysis complement the direct and non-destructive methods used to study the organometallic compounds in situ in aqueous solutions. In the majority of the cases it is generally accepted that crystallization from aqueous solutions does not change the basic structures, compositions, or bonding orders. However, it is not always the case for bonding distances and angles. One has to be very careful when using crystallographic results to explain structures, reactions, catalysis, or kinetic behavior in solution. Incontestable and independent evidence is necessary that the structure of the studied compounds is the same in aqueous solution as in the solid state.

Computation completes more and more experimental results to explain, interpret, and predict structures, reactivities, and to contribute to the better understanding of reaction mechanisms. The powerful computers and software now available (density functional theory, ab-initio and molecular dynamic calculations, etc.) give calculated parameters closer and closer to the experimental values.

2.3.3 Effect of High Hydrostatic Pressure on Aqueous Organometallic Systems

Since the late 1960s pressure has become a common and important variable in the study of chemical kinetics and equilibrium [11–13]. High-pressure techniques have been developed for the majority of physicochemical methods (NMR, IR, and UV–visible spectroscopy, electrochemistry, etc.) [14, 15].

The pressure dependence of a rate constant, k, is given by Eq. (1) where $\Delta V^{\ddagger}$ is the volume difference between the volume of the transition state and the volume of the reactants. Generally the pressure dependence of $\ln k$ is linear,

$$RT \ln(k/k_0) = -\Delta V^{\ddagger} p \quad (1)$$

and the compressibility of the transition state can be neglected. Symmetric reactions like solvent exchange are especially suited to the study of the effect of pressure on the reaction rate, since the reaction volume, ΔV^0, is 0. Since there is no change formation or cancelation during this reaction, the activation volumes determined give direct information about the transition state. In the case of complex formation or non-symmetric ligand-exchange reactions, only the complete volume profile (activation volumes for the forward and for the reverse reaction, reaction volume) allow mechanistic conclusions.

Water exchanges on $[Ru(\eta^6\text{-}C_6H_6)(H_2O)_3]^{2+}$ and on $[Os(\eta^6\text{-}C_6H_6)(H_2O)_3]^{2+}$ [16] were studied by variable-pressure ^{17}O NMR spectroscopy. A home-built high-pressure ^{17}O probehead was used with a Bruker CXP-200 spectrometer up to 200 MPa [17]. The exchange rates (Eq. (2), M = Ru or Os) were in the slow-exchange domain on the NMR timescale, so the rate constants at different pressures were determined from the line broadening of the water signals of the diamagnetic complexes: low-frequency shifts at δ −73.4 ppm for M = Ru and δ −66.6 ppm for M = Os (the organometallic aquo ions were enriched previously in $H_2{}^{17}O$).

$$[M(\eta^6\text{-}C_6H_6)(H_2O)_3]^{2+} + 3\,H_2O^* \rightleftharpoons [M(\eta^6\text{-}C_6H_6)(H_2O^*)_3]^{2+} + 3\,H_2O \quad (2)$$

The slightly positive activation volume values for $[Ru(\eta^6\text{-}C_6H_6)(H_2O)_3]^{2+}$ and for $[Os(\eta^6\text{-}C_6H_6)(H_2O)_3]^{2+}$ show that the water exchange takes place via an interchange mechanism. The same technique, high-pressure ^{17}O NMR [18], was applied to study water exchange on $[Rh(Cp^*)(H_2O)_3]^{2+}$ and $[Ir(Cp^*)(H_2O)_3]^{2+}$ [19], where Cp* = pentamethylcyclopentadienyl anion; on $[Pt(L)(H_2O)]^+$ arylplatinum complex [20], where L = $C_6H_3(CH_2NMe_2)_2$-2,6, a terdentate *N,C,N'*-chelating monoanionic ligand, and on $[Ru(H_2O)_5(CH_2{=}CH_2)]^{2+}$ [21]. From the experimental results it was concluded that the water exchange in the triaqua species follows an interchange mechanism with the M–O(water) bond breaking being rate-controlling. In the case of $[Pt(L)(H_2O)]^+$ the large negative activation volume ($\Delta V^{\ddagger}_{kex} = -9.2 \pm 1.0\ cm^3\ mol^{-1}$) indicates an associative model of activation for the water exchange. The positive volumes obtained for the axial and equatorial water exchange reactions on $[Ru(H_2O)_4^{eq}(H_2O)^{ax}(CH_2{=}CH_2)]^{2+}$ are in accordance with a dissociative activation mode for the exchange process.

A high-pressure stopped-flow technique [22] was applied to study the complex formation reactions between $[Pt(L)(H_2O)]^+$ and N_3^- or tmtu, where L = $C_6H_3(CH_2NMe_2)_2$-2,6 [23] or *N,N*-dimethylbenzylamine and pyridine-3-sulfonic acid [24], tmtu = *N,N,N',N'*-tetramethylthiourea. The large negative activation volumes found for the forward and reverse reactions ($\Delta V^{\ddagger}$ values are between −10.1 and $-12.4\ cm^3\ mol^{-1}$) are evidence for the operation of an associative activation mode for the substitution.

Methyltrioxorhenium, one of the most versatile organometallic catalysts [25–28], hydrolyzes very slowly in dilute aqueous solution. This reaction was studied as a function of pressure by a spectrophotometric method [29]. The hydrolytic reaction (Eq. (3)) was followed at 268.5 and 227.0 nm, where the absorbance changes were maximal, and it was found the reaction has the experimental rate law described by Eq. (4).

$$CH_3ReO_3 + OH^- \longrightarrow CH_4 + ReO_4^- \qquad (3)$$

$$-\, d(CH_3ReO_3)/dt = k\,[OH^-][CH_3ReO_3] \qquad (4)$$

An UV–visible–near-IR high-pressure optical unit [30] equipped with pill-box cells [31] was used for the kinetic experiments. Here the observed small negative activation volume does not necessarily indicate an associatively activated reaction pathway. Indeed, the $\Delta V^{\neq}$ determined is a composite one, as it includes the activation volume of the rate-determining step and the unknown reaction volume. Attempts to perform the reverse reaction, the synthesis of CH_3ReO_3 from CH_4 (200 MPa) and ReO_4^-, failed, so the hydrolysis of CH_3ReO_3 seems to be irreversible.

The pressure dependences of formation, homolytic and heterolytic fission of chromium–carbon bonds in aqueous solutions were investigated for the $[(H_2O)_5CrR]^{2+}$ complexes, where R = different alkyl groups [32–34]. The reactions (Eqs. (5) and (6)) were followed by spectrophotometry [35], the alkyl radicals being generated in situ by pulse radiolysis [36].

$$[Cr(H_2O)_6]^{2+} + {\bullet}R \rightleftharpoons [(H_2O)_5CrR]^{2+} + H_2O \qquad (5)$$

$$[(H_2O)_5CrR]^{2+} + H_2O \longrightarrow [(H_2O)_5CrOH]^{2+} + RH \qquad (6)$$

The determined activation volumes for complex formation ($\Delta V^{\neq}/cm^3\,mol^{-1}$, between +3.4 and +6.3 for the different alkyl groups) and the large volumes of reaction support a dissociative interchange mechanism for $[(H_2O)_5CrR]^{2+}$ formation. Similar reactions were studied for the $[(H_2O)_2ML]$ transition metal complexes with the $\cdot CH_3$ radical, where L = nitrilotriacetate (nta) for M = Co(II) and Fe(II); 1,4,8,11-tetraazacyclotetradecane (cyclam) for M = Ni(II) [37–39]. The activation volumes calculated for the $[ML(H_2O)(CH_3)]$ complex formation, $\Delta V^{\neq}$, and the suggested mechanisms are in accordance with those found for the water exchange on these aquo ions.

The acidolysis reaction of dimethylmercury as a function of pressure [40] was studied in aqueous medium, in dilute HCl and HBr solutions (Eq. (7), X = Cl or Br) by a conductivity technique [41]. The volumes of activation, $\Delta V^{\neq}$, were measured to be −22.0 and −37.0 $cm^3\,mol^{-1}$ for X = Cl and Br, respectively.

$$(CH_3)_2Hg + HX \rightleftharpoons (CH_3)HgX + CH_4 \qquad (7)$$

As an explanation of the large negative values the authors suppose that the activated complexes are much more polar than the reactants and the solvation term contributing to the volume of activation is larger than the contributions arising from breakage or formation of covalent bonds on the reacting species in the transition state.

2.3.4 Aqueous Organometallics with Pressurized Gases

Reactions of organometallic compounds with pressurized gases in water are becoming more and more important for industrial applications (cf. Section 6.1). High pressure increases the dissolved gas concentration according to Henry's law. It has a twofold effect, thermodynamic and kinetic. The higher concentration of the dissolved compound shifts equilibrium toward product formation, increasing the yield. Higher concentrations accelerate the second- and higher-order reactions, and in the case of parallel reactions help to increase the formation of the desired product. Applications of pressurized gases in aqueous organometallic systems are described here.

The synthesis [42] of one of the simplest organometallic aquo ions, $[Ru(CO)(H_2O)_5]^{2+}$, was carried out under 5.0 MPa CO pressure in a 10 mm sapphire NMR tube [43] (Eq. (8)). The $[Ru(H_2O)_6]^{2+}$ was previously enriched

$$[Ru(H_2O)_6]^{2+} + CO \longrightarrow [Ru(CO)(H_2O)_5]^{2+} + H_2O \qquad (8)$$

with 10% $H_2{}^{17}O$, so the reaction was followed simultaneously by ^{13}C and ^{17}O NMR spectroscopy. During the reaction the disappearance of the ^{17}O signal of the $[Ru(H_2O)_6]^{2+}$ at -192 ppm and the appearance of two new resonances in a 1:4 ratio (at -29.3 ppm and -154.8 ppm) were observed, corresponding to the water molecules axial and equatorial to the CO in $[Ru(CO)(H_2O)^{ax}$-$(H_2O)_4^{eq}]^{2+}$. At the same time the ^{13}C NMR spectrum shows the growth of a new signal at 205.5 ppm due to the coordinated CO, next to the signal of the free CO (187.0 ppm). The FT-IR spectra [44] of $[Ru(CO)(H_2O)_5]^{2+}$, yellow in aqueous solution, shows the characteristic stretching frequency at 1971 cm^{-1} due to the coordinated CO (the free-CO frequency in water appears at 2134 cm^{-1}). Substitution of a second H_2O by CO in $[Ru(CO)(H_2O)_5]^{2+}$ was not detected even at 200 MPa CO pressure and after several days, as followed by spectrophotometry with a high-pressure UV–visible unit [30] or in a high-pressure FT-IR cell [44].

Several olefin complexes of the Ru(II) aquo ion [45–47] and of the other Ru complexes [48, 49] have been synthesized and characterized in the ring-opening metathesis polymerization (ROMP; cf. Section 6.13) or olefin isomerization reactions. The simplest olefin complex of Ru was also observed, isolated, and characterized under ethylene pressure [50]. A $H_2{}^{17}O$-enriched aqueous solution of $[Ru(H_2O)_6]^{2+}$ was pressurized with ethylene and mixed. The reaction (Eq. (9)) was followed by ^{1}H, ^{13}C, and ^{17}O NMR spectroscopy.

$$[Ru(H_2O)_6]^{2+} + CH_2{=}CH_2 \longrightarrow [Ru(CH_2{=}CH_2)(H_2O)_5]^{2+} + H_2O \qquad (9)$$

After 6 h the ^{17}O NMR spectra showed the complete loss of resonance at -192 ppm due to the hexaaqua Ru(II). At the same time new signals appeared corresponding to $[Ru(CH_2{=}CH_2)(H_2O)_4^{eq}(H_2O)^{ax}]^{2+}$. After 18 h of mixing the NMR spectra show the substitution of a second water molecule (Eq. (10)).

$$[Ru(CH_2{=}CH_2)(H_2O)_5]^{2+} + CH_2{=}CH_2 \longrightarrow [Ru(CH_2{=}CH_2)_2(H_2O)_4]^{2+} + H_2O \qquad (10)$$

The ^1H, ^{13}C and ^{17}O NMR signals characteristic for the *cis*-$[Ru(CH_2{=}CH_2)_2(H_2O)_4]^{2+}$ cation (Structure **1**) were observed.

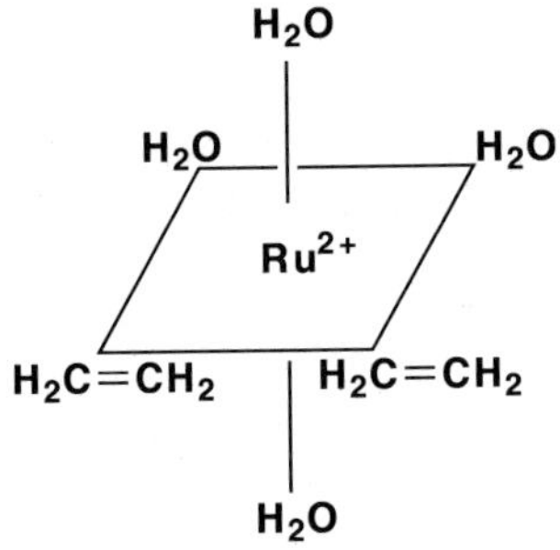

After three days on mixing, an organic phase was formed above the aqueous solution. The analysis of this new phase showed the presence of free butenes: *cis*-2-butene (18%), *trans*-2-butene (41%) and 1-butene (41%). This aqueous catalytic dimerization of ethylene is not stereoselective (Scheme 1), all possible butenes, except for 2-methylpropene, being formed. No higher oligomers were detected, probably due to the low solubility of butenes in aqueous solutions and to the lack of their coordination to Ru(II). Besides $[Ru(CH_2{=}CH_2)(H_2O)_5]^{2+}$ and *cis*-$[Ru(CH_2{=}CH_2)_2(H_2O)_4]^{2+}$, no more highly substituted Ru(II) complexes were isolated or detected.

The catalytic activity of the $HRh(CO)[P(m\text{-}C_6H_4SO_3Na)_3]_3$ complex [51] was checked in aqueous solution up to 20.0 MPa $CO{:}H_2$ (1:1) pressure [52, 53]. The high-pressure NMR spectra of $HRh(CO)[P(m\text{-}C_6H_4SO_3Na)_3]_3$ show the char-

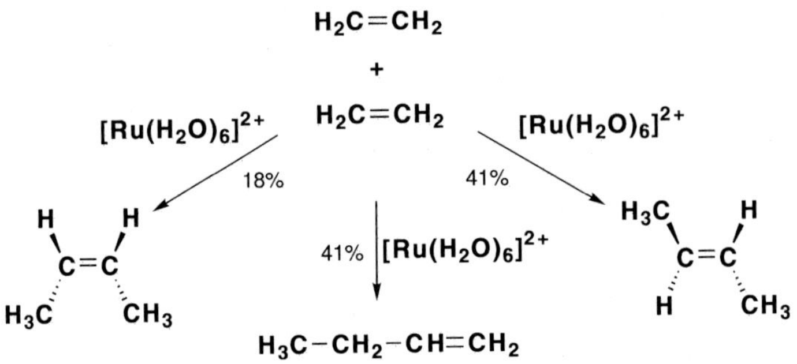

Scheme 1

acteristic chemical shifts and infrared resonances. This complex does not show any further substitution of phosphine ligands by CO. The catalytic activity in hydroformylation reactions of the similar $HRh(CO)(PPh_3)_2$ complex in toluene has been proven to be linked to the coordinatively unsaturated $HRh(CO)_2(PPh_3)$ and $HRh(CO)(PPh_3)_2$ species. The absence of the further-substituted complexes explains the lower catalytic activity and the higher selectivity towards *n*-aldehydes. The $HRh(CO)[P(m\text{-}C_6H_4SO_3Na)_3]_3$ complex is successfully applied as an aqueous catalyst in industrial hydroformylation processes [54–57].

Ligand substitution on $M(CO)_5L$ complexes, where M = Mo or W and L = $P(m\text{-}C_6H_4SO_3Na)_3$ or PPh_3, were studied under 3,4 MPa CO pressure in water and a water–THF (1:1) mixture [58]. The reactions (Eq. (11)) were followed in situ by infrared spectroscopy in a reactor cell [59] between 388 and 423 K.

$$M(CO)_5L + CO \longrightarrow M(CO)_6 + L \tag{11}$$

Under constant CO gas pressure the kinetics of the substitution reaction can be described by a first-order rate law. The calculated activation enthalpies and entropies for the $P(m\text{-}C_6H_4SO_3Na)_3$ ligand are quite similar to those determined for the analogous processes involving the unsulfonated PPh_3 ligand.

2.3.5 Concluding Remarks

Many organometallic compounds are astonishingly stable in protic media and in aqueous solution. There is an increasing interest in the use of water as solvent for industrial organic reactions. The main advantages are the ease of separation processes as well as economic and ecological considerations. In aqueous solution or in a two-phase system homogeneous catalytic organic syntheses involve organometallic compounds. Structural information on the catalysts leads to better understanding of the reaction mechanism, to improve the yield and the selectivity. Non-destructive analytical methods play a key role in this respect. Although a very great number of catalytic organic syntheses have been studied under high gas pressures in autoclaves, the majority of investigations have dealt only with the reaction products when the reaction was complete, after the release of pressure, and analyzed them mainly by chromatographic methods (HPLC, GC). As has been shown, the availability of non-destructive in situ methods allows these reactions to be followed in detail. As can be realized, multinuclear NMR plays a predominant role in this field. New organometallic aquo ions have thus been synthesized, isolated, and characterized.

Techniques applying high hydrostatic pressure have proven to be one of the most powerful methods to obtain information about the reaction mechanisms and about the transition state. These techniques were used to study water exchange on organometallic aquo ions by variable-pressure ^{17}O NMR. The activation volumes, ΔV^{+}, have shown whether bond breaking or bond making is the rate-determining step in the exchange reaction.

These examples confirm that water as a solvent offers a number of variations to the possibilities offered by organic solvents in organometallic reactions and underline the importance of the analytical methods to characterization of organometallic compounds in water. There is still a huge potential for the use of water as the solvent of choice for homogeneous or biphasic catalysis.

References

[1] B. Cornils, W. A. Herrmann (Eds.) *Applied Homogeneous Catalysis with Organometallic Compounds*, Vols 1 and 2, VCH, Weinheim, **1996**.

[2] I. T. Horváth, F. Joó (Eds.) *Aqueous Organometallic Chemistry and Catalysis*, NATO ASI Series 3, Kluwer, Dordrecht, **1995**.

[3] F. Joó, Á. Kathó, *J. Mol. Catal. A* **1997**, *116*, 3.

[4] B. Cornils, W. A. Herrmann, R. W. Eckl, *J. Mol. Catal. A* **1997**, *116*, 27.

[5] M. Beller, B. Cornils, C. D. Frohning, C. W. Kohlpaintner, *J. Mol. Catal. A* **1995**, *104*, 17.

[6] M. Gielen, R. Willem, B. Wrackmeyer (Eds.) *Advanced Applications of NMR to Organometallic Chemistry*, Wiley, Chichester, **1996**.

[7] A. T. Bell, A. Pines (Eds.) *NMR Technique in Catalysis*, Marcel Dekker, New York, **1994**.

[8] B. Schrader (Ed.) *Infrared and Raman Spectroscopy*, VCH, Weinheim, **1995**.

[9] H. A. Willis, J. H. van der Maas, R. G. J. Miller (Eds.) *Laboratory Methods in Vibrational Spectroscopy*, Wiley, **1987**.

[10] G. L. Geoffroy, M. S. Wrighton, *Organometallic Photochemistry*, Academic Press, New York, **1979**.

[11] A. E. Merbach, *Pure Appl. Chem.* **1987**, *59*, 161.

[12] R. van Eldik, T. Asano, W. J. le Noble, *Chem. Rev.* **1989**, *89*, 549.

[13] R. van Eldik, J. Jonas (Eds.) *High Pressure Chemistry and Biochemistry*, NATO ASI Series, D. Reidel, Dordrecht, **1986**.

[14] R. van Eldik (Ed.) *Inorganic High Pressure Chemistry: Kinetics and Mechanisms*, Elsevier, **1986**.

[15] W. B. Holzapfel, N. S. Isaacs (Eds.) *High-pressure Techniques in Chemistry and Physics*, Oxford University Press, Oxford, **1997**.

[16] M. Stebler-Röthlisberger, W. Hummel, P.-A. Pittet, H.-B. Bürgi, A. Lundi, A. E. Merbach, *Inorg. Chem.* **1988**, *27*, 1358.

[17] D. L. Pisaniello, L. Helm, P. Meier, A. E. Merbach, *J. Am. Chem. Soc.* **1983**, *105*, 4258.

[18] U. Frey, L. Helm, A. E. Merbach, *High Press Res.* **1990**, *2*, 237.

[19] L. Dadci, H. Elias, U. Frey, A. Hörnig, U. Koelle, A. E. Merbach, H. Paulus, J. S. Schneider, *Inorg. Chem.* **1995**, *34*, 306.

[20] U. Frey, D. M. Grove, G. van Koten (personal communication).

[21] N. Aebischer, E. Sidorenkova, M. Ravera, G. Laurenczy, D. Osella, J. Weber, A. E. Merbach, *Inorg. Chem.* **1997**, *36*, 6009.

[22] R. van Eldik, W. Gaede, S. Wieland, J. Kraft, M. Spitzer, D. A. Palmer, *Rev. Sci. Instr.* **1993**, *64*, 1355.

[23] M. Schmülling, A. D. Ryabov, R. van Eldik, *J. Chem. Soc., Dalton Trans.* **1994**, 1257.

[24] M. Schmülling, D. M. Grove, G. van Koten, R. van Eldik, N. Veldman, A. L. Spek, *Organometallics* **1996**, *15*, 1384.
[25] W. A. Herrmann, W. Wagner, U. N. Flessner, U. Volkhardt, H. Komber, *Angew. Chem., Int. Ed. Engl.* **1991**, *30*, 1636.
[26] W. A. Herrmann, R. W. Fischer, D. W. Marz, *Angew. Chem., Int. Ed. Engl.* **1991**, *30*, 1638.
[27] W. A. Herrmann, M. Wang, *Angew. Chem., Int. Ed. Engl.* **1991**, *30*, 1641.
[28] W. A. Herrmann, W. Scherer, R. W. Fischer, J. Blümel, M. Kleine, W. Mertin, R. Gruehn, J. Mink, H. Boysen, C. C. Wilson, R. M. Ibberson, L. Bachmann, M. Mattner, *J. Am. Chem. Soc.* **1995**, *117*, 3231.
[29] G. Laurenczy, F. Lukács, R. Roulet, W. A. Herrmann, R. W. Fischer, *Organometallics* **1996**, *15*, 848.
[30] D. T. Richens, Y. Ducommun, A. E. Merbach, *J. Am. Chem. Soc.* **1987**, *109*, 603.
[31] W. le Noble, R. Schlott, *Rev. Sci. Instr.* **1976**, *47*, 770.
[32] M. J. Sisley, W. Rindermann, R. van Eldik, T. W. Swaddle, *J. Am. Chem. Soc.* **1984**, *106*, 7432.
[33] K. Ishihara, T. W. Swaddle, *Can. J. Chem.* **1986**, *64*, 2168.
[34] R. van Eldik, W. Gaede, H. Cohen, D. Meyerstein, *Inorg. Chem.* **1992**, *31*, 3695.
[35] M. Spitzer, F. Gartig, R. van Eldik, *Rev. Sci. Instr.* **1988**, *59*, 2092.
[36] H. Cohen, D. Meyerstein, *Inorg. Chem.* **1974**, *10*, 2435.
[37] R. van Eldik, H. Cohen, D. Meyerstein, *Angew. Chem., Int. Ed. Engl.* **1991**, *30*, 1158.
[38] R. van Eldik, H. Cohen, A. Meshulam, D. Meyerstein, *Inorg. Chem.* **1990**, *29*, 4156.
[39] R. van Eldik, H. Cohen, D. Meyerstein, *Inorg. Chem.* **1994**, *33*, 1566.
[40] R. J. Maguire, S. Anand, *J. Inorg. Nucl. Chem.* **1976**, *38*, 1167.
[41] B. G. Oliver, W. A. Adams, *Rev. Sci. Instr.* **1972**, *43*, 830.
[42] G. Laurenczy, L. Helm, A. Ludi, A. E. Merbach, *Helv. Chim. Acta* **1991**, *74*, 1236.
[43] (a) C. D. Roe, *J. Magn. Reson.* **1985**, *63*, 388; (b) A. Cusanelli, U. Frey, D. T. Richens, A. E. Merbach, *J. Am. Chem. Soc.* **1996**, *118*, 5265.
[44] G. Laurenczy, F. Lukács, R. Roulet, *Anal. Chim. Acta* **1998**, *359*, 275.
[45] B. M. Novak, R. H. Grubbs, *J. Am. Chem. Soc.* **1988**, *110*, 7542.
[46] D. V. McGrath, R. H. Grubbs, *J. Am. Chem. Soc.* **1991**, *113*, 3611.
[47] T. Karlen, A. Ludi, *Helv. Chim. Acta* **1992**, *75*, 1604.
[48] S. T. Nguyen, L. K. Johnson, R. H. Grubbs, *J. Am. Chem. Soc.* **1992**, *114*, 3974.
[49] S. T. Nguyen, R. H. Grubbs, *J. Am. Chem. Soc.* **1992**, *115*, 9858.
[50] G. Laurenczy, A. E. Merbach, *Chem. Comm.* **1993**, 187.
[51] I. T. Horváth, R. V. Kastrup, A. A. Oswald, E. J. Mozeleski, *Catal. Lett.* **1989**, *2*, 85.
[52] I. T. Horváth, J. M. Millar, *Chem. Rev.* **1991**, *91*, 1939.
[53] I. T. Horváth, E. E. Ponce, *Rev. Sci. Instr.* **1991**, *62*, 1104.
[54] (a) B. Cornils, E. Wiebus, *Recl. Trav. Chim. Pays-Bas* **1996**, *115*, 211; (b) B. Cornils, E. Wiebus, *CHEMTECH* **1995**, *25*, 33.
[55] B. Cornils, E. G. Kuntz, *J. Organomet. Chem.* **1995**, *502*, 177.
[56] W. A. Herrmann, C. W. Kohlpaintner, *Angew. Chem., Int. Ed. Engl.* **1993**, *32*, 1524.
[57] E. G. Kuntz, *CHEMTECH* **1987**, *17*, 570.
[58] D. J. Darensbourg, C. J. Bischoff, *Inorg. Chem.* **1993**, *32*, 47.
[59] C. J. Bischoff, Ph. D. Dissertation, Texas A & M University, **1991**.

3

Catalysts for an Aqueous Catalysis

3.1 Variation of Central Atoms

Joseph S. Merola

3.1.1 Introduction

With little fear of contradiction, one can point to the discovery of the synthesis and structure of ferrocene as the birth of "modern" organometallic chemistry [1]. The decades of 1960, 1970, and 1980 saw huge efforts in the field of organometallic chemistry (which continue to this day) spurred on mostly by the promise of new, more active, and more selective catalysts. The organometallic chemist's laboratory was fitted with Schlenk equipment and glove-boxes since much (if not most) of this chemistry involved air- and/or moisture-sensitive compounds. Ironically, the first transition metal organometallic complex, Zeise's salt, $K[PtCl_3(C_2H_4)] \cdot H_2O$, was discovered nearly 200 years ago and is quite stable to moisture and indeed water-soluble [2].

This book is devoted to an entirely different view of organometallic chemistry: chemistry and catalysis in aqueous solution. Instead of fighting to exclude moisture, why not take advantage of the unique solvent and chemical properties of water and devise organometallic systems which are water-soluble and capable of catalyzing organic reactions in aqueous media? This particular section will examine the wide variety of different metal systems which have been studied in aqueous solution. In some cases, very active catalysts have been developed which operate in water. In other cases, new water-soluble organometallic compounds have been prepared which show great promise for future catalytic applications.

The way in which organometallic systems and their central atoms gain water-solubility can be divided into two very broad categories:

1. The attachment of water-soluble ligands to the transition metal leads to a complex which is water-soluble; and
2. The direct interaction of water with the metal center yields water-solubility.

3.1.2 Water-solubility through the Attachment of Water-soluble Ligands

3.1.2.1 Sulfonated Arylphosphine Ligands

A survey of the various types of water-soluble ligands used to make water-soluble organometallic compounds shows that the bulk of the approaches involve sulfonated aryl phosphines (cf. Section 3.2.1). The first water-soluble sulfonated phosphine, *m*-sulfophenyldiphenylphosphine (TPPMS), **1**, was reported in 1958 and complexes of Ag^+ and Cd^+ were made [3].

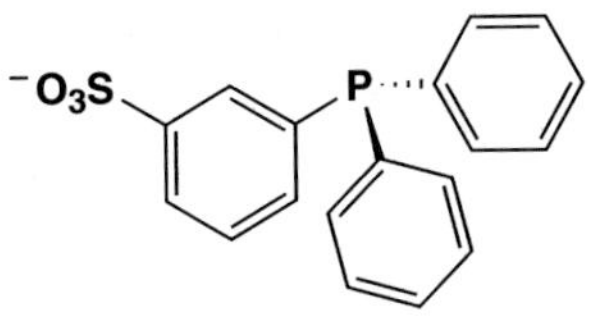

^-O_3S P

1

In 1977, Joó and co-workers described using this phosphine with Rh for an aqueous olefin hydrogenation system [4] (cf. Section 6.2). In 1978, Wilkinson and co-workers reported complexes of this same ligand with Ru, Rh, Pd, and Pt; hydrogenation and hydroformylation catalysis in water was also reported [5]. Joó et al. showed that the chemistry in water may sometimes be quite different from that in organic solvents [6]. For example, whereas dihydrido rhodium species predominate in organic media, monohydrido compounds form via the reaction shown in Eq. (1).

$$RhCl(TPPMS)_3 + H_2 \dashrightarrow HRh(TPPMS)_3 + H^+ + Cl^- \quad (1)$$

Novak and co-workers used TPPMS in the form of $Pd(TPPMS)_3$ as a catalyst for the synthesis of water-soluble poly(*p*-phenylene) derivatives via cross coupling of boronic acids and aryl halides in aqueous solution [7]. Casalnuovo had already shown that this particular catalyst was effective for aqueous alkylation and cross-coupling chemistry [8] (cf. Sections 6.4 and 6.10).

The singly sulfonated ligand does indeed impart water-solubility. However, with all three aromatic rings of triphenylphosphine sulfonated, the resulting tris(*m*-sulfophenyl)phosphine, TPPTS, is much more water-soluble. In turn, metal compounds with this ligand are also very water-soluble, and therefore much of the work in this area utilizes the TPPTS ligand. In 1974, Rhône-Poulenc began studies on using TPPTS in conjunction with rhodium to develop aqueous hydroformylation systems, a venture which resulted in commercial success [9]. This same ligand has also been an important part of some other commercial centures using Rh and Pd (cf. Sections 6.1 and 6.3) [9, 10].

It would appear that most of the research involving TPPTS deals with rhodium as the central metal. This is not particularly surprising since almost all of the investigations of catalysis with TPPTS systems focus on hydroformylation catalysis where Rh appears to be the metal of choice. Quite a number of reviews have been written on this subject [10–14]. An interesting variant of using a TPPTS-solubilized Rh system involves using these systems in a thin aqueous layer adsorbed onto a solid substrate, a mode of catalysis referred to as "supported aqueous-phase catalysis", SAPC (cf. Section 4.7). The first SAP catalyst reported was $HRh(CO)(TPPTS)_3$ in an aqueous layer on beads of controlled-pore-size silica. These systems were effective for the hydroformylation of a number of low-molecular-mass olefins [15, 16].

Herrmann and co-workers did much to extend the use of TPPTS as a ligand to many more systems beyond Rh. They have carried out extensive investigations into the synthesis and purification of TPPTS and have made numerous transition metal complexes, including those of Mn, Fe, Ru, Co, Rh, Ir, Ni, Pd, Pt, Ag, and Au [17–19]. Among the compounds made include the first homoleptic TPPTS compounds such as $Ni(TPPTS)_3$, $Pd(TPPTS)_3$, and $Pt(TPPTS)_4$. It is noteworthy that the coordination number for the metals in these latter complexes is in general lower than for the analogous triphenylphosphine compounds, indicating a somewhat larger cone angle for TPPTS compared with triphenylphosphine.

Darensbourg et al. have attached TPPTS to Cr, Mo, and W carbonyls yielding compounds of the general formula $M(CO)_5(TPPTS)$ and *cis*-$M(CO)_4(TPPTS)_2$. The kinetics of ligand substitution for these compounds was also reported [20, 21].

Hanson and co-workers synthesized the cobalt carbonyl dimer $Co_2(CO)_6(TPPTS)_2$ and showed that it was active for hexene hydroformylation in water [22]. In addition, Hanson, Davis et al. have extended the work on SAPC systems by using TPPTS complexes of Pt with $SnCl_2$ as well as Co for supported hydroformylation catalysis [23].

A number of groups have extended the sulfonation of arylphosphines to more complicated systems. Herrmann's group has developed a number of sulfonated bisphosphines such as BINAS, Structure **2** [24, 25].

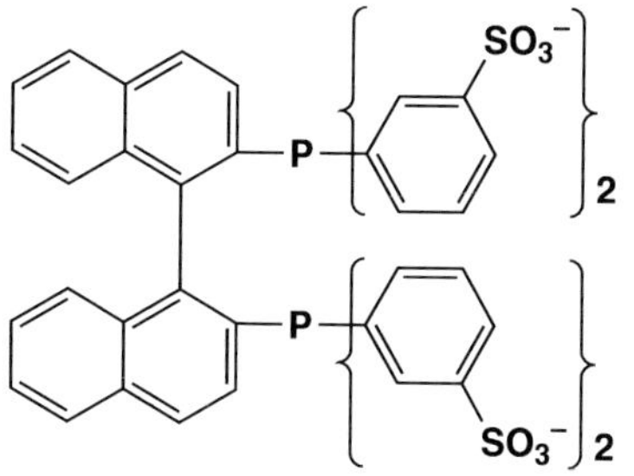

SO_3^- P 2 P SO_3^- 2

2

Using these ligands in combination with Rh produced a catalytic system for the asymmetric hydroformylation of styrene derivatives as well as a "superior" system for the hydroformylation of propylene. A sulfonated BISBIS derivative was also prepared by Herrmann et al. and was used with Rh for hydroformylation catalysis [26]. Hanson and co-workers have also been active in extending the area of sulfonated arylphosphines. Using a number of interesting routes, they have synthesized some unusual surfactant-like sulfonated phosphines and have used them attached to Rh for hydroformylation catalysis [27–29].

Although the "active metal center" is the focus of this review, it is interesting to note that Hanson et al. have shown that the nature and concentration of the "spectator ions" (those that balance the charge of the sulfonate groups) play a role in the catalytic activity of Rh-based hydroformylation systems. Increasing the concentration of monovalent cations (Li^+, Na^+, Cs^+) led to a decrease in hydroformylation activity but an increase in selectivity. On the other hand a trivalent cation (Al^{3+}) had a negative effect on selectivity [30].

3.1.2.2 Other Water-soluble Phosphine Ligands

Although sulfonated aryl phosphine catalysts are the subject of most of the research carried out to date in the area of aqueous catalysis, investigations into other types of water-solubilizing ligands are becoming more prevalent. Instead of using the anionic sulfonate group, a number of instances of using cationic ammonium groups for water-solubility have also been reported. One of the first of these was the AMPHOS, structure **3**, system where attachment to Rh led to water-soluble hydrogenation and hydroformylation systems [31–33]. Hanson and co-workers have also investigated a number of quaternary-ammonium-based systems with Rh for hydroformylation [34, 35].

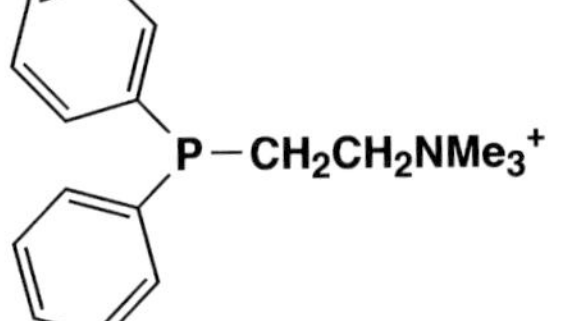

$P-CH_2CH_2NMe_3^+$

3

A number of groups were interested in developing some aliphatic water-soluble phosphines for aqueous catalysis, because of their greater basicity. Pringle et al. (following a report from a 1958 German patent) synthesized tris(hydroxymethyl)phosphine, THMP, structure **4**, from PH_3 and formaldehyde [36–38]. Complexes of THMP with Ni, Pd, and Pt showed that it behaves, in many

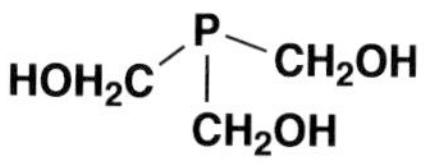
P
HOH2C
CH2OH
CH2OH

4

respects, in a manner similar to that of triethylphosphine. However, the presence of intramolecular hydrogen bonding leads to different properties for the metal complexes, most notably air-stability. Hydrogenation catalysis with these metal systems was investigated and they were shown to be quite active.

Pringle et al. have shown that the addition of PH_3 to aldehydes and ketones provides a general route to a number of hydroxy-substituted aliphatic phosphines which can be used to prepare water-soluble catalyst systems. Addition of formaldehyde to chelating phosphines such as 1,2-diphosphinoethane yield chelating ligands such as 1,2-bis[bis(hydroxymethyl)phosphino]ethane, structure **5** [36]. These ligands were attached to Pt and, again, the role of intramolecular hydrogen bonding was shown to be important in the structure and reactivity of these systems. Tyler et al. used **5** to intercept 19-electron intermediates in aqueous cyclopentadienyl (Cp) molybdenum chemistry (see below) and also reported on the structure of a Ni complex of that ligand [39].

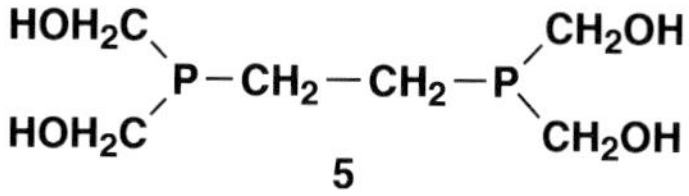
HOH2C
HOH2C
P–CH2–CH2–P
CH2OH
CH2OH

5

Tyler and co-workers also prepared a series of 1,2-bis[bis(hydroxyalkyl)-phosphino]ethane systems by the addition of 1,2-bis(phosphino)ethane to 3-buten-1-ol and 4-penten-1-ol. Ni, Ru, and Rh complexes of this ligand were prepared [40, 41].

Grubbs and co-workers had need of a bulky, water-soluble, aliphatic phosphine, so they developed some novel routes to dicyclohexylphosphine ligands starting with dicyclohexylphosphine–borane and cyclohexylphosphine ligands starting with cyclohexylphosphine–borane [42]. These ligands were used with Ru to create highly active, water-soluble, ring-opening metathesis polymerization (ROMP) catalysts, with complex **6** typifying the Ru carbene complexes synthesized in this study (cf. Section 6.13).

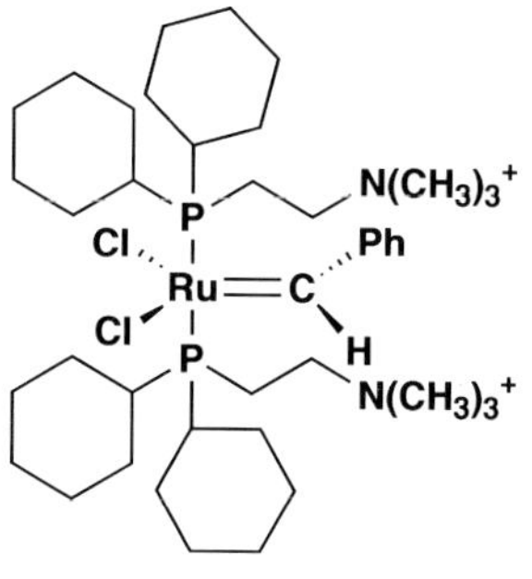
N(CH3)3+
P
Cl
Ph
Ru
C
Cl
P
H
N(CH3)3+

6

Tris(hydroxymethyl)phosphine also plays a role as a precursor to another water-soluble phosphine, 1,3,5-triaza-7-phosphaadamantane (PTA), structure **7**. This ligand was first prepared by Daigle [43] but its coordination chemistry has been extensively developed by Darensbourg et al. One of the first applications of this ligand was in a Mo carbonyl system to make $Mo(CO)_5(PTA)$ [44].

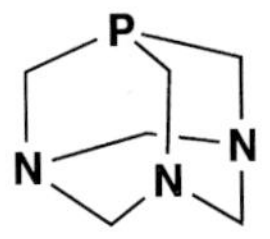

7

Darensbourg and co-workers went on to synthesize numerous Rh and Ru complexes with PTA. The N sites in PTA are also basic and some of these complexes could be isolated as both the neutral PTA adducts as well as the HCl salts. For example, they isolated both $RuCl_2(PTA)_4$, **8**, and $RuCl_2(PTA)_4 \cdot 2HCl$. Systems with PTA and Rh or Ru have been found to be active catalysts for the hydrogenation of aldehydes and olefins in water and also for the hydrogenation of unsaturated aldehydes [20, 45–49].

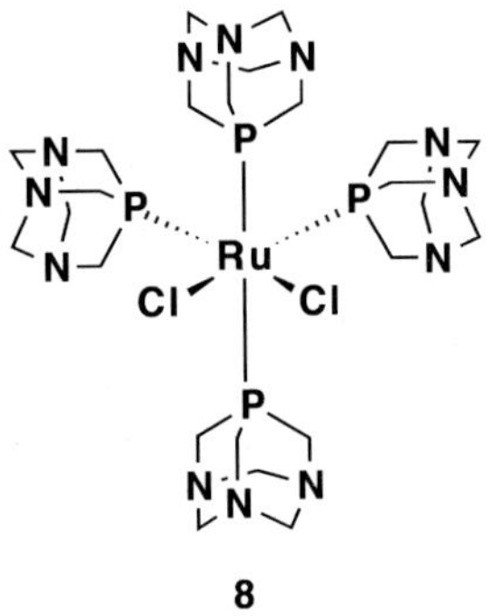

8

3.1.2.3 Nonphosphine-based Water-soluble Ligands

Water-soluble ligands are not restricted to phosphine-based systems. A number of cyclopentadienyl-based ligands have also been reported. The Tyler group has been investigating the generation of 19-electron species in aqueous solution with the hope of exploiting the highly reducing nature of these compounds [40]. A carboxylic acid substituted Cp was used to make a water-soluble tungsten carbonyl dimer, **9** [50]. The pK_a of the carboxylic acid in **9** was 4.5 and so the

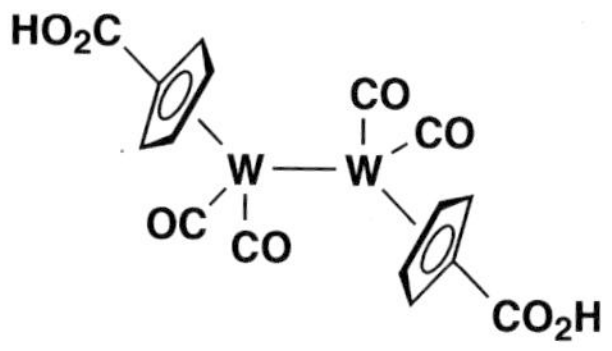

HO2C
CO
CO
W—W
OC CO
CO2H

9

complex was soluble in water at a pH of 6 or greater. In order to carry out studies in water at pH values lower than 6, Tyler et al. also developed the ammonium-substituted CpMo complex, which is soluble at a pH of 8 or lower [51]. They have shown that the photochemistry of these compounds in water is analogous to the photochemistry of the unsubstituted compounds in non-aqueous solution.

3.1.3 Intrinsically Water-soluble Metal Complexes

A second broad class of water-soluble compounds are those which are water-soluble, not as the result of a water-solubilizing ligand, but due to the direct interaction of the metal center with water (intrinsic water-solubility). A very recent example of this type of complex which has very exciting catalytic properties is $[Ru(OH_2)_6](tos)_2$, **10**, where tos = tosylate [52–55]. This compound is very water-soluble and is very active for ROMP catalysis.

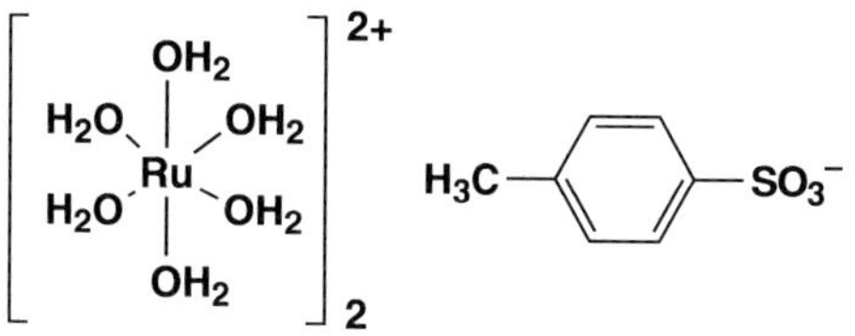

$[Ru(OH_2)_6]^{2+}$ $(H_3C{-}C_6H_4{-}SO_3^-)_2$

10

Grubbs et al. have studied ROMP catalysis with this compound and have also elucidated the mechanism by which this complex catalyzes olefin isomerization [56]. Grubbs was able to isolate and structurally characterize an olefin complex, **11**, derived from this Ru(II) complex and 3-pentenoic acid [57]. Bényei has investigated the use of **10** as a precursor to a number of other water-soluble organometallic catalysts [58].

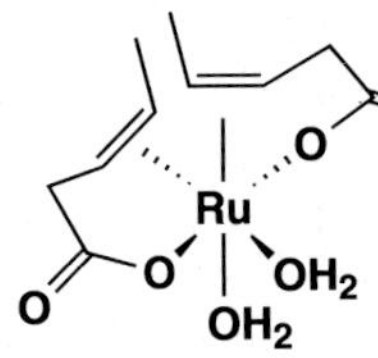

11

However, aqueous organometallic chemistry and catalysis goes back further than that. Quite a number of studies were reported by Darensbourg et al. involving anionic Group 6 metal carbonyl species involved in water-gas shift chemistry. Species such as $M(CO)_5COOH^-$, $M(CO)_5H^-$, and even $M(CO)_5CO_2^{2-}$, as well as dinuclear metal complexes were shown to play a role in water-gas shift catalysis with M = Cr, Mo, W [59–63].

Fish et al. have described the aqueous organometallic chemistry of the $Cp^*Rh(OH_2)_3^{2+}$ species, **12**, (Cp^* = pentamethylcyclopentadienyl) and investigated the structures present in solution as a function of pH [64]. They also showed that the $Cp^*Rh(OH_2)_3^{2+}$ fragment interacts with a number of nitrogen ligands, including amino acids [65].

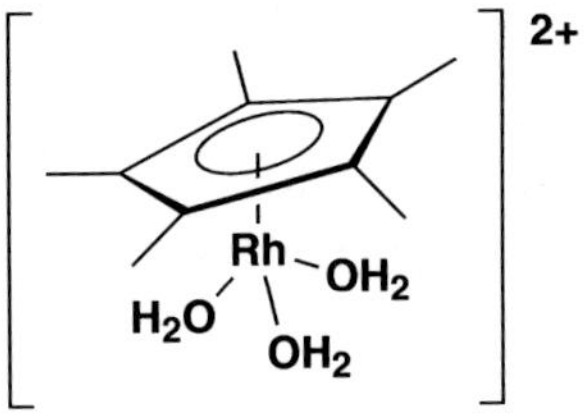

12

Merola et al. described the chemistry of $Cp^*Rh(\eta^2\text{-}O_2CCH_3)(\eta^1\text{-}O_2CCH_3)$, **13** [66]. Upon reaction with water, one molecule of water binds to the Rh and hydrogen-bonds intramolecularly to the two acetate ligands. In aqueous solutions, there is a rapid exchange at room temperature between free and bound water.

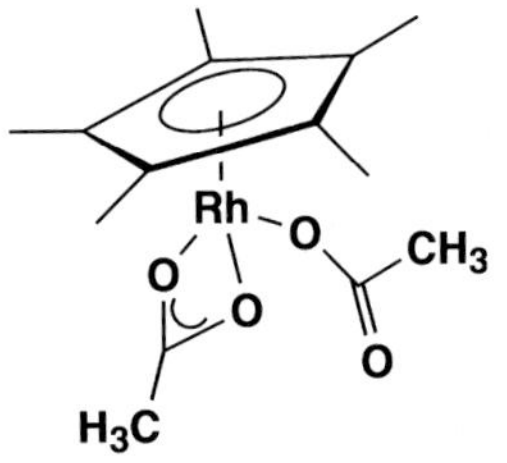

13

Hubbard and co-workers have reported on the aqueous chemistry of the $Cp^*Ru(NO)^{2+}$ fragment and isolated and structurally characterized both monoaquo and diaquo compounds [67].

Merola and co-workers have reported the synthesis and structural characterization of a very water-soluble Ir(I) complex, $[Ir(cod)(PMe_3)_3]Cl$, **14** (cod = 1,5-cyclooctadiene, PMe_3 = trimethylphosphine) [68]. Structure **14** is relatively basic and can be protonated quite readily in water to yield the dicationic $[IrH(cod)(PMe_3)_3]Cl_2$, which can generate chloroiridium compounds in aqueous solution with loss of cod [69].

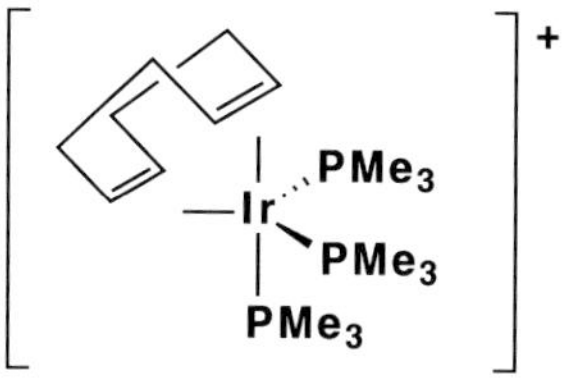

14

Compound **14** can also serve as a precursor to a water-soluble dihydridoiridium complex, *mer*-$(Me_3P)_3IrH_2Cl$, **15** [70]. Compound **15** is quite active for the hydrogenation of alkynes and alkenes in water with the aquo species, **16**, playing an important role in the catalysis. Alkenyl hydrido iridium complexes were isolated from the reaction between the dihydride **15** and alkynes in water. Merola and co-workers also described some water-soluble iridium complexes based on the chelating ligands bis(dimethylphosphino)ethane (dmpe) and various 1,2-ethylenediamines [69]. These compounds are all quite active for H_2 addition and for hydrogenation catalysis.

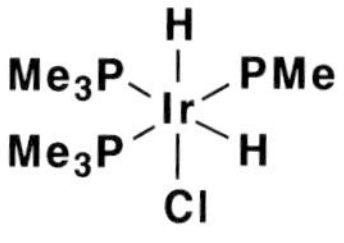

15 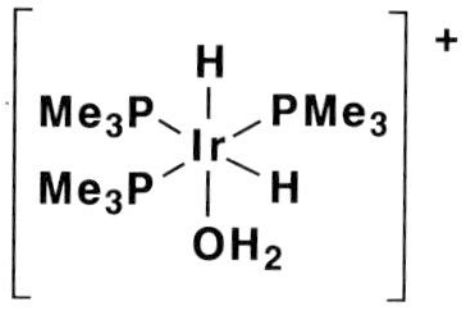 **16**

A very exciting report by Flood and co-workers involves rhodium complexes with 1,4,7-trimethyl-1,4,7-triazacyclononane (Cn). $CnRhCl_3$ was originally reported by Weighardt et al. and Flood and co-workers used this precursor to synthesize $CnRhMe_3$. From $CnRhMe_3$, using carefully controlled additions of HX (X = Cl, OTf, BF_4) (OTf = triflate, trifluoromethanesulfonate), they were able to synthesize $CnRhMe_2X$ and $CnRhMeX_2$. These two compounds dissolve in water to form cationic aquo compounds which undergo insertion of ethylene into the Rh–C bonds. At pH = 8.6, an aquohydroxorhodium complex is formed which is active for ethylene polymerization [71, 72].

Palladium-based coupling reactions have also been successfully transferred into aqueous media. Grotjahn et al. have investigated the intramolecular Heck reaction in aqueous media using Pd acetate. Grotjahn's group showed that aqueous media played an important role in favoring alkene insertion over β-hydride elimination to allow a double Heck reaction [73, 74]. Kiji also demonstrated the efficiency of aqueous solutions for the Heck reaction using Pd [75].

Finally, one of the more unusual aqueous systems reported to date is the use of methylrhenium trioxide (MTO), $MeReO_3$, for homogeneously catalyzed Diels–Alder reactions in water [76]. The Diels–Alder reactions were investigated in both organic solvents and water, and water was usually found to be the best solvent when the dienophiles were α,β-unsaturated ketones or aldehydes.

3.1.4 Conclusions

There is a large variety of metal centers which have been shown to have a rich organometallic chemistry in water. As researchers develop new water-soluble ligands and look at transition metal interactions with water more closely, this field will develop even further. Taking advantage of the unique properties of water will be a challenge for many years to come.

References

[1] G. Wilkinson, *J. Organomet. Chem.* **1975**, *100*, 273.
[2] P. I. Starosel'skii, Y. I. Solov'ev, *Koord. Khim.* **1978**, *4*, 1123.
[3] S. Ahrland, J. Chatt, N. R. Davies, A. A. Williams, *J. Chem. Soc.* **1958**, 264.
[4] F. Joó, Z. Toth, M. T. Beck, *Inorg. Chim. Acta* **1977**, *25*, L61–L62.
[5] A. F. Borowski, D. J. Cole-Hamilton, G. Wilkinson, *Nouv. J. Chim.* **1978**, *2*, 137.
[6] F. Joó, L. Nádasdi, A. Bényei, P. Csiba, A. Kathó, *NATO ASI Ser., Ser. B* **1995**, *5*, 23.
[7] T. I. Wallow, B. M. Novak, *J. Am. Chem. Soc.* **1991**, *113*, 7411.
[8] A. L. Casalnuovo, J. C. Calabrese, *J. Am. Chem. Soc.* **1990**, *112*, 4324.
[9] B. Cornils, E. Wiebus, *Chem. Ing. Techn.* **1994**, *66*, 916; *CHEMTECH* **1995**, *25*, 33.
[10] M. Beller, B. Cornils, C. D. Frohning, C. W. Kohlpaintner, *J. Mol. Catal. A*, **1995**, *104*, 17.
[11] W. A. Herrmann, C. W. Kohlpaintner, *Angew. Chem.* **1993**, *105*, 1588; *Angew. Chem., Int. Ed. Engl.* **1993**, *32*, 1524.
[12] M. Barton, J. D. Atwood, *J. Coord. Chem.* **1991**, *24*, 43.
[13] F. Joó, A. Kathó, *J. Mol. Catal. A: Chem.* **1997**, *116*, 3.
[14] D. M. Roundhill, *Adv. Organomet. Chem.* **1995**, *38*, 155.
[15] J. P. Arhancet, M. E. Davis, J. S. Merola, B. E. Hanson, *J. Catal.* **1990**, *121*, 327.
[16] J. P. Arhancet, M. E. Davis, J. S. Merola, B. E. Hanson, *Nature (London)* **1989**, *339*, 454.
[17] W. A. Herrmann, J. Kellner, H. Riepl, *J. Organomet. Chem.* **1990**, *389*, 103.

[18] W. A. Herrmann, J. A. Kulpe, W. Konkol, H. Bahrmann, *J. Organomet. Chem.* **1990**, *389*, 85.
[19] W. A. Herrmann, J. A. Kulpe, J. Kellner, H. Riepl, H. Bahrmann, W. Konkol, *Angew. Chem.* **1990**, *102*, 408.
[20] D. J. Darensbourg, C. J. Bischoff, *Inorg. Chem.* **1993**, *32*, 47.
[21] D. J. Darensbourg, C. J. Bischoff, J. H. Reibenspies, *Inorg. Chem.* **1991**, *30*, 1144.
[22] I. Guo, B. E. Hanson, I. Toth, M. E. Davis, *J. Organomet. Chem.* **1991**, *403*, 221.
[23] I. Toth, I. Guo, B. E. Hanson, *J. Mol. Catal. A: Chem.* **1997**, *116*, 217.
[24] W. A. Herrmann, C. W. Kohlpaintner, R. B. Manetsberger, H. Bahrmann, H. Kottmann, *J. Mol. Catal. A: Chem.* **1995**, *97*, 65.
[25] R. W. Eckl, T. Priermeier, W. A. Herrmann, *J. Organomet. Chem.* **1997**, *532*, 243.
[26] W. A. Herrmann, C. W. Kohlpaintner, H. Bahrmann, W. Konkol, *J. Mol. Catal.* **1992**, *73*, 191.
[27] T. Bartik, B. Bartik, B. E. Hanson, *J. Mol. Catal.* **1994**, *88*, 43.
[28] T. Bartik, B. Bartik, B. E. Hanson, I. Guo, I. Toth, *Organometallics* **1993**, *12*, 164.
[29] H. Ding, B. E. Hanson, T. Bartik, B. Bartik, *Organometallics* **1994**, *13*, 3761.
[30] H. Ding, B. E. Hanson, *J. Mol. Catal. A: Chem.* **1995**, *99*, 131.
[31] R. T. Smith, R. K. Ungar, L. J. Sanderson, M. C. Baird, *Organometallics* **1983**, *2*, 1138.
[32] R. T. Smith, M. C. Baird, *Inorg. Chim. Acta* **1982**, *62*, 135.
[33] R. T. Smith, R. K. Ungar, M. C. Baird, *Transition Met. Chem.* **1982**, *7(5)*, 288.
[34] I. Toth, B. E. Hanson, *Tetrahedron: Asymm.* **1990**, *1*, 895.
[35] I. Toth, B. E. Hanson, M. E. Davis, *Tetrahedron: Asymm.* **1990**, *1*, 913.
[36] P. G. Pringle, D. Brewin, M. B. Smith, K. Worboys, *NATO ASI Ser, Ser. 3* **1995**, *5*, 111.
[37] P. A. T. Hoye, P. G. Pringle, M. B. Smith, K. Worboys, *J. Chem. Soc., Dalton Trans.* **1993**, 269.
[38] J. W. Ellis, K. N. Harrison, P. A. T. Hoye, A. G. Orpen, P. G. Pringle, M. B. Smith, *Inorg. Chem.* **1992**, *31*, 3026.
[39] G. F. Nickarz, T. J. R. Weakley, W. K. Miller, B. E. Miller, D. K. Lyon, D. R. Tyler, *Inorg. Chem.* **1996**, *35*, 1721.
[40] D. R. Tyler, *NATO ASI Ser., Ser. 3* **1995**, *5*, 47.
[41] G. T. Baxley, W. K. Miller, D. K. Lyon, B. E. Miller, G. F. Nieckarz, T. J. R. Weakley, D. R. Tyler, *Inorg. Chem.* **1996**, *35*, 6688.
[42] B. Mohr, D. M. Lynn, R. H. Grubbs, *Organometallics* **1996**, *15*, 4317.
[43] D. J. Daigle, A. B. Pepperman, Jr., *J. Chem. Eng. Data* **1975**, *20*, 448.
[44] M. Y. Darensbourg, D. Daigle, *Inorg. Chem.* **1975**, *14*, 1217.
[45] F. Joó, L. Nádasdi, A. C. Bényei, D. J. Darensbourg, *J. Organomet. Chem.* **1996**, *512*, 45.
[46] D. J. Darensbourg, T. J. Decuir, J. H. Reibenspies, *NATO ASI Ser., Ser. 3* **1995**, *5*, 61.
[47] D. J. Darensbourg, N. W. Stafford, F. Joó, J. H. Reibenspies, *J. Organomet. Chem.* **1995**, *488*, 99.
[48] D. J. Darensbourg, F. Joó, M. Kannisto, A. Kathó, J. H. Reibenspies, *Organometallics* **1992**, *11*, 1990.
[49] D. J. Darensbourg, F. Joó, M. Kannisto, A. Kathó, J. H. Reibenspies, D. J. Daigle, *Inorg. Chem.* **1994**, *33*, 200.
[50] D. M. Schut, T. J. R. Weakley, D. R. Tyler, *New J. Chem.* **1996**, *20*, 113.
[51] A. Avey, D. R. Tyler, *Organometallics* **1992**, *11*, 3856.
[52] B. M. Novak, R. H. Grubbs, *J. Am. Chem. Soc.* **1988**, *110*, 7542.
[53] D. V. McGrath, B. M. Novak, R. H. Grubbs, *NATO ASI Ser., Ser. C* **1990**, *326*, 525.
[54] M. A. Hillmyer, C. Lepetit, D. V. McGrath, B. M. Novak, R. H. Grubbs, *Macromolecules* **1992**, *25*, 3345.
[55] M. A. Hillmyer, C. Lepetit, D. V. McGrath, R. H. Grubbs, *Polym. Prepr.* **1991**, *32(1)*, 162.

[56] D. V. McGrath, R. H. Grubbs, *Organometallics* **1994**, *13*, 224.
[57] D. V. McGrath, R. H. Grubbs, J. W. Ziller, *J. Am. Chem. Soc.* **1991**, *113*, 3611.
[58] A. C. Bényei, *NATO ASI Ser., Ser. 3* **1995**, *5*, 159.
[59] D. J. Darensbourg, M. Y. Darensbourg, R. R. Burch, Jr., J. A. Froelich, M. J. Incorvia, *Adv. Chem. Ser.* **1979**, *173*, 106.
[60] D. J. Darensbourg, A. Rokicki, M. Y. Darensbourg, *J. Am. Chem. Soc.* **1981**, *103*, 3223.
[61] D. J. Darensbourg, A. Rokicki, *ACS Symp. Ser.* **1981**, *152*, 107.
[62] D. J. Darensbourg, A. Rockicki, *Organometallics* **1982**, *1*, 1685.
[63] D. J. Darensbourg, R. L. Gray, C. Ovalles, *J. Mol. Catal.* **1987**, *41*, 329.
[64] M. S. Eisen, A. Haskel, H. Chen, M. M. Olmstead, D. P. Smith, M. F. Maestre, R. H. Fish, *Organometallics* **1995**, *Volume Date 1995, 14,* 2806.
[65] S. Ogo, B. H. Chen, M. M. Olmstead, R. H. Fish, *Organometallics* **1996**, *15*, 2009.
[66] P. M. Boyer, C. P. Roy, J. M. Bielski, J. S. Merola, *Inorg. Chim. Acta* **1996**, *245*, 7.
[67] A. Svetlanova-Larsen, C. R. Zoch, J. L. Hubbard, *Organometallics* **1996**, *15*, 3076.
[68] J. F. Frazier, J. S. Merola, *Polyhedron* **1992**, *11*, 2917.
[69] J. S. Merola, T. L. Husebo, K. E. Matthews, M. A. Franks, R. Pafford, P. Chirik, *NATO ASI Ser., Ser. 3* **1995**, *5*, 33.
[70] T. X. Le, J. S. Merola, *Organometallics* **1993**, *12*, 3798.
[71] L. Wang, T. C. Flood, *J. Am. Chem. Soc.* **1992**, *114*, 3169.
[72] L. Wang, R. S. Lu, R. Bau, T. C. Flood, *J. Am. Chem. Soc.* **1993**, *115*, 6999.
[73] D. B. Grotjahn, X. Zhang, *J. Mol. Catal. A: Chem.* **1997**, *116*, 99.
[74] D. B. Grotjahn, X. Zhang, *NATO ASI Ser., Ser. 3* **1995**, *5*, 123.
[75] J. Kiji, T. Okano, T. Kasegawa, *J. Mol. Catal.* **1995**, *97*, 73.
[76] Z. Zhu, J. H. Espenson, *J. Am. Chem. Soc.* **1997**, *119*, 3507.

3.2 Variation of Ligands

3.2.1 Monophosphines

Othmar Stelzer

3.2.1.1 General Features, Scope, and Limitations

A great deal of the development of catalysts with unprecedented activities and selectivities within the last decade has been achieved by a systematic modification of suitable phosphorus ligands bearing functionalities and elements of chirality in their backbones or peripheries. It was already recognized, however, in the early stages of this development that these homogeneous processes suffer from the difficulties of separating the catalysts from the products. This problem has been solved in an elegant manner using strongly hydrosoluble catalysts in immiscible aqueous/organic two-phase systems following a proposal by Manassen [1 a]. The water-solubility of the catalysts employed in aqueous two-phase catalysis can be achieved by appropriate modification of the phosphine ligands with polar groups such as SO_3^-, COO^-, NMe_3^+, OH, etc. First attempts to carry out transition metal catalyzed reactions using water-soluble phosphine ligands date back to 1973 [1 b]. The monosulfonated derivative of Ph_3P had already been synthesized in 1958 [1 c], and the well-known standard ligand TPPTS (trisulfonated triphenylphosphine) was reported by Kuntz in 1975 [2]. The development of new types of hydrosoluble phosphine ligands with "tailor-made" structures for highly active and selective two-phase catalysts is an ongoing challenge to chemists working in this field. Aspects of the topic under review have been covered in the literature by review articles [3–6] and monographs [7].

In this section catalysts containing monophosphines for aqueous-phase catalysis will be presented, the influence of ligand variation on catalyst activity being emphasized. Syntheses of hydrophilic monodentate phosphine ligands will be discussed very briefly. Reference to catalysts containing bidentate and tenside ligands (see Sections 3.2.2–3.2.4) is made only where appropriate.

3.2.1.2 Anionic Phosphines

3.2.1.2.1 Phosphines Containing Sulfonated Aromatic and Aliphatic Groups

Ligand Syntheses

Water-soluble phosphines of this type reported in the literature so far are collected as Structures **1**–**6**. Direct sulfonation of the neutral "mother phosphine" with oleum, introduced originally by Kuntz in 1975 [2] for the preparation of TPPTS (**1a**), is still the most important procedure; it has also been used for the syntheses of an extended series of TPPTS-type catalyst ligands, e.g., **1b** (R = Ph [8], C_6H_{13} [9]), **1c** (R = alkyl, *c*-Hex; R′ = Me, OMe [10]), **1d** (R = 4-F-C_6H_4 [11]), **1e** [12]. Sulfonated tris (ω-phenylalkyl)phosphines **5a** [13] and their *p*-phenylene analogs **5b** [14] as well as the bicyclic phosphine **4** [15] were obtained in an analogous manner. The kinetics of PPh_3 sulfonation have been investigated by Lecomte and Sinou [16].

The formation of phosphine oxides, which is a serious disadvantage inherent to this synthetic procedure [17], may be suppressed by addition of boric acid to the reaction mixtures [18]. Alternative synthetic routes have been developed using either nucleophilic phosphination of sulfonated fluorobenzenes, e.g., F-C_6H_4-2-SO_3K, F-C_6H_4-4-SO_3K, F-C_6H_3-2,4-$(SO_3K)_2$, in superbasic media (DMSO/KOH) or Pd-catalyzed P–C coupling of sulfonated bromo- or iodoaromatic compounds with PH_3 or primary and secondary phosphines. The scope of these synthetic strategies is rather broad, the *p*-isomer of TPPTS (**2a**) [19], higher and even secondary (R = H, R′ = Ph) sulfonated phosphines **2b**, **2c**, and **3** [20] being accessible. The phosphines **2b**, **2c** (R, R′ = Ph) have been obtained by reaction between Ph_2PK or Ph_2PCl with Li or K 4-chlorobenzenesulfonate or *o*-lithiated lithiumbenzenesulfonate, respectively [21 b, 22]. A multistage low-yield synthesis has been published for the methyl derivative **1b** (R = Me; $n = 1$) [23]. Alkylation of the alkali metal phosphides R_2PM (M = Li, Na, K) or $RPLi_2$ with sultones [24–26] or ω-haloalkyl sulfonates [27] affords the sulfonated benzyldiphenylphosphine **5c**, phenylsulfonatoalkylphosphines **6a**, **6c**, and the peralkyl derivative **6b** [22 b, 27, 28].

Catalysts Containing Sulfonated Aromatic Phosphines as Ligands

Ruthenium complexes of TPPTS and TPPMS [29–32] have been employed as catalysts or catalyst precursors for the hydrogenation of α,β-unsaturated carbonyl compounds in biphasic systems ([29–33], cf. Section 6.2). Dimeric structures **7**, **8** were assigned to the catalyst complexes $[RuX(\mu\text{-Cl})(TPPMS)_2]_2$ (X = H, Cl) on the basis of NMR studies [30]. The regioselectivity toward the production of α,β-unsaturated alcohols is greatly enhanced on going from the

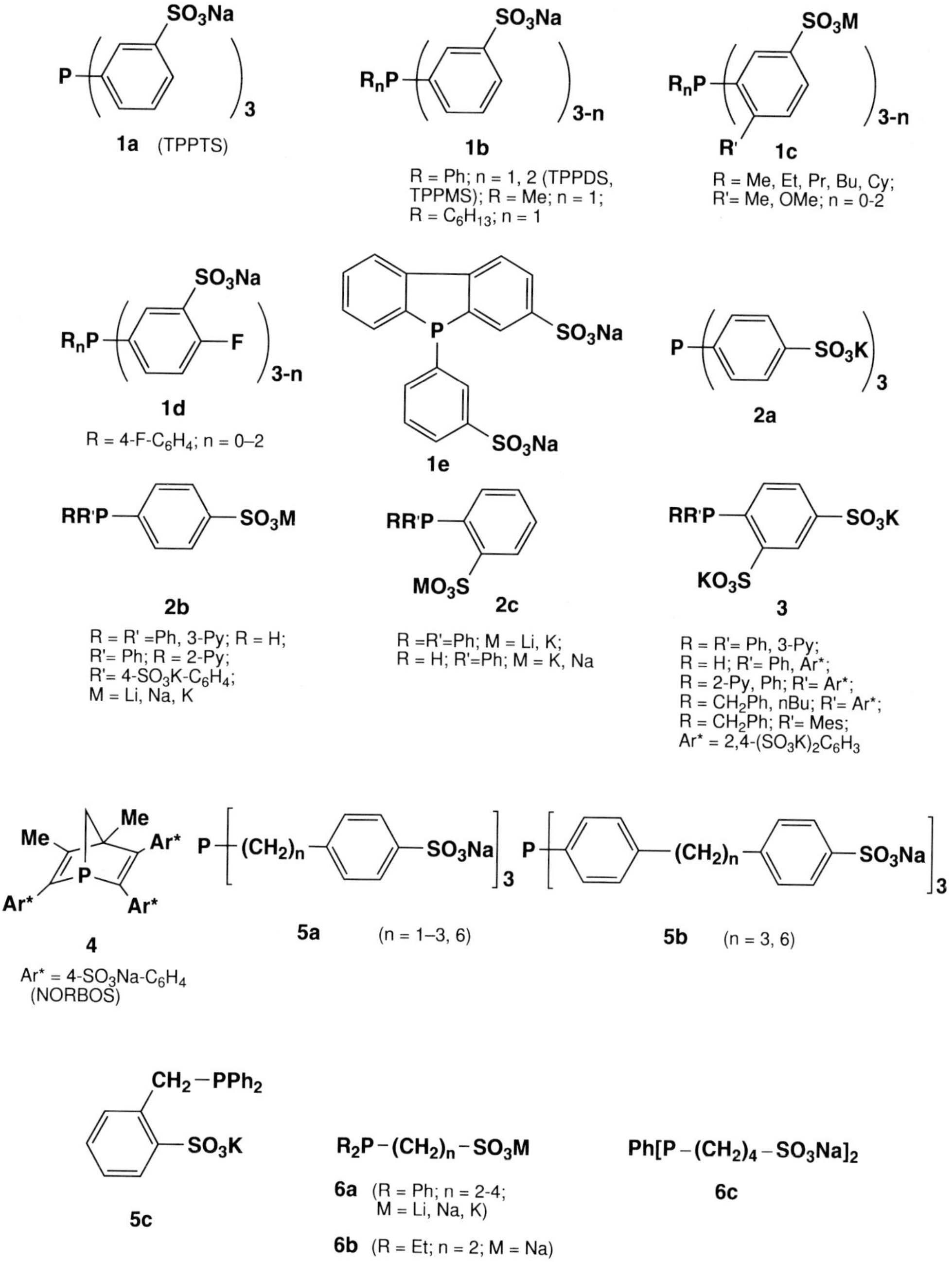

1a (TPPTS)

1b

R = Ph; n = 1, 2 (TPPDS, TPPMS); R = Me; n = 1; R = C_6H_{13}; n = 1

1c

R = Me, Et, Pr, Bu, Cy; R'= Me, OMe; n = 0-2

1d

R = 4-F-C_6H_4; n = 0–2

1e

2a

2b

R = R' =Ph, 3-Py; R = H; R'= Ph; R = 2-Py; R'= 4-SO_3K-C_6H_4; M = Li, Na, K

2c

R =R'=Ph; M = Li, K; R = H; R'=Ph; M = K, Na

3

R = R'= Ph, 3-Py; R = H; R'= Ph, Ar*; R = 2-Py, Ph; R'= Ar*; R = CH_2Ph, nBu; R'= Ar*; R = CH_2Ph; R'= Mes; Ar* = 2,4-$(SO_3K)_2C_6H_3$

4

Ar* = 4-SO_3Na-C_6H_4 (NORBOS)

5a (n = 1–3, 6)

5b (n = 3, 6)

5c

$R_2P-(CH_2)_n-SO_3M$

6a (R = Ph; n = 2-4; M = Li, Na, K)

6b (R = Et; n = 2; M = Na)

$Ph[P-(CH_2)_4-SO_3Na]_2$

6c

Water-soluble phosphines containing sulfonated aryl and alkyl side chains.

homogeneous medium (using Ph_3P as the ligand) to the biphasic system using the TPPMS or TPPTS catalysts $[RuCl(\mu\text{-}Cl)L_2]_2$ [30]. While TPPMS is acting as a surface-active agent, TPPTS behaves more like an electrolyte [29]. In acidic solution formation of phosphonium salts, e.g., **9**, was observed if the $RuCl_3 \cdot 3H_2O$/TPPTS catalyst system was used [32]. In Rh-catalyzed hydrogenation of activated olefins, phosphonium salt formation obviously promotes the formation of the catalytically active species [34a] by consuming free phosphine [34b]. This reaction may lead to severe phosphine loss and may account in part for the large excess phosphine requirement in related catalytic processes. The hydrogenation of unsaturated aldehydes can be directed to either the allylic alcohols or the saturated aldehydes by proper choice of the metal (Ru, Rh) [33]. TPPMS complexes of osmium, OsH_4L_3, $OsHCl(CO)L_2$ and $[OsCl(\mu\text{-}Cl)L_2]_2$, have also been found to be active catalysts for hydrogenation of α,β-unsaturated aldehydes [30]. A monomeric structure **10** was assigned to the hydrido complex $OsHCl(CO)L_2$. In Ru-catalyzed propionaldehyde hydrogenation the complexes $[RuCl(\mu\text{-}Cl)L_2]_2$, $RuH(X)L_3$ (X = Cl, I, OAc) and H_2RuL_4 (L = TPPTS) have been employed, the real catalyst being H_2RuL_3 [35]. Complexes prepared by ligand exchange between $RuCl_2(Ph_3P)_3$ and TPPTS have been used in aqueous solution for hydrogenation of D-glucose and D-mannose with molecular hydrogen and by transfer hydrogenation [36].

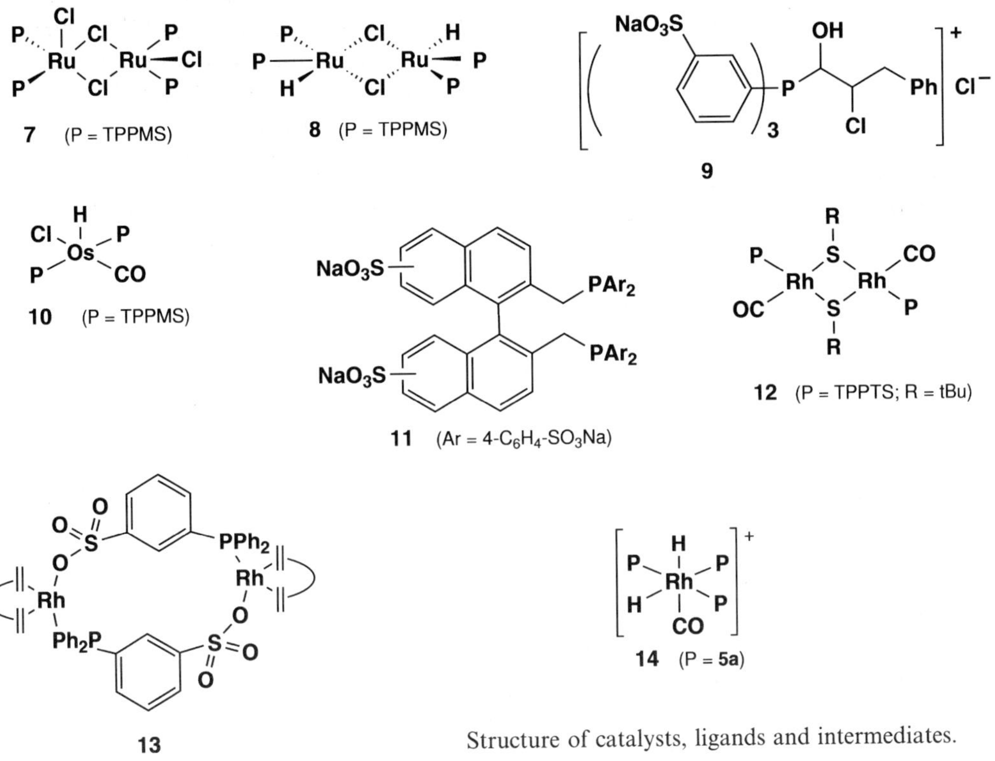

7 (P = TPPMS)

8 (P = TPPMS)

9

10 (P = TPPMS)

11 (Ar = 4-C_6H_4-SO_3Na)

12 (P = TPPTS; R = tBu)

13

14 (P = **5a**)

Structure of catalysts, ligands and intermediates.

The technically most important biphasic process is the Ruhrchemie–Rhône–Poulenc hydroformylation of propene using the in-situ Rh(I) catalyst $HRh(CO)(TPPTS)_3$ [6, 37]. Its formation from $Rh(CO)_2(acac)$ and TPPTS in a syngas atmosphere has been studied in detail [38, 39]. The BINAS-Na (**11**)/Rh catalyst showed an outstanding performance in propene hydroformylation [15]. Binuclear thiolato bridged rhodium complexes **12** have been used in 1-octene hydroformylation as precatalysts [41]. For details of the hydroformylation, cf. Section 6.1 [15, 40, 41].

Rh(I) catalysts containing TPPTS-type ligands with electron-withdrawing groups in the aromatic rings show higher *n*/*i* selectivity in hydroformylation reactions. Thus *para*-fluorinated derivatives of TPPTS and TPPDS (**1d**), which are weaker bases and stronger π-acids than TPPTS and TPPDS [42], in 1-hexene hydroformylation gave *n*/*i* selectivities of 93:7, compared with 86:14 for the non-fluorinated ligands [11]. The application of aqueous two-phase catalysis in hydroformylation of longer-chain olefins is hampered, however, in most cases by their low solubility in water. This problem has been overcome by using phosphine ligands like **5b**, showing a more pronounced surface-active character than TPPTS. Light-scattering experiments [43] on aqueous solutions of catalysts obtained from $(acac)Rh(CO)_2$ and **5b** ($n = 6$) (L/Rh ratio = 2:1) indicated the presence of aggregates (micelles) with a hydrodynamic radius of ca. 19 Å. In hydroformylation of *n*-octene an average TOF of 160 was measured for **5b**, compared with 90 for TPPTS, the *n*/*i* ratio (at L/Rh = 2:1) being much better [14]. This was attributed to micelle formation providing a hydrophobic pocket for binding the olefin. Reaction rate and selectivity increase with ionic strength of the solution, since added salt promotes the formation of micelles [44]. For a detailed discussion of stability and structure of complex catalysts containing highly sulfonated ligands, e.g., $HRh(CO)(TPPTS)_3$, the role of the "spectator" cations (Li^+, Na^+, K^+, Cs^+, Al^{3+}) has to be considered [45]. One of the 2^9 possible ligand conformations in $RhH(CO)(TPPTS)_3$ derived from the solid-state structure of $HRh(CO)(PPh_3)_3$ with the *meta*-SO_3^- groups being as far apart on average as possible is shown in Figure 1.

High ionic strengths stabilize the hydration sphere by minimizing the electrostatic repulsions between the sulfonate groups. As a consequence the dissociation energy for TPPTS with formation of the active species $\{HRh(CO)(TPPTS)_2\}$ will be increased, thus lowering the activity of the catalyst. In agreement with this reasoning a value of 30.6 or 22.4 kcal mol^{-1} has been calculated for the barrier of exchange at high or low ligand and complex concentration, respectively [44]. The decrease in phosphine dissociation rates in the complexes *cis*-$Mo(CO)_4L_2$ (L = $\{(Na^+\text{-kryptofix-221})_3[P\text{-}(C_6H_4\text{-}m\text{-}SO_3^-)_3]\} > PPh_3 \geq$ TPPTS) is in line with this, showing the stabilizing effect of the alkali-metal cations. For the tri-anionic cryptated and encapsulated TPPTS containing $P(C_6H_4\text{-}m\text{-}SO_3^-)_3$ a cone angle $\cong 30\%$ greater than that of PPh_3 was estimated [46] in agreement with the value of 170° as determined from the X-ray structure of $(Na^+\text{-kryptofix-221})_3[(CO)_5W\text{-}P(C_6H_4\text{-}m\text{-}SO_3^-)_3]$ [47]. Using PH_3 as the model phosphine, the PH_3 gas-phase dissociation energies for $HRh(CO)(PH_3)_3$ have been calculated at different theoretical levels, being 19.4

or 22.7 ($kcal\,mol^{-1}$) for HF or MP2 optimized geometries [48]. The agreement of the calculated with the experimental value for the PPh_3 complex (20 ± 1 $kcal\,mol^{-1}$) [49] might be the result of a fortuitous error cancellation.

The surface-active character of sulfonated phosphines, e.g., TPPMS (**1b**, $n = 2$), was already noted in 1977 by Wilkinson et al., who also commented on the participation of the SO_3^- groups in the coordination of these ligands to transition metals in catalyst complexes (**13**) [50]. Rh complexes of **1b** (TPPDS) have been employed as catalysts in hydroformylation of olefins in biphasic media [21 a]. The *p*-isomer **2b** (Ph_2P-C_6H_4-*p*-SO_3K) was, however, much more reactive in the presence of a surfactant (lauric acid) [21 b]. Rh catalysts using TPPMS have been employed in the Union Carbide process of hydroformylation of higher olefins combining a homogeneous catalytic process in polyalkene glycols and an aqueous bisphasic recovery of the catalyst [51, 52]. Problems in Rh/TPPTS-catalyzed hydroformylation arising from the insolubility of the olefin in water can be solved at least in part by addition of co-solvents [41, 53], quaternary phosphonium salts or by performing the reaction in micellar systems which can be created by addition of tensides to the organic/aqueous phase [54, 55] (cf. Sections 3.2.4 and 4.5). Alternatively supported aqueous-phase catalysis (SAPCs; cf. Section 4.7) may be employed, which, with two recent exceptions [56], has been limited to the use of TPPTS. Replacing TPPTS by the more basic weakly surface-active ligand **1b** ($R = C_6H_{13}$) provided no improvement in the activity of the SAP Rh catalyst in C_8–C_{12} alkene hydroformylation, however [9].

Catalysis of 1-octene hydroformylation in biphasic systems by $[Rh(cod)Cl]_2$-TPPTS can be dramatically enhanced by addition of lipophilic phosphines, e.g., Ph_3P. The rate enhancement by the "promoter ligands" is believed to be the result of an increasing local concentration of the catalytic species at the interface [57]. Mixed-ligand complexes $HRh(CO)(TPPTS)_{3-x}(Ph_3P)_x$ prepared separately from HRh(CO)(TPPTS) and Ph_3P are likely not to be catalyst precursors.

Rh(I) complexes such as $RhCl(TPPMS)_3 \cdot 4H_2O$, $Rh(cod)Cl(TPPMS) \cdot H_2O$, $[Rh(cod)(\mu_2\text{-}Ph_2P\text{-}C_6H_4\text{-}3\text{-}SO_2\text{-}O)]_2$ catalyze oligomerization of alkynes (propynoic acid, arylacetylenes); this has been studied only very recently in terms of biphasic catalysis. The Rh(I) complex $[Rh(cod)(\mu_2\text{-}Ph_2P\text{-}C_6H_4\text{-}3\text{-}SO_2\text{-}O)]_2$ already reported by Wilkinson et al. [50] promotes the conversion of phenylacetylene to 1,2,4- and 1,3,5-triphenylbenzene under biphasic conditions (toluene/water) [58].

Palladium complexes of TPPTS and TPPMS have been employed extensively as catalysts for carbonylation, hydroxycarbonylation, and C–C cross-coupling reactions (cf. Section 6.3). Hydroxycarbonylation of bromobenzene in biphasic medium using $Pd(TPPTS)_3$ as catalyst yields benzoic acid, which remains in the aqueous phase, thus avoiding the direct recycling of the catalyst [59]. The formation of $Pd(TPPTS)_3$ from $PdCl_2$ and TPPTS in aqueous solution has been studied in detail by ^{17}O, $\{^1H\}^{31}P$, and ^{35}Cl NMR spectroscopy. The complex $[PdCl(TPPTS)_3]^+Cl^-$ obtained initially is reduced by excess TPPTS, TPPTSO being formed. A more attractive synthesis of $Pd(TPPTS)_3$ involves the facile reduction of $[PdCl(TPPTS)_3]^+Cl^-$ with CO (Scheme 1) [60].

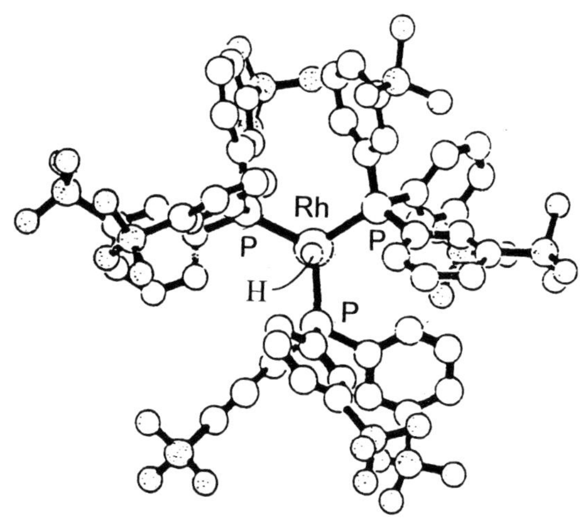

Figure 1. Representation of $HRh(CO)(TPPTS)_3$.

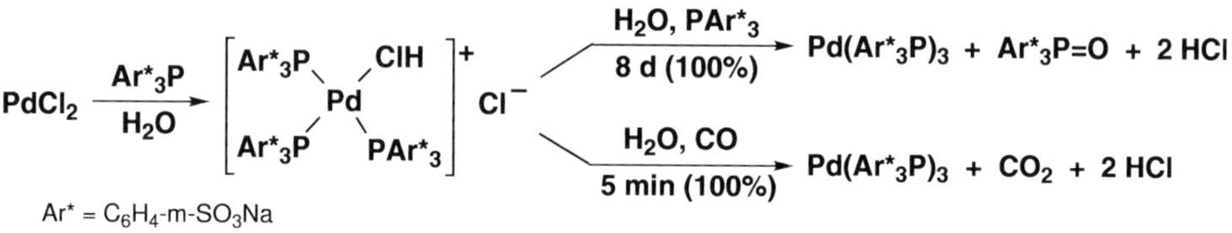

$$PdCl_2 \xrightarrow[H_2O]{Ar^*_3P} \left[\begin{matrix} Ar^*_3P & & ClH \\ & Pd & \\ Ar^*_3P & & PAr^*_3 \end{matrix} \right]^+ Cl^-$$

$$\xrightarrow[8\ d\ (100\%)]{H_2O,\ PAr^*_3} Pd(Ar^*_3P)_3 + Ar^*_3P{=}O + 2\ HCl$$

$$\xrightarrow[5\ min\ (100\%)]{H_2O,\ CO} Pd(Ar^*_3P)_3 + CO_2 + 2\ HCl$$

$Ar^* = C_6H_4\text{-m-}SO_3Na$

Scheme 1. Formation of $Pd(TPPTS)_3$.

Using $Pd(TPPTS)_3$ as a catalyst, 5-hydroxymethylfurfural could be carbonylated chemoselectively in water to yield 5-formylfuran-2-acetic acid. Replacement of TPPTS by ligands containing fewer SO_3Na groups (TPPDS, TPPMS, **1d**, $n = 1$) gives rise to a dramatic drop in catalytical activity [61]. The $Pd(OAc)_2$/TPPTS mixture widely applied as catalyst for sp–sp and sp^2–sp C–C coupling reactions without a Cu(I) promoter spontaneously affords a palladium(0) complex "$Pd(TPPTS)_n$" [62]. Using $PdCl_2(TPPMS)_2$ as catalyst, benzyl chloride could be carbonylated to form phenylacetic acid in 93% yield. The reaction is accelerated by addition of surfactants [63].

The x-ray structure of $Pd(TPPMS\text{-}K)_3(H_2O)_4$ obtained by reductive phosphination of K_2PdCl_4 with TPPMS has been determined [64a]. $Pd(TPPMS)_3$ is a highly efficient catalyst for C–C coupling reactions between aryl and heteroaromatic halides with aryl- and vinylboronic acids and alkynes in biphasic systems and single-phase aqueous media [64]. This catalyst system using either TPPMS or TPPDS is also active in Stille-type coupling reactions (cf. Section 6.4) between organohydroxotin complexes $K_n[RSn(OH)_{3+n}]$ or organotin trihalides and aryl halides [65, 66]. Water-soluble $ArPdL_2I$ complexes (L = TPPMS, **2b** ($R = R' = Ph$) [19, 21 b, 22 a], which may be envisaged as intermediates in these reactions, have been isolated [22 a]. Pd(0) complexes formed in the TPPTS/$Pd(OAc)_2$ system have been employed as catalysts for the syntheses of polynuclear aromatic compounds by cross-coupling reactions [67]. The same catalyst was employed for the substitution of allylic substrates with carbon- and heteronucleophiles under mild conditions [68].

Using a catalyst mixture of $Pd(OAc)_2$ and TPPTS, TPPMS or **1c** ($R' = Me$, $n = 0$) [69–72] in a molar ratio of about 1:5, telomerization of butadiene to octadienol could be achieved with high efficiency in micellar systems formed by addition of surfactants (cf. Section 6.7). Surprisingly, butadiene conversion and octadienol selectivities were not affected significantly by the hydrophilicity of the phosphines. Selectivity reached an unprecedented value of 97% with **1c** ($R' = Me$, $n = 0$), employing $C_{12}H_{25}NMe_2$ as surfactant precursor. This was interpreted in terms of the steric effects of the ligands, the cone angles increasing significantly within the series Ph_3P (145°), TPPMS (151°), TPPTS (166°), and **1c** ($R' = Me$, $n = 0$) (194°) [46]. The Pd/TPPTS catalyst has been very recently employed in the reductive carbonylation of nitroaromatic compounds in aqueous biphasic systems proceeding in a highly selective manner and tolerating reactive functional groups [73]. For the reduction of nitroaromatic compounds with molecular hydrogen the complex $Rh(CO)Cl(TPPTS)_2$ was used as catalyst [74].

Ni(0) complexes of TPPTS have been employed as catalysts for the hydrocyanation of dienes and unsaturated nitriles (cf. Section 6.5). Product linearity and catalyst lifetimes can be improved if the catalysis is performed in a xylene/water biphasic system by using TPPTS as co-catalyst [75]. The Ni(0)/TPPTS complexes employed may be obtained by electrochemical reduction of $Ni(CN)_2$ in water in presence of TPPTS [76].

Catalysts Containing Electron-donating Phosphines with Sulfonated Alkyl Side Chains

Rhodium(I) complexes $HRh(CO)L_3$ of alkali-metal phenylphosphinoalkylsulfonates (L = **6a**, $n = 3, 4$; **6c**) have been used as catalysts for hydroformylation of higher olefins (e.g., *n*-1-tetradecene) in methanolic solution. The catalyst could be recovered without loss of activity by extraction of the isolated product with water. Using the bis-sulfonated ligand **6c** instead of monosulfonated **6a** more than half of the aldehyde was reduced to the alcohol [25]. The electron-donating water-soluble phosphines **5a** have been used to generate aqueous catalysts with $Rh(CO)_2(acac)$ for hydroformylation of 1-octene, its conversion increasing with increasing *n*. On reaction of **5a** (L) with $Rh(CO)_2(acac)$ complexes Rh(acac)(CO)L are formed initially, which with dihydrogen give $HRh(CO)L_3$. These complexes, which are stable for L = TPPTS, decompose via oxidative addition of water to yield $[H_2Rh(CO)L_3]^+OH^-$, to which the dihydride structure **14** has been assigned [77]. Zwitterionic cobalt(I) complexes $Na_5[Co(CO)_3L_2]$ obtained from $Co_2(CO)_8$ and the electron-donating water-soluble phosphines **5a** (L) are precatalysts in aqueous biphasic and SAP systems, giving almost exclusively the aldehyde [78]. Palladium(II) complexes $PdCl_2L_2$ of **5c** (L) and **6a** (L) have been proposed as catalysts for the hydrodehalogenation of allyl and benzyl halides using potassium formate as hydrogen source in a biphasic water/heptane system. The selectivity (allylbenzene vs. propenylbenzene) was controlled to a larger extent by different surfactants than by variation of the ligands [79]. The complexes formed from bis(dibenzylideneacetone)-palladium and the electron-rich $Et_2PCH_2CH_2SO_3Na$ (**6b**) are effective catalysts for telomerization of conjugated dienes [28].

3.2.1.2.2 Phosphines Containing Carboxylated Aromatic Groups and Side Chains

Ligand Syntheses

Structures **15**–**23** are hydrophilic phosphine ligands bearing carboxylic groups. In contrast to their alkali-metal salts, the free acids show only moderate solubilities in water. The ligands **15**–**17** have been obtained by standard preparative methods comprising alkylation and arylation of alkali-metal organophosphides in organic solvents (THF, DME, dioxane) [80–82]. Improved synthetic procedures based on the nucleophilic phosphination of fluorobenzoic acids [83] or Pd-catalyzed P–C coupling reaction of bromo- and iodobenzoic acid with primary or secondary phosphines [20] are known. These methods are of broad applicability and can be used also for the syntheses of multiply functionalized phosphinocarboxylic acids (**20**, **22**) including the diphenylphosphinophenylacetic acids (**21**) and the novel phosphine ligands containing amino acid moieties (**23**) [84]. Phosphines of type **16** ($n = 2$) have been obtained by oxidation of

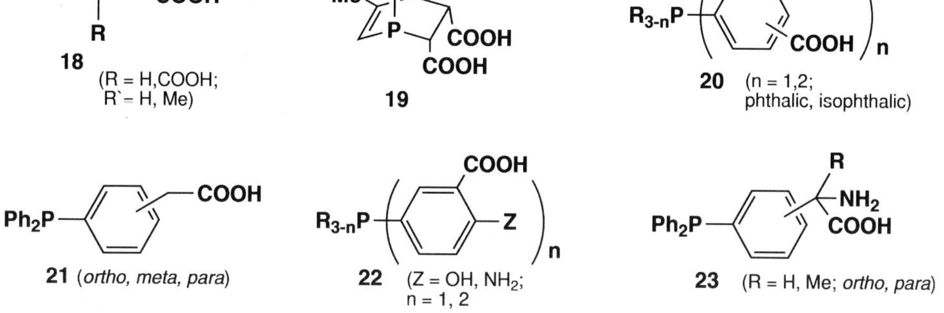

15 (R, R' = H, Me) **16** (n = 1-5, 7; M = H, Na) **17** (n = 1-3; *ortho, meta, para*)

18 (R = H,COOH; R' = H, Me) **19** **20** (n = 1,2; phthalic, isophthalic)

21 (*ortho, meta, para*) **22** (Z = OH, NH2; n = 1, 2) **23** (R = H, Me; *ortho, para*)

Carboxylated phosphines.

hydroxyalkylphosphine–borane complexes with chromic anhydride in sulfuric acid following by decomplexation with NEt_2H [85]. Schumann and co-workers published a multistage synthesis for the ligands **17** [86]. The novel bicyclic carboxylated phosphine **19** was obtained using the [4+2] dimer of 3,4-dimethyl-2*H*-phosphole as a starting material [87]. Free-radical addition of diphenylphosphine to unsaturated carboxylic esters followed by saponification gave the mono- and dicarboxylated phosphines **18** [88].

Catalysts Containing Carboxylated Phosphines as Ligands

Compared with their sulfonated analogs, phosphine ligands containing carboxylic acid moieties have been much less investigated as catalyst components, although some of them (e.g., Ph_2PCH_2COOH, **15**) have already been applied at an early stage in the Shell Higher-olefin Process (SHOP; cf. Section 7.1) [89], the first large-scale industrial biphasic but non-aqueous catalytic process.

The surface-active rhodium(I) complexes obtained by ligand exchange between $RhCl(Ph_3P)_3$ and long-chain ligands of type **16** were found to be effective catalysts for hydrogenation of olefins and polybutadiene in aqueous and non-aqueous biphasic media. Terminal olefins were hydrogenated much faster than internal ones. An unusual enhanced reaction rate was observed for the internal double bond in 2-pentene and 3-pentene nitriles [90]. Rh(I) complexes of the amphiphilic ligands of type **17** ($n = 1$, *para*) have shown to be active catalysts in 1-octene hydroformylation. They could be separated from the product by basic extraction into water and re-extraction after neutralization into a new batch of *n*-octene with toluene [91]. Preliminary experiments of van Leeuwen and co-workers [92] have shown, however, that this ligand in its protonated form with $Rh(CO)_2(acac)$ produces complexes that are insoluble in all common solvents and water at variable pH values. They were of low activity in hydroformylation of *n*-octene due to formation of polymeric rhodium carboxylate structures. The binuclear thiolato-bridged Rh(I) complex $[Rh(CO)(\mu\text{-SPh})(\mathbf{17})]_2$ containing the carboxylated ligand **17** ($n = 1$, *meta*) catalyzes trimerization of propynoic acid or phenylacetylene in aqueous or biphasic medium (toluene/water), trimesic (38%) and trimellitic acid (62%) or 1,2,4- and 1,3,5-triphenylbenzene (22%, 5%) being formed [58].

3.2.1.2.3 Phosphines Containing Phosphonated Aromatic Groups and Alkyl Side Chains

Ligand Syntheses

The water-soluble phosphonate-functionalized phosphines **24** (M = Na) have been obtained by halogen–metal exchange on bromophenyldiphenylphosphines with *n*BuLi followed by reaction with diethyl-chlorophosphate and subsequent hydrolysis [93]. An alternative more generally applicable synthesis of phosphines of type **24** comprises the nucleophilic phosphination of fluorophenylphosphonic ethyl esters with Ph_2PK [94]. ω-Alkylphosphonate-functionalized phosphines **25** are accessible by reaction of Li-CH_2-P(O)(O*i*Pr)$_2$ with $BrPPh_2$ ($n = 1$), alkylation of Ph_2PLi and PPh_2H with ω-bromoalkylphosphonates ($n = 2$) [27] or 3-bromopropyldiisopropylphosphonate in presence of KO*t*Bu ($n = 3$) [95], respectively. The synthesis of the bicyclic phosphines **26** and **27** is based on the [4 + 2] cycloaddition of ethynyl phosphonates or phosphonamides to 1-phenyl-3,4-dimethylphosphole [96]. Hydrolysis with HCl or de-ethylation with Me_3SiBr followed by hydrolysis gave the free acids in all cases; neutralization with NaOH yielded the sodium salts.

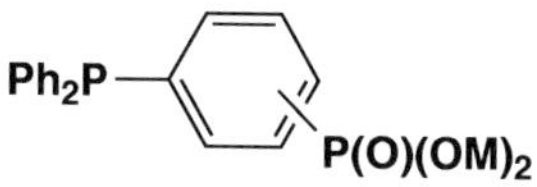

Ph_2P–C_6H_4–$P(O)(OM)_2$ **24** (M = Et, H, Na; *ortho, meta, para*)

$R_2P-(CH_2)_n-P(O)(OM)_2$ **25** (M = Me, Et, iPr, H, Na; R = Ph, Et; n = 1-3)

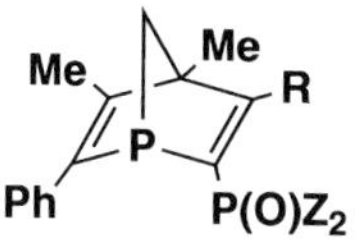

26 (Z = NEt_2, OEt, OH, ONa; R = Ph, H)

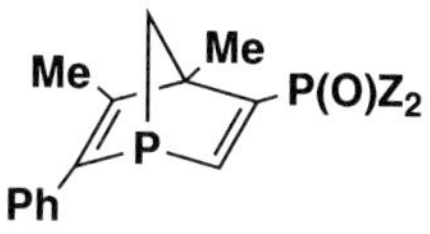

27 (Z = NEt_2, OH, ONa)

Phosphonated phosphines.

Catalysts Containing Phosphonated Phosphines as Ligands

The cationic palladium(II) complex $[Pd(\mathbf{24})_3Cl]^+$ of the *para*-isomer of **24** (M = Na) catalyzes the carbonylation of benzyl chloride in basic medium to give phenylacetic acid in high yields. The Pd(0) complex $[Pd(\mathbf{24})_3]$, formed by reduction of $[Pd(\mathbf{24})_3Cl]^+$ with CO, is asumed to be the catalytic species [93] (see Scheme 1). Palladium complexes of ligands related to **24** (M = Na) have also been employed in aqueous ethylene glycol phases as catalysts for Suzuki-type C–C cross-coupling reactions for the syntheses of substituted biphenyls [97].

The bicyclic ligands **26** were considered to be of interest as substitutes for TPPMS in the new oxo process developed by Union Carbide for the hydroformylation of higher olefins using *N*-methylpyrrolidone or polyalkylene glycols as solvents [7, 51, 52]. Rh(I) complexes $[Rh(\mathbf{26})_2]^+$ [96] showed, however, a very poor performance as catalysts in biphasic systems for hydrogenation and hydroformylations in contrast to non-functionalized 1-phosphanorbornadiene [98]. This was explained by formation of P,P(O) chelates blocking the catalytic cycle. This hypothesis was confirmed by the pronounced activity of analogous Rh(I) complexes obtained from **27** (Z = ONa) for which P(O) chelation is not possible for steric reasons.

3.2.1.3 Cationic Phosphines

3.2.1.3.1 Ligand Syntheses

The synthesis of AMPHOS, the prototype of cationic phosphines, known for almost 20 years, requires intermediate protection of the phosphorus in $Ph_2P\text{-}(CH_2)_2\text{-}NMe_2$ by oxidation or coordination to a transition metal before *N*-quaternization [99, 100]. In the case of the cyclohexyl analog **28** (R = Cy) and dicationic **30**, borane was used as the protecting group.

Thus, starting with $Cy_2P(H)BH_3$ and $CyP(H_2)BH_3$ base-assisted aminoalkylation, *N*-quaternization and *P*-deprotection with morpholine gave **28** or **30** (R = Cy) [101]. Further AMPHOS analogs could be obtained without using a protecting group by two-phase protected selective *N*-quaternization and subsequent *P*-alkylation of primary phosphines $H_2P\text{-}(CH_2)_n\text{-}NR'_2$ [102]. Phosphines **33** bearing terminal phosphonium groups are accessible by quaternization of PMe_3 with ω-chloroalkyl-diphenylphosphines or alkylation of $LiPPh_2$ with ω-halogenoalkylphosphonium salts $[X\text{-}(CH_2)_n\text{-}PMe_3]^+X^-$, respectively [103].

Ligands **31** and **32** containing guanidinium moieties constitute a novel type of cationic phosphine ligand showing extreme solubilities in water. The guanidinium groups were introduced into the corresponding aminoalkyl- and aminoarylphosphines by addition of cyanamides $R_2N\text{-}CN$ (R = H, Me) or 1*H*-pyrazole-1-carboxamidine [104].

$[R_2P\text{-}(CH_2)_n\text{-}NR'_3]^+ X^-$

28

R = Ph; R' = Me; n = 2 (AMPHOS), 3;
R = Cy; R' = Me; n = 2;
R = Me; R'$_3$ = Me_2Oct, Me_2Dod; n = 2;
R = H; R'$_3$ = Me_3, $Me_2(C_mH_{2m+1})$;
m = 6-8, 12, 16, 18; n = 2, 3;
X = Cl, Br, I, NO_3

$\{CyP\text{-}[(CH_2)_2\text{-}NMe_3]_2\}^{2+}\ 2\ Cl^-$

30

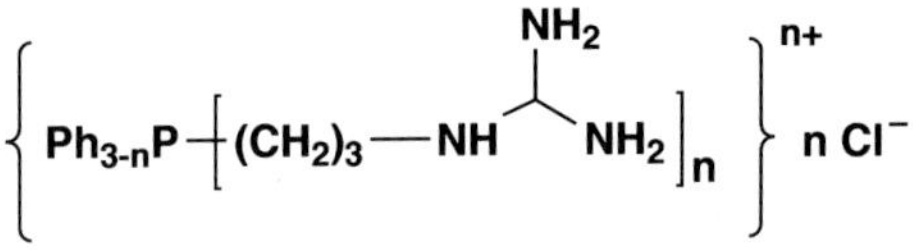

29

31 (n = 1, 2)

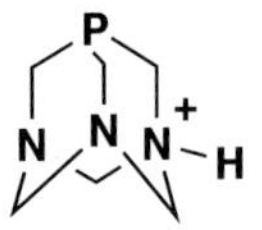

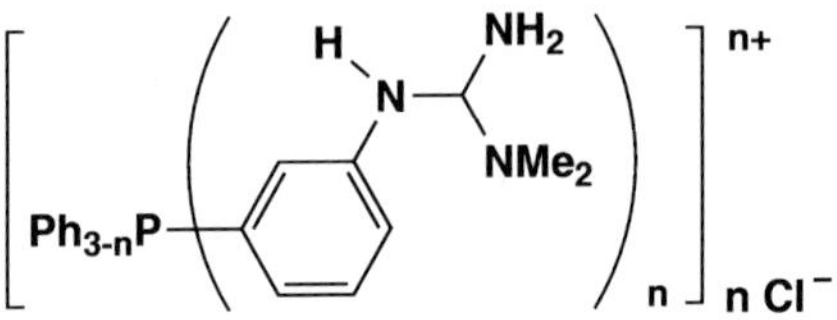

$[Ph_2P\text{-}(CH_2)_n\text{-}PMe_3]^+ X^-$

33 (n = 2, 3, 6, 10; X= Cl, PF_6, NO_3; PHOPHOS II, III, VI, X)

32 (n = 1-3)

Cationic phosphine ligands.

3.2.1.3.2 Catalysts Containing Cationic Phosphines as Ligands

Rhodium(I) and cobalt(0) complexes of AMPHOS, $[(nbd)Rh(AMPHOS)_2]^{3+}$ and $[Co(CO)_3(AMPHOS)]_2(PF_6)_2$, were already being used in the early 1980s for the hydrogenation and hydroformylation of maleic and crotonic acid or styrene and 1-hexene in water or aqueous biphasic systems [99, 105]. The lower effectiveness of the Co catalyst was attributed to the lighter metal's proclivity to oxidation and phosphine dissociation. Binuclear thiolato-bridged Rh(I) complexes, $[Rh(CO)(AMPHOS)(tBuS)]_2^{2+}2[BPh_4]^-$ are highly active catalysts for the hydrogenation of unsaturated alcohols and acids [106]. Water-soluble Rh(I) complexes of monoprotonated 1,3,5-triaza-7-phosphaadamantane (PTA), $[Rh(PTA)_2(PTAH)Cl]Cl$ (PTAH = **29**), have been used as catalysts for the regioselective reduction of unsaturated to saturated aldehydes with sodium formate in an aqueous biphasic system or dihydrogen in presence of ethanol as co-solvent [107]. The P-basicity of PTA and PTAH is believed to be similar to that of triarylphosphines, but the cone angles are smaller [108]. The cyclohexyl analog of AMPHOS (**28**, R = Cy) is less electron-donating than the corresponding ligand bearing an SO_3^- instead of an NMe_3^+ group. Their Ru complexes $L_2Cl_2Ru{=}CHPh$ (L = **28**) initiate ring-opening metathesis polymerization (ROMP) of 7-oxanorbornenes in water, methanol, and aqueous emulsions [101].

Palladium complexes of AMPHOS (**28**, $R'_3 = Me_3$ and Me_2H; X = Cl) are suitable catalysts for telomerization of butadiene and isoprene in biphasic systems in the presence of CH_3OH/CH_3ONa. Conversion rates and selectivities are altered only very little upon catalyst recycling, the results obtained for butadiene comparing favorably with those obtained for TPPTS [109]. A comparative study of the alkyl- and arylguanidinium phosphines **31** and **32** including TPPTS in the palladium-catalyzed Castro–Stephens C–C coupling reactions between *p*-iodobenzoate and (trifluoroacetyl)propargylamine has shown **32** to be of superior activity [104]. Copper(I) iodide promotes the conversion rates but is not vital to these reactions. The Pd complex formed from $Pd(OAc)_2$ and TPPTS barely exceeds the $Pd(OAc)_2$-catalyzed background reaction.

$^-OOC{-}C_6H_4{-}I$ + $H{-}C{\equiv}C{-}CH_2{-}NH{-}C(=O){-}CF_3$ $\xrightarrow[H_2O/CH_3CN]{[Cat]/base}$ $^-OOC{-}C_6H_4{-}C{\equiv}C{-}CH_2{-}NH{-}C(=O){-}CF_3$

[Cat] = $Pd(OAc)_2$/**32** (n = 2)

Rh complexes of composition $[Rh(nbd)(\mathbf{33})_2]^+X^-$ containing the phosphonium phosphines **33** (PHOPHOS II, III, VI, X) have been shown to be very active catalysts for the biphasic hydrogenation of *n*-hexene and maleic acid in water [103, 110]. A definite chain length effect was observed, the system where $n = 6$ being the most active. The biphasic systems containing longer chain ligands are not well behaved as catalysts since they are prone to formation of stable emulsions which are quite difficult to break.

3.2.1.4 Nonionic Water-soluble Phosphines

3.2.1.4.1 Ligand Syntheses

Water-solubilization of organometallic catalysts is preferably performed by introduction of anionic or cationic functional groups into the phosphine ligands. The concept of attaining water-solubility by incorporation of hydroxylic groups into the ligand periphery has attracted much less attention. Large-scale syntheses for **34** [111] based on K_2PtCl_4-catalyzed addition of aqueous formaldehyde to PH_3 or decomposition of commercially available $[P(CH_2OH)_4]^+Cl^-$ have been published recently [112]. Ligands of type **35a** have been obtained by addition of $Ph_{3-n}PH_n$ to ethylene glycol monoallyl ethers [113]. Chiral ligands of this type, e.g. **35b**, showing unprecented η^3-mode bonding to transition metals, were prepared by Mathieu and co-workers [114]. The hydroxyphenylphosphines **36** are accessible by multistage syntheses [115] or by making use of the Pd-catalyzed P–C coupling reaction employing iodophenols as starting materials [20]. Ethoxylation of mono-, di- and tris-*p*-hydroxytriphenylphosphines with ethylene oxide yields polyether-substituted triphenylphosphines (PETPPs) designed for use in micellar [54, 116] and thermally regulated [117] phase-transfer catalysis. Ligands whose water-solubility is inversely dependent on temperature were first reported by Bergbreiter et al. [118]. This subject will be discussed in more detail in Sections 3.2.4, 4.5, and 4.6.3.

Ligands incorporating the diphenylphosphine moiety into sugar structures have been reported by a number of authors [119]. This subject has been reviewed

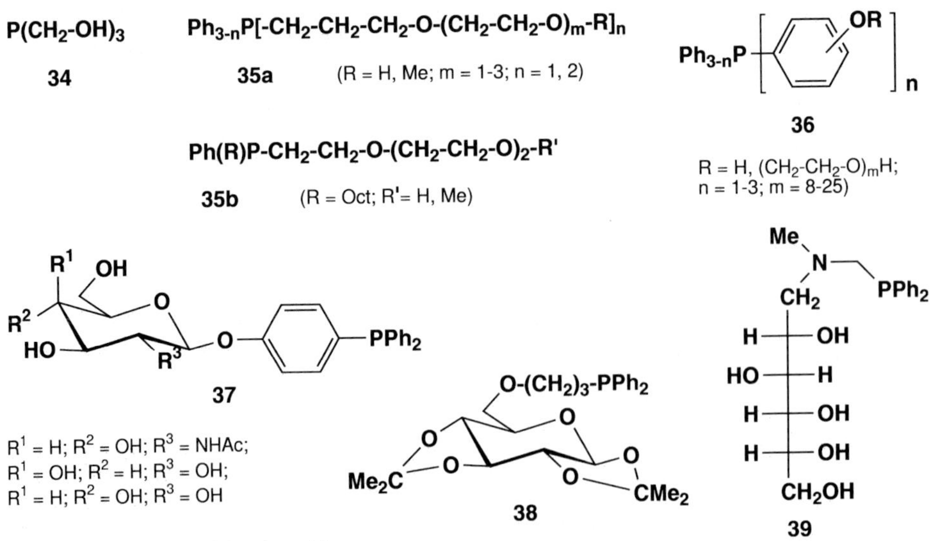

$P(CH_2\text{-}OH)_3$ **34**

$Ph_{3-n}P[\text{-}CH_2\text{-}CH_2\text{-}CH_2\text{-}O\text{-}(CH_2\text{-}CH_2\text{-}O)_m\text{-}R]_n$ **35a** (R = H, Me; m = 1-3; n = 1, 2)

$Ph(R)P\text{-}CH_2\text{-}CH_2\text{-}O\text{-}(CH_2\text{-}CH_2\text{-}O)_2\text{-}R'$ **35b** (R = Oct; R'= H, Me)

36 R = H, $(CH_2\text{-}CH_2\text{-}O)_mH$; n = 1-3; m = 8-25)

37 R^1 = H; R^2 = OH; R^3 = NHAc; R^1 = OH; R^2 = H; R^3 = OH; R^1 = H; R^2 = OH; R^3 = OH

38

39

Nonionic water-soluble phosphines.

by Blaser [120]. In most cases, however, the hydroxyl functions were fully (e.g., **38**) or at least in part protected, the solubilities of these ligands (especially those of the acetonides) in water being low [113]. By two-phase glycosidation of acetyl-protected halopyranoses with *p*-hydroxyphenyldiphenylphosphine at ambient temperature and subsequent O–Ac deprotection the aryl-*β*-*O*-glycosides of glucose, galactose, and glucosamine (**37**) could be obtained [121]. Temperature-dependent partition coefficients in the system water/ethanol/di-*n*-butyl ether indicate thermally reversible solvation for **37** ($R^1 = H$, $R^2 = OH$, $R^3 = NHAc$). The sugar phosphines **39** have been synthesized by a Mannich-type condensation involving hydroxymethyldiphenylphosphine and sugars which incorporate the amine function [122].

3.2.1.4.2 Catalysts Containing Neutral Phosphines as Ligands

Homoleptic Ni(0) and Pd(0) complexes of **34**, $M[P(CH_2OH)_3]_4$, have been reported [112]. The rhodium complexes, e.g., *trans*-$RhCl(CO)[P(CH_2OH)_3]_2$, are relatively active catalysts for the water-gas shift reaction [108]. The catalytic activities and *n*/*i* selectivities of Rh(I) complexes $HR(CO)_nL_m$ of 3-hydroxyphenyldiphenylphosphine (L = **36**; R = H; $n = 1$) and unmodified Ph_3P in *n*-octene hydroformylation in organic solvents (toluene) are comparable [92]. Within the new catalyst recycling approach, which is not solely restricted to water but can switch between an organic and aqueous phase by varying the pH of the system, the ligand **36** could not be employed, however, because of its weakly acidic character [91]. The Pd complexes obtained from the sugar phosphines **37** and $Pd(OAc)_2$ turned out to be catalytically more active in Heck-type and Suzuki-type C–C coupling reactions than those formed with TPPTS. This was attributed to the higher catalyst concentration of the organic phase in the case of **37**. The concept of thermally regulated phase-transfer catalysis has been successfully applied to the biphasic hydroformylation of *n*-1-dodecene using Rh(I) complexes of **36** (L; $R = (CH_2CH_2O)_mH$; $n = 1-3$) of composition $HRh(CO)L_3$ obtained by reaction of the ligands with $Rh(CO)_2(acac)$ in a syngas atmosphere [116, 117].

References

[1] (a) J. Manassen, *Catalysis: Progress in Research*, pp. 177, 183 (Eds.: F. Basolo, R. L. Burwell), Plenum Press, London, **1973**; (b) J. Chatt, G. J. Leigh, R. M. Slade, *J. Chem. Soc., Dalton Trans.* **1973**, 2021; (c) S. Ahrland, J. Chatt, N. R. Davies, A. A. Williams, *J. Chem. Soc.* **1958**, 276.

[2] Rhône–Poulenc Recherche (E. Kuntz), FR 2.314.910 (1975).

[3] M. Barton, J. D. Atwood, *J. Coord. Chem.* **1991**, *24*, 43; D. Sinou, *Bull. Soc. Chim. Fr.* **1987**, *3*, 480.

[4] W. A. Herrmann, C. W. Kohlpaintner, *Angew. Chem.* **1993**, *105*, 1588; *Angew. Chem., Int. Ed. Engl.* **1993**, *32*, 1524.

[5] B. Cornils, *Angew. Chem.* **1995**, *107*, 1709; *Angew. Chem., Int. Ed. Engl.* **1995**, *34*, 1574.
[6] B. Cornils, E. G. Kuntz, *J. Organomet. Chem.* **1995**, *502*, 177; B. Cornils, W. A. Herrmann, R. W. Eckl, *J. Mol. Catal. A: Chem.* **1997**, *116*, 27.
[7] B. Cornils, W. A. Herrmann, *Applied Homogeneous Catalysis with Organometallic Compounds* (Eds.: B. Cornils, W. A. Herrmann), VCH, Weinheim, **1996**.
[8] Ruhrchemie AG (H. Bahrmann, B. Cornils, W. Lipps, P. Lappe, H. Springer), Can. CA 1.247.642 (1988), *Chem. Abstr. 111*, 97503; Ruhrchemie AG (L. Bexten, B. Cornils, D. Kupies), DE 3.431.643 (1986).
[9] I. Tóth, I. Guo, B. E. Hanson, *J. Mol. Catal. A: Chem.* **1997**, *116*, 217.
[10] Hoechst AG (G. Albanese, R. Manetsberger, W. A. Herrmann), EP 704.450 (1996).
[11] B. Fell, G. Papadogianakis, *J. Prakt. Chem.* **1994**, *336*, 591.
[12] Hoechst AG (G. Albanese, R. Manetsberger, W. A. Herrmann, R. Schmid), EP 704.452 (1996).
[13] T. Bartik, B. Bartik, B. E. Hanson, I. Guo, I. Tóth, *Organometallics* **1993**, *12*, 164.
[14] H. Ding, B. E. Hanson, J. Bakos, *Angew. Chem.* **1995**, *107*, 1728; *Angew. Chem., Int. Ed. Engl.* **1995**, *34*, 1645; H. Ding, B. E. Hanson, T. Bartik, B. Bartik, *Organometallics* **1994**, *13*, 3761.
[15] W. A. Herrmann, C. W. Kohlpaintner, R. B. Manetsberger, H. Bahrmann, H. Kottmann, *J. Mol. Catal. A: Chem.* **1995**, *97*, 65.
[16] L. Lecomte, D. Sinou, *Phosphorus, Sulfur, Silicon* **1990**, *53*, 239.
[17] T. Bartik, B. Bartik, B. E. Hanson, T. Glass, W. Bebout, *Inorg. Chem.* **1992**, *31*, 2667.
[18] W. A. Herrmann, G. P. Albanese, R. B. Manetsberger, P. Lappe, H. Bahrmann, *Angew. Chem.* **1995**, *107*, 893; *Angew. Chem., Int. Ed. Engl.* **1995**, *34*, 811; Hoechst AG (G. Albanese, R. Manetsberger, W. A. Herrmann, C. Schwer), EP 704.451 (1996).
[19] Hoechst AG (O. Stelzer, K. P. Langhans, N. Weferling), DE 4.141.299 (1993); Hoechst AG (O. Stelzer, O. Herd, N. Weferling), EP 638.578 (1995); F. Bitterer, O. Herd, A. Heßler, M. Kühnel, K. Rettig, O. Stelzer, W. S. Sheldrick, S. Nagel, N. Rösch, *Inorg. Chem.* **1996**, *35*, 4103; O. Herd, A. Heßler, K. P. Langhans, O. Stelzer, W. S. Sheldrick, N. Weferling, *J. Organomet. Chem.* **1994**, *475*, 99.
[20] O. Herd, A. Heßler, M. Hingst, M. Tepper, O. Stelzer, *J. Organomet. Chem.* **1996**, *522*, 69.
[21] (a) Kuraray Corp. (Y. Tokitoh, N. Yoshimura), US 4.808.756 (1989); (b) Exxon Chemical Patents, Inc. (E. N. Suciu, J. R. Livingston, E. J. Mozeleski), US 5.300.617 (1994).
[22] (a) T. I. Wallow, F. E. Goodson, B. M. Novak, *Organometallics* **1996**, *15*, 3708; (b) Shell Internationale Research Maatschappij (J. A. van Doorn, E. Drent, P. W. N. M. van Leeuwen, N. Meijboom, A. B. van Oort, R. L. Wite), EP 280.380 (1993); Shell Internationale Research Maatschappij (E. Drent, D. H. L. Pello), EP 632.084 (1995).
[23] C. Larpent, H. Patin, N. Thilmont, J. F. Valdor, *Synth. Commun.* **1991**, *21*, 495.
[24] E. Paetzold, G. Oehme, B. Costisella, *Z. Chem.* **1989**, *29*, 447.
[25] S. Kanagasabapathy, Z. Xia, G. Papadogianakis, B. Fell, *J. Prakt. Chem.* **1995**, *337*, 446.
[26] E. Paetzold, A. Kinting, G. Oehme, *J. Prakt. Chem.* **1987**, *329*, 725.
[27] S. Ganguly, J. T. Mague, D. M. Roundhill, *Inorg. Chem.* **1992**, *31*, 3500.
[28] Snamprogetti S.p.A. (R. Patrini, M. Marchionna), EP 613.875 (1994).
[29] A. Andriollo, J. Carrasquel, J. Mariño, F. A. López, D. E. Páez, I. Rojas, N. Valencia, *J. Mol. Catal. A: Chem.* **1997**, *116*, 157.
[30] R. A. Sánchez-Delgado, M. Medina, F. López-Linares, A. Fuentes, *J. Mol. Catal. A: Chem.* **1997**, *116*, 167.
[31] M. Hernandez, P. Kalck, *J. Mol. Catal. A: Chem.* **1997**, *116*, 117.
[32] M. Hernandez, P. Kalck, *J. Mol. Catal. A: Chem.* **1997**, *116*, 131.
[33] J. M. Grosselin, C. Mercier, G. Allmang, F. Grass, *Organometallics* **1991**, *10*, 2126.

[34] (a) A. Bényei, J. N. W. Stafford, A. Kathó, D. J. Darensbourg, F. Joó, *J. Mol. Catal.* **1993**, *84*, 157; (b) D. J. Darensbourg, F. Joó, A. Kathó, J. N. White-Stafford, A. Bényei, J. H. Reibenspies, *Inorg. Chem.* **1994**, *33*, 175.
[35] E. Fache, F. Senocq, C. Santini, J. M. Basset, *J. Chem. Soc., Chem. Commun.* **1990**, 1776; E. Fache, C. Santini, F. Senocq, J. M. Basset, *J. Mol. Catal.* **1992**, *72*, 337.
[36] S. Kolaric, V. Sunjic, *J. Mol. Catal. A: Chem.* **1996**, *110*, 189.
[37] Hoechst AG (W. A. Herrmann, J. Kulpe, J. Kellner, H. Riepl), DE 3.840.600 (1990); DE 3.921.295 (1991); Rhône–Poulenc SA (E. Kuntz), DE 2.627.354 (1976).
[38] J. P. Arhancet, M. E. Davis, J. S. Merola, B. E. Hanson, *J. Catal.* **1990**, *121*, 327.
[39] I. T. Horvath, R. V. Kastrup, A. A. Oswald, E. J. Mozeleski, *Catal. Lett.* **1989**, *2*, 85.
[40] (a) H. Bahrmann, C. D. Frohning, P. Heymanns, H. Kalbfell, P. Lappe, D. Peters, E. Wiebus, *J. Mol. Catal. A: Chem.* **1997**, *116*, 35; (b) Hoechst AG (H. Bahrmann, P. Lappe, E. Wiebus, B. Fell, P. Hermanns), DE 19.532.394 (1997).
[41] F. Monteil, R. Queau, P. Kalck, *J. Organomet. Chem.* **1994**, *480*, 177.
[42] T. Allman, R. G. Goel, *Can. J. Chem.* **1982**, *60*, 716.
[43] Z. J. Yu, R. D. Neuman, *Langmuir* **1992**, *8*, 2074.
[44] H. Ding, B. E. Hanson, T. E. Glass, *Inorg. Chim. Acta* **1995**, *229*, 329; H. Ding, *Dissert. Abstr. Int.* **1996**, *56*, 6735.
[45] H. Ding, B. E. Hanson, *J. Mol. Catal. A: Chem.* **1995**, *99*, 131.
[46] D. J. Darensbourg, C. J. Bischoff, *Inorg. Chem.* **1993**, *32*, 47.
[47] D. J. Darensbourg, C. J. Bischoff, J. H. Reibenspies, *Inorg. Chem.* **1991**, *30*, 1144.
[48] R. Schmid, W. A. Herrmann, G. Frenking, *Organometallics* **1997**, *16*, 701.
[49] R. V. Kastrup, J. S. Merola, A. A. Oswald, *Adv. Chem. Ser.* **1982**, *196*, 43.
[50] A. F. Borowski, D. J. Cole-Hamilton, G. Wilkinson, *Nouv. J. Chim.* **1978**, *2*, 137.
[51] A. G. Abatjoglou, 209th ACS National Meeting, Anaheim, USA, **1995**.
[52] J. Haggin, *Chem. Eng. News* **1995**, 25.
[53] P. Purwanto, H. Delmas, *Catal. Today* **1995**, *24*, 135.
[54] (a) B. Fell, C. Schobben, G. Papadogianakis, *J. Mol. Catal. A: Chem.* **1995**, *101*, 179; (b) B. Fell, D. Leckel, C. Schobben, *Fat. Sci. Technol.* **1995**, *97*, 219.
[55] (a) Hoechst AG (H. Bahrmann, P. Lappe, B. Fell, D. Leckel, C. Schobben), EP 711.748 (1996); (b) Hoechst AG (H. Bahrmann, P. Lappe), EP 602.463 (1996).
[56] K. T. Wan, M. E. Davies, *J. Catal.* **1994**, *148*, 1; **1995**, *152*, 25.
[57] (a) R. V. Chaudhari, B. M. Bhanage, R. M. Deshpande, H. Delmas, *Nature (London)* **1995**, *373*, 501; (b) Council of Scientific and Industrial Interest, India (R. V. Chaudhari, B. M. Bhanage, S. S. Divekar, R. M. Deshpande), US 5.498.801 (1996).
[58] W. Baidossi, N. Goren, J. Blum, H. Schumann, H. Hemling, *J. Mol. Catal.* **1993**, *85*, 153.
[59] F. Monteil, P. Kalck, *J. Organomet. Chem.* **1994**, *482*, 45.
[60] G. Papadogianakis, J. A. Peters, L. Maat, R. A. Sheldon, *J. Chem. Soc., Chem. Commun.* **1995**, 1105.
[61] G. Papadogianakis, L. Maat, R. A. Sheldon, *J. Mol. Catal. A: Chem.* **1997**, *116*, 179; G. Papadogianakis, L. Maat, R. A. Sheldon, *J. Chem. Soc., Chem. Commun.* **1994**, 2659.
[62] C. Amatore, E. Blart, J. P. Genêt, A. Jutand, S. Lemaire-Audoire, M. Savignac, *J. Org. Chem.* **1995**, *60*, 6829.
[63] T. Okano, I. Uchida, T. Nakagaki, H. Konishi, J. Kiji, *J. Mol. Catal.* **1989**, *54*, 65; T. Okano, T. Hayashi, J. Kiji, *Bull. Chem. Soc. Jpn.* **1994**, *67*, 2339.
[64] A. L. Casalnuovo, J. C. Calabrese, *J. Am. Chem. Soc.* **1990**, *112*, 4324; T. Wallow, B. M. Novak, *Polymer Prep.* **1992**, *33*, 908.
[65] A. I. Roshchin, N. A. Bumagin, I. P. Beletskaya, *Tetrahedron Lett.* **1995**, *36*, 125.
[66] R. Rai, K. B. Aubrecht, D. B. Collum, *Tetrahedron Lett.* **1995**, *36*, 3111.

[67] Hoechst AG (S. Haber, N. Egger), WO 97/05151 (1997); Hoechst AG (S. Haber, J. Manero), EP 694.530 (1996); Hoechst AG (S. Haber, J. Manero, G. Beck, J. Lagouardat), EP 709.357 (1996).
[68] E. Blart, J. P. Genêt, M. Safi, M. Savignac, D. Sinou, *Tetrahedron* **1994**, *50*, 505; M. Safi, D. Sinou, *Tetrahedron Lett.* **1991**, *32*, 2025.
[69] E. Monflier, P. Bourdauducq, J. L. Couturier, J. Kervennal, A. Mortreux, *Appl. Catal. A: Gen.* **1995**, *131*, 167.
[70] Kuraray Co (Y. Tokito, K. Watanabe, N. Yoshimura), JP 6.321.828 (1994); *Chem. Abstr.* *122*, 213604.
[71] Elf Atochem S. A. (E. Monflier, P. Bourdauducq, J. L. Couturier), WO 95/30636 (1995).
[72] E. Monflier, P. Bourdauducq, J. L. Couturier, J. Kervennal, I. Suisse, A. Mortreux, *Catal. Lett.* **1995**, *34*, 201.
[73] Hoechst AG (M. Beller, A. Tafesh, C. Kohlpaintner, C. Naumann), EP 728.734 (1996); A. M. Tafesh, M. Beller, *Tetrahedron Lett.* **1995**, *36*, 9305.
[74] Hoechst AG (H. Bahrmann, B. Cornils, A. Dierdorf, S. Haber), WO 97/07087 (1997); Hoechst AG (H. Bahrmann, B. Cornils, A. Dierdorf, S. Haber), DE 19.529.874 (1997).
[75] Rhône–Poulenc Chimie (M. Huser, R. Perron), EP 650.959 (1995).
[76] Rhône–Poulenc Fiber et Resin Intermediates (A. Chamard, D. Horbez, M. Huser, R. Perron), EP 715.890 (1996).
[77] T. Bartik, B. Bartik, B. E. Hanson, *J. Mol. Catal.* **1994**, *88*, 43.
[78] T. Bartik, B. Bartik, I. Guo, B. E. Hanson, *J. Organomet. Chem.* **1994**, *480*, 15.
[79] E. Paetzold, G. Oehme, *J. Prakt. Chem.* **1993**, *335*, 181.
[80] W. Keim, R. P. Schulz, *J. Mol. Catal.* **1994**, *92*, 21.
[81] J. A. van Doorn, N. Meijboom, *Phosphorus, Sulfur, Silicon* **1989**, *42*, 211.
[82] J. E. Hoots, T. B. Rauchfuss, D. A. Wrobleski, *Inorg. Synth.* **1982**, *21*, 175.
[83] M. Hingst, M. Tepper, O. Stelzer, *Eur. J. Inorg. Chem.* **1998**, 73.
[84] M. Tepper, O. Stelzer, T. Häusler, W. S. Sheldrick, *Tetrahedron Lett.* **1997**, *38*, 2257.
[85] P. Pellon, *Tetrahedron Lett.* **1992**, *33*, 4451.
[86] (a) V. Ravindar, H. Hemling, H. Schumann, J. Blum, *Synth. Commun.* **1992**, *22*, 1453; (b) *idem*, *Synth. Commun.* **1992**, *22*, 841.
[87] F. Mercier, F. Mathey, *J. Organomet. Chem.* **1993**, *462*, 103.
[88] K. Heesche-Wagner, T. N. Mitchell, *J. Organomet. Chem.* **1994**, *468*, 99.
[89] (a) M. Peuckert, W. Keim, *Organometallics* **1983**, *2*, 594; (b) Shell Int. Res. (W. Keim, T. M. Shryne, R. S. Bauer, H. Chung, P. W. Glockner, H. van Zwet), DE 2.054.009 (1969); W. Keim, *Chem. Ing. Techn.* **1984**, *56*, 850; (c) A. Behr, W. Keim, *Arab. J. Science Eng.* **1985**, *10*, 377.
[90] D. C. Mudalige, G. L. Rempel, *J. Mol. Catal. A: Chem.* **1997**, *116*, 309.
[91] A. Buhling, P. C. J. Kamer, P. W. N. M. van Leeuwen, J. W. Elgersma, *J. Mol. Catal. A: Chem.* **1997**, *116*, 297.
[92] A. Buhling, P. C. J. Kamer, P. W. N. M. van Leeuwen, *J. Mol. Catal. A: Chem.* **1995**, *98*, 69.
[93] (a) T. L. Schull, J. C. Fettinger, D. A. Knight, *Inorg. Chem.* **1996**, *35*, 6717; (b) T. L. Schull, J. C. Fettinger, D. A. Knight, *J. Chem. Soc., Chem. Commun.* **1995**, 1487.
[94] P. Machnitzki, T. Nickel, O. Stelzer, C. Landgrafe, *Eur. J. Inorg. Chem.* **1998**, in press.
[95] S. Bischoff, A. Weigt, H. Mießner, B. Lücke, *National Meeting, Am. Chem. Soc., Div. Fuel Chem., Washington* **1995**, *40*, 114.
[96] S. Lelièvre, F. Mercier, F. Mathey, *J. Org. Chem.* **1996**, *61*, 3531.
[97] Hoechst AG (S. Haber, H. J. Kleiner), WO 97/05104 (1997).
[98] SNPE (F. Mathey, D. Neibecker, A. Brèque), FR 2.588.197 (1985).
[99] R. T. Smith, R. K. Ungar, L. J. Sanderson, M. C. Baird, *Organometallics* **1983**, *2*, 1138.
[100] R. T. Smith, M. C. Baird, *Transition Met. Chem.* **1981**, *6*, 197.

[101] B. Mohr, D. M. Lynn, R. H. Grubbs, *Organometallics* **1996**, *15*, 4317.

[102] F. Bitterer, S. Kucken, O. Stelzer, *Chem. Ber.* **1995**, *128*, 275; D. J. Brauer, J. Fischer, S. Kucken, K. P. Langhans, O. Stelzer, N. Weferling, *Z. Naturforsch. Teil B* **1994**, *49*, 1511.

[103] E. Renaud, R. B. Russell, S. Fortier, S. J. Brown, M. C. Baird, *J. Organomet. Chem.* **1991**, *419*, 403.

[104] (a) A. Heßler, O. Stelzer, H. Dibowski, K. Worm, F. P. Schmidtchen, *J. Org. Chem.* **1997**, *62*, 2362; (b) H. Dibowski, F. P. Schmidtchen, *Tetrahedron* **1995**, *51*, 2325.

[105] M. K. Markiewicz, M. C. Baird, *Inorg. Chim. Acta* **1986**, *113*, 95.

[106] H. Schumann, H. Hemling, N. Goren, J. Blum, *J. Organomet. Chem.* **1995**, *485*, 209.

[107] D. J. Darensbourg, N. White Stafford, F. Joó, J. H. Reibenspies, *J. Organomet. Chem.* **1995**, *488*, 99.

[108] F. P. Pruchnik, P. Smolenski, I. Raksa, *Pol. J. Chem.* **1995**, *69*, 5.

[109] G. Peiffer, S. Chhan, A. Bendayan, B. Waegell, J. P. Zahra, *J. Mol. Catal.* **1990**, *59*, 1.

[110] E. Renaud, M. C. Baird, *J. Chem. Soc., Dalton Trans.* **1992**, 2905.

[111] P. G. Pringle, M. B. Smith, *Platinum Met. Rev.* **1990**, *34*, 74.

[112] J. W. Ellis, K. N. Harrison, P. A. T. Hoye, A. G. Orpen, P. G. Pringle, M. B. Smith, *Inorg. Chem.* **1992**, *31*, 3026.

[113] T. N. Mitchell, K. Heesche-Wagner, *J. Organomet. Chem.* **1992**, *436*, 43.

[114] E. Valls, J. Suades, B. Donadieu, R. Mathieu, *J. Chem. Soc., Chem. Commun.* **1996**, 771.

[115] L. Maier, *Organic Phosphorus Compounds* (Eds.: G. M. Kosolapoff, L. Maier), pp. 1–288, Wiley Interscience, New York, **1972**.

[116] Z. Jin, Y. Yan, H. Zuo, B. Fell, *J. Prakt. Chem.* **1996**, *338*, 124.

[117] Z. Jin, X. Zheng, B. Fell, *J. Mol. Catal. A: Chem.* **1997**, *116*, 55.

[118] D. E. Bergbreiter, L. Zhang, V. M. Mariagnanam, *J. Am. Chem. Soc.* **1993**, *115*, 9295.

[119] (a) M. Yamashita, M. Kobayashi, M. Sugiura, K. Tsunekawa, T. Oshikawa, S. Inokawa, H. Yamamoto, *Bull. Chem. Soc. Jpn.* **1986**, *59*, 175; (b) T. H. Johnson, G. Rangerjan, *J. Org. Chem.* **1980**, *45*, 62; (c) D. Lafont, D. Sinou, G. Descotes, *J. Organomet. Chem.* **1979**, *169*, 87.

[120] H. U. Blaser, *Chem. Rev.* **1992**, *92*, 935.

[121] M. Beller, J. G. E. Krauter, A. Zapf, *Angew. Chem.* **1997**, *109*, 793; *Angew. Chem., Int. Ed. Engl.* **1997**, *36*, 793.

[122] G. M. Olsen, W. Henderson, *Proc. XVIIth Int. Conf. on Organometallic Chemistry,* Brisbane, **1996**, p. 168.

3.2.2 Diphosphines and Other Phosphines

Marcel Schreuder Goedheijt, Paul Kamer, Piet van Leeuwen

3.2.2.1 General

Tertiary phosphines represent the major class of ligands used in homogeneous catalysis to stabilize metal centers in low oxidation states (cf. Section 3.2.1). Recent developments in the field, for example, rhodium-catalyzed hydroformylation in the organic phase have shown that the selectivity of the reaction is greatly enhanced toward the production of linear aldehydes [1, 2] upon replacement of monodentate ligands for bidentate ligands, often at the expense of activity. The next logical step to deal with the problem of product separation is the conversion of chelating diphosphines into water-soluble derivatives. A useful strategy for obtaining water-soluble ligands for organometallic catalysis is the attachment of ionic or polar substituents. This is usually done by introducting sulfonate groups, either by direct sulfonation or at an earlier stage in the synthesis of the diphosphine. Other classes of water-soluble diphosphines include diphosphines with quaternized aminoalkyl or aminoaryl groups, diphosphines with hydroxyalkyl or polyether substituents, carboxylated diphosphines, and amphiphilic diphosphines (cf. Sections 3.2.2.2–3.2.2.7). Chiral diphosphines will not be discussed here (see Section 3.2.5).

3.2.2.2 Diphosphines – Introduction of Sulfonate Groups by Direct Sulfonation

As already outlined in Section 3.2.1, several water-soluble monophosphines, such as TPPTS, have been synthesized by controlled sulfonation in oleum. In the same way several disphosphines have been sulfonated (Structures **1**–**5**). However, only a small number of sulfonated diphosphines are known so far, including BINAS-8 (**1**) [3, 4] and BISBIS (**2**) [5], both of which contain an aromatic bridge between the two phosphino moieties, and the alkyl-bridged bidentates 1,2-bis[di(3-sulfonatophenyl)phosphino]ethane (**3**) [6] and 1,3-bis[di(3-sulfonatophenyl)phosphino]propane (**4**).

The precise control of the number and position of the sulfonate groups are still troublesome. The concentration of SO_3 and reaction temperature have a major effect on the degree of sulfonation [3]. The best results are obtained when the SO_3 concentration is $\pm 25-40\%$ and the temperature is not allowed to rise

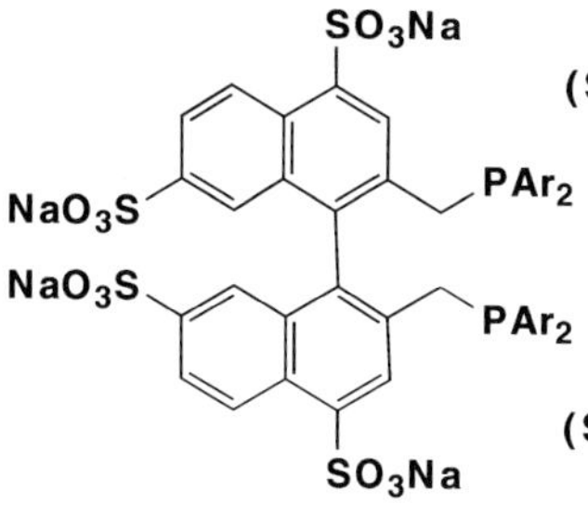

1 Ar = C_6H_4-*m*-SO_3Na

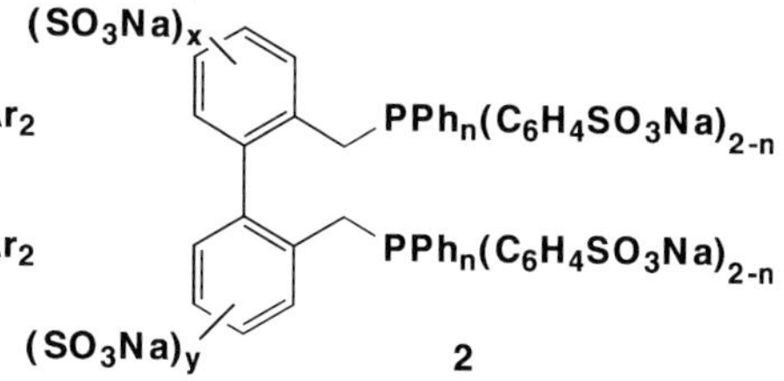

2

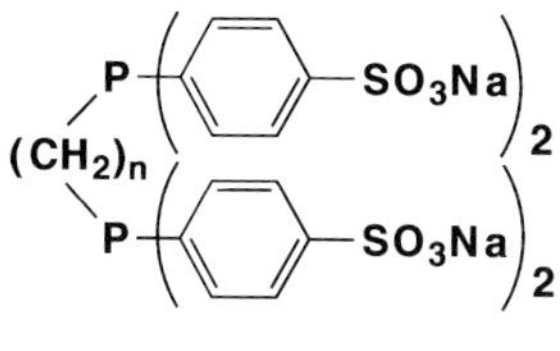

3 n = 2
4 n = 3

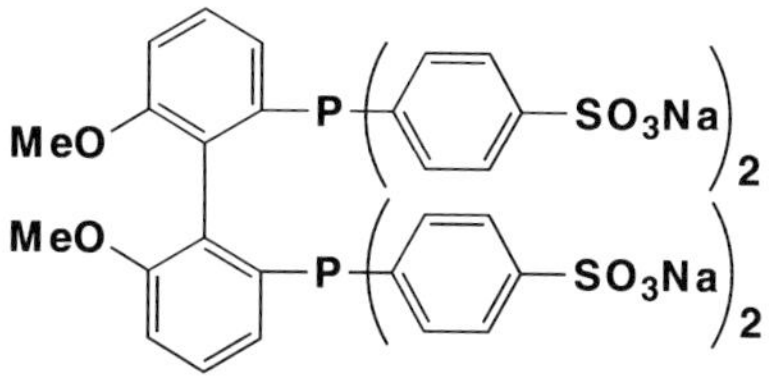

5

above room temperature. Oxidation of phosphorus may be avoided using a new method recently developed by Herrmann et al., which makes use of a superacidic medium derived from orthoboric acid and anhydrous sulfuric acid [8a]. In that manner, BINAP was sulfonated almost without formation of phosphine oxides (≤ 3 mol%). The isolation of the sulfonated ligands is often difficult, but by successive extraction with methanol, to remove inorganic salts, and subsequent filtration, the diphosphines are obtained in appreciable purity. The more economical method (and obligatory for industrial use) is extinction with tertiary amines [8b].

The diphosphines **1** and **2** were tested as ligands in the rhodium-catalyzed *biphasic* hydroformylation of propene. Both catalysts were found to exhibit higher activities and gave rise to higher *l*/*b* ratios [4] than TPPTS. Furthermore, it was shown that displacement of the *biphenyl* unit of **2** by a *binaphthyl* unit in **1** leads to an increase of the catalytic activity which was ascribed to electronic effects. In addition, the steric effect of the binaphthyl unit was believed to cause higher *l*/*b* ratios.

The alkyl-bridged diphosphine **3** was also tested as a ligand for hydroformylation in the *biphasic* process, but it was found to be not very active [9]. The use of previously described ligands in catalysis will be discussed in more detail in Chapter 6.

3.2.2.3 Introduction of Sulfonate Groups During Synthesis

Sulfonate groups can also be introduced during the course of the diphosphine synthesis. For example, as shown in Scheme 1, the surfactant sulfonate derivative of bis(diphenylphosphino)pentane is formed by reaction of $Li\{P[C_6H_4(CH_2)_3C_6H_5]_2\}$ with (*R*,*R*)-2,4-pentanediylditosylate and subsequent sulfonation with sulfuric acid [10]. Here, the phenyl group which is not attached directly to phosphorus is less deactivated for sulfonation.

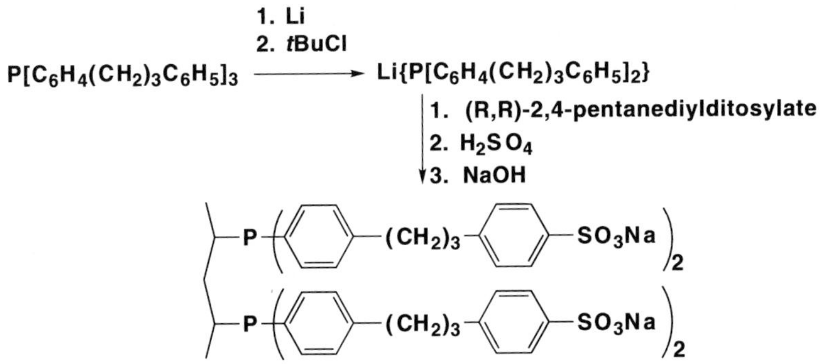

$P[C_6H_4(CH_2)_3C_6H_5]_3$ —(1. Li; 2. *t*BuCl)→ $Li\{P[C_6H_4(CH_2)_3C_6H_5]_2\}$ —(1. (R,R)-2,4-pentanediylditosylate; 2. H_2SO_4; 3. NaOH)→ **6**

Scheme 1. Synthesis of the surfactant sulfonate derivative of 2,4-bis(diphenylphosphino)-pentane (**5**).

The tetrasulfonated product **6** is formed in high yield after 6 h. The ^{1}H and ^{13}C NMR spectra were consistent with sulfonation in the *para* position; that is, only the AA′BB′ pattern of the signals expected for *para* substitution in an aromatic ring is observed in the ^{1}H NMR spectrum. Diphosphine **6** was used in the asymmetric hydrogenation of prochiral olefins in a two-phase system consisting of water and dichloromethane or ethyl acetate. The *ee* values varied from 18 to 75%. From dynamic light-scattering experiments it was shown that **6** forms aggregates with a radius of 25 Å in aqueous solutions that are 0.01 M in **6** and 0.25 M in NaCl, though its suggested surface-active character has only a minor influence on its catalytic activity. Furthermore, its activity decreased when a salt, such as Na_2HPO_4, was added, which is not surprising because there is no excess of phosphine present to form micelles at high ionic strength. Also, in other systems it has been observed that high ionic strength, inhibiting micelle formation, has the effect of decreasing the activity in two-phase system with a poorly water-soluble substrate.

In the same way as has been described for **6**, *chiral* tertiary arylphosphines, such as the atropisomeric tetrasulfonate (*R*)- or (*S*)-MeO-BIPHEP-TS **3** were prepared [11] (cf. Section 3.2.5).

Functionalized diphosphines which carry sulfonic acid groups in one or more side chains are also known. This class of compounds has mainly been developed by Whitesides and co-workers [12–14]. For example, acylation of bis[2-(diphenylphosphino)ethyl]amine, which was first prepared by Sacconi et al. [15], with *o*-sulfobenzoic anhydride in the presence of HCl afforded the water-soluble diphosphine **7** in high yield (Scheme 2) [14]. It was found that **7** is highly soluble in water, although the complex **7** · Rh(I)nbd$^+$Tf$^-$ (nbd = norbornadiene, Tf = triflate) was only sparingly soluble. This was ascribed to coordination of the sulfonic acid group to rhodium, resulting even in a decrease of activity when used as a homogeneous hydrogenation catalyst. Related effects have been observed with sulfonated triphenylphosphine complexes [16].

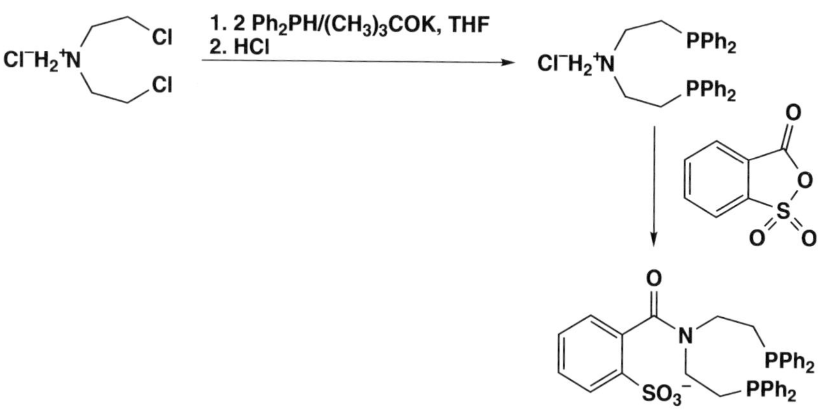

1. 2 $Ph_2PH/(CH_3)_3COK$, THF
2. HCl

$Cl^-H_2^+N$ Cl Cl → $Cl^-H_2^+N$ PPh_2 PPh_2

O O S O O

O N PPh_2 SO_3^- PPh_2

7

Scheme 2. Reaction of bis[2-(diphenylphosphino)ethyl]amine · HCl with *o*-sulfobenzoic anhydride; synthesis of **7**.

Another example published by Whitesides and co-workers is the preparation of diphosphine **8** using sodium taurinate as the sulfonate donor. Diphosphine **8** and its metal complexes appeared to be highly water-soluble, with a concentration in aqueous solution of ±0.3 M (pH 7.0, 25 °C) [13]. In the same way, several other diphosphines were synthesized.

The corresponding rhodium complexes L · Rh(I)nbd$^+$Tf$^-$ (L = diphosphine) were prepared *in situ* and appeared to form homogeneous solutions in water, although in some cases formation of micelles was observed, especially at higher

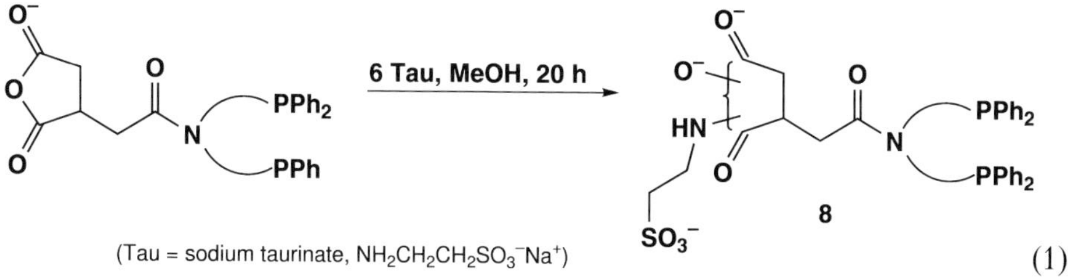

6 Tau, MeOH, 20 h

O^- O O O N PPh_2 PPh → O^- O^- HN O O N PPh_2 PPh_2 SO_3^- **8**

(Tau = sodium taurinate, $NH_2CH_2CH_2SO_3^-Na^+$) (1)

temperatures. Not only were these cationic rhodium complexes shown to be catalytically active in homogeneous hydrogenation reactions in water using several substrates, such as unsaturated carboxylic acids, but also they may catalyze the *water-gas shift* reaction [13].

3.2.2.4 Diphosphines with Quaternized Aminoalkyl and Aminoaryl Groups

Quaternization of nitrogen atoms of aminoalkyl or aminoaryl diphosphines opens up another route to water-soluble diphosphines (cf. Section 6.1.5), but until now only one (chiral) diphosphine has been synthesized by this method. Before quaternization, the phosphorous atom has to be protected either by oxidation (e.g., with hydrogen peroxide) or by coordination to a metal. Subsequent reduction or decomplexation then affords the water-soluble diphosphine. For example, Tòth and Hanson have synthesized the *chiral* diphosphine **9** with *p*-trimethylaniline substituents opposite phosphorus [17].

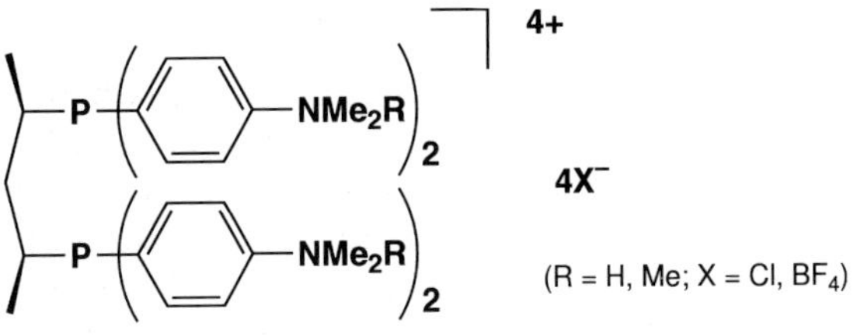

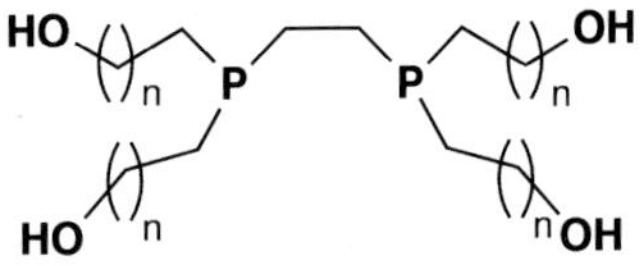

3.2.2.5 Diphosphines with Hydroxyalkyl or Polyether Substituents

Investigations into water-soluble hydroxyalkyl-substituted phosphines were first carried out by Chatt et al. [18]. Klötzer et al. reported the synthesis of 1,2-bis[di(hydroxymethyl)phosphino]ethane (**10**, $n = 0$; DHMPE) [19], and Boerner and co-workers synthesized several *chiral* bis(phosphines) bearing hydroxyl groups [20].

Recently, the synthesis of 1,2-bis(di(hydroxyalkyl)phosphino)ethane where alkyl is propyl (**11**, $n = 2$, DHPrPE), butyl (**12**, $n = 3$, DHBuPE) or pentyl (**13**, $n = 4$, DHPePE) was described [21 a]. They were synthesized by the well-known reaction [22] between an olefinic alcohol and 1,2-bis(phosphino)ethane under free-radical conditions in good yields [Eq. (2)].

$$H_2P\text{-}CH_2CH_2\text{-}PH_2 + 4\ CH_2{=}CH(CH_2)_nOH \xrightarrow[\text{MeOH, 60°C}]{\text{VAZO 67}} (HO(CH_2)_{n+2})_2P\text{-}CH_2CH_2\text{-}P((CH_2)_{n+2}OH)_2 \qquad (2)$$

VAZO 67: 2,2'azobis(2-methylbutyronitrile)

n = 0, 2, 3, 4

To demonstrate the potential use of these ligands in aqueous catalysis, several metal complexes were prepared and characterized. For example, the reaction of **11** with $NiCl_2$ in a 2:1 ratio in methanol gave an orange product identified as $Ni(DHPrPE)_2Cl_2$ [**14**; Eq. (3)]. The structure of **14** was confirmed by X-ray

$$NiCl_2 + 4\ DHPrPE \xrightarrow{\text{MeOH}} [NiCl(DHPrPE)_2]^+ + Cl^- \qquad (3)$$

14

crystallography (Figure 1). As expected, due to its cationic nature it is highly water-soluble (>0.5 M). In the same way, complexes similar to **14** were synthesized from **12** and **13** with $NiCl_2$. Both complexes were found to be highly water-soluble as well (>0.5 M), which was determined by monitoring the solubility of the ligand in the presence of $NiCl_2$. For example, the solubility of the DHPePE (**13**) ligand is ± 0.002 M in water, whereas the solubility in the presence of 0.5 equiv. of $NiCl_2$ increases the solubility of the ligand to >0.5 M. The solubility of the metal complexes in water can be attributed to the hydrophilic hydroxyl groups, which surround the outside of the complexes, but even more to the charges on the molecules.

Besides Ni(II) complexes, Ru(III), Rh(I) and Rh(II) [21 b] complexes were prepared with these ligands which also appeared to be highly water-soluble. Recently, the sodium sulfonate analog of **11**, DSPrPE in which OH is replaced by SO_3Na, was synthesized [21 b] and was shown to be highly soluble in water (1.5 M). Rh(I) complexes with DSPrPE were demonstrated to hydrogenate alkenes completely with a higher selectivity and activity than a similar system prepared with TPPTS.

Pringle and co-workers recently found that formaldehyde adds readily to primary diphosphines, without the need for a catalyst, to give bidentate, water-soluble diphosphine ligands with bis(hydroxymethyl)phosphino groups [23].

Substituting a phosphine with a polyether chain may also make the phosphine water-soluble. However, diphosphines of the type **15** (Structures **15**–**17**) are only soluble in water when $n > 15$ [24]. Other examples related to **15** are the

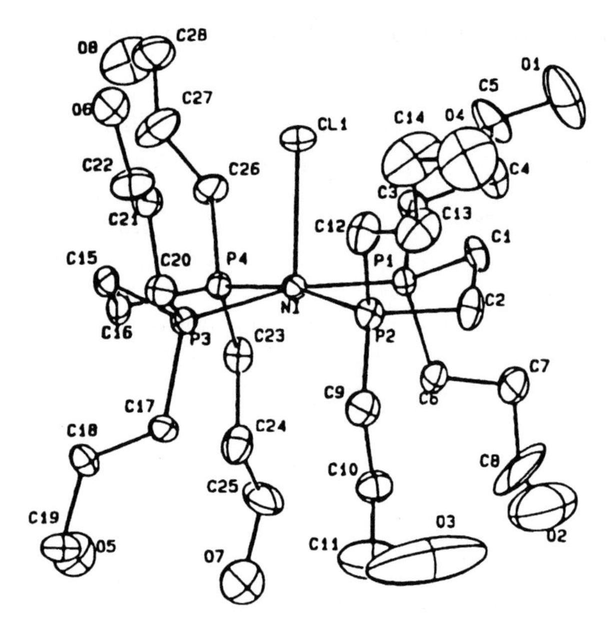

Figure 1. X-ray crystal structure of $Ni(DHPrPE)_2Cl_2$ (**14**). Reproduced by the permission of the American Chemical Society (© (1997) American Chemical Society).

class of compounds **16** (cf. Section 3.2.2.3; $n = 12, 16, 110$) and **17** ($n = 18$) [12, 25]. The number n gives the average value of the "degree of polycondensation". So far, these polyether-based diphosphines have mainly been used in asymmetric hydrogenation of prochiral substrates such as α-acetamidocinnamic acid where *ee* values vary from 11 to 91 % depending on the reaction medium and ligand used (cf. Section 4.6.3).

15

16

17

3.2.2.6 Carboxylated Diphosphines

Phosphines with carboxylic groups were some of the earliest water-soluble phosphines investigated [26]. They also prepared a diphosphine, the phosphine analog of ethylenediaminetetraacetic acid, which is obtained as a monohydrate of the tetrasodium salt (**18**) [26]. Jegorov and Podlahová recently published a short review on the catalytic uses of these carboxyalkylphosphines [27].

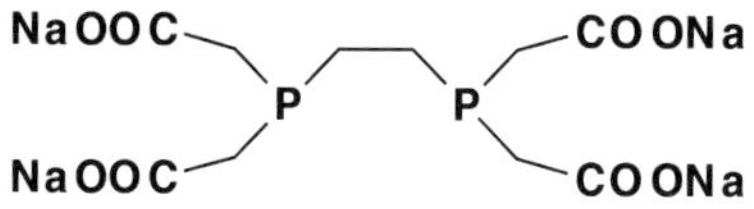

NaOOC, NaOOC, P, P, COONa, COONa

18

Van Doorn developed a water-soluble diphosphine based on 2,3-bis(diphenylphosphino)maleic anhydride, which was converted into the biscarboxylic acid **19** with sodium hydroxide (Scheme 3) [28a]. The compound was also described by Tyler and co-workers in 1993 [28 b].

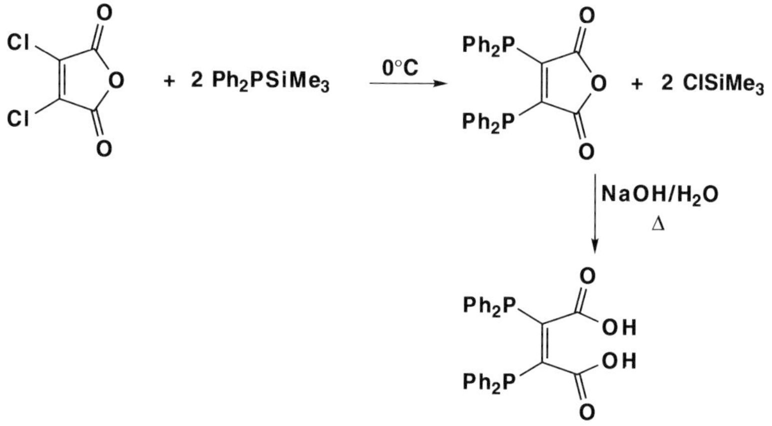

Cl, Cl, O, O, O + 2 Ph_2PSiMe_3 —0°C→ Ph_2P, Ph_2P, O, O, O + 2 $ClSiMe_3$

$NaOH/H_2O$, Δ

Ph_2P, Ph_2P, O, OH, OH, O

19

Scheme 3. Synthesis of the water-soluble diphosphine **19** based on 2,3-bis(diphenylphosphino)maleic anhydride.

An advantage over other methods for the preparation of water-soluble phosphines is that diphosphine **19** could easily be purified by extraction followed by crystallization from diethyl ether. The structure of **19** was confirmed by X-ray crystallography. The ligand is readily soluble in aqueous solution at pH 5 or higher. For example, the solubility is >1 M at pH 7. At this pH the ligand is deprotonated. To demonstrate the utility of **19** in aqueous organometallic chemistry, solutions of several water-soluble complexes, such as $(cpCOOH)_2W_2(CO)_6$, were irradiated in the presence of the ligand. This resulted in the disproportionation of the metal complex [Eq. (4)], which proceeds analogously to the reaction of $cp_2M_2(CO)_6$ (M = Mo or W) and dppe (1,2-bis(diphenylphosphino)ethane) in benzene solution [29]. So far, the ligand has not been used for catalytic purposes.

$$[(CpCOO^-)_2W_2(CO)_6]^{2-} + L_2^{2-} \xrightarrow[H_2O,\ pH\ 10]{h\upsilon} [(CpCOO^-)_2W^{II}(CO)_2(L_2^{2-})]^{2-} + [(CpCOO^-)_2W^{0}(CO)_3]^{2-} + CO \quad (4)$$

L_2^{2-} = deprotonated **19**

3.2.2.7 Amphiphilic Diphosphines

Another approach to water-soluble phosphines with the emphasis on metal recycling was reported by van Leeuwen and co-workers [30]. They have synthesized a number of diphosphines that, when coordinated to rhodium, form complexes having an amphiphilic character. The ligands synthesized are based on BISBI and Structures **20–22**, and hydroformylation (for example) can be

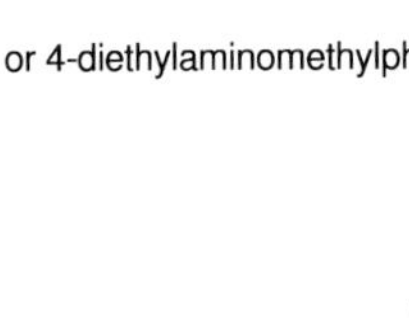

XPhP PPhX

20

X = Py or 4-diethylaminomethylphenyl

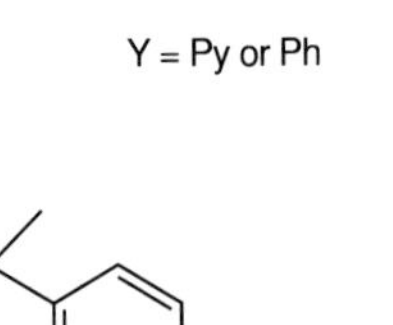

YPhP PPhY

21

Y = Py or Ph

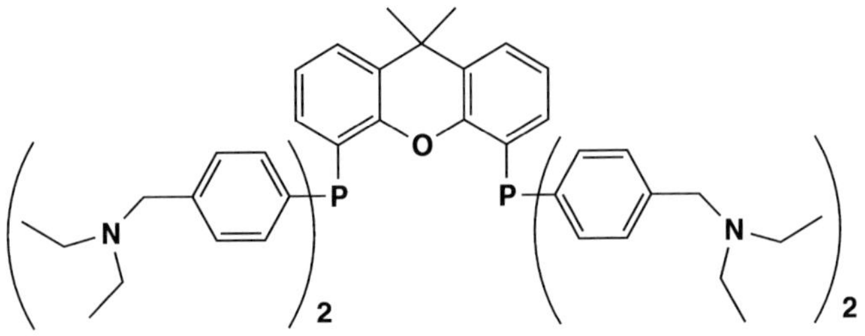

22

conducted in a homogeneous (organic) phase [30a]. After it has been used in the hydroformylation of olefins the catalyst can be removed by acidic extraction. It was established that these novel diphosphines form active and highly selective catalysts. This amphiphilic approach, i.e., rhodium recycling based on the extraction and re-extraction principle, will be discussed in more detail in Section 7.4.

3.2.2.8 Other Phosphines

Water-soluble mono- and diphosphines represent the major class of phosphines used in aqueous-phase homogeneous catalysis. However, some new types of water-soluble phosphines have recently been developed. Some of the latest include phosphines containing sugar substructures [31] or phosphonate chains [32], and chiral sulfonated phosphines for the asymmetric hydrogenation of dehydropeptides [33].

A different class was developed by Mathey et al. [34], who described a simple route for the synthesis of bicyclic phosphines with phosphorus atoms at the bridgehead from transient 2*H*-phospholes (Scheme 4) and water-soluble analogues **23**. These phosphines showed a significant activity in the rhodium-catalyzed hydrogenation [34] and hydroformylation [35] of olefins. In view of the growing academic and industrial interest in bisphasic catalysis, water-soluble versions of these phosphanorbornadienes represent a further logical development, if cheap enough. Mathey et al. extended the synthesis to water-soluble bicyclic phosphines starting from 3,4-dimethyl-2*H*-phosphole, which was reacted with maleic anhydride to give the 2*H*-phosphole-maleic anhydride [4+2] *endo* cycloadduct. Upon treatment with sodium hydroxide, this cycloadduct was converted into the sodium salt (**23**) of 3,4-dimethyl-1-phospha-2-norbornene-5,6-dicarboxylic acid, which proved to be very soluble in water (>300 g/L) [Eq. (5)]. From NMR experiments it was shown that both carboxylate groups are on the *endo* side. The diacid, obtained when **23** is treated with 3 M hydrochloric acid, slowly loses water to regenerate the anhydride. Its coordination chemistry is currently under investigation.

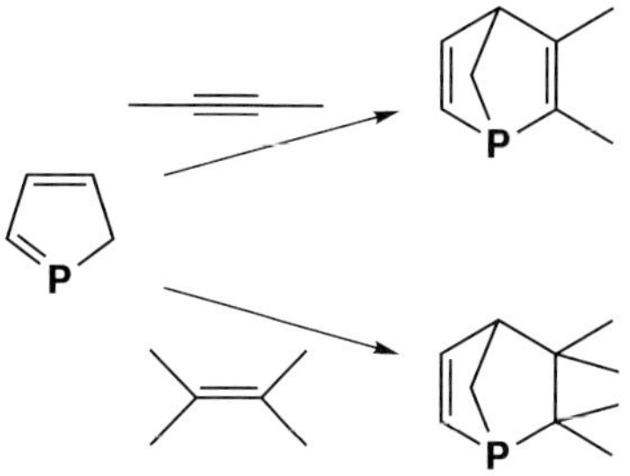

Scheme 4. Bicyclic phosphines based on 2H-phospholes.

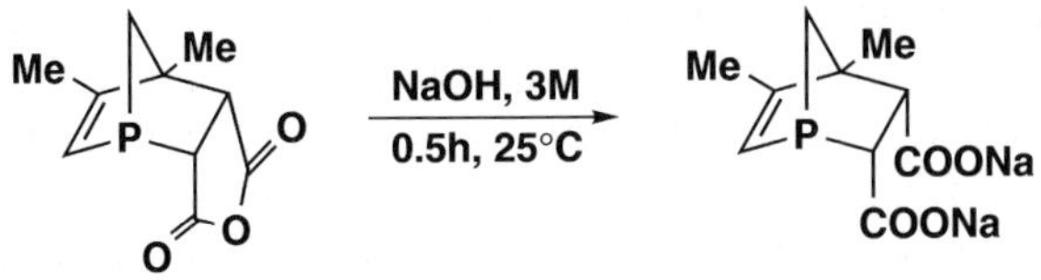

(5)

23

In line with the so-called NORBOS ligand, which shows an outstanding activity in the biphasic hydroformylation of propene [4], Mathey and co-workers developed several phosphanorbornadienes functionalized with phosphonate or phosphonamide groups [36]. For example, reaction of 2-phenyl-3,4-dimethyl-5*H*-phosphole, which was obtained by thermal isomerization of 1-phenyl-3,4-dimethylphosphole, with various (phenylethynyl)phosphonic acid derivatives produced phosphanorbornadienes having a phosphoryl group at C_2 (Scheme 5). Upon acid hydrolysis they could be converted into the correspond ing phosphonic acid, of which the sodium salt (**24**) (after treatment with sodium hydroxide) is moderately soluble in water (20 g/L). Because of this low solubility, **24** was potentially of some interest as a substitute for the sodium salt of triphenylphosphinemonosulfonic acid, which is used in the proposed new oxo process developed by Union Carbide for the hydroformylation of higher olefins [37].

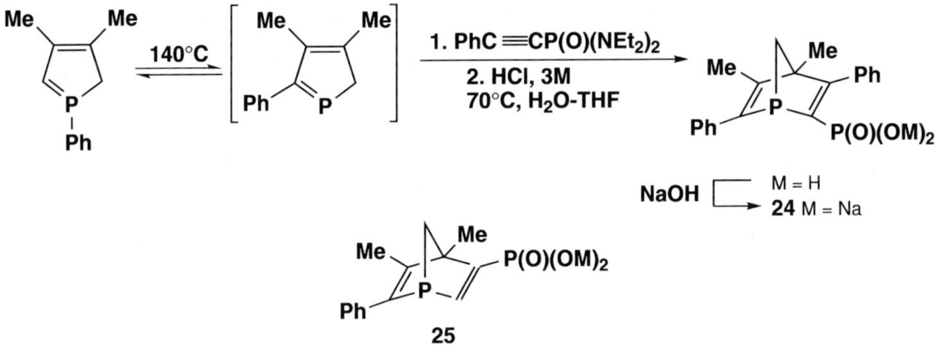

Scheme 5. Synthesis of phosphanorbornadienes having a phosphoryl group.

The catalytic activity of the functionalized phophanorbornadienes, various $[RhL_2]^+$ complexes, was used in the hydrogenation of 1-methylcyclohexene. However, the results were poor, which contrasts with the high activity observed for the nonfunctional 1-phosphanorbornadiene [38]. Poor results were also observed in the hydroformylation of 1-hexene catalyzed by $[Rh(CO)_2Cl]_2^+ \cdot 2\,L$, where L is **24**. These results were ascribed to a detrimental effect of the α-P(O) groups (formation of P,P(O) chelates), which block the catalytic cycle. This hypothesis was confirmed by using **25** (see Scheme 5) as a ligand, since in this case the hydrogenation proceeds quantitatively. The improvement is less impressive for the hydroformylation of 1-hexene, although it was found that **25** gave higher yields and rates than **24**. As expected, in this case the potential P(O) chelation has a less significant effect on the catalysis.

Besides the well-known applications of water-soluble phosphines, e.g., in hydroformylation, water-soluble catalysts may have significant advantages for electrochemical processes in which the much lower resistance of aqueous solutions compared with that or organic solutions would improve the energy efficiency of the process. It is known that the electrochemical reduction of carbon dioxide to carbon monoxide in acidic acetonitrile or DMF solutions is catalyzed by [Pd(triphosphine)(solvent)]$(BF_4)_2$ [38]. These complexes exhibit interesting catalytic properties, such as high turnover rates and high selectivities (under appropriate reaction conditions). To study the effect of an aqueous environment on the rates and selectivities of these catalysts, various water-soluble triphosphine complexes of the type [Pd(triphosphine)(CH_3CN)]$(BF_4)_2$ were prepared by DuBois et al. [38]. Hydroxyl, phosphonate, or amines were used as water-solubilizing groups. The synthetic strategy used extends previous work on monodentate [31] and bidentate [39] ligands. As concluded from kinetic studies carried out in DMF solutions, the mechanism for the carbon dioxide reduction is the same as that proposed for complexes without water-soluble functional groups. In addition, the ligands may have applications in other catalytic processes in which water-solubility is an important feature.

Bianchini et al. [40] synthesized another water-soluble triphosphine, $NaO_3S(C_6H_4)CH_2C(CH_2PPh_2)_3$, the so-called SULPHOS (**26**; Scheme 6). This ligand is a TRIPHOS ligand [41] with a hydrophilic tail attached to the bridgehead carbon atom. It has been developed to facilitate catalyst separation in biphasic but aqueous systems. The synthesis involves the treatment of benzyltris(chloroethyl)methane with concentrated sulfuric acid at 100 °C, which results in the regioselective *para* sulfonation of the phenyl ring. Reaction of $NaO_3S(C_6H_4)CH_2C(CH_2Cl)_3$ with $KPPh_2$ in DMSO at 100 °C gives **26**. The complexes (**26**)Rh(cod) and (**26**)$Rh(CO)_2$ were used in biphasic catalysis as hydrogenation and hydroformylation catalysts respectively. Typical of the (**26**)$Rh(CO)_2$ catalyst system is that when methanol is used as a solvent, mainly alcohols are formed during the hydroformylation of higher olefins (*e.g.*,

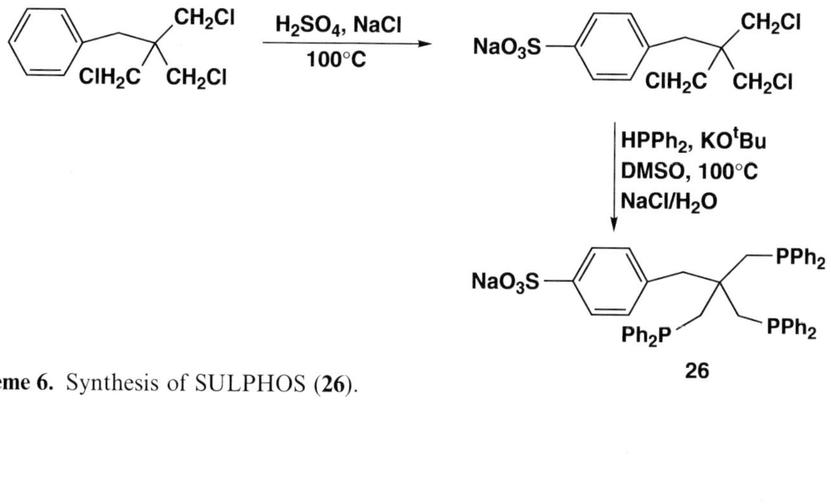

CH_2Cl ClH_2C CH_2Cl — H_2SO_4, NaCl, 100°C → NaO_3S … CH_2Cl ClH_2C CH_2Cl

$HPPh_2$, KO^tBu
DMSO, 100°C
$NaCl/H_2O$

NaO_3S … PPh_2 Ph_2P PPh_2

26

Scheme 6. Synthesis of SULPHOS (**26**).

1-hexene) whereas rhodium catalysts usually only produce aldehydes (*e.g.*, the Rh/TRIPHOS system [41 a]).

Recently, Katti et al. developed the new water-soluble triphosphine $PhP[CH_2CH_2P(CH_2OH)_2]$ (**27**; Scheme 7) [42]. This triphosphine, upon interaction with $[Rh(cod)Cl]_2$ under biphasic (water/dichloromethane) conditions, produces a water-soluble rhodium(I) complex in which the rhodium center is tripodally coordinated *via* the PPh and $P(CH_2OH)_2$ functionalities, as established by NMR spectroscopy. The presence of PPh and $P(CH_2OH)_2$ groups of disparate basicities makes it unusual in comparison with the traditional triphosphines (*e.g.*, TRIPHOS: $PhP(CH_2CH_2PPh_2)_2$). They suggested that the different basicities in **27** may aid the development of catalytically useful transition metal complexes in which the weaker of the two different M–P bonds may be reversibly cleaved in the presence of a substrate molecule.

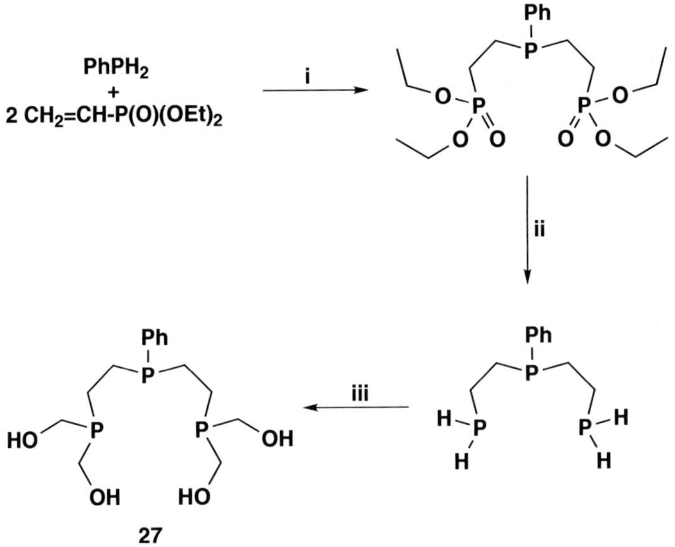

$PhPH_2$ + 2 $CH_2{=}CH\text{-}P(O)(OEt)_2$ —i→ ... —ii→ ... —iii→ **27**

Scheme 7. Synthesis of water-soluble triphosphine **27**. Reagents: (i) $KOBu^t$, thf; (ii) $LiAlH_4$, Et_2O; (iii) HCHO, EtOH.

The following topics have not been considered in this section: metal complexes of sulfonated 2,2′-bipyridine [43] and alizarin [44], water-soluble porphyrins and cyclopentadienyl ligands [45], and metal complexes of sulfonated phenanthroline derivatives [46]. Also *P,N*-bidentate phosphines [47], of which coordination properties await further study, have not been discussed here.

In conclusion, monophosphines are still the most widely investigated and applied water-soluble phosphines (*e.g.*, TPPTS in the Ruhrchemie–Rhône-Poulenc process [48]). However water-soluble diphosphines and other phosphines, such as triphosphines, are currently under investigation; chelating diphosphines, for example, show significantly improved regioselectivity in hydroformylation and may have other advantages over monophosphines. The combination of recent successes of newly developed diphosphines and the success of the Ruhrchemie–Rhône-Poulenc process have led to the development of a range of new water-soluble di- and triphosphines. They might be applicable if cheap enough under industrial conditions [49].

References

[1] M. E. Broussard, R. Juma, S. G. Train, W. J. Peng, S. A. Laneman, G. G. Stanley, *Science* **1993**, *260*, 1784.

[2] G. Süss-Fink, *Angew. Chem.* **1994**, *106*, 71; *Angew. Chem., Int. Ed. Engl.* **1994**, *33*, 67.

[3] Hoechst AG (H. Bahrmann, P. Lappe, W. A. Herrmann, R. Manetsberger, G. Albanese), DE 4.321.512 (1993).

[4] W. A. Herrmann, C. W. Kohlpaintner, R. B. Manetsberger, H. Bahrmann, H. Kottmann, *J. Mol. Catal.* **1995**, *97*, 65.

[5] W. A. Herrmann, C. W. Kohlpaintner, H. Bahrmann, W. Konkol, *J. Mol. Catal.* **1992**, *73*, 191.

[6] T. Bartik, B. Bunn, B. Bartik, B. E. Hanson, *Inorg. Chem.* **1994**, *33*, 164.

[7] Z. Jiang, A. Sen, *Macromolecules* **1994**, *27*, 7215.

[8] (a) W. A. Herrmann, G. P. Albanese, R. B. Manetsberger, P. Lappe, H. Bahrmann, *Angew. Chem., Int. Ed. Engl.* **1995**, *34*, 811; (b) Ruhrchemie AG (R. Gärtner, B. Cornils et al.), DE 3.235.029 and 3.235.030 (1982).

[9] T. Bartik, B. Bunn, B. Bartik, B. E. Hanson, *Inorg. Chem.* **1994**, *33*, 164.

[10] H. Ding, B. E. Hanson, J. Bakos, *Angew. Chem., Int. Ed. Engl.* **1995**, *34*, 1645.

[11] R. Schmid, E. A. Broger, M. Cereghetti, Y. Crameri, J. Foricher, M. Lalonde, R. K. Müller, M. Scalone, G. Schoettel, U. Zutter, *Pure Appl. Chem.* **1996**, *68*, 131.

[12] M. E. Wilson, R. G. Nuzzo, G. M. Whitesides, *J. Am. Chem. Soc.* **1978**, *100*, 2269.

[13] R. G. Nuzzo, D. Feitler, G. M. Whitesites, *J. Am. Chem. Soc.* **1979**, *101*, 3683.

[14] R. G. Nuzzo, S. L. Haynie, M. E. Wilson, G. M. Whitesides, *J. Org. Chem.* **1981**, *46*, 2861.

[15] L. Sacconi, R. Marassi, *J. Chem. Soc. A* **1968**, 2997.

[16] A. F. Borowski, D. J. Cole-Hamilton, G. Wilkinson, *Nouv. J. Chim.* **1978**, *2*, 137.

[17] I. Tóth, B. E. Hanson, *Tetrahedron: Asymmetry* **1990**, *1*, 895.

[18] J. Chatt, G. L. Leigh, R. M. Slade, *J. Chem. Soc., Dalton Trans.* **1973**, 2021.

[19] D. Klötzer, P. Mäding, R. Munze, *Z. Chem.* **1984**, *24*, 224.

[20] J. Holz, A. Boerner, A. Kless, S. Borns, S. Trinkhaus, R. Seleke, D. Heller, *Tetrahedron: Asymmetry* **1995**, *6*, 1973.

[21] (a) G. T. Baxley, W. K. Miller, D. K. Lyon, B. E. Miller, G. F. Nieckarz, T. J. R. Weakley, D. R. Tyler, *Inorg. Chem.* **1996**, *35*, 6688; (b) G. T. Baxley, T. J. R. Weakley, W. K. Miller, D. K. Lyon, D. R. Tyler, *J. Mol. Catal. A* **1997**, *116*, 191.

[22] L. Lavenot, M. H. Bortoletto, A. Roucoux, C. Larpent, H. Patin, *J. Organomet. Chem.* **1996**, *509*, 9 and references cited therein.

[23] P. G. Pringle, D. Brewin, M. B. Smith, K. Worboys in *Aqueous Organometallic Chemistry and Catalysis* (Eds.: I. T. Horváth, F. Joó), Kluwer, Dordrecht, **1995**, p. 111.

[24] Y. Amrani, D. Sinou, *J. Mol. Catal.* **1984**, *24*, 231.

[25] Y. Amrani, D. Sinou, *J. Mol. Catal.* **1986**, *36*, 319.

[26] J. Podlahová, J. Podlaha, *Collect. Czech. Chem. Commun.* **1980**, *45*, 2049.

[27] A. Jegorov, J. Podlaha, *Catal. Lett.* **1991**, *9*, 9.

[28] (a) J. A. van Doorn, *Thesis*, University of Amsterdam, **1991**, p. 31; (b) A. Avey, D. M. Schut, T. J. R. Weakley, D. R. Tyler, *Inorg. Chem.* **1993**, *32*, 233.
[29] A. Avey, S. C. Tenhaeff, T. J. R. Weakley, D. R. Tyler, *Organometallics* **1991**, *10*, 3607 and references cited therein.
[30] (a) A. Buhling, P. C. J. Kamer, P. W. N. M. van Leeuwen, *J. Mol. Catal. A* **1995**, *98*, 69; (b) A. Buhling, J. W. Elgersma, S. Nkrumah, P. C. J. Kamer, P. W. N. M. van Leeuwen, *J. Chem. Soc., Dalton Trans.* **1996**, 2143; (c) A. Buhling, P. C. J. Kamer, P. W. N. M. van Leeuwen, J. W. Elgersma, K. Goubitz, J. Fraanje, *Organometallics* **1998**, in press.
[31] T. N. Mitchell, K. Heesche-Wagner, *J. Organomet. Chem.* **1992**, *436*, 43.
[32] S. Ganguly, J. T. Mague, D. M. Roundhill, *Inorg. Chem.* **1992**, *31*, 3500.
[33] M. Laghmari, D. Sinou, A. Masdeu, C. Claver, *J. Organomet. Chem.* **1992**, *438*, 213.
[34] F. Mathey, D. Neibecker, A. Brèque, *Chem. Abstr.* **1987**, *107*, 219468v.
[35] D. Neibecker, R. Réau, *Angew. Chem., Int. Ed. Engl.* **1989**, *28*, 500.
[36] S. Lelièvre, F. Mercier, F. Mathey, *J. Org. Chem.* **1996**, *61*, 3531.
[37] J. Haggin, *Chem. Eng. News* **1995**, 25.
[38] A. M. Herring, B. D. Steffey, A. Miedaner, S. A. Wander, D. L. DuBois, *Inorg. Chem.* **1995**, *34*, 1100.
[39] R. B. King, W. F. Master, *J. Am. Chem. Soc.* **1977**, *99*, 4001.
[40] C. Bianchini, P. Frediani, V. Sernau, *Organometallics* **1995**, *14*, 5458.
[41] (a) C. Bianchini, A. Meli, M. Peruzzini, F. Vizza, P. Frediani, J. A. Ramirez, *Organometallics* **1990**, *9*, 226; (b) C. Bianchini, A. Meli, M. Peruzzini, F. Vizza, F. Zanobini, *Coord. Chem. Rev.* **1992**, *120*, 193; (c) V. Sernau, G. Huttner, M. Fritz, L. Zsolnai, O. Walter, *J. Organomet. Chem.* **1993**, *453*, C23.
[42] C. J. Smith, V. S. Reddy, K. V. Katti, *J. Chem. Soc., Chem. Commun.* **1996**, 2557.
[43] (a) W. A. Herrmann, W. R. Thiel, J. G. Kuchler, *Chem. Ber.* **1990**, *123*, 1953; (b) W. A. Herrmann, W. R. Thiel, J. G. Kuchler, J. Behm, E. Herdtweck, *Chem. Ber.* **1990**, *123*, 1963.
[44] (a) L. Vigh, F. Joó, *FEBS Lett.* **1983**, *162*, 423; (b) L. Vigh, Z. Gombos, F. Joó, *ibid.* **1985**, *191*, 200.
[45] U. Siemeling, *J. Organomet. Chem.* **1992**, *429*, C14.
[46] G. Schmid, B. Morun, J. O. Malm, *Angew. Chem.* **1989**, *101*, 772; *Angew. Chem., Int. Ed. Engl.* **1989**, *28*, 778.
[47] (a) B. Assmann, K. Angermaier, H. Schmidbaur, *J. Chem. Soc., Chem. Commun.* **1994**, 941; (b) B. Assmann, K. Angermaier, M. Paul, J. Riede, H. Schmidbaur, *Chem. Ber.* **1995**, *128*, 891.
[48] (a) E. G. Kuntz, *CHEMTECH* **1987**, *17*, 570; (b) B. Cornils, J. Falbe, *Proc. 4th Int. Symp. Homogeneous Catalysis,* Leningrad, Sept. **1984**, p. 487; (c) H. Bach, W. Gick, E. Wiebus, B. Cornils, *Prepr. Int. Symp. High-Pressure Chem. Eng.*, Erlangen, Sept. **1984**, p. 129; (d) H. Bach, W. Gick, E. Wiebus, B. Cornils, *Prepr. 8th ICC,* Berlin, **1984**, Vol. V, p. 417; *Chem. Abstr.* **1987**, *106*, 198 051; cited also in A. Behr, M. Röper, *Erdgas Kohle* **1984**, *37(11)*, 485; (e) H. Bach, W. Gick, E. Wiebus, B. Cornils, *1st IUPAC Symp. Org. Chemistry,* Jerusalem, **1986**, Abstracts, p. 295.
[49] (a) E. Wiebus, B. Cornils, *Chem.-Ing.-Techn.* **1994**, *66*, 916; (b) B. Cornils, E. Wiebus, *CHEMTECH* **1995**, *25*, 33; (c) E. G. Kuntz, *CHEMTECH* **1987**, *17*, 570.

3.2.3 Ligands or Complexes Containing Ancillary Functionalities

Philippe Kalck, Michel Dessoudeix

Since the early preparation of TPPMS by Chatt and co-workers [1] in 1958 and the spectacular properties of TPPTS discovered by Kuntz in 1975 [2, 3], many complexes have been prepared, most of them in situ, containing in addition to the water-soluble phosphines other ligands required for catalysis, such as hydride, carbon monoxide, alkene, etc.

This section deals with complexes containing a main phosphorus ligand and ancillary organo- or water-soluble groups which make it possible either to tune the metal coordination sphere or to graft the complex onto various supports.

3.2.3.1 Complexes Containing at Least Two Classical Functionalities

Because of their great capacity of water-solubility, a large proportion of the research has been directed toward the sulfonated phosphines. Nevertheless, it is obvious that other hydrophilic functionalities can be introduced on phosphines. Adapting the kind and the number of hydrophilic groups, it is thus possible to control the water-solubility of the complexes better.

3.2.3.1.1 Phosphanorbornadiene Phosphonate

As an introduction, it is of interest to report the special influence of the position of a functionality, taking as an example the phosphonation of 1-phosphanorbornadiene **1** [4]. This molecule has shown very good properties as a ligand for transition-metal catalyzed hydrogenation [5] and hydroformylation [6] of alkenes. Therefore, the influence of the substituents on the water-solubility has been studied. For this purpose, a phosphonate group was introduced, leading to ligand **2**. This ligand is quantitatively extracted from water-saturated toluene and it is a good substitute for the sodium salt of triphenylphosphinomonosulfonic acid, which is operating the new hydroformylation process of higher olefins developed by Union Carbide [7]. That molecule was used in the hydrogenation of (*Z*)-α-(*N*-acetamido)cinnamic acid. The results were poor compared with those obtained with the unsubstituted phosphanorbornadiene **1** [5]. In order to find a basis for the hypothesis that the inhibition of the catalytic cycle was due to P,P(O) chelate formation, molecule **3** was synthesized. Although the solubility of the sodium salt of molecule **2** was only 20 g/L, the solubility of the corresponding salt of **3** was 230 g/L. Used in catalysis, it allowed the quantitative hydrogenation of the substrate.

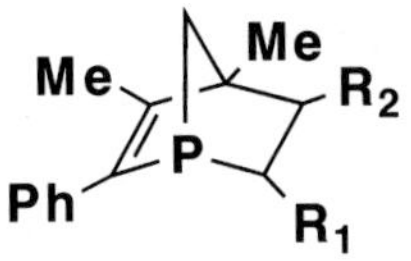

1 $R_1 = R_2 = Ph$
2 $R_1 = -P(O)(ONa)_2$, $R_2 = Ph$
3 $R_1 = H$, $R_2 = -P(O)(ONa)_2$

3.2.3.1.2 Complexes Containing at Least Two Functionalities

Recently, Patin and co-workers reported the synthesis under mild conditions of a series of ligands obtained from the mixture of Ph_2PH and an activated alkene in the presence of a small amount of tetraethylammonium hydroxide [8], and observed, when the olefin is very hygroscopic, that some phospine oxide is produced. During complexation with rhodium, oxidation is avoided using an Rh(I) dimer as a precursor. The synthetic reaction is shown in Scheme 1 and the various complexes obtained are listed in Table 1.

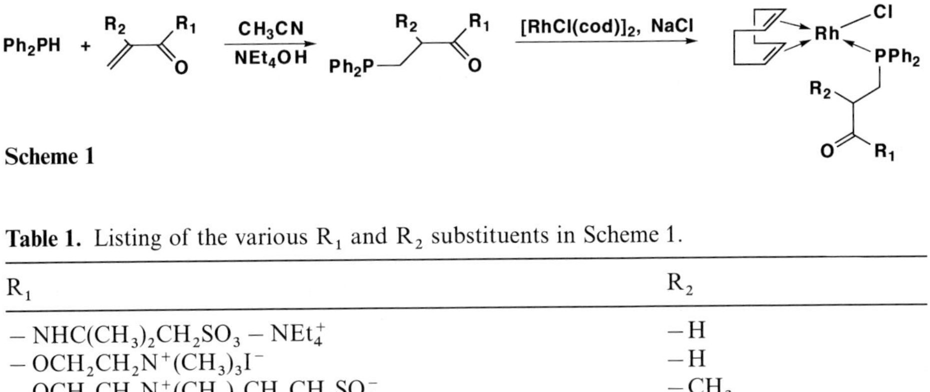

Scheme 1

Table 1. Listing of the various R_1 and R_2 substituents in Scheme 1.

R_1	R_2
$-NHC(CH_3)_2CH_2SO_3-NEt_4^+$	$-H$
$-OCH_2CH_2N^+(CH_3)_3I^-$	$-H$
$-OCH_2CH_2N^+(CH_3)_2CH_2CH_2SO_3^-$	$-CH_3$
$-OCH_3$	$-CH_2CO_2CH_3$
$-NHC(CH_2OH)_3$	$-H$
$-OCH_2CH_2N(CH_3)_2$	$-H$
$-OH$	$-CH_2CO_2H$

Generally, polydentate phosphines are synthesized by sequential introduction of phosphine units into a precursor [9]. Nevertheless, Whitesides and co-workers proposed a synthetic route by which diphosphine **4** is obtained in situ, as shown in Eq. (1) [10–12].

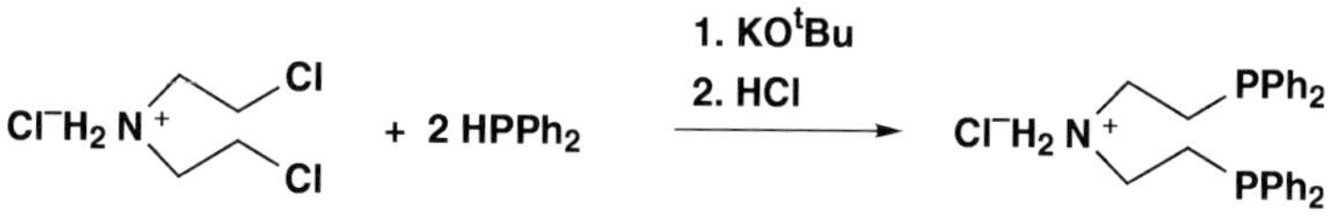

$$Cl^{-}H_2N^{+}(CH_2CH_2Cl)_2 + 2\,HPPh_2 \xrightarrow[2.\ HCl]{1.\ KO^tBu} Cl^{-}H_2N^{+}(CH_2CH_2PPh_2)_2 \quad \mathbf{4} \tag{1}$$

From this molecule, it became possible to synthesize easily various ligands containing complex frameworks, and different functionalities. Once acylated by acid chlorides, like molecules **5**, **6**, and **7** of Scheme 2, it reacts with sodium taurinate, $H_2NCH_2CH_2SO_3Na$. Complex **4** can also condense with sulfonic acid **8** in order to obtain **9**.

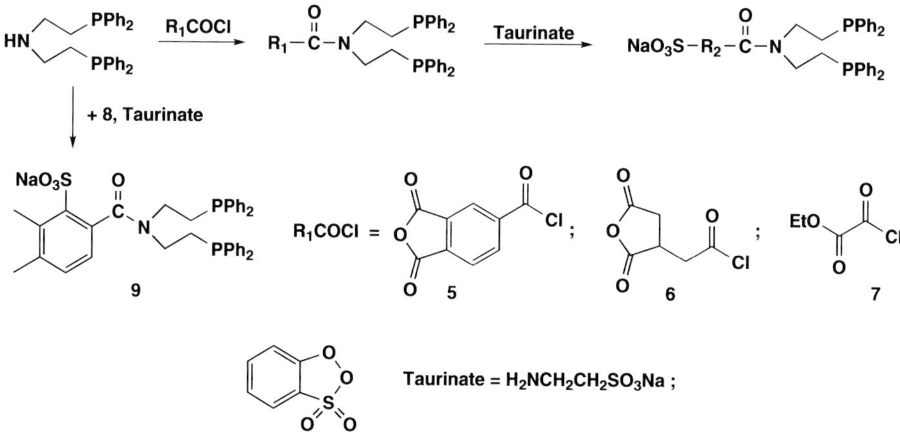

R_1COCl; Taurinate; + 8, Taurinate; R_1COCl = 5; 6; 7; Taurinate = $H_2NCH_2CH_2SO_3Na$;

Scheme 2

To extend Whiteside's method, it is possible to synthesize ligands containing hydroxyl, ether, or carboxyl groups [13], e.g. acid chlorides or anhydrides containing poly(ethylene glycol), poly(ether) or poly(hydroxyalkyl) chains.

In Scheme 3 [12], the use of this method to synthesize water-soluble derivatives of the DIOP ligand is described [14, 15].

Smith and co-workers reported the synthesis of a new triphosphine ligand via formylation of $PhP(CH_2CH_2PH_2)_2$ [16, 17]. From $[Rh(cod)Cl]_2$ (cod = cyclooctadiene) or $[Pt(cod)Cl_2]$, the authors formed tripodally-coordinated complexes of Rh(I) and Pt(II) (**10** and **11** in Scheme 4 [16]).

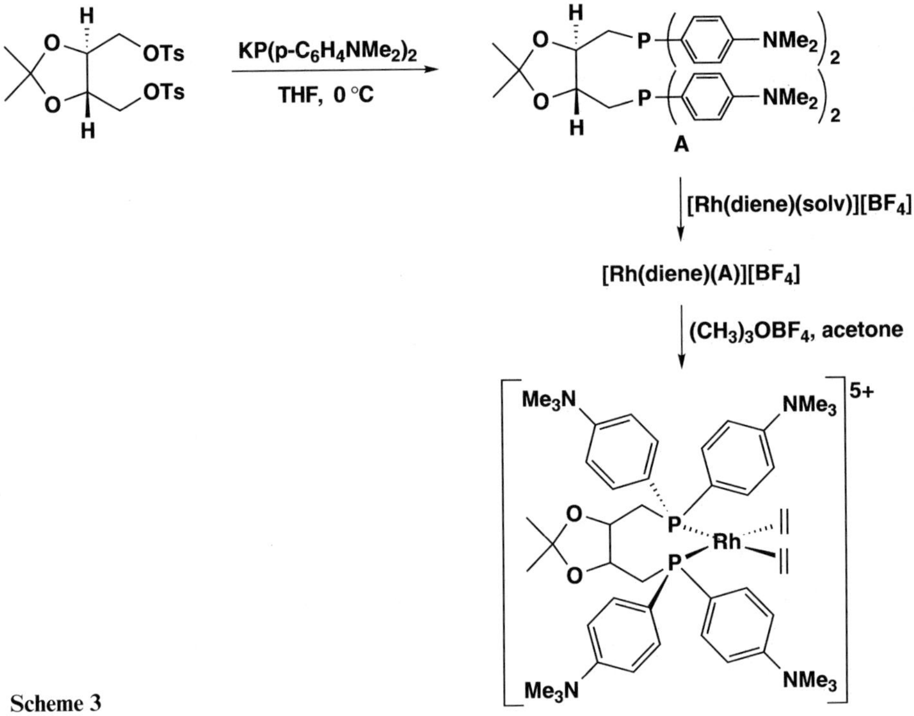

OTs
OTs
KP(p-C6H4NMe2)2
THF, 0 °C
NMe2
NMe2
A
[Rh(diene)(solv)][BF4]
[Rh(diene)(A)][BF4]
(CH3)3OBF4, acetone
Me3N
NMe3
5+
Rh
Me3N
NMe3

Scheme 3

3.2.3.2 Cationic Complexes

There are a few complexes containing more or less hydrophilic ligands which induce very high solubilities in water. For example, Sreenivasa Reddy et al. studied new ligands as in structures **12** and **13** [18–20]. The authors studied and characterized different complexes of palladium, rhenium, and platinum.

Water-solubility does not always depend solely upon the water-solubility of the functionalized ligand itself; it can also arise from specific water–metal interactions. The literature reports cationic as well as zwitterionic complexes [21–24]. These systems are not yet well known and some complexes can behave in an unexpected way. Thus, it has been reported that the SULPHOS–rhodium complex **14** is not soluble in water but it is soluble in light alcohols or in 1:1 (v/v) alcohol–water mixtures even if each molecule is a zwitterion [21]. In the same way, complexes such as **15** with limited water-solubility and stability have been characterized and their reactivity has been studied [23, 24]. Even if authors did not mention it, the low solubility in water seems not to be a drawback for catalytic purposes.

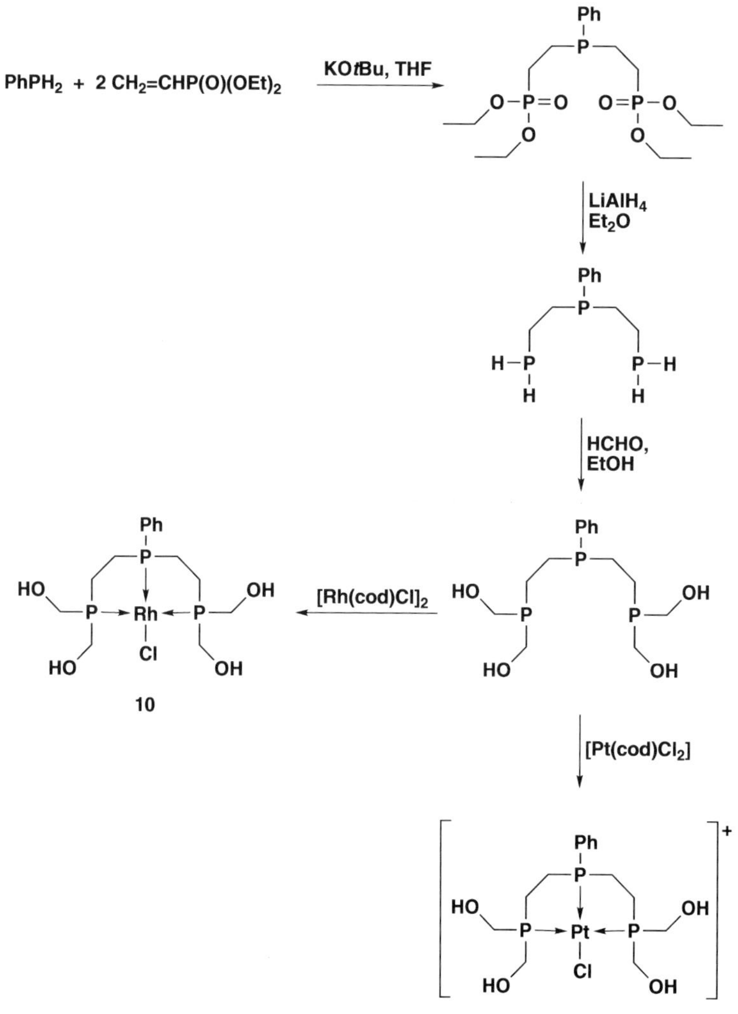

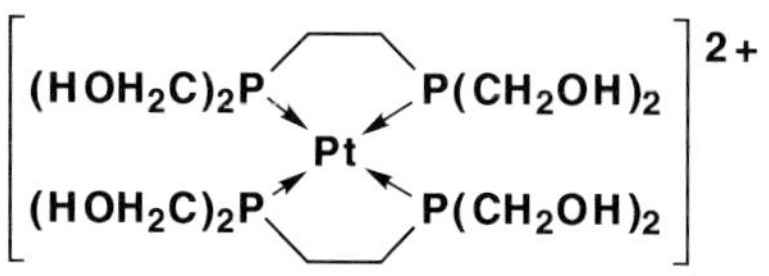

$PhPH_2 + 2\ CH_2{=}CHP(O)(OEt)_2$ → KO*t*Bu, THF

↓ $LiAlH_4$, Et_2O

↓ HCHO, EtOH

← $[Rh(cod)Cl]_2$

10

↓ $[Pt(cod)Cl_2]$

11

12

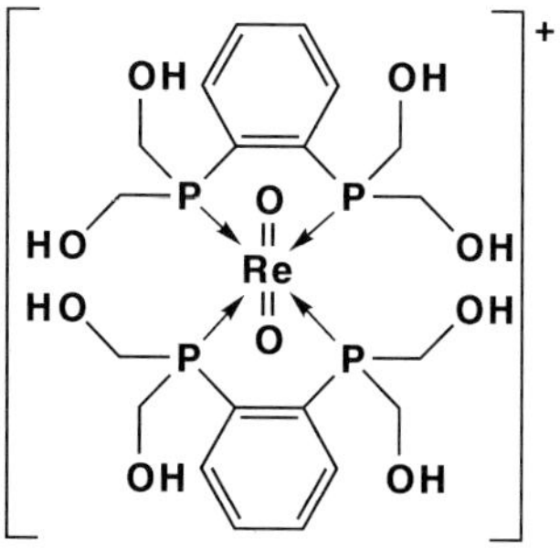

13

Scheme 4

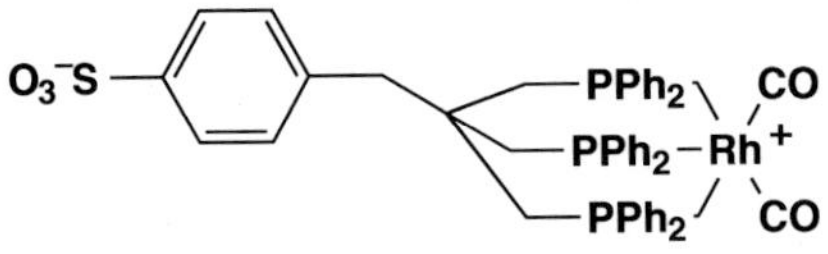

14

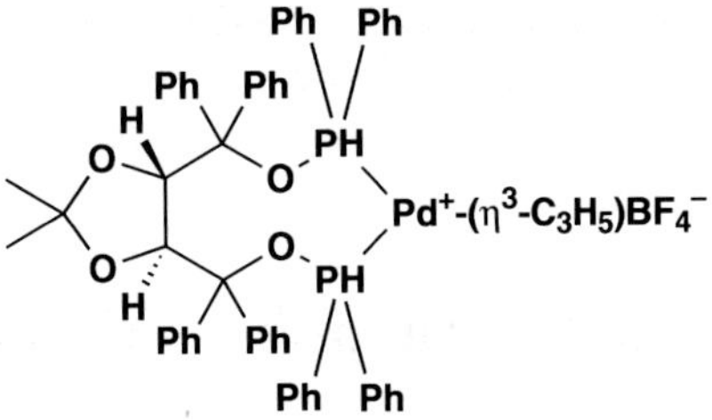

15

The chemistry of complexes containing cationic iridium has been studied by Merola and co-workers. Among all the complexes described in the literature, they reported the complex *cis,mer*-$(Me_3P)_3IrH_2Cl$, whose aqueous chemistry looks very rich [25]. It resulted from the oxidative addition of H_2 to $[Ir(cod)(PMe_3)_3]Cl$ [26]. From ^1H and ^{31}P NMR measurements, they showed that two different species are formed upon dissolution in water. One species is the cationic aquo complex [*mer,cis*-$(Me_3P)_3IrH_2(OH_2)$] **17** and the other one is the dinuclear cation **18** (Scheme 5).

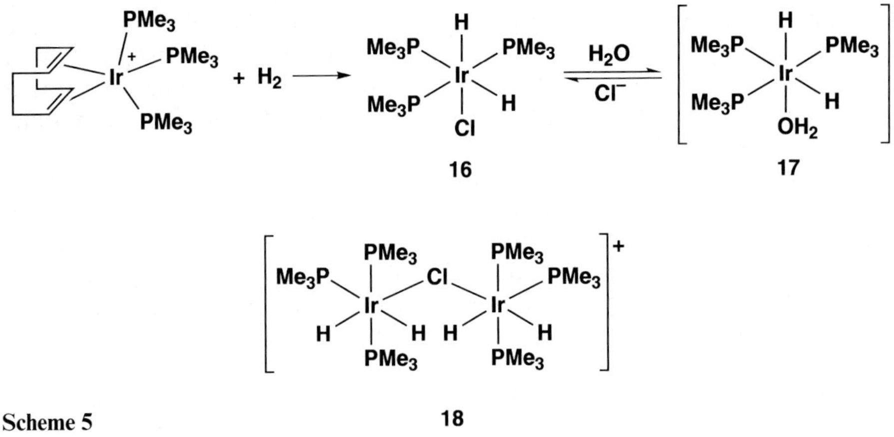

16

17

Scheme 5

18

It was also demonstrated that both **16** and **17** are in equilibrium, which involves loss or gain of chloride ligand. Compound **16** is very reactive toward alkenes and alkynes in water, whereas it is absolutely unreactive in organic solvents. The reaction which takes place with alkynes is shown in Eq. (2).

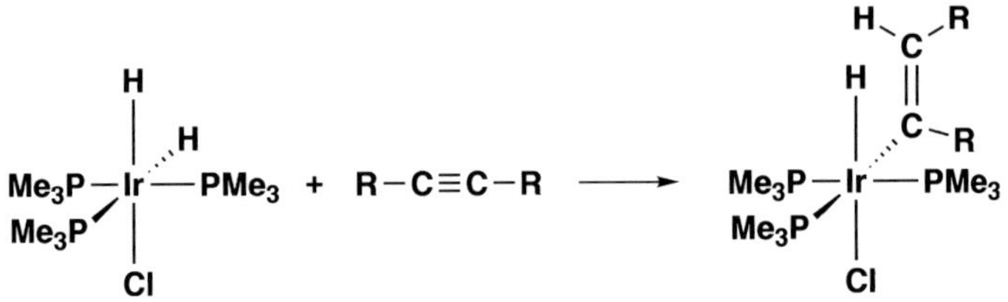

(2)

In catalysis, **16** showed very good activity for the hydrogenation of alkynes and alkenes to alkanes or ketones to alcohols (cf. Section 3.1).

3.2.3.3 Immobilization on Silica Supports

In order to combine all the advantages of both homogeneous (great activity and high selectivity) and heterogeneous catalysis (ease in separating catalyst from the medium), much work has been done with the aim of binding the active metal centers to organic polymers or inorganic oxides. Silica is certainly the most widely studied support in surface organometallic chemistry because it resists elevated temperatures and most solvents. These properties contribute greatly to enhance the lifetime and deactivation resistance of the catalyst. It is convenient to include the work done using silica with a thin film of water on its surface and generally called supported aqueous-phase catalysis (SAPC).

In 1992, Mercier and co-workers studied the chemoselective hydrogenation of several α,β-unsaturated aldehydes using, among other methods, SAPC [27, 28]. The catalysts were synthesized by adding a solution of a Ru-TPPTS complex to silica gel, resulting in two catalysts: $RuCl_2(TPPTS)_3/SiO_2$ and $RuH_2(TPPTS)_4/SiO_2$. Like Davis et al. [29–31], the authors assumed that the immobilization is due to strong interactions between the sulfonated groups and the silanol functions inside a thin surface water film (cf. Section 6.14).

The catalyst immobilized in this way is often leached from the support during the reaction because of the oxidation of the ligand and above all because of insufficiently strong bonds or interactions [32–38]. There are two synthetic routes to obtain strong immobilization: (a) silica was phosphinated, then treated by the appropriate precursor although phosphine exchange was described as tedious and difficult to control, or (b) the complex was first coordinated by a bifunctional phosphine, then linked to silica [36]. This method allows good characterization of the complex before linking. In this latter method, one uses the weakly acidic properties of a silica surface and the silane condensation reactions of donor molecules containing terminal trialkoxy groups. Because of the flat surface of silica, the ligands act as anchors leading to a robust, densely packed, single layer of transition metal centers maintained either in water or in the interfacial region and soluble in the reaction medium. Behringer started from the monoalkoxy phosphine $PPh_2(C_6H_4)SiMe_2OEt$, which forms just one siloxane bridge, to graft a dicarbonylnickel onto a support [Eq. (3)] [35].

One can exclusively obtain species **19** grafted by the two phosphine chains when DIPHOS in very dilute solutions of toluene is added dropwise to a silica suspension at room temperature. Petrucci used the same approach but started from a trialkoxyphosphine [Eq. (4)] [37], or reacted an amine with a terminal alcohol containing a phosphine [37, 38].

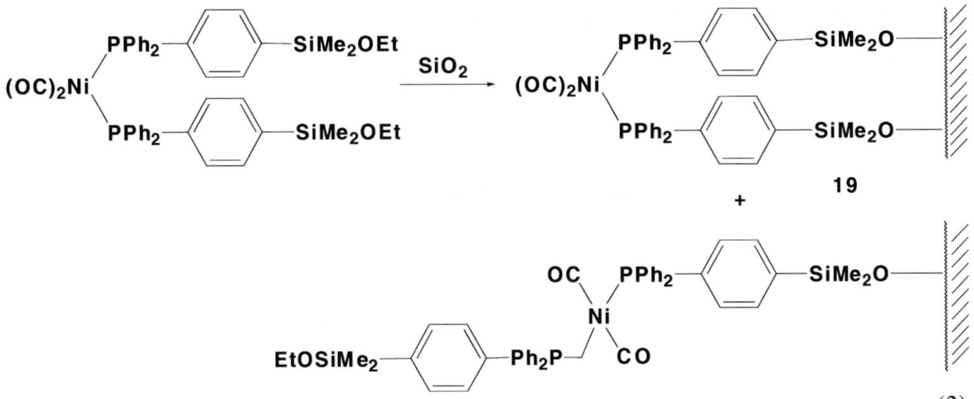

$(OC)_2Ni(PPh_2C_6H_4SiMe_2OEt)_2$ $\xrightarrow{SiO_2}$ **19** + mixed complex

(3)

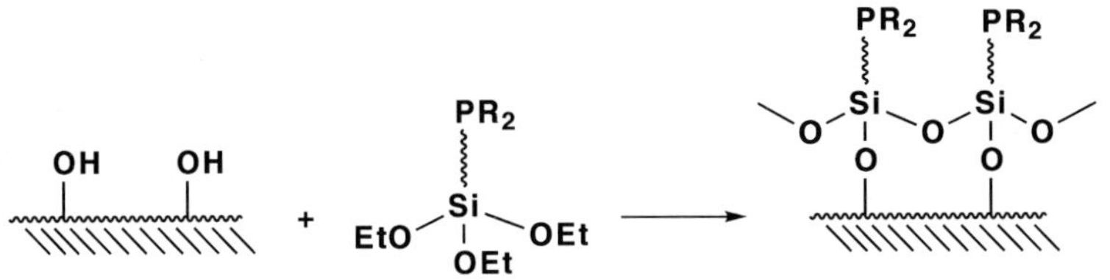

(4)

An example is provided by complex **20**. Placed in an acidic medium, the reaction can easily be carried out in biphasic conditions.

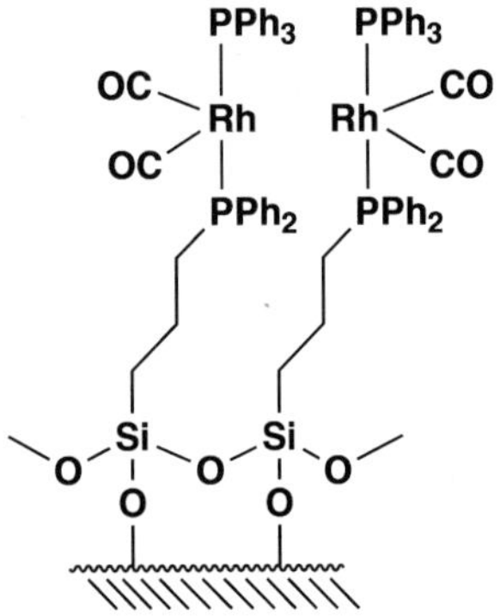

20

3.2.3.4 Macromolecular Ligands or Supports

3.2.3.4.1 Polymers

One of the very first methods of heterogenizing the catalyst in a homogeneous catalytic reaction was by using of polymers. Thus, thanks to these supports, the catalyst acquires the property of insolubility while maintaining its catalytic performance [39–42]. Some authors synthesized phosphonated resins, such as polystyrene, and used them as a ligand in several rhodium and platinum complexes. Thus, hydrogenation [43, 44], hydrosilylation [45], and hydroformylation of olefins were catalyzed.

Some authors have described the use of the previously seen framework [39]. The grafting of the DIOP ligand [Eqs. (**5–7**)] onto a Merrifield resin was performed by treatment of the insoluble aldehyde **21** with the (+)-diol **22** of the DIOP, leading to the ditosylate **23**. This product was then treated with lithium diphenylphosphide, affording the desired phosphinated resin **24**, which contains 0.5 meq/g of phosphorus functions. The reaction with $[RhCl(C_2H_4)_2]_2$ gave an insoluble complex that may be used further in catalysis.

CH_2Cl —DMSO→ CHO **21** (5)

(6) → **22**

21 + **22** → **23** —$LiPPh_2$→ **24** (7)

The second procedure is related to a copolymerization reaction [40]. Reacting (−)-1,4-ditosylthreitol with 4-vinylbenzaldehyde afforded a styryl monomer [Eq. (8)]. When this monomer (8 %, w/w) copolymerized at 70 °C with hydroxyethyl methacrylate (HEMA) in the presence of azobisisobutyronitrile (AIBN), the polymer formed was then treated with sodium diphenylphosphide (Scheme 6). The rhodium complex was obtained by reaction with $[RhCl(C_2H_4)_2]_2$.

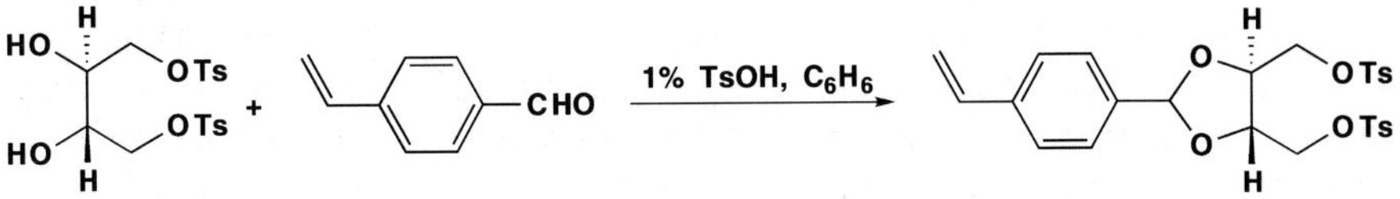

1% TsOH, C_6H_6 (8)

1% AIBN, C_6H_6

1) $NaPPh_2$
2) H_2O

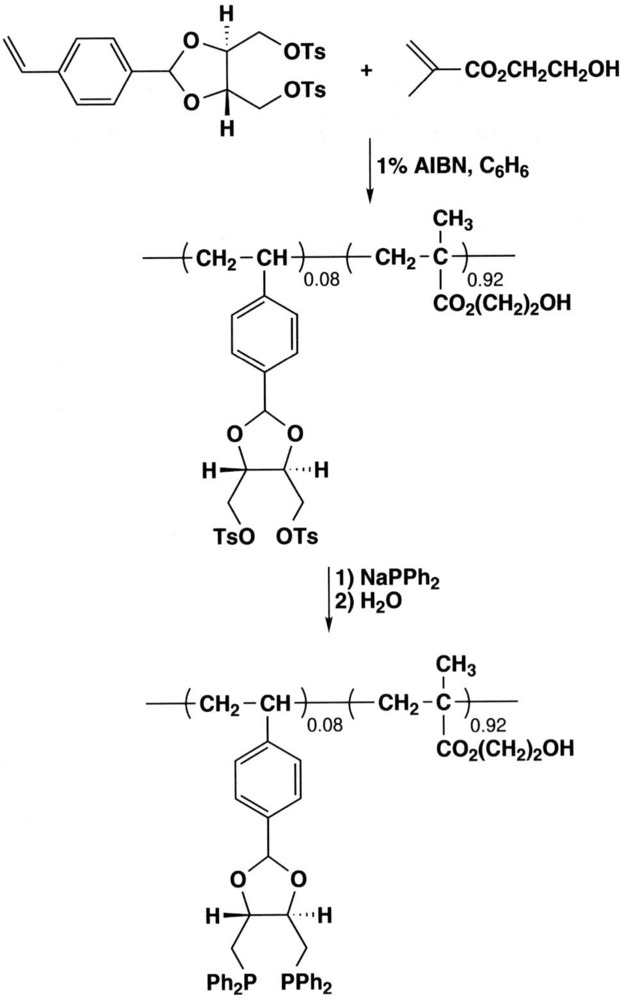

Scheme 6

Although these authors used the supported catalyst only in benzene for the grafted polymer, and in polar solvents for the copolymer, it is obvious that these catalysts can be used in water: the copolymer is indeed insoluble but the active part is water-soluble.

Recently, Malmström et al. synthesized water-soluble metal phosphine complexes based on water-soluble polymers [41]. In order to have solubility in both an acidic and a basic medium, they prepared two different water-soluble polymers. For the first, they made methyl [4-(diphenylphosphino)benzyl]amine (PNH) react with poly(acrylic acid) (PAA) using dicyclocarbodiimide (DCC) as the coupling agent, under strict exclusion of oxygen (**25**). For the second, they reacted (4-carboxyphenyl)diphenylphosphine with polyethylene imine (PEI) at room temperature (**26**). The reduction by sodium borohydride was made in situ, followed by the addition of methanesulfonic acid and diethyl ether. Then, the methanesulfonic salt of phosphinated polyethylenimine was precipitated.

25 PAA-PNH

26 PEI-PNH

Table 2. Comparison of various phosphines grafted onto polymers [41, 42].

Polymer	PAA–PNH	PEI–PNH	PAA–PPM
COO^-/P or N/P ratio	5	7	5
P or N content [%]	3.2	2.6	3.6
Water-solubility [$mg L^{-1}$]	165	260	115
pH dependence	pH $\geq$ 7	pH $\leq$ 6	–

The main characteristics of both polymer-based phosphines are listed in Table 2. In both cases, water-solubility can be varied changing the phosphine groups to carbonyl or nitrogen ratios. To warrant a high solubility, these ratios were kept at ca. 5 for PAA and ca. 7 for PEI during all the experiments. More recently, an extension of this concept to chiral ligands was reported. (2*S*,4*S*)-4-diphenylphosphino-2-diphenylphosphinomethylpyrrolidine (PPM) was acylated by PAA, using DCC [42]. Results are given in Table 2.

The reaction of an aqueous solution of PAA–PPM (**27**) with $[Rh(nbd)_2][CF_3SO_3]$ (nbd, norbornadiene) led to the *cis*-phosphino complex $[(PAA\text{–}PPM)Rh(nbd)][CF_3SO_3]$. Stirring an aqueous solution of this complex for 10 min under an H_2 atmosphere gave the solvato complex $[(PAA\text{–}PPM)Rh(H_2O)_2]^+$. $[Rh(nbd)_2][CF_3SO_3]$ was used for the hydrogenation of (*Z*)-2-acetamidocinnamic acid in water and in water–ethyl acetate. Results are given in Table 3.

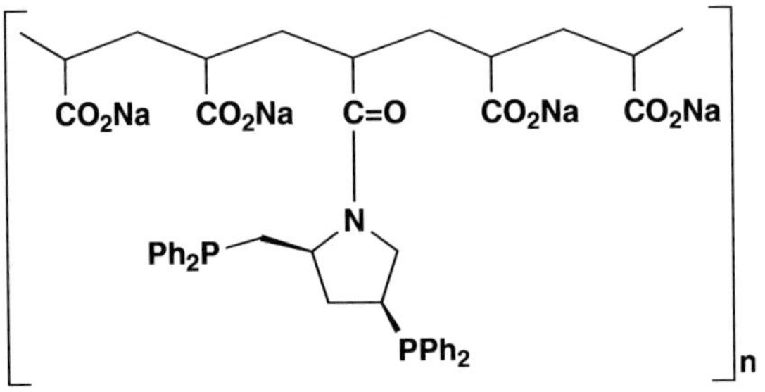

27 **PAA-PPM**

Table 3. Enantiomeric excesses obtained in various solvents for the hydrogenation of (*Z*)-2-acetamidocinnamic acid [42].

Solvent	pH	*t* [min]	Rh/alkene	Yield (%)	*ee* [%]
H_2O	8.26	120	1:35	>97	56
H_2O/EtOAc	7.00	50	1:83	>97	74
MeOH	–	–	1:100	100	93

3.2.3.4.2 Dendrimers

Most of the phosphines used in enantioselective catalysis possess a chiral backbone terminated by two diphenylphosphino groups. The chiral information is transmitted from the ligands to the metallic center via ordering of benzene rings [46]. Owing to the size of these groups, a long-distance effect is forbidden in spite of their great effectiveness. Therefore, it is of interest to use the dendrimers model in order to increase the size of the optically active phosphines. Dendrimers are spherical polymers with a highly symmetrical structure [47–49]. Thanks to internal cavities and functional end groups, dendrimers possess unusual properties. Grafting terminal P(III) phosphine groups at the periphery of the P(V) dendrimer, it becomes possible to cover its surface with metal complexes.

To ensure a strong chelation effect on the metal center, the backbone of the ligand must contain a PCH_2CH_2P framework. The dendrimer is built up by successive branchings, the final layer containing an enantiomerically pure compound; as this compound can easily be changed, such a method appears to be very flexible. Because of the space-filling nature of the ligands, chiral information is forced toward the center of the catalyst, which was called by Brunner the "pocket of the catalyst," i.e., the place where the enantioselective reaction occurs.

From a menthyl-based diphenyl ligand, a complex containing the $Rh(cod)^+$ residue was, for instance, easily obtained. This complex is soluble in toluene, pentane, and hexane; the ion pair Rh^+/PF_6^- is surrounded by a kind of "membrane" of menthyl groups whose hydrophobic sides are oriented toward the exterior, leading to the solubility in hydrocarbons. The presence of the Rh^+ cation and the PF_6^- anion suggests to us that this complex should be quite suitable for a biphasic catalysis.

3.2.3.4.3 Peptides

For several years, the use of peptides in organometallic chemistry has been a growing area. The interest in peptides stems from their secondary and tertiary structures, which is why all methods using these systems must involve the incorporation of just one functionality. The synthesis of peptides containing phosphine ligands opens new prospects for transition metal coordination chemistry. Indeed, it becomes possible to have a better control over the reactivity of the metal and to fix it in various membranes [50, 51].

During the synthetic procedure, it is fitting to prevent formation of the unwanted phosphine oxide by converting phosphine to phosphine sulfide, this method making it possible to protect and deprotect the phosphine easily. Reduction by Raney nickel [50] appeared inappropriate for the reduction of a large number of ligands [51]; then a new homogeneous reducing method was reported in which the sulfide-containing peptides remain attached to the polymer on which they are synthesized [52]. Phosphine sulfide is methylated by trifluoromethanesulfonate to give the phosphonium salt, which is treated with tris(dimethylamino)phosphine in order to remove the sulfur atom, giving the expected bis(phosphine) (Scheme 7) [51]. It is assumed that, as the peptides possess an α-helical secondary structure [53] and the phosphine-containing amino acids lie in the *i* and $i+4$ positions on the same side of the α-helical structure, the phosphines are able to bind a metal. This bis(phosphine) is treated with $[Rh(nbd)_2][ClO_4]$ to give the desired complex, whose structure is not yet well known: it may be an α-helical peptide coordinated to rhodium by two phosphorus atoms, or one rhodium atom linked to two different peptides, or both structures at the same time. The authors assume that it is probably the last possibility which exists in the system.

This complex was tested as a catalyst precursor for the hydrogenation of an enamide. This experiment showed a 100% conversion in spite of a low *ee* ranging from 4 to 9%.

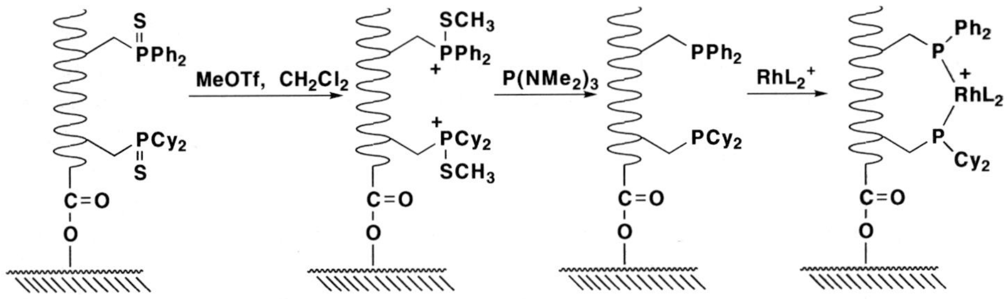

Scheme 7

3.2.3.4.4 Sugars

A very unexpected ligand was synthesized from sugars for asymmetric catalysis by Shi and co-workers [54]. They bound the phosphorus atom of a phosphine to a carbon atom in a system of pyrano rings (see **28** and **29**). The aim was to coordinate this chiral ligand with gold for medicinal applications. They reported synthesis and characterization of a series of gold(I) complexes whose general formula is [Au(P′)X], where P′ is for example methyl 4,6-*O*-benzylidene-2-deoxy-2-(diphenylphosphino)-α-D-altropyranoside (2-MBPA, **28**) and X is a chlorine atom or pyridine-2-thiol. The authors assumed that the complex [Au(2-MBPA)Cl], which presents a chelate structure where the hydroxyl oxygen O(3) is bound to gold, gives rise to the release of HCl and then to the hydrolysis of the molecule. They also showed by crystallographic measurements that, for the phosphine ligands, the donor ability toward gold(I) decreases in the order DBP > DMPP > 2-MBPA > PPh_3 > PCy_3.

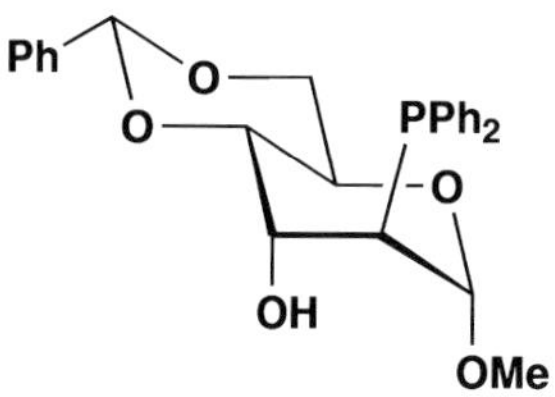

28 2-MBPA

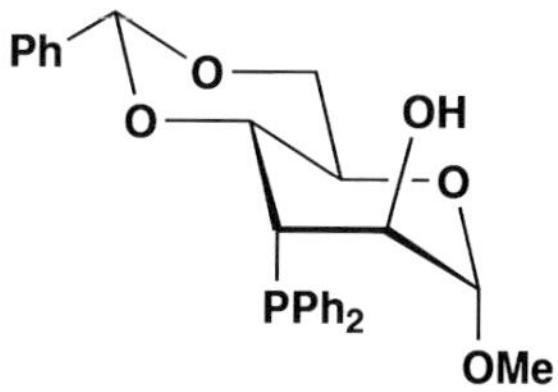

29 3-MBPA

3.2.3.5 Ligands not Containing Phosphorus

3.2.3.5.1 Nitrogen-containing Ligands

The literature about ligands not containing phosphorus is scarce. Most of the articles report nitrogen-containing ligands, especially substituted bipyridine and ethylene diamine *N*,*N*,*N'*,*N'*-tetraacetic acid (EDTA), and were recently reviewed [12]. The most commonly used functionalities are sulfonate and carboxylic acid groups. The first functionalization consists of various mono- and disulfonated bipyridines, like the structures **30**–**32**. Formic or acetic acid is usually introduced on bipyridine (**33**), aniline and above all ethylenediamine. The most commonly used ligand in this chemistry is EDTA, and the synthesis of EDTA-containing complexes has been detailed in a recent review [55]. The chemistry of EDTA–ruthenium species was widely studied, especially by Taqui Khan and co-workers [56, 57]. They investigated the hydroformylation reaction catalyzed by $[Ru(EDTA)]^-$ and observed a poor selectivity for allyl alcohol but a satisfactory linearity starting from 1-hexene. They showed that the rate-determining step was the hydride transfer to the coordinated alkene.

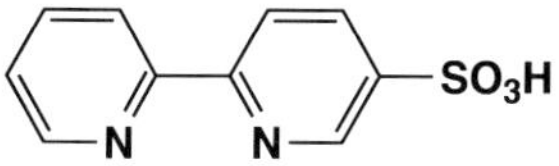

30

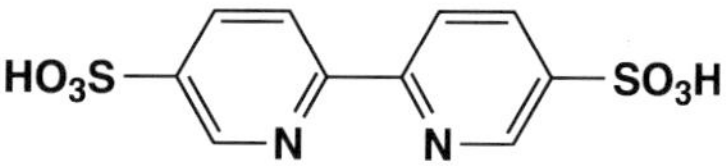

31

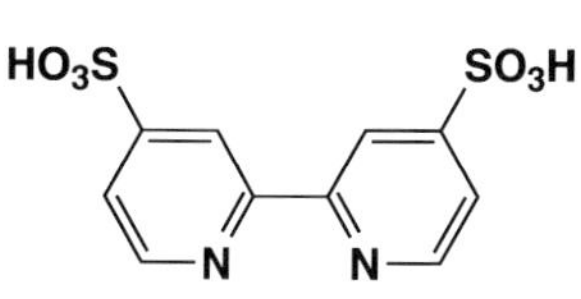

32

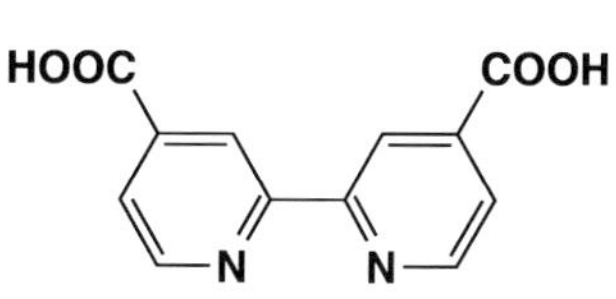

33

3.2.3.5.2 Sulfur-containing Ligands

Studying iron models for the FeMo cofactor of FeMo nitrogenases, Sellmann et al. isolated iron complexes with sulfur ligands [58, 59]. Generally, complexes such as **34**–**36** are only soluble in organic solvents like CH_2Cl_2, THF, DMSO, or DMF.

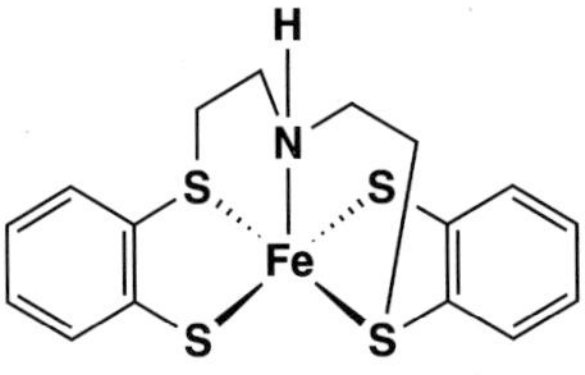

34

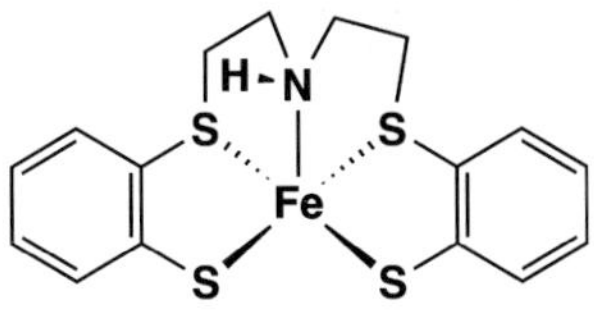

35

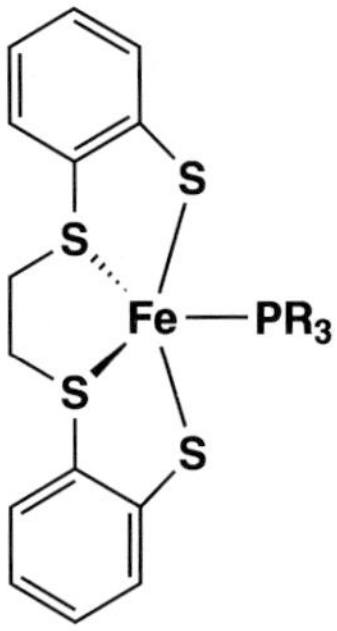

36

They succeeded in rendering them water-soluble by introducing carboxylic acid functions. These iron or ruthenium complexes become water-soluble when the carboxylic groups are deprotonated. Various complexes of iron and ruthenium can be obtained according to the method developed. An example is given in Eq. (9) [58].

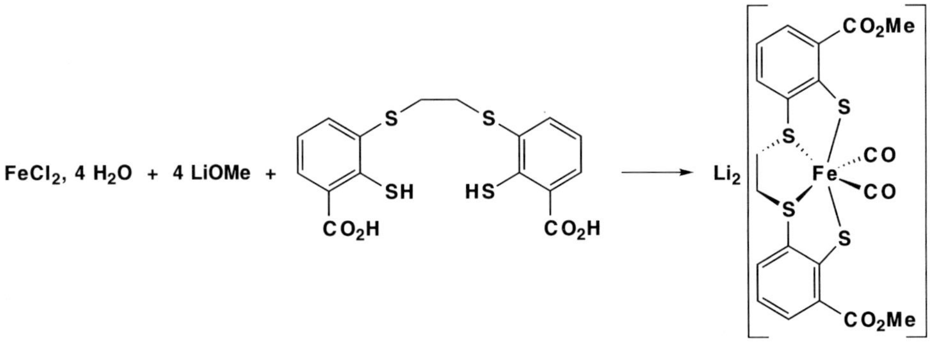

$FeCl_2$, 4 H_2O + 4 LiOMe + [ligand] → Li_2[Fe(CO)$_2$(ligand)] (9)

The dinuclear rhodium complex containing the water-soluble TPPTS ligand and the bridging *t*-butylthiolato ligands $[Rh_2(\mu\text{-}S^tBu)_2(CO)_2(TPPTS)_2]$ was shown to be an active catalyst of the hydroformylation of alkenes [60]. This complex can also use the CO/H_2O couple to combine the water-gas shift reaction and the hydroformylation reaction under moderately acidic conditions [61]. This compound and a cationic complex containing two aminothiolato bridges, $[Rh_2(\mu\text{-}SCH_2CH_2CH_2NHMe_2)_2(CO)_2(TPPTS)_2]Cl_2$, catalyze more or less rapidly the hydroformylation of heavy alkenes in the presence of ethanol as co-solvent in order to increase the transfer between the organic and aqueous phases [62].

References

[1] S. Ahrland, J. Chatt, N. R. Davies, A. A. Williams, *J. Chem. Soc.* **1958**, 276.
[2] Rhône-Poulenc Industries (E. G. Kuntz), FR 2.314.910 (1975).
[3] E. G. Kuntz, *CHEMTECH* **1987**, 570.
[4] S. Lelièvre, F. Mercier, F. Mathey, *J. Org. Chem.* **1996**, *61*, 3531.
[5] SNPE (F. Mathey, D. Neibecker, A. Brèque) FR 2588.197 (1987).
[6] D. Neibecker, D. Reau, *Angew. Chem., Int. Ed. Engl.* **1989**, *28*, 500.
[7] J. Haggin, *Chem. Eng. News* **1995**, *17*, 25.
[8] L. Lavenot, M. H. Bortoletto, A. Roucoux, C. Larpent, H. Patin, *J. Organomet. Chem.* **1996**, *509*, 9.
[9] B. W. Bangerter, R. P. Beatty, J. K. Kouba, S. S. Wreford, *J. Org. Chem.* **1977**, *42*, 3247.
[10] R. G. Nuzzo, D. Feitler, G. M. Whitesides, *J. Am. Chem. Soc.* **1979**, *101*, 3683.
[11] R. G. Nuzzo, S. L. Haynie, M. E. Wilson, G. M. Whitesides, *J. Org. Chem.* **1981**, *46*, 2861.
[12] Ph. Kalck, F. Monteil, *Adv. Organomet. Chem.* **1992**, *34*, 219, and references therein.
[13] M. E. Wilson, R. G. Nuzzo, G. M. Whitesides, *J. Am. Chem. Soc.* **1978**, *100*, 2269.
[14] Y. Amrani, Ph. D. Thesis, University of Lyon, Lyon, **1986**.
[15] D. Sinou, Y. Amrani, *J. Mol. Catal.* **1986**, *36*, 319.
[16] C. J. Smith, V. Sreenivasa Reddy, K. V. Katti, *Chem. Commun.* **1996**, 2557.
[17] R. T. Smith, M. C. Baird, *Inorg. Chim. Acta* **1982**, *62*, 135.
[18] V. Sreenivasa Reddy, K. V. Katti, C. L. Barnes, *Inorg. Chim. Acta* **1995**, *240*, 367.
[19] V. Sreenivasa Reddy, D. E. Berning, K. V. Katti, C. L. Barnes, W. A. Volkert, A. R. Ketring, *Inorg. Chem.* **1996**, *35*, 1753.
[20] V. Sreenivasa, K. V. Katti, C. L. Barnes, *J. Chem. Soc., Dalton Trans.* **1996**, 1301.
[21] C. Bianchini, P. Frediani, V. Sernau, *Organometallics* **1995**, *14*, 5458.
[22] C. Bianchini, A. Meli, *J. Chem. Soc., Dalton Trans.* **1996**, 801.
[23] W. J. Peng, S. G. Train, D. K. Howell, F. R. Fronczek, G. G. Stanley, *Chem. Commun.* **1996**, 2607.
[24] D. Seebach, E. Devaquet, A. Ernst, M. Hayakawa, F. N. M. Kühnle, W. B. Schweizer, B. Weber, *Helv. Chim. Acta* **1995**, *78*, 1636.
[25] J. S. Merola, T. L. Husebo, K. E. Matthews, M. A. Franks, R. Pafford, P. Chirik, in *Aqueous Organometallic Chemistry and Catalysis* (Eds.: I. T. Horváth, F. Joó) **1995**, 33, Kluwer, Dordrecht.
[26] T. X. Le, J. S. Merola, *Organometallics* **1993**, *12*, 3798.
[27] E. Fache, C. Mercier, N. Pagnier, B. Despeyroux, P. Panster, *J. Mol. Catal.* **1993**, *79*, 117.

[28] G. Allmang, F. Grass, J. M. Grosselin, C. Mercier, *J. Mol. Catal.* **1991**, *66*, L27.
[29] M. E. Davis, *CHEMTECH* **1992**, 498.
[30] (a) J. P. Arhancet, M. E. Davis, J. S. Merola, B. E. Hanson, *Nature (London)* **1989**, *339*, 454; (b) J. P. Arhancet, M. E. Davis, J. S. Merola, B. E. Hanson, *J. Catal.* **1990**, *121*, 327; (c) J. P. Arhancet, M. E. Davis, B. E. Hanson, *J. Catal.* **1991**, *129*, 94.
[31] (a) I. Guo, B. E. Hanson, I. Tóth, M. E. Davis, *J. Mol. Catal.* **1991**, *70*, 363; (b) I. Guo, B. E. Hanson, I. Tóth, M. E. Davis, *J. Organomet. Chem.* **1991**, *403*, 221; (c) T. Horváth, *Catal. Lett.* **1990**, *6*, 43.
[32] DeQuan Li, B. I. Swanson, J. M. Robinson, M. A. Hoffbauer, *J. Am. Chem. Soc.* **1993**, *115*, 6975.
[33] L. Bemi, H. C. Clark, J. A. Davies, C. A. Fyfe, R. E. Wasylishen, *J. Am. Chem. Soc.* **1982**, *104*, 438.
[34] T. Shido, T. Okazaki, M. Ichikawa, *J. Catal.* **1995**, *157*, 436.
[35] K. D. Behringer, J. Blümel, *Chem. Commun.* **1996**, 653.
[36] J. Blümel, *J. Am. Chem. Soc.* **1995**, *117*, 2112.
[37] M. G. L. Petrucci, A. K. Kakkar, *Adv. Mater.* **1996**, *8*, 251.
[38] M. G. L. Petrucci, A. K. Kakkar, *J. Chem. Soc., Chem. Commun.* **1995**, 1577.
[39] W. Dumont, J. C. Poulin, T. P. Dang, H. B. Kagan, *J. Am. Chem. Soc.* **1973**, *95*, 8295.
[40] N. Takaishi, H. Imai, C. A. Bertelo, J. K. Stille, *J. Am. Chem. Soc.* **1976**, *98*, 5400.
[41] T. Malmström, H. Weigl, C. Andersson, *Organometallics* **1995**, *14*, 2593.
[42] T. Malmström, C. Andersson, *Chem. Commun.* **1996**, 1135.
[43] R. H. Grubbs, L. C. Croll, *J. Am. Chem. Soc.* **1971**, *93*, 3062.
[44] J. P. Collmann, L. J. Hegedus, M. P. Cooke, J. R. Norton, G. Dolcetti, D. N. Marquardt, *J. Am. Chem. Soc.* **1972**, *94*, 1789.
[45] M. Capka, P. Svoboda, M. Creny, J. Hetflejs, *Tetrahedron Lett.* **1971**, 4787.
[46] H. Brunner, *J. Organomet. Chem.* **1995**, *500*, 39.
[47] C. Galliot, D. Prévoté, A. M. Caminade, J. P. Majoral, *J. Am. Chem. Soc.* **1995**, *117*, 5470.
[48] M. Slany, M. Bardaji, M. J. Casanove, A. M. Caminade, J. P. Majoral, B. Chaudret, *J. Am. Chem. Soc.* **1995**, *117*, 9764.
[49] D. Prévoté, C. Galliot, A. M. Caminade, J. P. Majoral, *Heteroatom Chem.* **1995**, *6*, 313.
[50] S. R. Gilbertson, G. Chen, M. McLoughlin, *J. Am. Chem. Soc.* **1994**, *116*, 4481.
[51] S. R. Gilbertson, X. Wang, G. S. Hoge, C. A. Klug, J. Schaefer, *Organometallics* **1996**, *15*, 4678.
[52] J. Omelanczuk, M. Mikolajczyk, *Tetrahedron Lett.* **1984**, *25*, 2493.
[53] M. R. Ghadiri, C. Soares, C. Choi, *J. Am. Chem. Soc.* **1992**, *114*, 825.
[54] J. C. Shi, X. Y. Huang, D. X. Wu, Q. T. Liu, B.-S. Kang, *Inorg. Chem.* **1996**, *35*, 2742, and references therein.
[55] M. M. Taqui Khan, *Platinum Metals Rev.* **1991**, *35*, 70.
[56] M. M. Taqui Khan, D. Chaterjee, M. Bala, K. N. Bhatt, *J. Mol. Catal.* **1992**, *73*, 265.
[57] M. M. Taqui Khan, S. B. Halligudi, S. H. R. Abdi, *J. Mol. Catal.* **1988**, *48*, 325.
[58] D. Sellmann, T. Becker, F. Knoch, *Chem. Ber.* **1996**, *129*, 509.
[59] D. Sellmann, W. Soglowek, F. Knoch, G. Ritter, J. Dengler, *Inorg. Chem.* **1992**, *31*, 3711.
[60] Ph. Kalck, P. Escaffre, F. Serein-Spirau, A. Thorez, B. Besson, Y. Colleuille, R. Perron, *New J. Chem.* **1988**, *12*, 687.
[61] P. Escaffre, A. Thorez, Ph. Kalck, *New J. Chim.* **1987**, *11*, 601.
[62] F. Monteil, R. Quéau, Ph. Kalck, *J. Organomet. Chem.* **1994**, *480*, 177.

3.2.4 Tenside Ligands

Georgios Papadogianakis, Roger A. Sheldon

3.2.4.1 Introduction

Organometallic catalysis in aqueous/organic two-phase systems combines the inherent advantages of homogeneous catalysts (high activity and selectivity) with the facile catalyst separation which is the great advantage of heterogeneous systems, thus affording both economic and environmental benefits [1]. The highly water-soluble $RhH(CO)(TPPTS)_3$ complex catalyst (TPPTS = triphenylphosphine trisulfonate), for example, is applied in the Ruhrchemie–Rhône-Poulenc (RCH/RP) biphasic process for the hydroformylation of propene with a capacity of 400000 t/a (debottle-necked) butyraldehyde (cf. Section 6.1.3.1). Other lower olefins such as 1-butene can also be hydroformylated with acceptable rates according to the RCH/RP process [2]. However, Rh/TPPTS catalysts exhibit low catalytic activity in the hydroformylation of higher olefins, due to the much lower solubility of such olefins in water. The reaction rates decrease dramatically with increasing C-number of the olefin. For example, 1-hexene is hydroformylated with conversions up to 22% [3] whereas the highly water-immiscible 1-tetradecene gives only traces of C_{15}-aldehydes [4].

Several concepts have been suggested to increase the rates in aqueous-phase catalytic conversion of higher substrates such as addition of conventional surfactants [3, 5] (cf. Sections 4.5 and 6.1.5), counter (inverse)-phase transfer catalysis using β-cyclodextrins [6] (cf. Section 4.6.1), addition of promoter ligands, e.g. PPh_3 [7], or co-solvents (cf. Section 4.3). However, addition of "foreign compounds" militates against the facile catalyst separation and purification of the products and increases the costs as well.

A particularly elegant approach to circumventing the solubility problem is to generate transition metal complexes from tenside phosphines which combine both the inherent properties of a ligand (appropriate steric and electronic environment) and a surfactant in one molecule, and to use them as catalysts in micellar systems.

3.2.4.2 Tenside Phosphines and Amines

The term "tenside" is synonymous with surfactant (surface-active agent), amphiphilic or amphipathic; it is *not* synonymous with detergent (often used interchangeably with "surfactant"), which is a substance capable of cleaning and contains, inter alia, surfactants. A surfactant is composed of a nonpolar hydrophobic (lipophilic) region, usually an elongated alkyl group called the tail, and a polar hydrophilic (lipophobic) portion (the head). It therefore has both pronounced hydrophobic and hydrophilic properties. Depending on the charges of their polar head groups, tensides are conveniently divided into anionic, cationic, zwitterionic (amphoteric), or nonionic. Surfactants are adsorbed at the interfaces of aqueous/organic two-phase systems, leading to a reduction of the interfacial tension. The term "hydrophile–lipophile balance" (HLB), according to Griffin [8], expresses the relative simultaneous attraction of a surfactant in the aqueous and nonpolar organic phases. The HLB value is an empirical number in the range from 1 (oleic acid) to 40 (sodium dodecylsulfate, SDS) where the low values indicate solubility in the nonpolar solvents and the high values solubility in water [8, 9c]. When surfactants are dissolved in water they have the characteristic property of assembling into molecular aggregates called micelles (cf. Section 4.5) above a certain concentration termed the critical micelle concentration (CMC) [9]. Structures **1**–**21** depict the tenside ligands used so far in organometallic catalysis in aqueous media. Surfactant phosphines containing sulfonate or polyether functionalities are the tenside ligands most frequently used to impart surface activity to transition metal complexes applied as catalysts in hydroformylation or hydrogenation reactions. Tenside phosphines bearing carboxylic, phosphonium, or phosphate moieties have also been used as ligands in hydrogenation reactions. Furthermore, another class of surfactant ligands, namely tenside amines (e.g. **20** and **21**) have been applied as modifiers in transition-metal-catalyzed hydrolysis reactions [10].

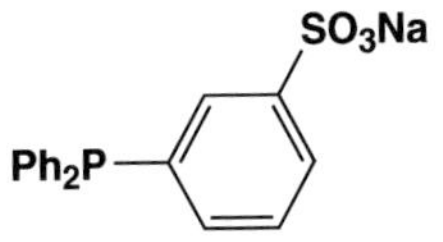

Ph_2P–C_6H_4–SO_3Na

1

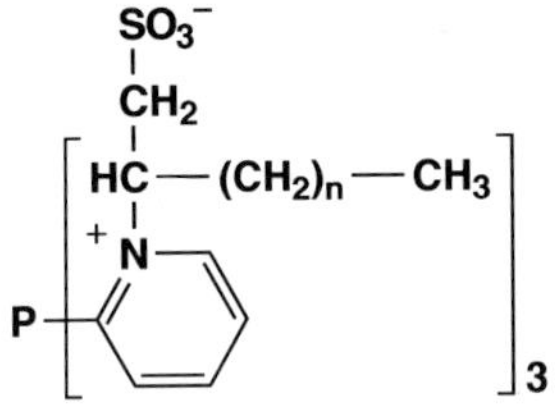

$P[C_5H_4N^+–CH(CH_2SO_3^-)–(CH_2)_n–CH_3]_3$

n = 0, 3, 5, 7, 9, 11

2

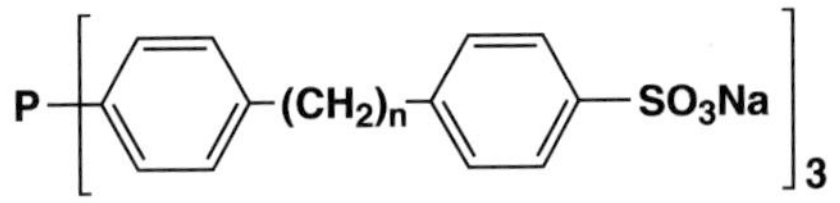

$P[C_6H_4–(CH_2)_n–C_6H_4–SO_3Na]_3$

n = 3, 6

3

$P[(CH_2)_n–C_6H_4–SO_3Na]_3$

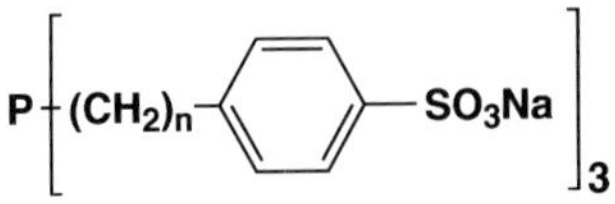

n = 1, 2, 3, 6

4

(menthyl)$P[(CH_2)_8–C_6H_4–SO_3Na]_2$

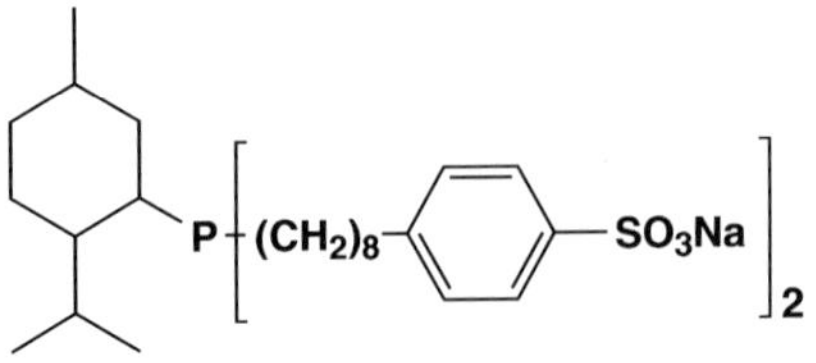

5

$R_2P(CH_2)_nSO_3Na$

R = Ph, menthyl, cyclohexyl
n = 2, 3, 4, 5

6

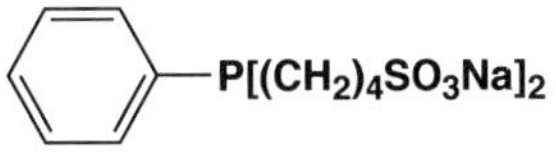

C_6H_5–$P[(CH_2)_4SO_3Na]_2$

7

$P[C_6H_4(CH_2)_3C_6H_4SO_3Na]_2$ / $P[C_6H_4(CH_2)_3C_6H_4SO_3Na]_2$

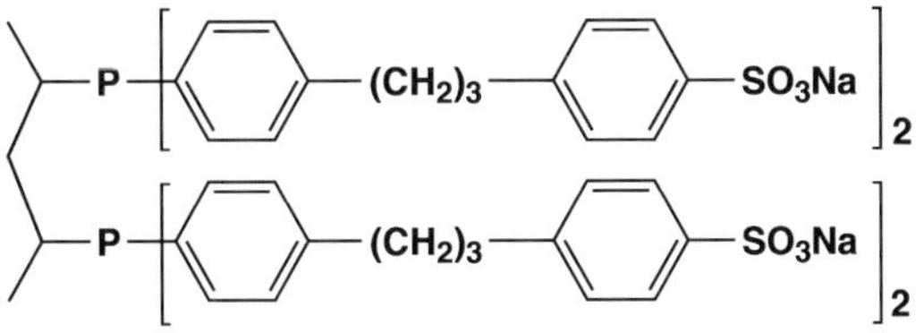

8

$Ph_nP[(CH_2)_x(CH_2CH_2O)_mR]_{3-n}$

n = 0, 1, 2; x = 0, 1; m = 1, 2, 3, 12, 16; R = Me, Bu

9

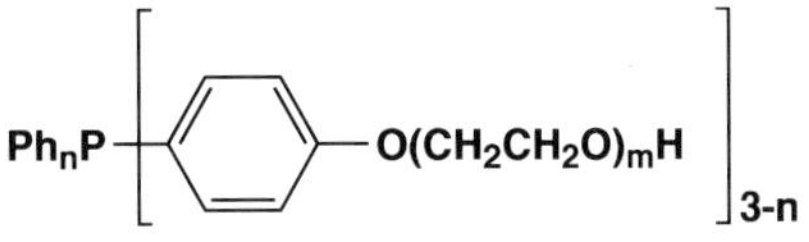

$Ph_nP[C_6H_4O(CH_2CH_2O)_mH]_{3-n}$

n = 0, 1, 2; m = 3–25

10

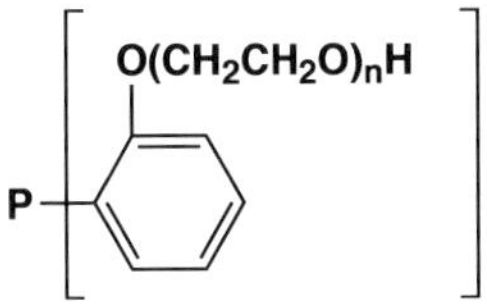

$P[C_6H_4O(CH_2CH_2O)_nH]_3$

n = 3–14

11

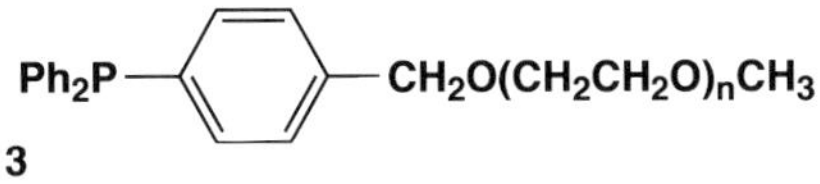

Ph_2P–C_6H_4–$CH_2O(CH_2CH_2O)_nCH_3$

12

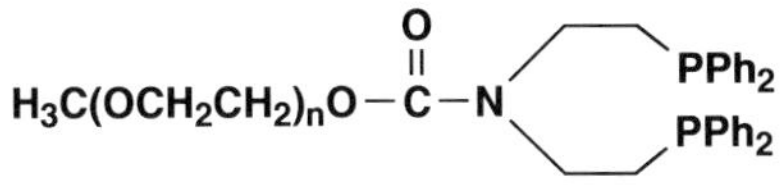

$H_3C(OCH_2CH_2)_nO-C(=O)-N$ (PPh₂, PPh₂)

n = 12, 16, 110

13

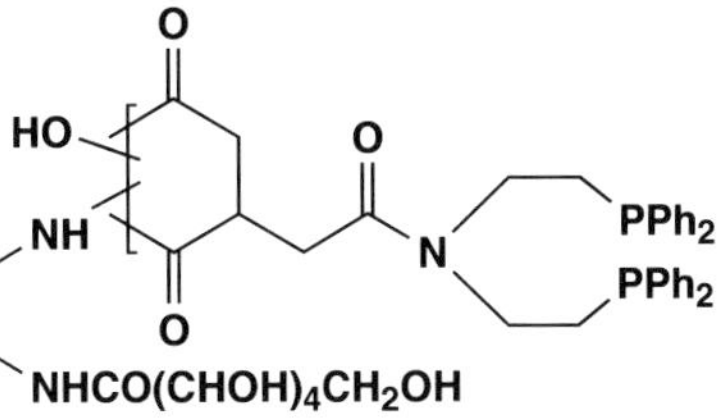

HO, NH, O, O, O, N, PPh_2, PPh_2, $NHCO(CHOH)_4CH_2OH$

14

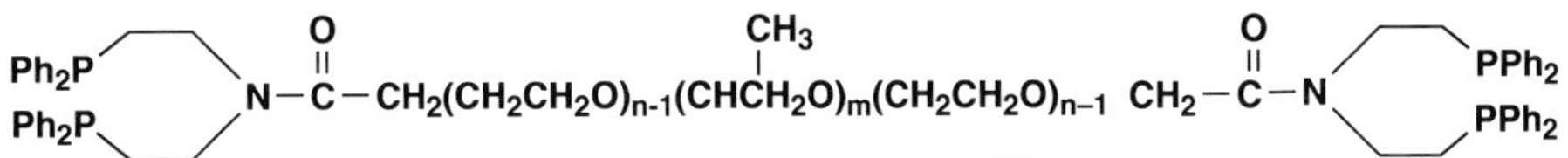

Ph_2P, Ph_2P $N-C(=O)-CH_2(CH_2CH_2O)_{n-1}(CH(CH_3)CH_2O)_m(CH_2CH_2O)_{n-1}\ CH_2-C(=O)-N$ PPh_2, PPh_2

n = 11; m = 34

15

$CH_3(OCH_2CH_2)_nOCH_2CH$ (O, O) PPh_2, PPh_2

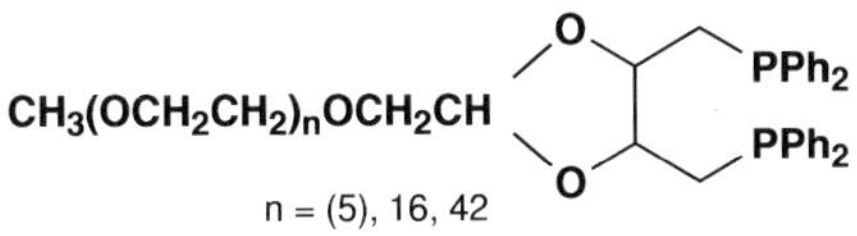

n = (5), 16, 42

16

$Ph_2P(CH_2)_nCOOH$

n = 1, 2, 3, 4, 5, 7, 9, 11

17

$[Ph_2P(CH_2)_nPMe_3]^+$

n = 2, 3, 6, 10

18

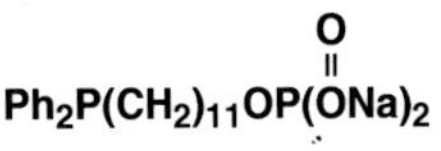

$Ph_2P(CH_2)_{11}OP(O)(ONa)_2$

19

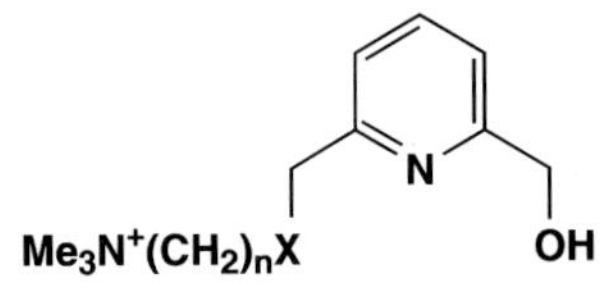

$Me_3N^+(CH_2)_nX$ OH

n = 2, 10, 11; X = CONH, S

20

R' R''

R–N N

CH_2OH

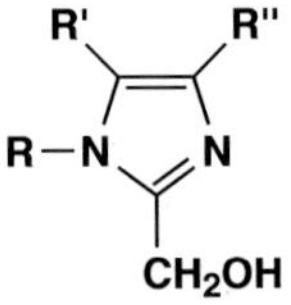

R = $C_{16}H_{33}CH(OSO_3Na)CH_2$, $C_{12}H_{25}$

R'= H, CH_2OH; R''= H, $CH_2S(CH_2)_2SO_3Na$

21

3.2.4.3 Hydroformylation Reactions Catalyzed by Transition Metal Surfactant–Phosphine Complexes

The fact that water-soluble sulfonated phosphines may combine the properties of a ligand and a surfactant in the same molecule was first mentioned in 1978 by Wilkinson et al. [11] in their study of the hydroformylation of 1-hexene using rhodium and ruthenium catalysts modified with TPPMS (triphenylphosphine monosulfonate; **1**) in an aqueous/organic two-phase system. The authors [11] noted that under all conditions employed using Rh/TPPMS catalysts some orange color was present in the organic phase after the reaction, indicating leaching of rhodium, probably due to the low HLB value of TPPMS. Quite recently, surface tension measurements of TPPMS and TPPTS (the corresponding trisulfonate) in a water/toluene two-phase system provided evidence for surface activity of TPPMS which gives rise to a lower interfacial tension between the two phases (15.05 dyn cm^{-2}) whereas TPPTS behaves as an electrolyte (51.23 dyn cm^{-2}) [12]. Rh/TPPMS catalysts generated from the potassium salt of TPPMS were active in the biphasic hydroformylation of 1-hexene and 1-dodecene to afford aldehyde yields as high as 86% and 66%, respectively [13]. However, the authors have not discussed the question of the carry-over of rhodium from the aqueous to the organic phase [13].

The first water-soluble system specifically designed to combine both functions of a ligand and a surfactant in one molecule and applied in transition-metal-catalyzed conversions of highly water-insoluble substrates in micellar systems is the zwitterionic tenside trisulfoalkylated tris(2-pyridyl)phosphine, **2** ($n = 0, 3, 5, 7, 9, 11$) [4, 14]. Turnover frequencies (TOF) up to 340 h^{-1} were achieved in the micellar hydroformylation of 1-tetradecene to pentadecanals, according to Eq. (1), using Rh/**2** catalysts at 125 °C, by fine tuning of the hydrophilic/lipophilic properties of the tenside system **2** [4, 14]. In sharp contrast, Rh/TPPTS catalysts gave only traces of pentadecanals under the same biphasic conditions.

1-tetradecene + CO / H_2 $\xrightarrow{Rh / 2}$ n-pentadecanal (CHO) + iso-pentadecanal (CHO) (1)

The conversions in the micellar Rh/**2**-catalyzed hydroformylation of 1-tetradecene increased with increasing length of the nonpolar hydrocarbon tail of the surfactant **2** up to a maximum for 8 C-atoms ($n = 5$) achieving a conversion of 79% within 3 h. A further increase in nonpolar chain length resulted in a decrease in conversion (C_{10} and C_{12} gave 72% and 39%, respectively). With high HLB values, i.e., a shorter nonpolar tail in the range of 3 to 10 C-atoms ($n = 0-7$) the Rh/**2** catalyst was quantitatively recovered by simple phase separation. In contrast, at lower HLB values, namely a longer tail with 12 or 14 C-atoms ($n = 9$, 11), very stable emulsions were formed which did not break into a desired two-phase system even after standing for one year at room temperature.

In order to rationalize the effect whereby the activity in the Rh/**2**-catalyzed hydroformylation of 1-tetradecene goes through a maximum as a function of the tail length of the surfactant **2**, the model of a simplified spherical (Hartley) ionic micelle [9a–c] (Figure 1) was proposed [14, 15]. The core of the micelle is probably composed of the hydrophobic tail of the tenside phosphine **2** where 1-tetradecene is solubilized (Figure 1, stippled part).

Surrounding the core is the Stern layer where the charged head groups (SO_3^-) of the surfactant **2** are located together with the counterions (Na^+) in a compact region a few angstroms wide. The rhodium atom of the catalyst is probably located on the polarity gradient between the Stern layer and the core of the micelle. The situation of the Rh should depend on the HLB value, i.e., on the tail length of the system **2**. The surfactant **2** with eight C-atoms ($n = 5$) apparently possesses the optimum length for solubilizing 1-tetradecene efficiently, and therefore maximum activity is observed in the hydroformylation reaction. In contrast, when $n < 5$ the ligand **2** is probably too short to solubilize the olefin and when $n > 5$ the Rh is probably located far from the polarity gradient between the Stern layer and the core of the micelle, which gives rise to a drop in the catalytic hydroformylation activity [4, 14]. The Rh/**2**-catalyzed hydroformylation of 1-tetradecene may, alternatively, proceed in "wet micelles" [16], which are water-permeated, porous micellar structures, or in reversed (inverse) micelles [9a, 17, 18], where the polar head groups of **2** form an aqueous interior while the hydrophobic tails are in contact with the olefin exterior bulk phase.

The anionic tenside phosphines **3**, **4**, and **5** [19–26] were used as ligands to impart surfactant properties to rhodium and cobalt catalysts in hydroformylation reactions of higher olefins in aqueous media. The aggregation of the system **3** ($n = 3$, 6) and **5** was investigated with dynamic light-scattering experiments at different ionic strengths of the solution by addition of NaCl [19]. It was assumed

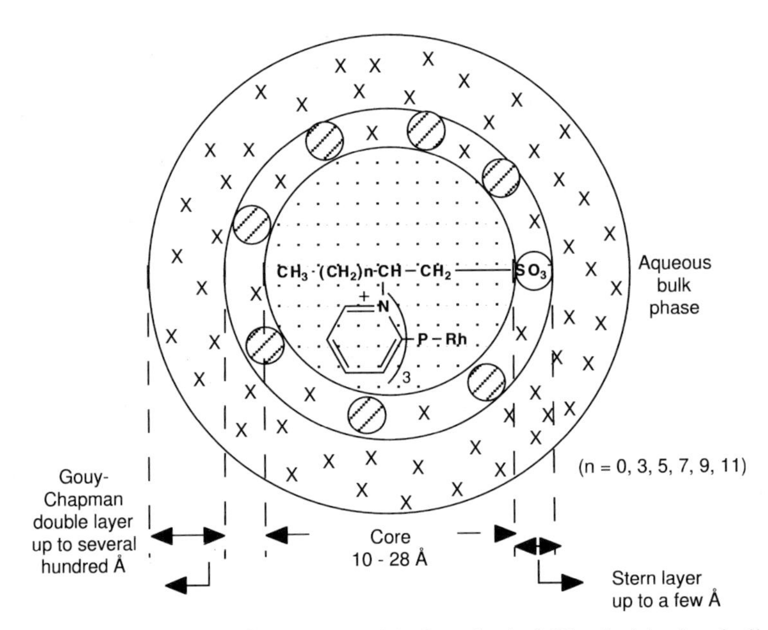

Figure 1. Representation of a simplified model of a spherical (Hartley) ionic micelle containing the Rh/**2** catalyst. The solubilized 1-tetradecene in the core (stippled area), the tail of the tenside ($CH_3(CH_2)_nCHCH_2$–), the head (SO_3^-), the counter ions (Na^+, OH^-, depicted as X) schematically indicate their relative locations and not the relationship to their molecular size, distribution, number, or configuration.

that if the relationship of the SO_3^- groups to the P-atom of the system **3** is pyramidal, then a tetrahedral array of four ligands **3** form a small spherical micelle with the P-atom comprising its core [19, 27]. Hydroformylation reactions of 1-octene were carried out using rhodium catalysts modified with the surfactant ligand **3** ($n = 3, 6$) dissolved in an aqueous methanol solution ($MeOH/H_2O = 1:1$) to give C_9-aldehydes and nonanols (alcohols up to 10%) [19, 20]. At a ligand/Rh ratio of 10:1 the TOF obtained with **3** ($n = 3$) was 335 h^{-1} and with **3** ($n = 6$) 360 h^{-1}. Under the same conditions using Rh/TPPTS catalysts, however, the TOF obtained was 260 h^{-1} [20]. When this biphasic reaction occurred in water instead of aqueous methanol the reaction rates were clearly diminished, using both Rh/**3** and Rh/TPPTS catalysts [19, 22]. Rhodium catalysts derived from the electron-donating surfactant ligand **4** have been used in the hydroformylation of 1-octene and in the presence of a conventional tenside, namely the sodium salt of dodecylbenzenesulfonic acid (0.5 wt.%) in a 1-octene/nonane/methanol/water (60:34:50:56) mixture [23]. At low ligand/Rh ratios in the range 2–4, the conversion increased with increasing number of methylene groups in the ligand **4**; with Rh/**4** ($n = 6$) the conversion of 1-octene achieved was 85% (TOF = 28 h^{-1}) [23].

Cobalt complexes generated from the tenside ligand **4** ($n = 1, 2, 3, 6$), namely $Na_5[Co^+(CO)_3(\mathbf{4})_2]^{5-}$, have been used as catalysts for the biphasic hydroformylation of 1-hexene and 1-octene [24]. The products obtained were almost exclusively aldehydes (4–38%) and there was very little (0.4–3%) or no alcohol formation [24]. This contrasts with the expected products for the phosphine-modified cobalt hydroformylation catalysts in conventional organic solvents which are alcohols. The *n*/iso ratios of the aldehydes using the $Na_5[Co^+(CO)_3(\mathbf{4})_2]^{5-}$ catalysts was low (1.1–2.5) [24] and never approached that expected for a phosphine-modified cobalt catalyst in organic media (7.3) [1c]. However, the reaction rates and the selectivity are dependent on the ligand **4**/Co ratio and no leaching of cobalt was observed into the organic phase [24].

Rhodium complexes modified with the tenside chiral phosphine **5** were used as catalysts in the hydroformylation of styrene, according to Eq. (2), in an aqueous/organic two-phase system [19, 26]. The TOFs achieved with the surfactant catalyst Rh/**5** were higher (245 h^{-1}) compared with the Rh/TPPTS system (100 h^{-1}) [26]. The *n*/iso ratios of the aldehydes were about 0.6 with Rh/TPPTS and ca. 0.4 with the Rh/**5** system. Although the phosphine **5** is chiral, virtually no optical activity was observed in the phenylisopropanal product.

Styrene + CO / H_2 —Rh / 5→ n- (CHO) + iso- (CHO) (2)

Phenylpropanal

The sulfonated phosphine system **6** [28–30] was also described as possessing surfactant properties [28]. However, **6** and **7** have only been applied in the hydroformylation of 1-tetradecene in methanol followed by biphasic separation of the catalyst after treatment with water [30].

Rhodium complexes generated from the poly(ethylene glycol)-functionalized phosphine **9** ($n = 1$, $x = 0$, R = Me, Bu), which should behave as a nonionic surfactant and be able to induce micelle formation, have been used as catalysts in the hydroformylation of 1-dodecene in an aqueous/organic two-phase system [31]. The conversion of 1-dodecene was 80% and the *n*/iso ratio 60:40, with no carry-over of the rhodium catalyst into the organic phase. The Rh/**9** ($n = 1$, $x = 0$, R = Me, Bu) catalyst remained active after one recycle step [31].

A novel concept in the field of hydroformylation of higher olefins consists in the use of rhodium catalysts modified with the nonionic surfactant phosphines **10** and **11**, which become organic-soluble on heating above a critical-temperature cloud point (T_p) and on cooling back to room temperature become water-soluble again, providing for higher rates and a quantitative catalyst separation by simple decantation [32] (cf. Section 4.6.3).

The concept of organometallic catalysis in micellar systems is compared by several writers [15, 33, 34] with the concept of heterogeneous catalysis on solid surfaces since the solubilization of the reactants in the core of the micelle containing catalytically active sites can be compared with adsorption on surfaces and the number of micelles to surface area (cf. Section 4.5).

3.2.4.4 Hydrogenation Reactions Catalyzed by Transition Metal Surfactant–Phosphine Complexes

Biphasic hydrogenation reactions of cyclohexene using Rh/TPPMS catalysts in a dimethylacetamide/H_2O mixture demonstrated that this reaction does not take place at the interface. Thus, doubling of the volume of the aqueous phase gave rise to an increase in the reaction rate. If this reaction were an interfacial reaction, the rates would not depend on the volume of the aqueous phase [35]. In contrast, the rates in the Ru/TPPMS-catalyzed biphasic hydrogenation of cyclohexene [36] were dependent on the rate of stirring (0–1000 rpm), indicating interfacial catalysis.

Rhodium complexes generated from the water-insoluble carboxylated surfactant phosphine **17** ($n = 3, 5, 7, 9, 11$) were used as catalysts in the micellar hydrogenation of α- and cyclic olefins, such as 1-octene, 1-dodecene, and cyclohexene, in the presence of conventional cationic or anionic tensides such as cetyltrimethylammonium bromide (CTAB) or SDS and co-solvents, e.g., dimethyl sulfoxide [15]. After the reaction the catalyst was separated from the organic products by decantation and recycled without loss in activity. There is a critical relationship between the length of the hydrocarbon chain of the ligand **17** and the length and nature of the added conventional surfactant, for obtaining maximum reactivity. For example, maximum efficiency was observed in the micellar hydrogenation of cyclohexene at 50 °C using rhodium catalysts modified with the ligand **17** possessing a C_5 or C_7 chain in the presence of the anionic surfactant SDS. In contrast, using the cationic tenside CTAB, the ligand with a C_5 chain was almost inactive but became active again in the presence of dodecyltrimethylammonium bromide. However, the C_9 ligand shows the same behavior with anionic (SDS) and cationic (CTAB) tensides [15]. Rh/**17** catalysts exhibit higher rates with linear than with cyclic olefins.

Rhodium catalysts modified with the sodium salt of the surfactant phosphines **17** ($n = 5, 7$) exhibited high catalytic activity in the hydrogenation of higher α-olefins such as 1-octene and 1-decene at 50 °C in an aqueous/organic two-phase system without any addition of conventional tensides [37]. Using the ligand **17** with a chain length of six C-atoms ($n = 5$) the rhodium catalysts were more efficient than with a tail of eight C-atoms ($n = 7$). For example, the TOFs achieved in the hydrogenation of 1-octene were 1880 h^{-1} with the Rh/**17** ($n = 5$) compared with 460 h^{-1} with the Rh/**17** ($n = 7$) catalyst [37]. With both catalysts the α-olefins were hydrogenated much faster than internal olefins. Unusually high reaction rates (up to TOF = 7920 h^{-1}) were obtained in the hydrogenation of the internal double bonds in *cis*-2-pentenenitrile (**22**) and *trans*-3-pentenenitrile (**23**), both giving valeronitrile (**24**) when catalyzed by the sodium salt of the surfactant Rh/**17** system [Eq. (3) and Table 1] [37].

$$\mathbf{22} + \mathbf{23} + H_2 \xrightarrow{Rh\,/\,17} \mathbf{24} \quad (3)$$

Table 1. Rh/**17** (n = 5,7)-catalyzed hydrogenation of **22** and **23**.

Substrate	Rh/**17**	TOF (h^{-1})
22	n = 5	7920
22	n = 7	6800
23	n = 5	1980
23	n = 7	1760

That the hydrophobic chain length of tenside ligands has a significant effect on the catalyst activity in micellar systems was further demonstrated in the biphasic hydrogenation of 1-hexene using rhodium catalysts generated from the cationic surfactant phosphonium phosphines **18** (n = 2, 3, 6, 10) [38]. The Rh/**18** with a chain length of six C-atoms was the most active catalyst; ca. 90% of 1-hexene were hydrogenated (with 10% isomerization) at 25 °C. The catalytic activity of the Rh/**18** system decreased with the chain length n in the order $6 > 10 > 3 > 2$ [38 a]. Using Rh/**18** catalysts with longer-chain ligands, very stable emulsions were formed which were quite difficult to break [38 a].

Hydrogenation of unsaturated membrane lipids, for modulation of membrane fluidity, is facilitated by water-soluble catalysts in aqueous media, thus rendering organic solvents such as THF or DMSO, which are necessary for penetration of the membrane with conventional organic-soluble complex catalysts, superfluous (cf. Section 6.8). For example hydrogenation of dioleoylphosphatidylcholine dispersions catalyzed by rhodium complexes modified with the surfactant phosphate phosphine ligand **19** and TPPMS were carried out at 37 °C and 1.2 atm H_2 in aqueous media [39]. Under these mild reaction conditions, isomerization of the oleoyl group to an elaidoyl moiety is followed by hydrogenation of the elaidoyl functionality. Rhodium catalysts modified with either **19** or TPPMS exhibit similar behavior in this reaction [39].

Whitesides and co-workers [40] synthesized the nonionic surfactant phosphines **9** ($n = 2$; $x = 0$; $m = 12, 16$; R = Me), **12**, **13**, **14** and used them as ligands in rhodium-catalyzed hydrogenation of water-immiscible starting materials such as cyclohexene in a two-phase system.

The inverse temperature dependence of the water-solubility of nonionic tenside phosphines at the T_p was applied by Bergbreiter et al. [41] in the hydrogenation of allyl alcohol using water-soluble rhodium catalysts modified with the "smart ligand" **15** in aqueous media. In this case, on heating the sample to 40–50 °C the reaction stopped but on cooling to 0 °C hydrogenation was resumed in the aqueous phase (cf. Section 4.6.3).

The anionic tenside chiral diphosphine **8** was used to generate rhodium catalysts for the hydrogenation of the prochiral olefin (*Z*)-2-(*N*-acetamido)cinnamic acid methyl ester in micellar ethyl acetate/H_2O two-phase systems [42]. The yield (100 %) and enantiomeric excess (69 %) were considerably higher than those observed with the tetrasulfonated bis(diphenylphosphino)pentane (32 % yield and 20 % *ee*) and the reaction time was shorter (1.5 versus 20 h) [42].

The asymmetric diphosphine **16**, used in rhodium-catalyzed enantioselective hydrogenation reactions [43], should exhibit surfactant properties due to the poly(ethylene glycol) functionality. The behavior of **16** in water (insoluble when $n = 5$ and water-soluble with $n = 16$) is typical for nonionic surfactant diphosphines and analogous to that of **13**, which shows increasing water-solubility with increasing molecular mass [40 b].

3.2.4.5 Concluding Remarks and Future Prospects

Organometallic catalysis in micellar systems using transition metal complexes generated from tenside ligands is a particularly attractive approach for circumventing the solubility problem of highly water-immiscible starting materials encountered in catalysis in aqueous media. Higher reaction rates combined with facile catalyst separation and recycling can be achieved by fine tuning of the HLB value of the surfactant ligand. Hence, we conclude that catalysis in micellar microheterogeneous systems has considerable potential for further applications in the synthesis of fine chemicals.

References

[1] (a) B. Cornils, W. A. Herrmann in *Applied Homogeneous Catalysis with Organometallic Compounds* (Eds.: B. Cornils, W. A. Herrmann), VCH, Weinheim, **1996**, Vol. 2, Chapter 3, p. 577; (b) I. T. Horváth, F. Joó in *Aqueous Organometallic Chemistry and Catalysis* (Eds.: I. T. Horváth, F. Joó), Kluwer, Dordrecht, **1995**, p. 1; (c) G. Papadogianakis, R. A. Sheldon, *New J. Chem.* **1996**, *20*, 175.

[2] H. Bahrmann, C. D. Frohning, P. Heymanns, H. Kalbfell, P. Lappe, D. Peters, E. Wiebus, *J. Mol. Catal. A: Chem.* **1997**, *116*, 35.

[3] (a) Ruhrchemie AG (H. Bahrmann, B. Cornils, W. Konkol, W. Lipps), DE 3.412.335 (1985); (b) Ruhrchemie AG (H. Bahrmann, B. Cornils, W. Konkol, W. Lipps), DE 3.420.491 (1985).

[4] B. Fell, G. Papadogianakis, *J. Mol. Catal.* **1991**, *66*, 143.

[5] (a) B. Fell, Ch. Schobben, G. Papadogianakis, *J. Mol. Catal. A: Chem.* **1995**, *101*, 179; (b) Hoechst AG (H. Bahrmann, P. Lappe), EP 602.463 (1994); (c) G. Oehme, I. Grassert, N. Flach, in [1 b], p. 245.

[6] E. Monflier, G. Fremy, Y. Castanet, A. Mortreux, *Angew. Chem.* **1995**, *107*, 2450; *Angew. Chem., Int. Ed. Engl.* **1995**, *34*, 2269.

[7] (a) R. V. Chaudhari, B. M. Bhanage, R. M. Deshpande, H. Delmas, *Nature (London)* **1995**, *373*, 501; (b) Council of Scientific & Industrial Interest (R. V. Chaudhari, B. M. Bhanage, S. S. Divekar, R. M. Deshpande), US 5.498.801 (1996).

[8] W. C. Griffin, *Kirk-Othmer's Encycl. Chem. Technol.* **1979**, 3rd ed., Vol. 8, p. 900.

[9] (a) J. H. Fendler, E. J. Fendler, *Catalysis in Micellar and Macromolecular Systems*, Academic Press, New York, **1975**, pp. 22, 31, 316, 339; (b) E. J. Fendler, J. H. Fendler, *Adv. Phys. Org. Chem.* **1970**, *8*, 271; (c) Y. Moroi, *Micelles: Theoretical and Applied Aspects*, Plenum, New York, **1992**, p. 41; (d) C. Tanford, *The Hydrophobic Effect: Formation of Micelles and Biological Membranes*, Wiley, New York, **1980**, p. 42.

[10] (a) R. Fornasier, P. Scrimin, P. Tecilla, U. Tonellato, *J. Am. Chem. Soc.* **1989**, *111*, 224; (b) P. Scrimin, U.Tonellato in *Surfactants in Solution* (Eds.: K. L. Mittal, D. O. Shah), Plenum Press, New York, **1991**, Vol. 11, p. 349; (c) K. Ogino, Y. Tokuda, T. Nakai, W. Tagaki, *Mem. Fac. Eng., Osaka City Univ.* **1994**, *35*, 187; (d) W. Tagaki, K. Ogino, T. Fujita, T. Yoshida, K. Nishi, Y. Inaba, *Bull. Chem. Soc. Jpn.* **1993**, *66*, 140.

[11] A. F. Borowski, D. J. Cole-Hamilton, G. Wilkinson, *Nouv. J. Chim.* **1978**, *2*, 137.

[12] A. A. Andriollo, J. Carrasquel, J. Mario, F. A. Lopez, D. E. Paez, I. Rojas, N. Valencia, *J. Mol. Catal. A: Chem.* **1997**, *116*, 157.

[13] Y. Y. Yan, H. P. Zuo, Z. L. Jin, *J. Nat. Gas Chem.* **1996**, *5*, 161.

[14] G. Papadogianakis, Ph. D. Thesis, Rheinisch-Westfälische Technische Hochschule Aachen, Germany, **1990**, p. 65.

[15] Y. Dror, J. Manassen, *Stud. Surf. Sci. Catal.* **1981**, *7* (Pt. B, New Horiz. Catal.), 887.

[16] F. M. Menger, C. E. Mounier, *J. Am. Chem. Soc.* **1993**, *115*, 12222 and references therein.

[17] M. J. H. Russell, *Plantinum Met. Rev.* **1988**, *32*, 179.

[18] C. M. Starks, C. L. Liotta, M. Halpern, in *Phase-transfer Catalysis: Fundamentals, Applications, and Industrial Perspectives*, Chapman, New York, **1994**, p. 9.

[19] B. E. Hanson, H. Ding, T. Bartik, B. Bartik, in [1 b], p. 149.

[20] H. Ding, B. E. Hanson, T. Bartik, B. Bartik, *Organometallics* **1994**, *13*, 3761.

[21] H. Ding, B. E. Hanson, *J. Chem. Soc., Chem. Commun.* **1994**, 2747.

[22] H. Ding, B. E. Hanson, T. E. Glass, *Inorg. Chim. Acta* **1995**, *229*, 329.

[23] T. Bartik, B. Bartik, B. E. Hanson, *J. Mol. Catal.* **1994**, *88*, 43.

[24] T. Bartik, B. Bartik, I. Guo, B. E. Hanson, *J. Organomet. Chem.* **1994**, *480*, 15.

[25] T. Bartik, B. Bartik, B. E. Hanson, I. Guo, I. Tóth, *Organometallics* **1993**, *12*, 164.

[26] T. Bartik, H. Ding, B. Bartik, B. E. Hanson, *J. Mol. Catal. A: Chem.* **1995**, *98*, 117.

[27] J. Haggin, *Chem. Eng. News* **1994**, *72(41)*, 28.

[28] Akademie d. Wissenschaften d. DDR (G. Oehme, E. Paetzold, A. Kinting), DD 259.194 (1988); *Chem. Abstr.* **1988**, *111*, 58020h.

[29] E. Paetzold, A. Kinting, G. Oehme, *J. Prakt. Chem.* **1987**, *329*, 725.

[30] S. Kanagasabapathy, Z. Xia, G. Papadogianakis, B. Fell, *J. Prakt. Chem./Chem.-Ztg.* **1995**, *337*, 446.

[31] E. A. Karakhanov, Y. S. Kardasheva, A. L. Maksimov, V. V. Predeina, E. A. Runova, A. M. Utukin, *J. Mol. Catal. A: Chem.* **1996**, *107*, 235.

[32] (a) Z. Jin, X. Zheng, B. Fell, *J. Mol. Catal. A: Chem.* **1997**, *116*, 55; (b) Z. Jin, Y. Yan, H. Zuo, B. Fell, *J. Prakt. Chem./Chem.-Ztg.* **1996**, *338*, 124; (c) Y. Y. Yan, H. P. Zhuo, B. Yang, Z. L. Jin, *J. Nat. Gas Chem.* **1994**, *3*, 436; (d) Y. Y. Yan, H. P. Zuo, Z. L. Jin, *Chin. Chem. Lett.* **1996**, *7*, 377; (e) Y. Yan, H. Zhuo, Z. Jin, *Fenzi Cuihua* **1994**, *8*, 147; *Chem. Abstr.* **1994**, *121*, 111875a.
[33] P. A. Chaloner, M. A. Esteruelas, F. Joó, L. A. Oro, in *Homogeneous Hydrogenation* (Eds.: R. Ugo, B. R. James), Catalysis by Metal Complexes, Kluwer, Dordrecht, **1994**, Vol. 15, Chapter 5, p. 183.
[34] G. Oehme, E. Paetzold, R. Selke, *J. Mol. Catal.* **1992**, *71*, L1.
[35] Y. Dror, J. Manassen, *J. Mol. Catal.* **1977**, *2*, 219.
[36] A. Andriollo, A. Bolivar, F. A. Lopez, D. E. Paez, *Inorg. Chim. Acta* **1995**, *238*, 187.
[37] D. C. Mudalige, G. L. Rempel, *J. Mol. Catal. A: Chem.* **1997**, *116*, 309.
[38] (a) E. Renaud, R. B. Russell, S. Fortier, S. J. Brown, M. C. Baird, *J. Organomet. Chem.* **1991**, *419*, 403; (b) I. Kovacs and M. C. Baird, *J. Organomet. Chem.* **1995**, *502*, 87.
[39] F. Farin, H. L. M. van Gaal, S. L. Bontinger, F. J. M. Daemen, *Biochim. Biophys. Acta* **1982**, *711*, 336.
[40] (a) R. G. Nuzzo, D. Feitler, G. M. Whitesides, *J. Am. Chem. Soc.* **1979**, *101*, 3683; (b) R. G. Nuzzo, S. L. Haynie, M. E. Wilson, G. M. Whitesides, *J. Org. Chem.* **1981**, *46*, 2861; (c) Anon., *CHEMTECH* **1978**, *8*, 43; (d) M. E. Wilson, R. G. Nuzzo, G. M. Whitesides, *J. Am. Chem. Soc.* **1978**, *100*, 2269.
[41] D. E. Bergbreiter, L. Zhang, V. M. Mariagnanam, *J. Am. Chem. Soc.* **1993**, *115*, 9295.
[42] H. Ding, B. E. Hanson, J. Bakos, *Angew. Chem.* **1995**, *107*, 1728; *Angew. Chem., Int. Ed. Engl.* **1995**, *34*, 1645.
[43] Y. Amrani, D. Sinou, *J. Mol. Catal.* **1984**, *24*, 231.

3.2.5 Chiral Ligands

Wolfgang A. Herrmann, Robert W. Eckl

3.2.5.1 Introduction

The growing demand for chiral precursors and products has reinforced academic efforts to control enantioselectivity in homogeneous catalysis. The application of chiral ligands coordinated to catalytically active transition metals has proven to be the most successful way to achieve such asymmetric transformations. A plethora of chiral ligands have been synthesized and applied in catalysis in the past three decades. In aqueous-phase catalysis the advantage of catalyst recovery becomes even more significant if, besides precious metals, costly ligands are employed and recycled. Further support for the efforts toward asymmetric aqueous phase catalysis may be seen in the very limited success of heterogeneous catalysts in enantioselective reactions since the great number of catalytically active species in most cases leads to a drop in selectivity.

The vast majority of chiral ligands for asymmetric catalysis in a homogeneous phase consists of tertiary phosphines, and this is also true for aqueous-phase catalysis. This section concentrates on the different approaches to achieve solubilization of these ligands in water.

3.2.5.2 Sulfonated Chiral Phosphines

(Di)phosphines containing chiral backbones equipped with diphenylphosphinosubstituents are the most successful and best-investigated chiral ligands in asymmetric homogeneous catalysis. Thus, a variety of chiral water-soluble ligands were prepared by direct sulfonation of these phosphorus ligands under conditions similar to those for the synthesis of achiral sulfonated phosphines.

Sinou et al. prepared several water-soluble derivatives of chiral 1,2-, 1,3-, and 1,4-diphosphines. The sulfonated counterparts **1–4**, respectively, of (*S,S*)-CHIRAPHOS, [1] (*S,S*)-BDPP, (*S,S*)-CYCLOBUTANEDIOP, and *R*-PROPHOS [2] are shown. Except for the latter case, all diphosphines can be obtained tetrasulfonated by treating the precursors with sulfuric acid containing 20% SO_3 for 2–5 days with subsequent neutralization and workup [2]. Unexpectedly, sulfonation of *R*-PROPHOS could only be extended to give a mixture of tetrasulfonates **4a** (55%), trisulfonates **4b** (35%), and phosphine oxides (10%) under optimized conditions. Nevertheless, yields of highly water-soluble products are generally quantitative and analysis was carried out mainly by $^{31}P\{^{1}H\}$NMR and a special HPLC technique referred to as "soap chromatography" [3]. Phosphines **1–4** were successfully applied in asymmetric hydrogenation of carbon–carbon, carbon–oxygen, and carbon–nitrogen double bonds under two-phase conditions with up to 89% *ee* (cf. Section 6.2). In almost all cases re-use of the catalysts did not lower the enantioselectivity of the products obtained.

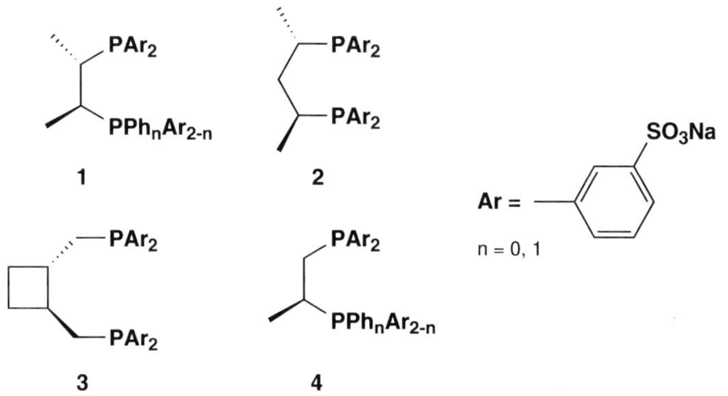

PAr_2 / PPh_nAr_{2-n} **1**; PAr_2 / PAr_2 **2**; PAr_2 / PAr_2 **3**; PAr_2 / PPh_nAr_{2-n} **4**

Ar = 3-(SO_3Na)phenyl; n = 0, 1

Some striking features were observed in the performance of these ligands, depending on the degree of sulfonation. Mono-, di- and trisulfonation of the diphosphines mentioned above gives rise to formation of diastereomers because of different configurations of the phosphorus atoms. Hence, the use of tetrasulfonated derivatives should give the highest enantioselectivity. However, a mixture of mono-, di- and trisulfonated (*S*,*S*)-BDPP gave higher optical yields compared with fourfold-sulfonated **2**. On the other hand there was no change in enantioselectivity between a mixture of partially sulfonated (*S*,*S*)-CHIRAPHOS and tetrasulfonated **1** when compared in the same reaction.

The dependence of the degree of sulfonation was also studied by the Vries and co-workers [4]. A remarkable effect was observed in the hydrogenation of acetophenone *N*-benzylimine. While the use of monosulfonated (*S*,*S*)-BDPP in an EtOAc/H_2O mixture gave the corresponding amine with 94% *ee* in a very fast reaction, only 2% *ee* was achieved when the disulfonated ligand was applied. Furthermore, application of the unsulfonated phosphine in the same solvent mixture gave no reaction at all and in homogeneous methanol solution only 65% *ee* could be attained. As was shown by HPLC and NMR analysis, the monosulfonated ligand and its Rh complexes contained equal amounts of two epimers and the high enantioselectivity was attributed to a single kinetically superior species formed during catalysis. On the other hand, the reaction probably takes place in the organic phase, since the same optical yield was found in a run that was performed in neat ethyl acetate.

BINAP, one of the most outstanding ligands in asymmetric hydrogenation, was also converted into a water-soluble diphosphine by direct sulfonation. Davis et al. reported that treatment of a solution of (*R*)-BINAP in concentrated sulfuric acid with fuming sulfuric acid (40 wt% SO_3) gives a mixture of a tetra- and a penta- or hexasulfonated compound [5]. The major species thereof (85%) was assigned to be the tetrasulfonated derivative **5**. The by-product was assumed to carry one or two extra sulfonate group(s) on the naphthyl ring(s) but their presence was believed not to affect the enantioselectivity since interaction of the four phenyl groups of a diphosphine with the substrate is generally presumed to control chiral recognition.

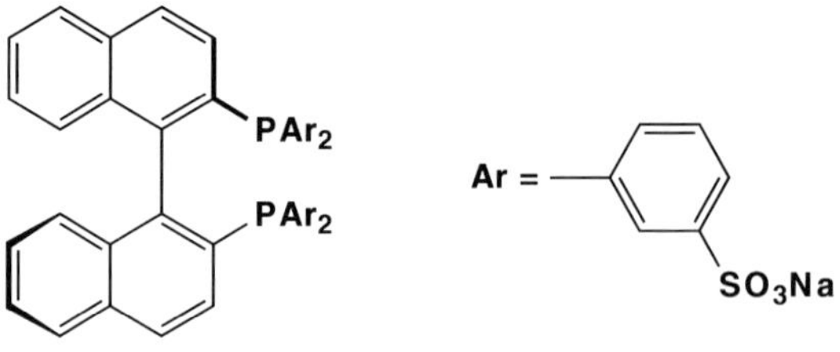

PAr2
PAr2
Ar =
SO3Na
5

Indeed catalysts prepared from several batches of sulfonated ligand with different ratios of major to minor species showed similar activity and selectivity in Rh-catalyzed hydrogenation of 2-acetamidoacrylic acid. No loss in enantioselectivity was observed in comparison with reactions using the unsulfonated ligand in organic solvents. Ruthenium(II) complexes of sulfonated BINAP proved to be even more efficient. 2-Acetamidoacrylic acids and methylenesuccinic acid can be hydrogenated with up to 85 % and 90 % *ee* in alcoholic solvents [6]. Although the optical yield decreased in most cases when the reaction was performed in neat water, some of the substrates were reduced with almost identical enantioselectivities compared with the unsulfonated parent system. The ligand was also applied in Ru-catalyzed, asymmetric hydrogenation of 2-(6′-methoxy-2′-naphthyl)acrylic acid to (*S*)-naproxen [7]. Impregnation with [Ru(BINAP-4-SO_3Na)(benzene)Cl]Cl of a controlled-pore glass (CPG) yields an efficient SAP catalyst whose activity and enantioselectivity were found to be very sensitive to the amount of water present on the surface. The limit in enantioselectivity of the hydrated SAP catalyst was restricted to about 77 % *ee*, which resembles the performance of the organometallic catalyst in neat water (80 % *ee*). In anhydrous methanol, however, enantiomeric excesses up to 90 % can be achieved. An explanation for this effect can be found in the rapid hydrolysis of the ruthenium–chloro bond in the presence of water, which was shown to be responsible for the decrease in optical yield. As a consequence, highly polar ethylene glycol was successfully used as a substitute for water in the SAPC system. Thus, enantioselectivities could be extended to about 95 % *ee* at a reaction temperature of 3 °C and proved to be dependent on the ethylene glycol loading in the triphasic system. With ethyl acetate as the organic phase, reasonably high rates were observed and recycling of the catalyst is possible without leaching of ruthenium at a detection limit of 32 ppb.

Another sulfonated derivative of BINAP was mentioned in a patent of Takasago International Corporation [8]. Cationic complexes of ruthenium and of iridium with this ligand – sulfonated on the 5-and 5′-positions of the naphthyl rings – are claimed to affect asymmetric hydrogenation of olefins, ketones, and imines.

The diphosphine ligand NAPHOS (2,2′-bis(diphenylphosphinomethyl)-1,1′-binaphthalene) proved to be an efficient ligand for hydroformylation reactions due to its large "natural bite angle" [9]. Direct sulfonation of enantiopure *S*-NAPHOS with oleum with addition of boric acid [10] yields the eightfold-sulfonated species *S*-(−)BINAS-8, **6** (80–90 % yield), accompanied with small

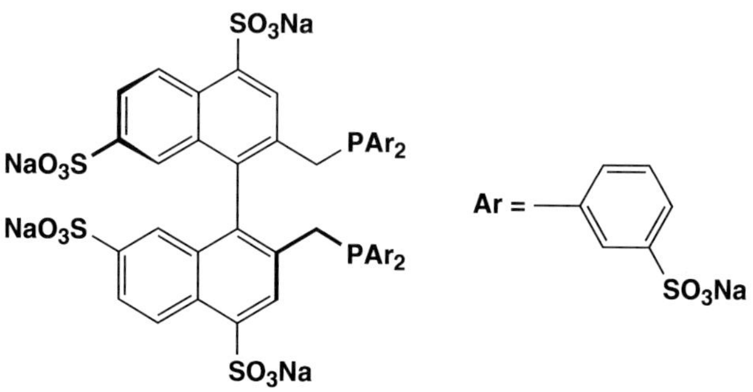

6

amounts of seven- and sixfold-functionalized derivatives as highly water-soluble sodium salts. Application in Rh-catalyzed two-phase hydroformylation of styrene furnishes the corresponding branched aldehyde with good regioselectivity but optical yields are lower (18% *ee*) than in the conventional monophase technique (34% *ee*) [11].

Diphosphine diversity was also exploited by atropisomeric biphenyl ligands. An interesting approach toward water solubilization was disclosed for one of these BINAP-type diphosphines (Scheme 1, MeOBIPHEP-S) [12] (cf. Section 3.2.1). The indolylsulfonyl group withstands the Grignard reaction and phosphine oxide reduction and is readily cleaved under mild alkaline conditions to yield the fourfold *p*-sulfonated diphosphine.

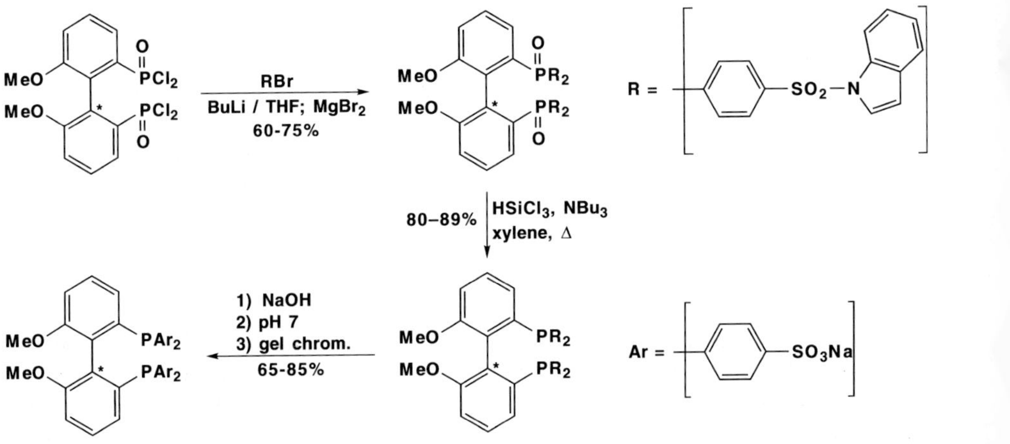

MeO, MeO, $P(O)Cl_2$, $P(O)Cl_2$ — RBr, BuLi / THF; $MgBr_2$, 60-75% → MeO, MeO, $P(O)R_2$, $P(O)R_2$ — 80–89%, $HSiCl_3$, NBu_3, xylene, Δ → MeO, MeO, PR_2, PR_2 — 1) NaOH, 2) pH 7, 3) gel chrom., 65-85% → MeO, MeO, PAr_2, PAr_2

R = C_6H_4–SO_2–N(indolyl)

Ar = C_6H_4–SO_3Na

Scheme 1

Promising results were achieved in hydrogenation experiments with Rh(I) and Ru(II) complexes derived from MeOBIPHEP-S. Hydrogenation of methyl acetoacetate and geraniol was accomplished with 93% and 98% *ee* in methanol and in an ethyl acetate/water two-phase system, respectively. Moreover, reduction of triethylammonium salts of two different unsaturated acids in water was performed with high S/C ratios (1000 to 10000:1) and with *ee* values up to 99%.

A novel access to chiral sulfonated phosphines was recently described [13]. The method is based on the acylation of chiral hydroxyphosphines with commercial *o*-sulfobenzoic anhydride and was demonstrated on a number of diphosphines based on the DIOP skeleton [Eq. (1)]. As the reaction was shown to proceed smoothly with *n*-BuLi in THF at low temperatures, it may be applied to other acid-sensitive precursors containing one or more OH groups. Although no asymmetric catalytic reactions have been reported yet, the chelating properties of the diphosphines were studied by preparation of cationic rhodium complexes. The presence of the *o*-sulfobenzoate group is thought not to influence the catalytic properties.

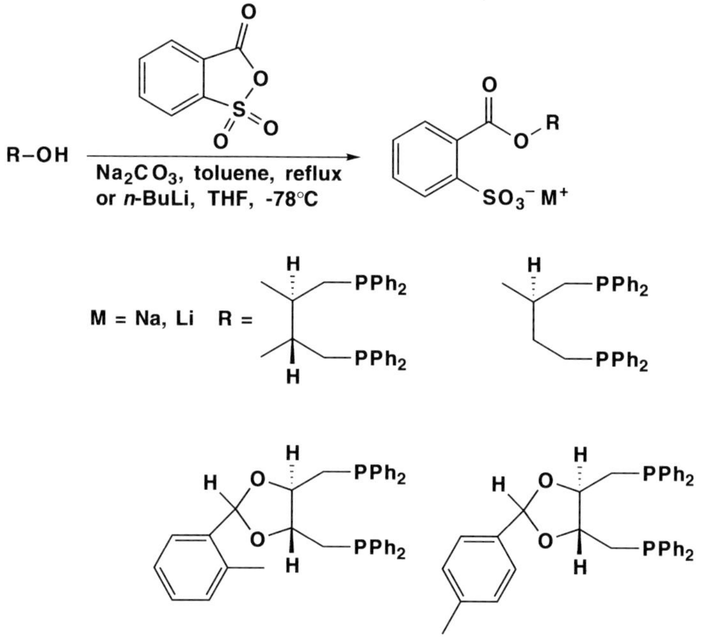

R–OH → (Na_2CO_3, toluene, reflux or *n*-BuLi, THF, -78°C) SO_3^- M^+

M = Na, Li R = PPh$_2$ PPh$_2$ (1)

Stelzer and co-workers reported a number of chiral water-soluble secondary phosphines [14], prepared by nucleophilic phosphination of primary phosphines with fluorinated aryl sulfonates in the superbasic medium DMSO/KOH. Further reaction with alkyl halides gives bidentate tertiary phosphines with P-chirality, but only racemic versions have been reported so far. Hanson et al. introduced so-called surface-active phosphines into asymmetric aqueous-phase catalysis. One of the main problems inherent to two-phase catalysis is the often very low miscibility of the substrates in the aqueous phase. Insertion of long alkyl chains between phoshorus atoms and phenyl groups in sulfonated phosphine ligands has been proven to increase reaction rates in the Rh-catalyzed hydroformylation of 1-octene [15]. This concept was extended to a number of chiral ligands, i.e., the monodentate bis(8-phenyloctyl) (1*R*,3*R*,4*S*-(−)menthyl) phosphine and the corresponding *p*,*p*-disulfonated phosphine by sulfonation under mild conditions. The latter ligand was applied in two-phase hydroformylation of styrene; higher activity was achieved compared with a TPPTS catalyst under similar conditions. Nevertheless, virtually no enantioselectivity was observed with this monodentate phosphine.

The preparation of chelating surface-active diphosphines proved to be more successful. The BDPP analogue *R*,*R*-(**7**) was synthesized by lithiation of tris[*p*-(3-phenylpropyl)phenyl]phosphine, reaction with (*R*,*R*)-2,4-pentanediylditosylate and subsequent sulfonation with concentrated sulfuric acid [16]. Hydrogenation of the prochiral olefin (*Z*)-2-(*N*-acetamido)cinnamic acid ester in a two-phase system (EtOAc/H_2O) with Rh complexes derived from *R*,*R*-(**7**)

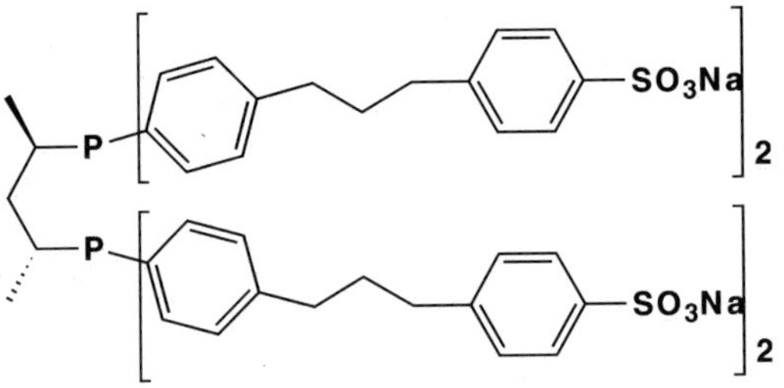

7

was accomplished with identical selectivity compared with the unsulfonated BDPP and improved reactivity compared with BDPPTS (**2**).

A surface-active equivalent of BINAP was also prepared recently [17]. According to the new preparative route to BINAP that was previously developed, di[*p*-(3-phenylpropyl)phenyl]phosphine was coupled with the ditriflate of 2,2′-binaphthol (2,2′-bis(trifluoromethanesulfonyloxy)-1,1′-binaphthalene) in a Ni-catalyzed reaction. Again, sulfonation of the diphosphine thus obtained with concentrated H_2SO_4 yields a water-soluble ligand tetrasulfonated on the terminal phenyl groups in the *para* position. Only the racemic version of the diphosphine has been synthesized so far; in biphasic hydroformylation of 1-octene its activity was equal to or less than that of TPPTS.

3.2.5.3 Other Water-soluble Chiral Ligands

Apart from sulfonate groups, it was mainly quaternary ammonium groups that were introduced in the preparation of water-soluble chiral phosphines. Toth and Hanson synthesized a number of tetra-amine functionalized diphosphines and achieved a very high water-solubility by methyl quaternization of their rhodium complexes. As outlined in Scheme 2 a new *p*-dimethylamino derivative of DIOP

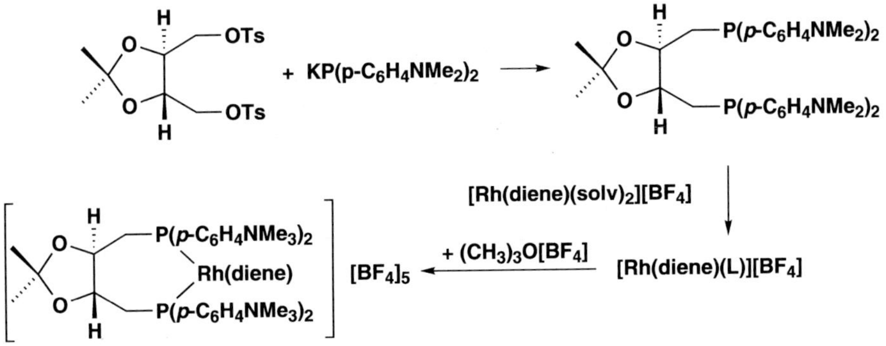

H
O
OTs
O
OTs
H
+ KP(p-C6H4NMe2)2
P(p-C6H4NMe2)2
P(p-C6H4NMe2)2
[Rh(diene)(solv)2][BF4]
+ (CH3)3O[BF4]
[Rh(diene)(L)][BF4]
[BF4]5
P(p-C6H4NMe3)2
Rh(diene)
P(p-C6H4NMe3)2

Scheme 2

was prepared by standard procedures and quaternized with Meerwein's salt, $[(CH_3)_3O][BF_4]$, after complexation to rhodium [18]. Coordination of the ligand prevents alkylation of the phosphorus atoms.

In the same way, cationic complexes of amine derivatives of BDPP and CHIRAPHOS were prepared which showed unlimited solubility in water and negligible solubility in common organic solvents. They were used in asymmetric hydrogenation of dehydroamino acids and provided modest to high enantioselectivities [19]. The presence of the dimethylamino group in the DIOP derivative resulted in a reversal in the observed dominant product antipode, which was attributed to a change in the preferred ligand conformation.

Carboxylate groups were used in many cases to achieve water-solubilization of simple achiral phosphines. With the synthesis of the water-soluble polymer shown in Eq. (2) this methodology was extended into the field of chiral ligands [20]. Acylation of the diphosphine (2*S*,4*S*)-4-diphenylphosphino-2-diphenylphosphinomethylpyrrolidone (PPM) with poly(acrylic acid) (PPA) yields the hydrophilic macroligand PAA-PMM whose Rh complexes were used in biphasic hydrogenation of (*Z*)-2-(*N*-acetamido)cinnamic acid with moderate enantioselectivities.

$[CO_2H\ CO_2H\ CO_2H\ CO_2H\ CO_2H]_n$ + Ph_2P … PPh_2 (N–H)

1) DCC
2) Na_2CO_3

CO_2^- CO_2^- C=O CO_2^- CO_2^- … N … Ph_2P … PPh_2 $]_n$ $[Na^+]_{4n}$

(2)

Various research groups took advantage of the chirality and hydrophilicity of sugar backbones. While investigating the influence of tensides and micelles upon asymmetric hydrogenation of enamide substrates in water, Oehme, Selke, and co-workers synthesized the water-soluble carbohydrate phosphinite complex **8** [21]. The compound provided increased enantioselectivity in the presence of a variety of surfactants and polymerized micelles. The catalytic results suggest that the phosphorus–oxygen bonds in the phosphinite are stable toward hydrolysis.

Reetz et al. prepared β-cyclodextrin-modified diphosphine 9 and a series of derivatives by phosphinomethylation of the corresponding amino-substituted cyclodextrin precursors [22]. The ligands were used in biphasic hydroformyla-

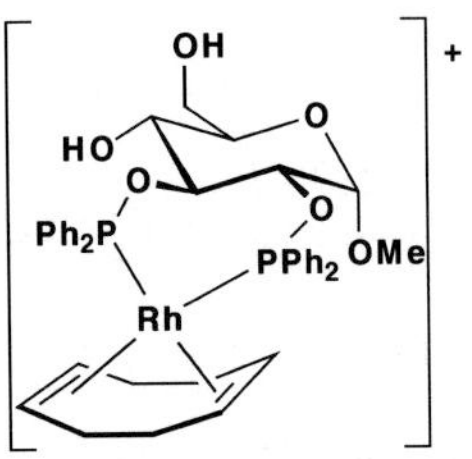

8 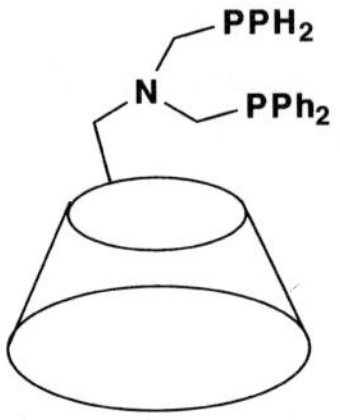 9 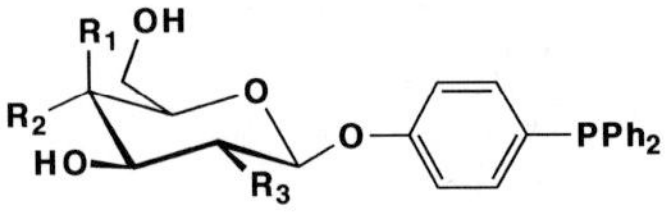 10

tion of 1-octene and competitive hydrogenation of 1-alkenes with rather poor results. Although no asymmetric catalysis has been attempted yet, the chiral cavity in the backbone might achieve molecular recognition of certain prochiral substrates and thus provide enantioselectivity in the products.

Some other nonionic ligands based on carbohydrates were synthesized by Beller et al. [23]. Glycosidation of glucose, galactose, and glucosamine with *p*-hydroxyphenyldiphenylphosphine yields water-soluble phosphines of the general type **10** (R_1, R_2 = H or OH, R_3 = NHAc or H), which were applied in two-phase Heck and Suzuki reactions. Due to the monodentate character and the remote chiral centers, asymmetric induction with this type of ligand should be negligibly low.

3.2.5.4 Conclusion

It is evident by now that enantioselective catalysis in water is a potentially rich area of research in aqueous-phase catalysis. Progress in this field is strongly dependent on the design and synthesis of tailormade, water-soluble ligands. Despite the fact that in most cases enantioselectivity in biphasic systems is decreased, some of the results mentioned are very encouraging with respect to activity, selectivity, and catalyst recovery. Beyond that, the solvent water with its intrinsic properties often makes it possible to follow different reaction pathways.

References

[1] F. Alario, Y. Amrani, Y. Colleuille, T. P. Dang, J. Jenck, D. Morel, D. Sinou, *J. Chem. Soc., Chem. Commun.* **1986**, 202.
[2] Y. Amrani, L. Lecomte, D. Sinou, J. Bakos, I. Toth, B. Heil, *Organometallics* **1989**, *8*, 542.
[3] L. Lecomte, J. Triolet, D. Shinou, J. Bakos, B. Heil, *J. Chromatogr.* **1987**, *408*, 416.
[4] C. Lensink, J. G. de Vries, *Tetrahedron: Asymm.* **1992**, *3*, 235.

[5] K. T. Wan, M. E. Davis, *J. Chem. Soc., Chem. Commun.* **1993**, 1262.
[6] K. T. Wan, M. E. Davis, *Tetrahedron: Asymm.* **1993**, *4*, 2461.
[7] K. T. Wan, M. E. Davis, *J. Catal.* **1994**, *148*, 1.
[8] Takasago Int. Corp. (T. Ishizaki, H. Kumobayashi), EP 544.455-A1 (1991).
[9] W. A. Herrmann, C. W. Kohlpaintner, T. Priermeier, R. Schmid, *Organometallics* **1995**, *14*, 1961.
[10] W. A. Herrmann, G. P. Albanese, R. B. Manetsberger, H. Bahrmann, P. Lappe, *Angew. Chem.* **1995**, *107*, 893; *Angew. Chem., Int. Ed. Engl.* **1995**, *34*, 811.
[11] W. A. Herrmann, R. W. Eckl, T. Priermeier, *J. Organometal. Chem.* **1997**, *532*, 243.
[12] R. Schmid, E. A. Broger, M. Cereghetti, Y. Crameri, J. Foricher, M. Lalonde, R. K. Müller, M. Scalone, G. Schoettel, U. Zutter, *Pure Appl. Chem.* **1996**, *68*, 131.
[13] S. Trinkhaus, J. Holz, R. Selke, A. Börner, *Tetrahedron Lett.* **1997**, *38*, 807.
[14] F. Bitterer, O. Herd, A. Hessler, M. Kühnel, K. Rettig, O. Stelzer, W. S. Sheldrick, S. Nagel, N. Rösch, *Inorg. Chem.* **1996**, *35*, 4103.
[15] T. Bartik, B. Bartik, B. E. Hanson, *J. Mol. Catal.* **1994**, *88*, 43.
[16] H. Ding, B. E. Hanson, J. Bakos, *Angew. Chem.* **1995**, *107*, 1728; *Angew. Chem., Int. Ed. Engl.* **1995**, *34*, 1645.
[17] H. Ding, J. Kang, B. E. Hanson, C. W. Kohlpaintner, *J. Mol. Catal. A: Chem.* **1997**, *124*, 21.
[18] I. Toth, B. E. Hanson, *Tetrahedron: Asymm.* **1990**, *1*, 895.
[19] I. Toth, B. E. Hanson, *Tetrahedron: Asymm.* **1990**, *1*, 913.
[20] T. Malmström, C. Andersson, *J. Chem. Soc., Chem. Commun.* **1996**, 1135.
[21] A. Kumar, G. Oehme, J. P. Roque, M. Schwarze, R. Selke, *Angew. Chem.* **1994**, *106*, 2272; *Angew. Chem., Int. Ed. Engl.* **1994**, *33*, 2197.
[22] M. T. Reetz, S. R. Waldvogel, *Angew. Chem.* **1997**, *109*, 870; *Angew. Chem., Int. Ed. Engl.* **1997**, *36*, 865.
[23] M. Beller, J. G. E. Krauter, A. Zapf, *Angew. Chem.* **1997**, *109*, 793; *Angew. Chem., Int. Ed. Engl.* **1997**, *36*, 793.

3.2.6 Alternative Catalyst Concepts

Wolfgang A. Herrmann, Claus-Peter Reisinger

Apart from the aforementioned concepts for water-soluble catalysts, there are some new trends for ligands and catalysts that promise to facilitate their utilization in water. These concepts include nonionic donor ligands, water-soluble polymers, and inorganic polyoxometallates as redox-active ligands and/or complex ions in homogeneous acqueous-phase catalysis.

3.2.6.1 Amines as Ligands

Neutral amine ligands are often applied with transition metals in higher oxidation states in polar solvents. This concept combined with the resistance of amines against oxidative conditions led to the development of novel bleaching catalysts (Structures **1**–**3**) for washing powders by Unilever [1].

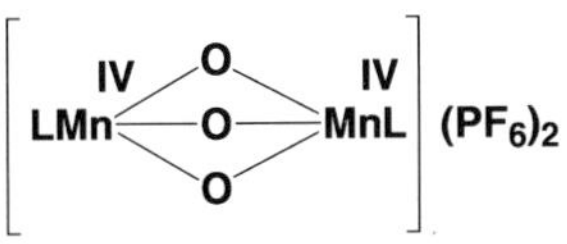

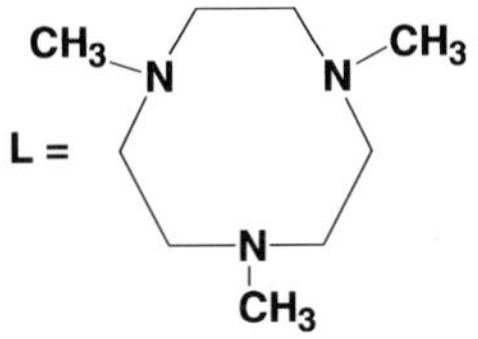

$[L\overset{IV}{Mn}(\mu\text{-}O)_3\overset{IV}{Mn}L](PF_6)_2$ **1**, L = 1,4,7-trimethyl-1,4,7-triazacyclononane (CH_3–N)

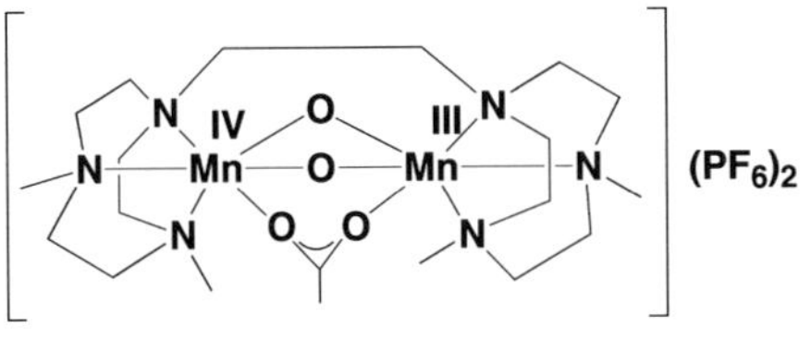

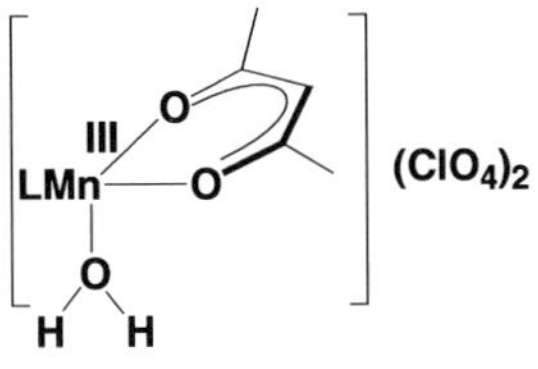

2 $(PF_6)_2$; **3** $(ClO_4)_2$

This class of binuclear compounds [2] with tripodal 1,4,7-trimethyl-1,4,7-triazacyclononane ligands [3] was originally reported by Wieghardt and co-workers [4]. The manganese(IV) complexes exhibit rather unusual structural and physical properties [5], combined with superior bleaching activity at room temperature (rt). The oxidizing agents are either hydrogen peroxide or peroxyacetic acid, generated by tetraacetylethylenediamine (TAED) in European detergent powders. Unligated "free" manganese salts show high instability as detergents, resulting in the formation of brown stains on textiles due to precipitation, while oxo- and acetato-bridged manganese complexes with 1,4,7-trimethyl-1,4,7-triazacyclononane combined with hydrogen peroxide show the highest activity at pH 10.

Bleaching is a complex process of different oxidation reactions involving manifold substrates, such as arenes and fatty acids. Thus, investigations of the epoxidation of 4-vinylbenzoic acid (a water-soluble alkene) with $H_2{}^{18}O_2$, as a model reaction, gave a product with 100% ^{18}O, suggesting that the oxygen of the epoxide is derived from hydrogen peroxide, and not from water. The optimum pH for epoxidation was established to be around 8, whereas the optimum pH for bleaching of tea-stained test cloths (commercially available as BC-1 cloths) was found to be around pH 10. Thus it appears that catalytic epoxidation and fabric bleaching involve different mechanisms [6].

A comparative study of the bleaching performance of complexes **1**–**3** with hydrogen peroxide as a function of pH revealed that compounds **1** and **3** show higher activities at $pH < 9.5$ and compound **2** demonstrates much better results at $pH > 9.5$. This difference in behavior appears to be related to the generally higher stability of the mixed-valence Mn species under catalytic conditions [6] and it may suggest that catalytic fabric bleaching involves mononuclear species at $pH < 9.5$ and binuclear species above this pH.

3.2.6.2 Water-soluble Polymers as Ligands

Biphasic catalysis relies on the transfer of organic substrates into the aqueous phase containing the catalyst. Therefore, studies have focused on improving the affinities between the two phases [7]. Addition of water-soluble organic ligands, such as TPPTS, increase the concentration of the catalyst at the interface, which enhances the reaction rate significantly. This ligand concept is also applied for a water-soluble, polymer-anchored, rhodium catalyst whose structure includes a lipophilic rhodium center and a hydrophilic long chain for the catalytic hydroformylation of higher olefins [8]. The appropriate ligand can be synthesized by controlled oxidation of poly(vinyl alcohol) (PVA) using sodium hypochlorite to afford lipophilic poly(methylene ketone) (PMK) [9]. Under basic conditions, tautomerization of PMK gives the enolate which chelates metal ions such as rhodium and nickel [9]. Thus, oxidation of PVA yields poly(vinyl alcohol-co-methylene ketone) composed of both lipophilic and hydrophilic components. The strong chelating ability of the PMK enolate hinders leaching of rhodium into the organic phase, while the unoxidized hydroxyl groups promote the dissolution of the catalyst in aqueous solution.

The rhodium complex, reported by Chen and Alper [8] was synthesized by addition of poly(enolate-*co*-vinyl alcohol-*co*-vinyl acetate) (PEVV) to $RhCl_3$ in water, and can be represented as Structure **4**.

Hence, treatment of 1-octene with 1:1 CO/H_2 and 0.2 mol% Rh–PEVV resulted in 49% conversion and 98% selectivity for hydroformylation with a linear/branched (*n*/iso) ratio of 1.14. Raising the reaction temperature to 90 °C affords an excellent TOF of 5.46×10^{-5} and improves the selectivity for the

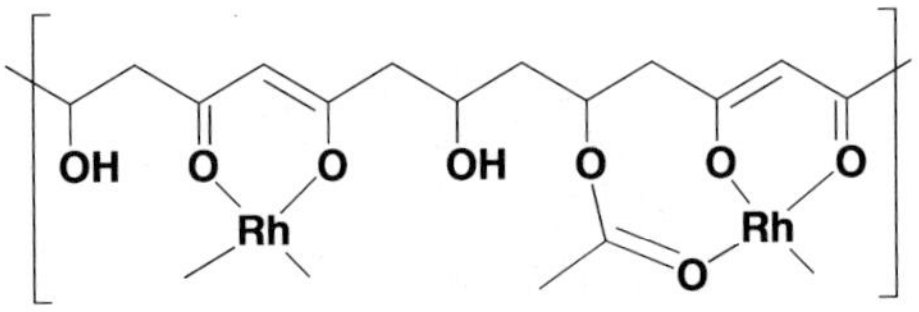

4

linear aldehyde. However, the rate of isomerization of 1-octene to 2-octene becomes greater than the hydroformylation reaction rate, resulting in lower conversions (25–43%). Changing the CO/H_2 ratio affected the rates of *n*/iso aldehyde, with a ratio of 3:1 being the most selective for the linear product up to 2.96. Re-use of the aqueous phase (three times) did not diminish the activity and selectivity for the hydroformylation reaction, whereas the recycled organic phase showed no catalytic activity, indicating at least minor rhodium leaching to the organic phase.

3.2.6.3 Polyoxometallates

In the field of catalysis with heteropolyacids (HPAs) and related polyoxometallate systems, large-scale processes such as oxidation of methacrolein, hydration of olefins, and polymerization of tetrahydrofuran have been developed and commercialized [10].

They have outstanding properties which are of great value for catalysis, such as strong Brønsted acidity [11], ability to catalyze reversible redox reactions under mild conditions [12], and high solubility in water [13]. In most applications, they are used as acid, redox, and bifunctional catalysts in homogeneous and heterogeneous systems.

HPAs are polyoxometallates incorporating anions (heteropolyanions) having metal–oxygen octahedra as the basic structural units [14]. Among a wide variety of HPAs, those belonging to the so-called Keggin series have the most importance for catalysis, being the most stable and easily accessible. They include heteropolyanions (HPANs) $XM_{12}O_{40}^{x-8}$, where X is the central atom (Si^{4+}, P^{5+}, etc.), x is the oxidation state, and M is the metal ion (Mo^{6+}, W^{6+}, V^{5+}, etc.). These HPANs are composed of a central tetrahedron XO_4 surrounded by 12 edge-sharing metal–oxygen octahedra MO_6 (Figure 1) [15].

Despite the utilization of polyoxometallates as Brønsted acid catalysts in organic synthesis [16], the most important application is the palladium-catalyzed Wacker oxidation of ethylene to acetaldehyde in aqueous phase. Under standard conditions ($PdCl_2$, $CuCl_2$, O_2, HCl), chlorine ions are corrosive and produce chlorinated by-products (mainly from $CuCl_2$); these conditions are not suitable for the oxidation of higher olefins, such as 1-butene to methyl ethyl ketone. For this reason HPANs with $PdSO_4$ are applied, instead of

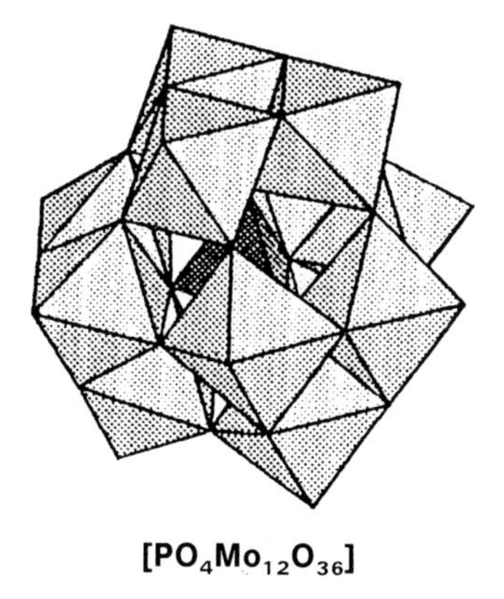

Figure 1. Structural pattern of polyoxymolybdates (HPANs).

$PdCl_2/CuCl_2/HCl$ as a chlorine-free reoxidizing agent with molecular oxygen. Mixed molybdovanadophosphoric heteropolyanions $[PMo^{6+}_{12-n}V^{5+}_{n}O_{40}]^{(3+n)-}$ ($n = 2$–6) [17] and dodecamolybdophosphoric homopolyanion $[PMo_{12}O_{40}]^{3-}$ [18] are used preferentially for the oxidation of ethylene and 1-butene with superior selectivity. The reaction proceeds in an acidic aqueous solution at a pH of 0.5–2.0 and can be represented by Eqs. (1–3):

$$C_2H_4 + Pd^{2+} + H_2O \longrightarrow CH_3CHO + Pd^0 + 2\,H^+ \quad (1)$$

$$Pd^0 + HPAN + 2\,H^+ \longrightarrow Pd^{2+} + H_2(HPAN) \quad (2)$$

$$H_2(HPAN) + 0.5\,O_2 \longrightarrow HPAN + H_2O \quad (3)$$

In general, homogeneous catalysts based on HPANs consist of complex equilibrium mixtures of polyanions of different compositions with products of the degradative dissociation of the heteropolyanion. All these species can function as the active form or as ligands of a transition metal complex [19]. The HPAN + Pd(II) and HPAN + Rh(I) systems are also used in carbonylation, hydroformylation, and hydrogenation reactions [20]. Other redox systems based on HPANs are also known. Their second component is Tl(III)/Tl(I) [21], Pt(IV)/Pt(II) [22], Ru(IV)/Ru(II), or Ir(IV)/Ir(III) [23].

References

[1] (a) T. L. F. Farre et al., EP 458.397 (1991) and 458.398 (1991); (b) J. Oakes, EP 127.910 (1984) and EP 145.091 (1985); (c) R. Hage, J. E. Iburg, J. Kerschner, J. H. Koek, E. L. M. Lempers, R. J. Martens, U. S. Racherla, S. W. Russell, T. Swarthoff, M. R. P. van Vliet, J. B. Warnaar, L. van der Wolf, B. Krijnen, *Nature (London)* **1994**, *369*, 637.

[2] (a) K. Wieghardt, *Angew. Chem.* **1989**, *101*, 1179; *Angew. Chem., Int. Ed. Engl.* **1989**, *28*, 1153; (b) G. Christou, *Acc. Chem. Res.* **1989**, *22*, 328; (c) G. W. Brudwig, R. H. Crabtree, *Prog. Inorg. Chem.* **1989**, *37*, 99; (d) C. G. Dismukes, *Bioinorganic Catalysis* (Ed.: J. Reedijk), Marcel Dekker, New York, **1993**; (e) H. H. Thorp, G. W. Brudwig, *New J. Chem.* **1991**, *15*, 479; (f) L. Que Jr., A. E. True, *Prog. Inorg. Chem.* **1990**, *38*, 97; (g) *Manganese Redox Enzymes* (Ed.: V. L. Pecoraro), VCH, New York, **1992**; (h) V. L. Pecoraro, M. J. Baldwin, A. Gelasco, *Chem. Rev.* **1994**, *94*, 807.

[3] T. J. Atkins, J. E. Richman, W. F. Oettle, *Org. Synth.* **1978**, *58*, 86.

[4] K. Wieghardt, U. Bossek, B. Nuber, J. Weiss, J. Bonvoisin, M. Corbella, S. E. Vitols, J. J. Giererd, *J. Am. Chem. Soc.* **1988**, *110*, 7398.

[5] R. Hage, B. Krijnen, J. B. Warnaar, F. Hartl, D. J. Stufkens, T. L. Snoeck, *Inorg. Chem.* **1995**, *34*, 4973.

[6] R. Hage, J. E. Iburg, J. Kerschner, J. H. Koek, E. L. M. Lempers, R. J. Martens, U. S. Racherla, S. W. Russell, T. Swarthoff, M. R. P. van Vliet, J. B. Warnaar, L. van der Wolf, B. Krijnen, *Nature (London)* **1994**, *369*, 637.

[7] (a) B. Cornils, *Angew. Chem.* **1995**, *107*, 1709; *Angew. Chem., Int. Ed. Engl.* **1995**, *34*, 1575; (b) R. V. Chaudhari, B. M. Bhanage, R. M. Deshpande, H. Delmas, *Nature (London)* **1995**, *373*, 501.

[8] J. Chen, H. Alper, *J. Am. Chem. Soc.* **1997**, *119*, 893.

[9] S. J. Huang, I.-F. Wang, E. Quinga, *Modification of Polymers* (Eds.: C. E. Canahes, J. E. Moore Jr.), Plenum Press, New York, **1983**.

[10] M. Misono, N. Nojiri, *Appl. Catal.* **1990**, *64*, 1.

[11] I. V. Kozhevnikov, *Russ. Chem. Rev.* **1987**, *56*, 811.

[12] (a) K. I. Matveev, *Kinet. Katal.* **1977**, *18*, 862; (b) I. V. Kozhevnikov, K. I. Matveev, *Russ. Chem. Rev.* **1982**, *51*, 1075.

[13] M. Misono, *Catal. Rev. Sci. Eng.* **1988**, *30*, 339.

[14] (a) M. T. Pope, A. Müller, *Angew. Chem.* **1991**, *103*, 56; *Angew. Chem., Int. Ed. Engl.* **1991**, *30*, 34; (b) *Polyoxometalates; From Platonic Solids to Anti-retroviral Activity* (Eds.: M. T. Pope, A. Müller), Kluwer, Dordrecht, **1994**.

[15] J. F. Keggin, *Proc. R. Soc. London Ser. A* **1934**, *75*, 144.

[16] I. V. Kozhevnikov, *Catal. Rev. – Sci. Eng.* **1995**, *37*, 311.

[17] K. I. Matveev, *Kinet. Katal.* **1977**, *18*, 862.

[18] S. F. Davison, B. E. Mann, P. M. Maitlis, *J. Chem. Soc., Dalton Trans.* **1984**, 1223.

[19] (a) J.-M. Brégeault (Ed.), *Catalyse Homogène Par Les Complexes Des Métaux de Transition*, Masson, Paris, **1992**; (b) I. V. Kozhevnikov, *Russ. Chem. Rev.* **1993**, *62*, 473.

[20] (a) Y. Izumi, Y. Tanaka, K. Urabe, *Chem. Lett.* **1982**, 679; (b) K. Urabe, Y. Tanaka, Y. Izumi, *Chem. Lett.* **1985**, 1595; (c) Z. Ainbinder, G. W. Parshall, US 4.386.217 (1983); *Chem. Abstr. 99*, 121789; (d) A. R. Siedle, C. G. Markell, P. A. Lyon, K. O. Hodson, A. L. Roe, *Inorg. Chem.* **1987**, *26*, 219.

[21] (a) T. Gorodetskaya, I. V. Kozhevnikov, K. I. Matveev, *React. Kinet. Catal. Lett.* **1981**, *16*, 17; (b) T. Gorodetskaya, I. V. Kozhevnikov, K. I. Matveev, USSR 893.058 (1982); *Byull. Izobret.* **1982**, 47.

[22] Y. V. Geletii, A. E. Shilov, *Kinet. Katal.* **1983**, *24*, 486.

[23] L. N. Arzamaskova, A. V. Romanenko, Y. I. Ermakov, *Kinet. Katal.* **1980**, *21*, 1068.

4

Catalyses in Water as a Special Unit Operation

4.1 Fundamentals of Biphasic Reactions in Water

Peter Claus, Manfred Baerns

4.1.1 Introduction

Catalytic biphasic reactions, including their aqueous variants, are widely used for the catalytic synthesis of organic products. Catalysts, various types of water-soluble ligands, synthetic uses, industrial applications, and their advantages over conventional homogeneously catalyzed reactions are well documented (e.g. [1–3]). However, knowledge of the physicochemical fundamentals of catalytic biphasic reactions, their kinetics, and mass-transfer processes related to reactions where a gas phase and a second (aqueous) or even a third (organic) phase are present lags behind the successful development of industrial processes such as, for example, the hydroformylation of propene using a water-soluble Rh–TPPTS catalyst (cf. Section 6.1).

Only for reactions that are usually homogeneously catalyzed in the liquid phase, and carried out in the absence of a second or even third phase, i.e., a gas or an immiscible liquid, are the procedures known required for kinetic analysis (e.g. [4–7]). In two-phase systems in which the catalytic reaction takes place in the liquid phase between a liquid reactant and gaseous reactants the quantitative analysis can be more complicated because the gaseous reactions have to be transferred over the gas/liquid boundary layer into the liquid phase. In this situation the reaction engineering prediction of the reactor performance can be performed easily as long as the rate of transfer of the gaseous reactants into the liquid phase is fast compared with the intrinsic catalytic reaction. Under these circumstances it can usually be assumed that the liquid-phase concentrations of the gaseous reactants correspond to gas/liquid thermodynamic equilibrium. By contrast, if the assumption of the rate-limiting catalytic reaction is not valid, reaction engineering modeling is no longer trivial. The situation may become even more complicated if a second liquid phase (e.g., water) is present: either this second phase may serve as the reaction space containing the catalyst while the product as well as a part of the reactants exist in the first liquid phase, or it may function as a solvent into which a desired intermediate is extracted from the reacting liquid phase. Thus, for a kinetic analysis of a homogeneous catalytic reaction in a multiphase chemical reactor it is necessary to combine several pieces of information (i.e., about intrinsic kinetics of the homogeneous catalytic reaction, mass transfer between the phases, the effect of the hydrodynamic conditions on mass transfer, and hydrodynamics within the reactor affecting the residence time distribution in continuous operation) in a suitable reactor model.

In this section, the required of kinetics and the mode of operation of biphasic reactions in water are considered, with the emphasis on problems which follow from chemical engineering aspects of gas/liquid/liquid reactions.

4.1.2 Gas/Liquid-phase Reactions

In homogenously catalyzed gas/liquid-phase reactions the overall reaction rate is determined by the actual chemical reaction rate and by mass transfer processes [1 b]. Depending on the magnitude of the rates of the catalytic reaction and of the transfer rate of the gaseous reactants, severe concentration gradients may exist near the gas/liquid interface. These phenomena are shown in Figure 1 for the reaction

$$A_{1,g} + A_{2,l} \rightarrow P_1 \tag{1}$$

As can be easily derived from the concentration pattern, the reaction takes place either mainly in the bulk of the well-mixed liquid phase or in the liquid-phase boundary layer. In reactions which occur in the bulk of the liquid phase, the concentration of gaseous educts decreases only within the interfacial layer (thickness δ) to the concentration $c_{A_1,l}$ by molecular diffusion processes. Only in the case of mass transport processes that are fast relative to the reaction rate is the latter proportional to the $c_{A_1,l}$ in the liquid phase. If the catalytic reaction is fast enough a "reaction surface" may develop within the boundary layer which may even move into the interface itself and, thus, neither the bulk of the liquid nor the liquid-phase boundary layer is utilized any more for the reaction.

From a chemical point of view it is obvious that not only the overall rate but also the selectivity towards a desired product is affected by these phenomena if a complex reaction network exists. If concentration gradients near the gas/liquid interphase are detrimental to good selectivity, they have to be avoided. This can be done by increasing the rate of transfer of the reactants or by lowering the rate of the catalytic reaction; the former is achieved by engineering means such as increasing the interfacial area per unit volume or/and increasing the rate of transfer per unit interfacial area by influencing the fluid dynamics (e.g., by stirring). The rate of the catalytic reaction may be reduced by decreasing the concentration of the catalyst, by diluting the reactants, or by lowering the temperature.

To derive the overall kinetics of a gas/liquid-phase reaction it is required to consider a volume element at the gas/liquid interface and to set up mass balances including the mass transport processes and the catalytic reaction. These balances are either differential in time (batch reactor) or in location (continuous operation). By making suitable assumptions about the hydrodynamics and, hence, the interfacial mass transfer rates, in both phases the concentration of the reactants and products can be calculated by integration of the respective differential equations either as a function of reaction time (batch reactor) or of

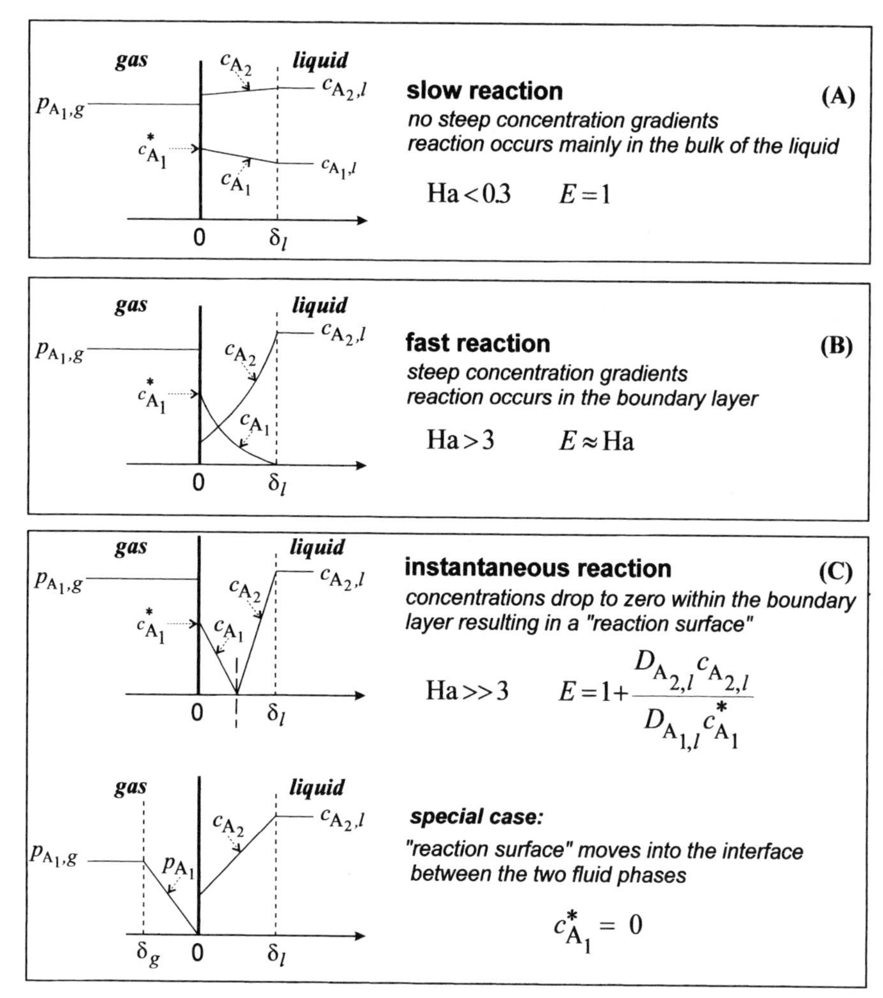

Figure 1. Concentration profiles in gas and liquid for a chemical reaction influenced by mass transfer in the liquid phase at various ratios of the reaction rate to the rate of mass transfer. D = binary diffusions coefficient.

location (continuously operated reactor). In continuous operation certain simplifications in setting up the balances are possible if one or all of the phases are well mixed, as in a continuous stirred tank reactor; thereby the mathematical treatment is significantly simplified.

Therefore, it is necessary to determine the influence of mass transfer to or from the above-mentioned interfaces on the conversion, leading to expressions for the flux of a reactant across the interface and for the overall reaction rate. After balancing the disappearance of the components A_1 and A_2 at, for instance, the gas/liquid interface by analogy with the treatment of the rate of chemical reactions and pore diffusion in heterogeneous catalysis, the overall reaction rate is given by Eq. (2) [5]:

$$r_{\text{eff}} = a \frac{\text{Ha}}{\tanh \text{Ha}} \frac{p_{1,g} - \dfrac{H_1 c_{1,l}}{\cosh \text{Ha}}}{\dfrac{RT}{k_{1,g}} + \dfrac{H_1}{k_{1,l}} \dfrac{\tanh \text{Ha}}{\text{Ha}}} \tag{2}$$

For analysis of such coupled fluid/fluid systems it is useful to distinguish between three regimes of the reaction rate (see Figure 1) which are characterized by different values of the Hatta number Ha (Eqs. (3) and (4)) and the enhancement factor E (see below).

$$\text{Ha} = \frac{1}{k_{1,l}} \sqrt{\frac{2}{n+1} k_{n,m} D_{A_{1,c}} c_{A_1}^{n-1} \bar{c}_{A_2}^{m}} \tag{3}$$

$$\text{Ha} = \frac{1}{k_{1,l}} \sqrt{k D_{1,l}} \qquad \text{(first-order reaction)} \tag{4}$$

The mass exchange rate between two phases during the course of the chemical reaction is compared with that for purely physical absorption. The ratio of these two rates (Eq. 4) is known as the enhancement factor E for mass transfer on the liquid side during the course of a chemical reaction:

$$E = \frac{J_{\text{A (with reaction)}}}{J_{\text{A (without reaction)}}} = \frac{\dfrac{\text{Ha}}{\tanh \text{Ha}} \left[1 - \dfrac{c_{1,l}}{c_1^*} \dfrac{1}{\cosh \text{Ha}} \right] k_{1,l} c_1^*}{k_{1,l} (c_1^* - c_{1,l})} \tag{5}$$

For slow reactions (Ha < 0.3) the rate of mass exchange through the fluid/fluid interface is not enhanced by the chemical reaction, which mainly takes place in the bulk of the reaction (catalytic) phase, and E becomes approximately 1 [5].

Under conditions where $0.3 <$ Ha < 3, the rate of mass exchange is enhanced by the chemical reaction ($E > 1$), and in the case of Ha > 3, A_1 and A_2 react so fast that the reaction proceeds only in the boundary layer ($E =$ Ha). Thus, equation (2) for the overall rate of reactions is reduced to Eq. (6).

$$r_{\text{eff}} \approx a \frac{p_{1,g}}{\dfrac{RT}{k_{1,g}} + \dfrac{H_1}{k_{1,l} \text{Ha}}} \tag{6}$$

A theoretical analysis to evaluate the mass transfer effects quantitatively for a hyperbolic form of the intrinsic kinetics of a homogeneous catalytic reaction has been developed [8], covering all the regimes of a gas/liquid reaction in the presence of a homogeneous catalyst, and thus leading to quantitative prediction of mass transfer effects on the kinetics of this operation mode. Assuming a gaseous reactant A, an organic liquid-phase reactant B and a homogeneous catalyst C, two cases have to be considered, namely that the reaction is assumed to occur either in bulk liquid, or completely in the film. For the second case the concept of a generalized Hatta number is used to obtain an approximate analytical solution for the enhancement factor. Plots of E against Ha at various values of the parameters are given [8]. In the former case a transition in the regimes of absorption with changes in Ha is indicated, which, with respect to the homogeneously catalyzed reaction, reflects a change in the concentration of the catalyst.

4.1.3 Gas/Liquid/Liquid-phase Reactions

In gas/liquid reactions the liquid phase contains the homogeneous catalyst together with the liquid reactant and the dissolved gaseous reactant. To perform gas/liquid/liquid reactions in the biphasic (liquid/liquid) mode it is essential that the solubility of the homogeneous catalyst in one of the two liquid phases is negligible. By using water-soluble ligands, homogeneous metal complex catalysts are dissolved in the aqueous phase where the catalytic reaction can take place (see below). In these aqueous biphasic liquid systems, different situations are possible which would have to be considered in any quantitative treatment. On the one hand, for example, the gaseous reactants are dissolved in the catalyst-containing aqueous phase where they react to form an immiscible product which is only soluble in the organic liquid phase. The gaseous reactants are transferred into the catalyst-containing phase either directly from the gas phase and/or from the liquid phase, in which they may be soluble, too. On the other hand, the gaseous reactants are soluble in only one liquid phase. This requires that the catalytic reaction takes place at the liquid/liquid interface only, i.e., the products, again soluble only in one of the two phases, are formed in the boundary layer or in the phase boundary interface for the case of an instantaneous reaction ($\text{Ha} \gg 3$).

These processes are still not fully understood and need further elucidation. In the case of the homogeneously catalyzed hydroformylation of olefins according to Ruhrchemie/Rhône-Poulenc, the reaction products are completely insoluble in the aqueous phase which contains the water-soluble catalyst ($[HRh(CO)(TPPTS)_3]$). By contrast, if the homogeneous catalyst is not soluble in water but in an organic liquid, the biphasic homogeneously catalyzed reaction may also occur in the organic phase. For instance, in the case of biphasic hydroformylation of allyl alcohol (Eq. 7), the educt and the reaction products are water-soluble but the catalyst is present in the organic phase [9]. Furthermore, in some cases the second liquid phase is formed during the catalytic reaction itself (e.g., in the SHOP process of Shell [1]).

$$\text{CH}_2{=}\text{CHCH}_2\text{OH} + CO/H_2 \xrightarrow[n\text{-heptanol}/H_2O]{HRh(CO)(PPh_3)_3} \text{OHC(CH}_2)_3\text{OH} \tag{7}$$

In all the above-mentioned cases conversion can only take place when the components are transferred to the catalytic phase or at least to the interface in which the reaction proceeds. The transport from one phase to the other(s) requires a driving force, i.e., the existence of concentration gradients. The principal steps of a homogeneously catalyzed gas/liquid/liquid reaction (8) are shown schematically in Figure 2, where the reaction product P_1 is formed by the reaction between a gaseous reactant A_1 and reactant A_2 from the organic liquid phase in presence of a second phase, i.e., an aqueous liquid phase which contains the water-soluble catalyst. Moreover, it is assumed that both liquid phases are immiscible. In such a scenario two cases can be distinguished: aqueous droplets containing the dissolved catalyst are dispersed in a continuous organic liquid phase, and the organic liquid phase is dispersed in a continuous aqueous phase which contains the water-soluble catalyst.

$$A_{1,g} + A_{2,l} \xrightarrow[\text{aqueous biphase}]{\text{homogeneous catalyst}} P_1 \tag{8}$$

Several steps influencing the overall rate of the reaction and the selectivity of a desired product have to be considered; they are outlined in Table 1. It is important to note that steps (e) and (f) cannot be separated from each other in either case because, in general, the transport from the interfaces occurs *simultaneously* with the catalytic reaction.

From the above qualitative discussion it can be clearly derived that at least four important factors, namely interphase mass transfer, solubility, thermodynamic phase equilibria, and intrinsic kinetics must be considered during quantitative analysis of gas/liquid/liquid reactions (Figure 3).

Thus, in the scenario shown above the various equations for the overall rate may be obtained, depending themselves on, for example, the types of catalyst and water-soluble ligands, solvents, surface-active compounds, phase equilibrium properties and the extent of gas/liquid and liquid/liquid resistances.

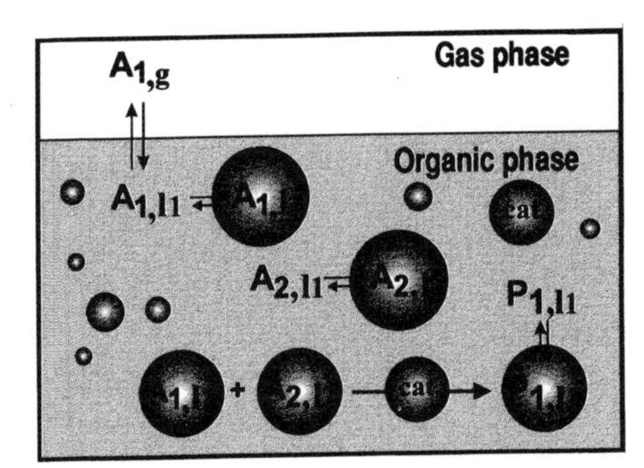

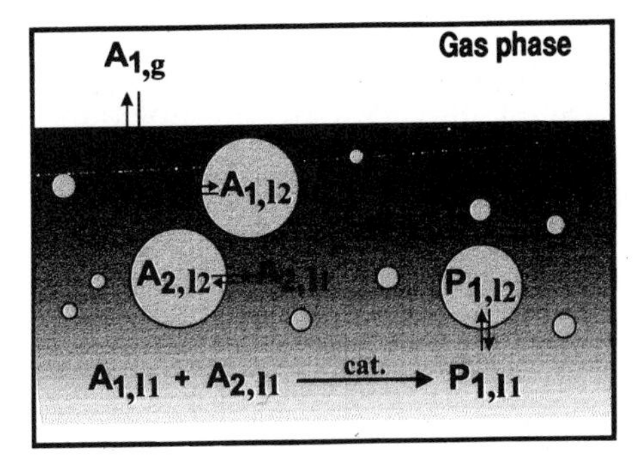

Figure 2. Two features of the principal steps of mass transfer and chemical reaction during a homogeneously catalyzed gas/liquid/liquid reaction (aqueous biphasic mode). A_1 = gaseous reactant; A_2 = liquid reactant; P_1 = reaction product. Aqueous droplets containing the dissolved catalyst are dispersed in a continuous organic liquid phase (left); the organic liquid phase is dispersed in a continuous aqueous phase which contains the water-soluble catalyst (right).

Table 1. Dependence of sequence of steps in an aqueous biphasic catalytic reaction on the mode of operation.

Step	Case I: Aqueous droplets containing the dissolved catalyst are dispersed in a continuous organic liquid phase	Case II: Organic liquid phase is dispersed in a continous aqueous phase which contains the catalyst
(a)	Transport of a gaseous reactant (A_1) from the bulk of the gas phase to the gas/organic liquid interface	Transport of a gaseous reactant (A_1) from the bulk of the gas phase to the gas/aqueous catalyst interface
(b)	Transport of A_1 through the gas/organic liquid interface	Transport of A_1 through the gas/aqueous catalyst interface
(c)	Transport of A_1 into the bulk of the organic liquid	Transport of A_1 into the bulk of the aqueous catalyst phase[a)]
(d)	Transport of both dissolved A_1 and liquid reactant (A_2) from the organic phase to the organic/aqueous interface	Transport of liquid reactant (A_2) from organic droplets to the organic/aqueous interface
(e)	Transport of A_1 and A_2 from the organic/aqueous interface to the aqueous catalyst phase	Transport of A_2 from the organic/aqueous interface to the aqueous catalyst phase
(f)	Homogeneously catalyzed reaction of dissolved A_1 and A_2 to product (P) in the aqueous phase	Homogeneously catalyzed reaction of dissolved A_1 and A_2 to products (P) in the aqueous phase
(g)	Transport of water-immiscible P from the aqueous to the organic phase	Transport of water-immiscible P from the aqueous to the organic liquid phase

[a)] In the presence of an organic liquid, further transport to the gas/organic liquid interphase and into the bulk of the organic liquid.

The droplet size of the dispersed aqueous and organic liquid phase is affected by the liquid properties and the reactor type. For instance, in the case of reaction (7) given above between $A_{1,g}$ and $A_{2,l}$ in a homogeneous catalyst-containing aqueous phase, the enhancement of gas-to-water mass transfer rates by a dispersed organic phase can be described with a new mass transfer theory without any additional parameter adjustment, the so-called film variable holdup (FVH) model [10]. This model takes into account the distribution of organic and continuous aqueous phase near the gas/liquid interface and explains quantitatively the influence of hold-up, droplet diameter, and permeability of the organic phase on the observed enhancement. Experimental data have been presented on the enhancement of O_2 mass transfer into an aqueous Na_2SO_3 solution in a stirred cell, due to the presence of a dispersed liquid 1-octene phase. Also the experimental data for O_2 mass transfer enhancement in hexadecane and for CO_2 mass transfer enhancement due to toluene droplets can be reasonably well described; this indicates that in different liquid/liquid systems the dispersed phase distribution is similar for different organic droplets in water.

The very low solubility of organic reactants in the catalytic phase often gives rise to a drastic decrease of the effective reaction rate. This drawback of gas/liquid/liquid reactions has been overcome by adding a co-solvent. In the case of

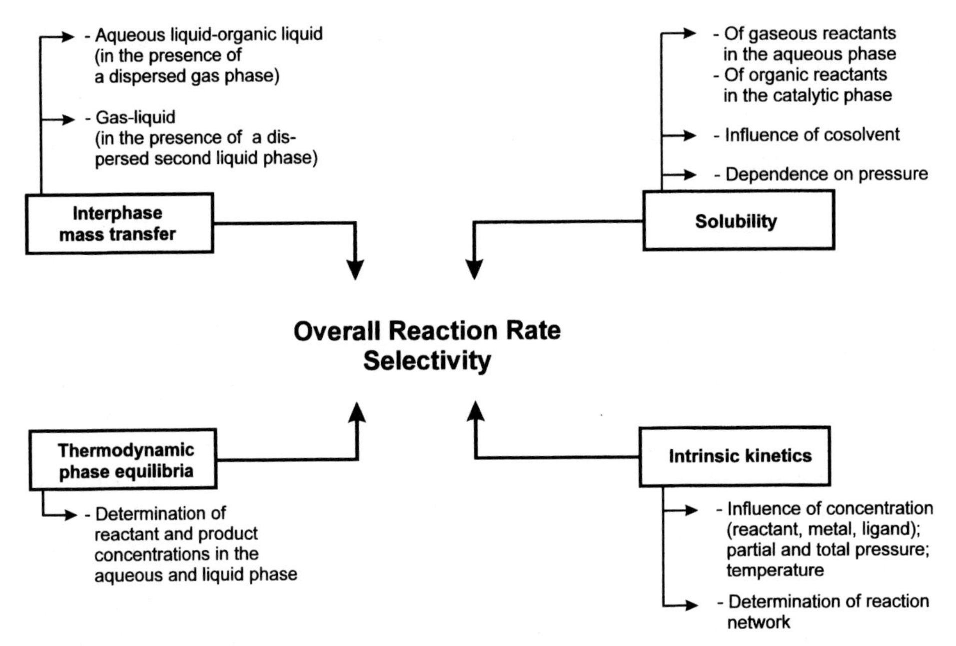

Figure 3. Factors controlling gas/liquid/liquid reactions (aqueous biphasic mode).

1-octene hydrogenation which was performed in semibatch operation and by using $[RhCl(COD)_2]$/TPPTS as water-soluble catalyst, the water/1-octene co-solvent equilibrium was estimated by a combination of the UNIFAC group distribution method and the UNIQUAC equation in order to select a convenient co-solvent [11]. Ethanol, *n*-propylamine, or ethylene glycol enhances the 1-octene concentration in the aqueous phase considerably without losing catalyst in the organic phase. Furthermore, it was shown that the hydrogen solubility in the aqueous catalytic phase increases according to Henry's law (9), with

$$p_i^* = H_i c_i^* \tag{9}$$

increasing hydrogen pressures up to 10 MPa. Then, initial kinetics, derived without any gas/liquid and liquid/liquid mass transfer limitations, showed first-order reaction rates with respect to dissolved hydrogen and 1-octene in the aqueous phase. Finally, by measuring time-dependent concentrations together with the thermodynamic liquid/liquid model, which makes it possible to calculate the concentrations of reactants and products in the aqueous catalytic and organic phases, complete kinetics of parallel hydrogenation and isomerization were analyzed [11]. Several of the factors of Figure 3 controlling the activity and selectivity of the biphasic selective hydrogenation of α,β-unsaturated aldehydes to allylic alcohols, for instance, 3-methyl-2-butenaldehyde to 3-methyl-2-buten-1-ol (Eq. 10) with ruthenium-sulfonated phosphine catalysts were investigated [12], such as the effect of agitation speed and the influence of aldehyde, ligand, and metal concentrations. Under optimized reaction conditions, where gas/liquid mass transfer was not rate-determining, the kinetic equation (Eq. 11) was found to apply. A zero-order dependence with respect to the concentration of the α,β-unsaturated aldehyde was found.

$$H_2 + \text{(3-methyl-2-butenal)} \xrightarrow{RuCl_3/TPPTS} \text{(3-methyl-2-buten-1-ol)} \tag{10}$$

$A_{1,g}$ $A_{2,l}$ P_1

$$r = k\, c_{Ru}\, c_{H_2} \tag{11}$$

Gas/liquid/liquid reaction engineering was studied by Purwanto and Delmas [13] during hydroformylation of 1-octene by $[Rh(COD)Cl_2]$/TPPTS catalyst in a batch reactor at pressures between 1.5 and 2.5 × 10^3 kPa and temperatures of 333 and 343 K. As also shown for olefin hydrogenation [11], the concentration of 1-octene in the aqueous phase was increased by using a co-solvent (e.g., ethanol), which gave rise to an enhancement of the reaction rate of hydroformylation [13, 14]. The effect of cosolvent addition on gas/liquid and liquid/liquid equilibria was studied both experimentally and theoretically by UNIFAC simulations. Kinetic studies showed that the reaction is first-order with respect to 1-octene and catalyst concentrations. The reaction rate was enhanced by hydrogen partial pressure and at low CO partial pressures, whereas inhibition was observed at high CO pressures. This dependence on CO partial pressure was typical of hydroformylation kinetics in homogeneously catalyzed reactions with a Rh complex catalyst. A semi-empirical kinetic model (Eq. 12) based on initial reaction rates obtained during this study was used to describe the overall rate of the 1-octene hydroformylation:

$$r = \frac{k\, c_{octene}\, c_{H_2}\, c_{CO}\, c_{cat}}{[1 + K_{H_2}\, c_{H_2}]\,[1 + K_{CO}\, c_{CO}]^2} \tag{12}$$

The parameter values obtained by an optimization routine are given in Table 2. Furthermore, it was observed that, if PPh_3 was added to the organic liquid phase (toluene) during hydroformylation, the rate was increased by a factor up to 50 [15].

The kinetics of the above-mentioned biphasic hydroformylation of allyl alcohol (see Eq. 7) was described by the rate Eq. (13) [9], which shows the inhibition of the reaction rate by the partial pressure of carbon monoxide.

$$r = \frac{k\, c_{C_3H_5OH}\, c_{H_2}\, c_{cat}}{[1 + K_{CO}\, c_{CO}]^3} \tag{13}$$

Table 2. Parameters of equation (12) obtained for the kinetics of hydroformylation of 1-octene at two temperatures [13].

Parameter	$T = 333$ K	$T = 343$ K
$k \times 10^{-5}$ [$m^9\ kmol^{-3}\ s^{-1}$]	1.571	7.441
$K_{H_2} \times 10^{-2}$ [$m^3\ kmol^{-1}$]	1.967	2.185
$K_{CO} \times 10^{-2}$ [$m^3\ kmol^{-1}$]	1.133	1.886

4.1.4 Aqueous Biphasic Systems: Where the Reaction Takes Place

With respect to the location where homogeneously catalyzed biphasic reactions occur, two main conclusions can be drawn from the experimental observations sketched above: in such cases where the reaction rate is accelerated by co-solvents and other solubility promoters (e.g., EtOH), it has been concluded that the bulk of the liquid is the reaction place [9, 11]. This seems to be confirmed by the observed decrease in reactivity of olefins as starting compounds for hydroformylation with their decreasing solubility [1, 19]. By contrast, the increase in the reaction rate and control of the selectivity in biphasic system observed with the use of surfactants or other surface-active compounds [16–18] (for "promotor ligand" see [15]) indicate an increase in the liquid/liquid interfacial area, and it has therefore been speculated that the latter seems to be the reaction site [15].

To answer the fundamental question of where the reaction takes place in RCH/RP's process, Cornils and co-workers [19] conducted relatively simple batch calorimetric experiments at various pressures, temperatures, stirrer speeds, and concentrations, using catalyst solutions from the large-scale plants at Oberhausen, together with reaction modeling. Two kinetic models were tested, namely one for a reaction in the bulk water phase (model 1) and another for a reaction at the phase interface (model 2) (Figure 4).

By comparison between the calculated and measured pressure and heat flux vs. time curves it was shown that the site of this hydroformylation reaction could not be the bulk of the liquid. Only the assumption of a reaction in the liquid boundary layer at the gas/liquid interface gave satisfactory agreement of the data under all experimental conditions. Thus, on this basis scale-up rules for the aqueous biphasic hydroformylation and appropriate kinetic models can be developed for optimal reactor design. The principle of both models applied to the general equation (Eq. 8) is shown in Figure 5.

However, reaction modeling will be more complex, for instance in the case of hydroformylation of higher olefins, if co-solvents or surfactants, i.e., additional liquid phase(s), are present in the reaction mixture.

In the case of another biphasic variant, namely phase-transfer-catalyzed reactions (cf. Section 4.6.1; e.g., [20]), mass transfer rates of ionic intermediates between the aqueous and organic phases, their phase and partition equilibria, as well as the reaction rate in the organic phase, have to be analyzed to model the overall reaction rate. Finally, the dynamics of liquid/liquid phase transfer reactions, which are a part of multiphase reactions as shown above, can be described by a new phase-plane model [21] based on the two-film theory.

In future, a complete quantitative analysis on the basis of chemical reaction engineering principles of homogeneously catalyzed gas/liquid/liquid reactions is needed to improve known aqueous biphasic reactions as well as to find new, highly active and selective homogeneous catalysts for organic synthesis.

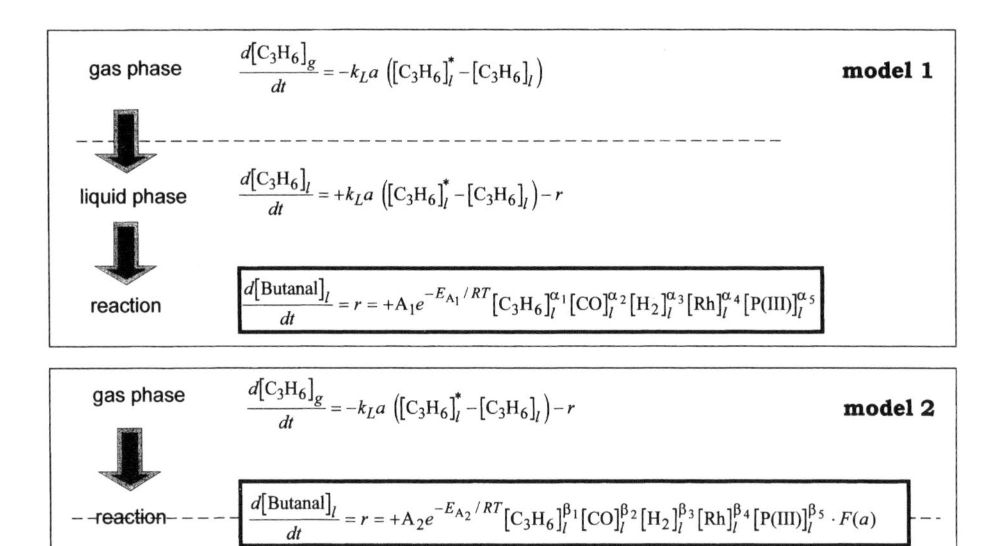

Figure 4. Kinetic models for the RCH/RP process by Cornils and co-workers [19].

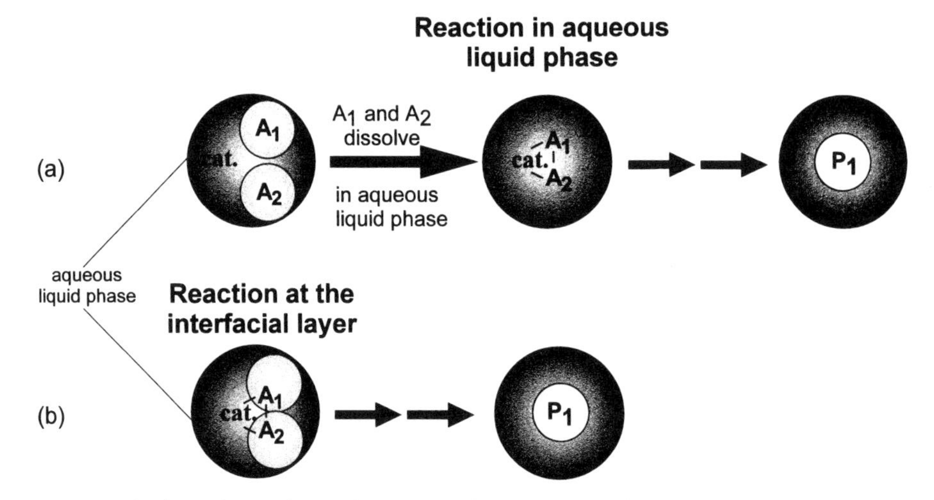

Figure 5. Principle of two alternative models for aqueous biphasic reactions. (a) Reaction in the bulk of the aqueous liquid phase (must be preceded by diffusion from the gas into the liquid phase). (b) Reaction in the interphase region.

Notation

a	interfacial area per unit volume [m^{-1}]	p_i	partial pressure [Pa]
c_i	concentration of species i [kmol m^{-3}]	r	reaction rate [kmol m^{-3} s^{-1}]
c_i^*	concentration of species i at phase equilibria [kmol m^{-3}]	r_{eff}	effective reaction rate [kmol m^{-3} s^{-1}]
D_{ik}	binary diffusion coefficient [m^2 s^{-1}]	R	gas constant [8.314 J mol^{-1} K^{-1}]
E	enhancement factor [–]	T	temperature [K]
H_i	Henry coefficient [$H_i = p_{A_1,g} c_{A_1}$]	δ	film thickness [m]
Ha	Hatta number [–]	λ	distance of reaction plane from interface [m]
J_i	diffusion flow density [mol m^{-2} s^{-1}]	F	fluid
k	rate constant (dimensions depend on kinetics) [$m^{3(n-1)}$ $kmol^{-(n-1)}$ s^{-1}]	g	gas
k_i	mass transfer coefficient [m s^{-1}]	i	component (e.g., $i = 1, 2$ or A, B ...)
n, m	reaction orders [–]	l	liquid

References

[1] (a) B. Cornils, W. A. Herrmann, *Applied Homogeneous Catalysis with Organometallic Compounds, Vol. 1,* VCH, Weinheim, **1996**, p. 2; (b) ibid., Vol. 2, p. 684.
[2] W. A. Herrmann, B. Cornils, *Angew. Chem.* **1997**, *109*, 1074.
[3] W. A. Herrmann, C. W. Kohlpaintner, *Angew. Chem.* **1993**, *105*, 1588.
[4] K. R. Westerterp, W. P. M. van Swaaij, A. A. C. M. Beenackers, *Chemical Reactor Design and Operation*, John Wiley, New York, **1984**.
[5] M. Baerns, H. Hofmann, A. Renken, *Chemische Reaktionstechnik,* Lehrbuch der Technischen Chemie I, George-Thieme-Verlag, Stuttgart, **1987**.
[6] J. M. Smith, *Chemical Engineering Kinetics*, 3rd ed., McGraw-Hill Chemical Engineering Series, New York, **1981**.
[7] O. Levenspiel, *Chemical Reaction Engineering,* 2nd ed., John Wiley, New York, **1972**.
[8] R. V. Chaudari, in *Frontiers in Chemical Reaction Engineering:* Proc. Int. Chem. React. Eng. Conf., New Delhi, Vol. 1 (Ed.: L. K. Doraiswamy), Wiley, New York, **1984**, p. 291.
[9] R. M. Deshpande, S. S. Divekar, B. M. Bhanage, R. V. Chaudari, *J. Mol. Cat.* **1992**, *75*, L19.
[10] C. J. van Ede, R. van Houten, A. A. C. M. Beenackers, *Chem. Eng. Sci.* **1995**, *50*, 2911.
[11] I. Hablot, J. Jenck, G. Casamatta, H. Delmas, *Chem. Eng. Sci.* **1992**, *47*, 2689.
[12] J. M. Grosselin, C. Mercier, G. Allmang, F. Grass, *Organometallics* **1991**, *10*, 2126.
[13] P. Purwanto, H. Delmas, *Catal. Today* **1995**, *24*, 135.
[14] R. V. Chaudari, A. Bhattacharya, B. M. Bhanage, *Catal. Today* **1995**, *24*, 123.
[15] R. V. Chaudari, B. M. Bhanage, R. M. Deshpande, H. Delmas, *Nature* (*London*) **1995**, *373*, 501.
[16] C. Larpent, F. Brise LeMenn, H. Patin, *New J. Chem.* **1991**, *15*, 361.
[17] H. Ding, B. E. Hanson, *J. Mol. Catal.* **1995**, *99*, 131.
[18] T. Bartik, H. Ding, B. Bartik, B. E. Hanson, *J. Mol. Catal.* **1995**, *98*, 117.
[19] O. Wachsen, K. Himmler, B. Cornils, *Catal. Today* **1998**, in press.
[20] Y. Lee, M. Yeh, Y. Shih, *Ind. Eng. Chem. Res.* **1995**, *34*, 1572.
[21] H. Wu, *Ind. Eng. Chem. Res.* **1993**, *32*, 1323.

4.2 Technical Solutions

Arno Behr

4.2.1 Reaction Systems

A general problem in homogeneous reactions with organometallic catalysts is the question of how to separate the catalysts from the products after the reaction is completed. In aqueous-phase catalysis this problem can become the most critical point in the technical realization of a process. One advantage of aqueous-phase chemistry is the fact that often more or less nonpolar starting chemicals are used which form a second organic phase. That means that the reaction system consists of two immiscible or partly miscible phases which can be used for the separation of catalyst and products. To obtain a better and more complete insight into this separation technique, the general Eq. (1) will be considered, in which the starting materials A and B react to give the products C and D. This reaction is carried out in the presence of the solvent water (or a similar strongly polar solvent or solvent mixture) and with the help of an organometallic catalyst, which may be polar or nonpolar.

$$A + B \xrightarrow[\text{water}]{\text{cat.}} C + D \qquad (1)$$

In Table 1 different variations are listed, depending on whether the starting compounds A and B or the products C and D and the catalyst are soluble in the aqueous phase (a) or in the nonpolar organic phase (o). In the last column of this Table a first evaluation is made of the possibility of separating the catalyst from the product chemicals (which may be mixed with unreacted starting compounds).

Entries 1 and 6 of Table 1 are the two cases of "homophase" homogeneous catalysis, either in aqueous or in organic solution. The evaluation (in view of catalyst separation) is negative, because in accordance with the definition of homogeneous reactions the catalyst is dissolved in the reaction medium and cannot be separated directly from the products.

Entries 2, 3, 5, 8, 10, and 11, however, are judged positive, because neither product (C or D) dissolves the catalyst. These estimations do not guarantee that all of these types of reactions can be carried out in a reasonable manner, however.

Entry 2, for example, comprises the reaction of aqueous-soluble compounds using a nonpolar catalyst. It is questionable whether this reaction really works

Table 1. General concept of aqueous-phase organometal-catalyzed reactions (Eq. 1).

Entry	Solubility[a]					
	A	B	C	D	Catalyst	Evaluation[b]
1	a	a	a	a	a	–
2	a	a	a	a	o	+
3	a	a	o	o	a	+
4	a	a	o	o	o	–
5	o	o	o	o	a	+
6	o	o	o	o	o	–
7	o	o	a	a	a	–
8	o	o	a	a	o	+
9	o	a	a	a	a	–
10	o	a	a	a	o	+
11	o	a	o	o	a	+
12	o	a	o	o	o	–
13	a	a	o	a	a	–
14	a	a	o	a	o	–
15	o	o	o	a	a	–
16	o	o	o	a	o	–

[a] a = soluble in aqueous phase; o = soluble in organic phase.
[b] – Low possibility of product separation; + good possibility of product separation.

because of the bad distribution of the catalyst in the reaction system. In this case it can be better to add a further nonpolar solvent which dissolves the catalyst, thus forming a true liquid–liquid two-phase system (LLTP).

The opposite of entry 2 is the phase combination of entry 5, a fully organic reaction in the presence of an aqueous phase which contains the catalyst. This type of reaction system is the most often used for the technical realization of aqueous-phase organometallic-catalyzed reactions, for instance in the oligomerization of ethylene using the SHOP process (cf. Section 7.1) or in the Ruhrchemie/Rhône-Poulenc process (cf. Section 6.1.1) of propene hydroformylation (see Section 4.2.2).

An interesting variant of entry 5 is entry 11, in which one of the educts is a polar and the other a nonpolar reactant yielding organic products. Once again the catalyst is in the aqueous phase and thus easily separated from the products. This technique is used in the telomerization of butadiene with water (Kuraray process; cf. Section 6.7) and the telomerization of butadiene with ammonia (see Section 4.2.2).

As already discussed with entry 2, all the aqueous-phase reactions of Table 1 can in principle be carried out as two-phase reactions by adding a second, nonpolar, solvent. The question of catalyst separation is not influenced by this solvent addition, but a second solvent can effectively control the selectivity of the reaction.

If a consecutive reaction (Eq. 2) is investigated, the initially formed organic product C will enter the organic solvent phase almost completely, thus disabling the consecutive reactions yielding products D or E. Consequently by this technique product C is formed with high selectivity.

$$A + B \xrightarrow{\text{cat.}} C \xrightarrow[+\,A]{\text{cat.}} D \xrightarrow[+\,A]{\text{cat.}} E \qquad (2)$$

4.2.2 Technical Realization: Variations

Table 2 gives a brief overview of the industrially realized aqueous-phase organometallic-catalyzed reactions, together with some related reactions which are still under development. The list includes not only water systems but also systems with polar solvents which have much in common with the aqueous-phase reactions.

The most important process worldwide is the "Shell higher-olefin process" (SHOP) in which ethylene is oligomerized to higher-molecular-mass, linear, α-olefins. The nickel-catalyst, containing a phosphorus/oxygen chelate ligand, is dissolved in the polar solvent 1,4-butanediol, which is not miscible with the α-olefins. Two big plant with a total capacity of 1 Mio t y^{-1} are built in Geismar (USA) and Stanlow (UK).

The most important process with the solvent water is the hydroformylation of propene to butyraldehydes, known as the Ruhrchemie/Rhône-Poulenc process. This reaction is catalyzed by a rhodium complex containing the water-soluble ligand triphenylphosphane trisulfonate (TPPTS). The aldehydes are formed with an annual capacity of approx. 300000 t.

All other industrial applications are still on a smaller scale. The Kuraray telomerization of butadiene with water is carried out at 5000 t/y^{-1} (see Section 6.7); the Rhône-Poulenc reactions have been developed into bulk processes. The investigations at the Universities of Dortmund and Aachen (Germany) are still on the laboratory or miniplant scale.

All the examples given in Table 2 make use of the same decisive advantage to recycle the homogeneous transition metal catalyst by applying the LLTP system. However, the technical realization of the catalyst recycle can be carried out in very different ways [60]. There are a number of possibilities:

Products and catalyst must be separated. This can be done by methods which either separate the *product* or the *catalyst* from the residue.

The separation methods can be varied: simple *separation* of two liquid phases, *extraction* with an additional extractant, or chemical *treatment* with additional bases and acids.

In all cases a reaction and a separation step are combined. Reaction and separation can be done in the same unit, i.e., *simultaneously*, or in separate units, i.e. *successively*.

Table 2. Important examples of polar-phase organometallic-catalyzed reactions.

Company or University	Reaction[a)]	Catalyst	Solvent	Literature
Shell (SHOP)	Oligomerization of ethene (7.1)	Ni	1,4-Butanediol	[1–9]
Ruhrchemie/ Rhône-Poulenc	Hydroformylation of olefins (6.1)	Rh	Water	[10–20]
Rhône-Poulenc	Co-oligomerization with myrcene (6.10)	Rh	Water	[21–23]
Rhône-Poulenc	Hydrogenation of unsaturated aldehydes (6.2)	Ru	Water	[24–27]
DuPont	Hydrocyanation of pentenenitrile (6.5)	Ni	Ionic liquids	[28]
Kuraray	Telomerization of butadiene with water (6.7)	Pd	Water/sulfolane	[29–44, 75–79]
University of Dortmund	Telomerization of isoprene with water	Pd	Water	[45–48]
University of Dortmund	Telomerization of butadiene with CO_2	Pd	Water	[49–55]
University of Aachen	Telomerization of butadiene with NH_3	Pd	Water	[56, 57]
University of Aachen	Telomerization of butadiene with phthalic acid	Pd	Acetonitrile	[58, 59]
University of Aachen	Dimerization of butadiene	Pd	DMSO	[58, 59]

[a)] Cf. Section indicated in parentheses.

In total – by combining all the variations listed above – there are 12 cases to be distinguished:

- simultaneous or successive reaction and product separation
- simultaneous or successive reaction and product extraction
- simultaneous or successive reaction and product treatment
- simultaneous or successive reaction and catalyst separation
- simultaneous or successive reaction and catalyst extraction
- simultaneous or successive reaction and catalyst treatment

In the following, these variants will be discussed in more detail, always looking at the same general reaction [Eq. (1)]. In this discussion some basic flow schemes will be presented which contain in some cases two different solvents, one polar (e.g., water) and one nonpolar. For reasons of simplicity in these cases it is assumed that the expected products are organic nonpolar compounds, that means we consider only entries 3–6, 11 and 12 in Table 1. Of course, analogous considerations apply if the products are polar compounds, i.e., in the case of entries 1, 2, and 7–10 in Table 1. In these cases the terms "polar" and "nonpolar" in the flow scheme have to be exchanged.

4.2.2.1 Reaction with Product Separation

The easiest technique to combine catalysis in a polar medium with product separation is shown in Figure 1. Both operations are done in the same unit at the same time. The nonpolar product phase is deposited from the polar catalyst phase and can be separated at the top of the reaction column. The SHOP oligomerization of ethene works in this way. The catalytic phase, consisting of 1,4-butanediol and nickel catalyst, always remains in the reaction unit. In the technical plant the reaction takes place not in only one reactor but in a series of tanks. This is so that the heat of reaction may be removed by water-cooled heat exchangers which are placed between the different reactor tanks. The flow scheme of the SHOP process is shown in Section 7.1.

A variant of this technique is "supported aqueous-phase catalysis" (SAPC), in which the polar catalyst phase is heterogenized on a solid support [61–68]. The principle of this technique is shown in Figure 2. The organometallic complex, e.g., rhodium with triphenylphosphine sulfonate ligands, is dissolved in an aqueous solution which is fixed onto a hydrophilic support, e.g., controlled-pore glasses on surface-modified silica. The advantage of this method lies in the easy

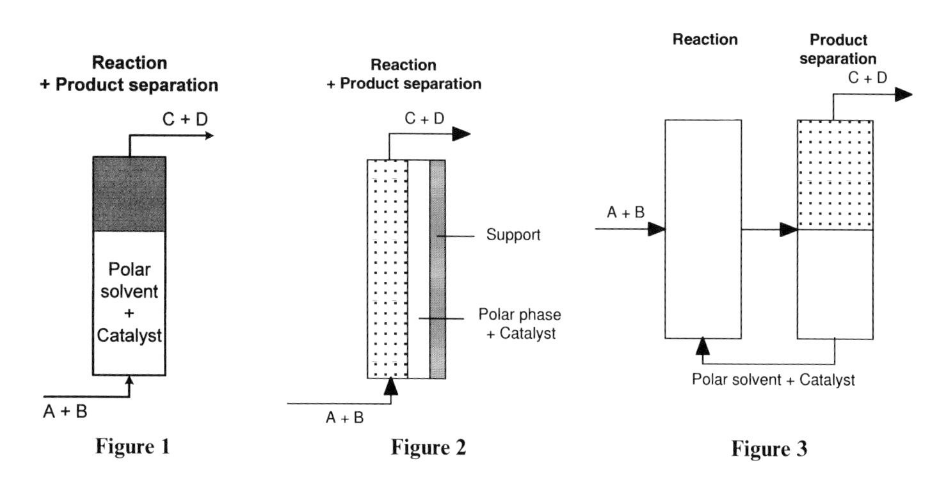

Figure 1. Simultaneous reaction and product separation (two liquid phases).
Figure 2. Simultaneous reaction and product separation (liquid nonpolar phase, heterogenized polar phase).
Figure 3. Successive reaction and product separation.

handling of the homogeneous catalyst, very similarly to a solid heterogeneous catalyst. However, no industrial application of SAPC is known so far (cf. Section 4.7).

Figure 3 shows a technical modification of the general principle illustrated in Figure 1; while in Figure 1 the two liquid phases are separated during the reaction, but in Figure 3 these steps are divided into two different units. As in the Ruhrchemie/Rhône-Poulenc process of propene hydroformylation, in this industrial application the reaction steps and the separation step are divided. The reaction takes place in a continuously stirred tank reactor, while the phase separation is carried out in a decanter.

4.2.2.2 Reaction and Product Extraction

The appropriate processes described so far use only an aqueous catalyst solution, and no additional organic solvent is necessary. In the processes described in this section an organic nonpolar solvent is used as product extractant. Hydrocarbons, chlorinated hydrocarbons, or ethers are often chosen as nonpolar solvents.

Figure 4 shows simultaneous reaction and product extraction, also called "in-situ extraction." The reaction of A and B runs in the reactor which contains the polar (e.g., aqueous) catalyst phase. The nonpolar extractant absorbs the organic products which are separated from the polar catalyst phase in the following separation step. The procedure is more costly because a further distillation is necessary in a third unit to separate the products from the low-boiling extractant, which is then recycled to the reactor.

For this technique some examples are given in the literature. Baird [69, 70] describes the hydrogenation and the hydroformylation of olefins applying this two-phase system. The aqueous phase contains the soluble rhodium catalyst

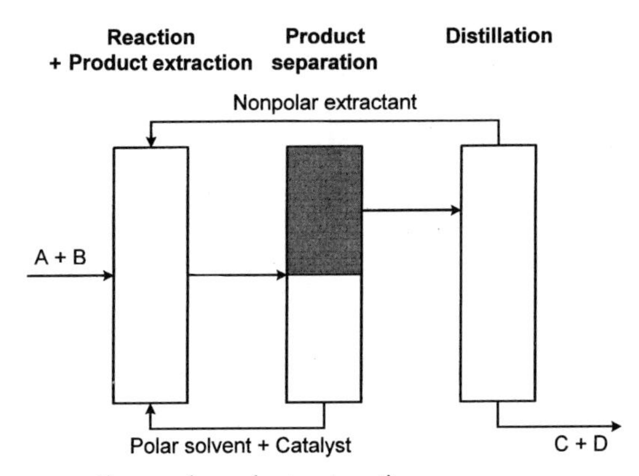

Figure 4. Simultaneous reaction and product extraction.

which is formed from norbornadienylrhodium chloride and AMPHOS nitrate ($Ph_2PCH_2CH_2NMe_3^+NO_3^-$); the organic phase contains solvents such as methylene chloride, diethyl ether, or pentane, or the olefin, e.g., 1-hexene. This system is very favorable because only traces of rhodium (0.25 ppm) move into the organic layer, and the aqueous catalyst phase could be re-used in the reaction with little or no loss of activity, even after exposure to air.

An interesting new example is the two-phase telomerization of butadiene with ammonia, yielding octadienylamines [57]. When this reaction is carried out in homogeneous one-phase solution, a great amount of primary, secondary and tertiary amines is formed [56]. With the two-phase technique using water, and toluene or pentane as the second phase, the consecutive reactions (Eq. 2) can be almost completely avoided, and the primary octodienylamines are the only main products. This reaction has not yet been realized industrially.

An important alternative is the successive reaction and product extraction procedure shown in Figure 5. First the reaction of A and B is carried out in a single polar homogeneous phase containing the catalyst. Downstream, the products are extracted with a nonpolar solvent or solvent mixture. In the third unit,the distillation, the low-boiling extractant is distilled off and recycled to the extraction unit. Two examples will demonstrate the practicability of this concept.

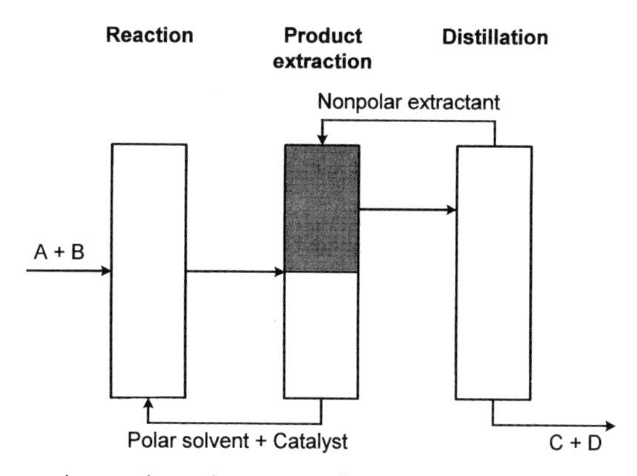

Figure 5. Successive reaction and product extraction.

Butadiene can be dimerized to 1,3,7-octatriene in the presence of the catalyst $Pd(PPh_3)_4$ and the alcohol $PhC(CF_3)_2OH$ (Eq. 3). This reaction is carried out favorably in the solvent acetonitrile. Then the product octatriene is extracted with isooctane while the catalyst and the alcohol remain in the acetonitrile phase, which is recycled to the reactor. Finally the isooctane is distilled off from the dimer. In a continuously working miniplant, 3 kg of octadiene/g Pd could be produced [58, 59].

$$2\ \text{butadiene} \xrightarrow[\text{CH}_3\text{CN}]{[\text{Pd}],\ \text{PhC(CF}_3)_2\text{OH}} \text{1,3,7-octatriene} \quad (3)$$

The second example is the telomerization of phthalic acid with butadiene yielding bis(octadienyl) phthalates, which can then be hydrogenated to the bisoctyl phthalates which may be used as plastic softeners. The catalyst is formed from palladium bis(acetylacetonate) and tris(*p*-methoxyphenyl)phosphite in the polar solvent dimethyl sulfoxide. Once again the extraction can be carried out with isooctane [58, 59].

A third example is the Kuraray process [29–44, 75–79].

The method shown in Figure 5 only works if the polar solvent (e.g., water) dissolves the catalyst completely and the nonpolar extractant dissolves only the product. Solvent and extractant must form a perfectly separable biphasic system. However, there are extractants which of course dissolve the products in good yields, but which dissolve also the solvent of the reaction. In that case, no proper biphasic system can appear. The solution of this problem is the – costly – distillation of the solvent before the extraction step.

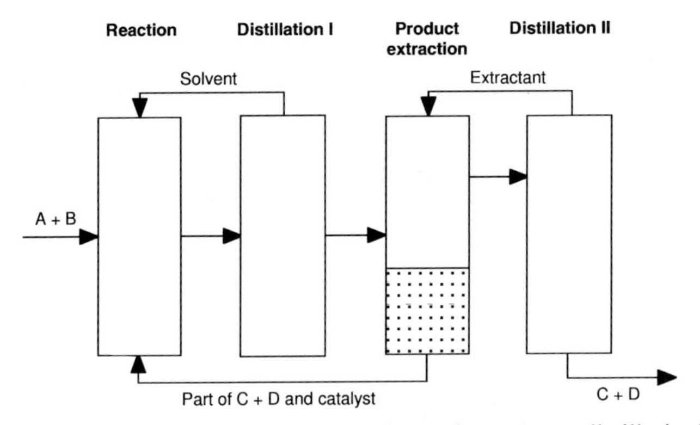

Figure 6. Successive reaction and product extraction (after solvent distillation).

In Figure 6 this principle is demonstrated for the telomerization of butadiene with carbon dioxide yielding a δ-lactone (Eq. 4). The reaction is carried out in a homogeneous acetonitrile solution using a palladium catalyst. After distillation of the acetonitrile in the second unit, the product/catalyst mixture is treated with the extractant, 1,2,4-butanetriol, which dissolves the product but not the catalyst [55]. The catalyst is then recycled to the reactor in a small amount of the liquid product. The main quantity of the lactone is separated from the extractant by a second distillation step.

$$2 \quad + \; CO_2 \xrightarrow{Pd} \qquad (4)$$

4.2.2.3 Reaction and Product Treatment

An interesting concept is to synthesize an organic nonpolar product with the help of a nonpolar organometallic catalyst and to convert this nonpolar product into a polar one. In sum, the catalyst and the product phase can be separated (entry 8 in Table 1), thus enabling the required catalyst recycle. An example of this chemical conversion is the synthesis of a nonpolar long-chain carboxylic acid which is then transformed into the polar carboxylate salt by addition of an aqueous base solution. This product treatment can be performed principally during the reaction (Figure 7), thus absorbing the product continuously into the aqueous phase. In a second unit the polar product, e.g., the carboxylate salt, is reconverted into the nonpolar product, the carboxylic acid, and the product and the aqueous salt solution can be separated in a decanter. This salt solution is one of the disadvantages of this and related separation techniques using chemical product treatment.

An alternative to the simultaneous reaction and product conversion shown in Figure 7 is the successive reaction and product treatment in Figure 8. This

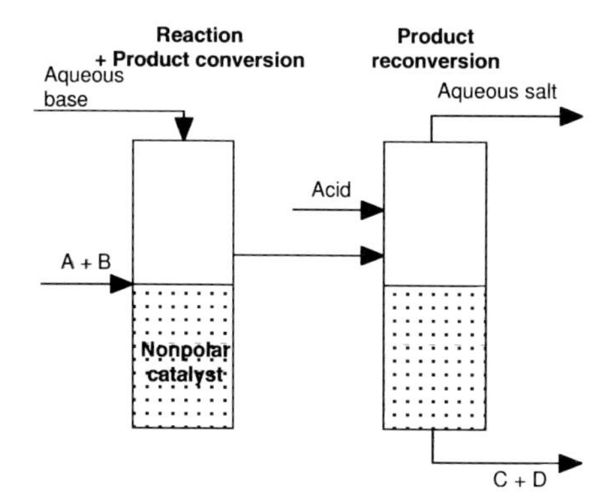

Figure 7. Simultaneous reaction and product treatment.

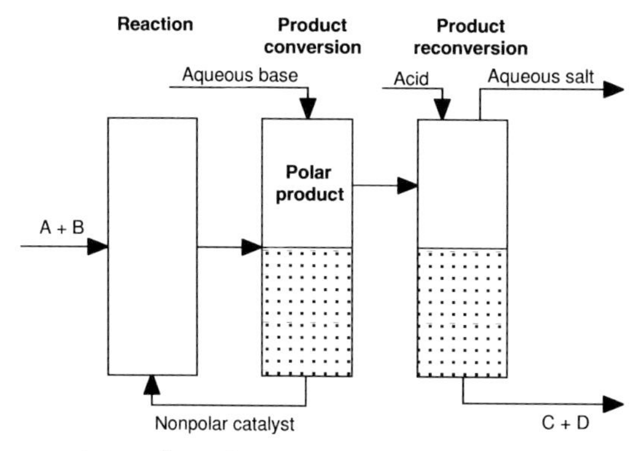

Figure 8. Successive reaction and product treatment.

sequence of three units comprises the reactor for the homogeneously catalyzed reaction, the separator of polar product and nonpolar catalyst phase, and the separator for nonpolar product and aqueous salt solution. An example for this case will be given in Section 6.12 (Fat Chemicals).

4.2.2.4 Reaction and Catalyst Separation

In Section 4.2.2.1 cases are described in which the catalyst stays in the reactor because of the liquid–liquid two-phase technique (LLTP). In other cases only one liquid phase occurs and attention is turned especially to the removal of the catalyst. This removal can be achieved by using the membrane technique shown in Figures 9 and 10. Special polyamide and polyimide membranes are able to separate transition metal complexes from smaller organic molecules [71–73]. The separation by membrane will be most effective if the difference in size between complex and organic molecules is as large as possible. Therefore, this

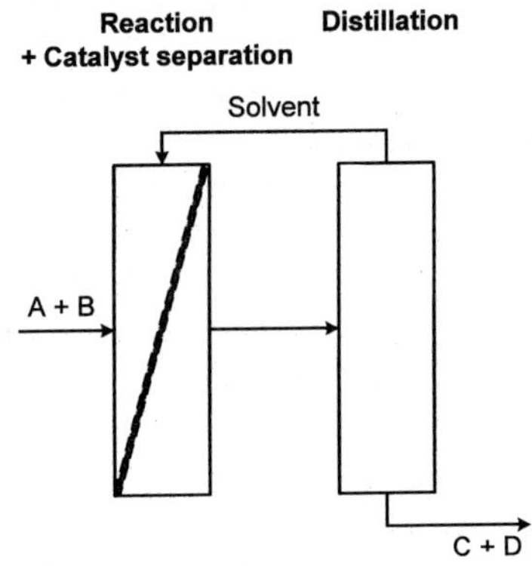

Figure 9. Simultaneous reaction and catalyst separation (via membrane).

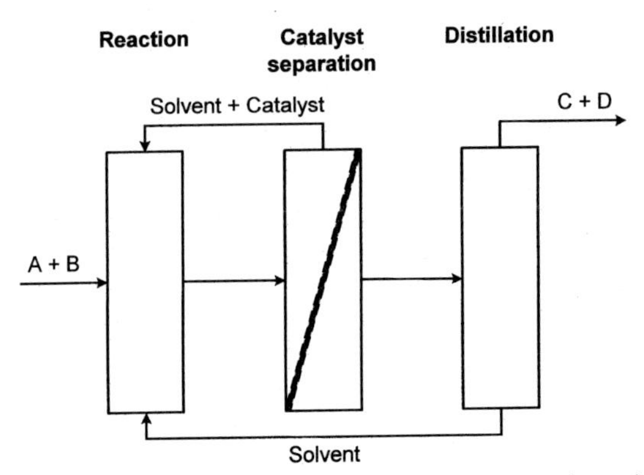

Figure 10. Successive reaction and catalyst separation (via membrane).

technique is particularly useful if the reaction is catalyzed by complexes with macromolecular ligands [74] (cf. Section 4.4).

In Figure 9 the simultaneous reaction and catalyst separation by membrane has been described. In one part of the reactor a catalyst is dissolved which cannot pass the membrane which is installed in the reactor. Here the starting chemicals A and B form the products C and D in a homogeneous catalyst solution. The products are able to pass through the membrane, perhaps together with a certain amount of the solvent. In the second unit, the distillation step, this solvent is recycled to the reactor and the products are isolated at the bottom of the distillation column.

Figure 10 shows a similar arrangement, but reaction and catalyst separation now occur in different units. The reaction is succeeded by the membrane separation step, which is then followed by the distillation unit. The products C and D leave the distillation at the top of the column and the solvent leaves at the bottom. Of course, the contrary is also conceivable, depending of the boiling points of products and solvent.

Besides catalyst separation by membranes, there is also the possibility of separating the catalyst via filtration after the catalyst has been precipitated. One interesting application is given by Fell and co-workers [83]. They describe the hydroformylation of higher-molecular-mass olefins with a rhodium/phenylsulfonatoalkylphosphine catalyst which is soluble both in methanol and in water. Figure 11 shows the principle of the separation: the reaction is first carried out in homogeneous methanol solution. After distillation of the methanol, the rhodium catalyst is precipitated and filtered off (or centrifuged) in a third unit. The products C and D, in the given example the higher-molecular-mass aldehydes, pass the filter. In a fourth unit the catalyst solution is prepared again by dissolving the solid catalyst in the solvent methanol.

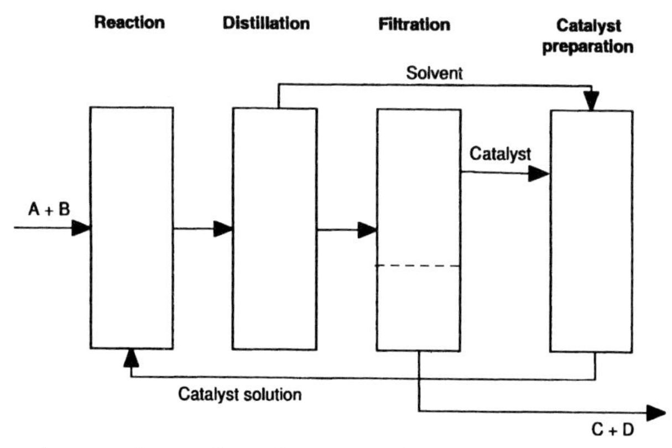

Figure 11. Successive reaction and catalyst separation (via solvent distillation and catalyst filtration).

4.2.2.5 Reaction and Catalyst Extraction

In Section 4.2.2.2 some processes are described in which the organic products are separated – simultaneously or successively – by extraction with a nonpolar extractant. In the cases presented in this section, not the product but the catalyst will be extracted and recycled to the reactor. Theoretically, this can be done simultaneously in the reactor, as shown in Figure 12. However, this arrangement is highly unfavorable, because it means that the catalyst is taken away from the reaction medium during the catalytic conversion. The successive variant, shown in Figure 13, makes much more sense: the reaction is first carried out in the reaction unit, then the catalyst is extracted by an extractant which does not dissolve the products C and D. Thus, the products can be removed, and the extractant and the catalyst solution are separated in a distillation step.

The crucial point of this method is the complete separation of the extractant (containing the catalyst) and the products. One example, where this problem could be solved, is shown in Figure 14, the hydroformylation process of higher olefins proposed by Union Carbide at the beginning of the 1990s [80–82]. The

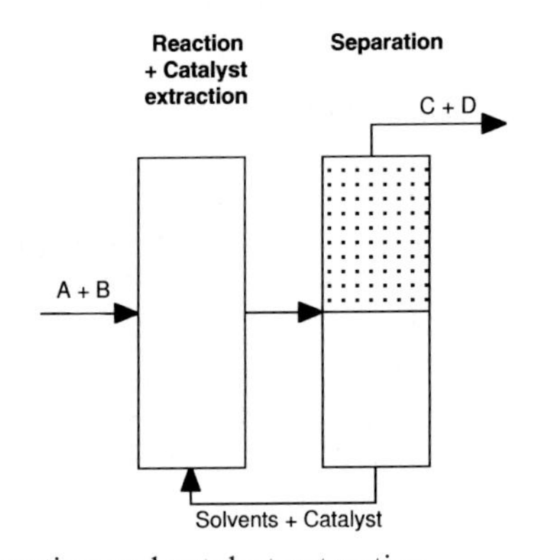

Figure 12. Simultaneous reaction and catalyst extraction.

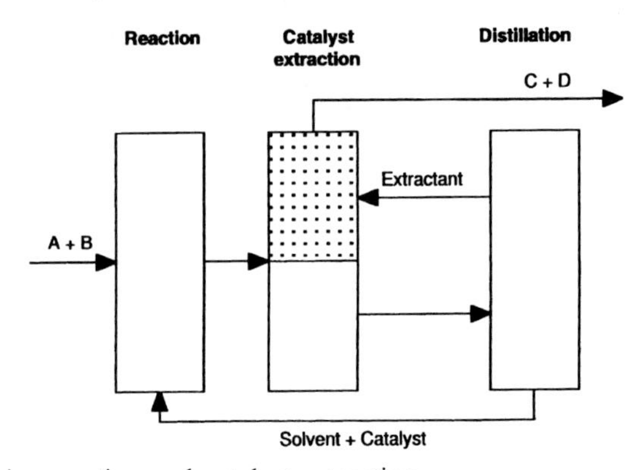

Figure 13. Successive reaction and catalyst extraction.

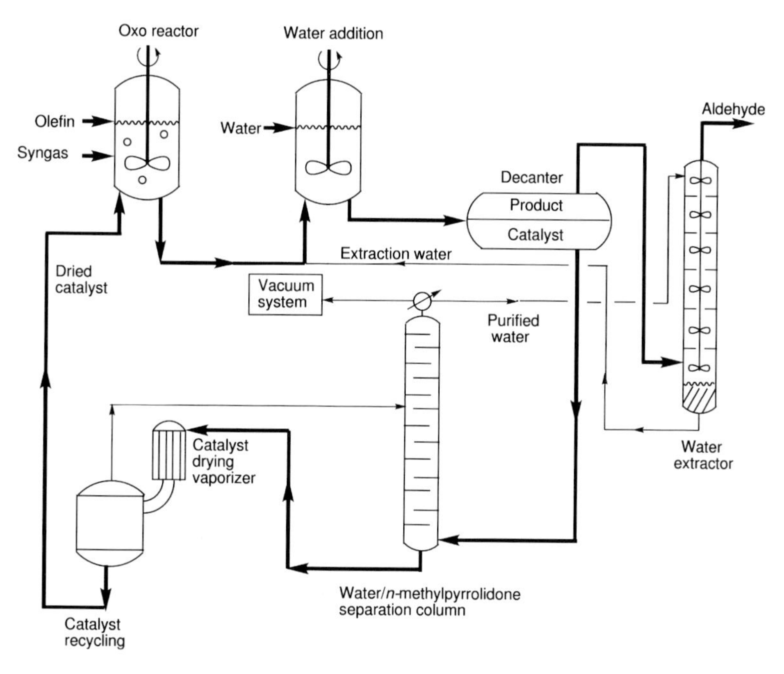

Figure 14. Hydroformylation of higher olefins [81].

reaction is carried out as a homogeneous catalytic process using the solvent *N*-methylpyrrolidone (NMP). The catalyst is then extracted by addition of water: the rhodium catalyst containing the water-soluble ligand monosulfonated triphenylphosphine (TPPMS) goes almost completely into the NMP/water phase, and the nonpolar aldehydes form the second phase. In this way C_7- and C_{15}-aldehydes may be produced with the final effluent being lower than 20 ppb in rhodium. However, the expenditure to regenerate the catalyst solution is

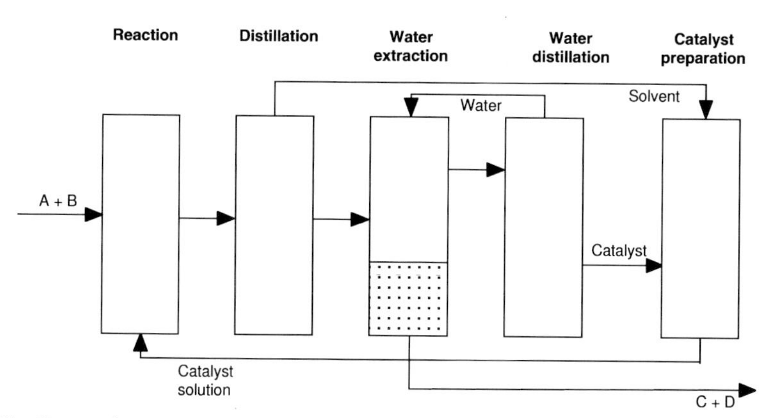

Figure 15. Successive reaction and catalyst extraction (after solvent distillation).

considerable: first, the water is partly distilled from the catalyst solution in vacuo, then a further catalyst-drying vaporizer is required until the catalyst can be recycled to the oxo reactor.

Another way to hydroformylate higher olefins and to extract the catalyst afterwards is described by Fell and co-workers [83–85]. They use a similar method to that described in Figure 11, but the catalyst is not precipitated and filtered off, but it is extracted with water.

The general flow scheme for this alternative is shown in Figure 15. After the homogeneous reaction, catalyzed for instance by a rhodium catalyst containing triphenylphosphine monosulfonic acid as complex ligand, the solubilizer methanol is distilled off. The catalyst system now becomes insoluble and is separated by extraction with water in the third unit. The products C and D, in this case the aldehydes, can be separated as the second liquid phase. After evaporation of the aqueous catalyst solution to dryness (unit 4) the catalyst is dissolved in the solvent methanol for a new reaction step (unit 5).

4.2.2.6 Reaction and Catalyst Treatment

In Section 4.2.2.3 examples are given where the product is treated with bases and acids to enable the separation of the product. In this section analogous cases will be considered in which, however, the catalyst is treated in such a way that it can be removed and recycled.

The principal flow scheme of a simultaneous reaction and catalyst treatment is shown in Figure 16. However, once again this technique seems not to be favorable, because during the conversion the nonpolar catalyst is removed from the organic reaction phase by the addition of a base.

It makes more sense to separate the two units as shown in Figure 17, the successive reaction and catalyst treatment. First, the reaction is carried out as

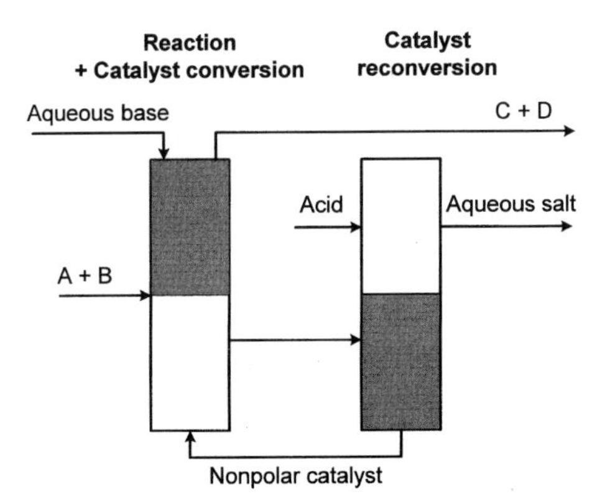

Figure 16. Simultaneous reaction and catalyst treatment.

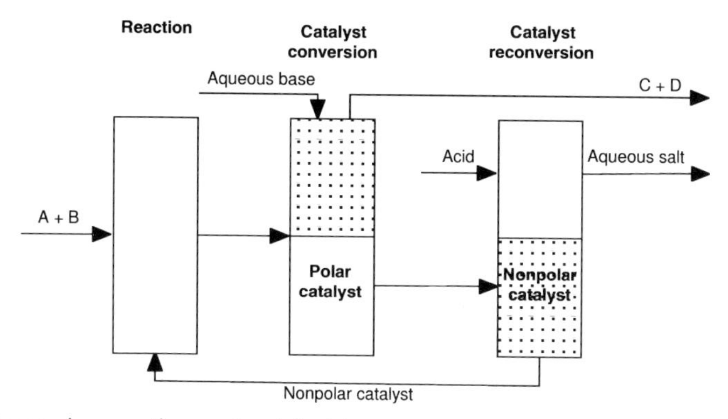

Figure 17. Successive reaction and catalyst treatment.

a homogeneously catalyzed conversion using a nonpolar solvent and a nonpolar catalyst. In the second unit, the nonpolar catalyst is converted into a polar one by adding an aqueous base. Finally, in the third unit, the polar catalyst is reconverted into a nonpolar one which can be recycled to the reactor. The main question remains, of how a catalyst can be converted from nonpolar to polar and vice versa. The answer is to use metal complexes with anionic ligands which contain a cationic counterion such as ammonium. When it has some higher-molecular-mass alkyl substituents, both this alkylammonium cation as well as the whole complex will be soluble in a nonpolar solvent such as toluene. If after the reaction an aqueous base such as a sodium hydroxide solution is added in excess, the ammonium counterion will be exchanged by the sodium cation, thus yielding a water-soluble complex (cf. Section 6.1.5). Of course, this process can be reversed by adding an acid and new ammonium ions.

4.2.3 Reaction Engineering Aspects

The reaction engineering aspects of liquid–liquid reactions have been well studied ([86–88], cf. Section 4.1). The performance of these reactions depends on the hydrodynamics of the dispersion, the mixing of the two fluid phases, the interface mass-transfer steps, the phase equilibria and kinetics of the reactions involved.

Some of the examples presented in Section 4.2.2 are, however, not only restricted to the two liquid phases discussed, but contain in addition a further gas phase [89]. For instance, in hydroformylation a water-gas phase exists, in oligomerizations an ethylene phase and in telomerizations a butadiene phase. For these gas–liquid–liquid systems there is so far only a limited amount of published information on the reaction engineering aspects [90]. One exception is the study of 1-octene hydrogenation using a rhodium/TPPTS catalyst [91], in which both thermodynamics and kinetics have been investigated in some detail. The same group also studied the hydroformylation of 1-octene [92].

Some further special technical aspects should be mentioned. The intensive mixture of the two liquid phases is an important condition for obtaining high reaction rates. This mixing can be achieved in bubble columns, tray columns or in stirred-tank reactors. In the few publications on industrially realized two-phase reactions the stirred tank reactor is always cited, but without detailed information on the stirring device. One further possible way to increase the mass transfer between the two liquid phases is by the influence of sonification. Cornils et al. applied this technique in the hydroformylation of hexene or diisobutene and found a considerable increase in the turnover numbers [93]. Another possibility for increasing the mass transfer may be by the use of microemulsions and micellar systems [94], which can be reached by addition of certain surfactants. This aspect is discussed in Sections 3.2.4 and 4.5. The separation of catalyst compounds in two-phase systems in combination with membranes has been studied recently by Müller and Bahrmann [95].

References

[1] Shell International Research (W. Keim, T. M. Shryne, R. S. Bauer, H. Chung, P. W. Glockner, H. van Zwet), DE-P 2.054.009 (1969).
[2] Shell Development Co. (W. Keim, H. van Zwet, R. Bauer), US 3.644.564 (1972).
[3] W. Keim, F. H. Kowaldt, R. Goddard, C. Krüger, *Angew. Chem.* **1978**, *90*, 493; *Angew. Chem., Int. Ed. Engl.* **1978**, *17*, 466.
[4] W. Keim, A. Behr, B. Limbäcker, C. Krüger, *Angew. Chem.* **1983**, *95*, 505; *Angew. Chem., Int. Ed. Engl.* **1983**, *22*, 503.
[5] W. Keim, *Chem.-Ing.-Tech.* **1984**, *56*, 850.
[6] A. Behr, W. Keim, *Arabian J. Sci. Eng.* **1985**, *10*, 377.
[7] W. Keim, A. Behr, B. Gruber, B. Hoffmann, F. H. Kowaldt, U. Kürschner, B. Limbäcker, F. P. Sistig, *Organometallics* **1986**, *5*, 2356.
[8] W. Keim, *New J. Chem.* **1987**, *11*, 531.
[9] W. Keim, *J. Mol. Catal.* **1989**, *52*, 19.
[10] Rhône-Poulenc (E. G. Kuntz), DE 2.627.354 (1976).
[11] B. Cornils, J. Falbe, *Proc. 4th Int. Symp. Homogeneous Catalysis, Leningrad* **1984**, 487.
[12] Ruhrchemie A.G. (B. Cornils, W. Konkol, H. Bach, G. Dämbkes, W. Gick, E. Wiebus, H. Bahrmann), DE 3.415.968 A1 (1984).
[13] Ruhrchemie A.G. (B. Cornils, W. Konkol, H. Bach, W. Gick, E. Wiebus, H. Bahrmann, H.-D. Hahn), DE 3.546.123 A1 (1985).
[14] E. G. Kuntz, *CHEMTECH* **1987**, 570.
[15] E. Wiebus, B. Cornils, *Chem.Ing.Tech.* **1994**, *66*, 961.
[16] B. Cornils, E. Wiebus, *Recl. Trav. Chim. Pays-Bas* **1996**, *115*, 211.
[17] B. Cornils, E. Wiebus, *CHEMTECH* **1995**, 33.
[18] B. Cornils, E. G. Kuntz, *J. Organometal. Chem.* **1995**, *502*, 177.
[19] E. Wiebus, B. Cornils, *Hydrocarbon Process.* March **1996**, 63.
[20] B. Cornils, W. A. Herrmann, R. W. Eckl, *J. Mol. Catal. A: Chem.* **1997**, *116*, 27.

[21] Rhône-Poulenc, EP 44.771 (1980), EP 441.708 (1991).
[22] C. Mercier, G. Mignani, M. Aufrand, G. Allmang, *Tetrahedron Lett.* **1991**, *32*, 1433.
[23] C. Mercier, P. Chabardes, *Pure Appl. Chem.* **1994**, *66*, 1509.
[24] Rhône-Poulenc, EP 362.037 (1987), EP 320.339 (1987).
[25] J. M. Grosselin, C. Mercier, *J. Mol. Catal.* **1990**, *63*, L26.
[26] G. Allmang, F. Grass, J. M. Grosselin, C. Mercier, *J. Mol. Catal.* **1991**, *66*, L27.
[27] J. M. Grosselin, C. Mercier, *Organometallics* **1991**, *10*, 2126.
[28] DuPont, *Proc. 10th Int. Symp. Homogeneous Catalysis, Princeton, USA,* August **1996**.
[29] K. E. Atkins, W. E. Walker, R. M. Manyik, *Chem. Commun.* **1971**, 330.
[30] K. E. Atkins, R. M. Manyik, G. L. O'Connor, GB 1.307.101 (1970), DE 2.018.054 (1970), FR 2.045.369 (1970), US 816.792 (1969).
[31] ICI (W. T. Dent), GB 1.354.507 (1970).
[32] Esso (M. G. Romanelli, R. J. Kelly), DE 2.011.163 (1970).
[33] Esso (M. G. Romanelli), US 3.670.032 (1972).
[34] Toray Ind. (T. Mitsuyasu, J. Tsuji), J 73 78.107 (1973).
[35] Kureha Chem. Ind. (S. Enomoto, H. Takita, S. Wada, Y. Mukaida, M. Yanaka), JP 49/35.603 B4 (1974).
[36] Toray Ind. (T. Tsuji, T. Mitsuyasu), JP 49/35.603 B4 (1974).
[37] Mitsubishi Chem. Ind. (T. Onoda, H. Wada, T. Sato, Y. Kasori), JP 53/147.013 A2 (1978).
[38] Kuraray Ind. (N. Yoshimura, M. Tamura), GB 2.074.156 (1981).
[39] Kuraray Ind. (N. Yoshimura, M. Tamura), FR 82 02.520 (1982).
[40] Kuraray Ind. (N. Yoshimura, M. Tamura), US 4.356.333 (1982).
[41] Kuraray Ind. (N. Yoshimura, M. Tamura), US 4.417.079 (1983).
[42] Kuraray Ind. (T. Maeda, Y. Tokitoh, N. Yoshimura), US 4.927.960 (1990), US 4.992.609 (1991), US 5.100.854 (1991).
[43] Kuraray Ind. (Y. Tokitoh, N. Yoshimura), US 5.057.631 (1991).
[44] Kuraray Ind. (Y. Tokitoh, T. Higashi, K. Hino, M. Murasawa, N. Yoshimura), US 5.118.885 (1992).
[45] W. Keim, A. Behr, H. Rzehak, *Tenside Detergents* **1979**, *16*, 113.
[46] A. Behr, *Aspects of Homogeneous Catalysis,* Vol. 5 (Ed.: R. Ugo), Reidel, Dordrecht **1984**, pp. 3–73.
[47] A. Behr, *Industrial Application of Homogeneous Catalysis* (Eds.: A. Mortreux, F. Petit), Reidel, Dordrecht, **1988**, pp. 141–175.
[48] A. Behr, T. Fischer (in preparation).
[49] A. Behr, K.-D. Juszak, W. Keim, *Synthesis* **1983**, *7*, 574.
[50] A. Behr, K.-D. Juszak, *J. Organometal. Chem.* **1983**, *255*, 263–268.
[51] A. Behr, R. He, K.-D. Juszak, C. Krüger, Y.-H. Tsay, *Chem. Ber.* **1986**, *119*, 991.
[52] A. Behr, *Angew. Chem.* **1988**, *100*, 681; *Angew. Chem., Int. Ed. Engl.* **1988**, *27*, 661.
[53] A. Behr, *Carbon Dioxide Activation by Metal Complexes,* VCH, Weinheim, **1988**.
[54] A. Behr, *Aspects of Homogeneous Catalysis,* Vol. 6 (Ed.: R. Ugo), Reidel, Dordrecht, **1988**, pp. 59–96.
[55] A. Behr, M. Heite, in preparation.
[56] J. Tsuji, M. Takahashi, *J. Mol. Catal.* **1981**, *10,* 107.
[57] T. Prinz, W. Keim, B. Drießen-Hölscher, *Angew. Chem.* **1996**, *108*, 1835; *Angew. Chem., Int. Ed. Engl.* **1996**, *35*, 1708.
[58] A. Durocher, W. Keim, P. Voncken, *Erdöl und Kohle, Erdgas, Petrochemie, Compendium* **1975/76**, pp. 347–354.
[59] W. Keim, A. Durocher, P. Voncken, *Erdöl, Kohle, Erdgas, Petrochem.* **1976**, *29*, 31.
[60] A. Behr, W. Keim, *Erdöl, Erdgas, Kohle* **1987**, *103*, 126.
[61] I. T. Horváth, *Catal. Lett.* **1990**, *6*, 43.

[62] J. P. Arhancet, M. E. Davis, J. S. Merola, B. E. Hanson, *J. Catal.* **1990**, *121*, 327.
[63] M. E. Davis, *CHEMTECH* **1992**, *22*, 498.
[64] J. P. Arhancet, M. E. Davis, B. E. Hanson, *Catal. Lett.* **1991**, *11*, 129.
[65] J. P. Arhancet, M. E. Davis, B. E. Hanson, *J. Catal.* **1991**, *129*, 94.
[66] J. P. Arhancet, M. E. Davis, B. E. Hanson, *J. Catal.* **1991**, *129*, 100.
[67] J. P. Arhancet, M. E. Davis, J. S. Merola, B. E. Hanson, *Nature (London)* **1989**, *339*, 454.
[68] Virginia Tech. Intellectual Properties Inc. (M. E. Davis, J. P. Arhancet, B. E. Hanson) EP 372.615 (1990).
[69] R. T. Smith, R. K. Ungar, L. J. Sanderson, M. C. Baird, *Organometallics* **1983**, *2*, 1138.
[70] R. T. Smith, R. K. Ungar, M. C. Baird, *Transition Metal. Chem.* **1982**, *7*, 288.
[71] J. E. Ellis, GB 1.312.076 (1973).
[72] L. W. Gosser, US 3.853.754 (1974).
[73] L. W. Gosser, W. H. Knoth, G. W. Parshall, *J. Mol. Catal.* **1977**, *2*, 253.
[74] E. Bayer, V. Schurig, *Angew. Chem., Int. Ed. Engl.* **1975**, *14*, 493.
[75] E. Monflier, P. Bourdauducq, J.-L. Couturier, J. Kervennal, I. Suisse, J. A. Mortreux, *Catal. Lett.* **1995**, *34*, 201.
[76] E. Monflier, P. Bourdauducq, J.-L. Couturier, J. Kervennal, J. A. Mortreux, *Appl. Catal. A* **1995**, *131*, 167.
[77] E. Monflier, P. Bourdauducq, J.-L. Couturier, J. Kervennal, I. Suisse, J. A. Mortreux, *J. Mol. Catal. A* **1995**, *97*, 29.
[78] E. Monflier, P. Bourdauducq, J.-L. Couturier, FR 9.403.897 (1994).
[79] E. Monflier, P. Bourdauducq, J.-L. Couturier, US 5.345.007 (1994).
[80] J. Haggin, *Chem. Eng. News* April 17, **1995**, 25.
[81] A. G. Abatjoglou, 209th ACS National Meeting, Anaheim, USA, **1995**.
[82] Union Carbide Chem. (A. G. Abatjoglou, D. R. Bryant, R. R. Peterson), EP 350.922 (1990).
[83] Z. Xia, B. Fell, *J. Prakt. Chem.* **1997**, *339*, 140.
[84] S. Kanagasabapathy, Z. Xia, G. Papadogianakis, B. Fell, *J. Prakt. Chem.* **1995**, *337*, 446.
[85] Z. Xia, Ch. Schobben, B. Fell, *Fett/Lipid* **1996**, *98*, 393.
[86] G. Astarita, *Mass Transfer with Chemical Reaction*, Elsevier, Amsterdam, **1967**.
[87] L. L. Tavlarides, M. Stamatoudis, *Adv. Chem. Eng.* **1981**, *11*, 199.
[88] L. K. Doraiswamy, M. M. Sharma, *Heterogeneous Reactions,* Vol. 2, John Wiley, New York, **1984**.
[89] R. V. Chaudhari, A. Bhattacharya, B. M. Bhanage, *Catal. Today* **1995**, *24*, 123.
[90] D. Wachsen, K. Himmler, B. Cornils, *Catal. Today* **1998**, in press.
[91] I. Hablot, J. Jenck, G. Casamatta, H. Delmas, *Chem. Eng. Sci.* **1992**, *47*, 2689.
[92] P. Purwanto, H. Delmas, *Catal. Today* **1995**, *24*, 135.
[93] Ruhrchemie (B. Cornils, H. Bahrmann, W. Lipps, W. Konkol), DE 35 11.428 (1985).
[94] Enichem S.p.A. (L. Tinucci, E. Platone) EP 0.380.154 (1990).
[95] Th. Müller, H. Bahrmann, *J. Mol. Catal. A: Chem.* **1997**, *116*, 39.

4.3 Side Effects, Solvents, and Co-solvents

Brian E. Hanson

4.3.1 Introduction

The concept of biphasic catalysis requires that the catalyst and product phases separate rapidly to achieve a practical approach to the recovery and recycling of the catalyst. It is obvious that simple aqueous/hydrocarbon systems form two phases under nearly all operating conditions and thus provide rapid product–catalyst separation. Ultimately, however, the application of water-soluble catalysts is limited to low-molecular-mass substrates which have appreciable water-solubility. The problem is illustrated by the data in Table 1, which gives the solubility of some simple alkenes in water at room temperature [1]. Although hydrocarbon (alkene)-solubility in water increases at higher temperature, most alkenes do not have sufficient solubility to give practical reaction rates in catalytic applications. The addition of salts further decreases the solubility of hydrocarbons in water. Substrate solubility in water is a significant issue and it is no accident that so-far the practiced and proposed commercial applications of water-soluble catalysts for hydroformylation are limited to propene and butene.

Table 1. Alkene solubility in water at 298 K [1].

Alkene	Solubility [ppm]
1-Pentene	148
1-Hexene	50
2-Heptene	15
1-Octene	2.7[a]
2-Octene	2.4[a]
1-Decene	0.6[a]

[a] Data in 0.001 M HNO_3.

The addition of co-solvents to the aqueous phase has been investigated extensively as a means of improving the solubility of higher olefin substrates in the catalyst-containing phase. Although data are difficult to find for the appropriate ternary systems, the effect on substrate solubility of a simple additive such as methanol is illustrated by the solubility of hydrocarbons in methanol alone.

Table 2. Distribution of methanol and hydrocarbon (mole fraction) between the methanol-rich phase and the hydrocarbon-rich phase.

Hydrocarbon	T [K]	$X_{methanol}$ (in olefin-rich phase)	$X_{hydrocarbon}$ (in methanol-rich phase)
Hexane	287	0.215	0.188
Heptane	295	0.147	0.095
1-Heptene	295	0.140	0.142
Octane	300	0.136	0.063
Decane	295	0.090	0.028
1-dodecene	274	Not reported	0.013

Methanol forms a biphasic system with many simple hydrocarbons, as seen in Table 2. However, the hydrocarbon solubilities in methanol are orders of magnitude greater than in water alone [2]. The effect of methanol addition to aqueous solutions of hydrocarbons is then to increase the solubility of the hydrocarbon.

4.3.2 Hydroformylation

Biphasic reaction conditions can be achieved within a wide range of operating conditions with respect to co-solvents. The most common co-solvents are the lower alcohols; the purpose is to improve substrate solubility and as a consequence to increase reaction rate. Recent work with ethanol as a co-solvent shows that this is very effective at improving reaction rates [3]. It is estimated for example that the solubility of 1-octene in a 50:50 mixture of ethanol and water is 10^4 times greater than in water alone [3]. In a comparison of several co-solvents – ethanol, methanol, acetone, and acetonitrile – it was found that ethanol was the most effective at improving reaction rates in the two-phase hydroformylation of 1-octene [4]. Generally, though, the use of co-solvents in hydroformylation reactions with Rh/TPPTS catalysts is not advisable, because of diminished reaction selectivity and the possibility of acetal formation (see below).

When the alkali-metal cation of a sulfonated phosphine or phosphite is replaced by ammonium ion it is possible to devise applications in nonaqueous solvents. Alkylammonium salts of TPPTS were reported in the first patents of Kuntz [5] and catalytic examples were given in which water alone was the solvent. More recently Fell et al. prepared a series of sulfonated phosphites including the example shown below, **1** [6], and examined these in a variety of solvents for the hydroformylation of tetradecene. Catalytic results are reported with acetone and acetophenone as the solvents; the yield of linear aldehydes among all aldehyde products is as high as 88 %.

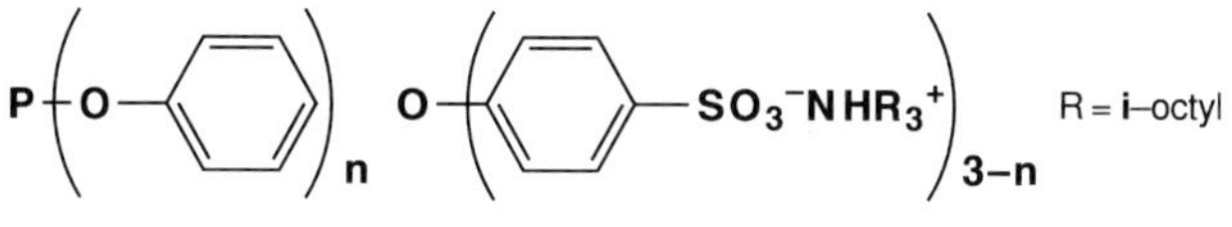

R = i–octyl

1

Chelating biphosphites (cf. Section 3.2.2) which apparently are hydrolytically stable, were also reported by Fell et al. [6]. Similar ionic phosphites were reported by Abatjoglou and Bryant; an example is shown below, compound **2** [7]. The bisphosphites are reported to be hydrolytically stable, but the applications shown for the ionic ligands are limited to nonaqueous solvents such as 2,2,4-trimethyl-1,3-pentanediol and poly(ethylene glycol) [7].

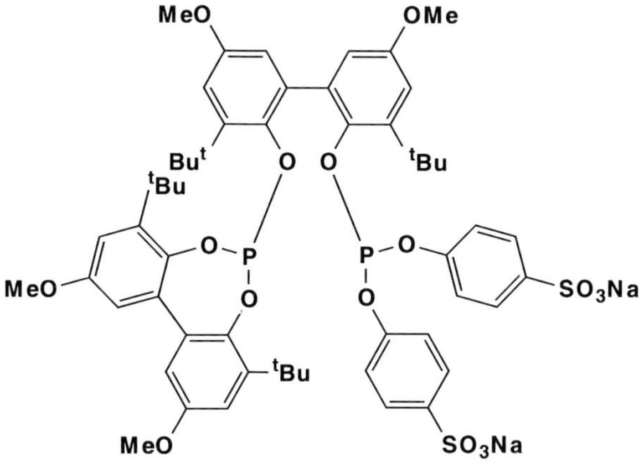

2

Monosulfonated triarylphosphines, for which TPPMS is the prototypical example, have good solubility in polar organic solvents. A variety of monosulfonated phosphines, including TPPMS, with several different metal cations as the counterion have been examined in the hydroformylation of propene, butene, and octene [8]. The solvent composition is reported to be a mixture of texanol and carbowax. The catalysts showed excellent selectivity to linear aldehydes, up to 96% depending on the reaction conditions [8].

Fell et al. describe the hydroformylation of tetradecene in methanol alone as the solvent. Since the standard ligand for two-phase catalysis, the sodium salt of TPPTS, is not soluble in methanol, the authors used the ligand $Ph_2P(CH_2)_4SO_3Na$. Good activities were observed although reaction selectivity to linear products was typically 70% and a relatively high proportion of alcohol products was observed [9]. Acetal formation is a possible side reaction when alcohols are used as the solvent for hydroformylation catalysis. The authors report that this side reaction is minimized when the solution is made alkaline by the addition of a strong base such as LiOH [9].

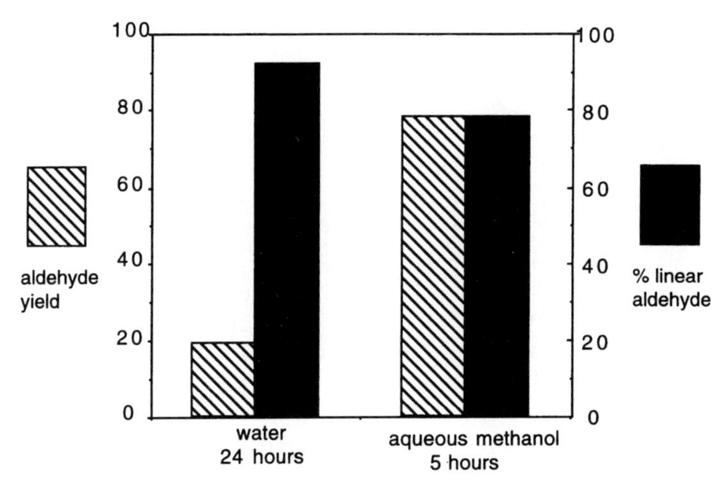

Figure 1. Comparison of 1-octene hydroformylation in water and aqueous methanol. The reaction time in water was 24 h; the reaction time in 50 % aqueous methanol was 5 h [10, 11].

As noted above, the addition of alcohols improves the reaction activity in the two-phase hydroformylation of higher olefins with Rh/TPPTS catalysts. Reaction selectivity, however, is diminished if the operating conditions are otherwise similar. This is illustrated in Figure 1, which compares the conversion and selectivity of 1-octene hydroformylation over the (acac)Rh$(CO)_2$/TPPTS catalyst in water alone and 50 % aqueous methanol as the solvent [10, 11]. Selectivity can be improved in the aqueous methanol system when ligands other than TPPTS are used, as for example with ionic phosphites of the type **1** cited above. These ligands differ significantly from TPPTS in that they are expected to be amphiphilic in character; the ionic portion is hydrophilic while the rest of the ligand is relatively hydrophobic.

A similar effect on activity and selectivity is observed upon addition of co-solvents to the dimeric rhodium hydroformylation catalyst, $[Ru_2(\mu\text{-S-}^t\text{Bu})_2(CO)_2(\text{TPPTS})_2]$ [12]. Specifically, there is an inverse relationship between solvent solvophobicity [13] and the log of the *n*/iso ratio of the aldehyde products for the hydroformylation of 1-octene. The catalyst is most selective to linear products in water.

In general, amphiphilic phosphines show both better activity and selectivity than TPPTS when compared in aqueous/solvent mixtures as the reaction medium. The series of phosphines, $P[(C_6H_4)(CH_2)_n(C_6H_4SO_3Na)]_3$, $n = 3, 6, 10$ has been compared with TPPTS for the hydroformylation of 1-octene in aqueous methanol [11, 14]. In the schematic representation of the ligand with $n = 10$ (compound **3**), hydrogen atoms are omitted for clarity.

Phosphine **3** has good solubility in both water and methanol. A comparison of **3** and TPPTS for the rhodium-catalyzed hydroformylation of 1-octene in aqueous methanol is shown in Figure 2 [14]. At all ligand/rhodium ratios studied, the amphiphilic phosphine shows both better activity and selectivity than

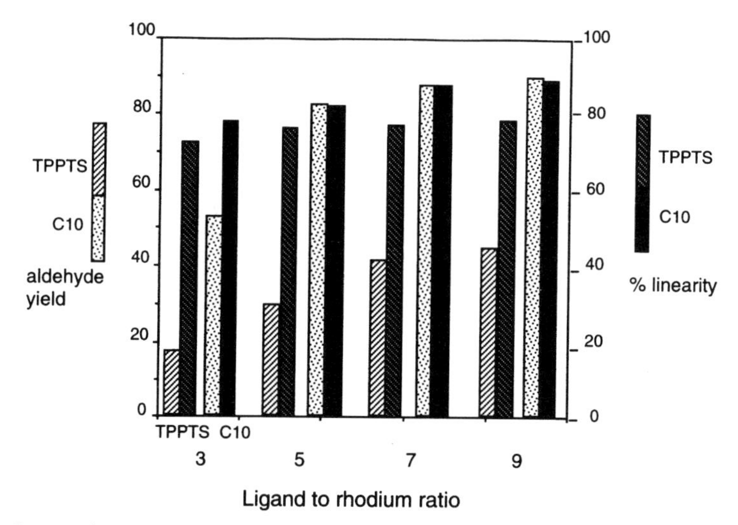

Figure 2. Comparison of TPPTS and C10-TS, **3**, for 1-octene hydroformylation under two-phase reaction conditions [14].

P
SO_3Na
SO_3Na
SO_3Na
3

TPPTS. The improved selectivity is attributed to diminished inter- and intraligand ionic repulsions with the greater distances between sulfonate groups in complexes of the amphiphilic ligands.

Aqueous-phase hydroformylation is dependent on both the pH of the aqueous solution and the ionic strength of the solution. Typical reaction conditions reported in patents show rhodium concentrations of approximately 0.05 M with the appropriate concentration of TPPTS to give the desired ligand/rhodium ratio. Buffers are typically added to control the pH. In the original patents the pH was buffered to approximately 7 [5]. As result of the added salts, the aqueous medium for catalysis has a very high solution ionic strength ("salt effect"). High solution ionic strength may play a role in stabilizing Rh/TPPTS complexes. In particular, $HRh(CO)(TPPTS)_3$ has a net charge of -9 due to the sulfonate groups on the TPPTS ligands. Stabilization of the ionic charge in this complex and in catalytic intermediates may contribute to the high selectivity of aqueous-phase catalysts toward linear products [15, 16]. This is consistent with the trend in selectivity with solvent solvophobicity noted above [12].

In more recent patents, examples are given that demonstrate that the formation of aldol condensation products and heavy ends is pH-dependent; specifically, condensation products are inhibited at relatively low pH. For example, propene hydroformylation at a reaction pH of 5.9 leads to a total of 1.2 wt.% C_8 and higher products. In contrast, when the reaction was repeated at a pH of 6.9, the C_8 and higher products formed from condensation reactions comprised 7.5 wt.% of the reaction mixture [17]. A higher proportion of linear products was observed at the lower pH as well. It has been noted also that the presence of CO_2 inhibits aldol condensation [18]. This may be a pH effect as well, since aqueous solutions of CO_2 are acidic.

4.3.3 Hydrogenations and Other Catalytic Reactions

Co-solvents also play a rate-enhancing role in two-phase hydrogenation reactions. In the two-phase hydrogenation of cyclohexene with a catalyst prepared in situ by the hydrogen reduction of $RhCl_3 \cdot 3H_2O$ in the presence of monosulfonated triphenylphosphine, TPPMS, the rate of hydrogenation was observed to be dependent on the co-solvent [19]. Lower alcohols, methanol plus ethanol, were superior to co-solvents such as acetone plus THF. At the co-solvent concentrations used, the system remained biphasic. The increase in activity was attributed to improved olefin solubility in the aqueous/co-solvent phase compared with water alone as the solvent. It was further observed that the addition of salts decreased the reaction rate, presumably by decreasing olefin solubility, while the addition of surfactants had little net effect on reaction rate.

In the deprotection of phenylacetic allyl esters over $Pd(OAc)_2$/TPPTS catalysts, it was noted that the reaction medium plays a role in determining reaction selectivity. The reaction is shown below in Eq. (1). When the reaction is performed homogeneously in CH_3CN/H_2O as the solvent, the allyl substituents with $R_1 = R_2 = H$; $R_1 = Ph$, $R_2 = H$; $R_1 = R_2 = Me$ are all substituted

[20, 21]; the first in the series is the fastest to react. In contrast, when the reaction takes place under biphasic conditions in C_3H_7CN/H_2O, only the simple allyl ester, $R_1 = R_2 = H$, is cleaved. When two allylic esters are incorporated into a single molecule the allyl ester can be selectively substituted under two-phase reaction conditions.

$$\text{PhCH}_2\text{C(O)O-CH}_2\text{CH=C(R}_1\text{)R}_2 \xrightarrow[\text{NEt}_2\text{H}]{\text{Pd(OAc)}_2\text{ / TPPTS}} \text{PhCH}_2\text{C(O)OH} + \text{Et}_2\text{N-CH}_2\text{CH=C(R}_1\text{)R}_2 \quad (1)$$

The alkylation of phenols with isoprene and isoprene derivatives can be catalyzed over rhodium complexes of TPPTS [22]. In the case where the isoprene derivative is myrcene, methanol is added to the reaction mixture. The apparent benefit is to increase reaction rates with the highly water-insoluble myrcene substrate. The alkylation of 2-naphthol with myrcene is shown in Eq. (2). In 2 h at 100 °C in aqueous methanol a 40% conversion of myrcene is observed (160 turnovers).

(2) [Scheme: 2-naphthol (OH) + myrcene → 1-alkylated 2-naphthol products (OH): 33% + 67%]

Similarly, the reaction of isoprenes with methyl acetylacetate can be accomplished over rhodium complexes of TPPTS [23]. The example with myrcene is shown in Eq. (3). In this case rates are also increased upon addition of methanol. For example, in water alone as the solvent 174 turnovers of myrcene are observed in 1 h at 90 °C, whereas 297 turnovers are observed under otherwise identical conditions in water/methanol (75:25). No difference in reaction selectivity is observed in the presence of methanol. In both cases the selectivity to the isomeric products shown in Eq. (3) is 99%.

[RhCl(cod)]$_2$
TPPTS, Na_2CO_3 (3)

References

[1] D. G. Shaw, M.-C. Hauclait-Pirson, G. T. Hefter, A. Maczynski (Eds.), *Hydrocarbons with Water and Sea Water*, IUPAC Solubility Data Series, Vols. 37 and 38, Pergamon Press, New York, **1989**.

[2] D. G. Shaw, A. Skrzecz, J. W. Lorimer, A. Maczynski (Eds.), *Alcohols and Hydrocarbons*, IUPAC Solubility Data Series, Vol. 56, Pergamon Press, New York, **1989**.

[3] P. Purwanto, H. Delmas, *Catal. Today* **1995**, *24*, 135.

[4] F. Monteil, R. Quéau, P. Kalck, *J. Organomet. Chem.* **1994**, *480*, 177.

[5] See, for example: Rhône-Poulenc (E. Kuntz), FR 2.314.910 (1975), US 4.248.802 (1981).

[6] B. Fell, G. Papadognianakis, W. Konkol, J. Weber, H. Bahrmann, *J. Prakt. Chem.* **1995**, *335*, 75.

[7] Union Carbide (A. G. Abatjoglou, D. R. Bryant), US 5.059.710 (1988).

[8] Union Carbide (A. G. Abatjoglou, D. R. Bryant), US 4.731.486 (1988).

[9] S. Kanagasabapathy, Z. Xia, G. Papadognianakis, B. Fell, *J. Prakt. Chem.* **1995**, *337*, 446.

[10] H. Ding, B. E. Hanson, T. E. Glass, *Inorg. Chim. Acta* **1995**, *229*, 329.

[11] H. Ding, B. E. Hanson, T. Bartik, B. Bartik, *Organometallics* **1994**, *13*, 3761.

[12] F. Monteil, R. Quéau, P. Kalck, *J. Organomet. Chem.* **1994**, *480*, 177.

[13] M. H. Abraham, P. L. Grellier, R. A. McGill, *J. Chem. Soc., Perkin Trans. II* **1988**, 339.

[14] H. Ding, B. E. Hanson, C. W. Kohlpaintner, US Patent Application Pending.

[15] H. Ding, B. E. Hanson, T. E. Glass, *Inorg. Chim. Acta* **1995**, *229*, 329.

[16] I. T. Horváth, R. V. Kastrup, A. A. Oswald, E. J. Mozeleski, *Catal. Lett.* **1989**, *2*, 85.

[17] Ruhrchemie AG (B. Cornils, W. Konkol, G. Dämbkes, W. Gick, W. Greb, E. Wiebus, H. Bahrmann), US 4.593.126 (1986).

[18] Ruhrchemie AG (B. Cornils, W. Konkol, H. Bach, G. Dämbkes, W. Gick, E. Wiebus, H. Bahrmann), DE 3.415.968 (1984).

[19] Y. Dror, J. Manassen, *J. Mol. Catal.* **1977**, *2*, 219.

[20] S. Lemaire-Audoire, M. Savignac, E. Blart, G. Pourcelot, J. P. Genêt, J.-M. Bernard, *Tetrahedron Lett.* **1994**, *35*, 8783.

[21] S. Lemaire-Audoire, M. Savignac, G. Pourcelot, J. P. Genêt, J.-M. Bernard, *J. Mol. Catal.* **1997**, *116*, 247.

[22] Rhône-Poulenc Industries (G. Mignani, D. Morel), US 4.594.460 (1986).

[23] Rhône-Poulenc Industries (D. Morel), US 4.460.786 (1984); FR 2.486.525 (1980).

4.4 Membrane Techniques

Helmut Bahrmann, Boy Cornils

Membranes, i.e., microporous separation layers in the form of thin films, can be used for selecting separation of substances (liquids) differing in particle sizes (e.g., colloids) and, under some circumstances, in molecular weight. For homogeneous catalysis, this means the possibility of separating reactants and reaction products from the catalysts (and possibly also their decomposition and aging products) [1]. In this respect, membrane techniques are employed simply as a substitute for the phase separation step in homogeneous aqueous catalysis. In conventional homogeneous catalysis, membrane and membrane processes are regarded as alternatives to techniques for the otherwise inherently difficult separation of catalyst and products, e.g., immobilizing the homogeneous species by "anchoring" it to supports ("heterogenization") [2, 3]. In this context the term "membrane reactor" refers to the fundamental part of this device: the membrane, acting either as a catalyst or a physical barrier. Thus, the following different applications may be distinguished:

- as a catalytically active membrane (catalytically active sites incorporated in or on the surface of membranes) – this describes catalytic membrane reactors (CMRs) [4];
- as "selective" membranes which allow a selective removal of product from the catalyst–reactant/product mixture;
- a membrane through which reactants and products can pass while the dissolved homogeneous catalyst is retained [2].

"Liquid membranes" [5] are in the transition to micellar systems (cf. Section 4.5).

In most cases the membrane material serves as a classification for reviews and surveys [6–9]. The "selectivity" of membranes is based on the pore size and various properties of the membrane material such as hydrophilicity or hydrophobicity, thus giving the ability, depending on the particle size, to recover soluble catalysts either by reverse osmosis or by ultrafiltration. In the case of ultrafiltration, the molecular weight of the catalyst has to be increased by suitable methods such as derivatization (cf. Section 6.1.5, re-immobilization [10, 11]) or coupling to soluble polymers (Section 7.5) [1]. Examples of the combination of various homogeneous catalytic processes and membrane techniques are compiled in [2, 11, 12]. In all cases, membrane separation of catalyst from product or of catalyst from reactant + product offers a gentle mechanical method without any thermal or chemical stress due to, for example, distillation or persistent effects of chemicals.

This is exactly the overall target when membrane techniques are combined with two-phase operations. However, since the basic task of separating catalyst from product has already be achieved by the phase separation (decantation) step in the two-phase processes, additional membrane processes can enhance the further work-up of the homogeneous catalyst, especially continuous, simultaneous, and in-situ workup ("makeup"). Membrane processes are particularly suitable here for separating the intact complex catalyst from its components (such as precursors, ligands, decomposition products (e.g., [13]), catalyst poisons) or accompanying substances (e.g., high-boiling condensation products, solvents, additives). Catalysts of the biphasic reaction can also be chemically modified (by coupling, introducing other cations, or tenside ligands [14], etc.; cf. Section 6.1.5) to alter their molecular weight and thus improve their ease of separation using a given membrane. In addition, Exxon patents (e.g., [15]) refer to the suitability of the membrane processes for removing the water-soluble catalyst in the presence of surfactants, which are reported to increase the reaction rate of reactants that are sparingly soluble in water (cf. Section 6.1.3.2).

The principal task of membrane separation components within two-phase aqueous homogeneous catalytic processes is therefore to perform fine separation of undesirable constituents of the catalyst or substances accompanying the catalyst, while the phase decantation has already performed the coarse separation of catalyst from product (or residual starting material) [16]. It is therefore generally useful to place the membrane separation processes in a catalyst recycling sidestream, and parallel to the main stream, as the schematic diagram in Figure 1, variant 1.1 shows. This means that the membrane treats only a portion of the entire catalyst stream to be recycled, and so is subjected to a lower loading and lasts longer.

In variant 1.2, which is the subject of patent claims by Exxon in particular [12, 15, 17, 18], the entire stream flows through the membrane stage, giving rise to considerably higher loadings of the membrane material. The membrane treat-

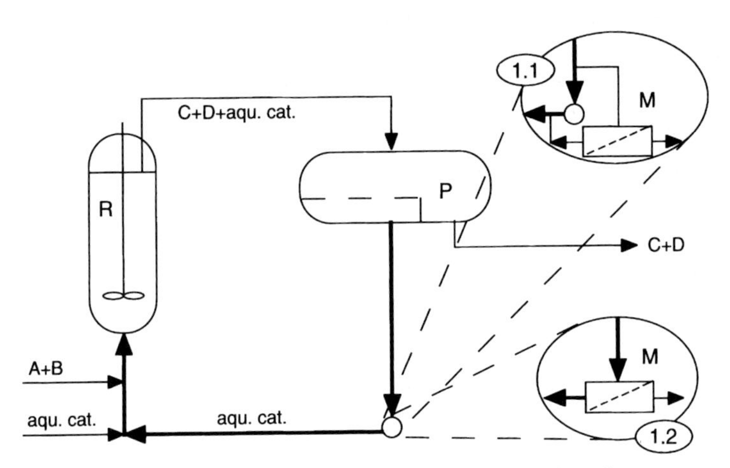

Figure 1. Membrane processes as a constituent of aqueous two-phase homogeneous catalytic methods for the reaction A + B → C + D (R = reactor; P = decanter; M = membrane separation): 1.1, sidestream membrane separation; 1.2, mainstream membrane separation.

ment of the mainstream of the organic phase may be advisable, if the preseparation by decantation is still convenient and only traces of the aqueous phase have to be separated.

In both variants, steps must be taken to ensure that the pressure and temperature conditions upstream and downstream of the membrane do not differ substantially from the standard conditions. Otherwise, deviations in conditions can lead to the complex catalysts breaking down or denaturing, resulting in recycled catalysts differing in behavior from fresh catalyst.

The relatively difficult experimental conditions explain why preference is given to membrane steps in technically and economically significant aqueous catalysis processes (e.g., oxosynthesis [12, 15–22]), and why it is chiefly industrial companies that have developed them (Hoechst [16, 22], Exxon [12, 15, 17, 18], BP [20], ICI [21], or Enichem [13]). Progress in this field is largely documented by patents. There is considerable scope for further development, because combination of membrane separation processes with homogeneous aqueous catalysis should be an interesting advance with improved activity and catalyst service life without the need for (costly) auxiliaries. A likely development is the connection of a simultaneous continuous makeup membrane stage in a sidestream to the circuit recirculating the aqueous catalyst solution stream.

Further developments are still being made in membrane materials and membrane modules. There are reports, in particular, of hydrophobic membranes made of polyolefins, crosslinked polyolefins, polyamides, polyaramids, poly(vinylidene fluoride), PTFE, polyimides, and suchlike.

References

[1] E. Bayer, V. Schurig, *CHEMTECH* **1976**, March, 212; *Angew. Chem., Int. Ed. Engl.* **1975**, *14*, 493.

[2] U. Kragl, C. Dreisbach, C. Wandrey, in *Applied Homogeneous Catalysis with Organometallic Compounds* (Eds.: B. Cornils, W. A. Herrmann), Vol. 2, VCH, Weinheim **1996**, p. 832.

[3] M.-Ch. Kuo, T.-Ch. Chou, *Ind. Eng. Chem. Res.* **1987**, *26*, 1140.

[4] T. T. Tsotsis, A. M. Champagnie, S. P. Vasileiadis, Z. D. Ziaka, R. G. Minet, *Sep. Sci. Technol.* **1993**, *29*, 397.

[5] E. T. Wolynic, D. F. Ollis, *CHEMTECH* **1974** (Feb.), 111.

[6] V. M. Gryaznov, *Platinum Met. Rev.* **1986**, *30*, 68.

[7] H. P. Hsieh, *Catal. Rev.–Sci. Eng.* **1991**, *33*, 1.

[8] G. Saracco, V. Specchia, *Catal. Rev.–Sci. Eng.* **1994**, *36*, 305.

[9] K. Keizer, V. T. Zaspalis, T. J. Burggraaf, *Mater. Sci. Monogr.* **1991**, *66D*, 2511.

[10] T. Müller, H. Bahrmann, *J. Mol. Catal. A* **1997**, *116*, 39.

[11] H. Bahrmann, M. Haubs, T. Müller, N. Schöpper, B. Cornils, *J. Organometal. Chem.* **1997**, 139–149.

[12] Exxon Chemical Patent Inc. (J. R. Livingston, E. J. Mozeleski, G. Sartori), WO 93/04.029 (1993).

[13] Enichem S.p.A. (N. Andriollo, G. Cassani, P. D'Olimpio, B. Donno, M. Ricci), EP 0.586.009 (1994).

[14] Hoechst AG (H. Bahrmann, B. Cornils, W. Konkol, W. Lipps), EP 0.163.234 (1984).
[15] Exxon Chemical Patent Inc. (J. H. Francis, J. R. Livingston, E. J. Mozeleski, J. G. Stevens), US 5.298.669 (1993).
[16] Hoechst AG (W. Greb, J. Hibbel, J. Much, V. Schmidt), EP 0.263.953 (1986).
[17] Exxon Chemical Patent Inc. (G. Sartori, E. J. Mozeleski, J. R. Livingston), US 5.288.818 (1994).
[18] Exxon Chemical Patent Inc. (J. R. Livingston, E. J. Mozeleski), EP 0.599.885 (1992).
[19] L. W. Gosser, W. H. Knoth, G. W. Parshall, *J. Mol. Catal.* **1977**, *2*, 253.
[20] BP Chemicals Ltd. (J. E. Ellis), GB 1.312.076 (1973).
[21] ICI Ltd. (W. Featherstone, T. Cox), GB 1.432.561 (1976).
[22] Hoechst AG (H. Bahrmann, M. Haubs, W. Kreuder, T. Müller), DE 3.842.819 (1988) and EP 0.374.615 (1988).

4.5 Micellar Systems

Günther Oehme

4.5.1 Introduction

Micelles, vesicles, and related species are supramolecular assemblies in a colloidal dimension [1]. This aggregation is based on amphiphilic molecules carrying a hydrophilic headgroup and a hydrophobic tail. Amphiphiles with a special structure can be surface-active agents (surfactants, detergents), which assemble in aqueous or nonpolar media [2]. A new proposal is to call surfactants "synkinons" (in analogy to synthons) and the aggregation by nonbonded forces "synkinesis" [3]. A simplified relationship between surfactant structure and the morphology of the aggregate is illustrated in Figure 1.

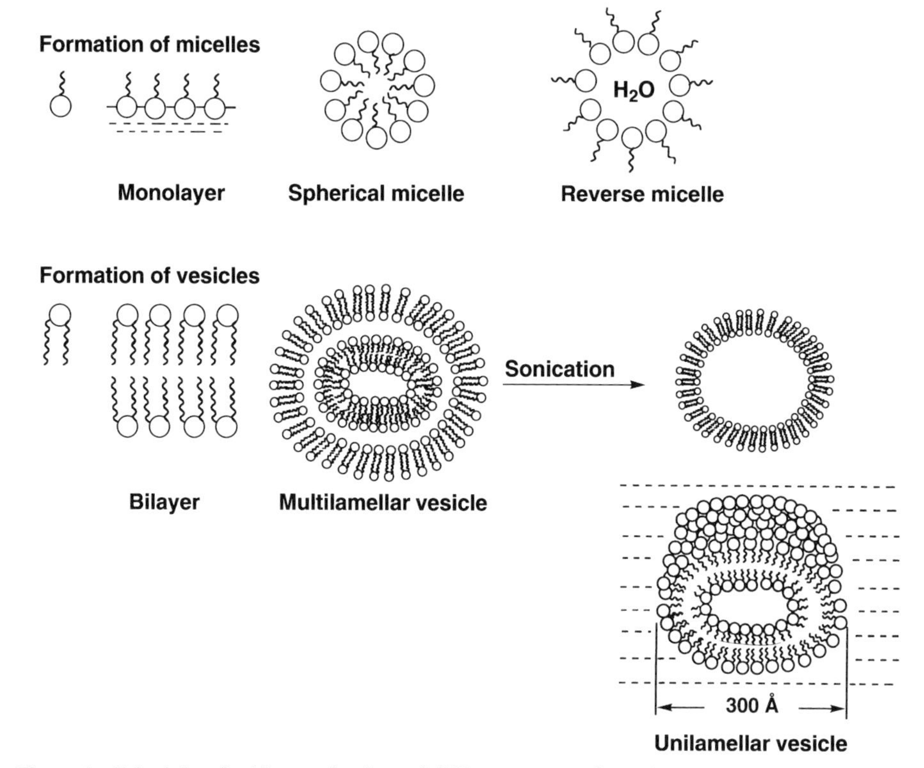

Figure 1. Principle of self-organization of different types of amphiphiles (all the sketches are idealized).

Aqueous micelles are thermodynamically stable and kinetically labile spherical assemblies. Their association–dissociation process is very fast and occurs within milliseconds. The actual order is less than shown in Figure 1. Driving forces for the formation of aqueous micelles or vesicles are the solvation of the headgroup and the desolvation of the alkyl chain ("hydrophobic effect"). Because of the rapid exchange of surfactants, the core of the micelle contains a small percentage of water molecules. Aqueous assemblies are preferentially stabilized by entropy, and reverse micelles by enthalpy [4]. The actual formation of micelles begins above a certain temperature (Krafft's point) and above a characteristic concentration (critical micelle concentration, CMC). Table 1 shows a selection of typical micelle-forming surfactants and their CMCs.

The hydrophilic "headgroup" is either charged (cationic, anionic, zwitterionic) or strongly polar (e.g., a polyether chain) and the hydrophobic (lipophilic) tail is usually a chain containing aromatic and aliphatic groups. Extremely hydrophobic properties were observed with perfluorinated alkyl groups [5].

Because of a very high polarity gradient between surface and core, micelles can enclose different organic species out of the surrounding aqueous phase [6]. This incorporation can be described by an equilibrium constant as a basis for kinetic treatments. Encapsulation of reactants in micelles often enhances or inhibits the reaction rates. Any reaction promotion has been called "micellar catalysis".

In an early review Morawetz [7] discussed three types of micellar catalysis:

- The surfactant forms the micelle and reacts as reagent.
- The interaction between the reacting species and the micelle influences the rate of reaction.
- The micelle carries catalytically active groups and acts as a catalyst.

Most reactions influenced by micelles are related to the second type [8].

Table 1. Typical micelle-forming surfactants and their CMC values.

Surfactant		CMC [mol L^{-1}]
alkyl chain–$O-SO_2-O^-Na^+$	(SDS)	8.1×10^{-3}
alkyl chain–$N^+(CH_3)_2-CH_3Br^-$	(CTABr)	9.2×10^{-4}
alkyl chain–$N^+(CH_3)_2$–(CH$_2$)$_3$–SO_3^-	(DDAPs)	1.2×10^{-3}
alkyl chain–$(OCH_2CH_2)_{20}OH$	(Brij 58)	7.7×10^{-5}

As mentioned above, surfactant micelles in water as medium create a special environment from high to low polarity. According to Brown et al. [9] the rate enhancement of organic reactions in micelles can be a combination of the following effects:

- There is a medium effect because of the lower dielectric constant in comparison with water.
- The reaction transition-state is stabilized by interaction with the polar headgroup.
- Reactants are concentrated by interacting with the micelle surface or due to incorporation. Thus the rate of bimolecular reactions should be increased. Surfaces of ionic surfactant micelles are more acid or more basic than the surrounding water phase.

Because of the hydrophobic core and a polar headgroup there are certain similarities between micellar and enzymic catalysis, but the lifetime of a micellar assembly is very short and there is no organization of catalytic and anchor groups as there is in enzymes. As a result, the enhancement of activity and enantioselectivity is much lower. Although formation and dissociation of micelles are very fast processes, some reactions occur with a high selectivity. Analogy between micellar catalysis and enzymic catalysis is also confirmed by a similar kinetic treatment. All kinetic models presuppose an interaction of the reactants with aggregated but not with dissociated surfactants. Detailed kinetic treatments are given in various reviews [10]. Certainly, there is no information about the morphology of surfactant aggregates in the presence of reactants. The existence of spherical micelles depends on the temperature, the concentration of surfactants, and the concentration of other solutes, e.g., salts. The transition from micelles to rods and even to vesicles of different shapes was observed. Nevertheless, it is very helpful to create micelles and most effects on the reaction rate and selectivity are observed near the CMC [11].

It is also noteworthy that micelle-forming surfactants may solubilize organic compounds sometimes in a very low concentration of the surfactant (still above the CMC). This embedding depends on the charge of surfactant and the charge of reactant. Only hydrophobic reactants may permeate into the hydrophobic core. Important for good solubilization properties is the hydrophile–lipophile balance (HLB) of the surfactant because sufficient water-solubility is required [12] (cf. Section 3.2.4). The transfer of a catalytic reaction into a micelle represents a case of microheterogenization.

4.5.2 Hydrolytic Reactions in Micelles

Nucleophilic reactions in micelles with water as reagent have been investigated as models of enzymic reactions. The enhancement of the reaction rate as well as the stereoselectivity of the reactions was studied. Typical substrates were activated esters of amino acids [13], carboxylic acids [14], and phosphoric acid [15], and typical catalysts were surface-active peptides with histidine as active component.

Kinetic resolution of racemic esters was determined. Brown et al. [16] and Moss et al. [17] gave explanations for the stereoselectivity. As a rule the effect of chiral inert micelles on the stereoselectivity is not important, but the use of chiral functional surfactants as catalysts sometimes gave unexpected effects. Surprisingly, nonfunctional amphiphiles as co-surfactants can improve the stereoselectivity enormously. One example given by Ueoka et al. [18] was that the saponification of D,L-*p*-nitrophenyl *N*-lauroylphenylalaninate with the tripeptide *Z*-Phe-His-Leu-OH as catalyst in assemblies of bis(tetradecyl)-dimethylammonium bromide ($2C_{14}Br$) gave almost pure L-*N*-dodecanoyl-phenylalanine upon addition of 7–20 mol% of the anionic surfactant sodium dodecylsulfate (SDS). A structural change of the vesicle into a rodlike or cylindrical micelle could be a possible explanation. Hydrolytic reactions are not favored in aqueous complex catalysis but there is an interest in finding models for the catalytic effects in metalloenzymes [19].

In a series of papers Tonellato, Tecilla, Scrimin, and co-workers [20] gave examples of highly active complexes with amphiphilic or hydrophilic ligands (Structures **5**–**8**) for the cleavage of various esters. Sometimes the rate is very drastically increased in comparison with the reaction in pure water (over a half-millionfold). These lipophilic ligands change into amphiphiles upon complexation with Ni(II), Cu(II), Zn(II), or Co(II), respectively. A comparison of amphiphiles and non-amphiphiles (methyl instead of dodecyl or hexadecyl chains) gave significant differences in rates and selectivity. The use of optically active ligands led to a moderate kinetic resolution of racemic α-amino acid esters [21]. Best results were obtained with bilayer-forming cationic co-surfactants in the gel state. In principle, the kinetic resolution of a racemic mixture of activated α-aminoesters appears feasible [22].

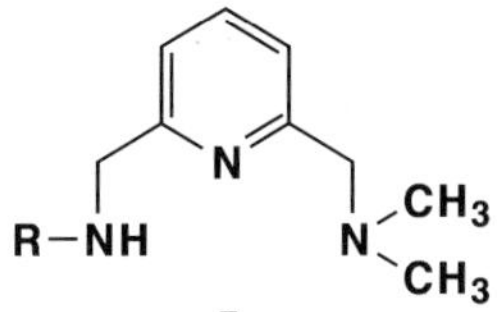

5 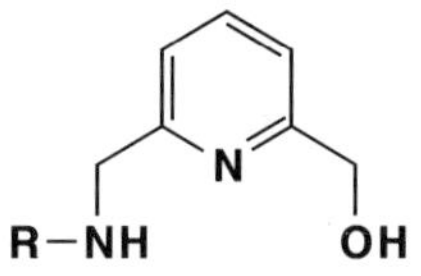 **6** 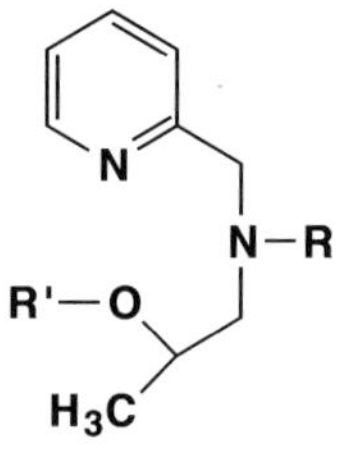 **7** 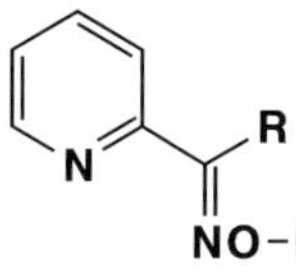 **8**

R = CH_3, C_nH_{2n+1}
R′ = H, CH_3
n = 8, 12, 13, 16

It is noteworthy that metallomicelles of Ni(II) complexes with long-chain *N*-alkylated ethylenediamine ligands catalyze the epimerization of aldoses in an aqueous dispersion [23]. The catalytic effect of metallosurfactants in enzyme-related reactions has been investigated by Nolte's working group [24], also carefully considering the assembly structure [25]. The wide field of artificial enzymes was recently reviewed by Murakami et al. [26].

4.5.3 Oxidation Reactions in Micelles

As pioneering work in micellar catalysis, Menger et al. [27] in 1975 described two reactions: the oxidation of piperonal to the corresponding carboxylic acid by means of $KMnO_4$ and the saponification of trichloromethylbenzene to benzoic acid, both in the presence of cetyltrimethylammonium bromide (CTABr). In these experiments the authors observed improved yields attributable to surfactant assemblies. The surfactant-supported oxidation of organic compounds with inorganic reagents has been developed into a useful method [28]. An enantioselective oxidation of 3,4-dihydroxy-L-phenylalanine catalyzed by an *N*-lauroyl-L-(or-D-)histidine-Cu(II) complex was submitted by Yamada et al. [29]. Using CTABr co-micelles, appreciable enantioselectivity was observed. Simon and co-workers [30] synthesized a series of amphiphilic copper(II)–amine complexes. These annelides exhibited oxygen activation [31] but no catalytic activity.

Shinozuka and co-workers [32] developed bleomycin model complexes bearing long alkyl chains. The multidentate amine ligand contains imidazole and shows high oxygen-activating capacity.

Recently, Rabion et al. [33] described a methane oxygen model, i.e., the oxidation of cyclohexane with *tert*-butyl hydroperoxide catalyzed by an iron complex in aqueous micelles. The favored surfactant was cetyltrimethylammonium hydrogen sulfate and the following complexes have been used: $[Fe_2O(\eta^1\text{-}H_2O)(\eta^1\text{-}OAc)(TPA)_2]^{3+}$ and $[Fe_2O(\eta^1\text{-}H_2O)(\eta^1\text{-}OAc)(BPIA)_2]^{3+}$ with TPA = tris[(2-pyridyl)methyl]amine and BPIA = bis[(2-pyridyl)methyl]-[2-(1-methylimidazolyl)methyl]amine. The investigated reaction does not occur in a biphasic aqueous system without surfactant.

4.5.4 Complex-catalyzed Hydrogenation in Micellar Media

Homogeneous hydrogenation is a typical example of the activation of small molecules by transition metals (cf. Section 6.2). The state-of-the-art has been reviewed in a series of articles and books [34]. A great number of transition metals are able to activate hydrogen but Co, Rh, Pd, Ru, and Pt are preferentially used. A special position in the asymmetric (enantioselective) hydrogenation is taken by rhodium and ruthenium complexes [35]. In the early investigation of homogeneous hydrogenation, water was strictly avoided as solvent except in experiments with the water-soluble complex of the potassium hydridopentacyanocobaltate, $K_3[HCo(CN)_5]$. By using a micellar medium, Reger and Habib [36] tried to stabilize this hydrido complex and even to influence the regioselectivity by the partial hydrogenation of conjugated dienes. The authors could observe stabilization of the catalyst but no alteration in the product distribution. Using the same catalyst but having atropic acid or the corresponding esters as substrate, Ohkubo et al. [37] found some acceleration in presence of micelle-forming reagents.

With the introduction of water-soluble phosphine ligands [38] and their application in complex catalysts in biphasic and phase-transfer systems, the influence of amphiphiles could be expected but the results were not really encouraging [39]. Quinn and Taylor [40] reported on selective hydrogenation in phospholipid bilayer membranes and Nuzzo et al. [41] described the rhodium complex of **9** as active exclusively in the presence of SDS.

HOOC–CH₂CH₂–C(=O)–N(CH₂CH₂PPH₂)(CH₂CH₂PPh₂)

9

It also seems to be noteworthy that in the presence of phospholipids the nitrogen reduction catalyzed by an artificial nitrogenase system is very selectively enhanced [42]. Ding et al. [43] gave examples of the use of chiral amphiphilic ligands in hydrogenation reactions but directed attention to their application in biphasic systems. Usually, the presence of water causes loss of activity and, in the case of enantioselective hydrogenation, loss of enantioselectivity [44].

Surprisingly, in the hydrogenation of unsaturated amino acid derivatives catalyzed by a chiral rhodium complex in water, these disadvantages are overcome by the addition of micelle-forming surfactants [45]. The mixture was solubilized by addition of a relatively small amount of surfactant, became yellow

Table 2. Hydrogenation with different types of surfactants (Rh/surfactant/substrate = 1:20:100).

$$\underset{\text{Ph}}{\overset{\text{H}}{}}\!\!>\!C{=}C\!<\!\!\underset{\text{NHCOCH}_3}{\overset{\text{COOCH}_3}{}} \xrightarrow[\text{cat: [Rh(cod)}_2]\text{BF}_4\ +\ \text{BPPM, surfactant}]{25^\circ\text{C, 0.1 MPa H}_2\text{, water}} \text{PhH}_2\text{C}{-}\overset{*}{\text{C}}\text{H}\!<\!\!\underset{\text{NHCOCH}_3}{\overset{\text{COOCH}_3}{}} \quad (1)$$

Surfactant	*t*/2 [min]	Optical yield [% *ee* (*R*)]
None: in water (methanol)	90 (2)	78 (90)
Anionic: sodium dodecyl sulfate (SDS)	6	94
Cationic: cetyltrimethylammonium hydrogen sulfate (CTA HSO_4)	5	95
Zwitterionic: N-dodecyl-N,N-dimethyl-3-ammonio-1-propanesulfonate	5	93
Nonionic: decaoxyethylene hexadecyl ether (Brij 56)	7	95

and consumed hydrogen [46]. Activity and enantioselectivity are comparable with the results obtained in methanol as standard solvent and in some cases the enantioselectivities were even higher than in organic solvents. Table 2 and Eq. (1) show selected examples with different amphiphiles.

All types of surfactants promote the reaction but only the hydrogen sulfate was active in the case of the cationic amphiphiles. No influence was found for the anions tetrafluoroborate or triflate. The most favorable chiral ligand was BPPM (**10**), described by Achiwa [47], but even DIOP (4.5-bis(diphenylphosphinomethyl)-2.2-dimethyl-1.3-dioxolan) [48] and Ph-β-glup-OH (phenyl 2.3-O-bis(diphenylphosphino)-β-D-glucopyranoside) [49] show the effect. The increase of activity and enantioselectivity depends on the concentration of the surfactant. There is a sudden increase near the CMC of the surfactant which might be an indication of the existence of micelles [11].

Ph_2P PPh_2 N CO–O*t*Bu

10

The enhancement of enantioselectivity cannot be an effect of solubility because water-soluble substrates gave enhanced activities but no enantioselectivity increase [50]. Acylation of PPM ((2*S*,4*S*)-4-diphenylphosphino-2-diphenylphosphinomethylpyrrolidine) with the chlorocarbonic ester of alkyl polyoxy-

Table 3. Chiral induction by optically active surfactants (for conditions see Table 2).

Surfactant	Catalyst: $[Rh(cod)_2]BF_4$ + BDPD[a)]	
	t/2 [min]	Optical yield [% *ee*]
Derived from amino acids:		
N-palmitoyl-L-Pro-Na	6	3 (*S*)
N-palmitoyl-D-Pro-Na	16	2 (*R*)
N-palmitoyl-L-Pro-L-Pro-Na	10	8 (*S*)
N-palmitoyl-D-Pro-D-Pro-Na	12	7 (*R*)
N-palmitoyl-D-Pro-L-Pro-Na	8	3 (*R*)
Derived from carbohydrates:		
Tetradecyl-β-D-maltoside	40	6 (*R*)
Dodecyl-β-D-maltoside	23	6 (*R*)
Derived from cholesterol:		
Cholesteryl (β-sodiumsulfonato)propionate	11	8.5 (*R*)

[a)] BDPD = $Ph_2P(CH_2)_4PPh_2$(bis(diphenylphosphino)butane).

ethylene ethers (Brij) leads to water-soluble or amphiphilic (that means in water as colloids dispersible) ligands. Complexation of these ligands to rhodium(I) leads to chiral metallomicelles which successfully catalyze the hydrogenation in methanol and even in water [51].

It is difficult to determine the location of the reactants in micelles. Both substrate and catalyst have to be located very close. Chiral induction by optically active surfactants, with the catalyst optically inactive, might indicate the location. Table 3 summarizes selected results with different types of chiral surfactants.

The induction is quite low, less than 10% *ee*, but the use of dipeptides containing two prolines of identical configuration causes amplification of the optical induction, whereas the acylated amino acid directs the induction when dipeptides with prolines of opposite configuration are being used [52]. This characteristic is true even for carbohydrate amphiphiles [53].

The most likely region for the reaction should be the transition between polar headgroup and hydrophobic chain ("palisade layer" [54]), which is also discussed by Monflier [55].

4.5.5 Carbon–Carbon Coupling Systems

One of the most important industrial reactions catalyzed by transition metal complexes is hydroformylation. Discussion of new aspects and developments are given in a review [56]. Milestones for this reaction are the development of water-soluble phosphines (Rhône-Poulenc [57]) and the first application as ligands by Ruhrchemie [58]. Gradually, a liquid/liquid biphasic system originated, which contained the catalytic acting complex in the water phase. It is difficult to work with higher olefins than propene because the water-solubility of educts and products is extremely low. Many proposals have been made in order to improve the solubility of the catalyst in the olefinic phase (cf. Section 6.1). It should be possible by addition of hydrophobic cations to anionic groups in the water-soluble ligands [59]. A hydroformylation in presence of micelle-forming surfactants was proposed by Dror and Manassen [60]. They did not find any indication of selectivity enhancement but there was clearly a stabilization of the catalyst. A comparable result was noted by Quinn and Taylor [40], who used phospholipid bilayers (vesicles) in water. The catalyst did not work in organic solvents but gave acceptable activities and regioselectivities in aqueous bilayer assemblies. Matsui and Orchin [61] investigated the stoichiometric reaction of $HMn(CO)_5$ with cyclopropenes which led to aldehyde and cyclopropane (Eq. 2).

$$\text{(cyclopropene; } H_3C\ CH_3) + HMn(CO)_5 \xrightarrow{CO} \underset{\mathbf{11}}{\text{(H, CHO, Ph, Ph, } H_3C\ CH_3\text{ cyclopropane)}} + \underset{\mathbf{12}}{\text{(H, H, Ph, Ph, } H_3C\ CH_3\text{ cyclopropane)}} + Mn_2(CO)_{10} \qquad (2)$$

In 1991 Fell and Papadogianakis provided an example of hydroformylation with surface-active phosphine ligands **13** [62].

$$P\left(\text{C}_5\text{H}_4\text{N}^+-\text{CH(R)}-\text{CH}_2\text{SO}_3^-\right)_3$$

13

Recently, Hanson and co-workers started a program of rhodium-catalyzed hydroformylation with various surface-active phosphine ligands (Structures **14**–**16**, [63]). In their ligand models the phosphorus is bound to the hydrophobic part.

$$P\left((CH_2)_x-C_6H_4-SO_3Na\right)_3 \qquad x = 1, 2, 3, 6$$

14

$$P\left(C_6H_4-(CH_2)_m-C_6H_4-SO_3Na\right)_3 \qquad m = 3, 6$$

15

$$P(\text{menthyl})\left((CH_2)_8-C_6H_4-SO_3Na\right)_2$$

16

As a new development, van Leeuwen and co-workers used ligands with amphiphilic properties in acid media for hydroformylation, but these phosphines are hardly able to form micelles [64].

Some other C–C bond coupling reactions in micellar systems should be mentioned here. Monflier et al. [65] described, in both papers and patents, the telomerization of 1,3-butadiene into octadienol in a micellar system by means of a palladium–phosphine catalyst. Water-soluble and amphiphilic phosphines have been used and the surfactants were widely varied. The authors have shown that the promoting effect of surfactants appeared above the CMCs of the surfactants, and conclude that micellar aggregates were present in the reaction mixture. Cationic, anionic, and nonionic surfactants gave this micellar effect but the combination of the highly water-soluble TPPTS and the surfactant dodecyldimethylamine hydrocarbonate was found to be best. A speculation about the location of reactants shows that the reaction probably occurs in the interface between the micellar pseudophase and water.

Surfactant-supported polymerization reactions are not included in this report.

4.5.6 Some Examples of Reactions in Reverse Micelles and Microemulsions

Reverse micelles are formed by association of polar headgroups of amphiphiles with colloidal drops of water in an organic medium. A favored surfactant seems to be AOT (di(2-ethylhexyl)sodium sulfosuccinnate) but SDS and tetraalkylammonium salts have also proved to be useful. Like aqueous micelles, reverse micelles exist in highly diluted systems.

A microemulsion is a multicomponent (3–4 components) system, e.g., water in hydrocarbon (water/oil) or hydrocarbon in water (oil/water), surfactant, and co-surfactant, and generally it exists only in small concentration ranges. Nevertheless, the capacity for reactants and variability of solubilization properties are high and of practical interest [66]. On the basis of microemulsions Menger and co-workers developed a method for an economical environmental cleanup of chemical warfare contamination [67]. As an example of organometallic catalysis in a microemulsion, Beletskaya [68] performed palladium-catalyzed C–C coupling reactions in aqueous medium with a very high content of surfactant.

An acceleration was observed in the hydrogen-transfer reaction from 1,2-cyclohexanedimethanol to (*E*)-4-phenyl-3-butene-2-one in toluene catalyzed by [RuCl((S)BINAP)(benzene)]Cl (2 mol%; BINAP = 2.2′-bis(diphenylphosphino)-1.1′-binaphthyl) and SDS (6 mol%) [69]. The surfactant is essential for the enhancement of catalytic activity and the authors proposed the existence of reverse micelles.

It was shown by Buriak and Osborn [70] that non-micelle-forming anions improved the enantioselectivity of an imine hydrogenation catalyzed by rhodium complexes in the same way as reverse micelles. Complexation of the sulfate or sulfonate anion with the catalyst appears to be responsible for the enhancement of the enantioselectivity. The very strong dependence of the product chirality on the structure of the anion is discussed.

Finally, a long-chain ephedrinium salt **17** as surfactant, should be mentioned.

OH
Ph Me
$N^{+}(Me)_2RBr^{-}$
R = ethyl, octyl, dodecyl, hexadecyl

17

Several ketones are reduced by $NaBH_4$ in a reaction of the type shown in Eq. (3). A maximum of 17.2% *ee* could be observed with R^1 = Ph and R = ^{t}Bu, which is noteworthy for optical induction from the micelle [71].

$$R_1C(=O)R_2 \xrightarrow[\text{chiral reverse micelle}]{NaBH_4} R_1-\overset{*}{C}H(OH)-R_2 \quad (3)$$

4.5.7 Perspectives

Micelles, vesicles, and other self-organized assemblies are interesting microreactors for a series of complex-catalyzed reactions. The solubilization of reactants in water, and the achievements in rate and selectivity enhancement, are sometimes unexpectedly high, but the main problem has been the separation of products, amphiphile, and catalyst after the reaction. One of the older arguments against the application of micelles is the low capacity for reactants, but there are some examples in which small concentrations of amphiphiles led to sufficient effects [72]. The challenge of catalyst recycling still remains. One solution should be the immobilization of micelle-like structures on polymers. Such polymers are well investigated and discussed as polysoaps in a series of reviews [73]. A first proposal to use polymers like micelles was made by Brown and Jenkins in 1976 [74].

For the first time Regen [75] used amphiphilic polymers in phase-transfer catalysis. Flach et al. [76] synthesized a variety of different polysoaps with organic and inorganic backbones for use in asymmetric hydrogenation. The simplest type is an admicelle [77]; this means that the surfactant (SDS) is adsorbed on alumina, thus forming a bilayer on its surface, which is stable against washing with water or other polar solvents. Best results have been observed with highly crosslinked organic ion exchange resins and inorganic ion exchanger. Polyether-surfactants bound to silica were also suitable as medium in the asymmetric hydrogenation of amino acid precursors with chiral rhodium complexes in water.

Thus, interesting new aspects in the use of surfactants in aqueous complex catalysis can be found, but at present there is no industrial application. On the other hand there is still potential, worth investigating in functional surfactants (amphiphilic complexes which are useful in both polar organic solvents and water), with surfactants with large HLBs which are consequently soluble in the water phase in biphasic systems with an extremely nonpolar organic phase (e.g. in hydroformylation and telomerization processes), and finally with immobilized micelle-like polymers as outlined above. In cases of unusual effects in activity and selectivity, industrial application might be advantageous despite the costly separation process.

References

[1] (a) W. M. Gelbert, A. Ben-Shaul, D. Roux (Eds.) *Micelles, Membranes, Microemulsions, and Monolayers,* Springer, New York, **1994**; (b) F. Vögtle, *Supramolekulare Chemie,* Teubner, Stuttgart, **1992**, Chapter 9.

[2] (a) J. H. Fendler, *Membrane Mimetic Chemistry,* Wiley, New York, **1982**; (b) F. M. Menger, *Angew. Chem.* **1991**, *103*, 1104; *Angew. Chem., Int. Ed. Engl.* **1991**, *30*, 1086.

[3] J.-H. Fuhrhop, J. Köning, *Membranes and Molecular Assemblies: The Synkinetic Approach,* The Royal Society of Chemistry, Cambridge, **1994**.

[4] (a) C. Tanford, *The Hydrophobic Effect: Formation of Micelles and Biological Membranes*, Wiley, New York, **1980**; (b) W. Blokzijl, J. B. F. N. Engberts, *Angew. Chem.* **1993**, *105*, 1610; *Angew. Chem., Int. Ed. Engl.* **1993**, *32*, 1545.
[5] E. Kissa, *Fluorinated Surfactants: Synthesis, Properties, Applications*, Marcel Dekker, New York, **1994**.
[6] S. D. Christian, J. F. Scamehorn (Eds.), *Solubilization in Surfactant Aggregates*, Marcel Dekker, New York, **1995**.
[7] H. Morawetz, *Adv. Catal.* **1969**, *20*, 341.
[8] (a) J. H. Fendler, E. J. Fendler, *Catalysis in Micellar and Macromolecular Systems*, Academic Press, New York, **1975**; (b) T. Kunitake, S. Shinkai, *Adv. Phys. Org. Chem.* **1980**, *17*, 435; (c) C. A. Bunton, G. Savelli, *Adv. Phys. Org. Chem.* **1986**, *22*, 213.
[9] J. M. Brown, S. K. Baker, A. Colens, J. R. Darwent in *Enzymic and Non-enzymic Catalysis* (Eds.: P. Dunnill, A. Wiseman, N. Blakebrough), Horwood, Chichester, **1980**, p. 111.
[10] (a) C. A. Bunton in *Kinetics and Catalysis on Microheterogeneous Systems* (Eds.: M. Grätzel, K. Kalyanasundaram), Marcel Dekker, New York, **1991**, p. 33; (b) M. A. Morini, P. C. Schulz, J. E. Puig, *Colloid Polym. Sci.* **1996**, *274*, 662; (c) M. L. Carreto, S. Rubio, D. Perez-Bendito, *Analyst (London)* **1996**, *121*, 33R.
[11] T. J. Broxton, J. R. Christie, A. J. Dole, *J. Phys. Org. Chem.* **1994**, *7*, 437.
[12] Y. Moroi, *Micelles, Theoretical and Applied Aspects*, Plenum Press, New York, **1992**.
[13] J. M. Brown, C. A. Bunton, *J. Chem. Soc., Chem. Commun.* **1974**, 969.
[14] J. Budka, F. Hampl, F. Liska, P. Scrimin, P. Tecilla, U. Tonellato, *J. Mol. Catal.* **1996**, *104*, L201.
[15] J. M. Brown, C. A. Bunton, S. Diaz, Y. Ihara, *J. Org. Chem.* **1980**, *45*, 4169.
[16] (a) J. M. Brown, C. A. Bunton, *J. Chem. Soc., Chem. Commun.* **1974**, 969; (b) J. M. Brown, R. L. Elliott, C. G. Griggs, G. Helmchen, G. Will, *Angew. Chem.* **1981**, *93*, 906; *Angew. Chem., Int. Ed. Engl.* **1981**, *20*, 890.
[17] (a) R. A. Moss, Y.-C. P. Chiang, Y. Hui, *J. Am. Chem. Soc.* **1984**, *106*, 7506; (b) R. Ueoka, Y. Matsumoto, R. A. Moss, S. Swarup, A. Sugii, K. Harada, J. Kikuchi, Y. Murakami, *J. Am. Chem. Soc.* **1988**, *110*, 1588.
[18] R. Ueoka, Y. Matsumoto, T. Yoshino, N. Watanabe, K. Omura, Y. Murakami, *Chem. Lett.* **1986**, 1743.
[19] (a) W. Tagaki, K. Ogino, *Top. Curr. Chem.* **1985**, *128*, 143; (b) Y. Murakami, J. Kikuchi, *Bioorg. Chem. Frontiers* **1991**, *2*, 73.
[20] P. Scrimin, U. Tonellato in *Surfactants in Solution* (Eds.: K. L. Mittal, D. D. Sha), Plenum Press, New York, **1991**, Vol. 11, p. 339; see also [14, 21].
[21] P. Scrimin, P. Tecilla, U. Tonellato, *J. Org. Chem.* **1994**, *59*, 4194.
[22] M. C. Cleij, P. Scrimin, P. Tecilla, U. Tonellato, *Langmuir* **1996**, *12*, 2956.
[23] S. Osanai, R. Yanagihara, K. Uematsu, A. Okumura, S. Yoshikawa, *J. Chem. Soc., Perkin Trans. 2* **1993**, 1937.
[24] M. C. Cleij, W. Drenth, R. J. Nolte, *J. Org. Chem.* **1991**, *56*, 3883.
[25] J. H. van Esch, M. Damen, M. C. Feitrs, R. J. M. Nolte, *Recl. Trav. Chim. Pays-Bas* **1994**, *113*, 186.
[26] Y. Murakami, J. Kikuchi, Y. Hiseada, O. Hayashida, *Chem. Rev.* **1996**, *96*, 721.
[27] F. M. Menger, J. V. Rhee, H. K. Rhee, *J. Org. Chem.* **1975**, *40*, 3803.
[28] (a) L. P. Panichewa, N. Y. Tretyakov, C. A. Jakovleva, A. Y. Yuffa, *Kinet. Katal.* **1990**, *31*, 96; *Chem. Abstr.* **1990**, *112*, 215 991; (b) M. Bressan, L. Forti, A. Morvillo, *J. Chem. Soc., Chem. Commun.* **1994**, 250; (c) S. K. Sahn, G. P. Panigrahi, *J. Ind. Chem. Soc.* **1996**, *73*, 576.
[29] K. Yamata, H. Shosenji, Y. Otsubo, S. Ono, *Tetrahedron Lett.* **1980**, *21*, 2649.
[30] (a) J. Simon, J. L. Moigne, *J. Mol. Catal.* **1980**, *7*, 137; (b) D. Markovitsi, J. Simon, E. Kraeminger, *Nouv. J. Chim.* **1981**, *5*, 141.

[31] J. Simon, J. LeMoigne, D. Markovitsi, J. Dayantis, *J. Am. Chem. Soc.* **1980**, *102*, 7247.
[32] (a) K. Shinozuka, H. Morishita, T. Yamazaki, Y. Sugiura, H. Sawai, *Tetrahedron Lett.* **1991**, *32*, 6869; (b) J. Kohda, K. Shinozuka, H. Sawai, *Tetrahedron Lett.* **1995**, *36*, 5575.
[33] A. Rabion, R. M. Buchanan, J.-L. Seris, R. H. Fish, *J. Mol. Catal.* **1997**, *116*, 43.
[34] As a selection: (a) H. Brunner in *Applied Homogeneous Catalysis with Organometallic Compounds* (Eds.: B. Cornils, W. A. Herrmann), VCH, Weinheim, **1996**, Vol. 1, p. 201; (b) P. A. Chaloner, M. A. Esteruelas, F. Joó, L. A. Oro, *Homogeneous Hydrogenation*, Kluwer, Dordrecht, **1994**.
[35] (a) H. Takaya, T. Ohta, R. Noyori in *Catalytic Asymmetric Synthesis* (Ed.: I. Ojima), VCH, Weinheim, **1993**, p. 1; (b) K. E. Koenig in *Asymmetric Synthesis* (Ed.: J. E. Morrison), Academic Press, Orlando, **1985**, Vol. 5, p. 71.
[36] (a) D. L. Reger, M. M. Habib, *J. Mol. Catal.* **1978**, *4*, 315; (b) D. L. Reger, M. M. Habib, *Adv. Chem. Ser.* **1979**, *173*, 43; (c) D. L. Reger, M. M. Habib, *J. Mol. Catal.* **1980**, *7*, 365; (d) D. L. Reger, A. Gabrielli, *J. Mol. Catal.* **1981**, *12*, 173.
[37] K. Ohkubo, T. Kawabe, K. Yamashita, S. Sakaki, *J. Mol. Catal.* **1984**, *24*, 83.
[38] (a) J. Joó, A. Kathó, *J. Mol. Catal.* **1997**, *116*, 3; (b) W. A. Herrmann, C. W. Kohlpaintner, *Angew. Chem.* **1993**, *105*, 1588; *Angew. Chem., Int. Ed. Engl.* **1993**, *32*, 1524; (c) F. Joó, Z. Toth, *J. Mol. Catal.* **1980**, *8*, 369.
[39] Y. Dror, J. Manassen, *J. Mol. Catal.* **1977**, *2*, 219.
[40] P. J. Quinn, C. E. Taylor, *J. Mol. Catal.* **1981**, *13*, 389.
[41] R. G. Nuzzo, S. L. Haynie, M. E. Wilson, G. M. Whitesides, *J. Org. Chem.* **1981**, *46*, 2861.
[42] L. P. Didenko, V. D. Makhaev, A. K. Shilova, A. E. Shilov, *Prepr. 4th Int. Symp. Homogen. Catal.*, Leningrad, **1984**, Vol. II, p. 102.
[43] H. Ding, B. E. Hanson, J. Bakos, *Angew. Chem.* **1995**, *107*, 1728; *Angew. Chem., Int. Ed. Engl.* **1995**, *34*, 1645.
[44] L. Lecomte, D. Sinou, J. Bakos, I. Toth, B. Heil, *J. Organomet. Chem.* **1989**, *370*, 277.
[45] G. Oehme, E. Paetzold, R. Selke, *J. Mol. Catal.* **1992**, *71*, L1.
[46] I. Grassert, E. Paetzold, G. Oehme, *Tetrahedron* **1993**, *49*, 6605.
[47] K. Achiwa, *J. Am. Chem. Soc.* **1976**, *98*, 8265.
[48] H. B. Kagan, T. B. Dang, *J. Am. Chem. Soc.* **1972**, *94*, 6429.
[49] R. Selke, *J. Organomet. Chem.* **1989**, *370*, 241.
[50] A. Kumar, G. Oehme, J. P. Roque, M. Schwarze, R. Selke, *Angew. Chem.* **1994**, *106*, 2272; *Angew. Chem., Int. Ed. Engl.* **1994**, *33*, 2197.
[51] S. Ziegler, Dissertation, Rostock, **1997**.
[52] I. Grassert, K. Schinkowski, D. Vollhardt, G. Oehme, to be published in *Chirality*.
[53] I. Grassert, V. Vill, G. Oehme, *J. Mol. Catal.* **1997**, *116*, 231.
[54] M. J. Rosen, *Surfactants and Interfacial Phenomena*, 2nd ed., Wiley, New York, **1989**.
[55] E. Monflier, P. Bourdauducq, J.-L. Couturier, J. Kervennal, A. Mortreux, *Appl. Catal. A: General* **1995**, *131*, 167.
[56] M. Beller, B. Cornils, C. D. Frohning, C. W. Kohlpaintner, *J. Mol. Catal.* **1995**, *104*, 17.
[57] E. G. Kuntz, *CHEMTECH* **1987**, *17*, 570.
[58] E. Wiebus, B. Cornils, *Chem.-Ing.-Techn.* **1994**, *66*, 916.
[59] (a) Johnson Matthey PLC (M. J. H. Russel, B. A. Murrer), BE 890.210 (1982); *Chem. Abstr.* **1982**, *97*, 23291; (b) Johnson Matthey PLC (M. J. H. Russel, B. A. Murrer), FR 2.489.308 (1982); *Chem. Abstr.* **1982**, *97*, 55308; (c) Ruhrchemie AG (H. Bahrmann, B. Cornils, W. Lipps, P. Lappe, H. Springer) DE 3.420.493 (1985); *Chem. Abstr.* **1987**, *106*, 33308.
[60] Y. Dror, J. Manassen, *Stud. Surf. Sci. Catal.* **1981**, *7*, 887.
[61] Y. Matsui, M. Orchin, *J. Organomet. Chem.* **1983**, *244*, 369.

[62] B. Fell, G. Papadogianakis, *J. Mol. Catal.* **1991**, *66*, 143.
[63] (a) T. Bartik, B. Bartik, B. E. Hanson, I. Guo, I. Toth, *Organometallics* **1993**, *12*, 164; (b) H. Ding, B. E. Hanson, *J. Chem. Soc. Chem. Commun.* **1994**, 2747; (c) T. Bartik, B. Bartik, B. E. Hanson, *J. Mol. Catal.* **1994**, *88*, 43; (d) H. Ding, B. E. Hanson, T. Bartik, B. Bartik, *Organometallics* **1994**, *13*, 3761; (e) T. Bartik, H. Ding, B. Bartik, B. E. Hanson, *J. Mol. Catal.* **1995**, *98*, 117.
[64] (a) A. Buhling, P. C. J. Kamer, P. W. N. M. van Leeuwen, *J. Mol. Catal.* **1995**, *98*, 69; (b) A. Buhling, J. W. Elgersma, S. Nkrumah, P. C. J. Kamer, P. W. N. M. van Leeuwen, *J. Chem. Soc., Dalton Trans.* **1996**, 2143; (c) A. Buhling, P. C. J. Kamer, P. W. N. M. van Leeuwen, J. W. Elgersma, *J. Mol. Catal.* **1997**, *116*, 297.
[65] (a) E. Monflier, P. Bourdauducq, J. L. Couturier, J. Kervennal, A. Mortreux, *J. Mol. Catal.* **1995**, *97*, 29; (b) E. Monflier, P. Bourdauducq, J.-L. Couturier, J. Kervennal, I. Suisse, A. Mortreux, *Catal. Lett.* **1996**, *34*, 201. See also [55].
[66] H.-J. Schwuger, K. Stickdorn, R. Schomäcker, *Chem. Rev.* **1995**, *95*, 849.
[67] (a) F. M. Menger, A. R. Elrington, *J. Am. Chem. Soc.* **1991**, *113*, 9621; (b) F. M. Menger, H. Park, *Recl. Trav. Chim. Pays-Bas* **1994**, *113*, 176.
[68] (a) I. P. Beletskaya, reported at *10th Int. Symp. Homog. Catal.,* **1996**, Princeton, NJ, USA; (b) I. P. Beletskaya, *Pure Appl. Chem.* **1997**, *69*, 471.
[69] K. Nozaki, M. Yoshida, H. Takaya, *J. Organomet. Chem.* **1994**, *473*, 253.
[70] J. M. Buriak, J. A. Osborn, *Organometallics* **1996**, *15*, 3161.
[71] Y. Zhang, P. Sun, *Tetrahedron: Asymm.* **1996**, *7*, 3055.
[72] (a) G. Oehme, I. Grassert, N. Flach in *Aqueous Organometallic Chemistry and Catalysis* (Eds.: I. T. Horvath, F. Joó), Kluwer, Dordrecht, **1995**, p. 245; (b) A. V. Cheprakov, N. V. Ponomareva, I. P. Beletskaya, *J. Organomet. Chem.* **1995**, *486*, 297.
[73] (a) A. Laschewsky, *Adv. Polym. Sci.* **1995**, *124*, 1; (b) C. M. Paleos (ed.) *Polymerization in Organized Media*, Gordon and Breach, Philadelphia, **1992**.
[74] J. M. Brown, J. A. Jenkins, *J. Chem. Soc., Chem. Commun.* **1976**, 458.
[75] S. L. Regen, *Angew. Chem.* **1979**, *91*, 464; *Angew. Chem., Int. Ed. Engl.* **1979**, *18*, 421.
[76] (a) H. N. Flach, I. Grassert, G. Oehme, *Macromol. Chem. Phys.* **1994**, *195*, 3289; (b) H. N. Flach, I. Grassert, G. Oehme, M. Capka, *Colloid Polym. Sci.* **1996**, *274*, 261.
[77] R. Sharma (Ed.), *Surfactant Adsorption and Surface Solubilization*, ACS Symp. Ser. No. 615, ACS, Washington DC, **1995**.

4.6 On the Borderline of Aqueous-phase Catalysis

4.6.1 Phase-transfer Catalysis

Eckehard Volker Dehmlow

4.6.1.1 General Overview, Fundamentals, and Definitions

Most organic substrates are hardly soluble in water, or miscible with it, whereas the appropriate reagents may be situated in an aqueous layer. Depending on their structure, organometallic catalysts can be present in either phase. The necessary transport often causes inefficient reaction rates. As detailed in other sections, the situation may be rectified by one of the following:

- use of a solvent mixture in which all species are (somewhat) soluble,
- increasing the interfacial area by fine dispersion, rapid stirring, emulsification, ultrasonification or the like,
- application of detergents and utilization of micellar processes, or
- modification of catalyst ligands to make the active species water-soluble.

Phase-transfer catalysis (PTC) has a different approach. In its archetypical version, the chemical reaction is located in the organic phase of a two-phase mixture. A catalytic amount of a quaternary ammonium salt QX is present. Catalyst cation Q^+ transfers an anionic species Y^- from the aqueous to the organic phase via extraction as a ion pair. The resultant ionized, but undissociated, salt QY is solvated much less in the organic layer than it would be under aqueous conditions. It therefore reacts very fast, and the anion Z^- (generated in the reaction) pairs with Q^+ to be transported to the interphase for a ion exchange of Z^- for Y^-. Regenerated QY returns to the depth of the organic phase for the next reaction cycle. Altogether the action of the phase-transfer (PT) catalyst is twofold, causing anion phase transfer and activation of the anion by removal of most of the hydration shell. Typically a limited number of water molecules are co-extracted with QY into an organic phase (Eqs. 1–3).

Organic phase: $$[QY] + R\text{-}Z \longrightarrow R\text{-}Y + [QZ] \quad (1)$$

Phase boundary: $$[QX] + Y^-_{(aq.)} \rightleftharpoons [QY] + X^-_{(aq.)} \quad (2)$$

$$[QZ] + Y^-_{(aq.)} \rightleftharpoons [QY] + Z^-_{(aq.)} \quad (3)$$

$Q^+ = NR_4^+$, PR_4^+, metal$^+$(crown), metal$^+$(cryptand)

This so-called "liquid/liquid phase-transfer catalysis" is what most chemists associate with the term: the extraction of a vital anionic species from the aqueous phase to the organic one, either by an onium salt or with the help of a crown ether or cryptand. Its application to metal organic catalysis will be described below. First it should be stressed, however, that the concept of "PTC" is much wider in scope and comprises numerous variants:

- The second phase can be a solid ("solid/liquid PTC") – or even a gas.
- Cations can be extracted from water to the organic phase by appropriate catalysts. This method is utilized in quite a number of organometallic applications.
- Various uncharged species (including metal oxides and Lewis acids) can be made organically soluble by complexation with some type of catalyst.
- Finally, extraction of the important reactive species can be executed in the opposite direction, from organic phase to water. This is called "inverse phase-transfer catalysis." Catalysts for such processes are mostly cyclodextrins or modified derivatives thereof. Relatively few applications of this type of PTC have been published. Whereas the present section is concerned only with the organic phase as the location of the proper chemical reaction, important contributions of inverse PTC toward organometallic catalysis are detailed in Section 4.6.2.

A comprehensive introduction to PTC may be found in recent monographs [1–3] and these as well as a symposium volume contain specific chapters on organometallic applications [4, 5]. Other short reviews are also available [6–8].

As stated above, anion extraction of the most common type (water to organic layer) is executed by quaternary ammonium or other onium cations or by crown-ether-masked metal cations. To be useful, at least two factors have to be met:

- The PTC cation must be sufficiently lipophilic. There is an increase in extraction constants E_{QX} (Eq. 4) of 10^{10} for any given QX when changing Q^+ from $N(CH_3)_4^+$ to $N(C_6H_{13})_4^+$. Tetrabutylammonium chloride or bromide, for instance, has a good balance of extractabilities for many organometallic applications.

$$E_{QX} = [QX]_{org} / [Q^+]_{aq} \cdot [X]^-_{aq} \tag{4}$$

- The original catalyst anion X^-, the reagent anion Y^-, and the anion Z^- generated in the reaction compete for the small catalytic concentration of Q^+ present. If Z^- is much more lipophilic than Y^- it will pair up with Q^+ and thereby encumber the catalytic effect ("catalyst poisoning"). Thus, the conversion will proceed properly only if the *relative* extractability of Y^- is better than that of Z^-. This condition is found in most organometallic applications. Extraction constants for many anions are tabulated in [1, 2]. These might help to get a feeling for new examples. QX values can vary by several powers of ten when Q^+ is kept constant and X^- is changed.

The kinetics of a PTC reaction can be very simple if anion transfer is fast relative to the intrinsic rate of the chemical reaction, and cases of pseudo first-order behavior are known. On the other hand, things can be very complicated if phase transfer is rate-limiting and special effects (e.g., competitive extraction, pH influences, complex stabilities, solubility limits, etc.) intervene. An extensive discussion of all possibilities and limiting rate laws has been published [1, Chapter 3].

Several fundamentally different ways of using PTC as a tool for organometallic catalysis can be distinguished, and some of these will be exemplified now with actual cases.

4.6.1.1.1 Solubilization by Complexation

Transition metal cations can be made organically soluble by complexation with a crown ether, a poly(ethylene glycol) or its dimethyl ether (an open crown) or tris(3,6-dioxaheptyl)amine (TDA-1, an open cryptand, **1**). TDA-1 is very hydrophilic and is most useful for the solubilization of solid salts. On the other hand, it also forms complexes with some metal carbonyls. Alternatively, a very lipophilic anion (for instance stearate) can make a salt "organic." Finally, some other special ligand (e.g., a bipyridine-*N,N'*-dioxide derivative) can be used. In all these cases positively charged species are brought into the organic phase for reaction.

$$N(CH_2CH_2OCH_2CH_2OCH_3)_3$$

1 (TDA-1)

4.6.1.1.2 Extraction of an Anionic Complex Salt

It can be shown that very many transition metal halides MX_n react with QX to form salts Q_mMX_{m+n}. For copper(I), for instance, the following simple salts and clusters have been characterized: NBu_4CuCl_2, NBu_4CuBr_2, $NBu_4Cu(CN)_2$ [9], $[NBu_4]_2Cu_2I_4$ [10], and $[MeNEt_3]_3Cu_6Br_9$ [11]. Other known ones include $NBu_4Ag(CN)_2$ [12] and NBu_4PdCl_3 [13].

Cerium(IV) presents an interesting case. The useful oxidant ceric ammonium nitrate (CAN) can be extracted from an aqueous solution as a complex salt $[NBu_4]_2[Ce(NO_3)_6]$ into hexane. Ceric ammonium sulfate, however, being a true double salt, cannot be transferred by a QX catalyst [14]. In an actual catalytic oxidation procedure, naphthalene is converted into naphthoquinone in a water/hexane two-phase system. CAN is added in a substoichiometric amount to the aqueous phase together with excess ammonium peroxidisulfate (the actual oxidant) and a small amount of $AgNO_3$ (catalyzing the reoxidation $Ce(III) \rightarrow Ce(IV)$) and NBu_4HSO_4 (PT catalyst). The water phase is acidified by sulfuric acid, and the oxidation proceeds at 55 °C. The reoxidation of Ce(III) occurs in the aqueous layer. It should be noted that there are three catalysts present in the system altogether: CAN, NBu_4HSO_4, Ag^+. It can be shown that the process becomes still faster and more efficient when a fourth catalyst, sodium dodecylsulfate as micellar catalyst, is added [14]!

4.6.1.1.3 Catalytic Transport from Aqueous Phase

In a third mechanistic possibility the transition metal catalyst remains in the organic phase all the time and is regenerated therein. Here the PT catalyst transports a different reagent to the reactive site from the water phase. A simple case in point is the oxidation of substituted (deactivated) toluenes to benzoic acids in dichloroethane/water by hypochlorite, catalyzed by organically dissolved RuO_4. OCl^- is extracted by NBu_4^+. No oxidation occurs in the absence of the PT catalyst, and the ruthenium precipitates as black RuO_2 [15]. Numerous, more complicated examples involve the reactions of metal carbonyls. In a reduction of nitrobenzene derivatives by carbon monoxide (1 bar), for instance, the catalytic reducing species $[HRu_3(CO)_{11}]^-$ is generated from $Ru_3(CO)_{12}$ at the interphase with 5 M NaOH. The PT catalyst ($PhCH_2NEt_3Cl$, "TEBA") carries it into the depths of the organic phase [16].

These principles of phase transfer catalysis allow manyfold applications in organometallic chemistry and, indeed, numerous studies have been published. The following sections concentrate on five subfields that have come to a certain maturity but quite a few other types have been explored.

4.6.1.2 Aqueous Organic-phase Heck and Other Cross Couplings Under Phase-transfer Catalysis Conditions

There is a limited number of publications on this subject as yet. It has been known for some time that the addition of tetrabutylammonium bromide or chloride to Pd(0) triphenylphosphine complexes enhances their stability and activates them for reactions at the same time [17]. It was tempting therefore to use the salts simultaneously as PT catalysts. Indeed, quite a number of such organometallic reactions can be accelerated in the presence both of salts QX and of *solid* inorganic bases such as potassium carbonate (solid/liquid PTC).

Zhang and Daves reported that water-containing solvents (H_2O/EtOH, 1:1) are more effective than conventionally used solvents (DMF, acetronitrile) in certain Heck couplings of iodoheterocycles and glycals or other cyclic enol ethers in the presence of NBu_4Cl [18]. This was corroborated by Jeffery for the coupling of iodobenzene and methyl acrylate. This reaction could even be performed in neat water with 0.05 equiv. of $Pd(OAc)_2$ with potassium carbonate or sodium carbonate as a base and 1 equiv. of a tetrabutylammonium salt as PT catalyst. Here it did not matter whether the NBu_4 counterion was chloride, bromide, or even hydrogensulfate [19]. The same group found, however, that even trace amounts of water brought in with the various tetrabutylammonium salts were detrimental to the same reaction when performed in DMF or acetonitrile solvents. In these cases the presence of molecular sieves as dehydrating agent was recommended [20].

Again in water as medium, styrene and acrylic acid were coupled with aryl bromides or iodides using K_2CO_3 as a base, NBu_4Br as a PT agent and $PdCl_2/P(o\text{-Tol})_3$ as an organometallic catalyst (Scheme 1). Isolated yields were mostly in the order of 70–90% [21].

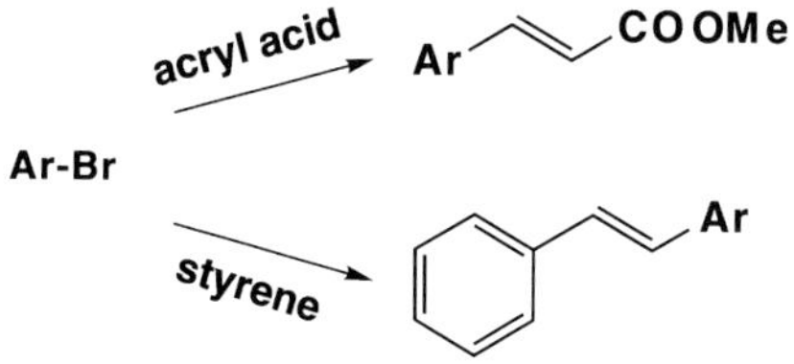

Ar-Br —acryl acid→ Ar–CH=CH–COOMe

Ar-Br —styrene→ Ph–CH=CH–Ar

Scheme 1

This methodology was extended very recently to the "ligandless" Suzuki type coupling of aryl bromide and substituted arylboronic acid in water [86].

Cross couplings of vinyl halides with alkynes to give enynes were also performed in benzene/10% aqueous sodium hydroxide in the presence of $Pd(PPh_3)_4$, CuI and TEBA at room temperature [22, 23]. This was extended to couplings of vinyl bromide and ethinyl thiophenes ([24], Scheme 2) and to the alternative combination of 2-halothiophene and propyne [25].

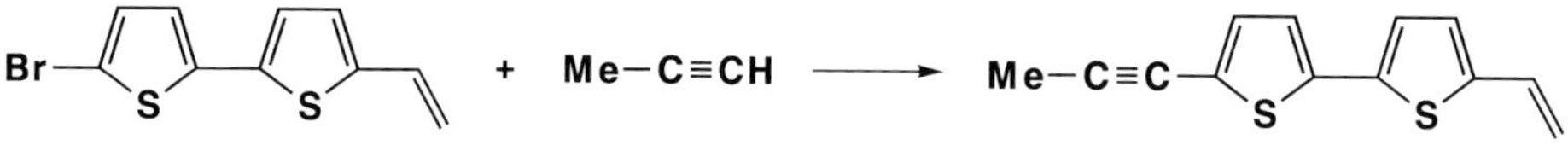

Br–(bithiophene)–CH=CH₂ + Me–C≡CH ⟶ Me–C≡C–(bithiophene)–CH=CH₂

Scheme 2

Rossi's group [26] pioneered a very elegant one-pot procedure for multistep couplings of two aryl or heteroaryl halides to a central acetylene unit that is partially masked in the beginning and becomes unmasked in due course: the first aryl halide is coupled with 2-methyl-3-butyn-2-ol at room temperature under PT conditions in benzene/5.5 M aqueous NaOH. TEBA is the PT agent, and a mixture of $Pd(PPh_3)_4$ and copper(I) iodide are the organometallic catalysts. The coupling product may be isolated, and the yield is often very high. More conveniently, however, the reaction mixture is charged thereafter with the second (different or identical) aryl halide and heated for up to 40 to 50 h at 80 °C. Fragmentation of acetone and a second coupling occur in moderate to good yields ([26], Scheme 3).

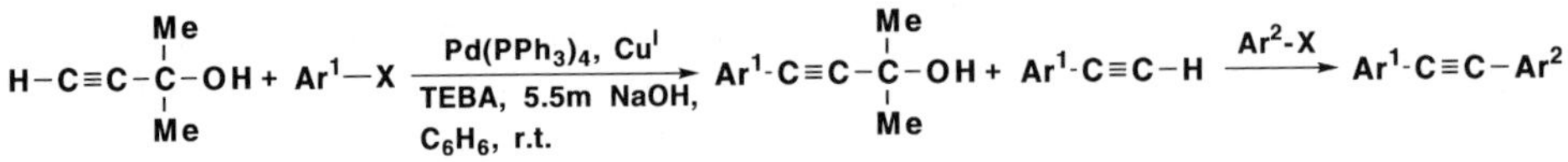

$$H{-}C{\equiv}C{-}C(Me)_2{-}OH + Ar^1{-}X \xrightarrow[\text{TEBA, 5.5m NaOH, } C_6H_6\text{, r.t.}]{Pd(PPh_3)_4,\ Cu^I} Ar^1{-}C{\equiv}C{-}C(Me)_2{-}OH + Ar^1{-}C{\equiv}C{-}H \xrightarrow{Ar^2{-}X} Ar^1{-}C{\equiv}C{-}Ar^2$$

Scheme 3

Applications of this method include among others the preparation of unsymmetrical diaryldiynes [27] and of 1,4-bis[2-(4′,4″-dialkoxyphenyl)ethenyl]-benzenes for polymer [28] and liquid-crystal synthesis [29]. Such one-pot "acetylene double couplings" are also possible between (substituted) 6-bromoazalenes and substituted bromobenzenes [30], and even macrocyclic polyene-polyynes have been made in this way ([31, 32], Scheme 4).

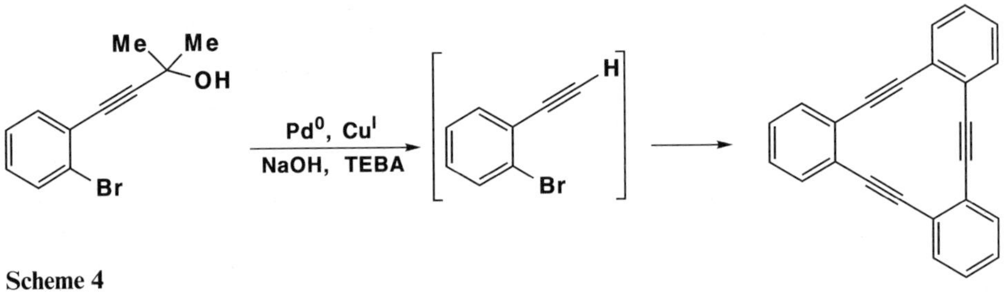

Pd^0, Cu^I; NaOH, TEBA

Scheme 4

4.6.1.3 Hydrogenations Mediated by Phase-transfer Catalysts

Very convenient aqueous/organic hydrogenations can be executed at room temperature with H_2 at atmospheric pressure with rhodium trichloride and Aliquat 336 (technical-grade methyltrioctylammonium chloride). QX extracts the rhodium salt from the aqueous layer as $Q^+ \; RhCl_4^-$. Hydrogenation occurs after an induction period. This can be shortened by pretreatment of the complex with a little H_2 and substrate before the proper hydrogenation [33]. It is believed that addition of a first substrate molecule to the rhodium takes place during this pretreatment. In the absence of the PT catalyst, metallic rhodium separates within minutes and the rate of hydrogenation is much lower. The metal complex is stabilized further by the addition of trialkylamines. The actual hydrogenation catalyst is homogeneous. It can be re-used for subsequent hydrogenations of different substrates. The structure of the active species is only speculative [34]. The extracted ion pair $Q^+ \; RhCl_4^-$ is somewhat hydrated and, as a matter of fact, a dried solution becomes totally inactive. Deuterium labeling indicates, however, that the water hydrogen atoms do not participate in the hydrogenation [33, 34].

Extensive work showed that this methodology can be applied to many hydrogenations: alkenes or alkynes [33], substituted benzene → substituted cylohexane [33, 34], deuterium exchange [34], selective reduction of polycyclic aromatics [35, 36], selective double-bond reduction in the presence of nitro groups (e.g., $PhCH{=}CHNO_2 \rightarrow PHCH_2CH_2NO_2$) [37], and reduction of double bonds in α,β-unsaturated esters and carbonyls [38, 39].

Rhodium(III) complexes $L_2Rh(H)Cl_2$ (L = tricyclohexylphosphine or triisopropylphosphine) were also shown to be excellently suited to a PTC-mediated hydrogenolysis of chloroarenes. The solvent mixture was 40% aqueous NaOH/toluene, and TEBA was the PT catalyst. Functional groups such as OR, CF_3, COR, COOH, and NH_2 were compatible with the $C{-}Cl \rightarrow C{-}H$ process [40].

Quite a different catalyst and technique were used by Wilkinson and co-workers for the hydrogenation of benzylideneaniline in water/diethyl ether. The water-soluble complex salt $[Rh(PPh_3)_2(cod)]PF_6$ was extracted by Triton X-100, a non-ionic detergent [41].

A unique application comprises a facile hydrodehalogenation (hydrogenolysis) of aromatic compounds in 15% KOH with a heterogeneous Pd/C catalyst and Aliquat 336 as PT catalyst [42]. The latter results in a marked rate enhancement in many cases. The ease of halide removal falls from iodide through chloride to fluoride. It is believed that the lipophilic Q^+ promotes the transport of the halide ion produced, from the Pd site to the alkaline aqueous phase.

Alper's group discovered a ring hydrogenation of variously substituted aromatics and heteroaromatics at 1 bar pressure in a mixture of a buffer (pH 7.4–7.6) and hexane. Acetophenone gave acetylcyclohexane as the major product; phenol led to cyclohexanone and cyclohexanol. Use was made of the dimer of chloro(1,5-hexadiene)rhodium as metal catalyst and cetyltrimethylammonium bromide or tetrabutylammonium hydrogensulfate as PT catalyst [43]. The constituents of the buffer were not important, only the pH applied. Reactions in the absence of the PT catalysts were much slower, but it is difficult to understand their role in the reaction apart from the stabilizing effect on the metal complex catalyst.

The hydrogenation catalyst $K_3[Co(CN)_5H]$ can be prepared easily from $CoCl_2$ and KCN with gaseous hydrogen. Its stability and usefulness is improved vastly by the presence of a PT catalyst for hydrogenation in benzene/dilute aqueous NaOH. TEBA and tetramethylammonium chloride are recommended by the Reger group as catalysts [44, 45], but more lipophilic onium salts might be even better. Japanese workers use dodecyldimethyl(α-methylbenzyl)-ammonium chloride [46]. The method allows transformations such as styrene $\rightarrow$ ethylbenzene [46], methyl sorbate $\rightarrow$ methyl hex-3-enoate and hex-2-enoate (ratio 3:1) [47], α,β-unsaturated ketones $\rightarrow$ saturated ketones [44, 45], and conjugated dienes $\rightarrow$ *E*-monoenes (main products) in 1,4-additions [44, 45].

4.6.1.4 Biphasic Transfer Hydrogenations

For the simplest form of such reactions, palladium on charcoal is the catalyst and aqueous formate the hydrogen donor. Interfacial transport of formate is facilitated by an organophilic counter ion, and for the present purpose a triethylammonium ion is mostly used [48, 49].

Applications include these conversions: alkyne → *Z*-alkene [48], aromatic nitrocompound → aniline derivative [48], aromatic ketone → hydrocarbon [48], and selective reduction of one nitro group in dinitroaromatic compounds [49]. A similar method was also applied for the reductive coupling of aryl halides or heteroaryl halides to yield biaryls or biheteroaryls, respectively [50, 51]. Aqueous sodium formate was used together with either a QX (cetyltrimethylammonium chloride, cetyltributylphosphonium bromide, TEBA), or a neutral or anionic surfactant instead. The alkyne → (*Z*)-alkene transfer hydrogenation was also performed with sodium phosphinate, NaH_2PO_2, as hydrogen donor, TEBA as phase-transfer agent, and Pd/C catalysts that were modified by Hg or Pb salt additions [52].

A somewhat different approach toward formate/QX transfer extraction and hydrogenation was taken by Blum and Sasson [53–56]. They started out by using $RuCl_2(PPh_3)_3$ for aldehyde → alcohol and unsaturated ketone → saturated ketone and $RhCl(PPh_3)_3$ for ketone → alcohol reductions at 20–80 °C [53–55]. A kinetic investigation showed a dependence on structure and concentration of the onium salt (among other factors). The observed activation energy suggested that the process is both chemical rate- and diffusion-controlled [54]. Hydrogenolysis of aryl bromides was possible under similar conditions with $PdCl_2(PPh_3)_2$ plus additional PPh_3 at 100 °C [56]. In these studies, the efficiency of various onium salts was compared: TEBA < PBu_4Br < NBu_4HSO_4 < Aliquat 336 < hexadecyltributylphosphonium bromide < tetrahexylammonium hydrogensulfate. Thus, the most lipophilic ammonium salt, predicted to be the best extractant for formate by phase-transfer catalysis theory, is the best suited agent.

There is yet another transfer/PTC hydrogenation method for the reduction of ketones to alcohols: here 2-propanol in dilute NaOH is the reductant, a QX is the phase-transfer agent, and $HFe_3(CO)_{11}^-$ (generated in situ from $Fe_3(CO)_{12}$) is the metal catalyst [57].

4.6.1.5 Aqueous/Organic-phase Oxidations Mediated by Metal and PT Catalysts

Aqueous hydrogen peroxide is one of the cheapest and most convenient oxidants. It can be extracted into organic media with an onium salt QX as a complex $[QX \cdots H_2O_2]$ [58]. Patents describe oxidations of alkenes in the additional presence of heavy metal oxides [59]. OsO_4, MoO_3, and H_2WO_4 give mainly *trans*-diols and epoxides; V_2O_5, Cr_2O_3, and TiO_2 lead mostly to allylic alcohols and α,β-unsaturated ketones. Styrene can be transformed into benzaldehyde with $RuCl_3/QX/H_2O_2$ at pH < 4 [60], and the side chains of aromatics are oxidized to ketones or tertiary alcohols under similar conditions [61]. In these experiments, the PT catalyst has a threefold role: extraction of both H_2O_2 and $RuCl_3$, and stabilization of Ru(III) against reduction. When the ratio $QX/RuCl_3$ is less than 8:1, metallic ruthenium is precipitated.

The situation becomes quite different when hydrogen peroxide is used to generate certain peroxo anions. As a matter of fact, many information exists on oxidations by aqueous H_2O_2 with homo- or hetero-polyanion redox carriers and a QX as carrier cation source. The anions ($PMo_{12}O_{40}^{3-}$; $PW_{12}O_{40}^{3-}$; $PW_4O_{16}^{3-}$; $Mo_7O_{24}^{6-}$; $PNiW_{11}O_{39}^{5-}$; $SiRu(H_2O)W_{11}O_{39}^{5-}$ and others) are formed in situ in some cases; in others they are brought into the reaction mixture as previously prepared catalyst salts. They are oxidized to peroxo anions in the aqueous layer (or at the interphase) prior to extraction.

The relevant literature in this subfield is too voluminous to be detailed here. Overviews are available [62], and only a few recent references to the newest publications are given below. Oxidations of the following types have been performed: alcohol → ketone [63]; aldehyde → acid; alkene → diol or epoxide [64–67]; alkene → aldehyde, acid; 1-alkyne → ketoaldehyde and acid (1 C-atom shorter); internal alkyne → α,β-epoxyketone; vic-diol → 1,2-diketone [68] or hydroxyketone [69]; amine → amine oxide [70]; aromatic amine → nitrosobenzene, nitrobenzene, azoxybenzene [71].

Aqueous sodium hypochlorite is another low-priced oxidant. Very efficient oxidative systems were developed which contain a *meso*-tetraarylporphyrinato-Mn(III) complex salt as the metal catalyst and a QX as the carrier of hypochlorite from the water phase to the organic environment. These reactions are of interest also as cytochrome *P*-450 models. Early experiments were concerned with epoxidations of alkenes, oxidations of benzyl alcohol and benzyl ether to benzaldehyde, and chlorination of cyclohexane at room temperature or 0 °C. A certain difficulty arose from the fact that the porphyrins were not really stable under the reaction conditions. Several research groups published extensively on optimization, factors governing catalytic efficiency, and stability of the catalysts. Most importantly, axial ligands on the Mn porphyrin (e.g., substituted imidazoles, 4-substituted piridines and their *N*-oxides), **2** increase rates and selectivities. This can be demonstrated most impressively with pyridine ligands directly tethered to the porphyrin [72]. Secondly, 2,4- and 2,4,6-trihalo- or 3,5-di-*tert*-butyl-substituted tetraarylporphyrins are more stable. Thirdly, the pH has a strong influence: at pH 12.7 only QOCl (extracted from the aqueous phase with Aliquat 336, for instance) is present as primary oxidant in the organic medium. At pH 9.5–10.5, however, HOCl is the oxidizing species in the nonaqueous environment. The reaction proceeds in this case without the presence of a phase-transfer catalyst [73]. Under optimal conditions the reactions can be over after a few minutes at 0 °C. Iron porphyrins proved to be less active for the present purpose. Regio- and stereochemistry of the Mn porphyrin/PT catalyst-mediated epoxidations of dienes and terpenes differed from those found for other epoxidation methods. A certain influence of the nature of the axial ligand on reaction results was also observed [74–76]. The actual oxidant is considered to be a Mn(V) oxo species, but the rate-determining step is still in doubt. Overviews of the extensive literature are available [2, 3, 77].

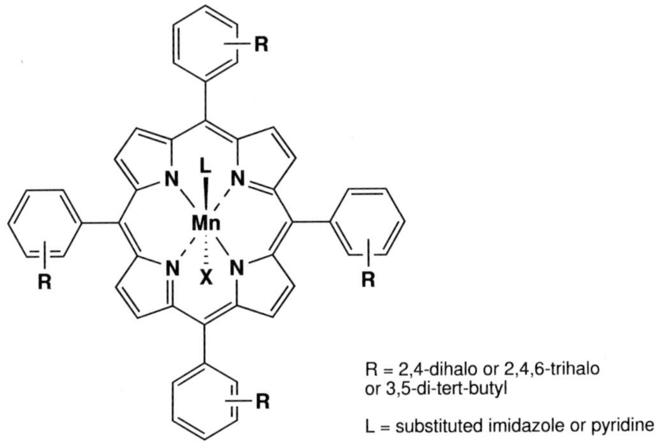

R = 2,4-dihalo or 2,4,6-trihalo
or 3,5-di-tert-butyl

L = substituted imidazole or pyridine

2

Potassium hydrogenperoxysulfate has been used alternatively to hypochlorite as primary oxidant in phosphate buffer (pH 6–7)/dichloromethane with the Mn porphyrin/tetraalkylammonium chloride system at room temperature. Very fast alkene epoxidations and hydroxylations of hydrocarbons (cyclohexane, adamantane, decalins, etc.) were observed. Methoxybenzenes were also oxidized [78].

4.6.1.6 Aqueous/Organic-phase Carbonylations

Carbonylations of a great many different compounds have been performed by CO, a catalytic amount of various metal carbonyls, 0.3–6 M NaOH, and a phase-transfer catalyst. Typical reactions occur at atmospheric pressure and often at room temperature. The literature in this field is very voluminous, but more or less extensive overviews are available up to the early 1990s [1–4, 6, 7], and further results are emerging continuously.

Benzyl, allyl, vinyl, and aryl, but also some aliphatic, halides have been transformed into carboxylic acids (Eq. 5). Under other conditions, couplings with CO to give symmetrical or unsymmetrical ketones were possible. Epoxides yielded unsaturated hydroxy acids, and α-oxo-butyrolactones (Eq. 6), couplings of alkynes and alkyl halides gave hydroxybutenolides (Eq. 7) and other more complex conversions could also be realized.

$$\text{Ar-X} + \text{Me-I} \xrightarrow[\text{C}_6\text{H}_6,\ \text{CO (1bar), 20 h, r.t.}]{\text{QBr, 50\% NaOH, Co}_2\text{(CO)}_8} \text{Ar-COOH} + \text{Ar-CO-Me} \quad (5)$$

$$\text{Ph(H)C(O)C(H)R (epoxide)} \xrightarrow[\text{C}_6\text{H}_6,\ \text{CO (1bar), Me=I, r.t., 24 h}]{\text{QBr, 0.5 } m \text{ NaOH, Co}_2\text{(CO)}_8} \text{(furanone: Ph, R, H, HO, O, O)} \quad (6)$$

$$\text{R—≡—H} + \text{Me-I} \xrightarrow[\text{C}_6\text{H}_6,\ \text{CO (1bar), r.t.}]{\text{QBr, 5 } m \text{ NaOH, Co}_2\text{(CO)}_8} \text{(butenolide: R, Me, OH, O, O)} \quad (7)$$

The carbonylation of a benzyl halide in the presence of iron pentacarbonyl to give a phenylacetic acid may serve to exemplify the interaction of a metal carbonyl, carbon monoxide, PT catalyst, aqueous sodium hydroxide, and the substrate [79]. $Fe(CO)_5$ is attacked by QOH at the interphase, and the species formed is extracted into the depths of the organic phase, where it reacts with CO and benzyl halide (Eqs. 8 and 9). This new anion **3** is the actual catalyst. It reacts with a second benzyl halide to give a non-ionic intermediate **4** (Eq. 10). By insertion of CO and attack of QOH, **4** is decomposed to the reaction product under regeneration of **3** (Eq. 11). Thus, the action of the PT catalyst is twofold. Firstly it transports the metal carbonyl anion. More important seems to be its involvement in the (rate-determining) decomposition step. A basically similar mechanism was proposed for cobalt carbonyl reactions [80], which have been modified somewhat quite recently (see below).

$$\text{Fe(CO)}_5 + 2\,\text{QOH} \longrightarrow \text{Q}_2[\text{Fe(CO)}_4] + \text{CO}_2 + \text{H}_2\text{O} \quad (8)$$

$$\underset{\mathbf{2}}{\text{Q}_2[\text{Fe(CO)}_4]} + \text{CO} + \text{ArCH}_2\text{X} \longrightarrow \underset{\mathbf{3}}{\text{Q}[\text{ArCH}_2\text{–CO–Fe(CO)}_4]} + \text{QX} \quad (9)$$

$$\text{Q}[\text{ArCH}_2\text{–CO–Fe(CO)}_4] + \text{ArCH}_2\text{X} \longrightarrow \underset{\mathbf{4}}{\text{ArCH}_2\text{–CO–Fe(CH}_2\text{R)(CO)}_4} + \text{QX} \quad (10)$$

$$\text{ArCH}_2\text{–CO–Fe(CH}_2\text{R)(CO)}_4 + \text{QOH} + \text{CO} \longrightarrow \text{ArCH}_2\text{COOH} + \text{Q}[\text{ArCH}_2\text{COFe(CO)}_4] \quad (11)$$

Most published PT-assisted carbonylation reactions were executed with $Co_2(CO)_8$, others with $Fe(CO)_5$ as indicated above, or with $Ru_3(CO)_{12}$, Pd complexes, $Ni(CN)_2$, and cobalt salts. The catalytically active species $Q[Ni(CN)(CO)_3]$ generated from nickel cyanide is much less hazardous than the very poisonous nickel carbonyl itself.

Research trends of the last few years highlight applications to more involved systems either from the substrate/product side or from the catalyst side. Furthermore, a deeper insight into underlying mechanism is intended. Thus, reductive carbonylation of dibromocyclopropanes was performed in toluene/5 M

KOH with syngas (CO/H_2, 3:1) at elevated temperature (90 °C) using a mixture of $CoCl_2$, KCN, and $Ni(CN)_2$ for the metal catalyst and PEG-400 as PT catalyst which was much more efficient than a quaternary ammonium catalyst [81]. 1,1-Dibromo-2-phenylcyclopropane furnished a 72% yield of 2-phenylcyclopropanecarboxylic acid (1:1 *cis*/*trans* mixture).

The same set of metal and PT catalyst and reaction conditions was used for a reductive carbonylation of 1-alkynes giving mostly branched acids [82] (Eq. 12).

$$\text{R–C}\equiv\text{C–H} \longrightarrow \text{R–CH(CH}_3\text{)–COOH} \qquad (12)$$

Poly(ethylene glycol)s (PEGs) substitute for onium salts as PT catalysts occasionally. There are cases where yields vary with the catalyst, but in one reaction the stereochemical outcome is influenced: nickel cyanide catalyzes double insertion of CO into alkynols (5 M NaOH, 95 °C) furnishing the acids shown in Eq. (13). When an NR_4X is the PT agent, the major products have *Z* geometry, and if PEG-400 is present *E* geometry prevails [83].

Alkyl(Me)C(OH)–C≡C–H → [Ni(CN)₂, catalyst; 5 *m* NaOH, toluene; 95°C, CO (1bar)] → Alkyl(Me)C=C(CH₂–COOH)(COOH) (E) + Alkyl(Me)C=C(COOH)(CH₂–COOH) (Z) (13)

Catalyst:	E	Z
NR_4X	5–15%	85–95%
PEG-400	65–90%	10–35%

Again using PEGs as PT catalysts, the Alper group reinvestigated the carbonylation of benzyl halides in the presence of $Co_2(CO)_8$. They were able to characterize and investigate η^1-benzyl, η^3-benzyl, and (η^1-phenylacetyl) cobalt carbonyls as intermediates, and arrived at an elaborate mechanistic cycle [84].

Another recent development is the preparation of polycycles that are difficult to make, by utilizing the extraction of rhodium chloride from water with Aliquat 336 for reductive ring-forming carbonylations (Eq. 14) [85].

R R → R R O and/or R R R R O O (14)

References

[1] Ch. M. Starks, Ch. L. Liotta, M. Halpern, *Phase Transfer Catalysis; Fundamentals, Applications, and Industrial Perspectives,* Chapman and Hall, New York, **1994**. Specifically Chapter 13: Phase transfer catalysis–transition metal cocatalyzed reactions, p. 594.
[2] E. V. Dehmlow, S. S. Dehmlow, *Phase Transfer Catalysis,* 3rd ed., VCH, Weinheim, New York, **1993**. Specifically Section 3.19: Organometallic PTC applications, p. 260.
[3] Y. Goldberg, *Phase Transfer Catalysis; Selected Problems and Applications,* Gordon and Breach, Yverdon, **1989**. Specifically Chapters 3 and 4: Phase transfer catalysis in organometallic chemistry, and Metal-complex catalysis under phase transfer conditions, p. 127.
[4] H. Alper, Phase transfer reactions catalyzed by metal complexes, in *Phase Transfer Catalysis; New Chemistry, Catalysts, and Applications* (Ed.: Ch. M. Starks), ACS Symposium Ser. No. 326, American Chemical Society, Washington, DC, USA, **1987**, Chapter 2, p. 8.
[5] R. A. Sawicki, *Triphase catalysis in organometallic anion chemistry,* in *Phase Transfer Catalysis; New Chemistry, Catalysts, and Applications* (Ed.: Ch. M. Starks), ACS Symposium Ser. No. 326, American Chemical Society, Washington, DC, USA, **1987**, Chapter 12, p. 143.
[6] H. Des Abbayes, *Israel J. Chem.* **1985**, *26*, 249.
[7] H. Alper, *Adv. Organomet. Chem.* **1981**, *19*, 183.
[8] J. F. Petrignani, *Chem. Met.–Carbon Bond* **1989**, *5*, 63.
[9] M. Nilsson, *Acta Chem. Scand. Ser. B* **1982**, *36*, 125.
[10] M. Asplund, S. Jagner, M. Nilsson, *Acta Chem. Scand. Ser. A* **1982**, *36*, 751.
[11] S. Andersson, S. Jagner, *Acta Chem. Scand. Ser. A* **1989**, *43*, 39.
[12] W. R. Mason, *J. Am. Chem. Soc.* **1973**, *95*, 3573.
[13] S. Cacchi, F. La Torre, D. Misiti, *Tetrahedron Lett.* **1979**, 4591.
[14] E. V. Dehmlow, J. K. Makrandi, *J. Chem. Res. (S)* **1986**, 32.
[15] Y. Sasson, G. D. Zappi, R. Neumann, *J. Org. Chem.* **1986**, *51*, 2880.
[16] H. Alper, S. Amaratunga, *Tetrahedron Lett.* **1980**, *21*, 2603.
[17] C. Amatore, M. Azzabi, A. Jutand, *J. Am. Chem. Soc.* **1991**, *113*, 8375.
[18] H.-C. Zhang, G. D. Daves, Jr., *Organometallics* **1993**, *12*, 1499.
[19] T. Jeffery, *Tetrahedron Lett.* **1994**, *35*, 3051.
[20] T. Jeffery, J.-C. Galland, *Tetrahedron Lett.* **1994**, *35*, 4103.
[21] N. A. Bumagin, V. V. Bykov, L. I. Sukhomlinova, T. P. Tostaya, I. P. Beletskaya, *J. Organomet. Chem.* **1995**, *486*, 259.
[22] R. Rossi, A. Carpita, M. G. Quirici, M. L. Gaudenzi, *Tetrahedron* **1982**, *38*, 631.
[23] R. Rossi, A. Carpita, *Tetrahedron* **1983**, *39*, 287.
[24] R. Rossi, A. Carpita, A. Lezzi, *Tetrahedron* **1984**, *40*, 2773.
[25] M. D'Auria, A. De Mico, F. D'Onofrio, G. Piancatelli, *Gazz. Chim. Ital.* **1986**, *116*, 747.
[26] A. Carpita, A. Lessi, R. Rossi, *Synthesis* **1984**, 571.
[27] S. A. Nye, K. T. Potts, *Synthesis* **1988**, 375.
[28] C. Pugh, V. Percec, *J. Polym. Sci., Part A, Polym. Chem.* **1990**, *28*, 1101.
[29] C. Pugh, V. Percec, *Mol. Cryst. Liq. Cryst.* **1990**, *178*, 193; *Chem. Abstr.* **1990**, *112*, 208389.
[30] D. Balschukat, E. V. Dehmlow, *Chem. Ber.* **1986**, *119*, 2272.
[31] C. Huynh, G. Linstrumelle, *Tetrahedron* **1988**, *44*, 6337.
[32] Y. Tobe, H. Matsumoto, K. Naemura, Y. Achiba, T. Wakabayashi, *Angew. Chem.* **1996**, *108*, 1924; *Angew. Chem., Int. Ed. Engl.* **1996**, *35*, 1800.
[33] J. Blum, I. Amer, A. Zoran, Y. Sasson, *Tetrahedron Lett.* **1983**, *24*, 4139.
[34] J. Blum, I. Amer, K. P. C. Vollhardt, H. Schwarz, G. Höhne, *J. Org. Chem.* **1987**, *52*, 2804.
[35] I. Amer, H. Amer, R. Ascher, J. Blum, Y. Sasson, K. P. C. Vollhardt, *J. Mol. Catal.* **1987**, *39*, 185.

[36] I. Amer, H. Amer, J. Blum, *J. Mol. Catal.* **1986**, *34*, 221.
[37] I. Amer, T. Bravdo, J. Blum, K. P. C. Vollhardt, *Tetrahedron Lett.* **1987**, *28*, 1321.
[38] J. Azran, O. Buchman, I. Amer, J. Blum, *J. Mol. Catal.* **1986**, *34*, 229.
[39] Henkel KG (A. Laufenberg, A. Behr, W. Keim), DE 3.841.698 (1989); *Chem. Abstr.* **1990**, *112*, 97724.
[40] V. V. Grushin, H. Alper, *Organometallics* **1991**, *10*, 1620.
[41] G. J. Longley, T. J. Goodwin, G. Wilkinson, *Polyhedron* **1986**, 1625.
[42] C. A. Marques, M. Selva, P. Tundo, *J. Org. Chem.* **1994**, *59*, 3830.
[43] K. R. Januskiewicz, H. Alper, *Organometallics* **1983**, *2*, 1055.
[44] D. L. Reger, M. M. Habib, D. J. Fauth, *Tetrahedron Lett.* **1979**, 115.
[45] D. L. Reger, M. M. Habib, D. J. Fauth, *J. Org. Chem.* **1980**, *45*, 3860.
[46] Fac. Eng. Kumamoto University (K. Yamashita, K. Ohkubo), Nippon Kagaku Kaishi **1984**, 505; *Chem. Abstr.* **1984**, *101,* 90102.
[47] D. L. Reger, M. M. Habib, *J. Mol. Catal.* **1980**, *7*, 365.
[48] J. R. Weir, B. A. Patel, R. F. Heck, *J. Org. Chem.* **1980**, *45*, 4926.
[49] M. O. Terpko, R. F. Heck, *J. Org. Chem.* **1980**, *45*, 4992.
[50] P. Bamfield, P. M. Quan, *Synthesis* **1978**, 537.
[51] G. R. Newkome, D. C. Pantaleo, W. E. Puckett, P. L. Ziefle, W. A. Deutsch, *J. Inorg. Nucl. Chem.* **1981**, *43*, 1529.
[52] R. A. W. Johnstone, A. H. Wilby, *Tetrahedron* **1981**, *37*, 3667.
[53] R. Bar, Y. Sasson, J. Blum, *J. Mol. Catal.* **1984**, *26*, 327.
[54] R. Bar, L. K. Bar, Y. Sasson, J. Blum, *J. Mol. Catal.* **1985**, *33*, 161.
[55] R. Bar, Y. Sasson, *Tetrahedron Lett.* **1981**, *22*, 1709.
[56] R. Bar, Y. Sasson, J. Blum, *J. Mol. Catal.* **1982**, *16*, 175.
[57] K. Jothimony, S. Vancheesan, *J. Mol. Catal.* **1989**, *52*, 301.
[58] E. V. Dehmlow, M. Slopianka, *Chem. Ber.* **1979**, *112*, 2765.
[59] Continental Oil Co. (D. R. Napier, C. M. Starks), US 3.992.432 (1976); *Derwent Abstr.* **1976**, 90477; (C. M. Starks, D. R. Napier), ZA 7.101.495 (1971); *Chem. Abstr.* **1972**, *76*, 153191.
[60] G. Barak, J. Dakka, Y. Sasson, *J. Org. Chem.* **1988**, *53*, 3553.
[61] Y. Sasson, G. Barak, *J. Chem. Soc., Chem. Commun.* **1988**, 637.
[62] P. 346 in [2]; p. 199 in [3].
[63] C. Venturello, M. Gambaro, *J. Org. Chem.* **1991**, *56*, 5924.
[64] L. J. Csanyi, K. Jaky, *J. Catal.* **1991**, *127,* 42.
[65] C. Aubry, G. Chottard, N. Platzer, J. M. Brégeault, R. Thouvenot, F. Chauveau, C. Huet, H. Ledon, *Inorg. Chem.* **1991**, *30*, 4409.
[66] Atochem. S.A. (P. Caubere, Y. Fort, A. Otar), EP 434.546 (1991); *Chem. Abstr.* **1991**, *115*, 280786.
[67] R. Neumann, A. M. Khenkin, *J. Org. Chem.* **1994**, *59*, 7577.
[68] T. Iwahama, S. Sakaguchi, Y. Nishiyama, Y. Ishii, *Tetrahedron Lett.* **1995**, *36*, 1523.
[69] Y. Sakata, Y. Ishii, *J. Org. Chem.* **1991**, *56*, 6233.
[70] A. J. Bailey, W. P. Griffith, B. C. Parkin, *J. Chem. Soc., Dalton Trans.* **1995**, 1833.
[71] S. Sakaue, T. Tsubakina, Y. Nishiyama, Y. Ishii, *J. Org. Chem.* **1993**, *58*, 3633.
[72] F. Montanari, M. Penso, S. Quici, P. Viganò, *J. Org. Chem.* **1985**, *50,* 4888.
[73] S. Banfi, F. Montanari, S. Quici, *J. Org. Chem.* **1989**, *54*, 1850.
[74] M.-E. De Carvalho, B. Meunier, *Nouv. J. Chim.* **1986**, *10*, 223.
[75] B. Meunier, E. Guilmet, M.-E. De Carvalho, R. Poilblanc, *J. Am. Chem. Soc.* **1984**, *106*, 6668.
[76] J. P. Collman, J. I. Brauman, B. Meunier, T. Hayashi, T. Kodadek, S. A. Raybuck, *J. Am. Chem. Soc.* **1985**, *107*, 2000.
[77] F. Montanari, S. Banfi, S. Quici, *Pure Appl. Chem.* **1989**, *61*, 1631.

[78] A. Robert, B. Meunier, *New J. Chem.* **1988**, *12*, 885.
[79] G. Tanguy, B. Weinberger, H. des Abbayes, *Tetrahedron Lett.* **1983**, *24*, 4005.
[80] H. des Abbayes, *New J. Chem.* **1987**, *11*, 535.
[81] V. V. Grushin, H. Alper, *Tetrahedron Lett.* **1991**, *32*, 3349.
[82] J.-T. Lee, H. Alper, *Tetrahedron Lett.* **1991**, *32*, 1769.
[83] Z. Zhou, H. Alper, *Organometallics* **1996**, *15*, 3282.
[84] C. Zucchi, G. Pályi, V. Galamb, E. Sámpar-Szerencsés, L. Markó, P. Li, H. Alper, *Organometallics* **1996**, *15*, 3222.
[85] Y. Badrieh, J. Blum, H. Schumann, *J. Mol. Catal.* **1994**, 231.
[86] D. Badone, M. Baroni, R. Cardamone, A. Ielmini, U. Guzzi, *J. Org. Chem.* **1997**, *62*, 7170.

4.6.2 Counter-phase Transfer Catalysis

Tamo Okano

4.6.2.1 Introduction

The phenomena accompanying interphase transfer of chemical materials are universally observed in biological systems. A number of carriers and various transportation methods are used and the selective transportation of materials contributes to controlling the biochemical reactions *in vivo*. In synthetic chemistry, however, the carriers as well as the methods are very limited. Phase-transfer catalysts (PTCs; cf. Section 4.6.1) such as crown ethers or onium salts are limited to the transportation of anions from an aqueous or solid phase into an organic phase; nevertheless, the PTCs contributed to the development of synthetic chemistry. The most important point is that these catalysts have enabled the biphasic reactions of lipophilic molecules with inexpensive inorganic salts and at the same time facilitated the separation of products.

Use of inorganic salts and easy separation are also important problems in catalytic reactions involving transition metal complexes. It is well known that the PTCs greatly accelerate the organic reactions catalyzed by anionic metal complexes which are water-soluble, owing to their easy extraction [1, 2] (Figure 1). However, the most important transition metal catalysts in homogeneous system consist of phosphine complexes, which are commonly lipophilic. Although PTCs are applicable to such catalytic reactions of hydrophobic substrates with inorganic salts [2a], they have no effect on the separation of the phosphine complexes and the products.

New tools for such biphasic reactions are inverse- or counter-phase transfer catalysts, which are able to transport lipophilic molecules from the organic phase into the aqueous phase. An advantage of the inverse- or counter-PTC is its applicability to reactions not only with ionic salts but also with non-ionic

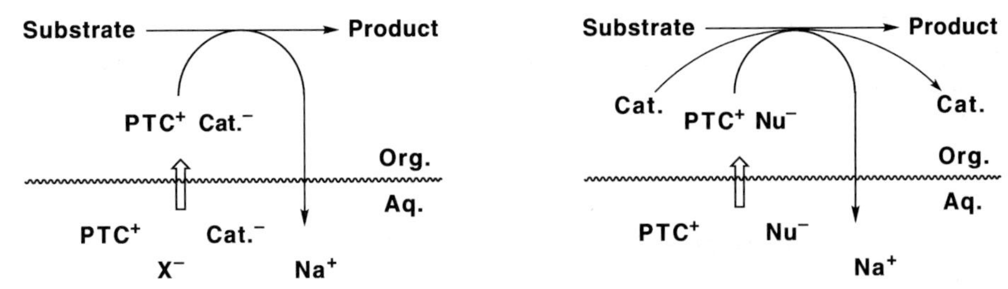

Figure 1. Mechanisms for metal-catalyzed reactions using phase-transfer catalysts: phase-transfer catalyst and anionic metal catalyst (left); phase-transfer catalyst and lipophilic metal catalyst (right).

reagents soluble in water. Such carriers were first reported by Mathias and Vaidya and by us in 1986 [3, 4], and three types of carriers are known at present. Mathias and Vaidya found that pyridine derivatives react with acid halides in the organic phase to form the pyridinium salts, which transfer into the aqueous phase [3]. This catalysis was named "inverse-phase transfer catalysis". As catalysis of this type, i.e., pyridine derivatives [5] and sulfides [6], are able to transport only alkyl and carboxylic acid halides, their application to organometallic reactions has not been reported.

The other type of inverse-PTCs are cyclodextrins, which are capable of forming water-soluble inclusion compounds with organic molecules with their lipophilic cavities. Owing to the absence of limitations of the functional groups of organic substrates on this transportation, the cyclodextrins are applicable to biphasic reactions using water-soluble metal catalysts (Figure 2). For example, Wacker oxidations [7, 8], deoxygenations [9, 10], and hydroformylations [11, 12] have been reported. However, inhibition of the catalytic activity due to complexations of metal catalysts with cyclodextrins is observed in some cases [8, 9, 11]. Another defect is a steric limitation on the transportation of the substrates due to the cavity size.

The inverse-PTCs with no steric limitations are hydrophilic phosphine complexes, which are able to transport the lipophilic substrates capable of forming complexes with transition metals. These hydrophilic phosphine complexes differ

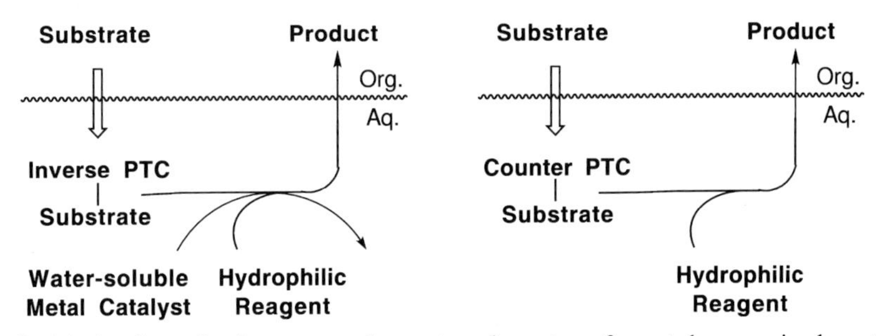

Figure 2. Mechanisms for inverse- and counter-phase transfer catalyses: mixed system of inverse PTC and water-soluble metal catalyst (left); counter-phase transfer catalyst (right).

essentially from the other inverse-PTCs in that they have the functions of both inverse-PTCs and transition metal catalysts by themselves. Therefore we named these catalysts "counter-phase transfer catalysts" [4].

4.6.2.2 Mechanism of the Counter-Phase Transfer Catalytic Reaction

It is difficult to verify the counter- or inverse-phase transfer catalysis strictly, because the catalyst more or less acts as a surfactant as well as a normal PTC [13]. However, it should be a positive proof of the counter-PTC to ascertain that the aqueous phase is where the products are formed.

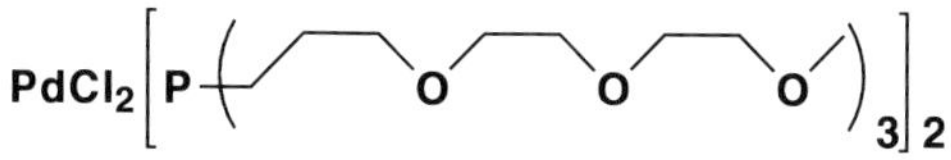

The reduction of allyl chlorides and acetates with sodium formate is efficiently catalyzed by hydrophilic phosphine complexes (Structures 1–3) under water–heptane two-phase conditions [4]. The catalytic activity of **1** is higher than that of $PdCl_2(PBu_3)_2$ (see Eq. (1) and Table 1). In order to ascertain the phase where the product is formed, the reduction of allyl acetate to form gaseous propene was carried out in a partitioned reactor as shown in Figure 3 (Reactor I). Part A is charged with an aqueous solution of catalyst and sodium formate, and a heptane solution of allyl acetate is placed in part B. The propene gas evolved

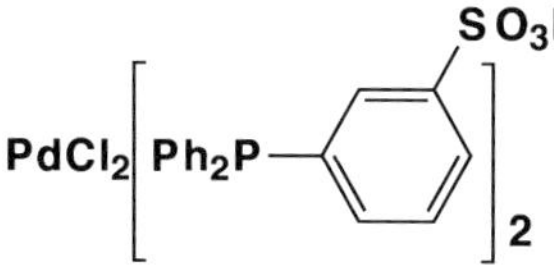

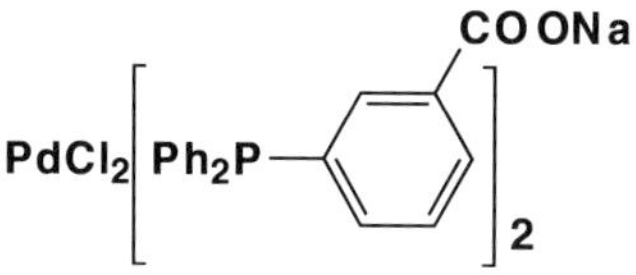

Table 1. Reduction of an allyl chloride (Eq. 1).

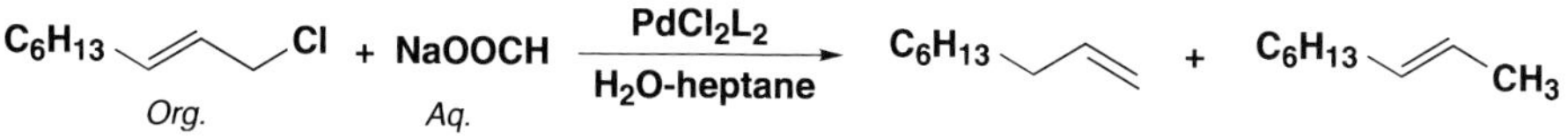

$$\underset{Org.}{C_6H_{13}CH_2CH{=}CHCl} + \underset{Aq.}{NaOOCH} \xrightarrow[H_2O\text{-heptane}]{PdCl_2L_2} C_6H_{13}CH_2CH{=}CH_2 + C_6H_{13}CH{=}CHCH_3 \quad (1)$$

Catalyst	Yield [%]	
	$C_6H_{13}CH_2CH{=}CH_2$	$C_6H_{13}CH{=}CHCH_3$
1	82	17
$PdCl_2[P(n\text{-}Bu)_3]_2$	20	3

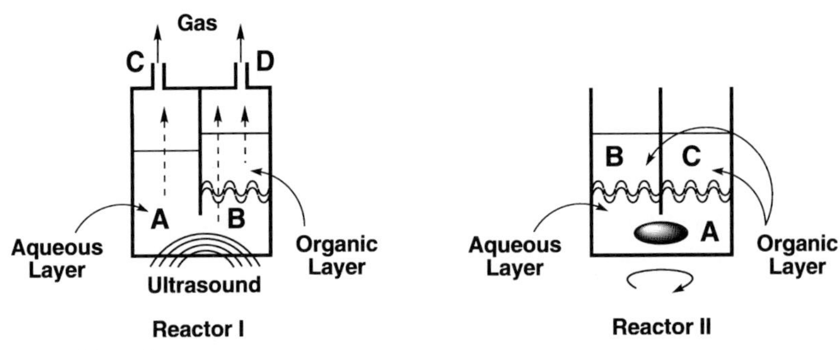

Figure 3. Partitioned reactors.

from part A of the aqueous layer is collected from outlet C, and the gas from the heptane layer, the interface, and part B of the aqueous layer is collected from outlet D. As the volume ratio of the aqueous solution in part A to that in part B is known, the amount of gas from each phase can be calculated. Reduction with a hydrophobic catalyst, $PdCl_2(PBu_3)_2$, causes evolution of the gas mostly from the heptane layer and/or the interface, whereas reaction with **1** leads to propene gas evolution mostly from the aqueous layer [see Eq. (2) and Table 2].

Table 2. Conversion of allyl acetate (Eq. 2).

$$\underset{Org.}{CH_2{=}CHCH_2OAc} + \underset{Aqu.}{NaOOCH} \xrightarrow[H_2O\text{-Heptane}]{PdCl_2L_2} \underset{Gas}{CH_2{=}CHCH_3} \uparrow \qquad (2)$$

Catalyst	Yield of C_3H_6 [%]	Source of C_3H_6 [aq.: org.]
$PdCl_2[P(n\text{-}Bu)_3]_2$	38	12:88
$PdCl_2[P(CH_2CH_2OCH_2CH_2OCH_2CH_2OCH_3)_3]_2$	55	98:2

Though the polyether phosphine is capable of acting as a normal PTC [14], this result indicates that the hydrophilic complex transports the allyl acetate from the heptane phase into the aqueous phase and catalyzes the reduction to propene with sodium formate in the aqueous phase [4] (see Scheme 1).

Palladium-catalyzed allylation of aldehydes with allyl compounds and stannous chloride in DMF or DMSO (Table 3) is a synthetically important reaction [15]. However, the normal PTCs might be inefficient for this class of reaction, because it is difficult to transport stannous chloride. The use of $PdCl_2(TPPMS)_2$ (**2**) enables us to achieve the carbonyl allylation under biphasic conditions [16]. Though **2** does not function as a normal PTC, its reactions give the allylated products in excellent yields even in a heptane–water two-phase system (see Table 3). One of the palladium-requiring steps in this allylation is the reaction of allyl compounds with stannous chloride to form allyl stannic compounds [15]. In the absence of aldehyde, the consumption rate of allyl chloride in the

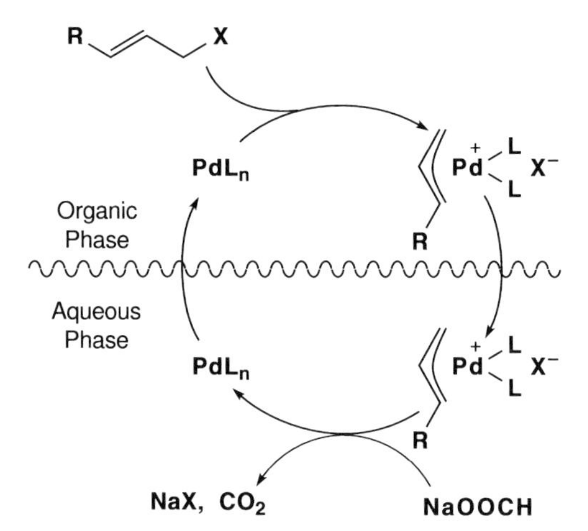

Scheme 1

Table 3. Biphasic allylation of aldehydes with allyl compounds and $SnCl_2$.

Catalyst [%]	Aldehyde	Allyl compound	Yield [%]
$PdCl_2(PPh_2\text{-}C_6H_4\text{-}SO_3Na)_2$	PhCHO	$CH_2{=}CHCH_2Cl$	98
	PhCHO	$CH_2{=}CHCH_2OH$	98
	p-$NaOOCC_6H_4CHO$	$CH_2{=}CHCH_2Cl$	95
	p-$NaOOCC_6H_4CHO$	$CH_2{=}CHCH_2OH$	93
$PdCl_2(P(C_6H_4\text{-}CH_3)_3)_2$	PhCHO	$CH_2{=}CHCH_2Cl$	5
	PhCHO	$CH_2{=}CHCH_2OH$	17
	p-$NaOOCC_6H_4CHO$	$CH_2{=}CHCH_2Cl$	16
	p-$NaOOCC_6H_4CHO$	$CH_2{=}CHCH_2OH$	8

$PdCl_2L_2$ (0.1 mmol): PhCHO: $CH_2{=}CHCH_2Cl$: $SnCl_2$ = 1:100:200:300, H_2O (10 mL), heptane (10 mL), 35 °C, 7 h.

$$\underset{Org.\,(Aq.)}{ArCHO} + \underset{Org.\,(Aq.)}{X\text{-}CH_2CH{=}CH_2} + \underset{Aqu.}{SnCl_2} \xrightarrow[H_2O\text{-heptane}]{PdCl_2L_2} Ar\text{-}CH(OH)\text{-}CH_2CH{=}CH_2 \qquad X = Cl, OH \tag{3}$$

heptane phase is much faster in the reaction with **2** than with $PdCl_2\{P(p\text{-}C_6H_4CH_3)_3\}_2$. This result suggests that the high efficiency of the hydrophilic catalyst is attributed to the fast transfer of allyl chloride from the organic phase into the aqueous phase via a hydrophilic π-allyl–palladium complex. In the presence of benzaldehyde (in part B, Figure 3), the carbonyl allylation with allyl chloride (in part C) in the partitioned Reactor II gives about half of the product in part B and another half in part C. This result indicates that the product is not formed in the organic phase or the interface but in the aquous phase. The proposed mechanism for the carbonyl allylation is depicted in Scheme 2.

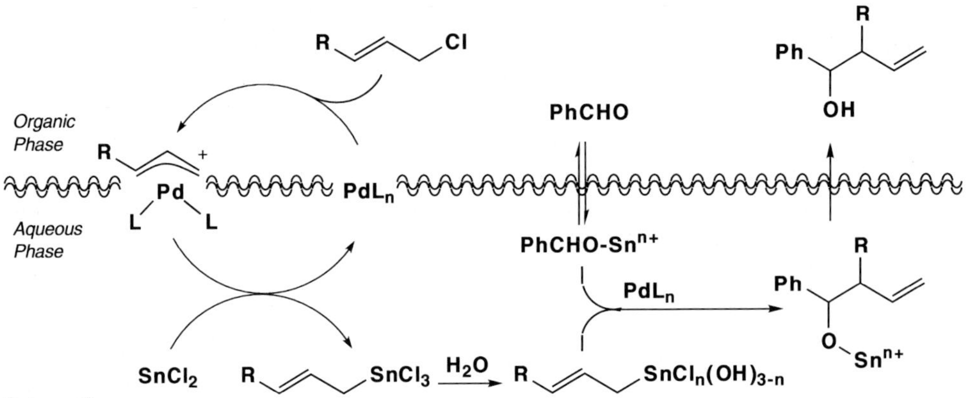

Scheme 2

The counter-phase transfer catalysis via π-allyl–palladium complexes exhibits an unusual solvent effect [16]. Both the reduction with sodium formate and the carbonyl allylation are faster in heptane than in toluene or anisole, whereas the efficiency of normal PTCs is generally higher in polar organic solvents than in nonpolar solvents. This inverse solvent effect is probably ascribed to a faster transfer of the π-allyl–palladium complexes from the heptane phase into the aqueous phase than that from the toluene or anisole phase. The use of nonpolar solvents is favorable for the separation of catalysts.

4.6.2.3 Counter-phase Transfer Catalytic Reactions

The biphasic reactions are essentially slow compared with the homogeneous reactions. However, some of the slow biphasic reactions are considerably improved by the use of the counter-PTCs, keeping the advantage of easy separation of the catalysts.

The biphasic carboxylation of allyl halides with NaOH and CO (Eq. 4) is quite fast even in the absence of PTCs; nevertheless, the use of normal or counter-PTCs is beneficial to the reaction [17, 18]. The selectivity as well as the

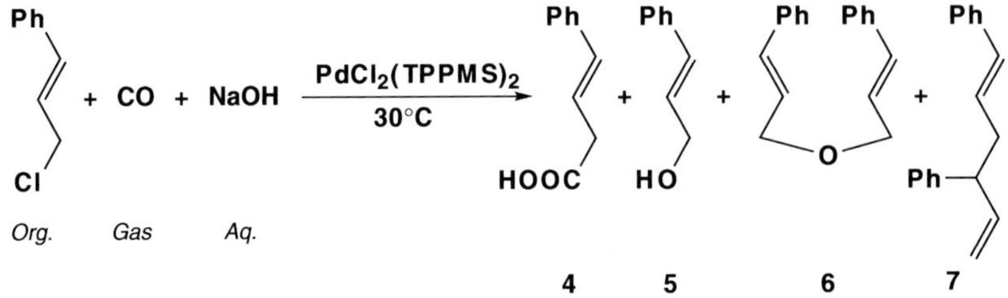

(4)

reactivity is improved by the use of **2** instead of $PdCl_2(PPh_3)_2$ in the biphasic carboxylation of cinnamyl chloride [19]. The solvent effect on the carboxylation with **2** is analogous to those on the reduction of allyl compounds and the carbonyl allylation as shown in Figure 4 [16]. The carboxylation in toluene or in chloroform forms considerable amounts of the dimer (**7**) (Table 4). The initial CO absorption rate in dioxane is the fastest, but large amounts of cinnamyl alcohol and the ether (**6**) are formed. This fact indicates that simple nucleophilic substitutions occurred significantly in a solvent that was miscible with water like dioxane. Though the carboxylation of cinnamyl chloride forms various products, simple acidification of the separated aqueous solution, which contains the sodium carboxylate and the catalyst, gives almost pure cinnamic acid, because the sulfonated phosphine complex is soluble even in the acidic aqueous solution.

The hydrophilic palladium complex (**2**) was also a good catalyst for the carboxylation of benzyl halides under heptane–water two-phase conditions [20]. Benzyl chloride and bromide give phenylacetic acid in high yields under

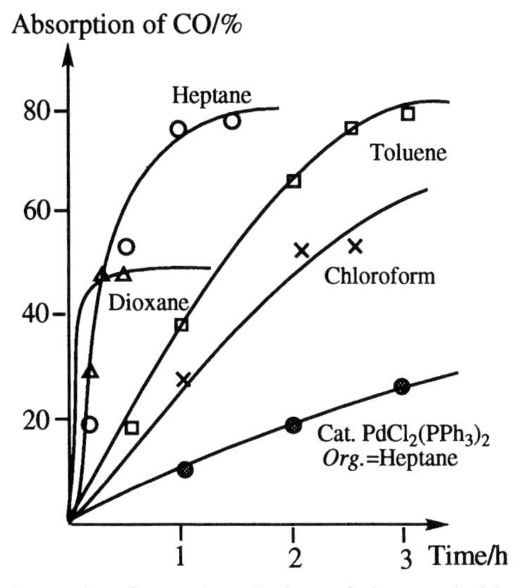

Figure 4. Rate of CO absorption for carboxylation of cinnamyl chloride.

Table 4. Biphasic carboxylation of cinnamyl chloride with CO and NaOH.

Organic phase	Time [h]	Conversion [%]	Yield [%]		
			4	**5 + 6**	**7**
Heptane	1.5	83	57	18	25
Toluene	4	95	53	7	40
Chloroform	2.5	95	40	18	40
Dioxane	0.5	66	40	45	15

$PdCl_2\{PPh_2(m\text{-}C_6H_4SO_3Na)\}_2$ (0.05 mmol): $PhCH_2=CHCH_2Cl$: NaOH = 1:100:1000, heptane (10 ml), H_2O (50 ml).

mild conditions (Eq. 5). However, the biphasic carboxylation with $PdCl_2(PPh_3)_2$ is very slow, and gives a considerable amount of benzyl alcohol. The addition of a normal PTC such as Bu_4NBr improves the catalytic activity of $PdCl_2(PPh_3)_2$ [16, 21]. However, the acceleration effect is poor, and the normal PTC is ineffective in retarding the formation of benzyl alcohol. Though the carboxylation with $PdCl_2(PPh_3)_2$ in ethanol is very fast, a large amount of benzyl alcohol is formed by the nucleophilic substitution. These results indicate that the water-soluble transition metal complex is very effective in retarding non-metal-catalyzed side reactions of organic substrates susceptible to nucleophilic substitution.

$$PhCH_2X + CO + NaOH \xrightarrow{PdCl_2(TPPMS)_2} PhCH_2COOH \quad (5)$$

The biphasic carboxylation shows an unexpected induction period. Interestingly, the addition of sodium phenylacetate shortens the induction period, and at the same time accelerates the reaction rate [22]. The other additives, such as sodium heptanesulfonate, Bu_4NBr, and n-$C_{18}H_{35}(OCH_2CH_2)_7OH$, are also effective. As the carboxylate and the sulfonate do not have the function of a normal PTC, these may as surfactants. The surfactants, however, have no effect on the reaction with the lipophilic catalysts $PdCl_2(PPh_3)_2$ and $Pd(PPh_3)_4$, suggesting that they do not act on the benzyl halides but on the hydrophilic catalyst.

Aryl iodides can also be carboxylated with NaOH and CO at atmospheric pressure in the presence of **2** under biphasic conditions, and give aryl carboxylic acids in high isolated yields [16] [Eq. (6) and Table 5]. The normal PTCs are not very effective in accelerating the carboxylation [23]. This may be ascribable to the poor extractability of hydroxy anion [24].

It is well known that the normal PTCs are effective for the biphasic cyanation of aryl halides with cyanide salts using hydrophobic catalysts [25]. However, a

Table 5. Biphasic carboxylation of phenyl iodide catalyzed by $PdCl_2L_2$

$$PhI + CO + NaOH \xrightarrow[H_2O\text{–organic solvent}]{PdCl_2L_2} PhCOOH \quad (6)$$

L	Organic phase	Isolated yield [%]
PPh_2–C_6H_4–SO_3Na (meta)	Heptane	88
	Toluene	84
	Anisole	81
	Dioxane	90
$P(C_6H_4$-$CH_3)_3$	Heptane	5
	Anisole	2
	Anisole + Bu_4NBr (1 mmol)	24

$PdCl_2L_2$ (0.1 mmol):L:PhI:NaOH = 1:5:100:250, CO (1 bar), H_2O (10 mL), organic solvent (10 mL), at 50 °C for 4 h.

simple application of hydrophilic catalysts results in failure. The cyanation with **2** requires $NaBH_4$ and zinc chloride (Eq. 7) [16]. These additives are necessary for reduction of the catalyst precursor and for prevention of deactivation of the catalyst by excess cyanide anion in the aqueous phase, respectively. As the use of zinc chloride in a Zn/CN molar ratio of more than 0.25:1 is required, the active cyanide source may be tetracyanozincate or zinc cyanide [26]. The efficiency of the counter-PTC in the heptane–water system exceeds that of the mixed catalyst system of lipophilic catalyst and normal PTC, though the cyanide anion is easily extractable by the normal PTCs (Table 6).

Carboxylated phosphine complex **3** is active for the cyanation, though this hydrophilic complex is inferior to **2** as a counter-PTC. As the cyanide anion is easily extractable, a phosphine complex having an ammonium group or a crown ether, which is insoluble in water [27], is also efficient for the cyanation. The

$$ArI + NaCN \xrightarrow[ZnCl_2\text{–}NaBH_4]{PdCl_2(TPPMS)_2} ArCN \quad (7)$$

Table 6. Biphasic cyanation in the presence of various phosphine complexes.

Phosphine (L)	Catalyst	Additive	PhCN [%]
P–(C₆H₄)–CH_3	$PdCl_2L_2$	$NaBH_4$	0
	$Pd_2(dba)_3$–4 L	Bu_4NBr (1 mmol)	4
	$Pd_2(dba)_3$–4 L	$n\text{-}C_7H_{15}SO_3Na$ (1 mmol)	1
Ph_2P–(benzo-crown ether)	$PdCl_2L_2$	$NaBH_4$	1
	$Pd_2(dba)_3$–8 L		25
Ph_2P–(C₆H₄)–$CH_2N^+Me_3Br^-$	$PdBr_2L_2$	$NaBH_4$	12
	$Pd_2(dba)_3$–4 L		24
Ph_2P–(C₆H₄)–COO^-Na^+	$PdCl_2L_2$	$NaBH_4$	22
Ph_2P–(C₆H₄)–$SO_3^-Na^+$	$PdCl_2L_2$	$NaBH_4$	99

$PdCl_2L_2$ (0.2 mmol) – $NaBH_4$ (0.4 mmol) or $Pd_2(dba)_3 \cdot CHCl_3$ (0.1 mmol) – L (0.4 mmol), NaCN (13 mmol), $ZnCl_2$ (5 mmol), heptane (10 mL), water (10 mL), reflux, 1 h.

catalytic efficiency is higher than that of the mixed system of lipophilic catalyst and normal PTC, owing to the proximity effect caused by the conjunction of metal catalyst and normal PTC. However, the hydrophilic complexes are superior to the doubly functional catalysts in the cyanation.

4.6.2.4 Concluding Remarks

Although the highly water-soluble catalysts such as polysulfonated phosphine complexes have the advantage of easy separation of catalyst from products, these catalytic reactions, especially using highly lipophilic substrates, are not so fast as in homogeneous systems [28, 29]. To improve the catalytic efficiency, the addition of inverse-PTCs, surfactants [30], and co-solvents [31] has been attempted. However, these additives are not beneficial to the separation. Recently, some of the water-soluble phosphine complexes with reduced hydrophilicity, including amphiphilic catalysts (see Section 7.4), have been recognized to be more effective than the highly hydrophilic catalysts. These low-hydrophilicity complexes are still easily and satisfactorily separable from the biphasic reaction mixtures [29, 32, 33]. Several mechanisms for these reactions have been proposed.

Biphasic hydroformylation with Rh-TPPTS (cf. Section 6.1) catalyst is markedly enhanced by the addition of PPh_3. The proposed mechanism is that the promoter ligand binds to the highly water-soluble catalyst, thereby increasing the catalyst concentration at the interface [29]. The use of phosphines having long alkyl chains and ionic groups also shows improved rates for the hydroformylation of liquid olefins [8, 32–35]. As sulfobetaine derivatives of tris(2-pyridyl)phosphine are surface-active, the high efficiency of their complexes is interpreted in the same way [32]. Some emulsification is observed in the biphasic hydroformylation using carboxylated phosphines, $PPh_2(CH_2)_nCO_2Na$ [33]. Dynamic light-scattering experiments show that sulfonated phosphines, $P[C_6H_4(CH_2)_n$-p-$C_6H_4SO_3Na]_3$, aggregate in aqueous salt solutions, though no stable emulsion is formed under biphasic reaction conditions [35]. The measured hydrodynamic radius for the catalyst solution is consistent with isolated molecules of the Rh–phosphine complex. The authors speculate that the two ligands of the Rh complex can mimic a small micelle wich provides a hydrophobic pocket for binding olefins near the Rh center. The sulfonated phosphines TPPMS [33] and TPPTS [35] do not form emulsions. In the counter-phase transfer catalytic reactions as mentioned above, obvious emulsification is not seen.

The aim of the most studies on the water-soluble phosphines is to find separable, active, and selective catalysts. The water-soluble catalysts are additionally useful for the reactions of hydrophobic substrates with inorganic salts. So far, amphiphilic solvents or mixed catalyst systems of normal PTCs and normal transition metal complexes have been used for some reactions. Though the mixed system enables the easy separation of inorganic salts from the reaction mixture, the separation of the metal catalyst is still difficult. Another problem of the mixed system is its slow rates, especially in reactions with poorly extractable anions such as the hydroxy anion. In the mixed system, the activated intermediate formed from an organic substrate and a metal complex reacts with the anionic reagent extracted by a normal PTC in the organic phase (see Figure 1). The concentrations of both species are ordinarily low, depending on the amounts of both catalysts, the reactivity of the metal catalyst, and the extractability of the anion. These conditions result in slow reaction rates and serious noncatalytic side reactions owing to a higher concentration of the substrate than that of the activated intermediate in the organic phase. These defects are diminished by means of chemical modification of the phosphine ligand with crowns or onium groups. However, neither the chemical binding nor the simple addition of phase-transfer catalysts has an effect on the separation of catalysts and products.

In such reactions as are described above, the counter-PTCs exhibit high efficiency and high selectivity, which arise from the concerted performance of their double function as an inverse-PTC and as a metal catalyst. The high efficiency is ascribed to the absence of any unnecessary transportation of substrates, because the carrier molecule itself is the metal catalyst. All of the transported substrates are able to react with reagents in the aqueous phase. The high selectivity is attributed to the specified transportation of the activated substrates by the metal catalyst from the organic phase into the aqueous phase in which the nucleophiles exist. The specific transportation is very effective in retarding the side reactions of any organic substrates susceptible to nucleophilic substitution, such as allyl halides and benzyl halides. The most important advantage is the easy separation of catalyst and products. Even if the product and the catalyst both exist in the aqueous layer, the product could be isolated, because the sulfonated phosphine complexes as well as polyether complexes have a high affinity for water over a wide range of pH. It is also beneficial to the separation that the catalytic efficiency of the counter-PTCs is higher in nonpolar solvents than in polar solvents.

Although counter-phase transfer catalysis has not yet attracted much attention, it is a very promising method for solving some of the problems of organometallic catalyses. Since transition metal complexes are capable of interacting with a wide variety of organic compounds, great development of counter-phase catalysis is expected.

References

[1] B. Cornils, W. A. Herrmann (Eds.), *Applied Homogeneous Catalysis with Organometallic Compounds,* VCH, Weinheim, **1996**, 2 Vols.

[2] (a) H. Alper, *Adv. Organomet. Chem.* **1981**, *19*, 183; J. F. Petrignani in *The Chemistry of the Metal–Carbon Bond* (Ed.: R. R. Hartly), Vol. 5, Wiley, New York, **1989**, Chapter 3; (b) H. Abbayea, A. Buloup, G. Tanguy, *Organometallics* **1983**, *2*, 1730; C. Zucchi, G. Palyi, V. Galamb, E. Sampar-Szerencses, L. Marko, P. Li, H. Alper, *Organometallics* **1996**, *15*, 3222.

[3] J. S. Mathias, R. A. Vaidya, *J. Am. Chem. Soc.* **1986**, *108*, 1093.
[4] T. Okano, Y. Moriyama, Y. Konishi, J. Kiji, *Chem. Lett.* **1986**, 1463.
[5] M. Yamada, Y. Watanabe, T. Sakakibara, R. Sudo, *J. Chem. Soc., Chem. Commun.* **1979**, 179; S. Rubinszlajn, M. Zeldin, W. K. Fife, *Macromolecules* **1990**, *23*, 4026; C. Kuo, J. Jwo, *J. Org. Chem.* **1992**, *57*, 1991; M. Wong, C. Ou, J. Jwo, *Ind. Eng. Chem. Res.* **1994**, *33*, 2034; S. Asai, H. Nakamura, W. Okada, M. Yamada, *Chem. Eng. Sci.* **1995**, *50*, 943; H. Nakamura, S. Asai, M. Yamada, *Chem. Eng. Sci.* **1996**, *51*, 1343.
[6] M. Takeishi, K. Se, N. Umeta, R. Sato, *Nippon Kagaku Kaishi* **1992**, 824.
[7] A. Harada, Y. Yu, S. Takahashi, *Chem. Lett.* **1986**, 2083; H. A. Zahalka, K. Januszkiewicz, H. Alper, *J. Mol. Catal.* **1986**, *35*, 249; E. Monflier, E. Blouet, Y. Barbaux, A. Mortreux, *Angew. Chem., Int. Ed. Engl.* **1994**, *33*, 2100; E. Monflier, S. Tilloy, G. Fremy, Y. Barbaux, A. Mortreux, *Tetrahedron Lett.* **1995**, *36*, 387.
[8] A. Benyei, F. Joó, *J. Mol. Catal.* **1990**, *58*, 151.
[9] H. A. Zahalka, H. Alper, *Organometallics* **1986**, *5*, 1909.
[10] J. Lee, H. Alper, *Tetrahedron Lett.* **1990**, *31*, 4104.
[11] J. R. Anderson, E. M. Campi, W. R. Jackson, *Catal. Lett.* **1991**, *9*, 55.
[12] E. Monflier, G. Fremy, Y. Castanet, A. Mortreux, *Angew. Chem., Int. Ed. Engl.* **1995**, *34*, 2269; E. Monflier, S. Tilloy, G. Fremy, Y. Castanet, A. Mortreux, *Tetrahedron Lett.* **1995**, *36*, 9481.
[13] C. C. Wasmser, J. A. Yates, *J. Org. Chem.* **1989**, *54*, 150; A. Bhattacharya, *Ind. Eng. Chem. Res.* **1996**, *35*, 645.
[14] T. Okano, M. Yamamoto, T. Noguchi, H. Konishi, J. Kiji, *Chem. Lett.* **1982**, 977; T. Okano, K. Morimoto, H. Konishi, J. Kiji, *Nippon Kagaku Kaishi* **1985**, 486.
[15] J. P. Takahara, Y. Masuyama, Y. Kurusu, *J. Am. Chem. Soc.* **1992**, *114*, 2577; Y. Masuyama, A. Hyakawa, Y. Kurusu, *J. Chem. Soc., Chem. Commun.* **1992**, 1102.
[16] T. Okano, in *Aqueous Organometallic Chemistry and Catalysis* (Ed.: I. T. Horváth, F. Joó), Kluwer, Dordrecht, **1995**, p. 97.
[17] F. Joó, H. Alper, *Organometallics* **1985**, *4*, 1775.
[18] J. Kiji, T. Okano, W. Nishiumi, H. Konishi, *Chem. Lett.* **1988**, 957.
[19] T. Okano, N. Okabe, J. Kiji, *Bull. Chem. Soc. Jpn.* **1992**, *65*, 2589.
[20] T. Okano, I. Uchida, T. Nakagaki, H. Konishi, J. Kiji, *J. Mol. Catal.* **1989**, *54*, 65.
[21] H. Alper, K. Hashem, J. Heveling, *Organometallics* **1982**, *1*, 775; H. Arzoumanian, G. Buono, M. Choukrad, J. F. Petrignani, *Organometallics* **1988**, *7*, 59.
[22] T. Okano, T. Hayashi, J. Kiji, *Bull. Chem. Soc. Jpn.* **1994**, *67*, 2329.
[23] L. Cassar, *Ann. N.Y. Acad. Sci.* **1980**, *333*, 208; V. V. Grushin, H. Alper, *Organometallics* **1993**, *12*, 1890.
[24] F. Montanari, D. Landini, F. Rolla, *Top. Curr. Chem.* **1982**, *101*, 147.
[25] K. Yamamura, S. Murahashi, *Tetrahedron Lett.* **1977**, 4429; L. Cassar, M. Foa, F. Montanari, G. P. Marinelli, *J. Organomet. Chem.* **1979**, *173*, 335; M. Prochazka, M. Siroky, *Coll. Czech. Chem. Commun.* **1983**, *48*, 1765; N. Sato, M. Suzuki, *J. Heterocyl. Chem.* **1987**, *24*, 1371; E. Piers, F. F. Fleming, *Can. J. Chem.* **1993**, *71*, 1867.
[26] D. M. Tschaen, R. Desmond, A. O. King, M. C. Fortin, B. Pipik, S. King, T. V. Verhoeven, *Syn. Commun.* **1994**, *24*, 887.
[27] T. Okano, M. Iwahara, T. Suzuki, H. Konishi, J. Kiji, *Chem. Lett.* **1986**, 1467; T. Okano, M. Iwahara, H. Konishi, J. Kiji, *J. Organomet. Chem.* **1988**, *346*, 267; T. Okano, N. Harada, J. Kiji, *Chem. Lett.* **1994**, 1057.
[28] W. A. Herrmann, C. W. Kohlpaintner, H. Bahrmann, W. Konkol, *J. Mol. Catal.* **1992**, *73*, 191.
[29] R. V. Chaudhari, B. M. Bhanage, R. M. Deshpande, H. Delmas, *Nature (London)* **1995**, *373*, 501.

[30] C. Larpent, F. Brise-LeMenn, H. Patin, *New J. Chem.* **1991**, *15*, 361.
[31] I. Hablot, J. Jenck, G. Casamatta, H. Delmas, *Chem. Eng. Sci.* **1992**, *47*, 2689; J. P. Genet, E. Blart, M. Savignac, S. Lemeune, J. Paris, *Tetrahedron Lett.* **1993**, *34*, 4189; F. Monteil, R. Queau, P. Kalck, *J. Organomet. Chem.* **1994**, *480*, 177; J. P. Genet, A. Linquist, E. Blart, V. Mouries, M. Savignac, M. Vaultier, *Tetrahedron Lett.* **1995**, *36*, 1443.
[32] B. Fell, G. Papadogianakis, *J. Mol. Catal.* **1991**, *66*, 143.
[33] D. C. Muudalige, G. L. Rempel, *J. Mol. Catal.* **1997**, *116*, 309.
[34] T. Bartik, B. Bartik, B. E. Hanson, *J. Mol. Catal.* **1994**, *88*, 43; H. Ding, B. E. Hanson, *J. Chem. Soc., Chem. Commun.* **1994**, 2747; H. Ding, B. E. Hanson, T. E. Glass, *Inorg. Chim. Acta* **1995**, *229*, 329.
[35] H. Ding, B. E. Hanson, T. Bartik, B. Bartik, *Organometallics* **1994**, *13*, 3761; H. Ding, B. E. Hanson, J. Bakos, *Angew. Chem., Int. Ed. Engl.* **1995**, *34*, 1645.

4.6.3 Thermoregulated Phase-transfer Catalysis

Zilin Jin, Xiaolai Zheng

4.6.3.1 Introduction

Over the past decade, increasing interest has been focused on the chemistry of water-soluble transition metal complexes and two-phase catalysis [1, 2]. One reason for this is the stimulating introduction of triphenylphosphine trisulfonate (TPPTS) and aqueous/organic two-phase systems to the rhodium-catalyzed hydroformylation of propene in Ruhrchemie/Rhône-Poulenc's process in 1984 [3]. However, the use of water as the second phase suffers from its drawbacks, especially when the water-solubility of the organic substrates proves too low. In this case, an undesirable reaction rate may arise due to a severe mass-transfer phenomenon. A variety of tentative approaches, including the addition of cosolvents, surfactants [4, 5], or "promoter ligands" [6], and the design of a "fluorous biphase system" (FBS) [7] and its variation [8], have been reported in order to ensure that the hydroformylation of extremely water-immiscible higher olefins may be smoothly carried out in the two-phase system. From the viewpoint of industrial development, however, the utilization of "foreign additives" will inevitably increase the difficulty of product separation, yet FBS would have to deal with the possible environmental impact of fluorinated solvents being employed.

Recently, a novel catalytic system based on the nonionic water-soluble phosphine-modified rhodium complexes has been successfully applied to the aqueous/organic two-phase hydroformylation of higher olefins [9, 10]. With polyoxyethylene chains as the hydrophilic group in the molecular structure, the ligands demonstrate a special property of inversely temperature-dependent water-solubility similar to that of the nonionic surfactants. As a result, the rhodium catalysts are soluble in the aqueous phase at a lower temperature and could transfer into the organic phase at a higher temperature. A catalyst which is capable of transferring between the aqueous phase and the organic phase in response to temperature changes is called a "thermoregulated phase-transfer catalyst" and a homogeneous catalytic process effected by a thermoregulated phase-transfer catalyst is called "thermoregulated phase-transfer catalysis" (TRPTC) [11]. Conspicuously, the concept of TRPTC as a "missing link" could not only provide a meaningful solution to the problem of catalyst/product separation, but also extricate itself from the limitation of low reaction rates of water-immiscible substrates. Thus, the scope of application of two-phase catalysis could be greatly broadened.

4.6.3.2 Thermoregulated Phase-transfer Catalysis and Nonionic Water-soluble Phosphines

It is well known that the water-solubility of nonionic surfactants with polyoxyethylene moieties as the hydrophilic group is based on the hydrogen bonds formed between polyether chains and water molecules. The solubility of this type of surfactant decreases with a rise in temperature, and their aqueous solutions will undergo an interesting phase separation process (a miscibility gap) on heating to a lower critical solution temperature – the "cloud point". A reasonable explanation attributes this phenomenon to the cleavage of hydrogen bonds. In addition, it is worth mentioning that such a process is a reversible one since the water-solubility could be restored on cooling to a temperature lower than the cloud point [12].

In view of the principles described above, a series of nonionic surface-active water-soluble phosphines **1**–**6** by introducing polyoxyethylene moieties to organophosphines [9–11] have been described. Results indicate that the ligands are appropriately soluble in water if the average number of ethylene oxide units per molecule exceeds 8 ($N \geq 8$). As shown in Table 1, the ligands reveal the clouding property pertaining to typical nonionic surfactants. Moreover, the cloud points could deliberately be controlled by adjusting the ratio of hydrophobic to hydrophilic groups in the molecule.

Several papers involving the synthesis of polyether-bound phosphines have been published previously. Okano et al. prepared tertiary phosphines **7** and **8** by Grignard reactions [14], and Harris et al. synthesized high-molecular-weight phosphines **9** and **10** through the functionalization of poly(ethylene glycols) (PEGs) [15]. Though not claimed by the authors, these ligands could hardly have a cloud point, since not enough hydrophobic groups are present in the molecules.

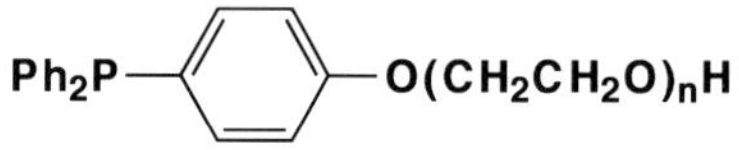

$Ph_2P-C_6H_4-O(CH_2CH_2O)_nH$ **1**

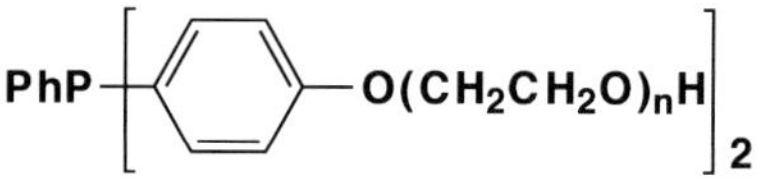

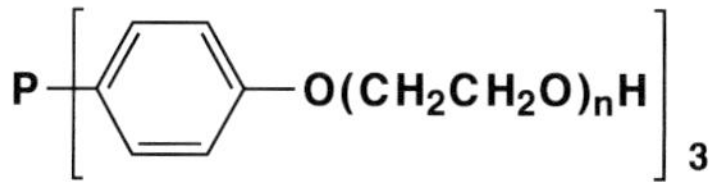

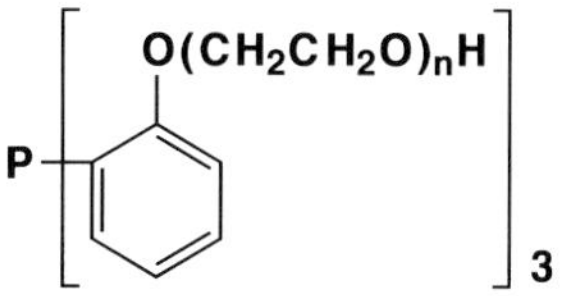

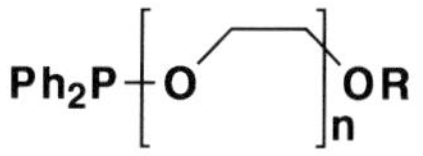

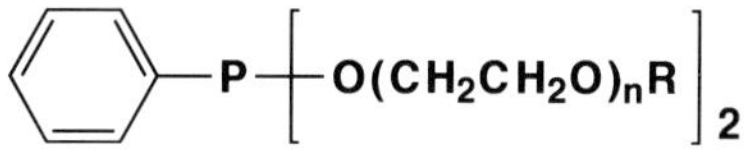

$PhP[C_6H_4-O(CH_2CH_2O)_nH]_2$ **2**

$P[C_6H_4-O(CH_2CH_2O)_nH]_3$ **3**

$P[C_6H_4-O(CH_2CH_2O)_nH]_3$ **4**

$Ph_2P[OCH_2CH_2]_nOR$ **5**

$PhP[O(CH_2CH_2O)_nR]_2$ **6**

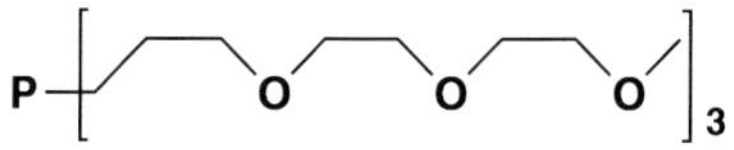

7

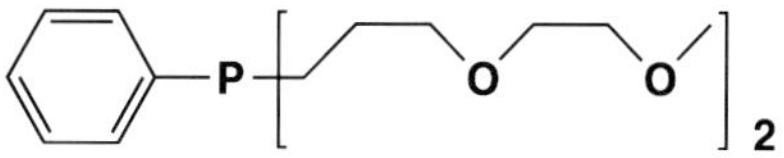

8

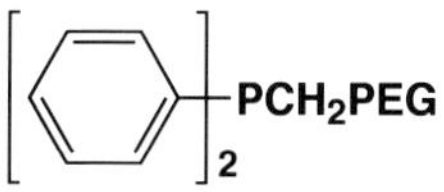

$[C_6H_5]_2PCH_2PEG$ **9**

$PhP(OCH_2CH_2PEG)_2$ **10**

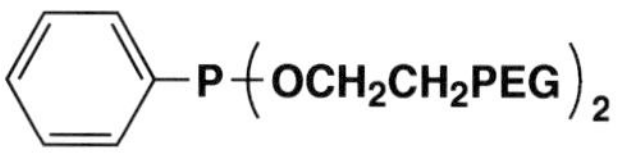

Table 1. Cloud points of nonionic water-soluble phosphines with polyoxyethylene moieties [11, 13].

Ligand	End group (R)	*n*	*N*[a)]	Cloud point [°C]
1	H	8	8	26
1	H	16	16	52
1	H	25	25	75
2	H	7	14	55
2	H	10	20	71
3	H	6	18	95
5	n-C_5H_{11}	16	16	48
5	n-C_8H_{17}	16	16	43
5	n-$C_{12}H_{25}$	16	16	39
6	n-C_5H_{11}	8	16	50
6	n-C_5H_{11}	16	32	66
6	n-$C_{12}H_{25}$	8	16	30
6	n-$C_{12}H_{25}$	16	32	60

[a)] *N* = average no. of ethylene oxide units per molecule of ligand.

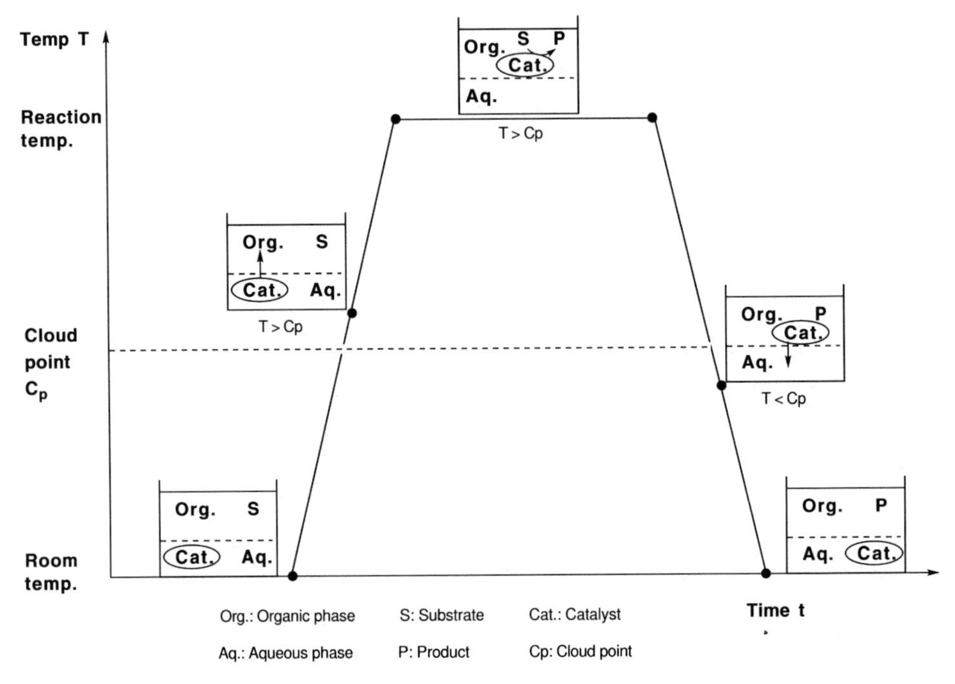

Figure 1. General principle of thermoregulated phase-transfer catalysis. The *mobile* catalyst transfers between the aqueous phase and the organic phase in response to temperature changes.

Based on the inversely temperature-dependent solubility of phosphines modified with polyoxyethylene chains, TRPTC has been proposed, and applied to the aqueous/organic two-phase reaction system [11]. The general principle of TRPTC is depicted in Figure 1.

The thermoregulated phase-transfer function of nonionic phosphines has been proved by means of the aqueous-phase hydrogenation of sodium cinnamate in the presence of Rh/**6** ($N = 32$, R = n-C_5H_{11}) complex as the catalyst [16]. As outlined in Figure 2, an unusual *inversely* temperature-dependent catalytic behavior has been observed. Such an anti-Arrhenius kinetic behavior could only be attributed to the loss of catalytic activity of the rhodium complex when it precipitates from the aqueous phase on heating to its cloud point. Moreover, the reactivity of the catalyst could be restored since the phase separation process is reversible on cooling to a temperature lower than the cloud point.

The analogous phenomenon was first reported by Bergbreiter et al. [17, 18]. In the presence of phosphorus-bonded block copolymers of ethylene oxide and propylene oxide as ligands ("smart ligands"), rhodium-catalyzed hydrogenation of maleic acid or allyl alcohol in the aqueous phase would show the same anti-Arrhenius reactivity.

Obviously, the existence of such a unusual temperature-dependent phenomenon provides fundamental support of TRPTC. By introducing an extra

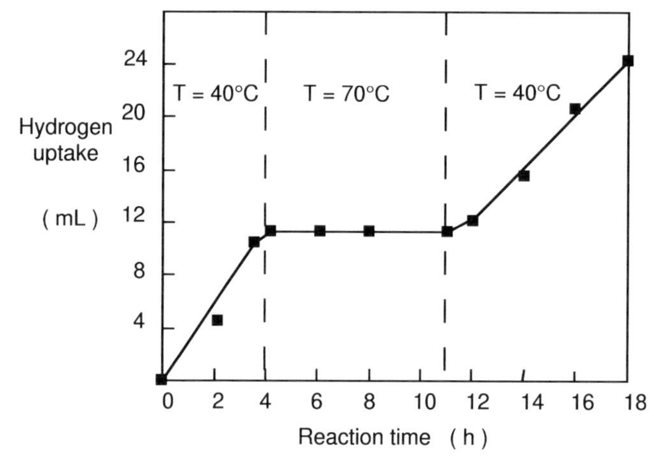

Figure 2. Atmospheric-pressure hydrogenation of sodium cinnamate using Rh/**6** ($N = 32$, $R = n\text{-}C_5H_{11}$) as the catalyst in water at different temperatures. The cloud point of the catalyst is 64 °C. Anti-Arrhenius kinetic behavior results due to the inversely temperature-dependent water-solubility of the nonionic phosphine [16].

organic phase containing a water-immiscible substrate into the reaction system, the catalyst being precipitated from the aqueous phase on heating to its low critical solution temperature would transfer into the organic phase, thus ensuring that the reaction takes place in the same – "homogeneous" – phase.

4.6.3.3 Hydroformylation of Higher Olefins Based on TRPTC

Hydroformylation of higher olefins with water-soluble catalysts is difficult to achieve (cf. Section 6.1.3.2). For example, the Rh/TPPTS complex, a perfect catalyst used in the two-phase hydroformylation of propene to generate butanol in the RCH/RP process, if applied to the hydroformylation of 1-hexene only gave conversions as low as 16–22% [4].

Thermoregulated phase-transfer catalysis, however, could be successfully put into effect for the hydroformylation of higher olefins in aqueous/organic two-phase media [11]. As shown in Table 2, various olefins have been converted to the corresponding aldehydes in the presence of nonionic phosphine-modified rhodium complexes as catalysts. An average turnover frequency (TOF) of 250 h^{-1} for 1-dodecene and 470 h^{-1} for styrene have been achieved. Even the hydroformylation of oleyl alcohol, an extremely hydrophobic internal olefin, would give a yield of 72% aldehyde [19]. In comparison, no reaction occurred if Rh/TPPTS complex was used as the catalyst under the same conditions.

Table 2. Two-phase hydroformylation of olefins catalyzed by nonionic phosphine-modified rhodium complexes [10, 19].

Olefin	Ligand	N[a]	P/Rh	P [Mpa]	T [°C]	Yield [%]	n/i	TOF [h^{-1}]
1-Hexene	**3**	18	5	5.0	100	85	2.0	–
1-Octene	**3**	18	5	5.0	100	88	2.8	–
1-Dodecene	**3**	18	5	5.0	100	84	1.8	250
Styrene	**1**	25	12	3.0	80	94	0.5	470
p-Chlorostyrene	**1**	25	12	3.0	80	92	0.6	460
p-Methoxystyrene	**1**	25	12	3.0	80	95	0.5	480
Cyclohexene	**1**	25	5	6.0	120	82	–	140
Oleyl alcohol	**1**	16	5	6.0	120	72	–	70

[a] N = average no. of ethylene oxide units per molecule of ligand.

The recycling effect of the thermoregulated phase-transfer catalyst was also examined. Aqueous phase containing the Rh/**3** ($N = 18$) catalyst after phase separation was re-used five times in the hydroformylation of 1-dodecene. Almost no loss in the reactivity was observed [10]. It should be pointed out that leaching of rhodium into the organic phase might difficultly be diminished to less than the ppm level by means of a single-phase separation. Fortunately, further washing of the organic phase with water could significantly reduce the rhodium loss to such an extent that the industrial requirement might be met.

Approaches in which surfactants [5, 20] and surface-active water-soluble phosphines [21, 22] are used to accelerate the reaction rate of hydroformylation in the two-phase system have been reported by several research groups. Surface-active materials tend to make it possible for hydrophobic higher olefins to *"enter"* the aqueous phase through micellar solubilization (cf. Section 4.5). As far as nonionic phosphines with polyoxyethylene chains as the hydrophilic group are concerned, micellar solubilization and other surface activities should hardly have played a decisive role at the temperature of hydroformylation (usually higher than 100 °C), since these ligands are designed to have a cloud point lower than 100 °C and would precipitate from the aqueous phase at those high hydroformylation temperatures.

The concept of TRPTC provides a reasonable explanation for the satisfactory catalytic reactivity of Rh/nonionic phosphine complexes in the case of the two-phase hydroformylation of higher olefins. At a temperature lower than the cloud point, a nonionic phosphine-modified rhodium catalyst would remain in the aqeous phase since the partition of the catalyst between water and a nonpolar aprotic organic solvent strongly favors the aqueous phase. On heating to a temperature higher than the cloud point, however, the catalyst loses its *hydrate shell*, transfers into the organic phase and then catalyzes the transformation of alkenes to aldehydes. As soon as the reaction is completed and the system cools to a temperature lower than the cloud point, the catalyst regains its hydrate shell and returns to the aqueous phase [11]. It deserves to be emphasized that the reaction site of TRPTC is the organic phase rather than the aqueous phase or the aqueous/organic interface. Thus, even the reaction of extremely water-immiscible substrates could be smoothly carried out.

It has been reported that the rates of hydroformylation decrease in the order 1-hexene > 1-octene > 1-decene in the classical aqueous/organic two-phase system, whereas the rates are almost identical in the homogeneous organic system [23]. Interestingly, in the hydroformylation of a mixture of equimolar 1-hexene and 1-decene in the presence of Rh/**1** ($N = 25$) complex as the catalyst, roughly the same reaction rates at various conversion levels have been observed [19]. This phenomenon further verified the conclusion that the organic phase is the reaction site of TRPTC.

4.6.3.4 Conclusion

Although the development of the aqueous/organic two-phase catalysis has only occurred more than 20 years since its emergence [24], crucial advantages of this methodology have been proved as it overcomes the immanent problem of catalyst/product separation associated with homogeneous catalysis. Compared with classical aqueous/organic two-phase catalysis, the process of TRPTC is more "*homogeneous*" to some extent because the catalyst and substrate would remain in the same organic phase at the reaction temperature. TRPTC is also quite different from FBS [5]. While the initial fluorous/organic biphase of FBS becomes a *single phase* at an appropriate higher temperature, TRPTC would maintain the aqueous/organic *two-phase* system throughout the reaction. It is the *mobile* catalyst that transfers between two phases without any additive in response to temperature changes.

The introduction of TRPTC to the aqueous/organic two-phase hydroformylation of higher olefins is free from the shortcomings of classical two-phase catalysis, in which the scope of application is more or less restrained by the water-solubility of the organic reactants. Obviously, the core of TRPTC is to use nonionic water-soluble phosphines with the property of clouding. Therefore, to design and prepare ligands with higher catalytic reactivity at lower cost will be a main topic in the scientific research and industrial exploitation of this strategy in the future. Just as Cornils remarks in this connection [25]: "since the agent responsible for the merger and subsequent separation of the phases is the approriately custom-designed ligand itself, there is no call for investing extra effort in the removal and recycling of a foreign additive, and this must therefore be regarded as a promising avenue for further exploration on a commercially realistic scale."

References

[1] (a) W. A. Herrmann, C. W. Kohlpaintner, *Angew. Chem.* **1993**, *105*, 1588; *Angew. Chem., Int. Ed. Engl.* **1993**, *105*, 1524; (b) W. A. Herrmann, B. Cornils, *Angew. Chem.* **1997**, *109*, 1074; *Angew. Chem. Int. Ed. Engl.* **1997**, *36*, 1048.

[2] (a) B. Cornils, W. A. Herrmann, R. W. Eckl, *J. Mol. Catal. A.: Chem.* **1997**, *116*, 27; (b) Special issue of *J. Mol. Catal.* **1997**, *116*.

[3] (a) B. Cornils, J. Falbe, *Proc. 4th Int. Symp. on Homogeneous Catalysis*, Leningrad, Sept. **1984**, p. 487; (b) H. Bach, W. Gick, E. Wiebus, B. Cornils, *Prepr. Int. Symp. High-Pressure Chem. Eng.*, Erlangen/Germany, Sept. **1984**, p. 129; (c) H. Bach, W. Gick, E. Wiebus, B. Cornils, *Prepr. 8th ICC*, Berlin **1984**, Vol. V, p. 417; *Chem. Abstr.* **1987**, *106*, 198051; cited also in A. Behr, M. Röper, *Erdgas und Kohle* **1984**, *37(11)*, 485; (d) H. Bach, W. Gick, E. Wiebus, B. Cornils, *Abstr. 1st IUPAC Symp. Org. Chemistry*, Jerusalem **1986**, p. 295; (e) E. Wiebus, B. Cornils, *Chem.-Ing.-Tech.* **1994**, *66*, 916; (f) B. Cornils, E. Wiebus, *CHEMTECH* **1995**, *25*, 33; (g) E. G. Kuntz, *CHEMTECH* **1987**, *17*, 570.

[4] Ruhrchemie AG (H. Bahrmann, B. Cornils), DE 3.420.491 (1985), DE 3.412.335 (1985).

[5] M. J. H. Russel, *Platinum Met. Rev.* **1988**, *32*, 179.

[6] R. V. Chaudhari, B. M. Bhanage, R. M. Deshpande, H. Delmas, *Nature (London)* **1995**, *373*, 501.

[7] (a) I. T. Horváth, J. Rabai, *Science* **1994**, *266*, 72; (b) B. Cornils, *Angew. Chem.* **1997**, *109*, in press.

[8] C. Bianchini, P. Frediani, V. Sernau, *Organometallics* **1995**, *14*, 5489.

[9] Y. Yan, H. Zuo, Z. Jin, *Fenzi Cuiha* **1994**, *8*, 147.

[10] Z. Jin, Y. Yan, H. Zuo, B. Fell, *J. Prakt. Chem.* **1996**, *338*, 124.

[11] Z. Jin, X. Zheng, B. Fell, *J. Mol. Catal. A: Chem.* **1997**, *116*, 55.

[12] N. Schonfeldt, *Surface-Active Ethylene Oxide Adducts*, Wissenschaftliche Verlagsges. mbH, Stuttgart, **1976**.

[13] Z. Jin, Y. Wang, X. Zheng, in *2nd Int. Symp. on Organometallic Chemistry and Catalysis*, Fuzhou, China, **1996**, IL10.

[14] T. Okano, M. Yamamoto, T. Noguchi, H. Konishi, J. Kiji, *Chem. Lett.* **1982**, 977.

[15] J. M. Harris, E. C. Struck, M. G. Case, M. S. Paley, M. Yalpani, J. M. van Alstine, D. E. Brooks, *J. Polym. Sci., Polym. Chem. Ed.* **1984**, *22*, 341.

[16] Y. Wang, Ph. D. Thesis, Dalian University of Technology, **1996**.

[17] D. E. Bergbreiter, V. M. Mariagnanam, L. Zhang, *Adv. Mater.* **1995**, *7*, 69.

[18] D. E. Bergbreiter, L. Zhang, V. M. Mariagnanam, *J. Am. Chem. Soc.* **1993**, *115*, 9295.

[19] X. Zheng, Z. Jin, unpublished results.

[20] B. Fell, D. Leckel, Ch. Schobben, *Fat.-Sci. Technol.* **1995**, *97*, 219.

[21] B. Fell, G. Papadogianakes, *J. Mol. Catal.* **1991**, *66*, 143.

[22] H. Ding, B. E. Hanson, T. Bartik, B. Bartik, *Organometallics* **1994**, *13*, 3761.

[23] I. T. Harvath, *Catal. Lett.* **1990**, *6*, 43.

[24] J. Manassen, in *Catalysis Progress in Research* (Ed.: F. Basolo, R. L. Burwell), Plenum Press, London, **1973**, p. 183.

[25] B. Cornils, *Angew. Chem.* **1995**, *107*, 1709; *Angew. Chem., Int. Ed. Engl.* **1995**, *34*, 1575.

4.7 Transitions to Heterogeneous Techniques (SAPC and Variations)

Mark E. Davis

4.7.1 Introduction

It is obvious that the diversity of catalytic chemistry that can be accomplished in aqueous phases by organometallic catalysts is burgeoning. Specific and newer details of the expanding field of catalysis in water can be found elsewhere in this book and in a special issue of the *Journal of Molecular Catalysts A* [1]. Here a concept is presented for converting known homogeneous catalytic systems that operate in aqueous media into heterogeneous analogues.

Comprehensive reviews of immobilization (heterogenization) techniques are available (e.g., [2]). In principle, immobilization may be achieved in one of several ways: physical adsorption or chemisorption of a metal complex onto a support; entrapment of metal complexes via in-situ synthesis within zeolites; dissolution of a metal complex in a nonvolatile solvent; and dissolution of a metal complex in a nonvolatile solvent that is adsorbed onto the surface of a support, i.e., as a supported liquid phase (SLP) [3]. The first three methods have not yet provided a commercially viable heterogeneous catalyst. In general, the immobilized systems never approach the combined activity/selectivity performance levels of their homogeneous counterparts and tend not to retain the metal complexes for a sufficiently long time ("leaching"). Dissolution of the metal complex has been successfully applied, as exemplified by the hydroformylations of propene of Union Carbide (hydrocarbon solvent) [4] and Ruhrchemie [5] after a laboratory development by Rhône-Poulenc (aqueous solvent) [6]. Both of these processes employ homogeneous catalysis; however, the reactants (propene, CO, H_2) and products (butanals) enter and leave as separate phases. The drawback of these immobilization methods is that they are not applicable for liquid-phase reactants/products that are miscible with the nonvolatile solvent phase.

4.7.2 The SAPC Concept of Immobilization

Although there is a vast literature on heterogenization of homogeneous catalysts, there is no successful commercialization of a solid catalyst that is an immobilized homogeneous catalyst. The primary reason for this is the lack of simultaneous high activity, high selectivity, and no leaching of active material. In order to convert a catalytic material into a commercially viable catalyst, these and other factors must all be optimized simultaneously. Thus, for an immobilized homogeneous catalytic material to have the opportunity to become a commercial catalyst, it must show high activity, selectivity, and enantioselectivity (if the reaction is chiral), without leaching.

A new immobilization method designed specifically to convert liquid-phase reactants has been developed [7, 8]. The catalytic materials consist of a thin film that resides on a high-surface-area support, such as controlled-pore glass or silica, and is composed of a hydrophilic liquid and a hydrophilic organometallic complex (see Figure 1).

Initially, water was used as the hydrophilic liquid and these catalysts are therefore denoted as supported aqueous-phase catalysts (SAPCs) [7–10]. Subsequently, we expanded this concept to other hydrophilic liquids such as ethylene glycol and glycerol [11]. Reactions of liquid-phase, hydrophobic organic reactants take place at the film–organic interface. SAP catalysis differs significantly from SLP catalysis in that the latter is used for gas-phase reactants whereas the former is specifically designed for liquid-phase substrates. Additionally, with SLP catalysis, the reaction proceeds homogeneously in the supported film while in SAP catalysis it occurs at the interface. Like many metals

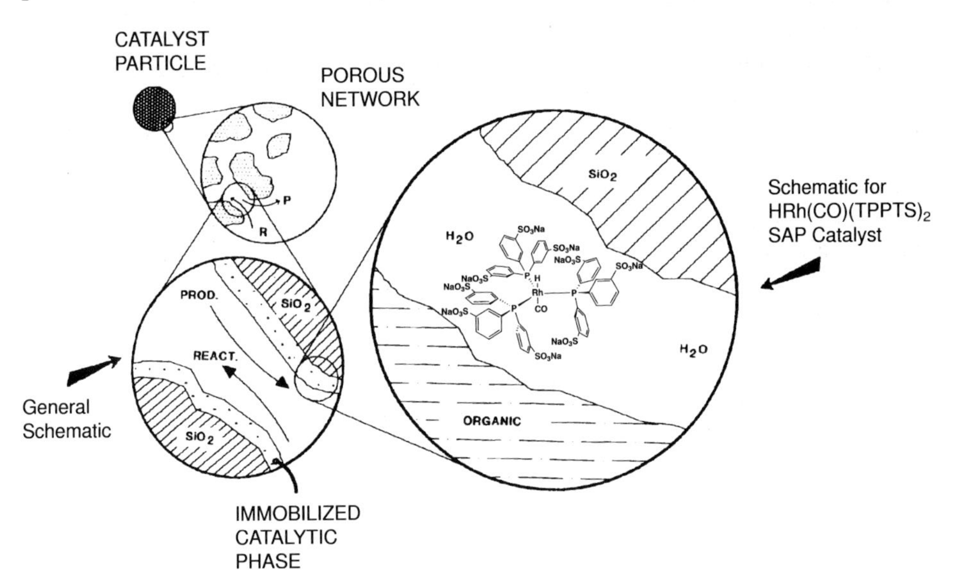

Figure 1. Schematic illustration of a supported aqueous-phase catalyst.

contained in biological systems, such as magnesium in chlorophyll or iron in hemoglobin, the hydrophilic catalysts contain ligands that ensure the hydrophilic properties while the environment at the metal center can remain essentially hydrophobic in character. Thus, the key is to impart hydrophilic properties to an organometallic complex that is known to be a homogeneous catalyst in organic media by modifying its ligands. The hydrophilic complex is supported on a hydrophilic solid to create a large interfacial area between the catalytic species and the organic reactants. The hydrophilicity of the ligands and the support creates interaction energies sufficient to maintain the immobilization. Thus, these catalysts are designed to conduct reactions efficiently at interfaces.

The proof of the concept of this type of catalysis was primarily obtained using hydroformylation as the test reaction. The commercial water-soluble rhodium hydroformylation catalyst (cf. Section 6.1.1) is used as the active organometallic complex. An SAP catalyst was created as follows: a controlled-pore glass, e.g., CPG-240 (mean pore size, 24.0 nm) was impregnated with $HRh(CO)[P(m\text{-}C_6H_4SO_3Na)_3]_3$ and $P(m\text{-}C_6H_4SO_3Na)_3$ (TPPTS). After the complex and excess ligand had been deposited, the water content of the solid was approximately 3 wt%. Water was then added to the catalyst. For details of the catalyst preparations and analyses, see [7, 8]. The rhodium-based SAP catalysts were used to hydroformylate liquid-phase olefins. For example, oleyl alcohol was hydroformylated with the SAP catalysts at 100 °C with 50 bar $CO + H_2$ ($CO:H_2$, 1:1). Extensive work was performed to show that rhodium is not leached into the organic phase. Elemental analysis of the organic phase for rhodium shows no detectable amounts, with a sensitivity of 1 ppb. Additionally, no reactivity is observed from the organic filtrate when it is tested for hydroformylation or hydrogenation activity (it is important to note that the filtrate must be obtained at the reaction temperature in order to eliminate the possibility of precipitation upon quenching the reaction medium). Since neither oleyl alcohol or its hydroformylation products are water-soluble, these reactivity results prove that rhodium is immobilized and that the reaction occurs at the organic–aqueous film interface. Finally, if the rhodium complexes are supported on various CPGs with different surface areas, the conversions obtained are proportional to the interfacial area in the reactor.

Horvath conducted several other interesting experiments with rhodium SAPCs (Table 1) [12]. Clearly, the water-solubility of the olefins does not limit the performance of the SAPCs since the TOFs (turnover frequencies) are essentially independent of olefin carbon number. This has been shown to be true also for carbon numbers as high as 17 [13]. Additionally, Horváth conducted experiments aimed at observing rhodium loss into the organic phase. He concluded that the SAPC does not leach catalytically active rhodium species under hydroformylation conditions. Another critical test for leaching was performed by Horvath. He conducted a 38 h continuous-flow experiment in a trickle-bed reactor and showed no loss of rhodium by elemental analysis. Thus, the combined data from all work shows that there is significant evidence to conclude that the SAPCs do remain immobilized and that the reaction occurs at the interface as designed.

Table 1. Hydroformylation of a 1:1:1 mixture of 1-hexene, 1-octene, and 1-decene with aqueous and organic soluble rhodium complexes and an SAPC (adapted from [12]).

Product	TOF[a)] [s^{-1}]		
	Biphasic[b)]	SAPC[c)]	Organic[d)]
Heptanals	0.0047	0.12	0.46
Nonanals	0.0014	0.12	0.50
Undecanals	0.0003	0.11	0.50

a) TOFs are estimated from conversions specified at various reaction times. Values are only estimates since the levels of conversion in some cases were very high.
b) $HRh(CO)[P(m\text{-}C_6H_4SO_3Na)_3]_3$ in water. T = 125 °C, P = 58 bar.
c) SAP catalyst containing $HRh(CO)[P(m\text{-}C_6H_4SO_3Na)_3]_3$ in water. Olefins were contained in hexane solvent. T = 100 °C, P = 51 bar.
d) $HRh(CO)[P(C_6H_5)_3]_3$ in hexane. T = 100 °C, P = 51 bar.

For long-term stability, the SAPC must remain assembled. To test for this type of stability, it was investigated whether the components can self-assemble. The rhodium complex $HRh(CO)(TPPTS)_3$, TPPTS and water were loaded into a reactor with cyclohexane and 1-heptene. The reactor was pressurized with approx. 70 bar H_2 + CO (CO:H_2, 1:1) and heated with stirring to 100 °C. A second experiment was carried out in a manner similar to the one previously described except that CPG-240 was added also. The components of the SAPC self-assemble to form an SAPC and carry out the hydroformylation reaction [13]. Upon termination of the reaction, the solid collected contained $HRh(CO)(TPPTS)_3$ and TPPTS. This test indicates that, under the conditions of the experiment, the individual components of the SAPC are more stable assembled in an SAPC configuration than separated. Therefore, the reverse, i.e., the separation of the solution and complex from the support, is not likely to happen under reaction conditions.

The water content of $HRh(CO)(TPPTS)_3$-based SAPCs has a great influence on their performance. For example, when 1-heptene is hydroformylated, the TOF increases by two orders of magnitude when the water content of the catalyst increases from approx. 2.9 wt.% to approx. 9 wt.% (Table 2).

What is interesting from these results is that, whereas the activity is dramatically affected by the water content, the selectivity does not change. Thus, at the low water content, the activity is low and the apparent activation energy (approx. 75 kJ mol^{-1}) is the same as that obtained from $HRh(CO)[P(C_6H_5)_3]_3$ in organic solvent. As the water level increases, the activity increases to almost that

Table 2. Hydroformylation of 1-heptene. P = 7 bar, H_2/CO = 1:1, T = 75 °C [13].

Catalyst	TOF [s^{-1}]	E_a [kJ mol^{-1}]
$HRh(CO)(TPPT)_3$–SAPC (≈2.9 wt.% H_2O)	0.0002	≈75
$HRh(CO)(TPPTS)_3$–SAPC (≈9 wt.% H_2O)	0.02	≈40
$HRh(CO)[P(C_6H_5)_3]_3$ (in toluene solvent)	0.08	≈75

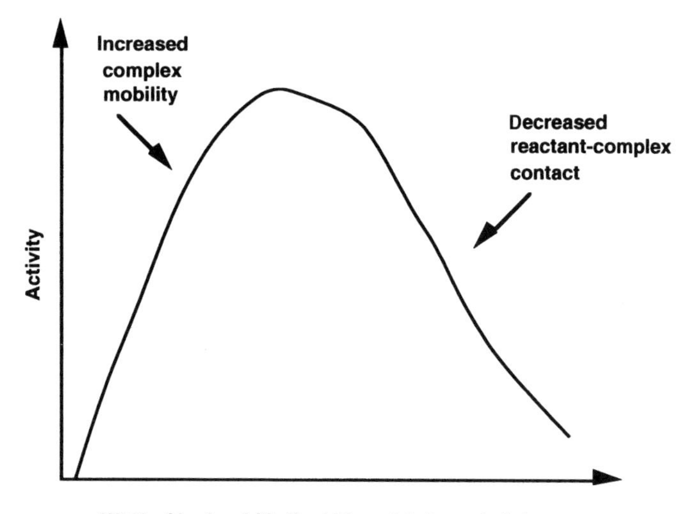

Figure 2. Schematic illustration of the activity of the immobilized catalyst as a function of the amount of hydrophilic liquid in the film.

obtained from $HRh(CO)[P(C_6H_5)_3]_3$ in toluene. It is surprising that the reaction rate is so high in view of the fact that the observed activity is most probably limited by mass transport (implied from observed activation energy). If too much water is added to the SAP catalyst the activity declines [8]. Thus, a bell-shaped curve describes the activity dependence on water (Figure 2).

This dependence has been noted by others as well [14, 15]. The increase in activity was suggested by use to be due to increase complex mobility (observed by NMR techniques) [7, 8]. ^{31}P NMR spin relaxation times do decline with increasing water content in SAP catalysts and the spin relaxation time at the optimal water loading is near that of the organometallic complex in water [16]. Thus, it is not unreasonable to suggest that the loss in activity as the catalyst is dried is due to the restrictions of the motion in the organometallic complex. As the water content becomes large, there is also a loss in activity. This decline is most probably due to losses in contacts between reactants and the organometallic complexes because of the lessening of interfacial area (the maximum water content would be that which fills the pore space and allows no contact with the hydrophobic organic phase). It is clear that one of the disadvantages of the SAPC concept is the sensitivity to water content in the film. Horváth reported that water is lost from the SAPC over the 38 h reaction conducted in the trickle-bed reactor and that the activity of the catalyst did vary over this time period [12]. Additionally, Fache et al. noticed that Ru–TPPTS–SAP catalysts lost activity upon recycling when hydrogenating α,β-unsaturated aldehydes [17]. In both of these studies, water-saturated solvents were not used. It is therefore expected that the water content and thus the activity of the SAP catalysts would vary with time on stream or batch number. For long-term stability, the water content of an SAPC must be carefully controlled.

Table 3. Catalyst types and reactions successfully used in the supported aqueous-phase configuration.

Catalyst type	Reaction	Ref.
Rh, TPPTS	Hydroformylation	[7, 8, 12–15, 18]
Co, TPPTS	Hydroformylation	[19]
Pt, Sn, TPPTS	Hydroformylation	[20]
Pd, Cu	Wacker oxidation	[21]
Pd, TPPTS	Allylic alkylation	[22]
Ru, TPPTS	Hydrogenation	[17]
Rh, TPPTS	Olefin isomerization	[23]
Pt, Sn, chiral ligand	Asymmetric hydroformylation	[24]
Rh, chiral ligand	Asymmetric hydrogenation	[24]
Ru, BINAP-4SO_3Na	Asymmetric hydrogenation	[10, 11]

The foregoing discussions show that the SAPC immobilization concept does reveal the desired properties of activity and selectivity with no catalyst leaching. Table 3 provides a summary of the catalytic materials/reactions reported using this immobilization technique.

4.7.3 Example of Rational Catalyst Design Strategy

After several years of work devoted to proving the concept of SAP catalysis, it was attempted to show that this class of heterogeneous catalyst could be prepared by design. In order to provide convincing evidence that this could be done, we chose to design and prepare a heterogeneous, asymmetric catalyst – which we believed to be the most difficult class of heterogeneous catalyst to prepare. Next, the successful design of such a catalyst is described.

It was known that ruthenium complexes of BINAP were efficient homogeneous catalysts for the asymmetric reduction of dehydronaproxen (**1**) to naproxen (**2**) (see Eq. 1) [25, 26].

This important and difficult reaction was used to test the design procedure. We followed the design steps listed in Table 4 except that initially we did not use ethylene glycol but rather water as the hydrophilic liquid for the immobilized film, i.e., an SAP-type catalyst. Implementation of the design sequence began by developing a water-soluble analogue of BINAP using direct sulfonation [27] and created an SAPC comprising a ruthenium(II)–sulfonated BINAP complex [10]. The catalyst revealed good activity and no leaching of the active component and gave an *ee* of approx. 70 %. For comparison, the *ee* values from homogeneous reactions using neat methanol as solvent were greater than 95 %. Thus, the

COOH
MeO
1

H_2/Ru catalyst (1)

Me
COOH
MeO
2 (S)–Naproxen

Table 4. Steps in a route to the rational design of immobilized homogeneous catalysts.

Step	Example (from [11])
1. Identify homogeneous reaction and catalytic material of interest.	Asymmetric hydrogenation to produce naproxen by Ru-BINAP catalytic material.
2. Convert hydrophobic ligand to hydrophilic ligand.	BINAP converted into sulfonated BINAP (BINAP-4SO_3Na).
3. Prepare hydrophilic catalytic material.	Ru–BINAP-4SO_3Na.
4. Immobilize hydrophilic catalytic material in a thin film of hydrophilic liquid on a high-surface-area hydrophilic solid.	Ru–BINAP-4SO_3Na in ethylene glycol film on a controlled-pore glass support.
5. Conduct reaction.	Synthesis of naproxen.

SAPC gave poorer *ee* values, and it was shown that this was due to the presence of water. Water hydrolyzed a Ru–Cl bond in the organometallic complex and lead to the reduction in *ee* [10]. Thus, the final step in the catalyst preparation was to replace water with another hydrophilic liquid that would preserve the Ru–Cl bond. This step was not anticipated and was therefore not included in the original design procedure. The substitution of water by ethylene glycol completed the catalyst construction (see Figure 3) and the reactivity results from this catalytic material are listed in Table 5. The heterogeneous, asymmetric catalyst produced naproxen in 100% yield with an *ee* of 96% and with no catalyst loss at approximately one-third the rate of homogeneous reaction.

In order to show that this procedure is not specific to the synthesis of naproxen, the asymmetric hydrogenation of β-ketoesters was investigated. Like the asymmetric hydrogenation to produce naproxen, the asymmetric hydrogenations of β-ketoesters by Ru–BINAP complexes were known to provide *ee* values above 95% [28, 29]. Using the procedure outline in Table 4 and armed with the knowledge that water should cause a loss in *ee* heterogeneous, asymmetric catalysts using Ru–BINAP-4SO_3Na complexes with ethylene glycol and

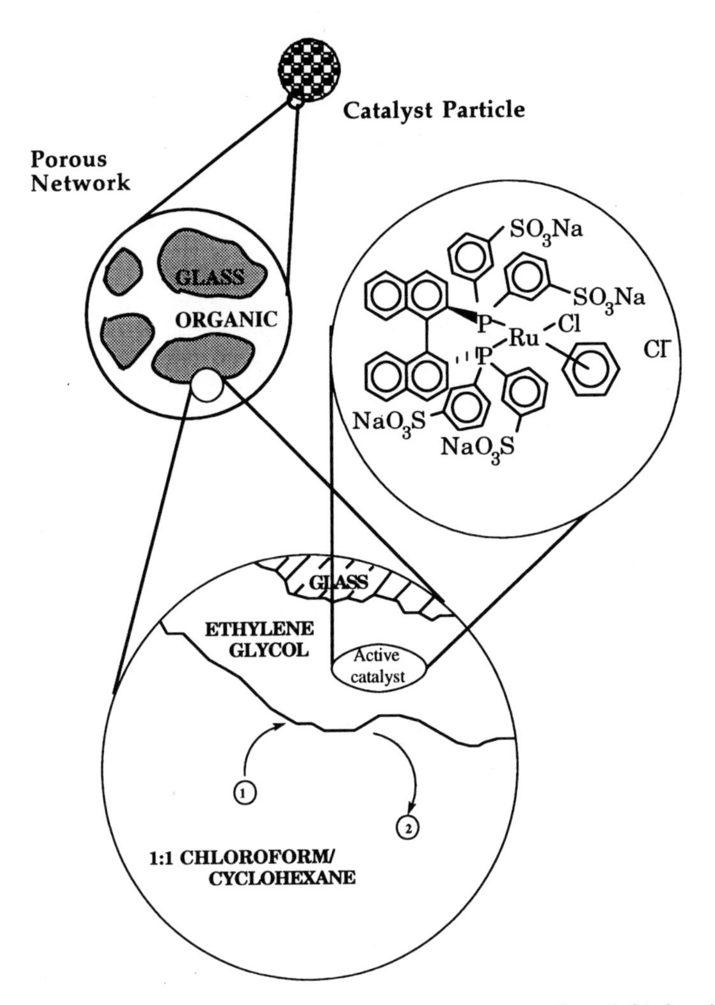

Figure 3. Schematic diagram of the designed heterogeneous catalyst (left) for the reaction shown in Eq. (1). (Adapted from [11]).

Table 5. Reaction results for the hydrogenation of **1** to yield naproxen (from [11]).[a)]

Solvent	TOF [h^{-1}]	*ee* [%]
1:1 $CHCl_3$/cyclohexane[b)] (heterogeneous catalyst)	40.7	88.4
1:1 $CHCl_3$/cyclohexane[b)] (heterogeneous catalyst)	–	95.7[c)]
MeOH (homogeneous catalyst)	131.0	88.2
MeOH (homogeneous catalyst)	–	96.1[d)]

a) Chiral catalytic material = [Ru(benzene) (BINAP-4SO_3Na)Cl]Cl.
b) No ruthenium found in the reaction filtrate at a detection limit of 32 ppb.
c) Reaction temperature = 276 K.
d) Reaction temperature = 277 K.

Table 6. Hydrogenation of ethyl butyrylacetate by homogeneous and heterogeneous catalysts at 85 °C and 60 bar H_2. The *ee* values are reported at 100% conversion (from [30]).

Catalysis	*ee* [%]
Homogeneous (methanol solvent)	97
Heterogeneous (ethylene glycol)	95
Heterogeneous (glycerol)	94
SAP (water)	83

glycerol were formulated (for comparison). Table 6 shows the results of hydrogenating ethyl butyrylacetate at 85 °C (Eq. 2). The heterogeneous catalysts gave *ee* values that were within experimental error (approx. 2% *ee*) of the *ee* obtained from the homogeneous reaction. Also and as now expected, the *ee* from the SAPC was lower.

O O OEt —H2 / Catalyst→ OH O OEt (2)

4.7.4 Suggested Reactions for Implementation of Design Concepts

The design strategy outlined here is general and should be applicable to the development of other chiral and achiral heterogeneous catalysts. For example, in the area of chiral hydrogenations two reaction chemistries that could greatly benefit from the use of a heterogeneous, asymmetric catalyst are: (1) the asymmetric hydrogenation of diketene; and (2) the asymmetric hydrogenation of imines. The efficient conversion of diketene **3** to its chiral lactone **4** (Eq. 3) could allow for low-cost production of polyhydroxybutyrate-type biodegradable polymers [28, 31].

3 —H2→ 4 (3)

Currently, these polymers are obtained via fermentation [31]. The Ru–BINAP type of catalysts have shown good selectivities and *ee* values for the asymmetric reduction of diketene [28, 29] and an immobilized version of the catalyst could be easily designed. The asymmetric reduction of imines to chiral amines has also been accomplished by homogeneous catalysts. Balkos et al. have hydrogenated imines with rhodium complexes of sulfonated bdpp ((−)(2*S*,4*S*)-2,4-bis(diphenylphosphino)pentane) with *ee* values exceeding 90 % [32]. Using this information the design of a heterogeneous catalyst for imine reduction is obvious and an important reaction for application of such a catalytic material would be the preparation of (*S*)-Metolachlor (Eq. 4). *S*-Metolachlor is an herbicide of which more than 10000 tons^{-1} are produced, making it the largest-scale product to be prepared by asymmetric hydrogenation.

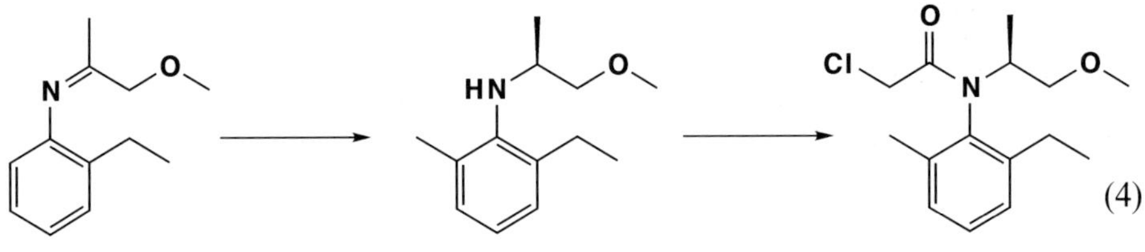

(4)

(*S*)–Metolachlor

Although the aforementioned examples involve only asymmetric hydrogenation, it is likely that the rational design strategy outlined here should provide a route to the design of a class of heterogeneous catalytic materials. However, skillful choice of the reaction system will be necessary. For example, reactions that provide large changes in hydrophilicity between reactants and products will most probably not be amenable to the immobilization procedure outlined here.

4.7.5 Outlook

Although SAP catalysts and their non-aqueous analogues have been reported only for less than a decade, it is clear that the reaction chemistries that can be accomplished in this configuration continues to burgeon. This method of immobilization has proven effective and as the realm of water-soluble organometallic catalysts expands, so can the field of SAPCs and its variants. It is hoped that economic proof will follow.

References

[1] *J. Mol. Catal. A* **1997**, *116*.
[2] R. R. Hartley, *Supported Metal Complexes,* Reidel, Dordrecht, **1985**.
[3] P. R. Rony, J. F. Roth, US 3.855.307 (1974).
[4] R. Fowler, H. Connor, R. A. Baehl, *CHEMTECH* **1976**, *6*, 772.
[5] (a) B. Cornils, J. Falbe, *Proc. 4th Int. Symp. Homogeneous Catalysis,* Leningrad, Sept. **1984**, p. 487; (b) H. Bach, W. Gick, E. Wiebus, B. Cornils, *Prepr. Int. Symp. High-Pressure Chem. Eng.,* Erlangen/Germany, Sept. **1984**, p. 129; (c) H. Bach, W. Gick,

E. Wiebus, B. Cornils, *Prepr. 8th ICC,* Berlin **1984**, Vol. V, p. 417; *Chem. Abstr.* **1987**, *106*, 198051; cited also in A. Behr, M. Röper, *Erdgas und Kohle* **1984**, *37(11)*, 485; (d) H. Bach, W. Gick, E. Wiebus, B. Cornils, *Abstr. 1st IUPAC Symp. Org. Chemistry*, Jerusalem, **1986**, p. 295; (e) E. Wiebus, B. Cornils, *Chem.–Ing.–Tech.* **1994**, *66*, 916.
[6] E. G. Kuntz, *CHEMTECH* **1987**, *17*, 570.
[7] J. P. Arhancet, M. E. Davis, J. S. Merola, B. E. Hanson, *Nature (London)* **1989**, *339*, 454.
[8] J. P. Arhancet, M. E. Davis, J. S. Merola, B. E. Hanson, *J. Catal.* **1990**, *121*, 327.
[9] M. E. Davis, *CHEMTECH* **1992**, 498.
[10] K. T. Wan, M. E. Davis, *J. Catal.* **1994**, *148*, 1.
[11] K. T. Wan, M. E. Davis, *Nature (London)* **1994**, *370*, 449.
[12] I. T. Horvath, *Catal. Lett.* **1990**, *6*, 43.
[13] J. P. Arhancet, M. E. Davis, B. E. Hanson, *J. Catal.* **1991**, *129*, 94.
[14] G. Frémy, E. Monflier, J. F. Carpentier, Y. Castanet, A. Mortreux, *Angew. Chem., Int. Ed. Engl.* **1995**, *34*, 1474.
[15] G. Frémy, E. Monflier, J. F. Carpentier, Y. Castanet, A. Mortreux, *J. Catal.* **1996**, *162*, 339.
[16] B. B. Bunn, T. Bartik, B. Bartik, W. R. Bebout, T. E. Glass, B. E. Hanson, *J. Mol. Catal.* **1994**, *94*, 157.
[17] E. Fache, C. Mercier, N. Pagnier, B. Desepeyroux, P. Panster, *J. Mol. Catal.* **1993**, *79*, 117.
[18] Y. Yuan, J. Xu, H. Zhang, K. Tsai, *Catal. Lett.* **1994**, *29*, 387.
[19] I. Guo, B. E. Hanson, I. Toth, M. E. Davis, *J. Organomet. Chem.* **1991**, *403*, 221.
[20] I. Guo, B. E. Hanson, I. Toth, M. E. Davis, *J. Mol. Catal.* **1991**, *70*, 363.
[21] J. P. Arhancet, M. E. Davis, B. E. Hanson, *Catal. Lett.* **1991**, *11*, 129.
[22] P. Schneider, F. Quignard, A. Choplin, D. Sinou, *New J. Chem.* **1996**, *20*, 545.
[23] J. P. Arhancet, Ph.D. Thesis, Blacksburg, VA, **1989**.
[24] I. Toth, I. Guo, B. E. Hanson, *J. Mol. Catal. A* **1997**, *116*, 217.
[25] R. Noyori, *Science* **1990**, *248*, 1194.
[26] A. S. C. Chan, S. A. Laneman, R. E. Miller, *Selectivity in Catalysis*, ACS Symp. Ser. No. 517, American Chemical Society, Washington DC, **1993**, pp. 27.
[27] K. T. Wan, M. E. Davis, *J. Chem. Soc., Chem. Commun.* **1993**, 1262.
[28] A. Akutagawa, *Appl. Catal. A* **1995**, *128*, 171.
[29] T. Ohta, T. Miyake, H. Takaya, *J. Chem. Soc., Chem. Commun.* **1992**, 1725.
[30] M. E. Davis, K. T. Wan, *Proc. Welch Found,* Proc. of the 40th Conf. **1997**, 45.
[31] E. Chiellini, R. Solaro, *Adv. Mater.* **1996**, *8*, 305.
[36] J. Balkos, A. Orosz, B. Heil, M. Laghmari, P. Lhoste, D. Sinou, *J. Chem. Soc., Chem. Commun.* **1991**, 1684.

5

Aqueous Catalysts for Environment and Safety

5.1 Water-soluble Organometallics in the Environment

Wolfgang A. Herrmann

5.1.1 Introduction

The focus of this book is on water as a solvent and reactant in organometallic chemistry. Although there are still a number of laboratory curiosities in this field, striking examples of industrial applications highlight the recent development and point toward a promising future. With the new recognition of aqueous organometallic chemistry, environmental aspects are coming into view that deserve further research or re-examination.

5.1.2 Organometallics in Nature: Cobalamines

With the possible exception of the redox enzymes cytochrome P-450 (iron) and the methanogenic bacterial cofactor F-430 (nickel) [1–3], the cobalamines are the only naturally occurring organometallics featuring σ-bonds between a transition metal (cobalt) and carbon ligands. Vitamin B_{12}, which can be found in the human body in amounts of 2–5 mg, is a derivative bearing a cyano group in place of the metal-attached ligand X (structure **1**). It is effective against several forms of anemic diseases of animals, e.g., "bush disease," "coast disease," and "salt sickness."

Vitamin B_{12} (**1a**) participates in the aqueous-phase biosynthesis of purine and pyrimidine bases, the reduction of ribonucleotide triphosphates, the conversion of methylmalonyl-coenzyme A to succinyl-coenzyme A, the biosynthesis of methionine from homocysteine, and the formation of myelin sheath in the nervous systems.

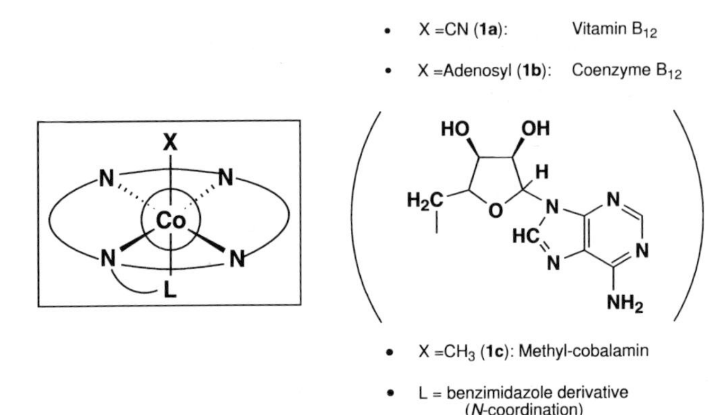

Methylcobalamin (**1 c**) can be isolated from microorganisms. Its largely covalent Co–CH_3 bond undergoes all three possible types of reactions, namely homolysis, carbonium-ion transfer, and carbanion transfer (Scheme 1), thus including reduction and oxidation of cobalt, respectively. Thermal degradation of cobalamin preferentially yields methane and ethane as radical-type reaction products [cf. Scheme 1, reaction (a)]. The carbonium ion $[CH_3]^+$ is transferred from **1 c** to sulfides and sulhydryl ions forming methylmercapto derivatives [cf. Scheme 1, reaction (b)]. The CH_3 group of **1 c** is available as an anion in the presence of metal cations [cf. Scheme 1, reaction (c)].

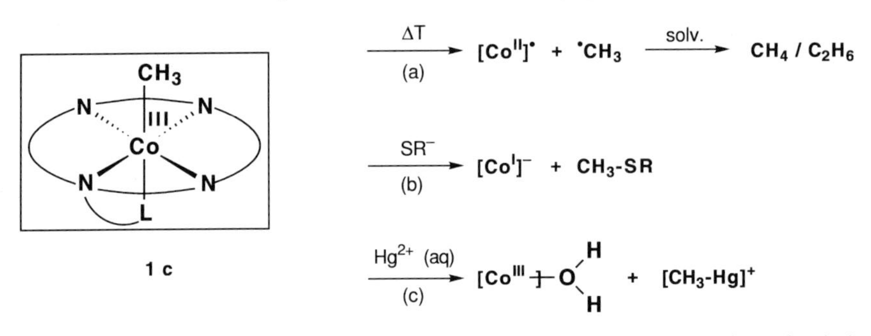

Scheme 1. Reaction of methylcobalamin in aqueous media; (a), (b) reduction of cobalt, (c) redox-neutral reaction (CH_3^- transfer: Co(III) → Hg(II)).

Coenzyme B_{12} catalyzes the 1,2-shift of alkyl groups (isomerase reaction) in hydrocarbons. An extensive chemistry was based on the reactivity aspects, where cobalamins acted as model systems, mainly in the work of Schrauzer et al. [5, 6].

Much of this chemistry depends on the redox processes. The strongly reducing, anionic cobalt(I) species, the so-called B_{12s} form, seems to play the central role in the catalytic reactions of the coenyme B_{12}. Methylcobalamin (**1 a**) is very similar to cyanocobalamine (**1 a**) in its geometry and ligand properties [5, 8].

There is a considerable demand for research on organometallic species generated in living organisms (cf. Section 6.8). Apart from the analogies with the chemistry of vitamin B_{12}, hardly any research has been concluded in this area.

5.1.3 Organomercury Compounds and Biological Methylation

Several organomercury compounds of formula CH_3HgX are water-soluble, which makes them very poisonous to organisms. They dissociate according to Eq. (1).

$$CH_3HgX \xrightarrow{(H_2O)} [CH_3Hg(H_2O)]^+ \; X^- \tag{1}$$

The biological methylation of mercury (e.g., from weathering, volcanism, fossil fuels, chloralkali electrolysis) is effected by microorganisms that utilize methylcobalamin (**1c**); see above.

The water-soluble organometallic **1c** is the *only natural product able to transfer the methyl group as a carbanion* (cf. Eq. 2). As a soft Lewis acid (in the classification of Pearson), the methylmercuronium ion $[CH_3Hg]^+$ is soluble in the presence of hard bases such as $[NO_3]^-$ or $[SO_4]^{2-}$. In turn, soft bases such as (organic) sulfides thus make it lipophilic and enable its uptake by living organisms.

$$CH_3\text{—[Cob]} \; + \; Hg^{2+} \xrightarrow{(H_2O)} H_2O\rightarrow\text{[Cob]}^+ \; + \; \underset{\mathbf{2}}{CH_3Hg^+} \tag{2}$$

Biological methylation is a process that converts inorganic, often harmless compounds, e.g., Hg^{2+}(aq), into highly toxic methylated species, e.g., CH_3Hg^+ [9]. The catastrophy of Minamata in Japan (1953–1960), with 55 people killed and more than 1200 poisoned, for the first time focused attention on the environmental consequences of water-soluble organometal species. In this particular case, it was mercury-containing waste water that had access to marine organisms. Fish, particularly, accumulates mercury, up to 250 ppb. In another case (Iraq, 1970–1971), the seed disinfectant ethylmercury *p*-toluenesulfonic anilide (**3**) had been applied to wheat and caused severe poisoning of consumers.

$$[C_2H_5Hg]^+ \left[H_3C\text{—}C_6H_4\text{—}S(=O)_2\text{—}NH\text{—}C_6H_5 \right]^-$$

3

Organomercury compounds derive their toxicity from their solubility in both aqueous and lipophilic systems. They primarily affect the central nervous system. The reversible ionic/covalent bonding in organomercury compounds distributes them in the body. Thus, water-soluble species such as **2** are converted in the stomach into lipophilic **4** (cf. Eq. 3) X = $[NO_3]^-$, etc., where it is then absorbed.

$$\underset{\mathbf{2}}{CH_3HgX} \; + \; Cl^- \xrightarrow{(H_2O)} \underset{\mathbf{4}}{CH_3HgCl} \; + \; X^- \tag{3}$$

Fixation to pyrimidine bases containing SH groups, such as uracil and thymine, is thought to explain the mutagenic effect of organomercurials.

5.1.4 Organoarsenic and Organotin Compounds

Other examples of toxic methylated metal ions are $[(CH_3)_2As]^+$ and $[(CH_3)_3Sn]^+$ [8]. The former is generated in rooms which were painted with dyes such as *Schweinfurter Grün*, a copper(II) acetate–arsenate(III), $Cu(CH_3CO_2)_2 \cdot 3\,Cu(AsO_2)_2$. Water-soluble stannonium compounds have been used widely as biocidal reagents, especially in wood treatment (e.g., antifouling chemicals). Note that tetramethyltin, $(CH_3)_4Sn$, is extremely poisonous because it reacts with water to form a water-soluble cation (cf. Eq. 4).

$$(CH_3)_4Sn + H_2O \longrightarrow [(CH_3)_3Sn]^+OH^- + CH_4 \qquad (4)$$

Organotin compounds find much application in agriculture. The advantage of the mostly used triorganyltin compounds is their excellent selectivity for lower organisms; in addition, their inorganic degradation products are not toxic. Examples of tin-based agrochemicals are $(n\text{-}C_4H_9)_3SnOH$ and $(C_6H_5)_3SnOAc$.

5.1.5 Perspectives

Water-soluble organometallic complexes should be developed further as reagents for alkyl- and aryl-group transfer processes. The example of cobalamin shows that nature exploits metal–carbon σ-bonds in aqueous systems for biomethylation processes.

References

[1] F. P. Guengerich, T. L. Macdonald, *Acc. Chem. Res.* **1984**, *17*, 9.
[2] R. R. Ortiz de Montellano, *Cytochrome P-450*, Plenum Press, New York, **1986**.
[3] G. Färber, W. Keller, C. Kratky, B. Jaun, A. Pfaltz, C. Spinner, A. Kobelt, A. Eschenmoser, *Helv. Chim. Acta* **1991**, *74*, 697.
[4] R. H. Abeles, D. Dolphin, *Acc. Chem. Res.* **1976**, *9*, 174.
[5] (a) G. N. Schrauzer, *Angew. Chem., Int. Ed. Engl.* **1976**, *15*, 417; (b) R. S. Young, *Cobalt in Biology and Biochemistry*, Academic Press, London, **1979**.
[6] G. H. Schrauzer, J. Kohnle, *Chem. Ber.* **1964**, *97*, 3056.
[7] (a) D. Lexa, J. M. Savéant, *J. Am. Chem. Soc.* **1978**, *100*, 3221; (b) D. Astruc, *Electron Transfer and Radical Processes in Transition Metal Chemistry*, VCH, Weinheim, **1995**.
[8] W. H. Lederer, R. J. Fensterheim (Eds.), *Arsenic: Industrial, Biomedical, Environmental Perspectives,* van Nostrand-Reinhold, New York, **1983**.

5.2 Environmental and Safety Aspects

Boy Cornils, Ernst Wiebus

5.2.1 Introduction

A comprehensive review [1] summarizes the environmental status of processes using catalytic conversion in water, and – more especially – several very recent comments highlight the main environmental features of Ruhrchemie/Rhône-Poulenc's (RCH/RP's) novel oxo process as a prototype of an aqueous biphasic technique [2]. Based on two-phase catalysis with water-soluble catalysts, this has now been used successfully for almost 15 years [3]. Astonishingly, in the early days of academic research (following far behind the industrial utilization; cf. Section 1) the importance of water as a "liquid support" of the thus immobilized homogeneous catalysts was underestimated and not undisputed.

Jiang and Sen, in a paper on water-soluble Pd(III) catalysts and their use in manufacturing CO/ethylene copolymers wrote [4]:

> "The use of water as a reaction medium in place of organic solvents is of great interest from the standpoint of environmentally benign synthesis of organics and polymers since the use of the former would significantly decrease harmful emissions, as well as cut costs associated with solvent recycling,"

whereas the organizers of a NATO Advanced Research Workshop on *Aqueous Organometallic Chemistry and Catalysis* (comprising the absolutely top level of academic researchers) took a more cautious view [5]:

> "Environmentally benign synthetic processes may also favor water as a solvent [...] though this question is not completely free of ambiguity."

This view, quoted from the concluding discussion of the NATO symposium [5c], is based on the disputed opinion that water as a solvent must be a bad thing because it has no inherent odor and therefore can disappear unnoticed (!) in the event of a leak. On the other hand, Papadogianakis et al. [1] mention

> "... Organometallic catalysis [...] affording both economic and environmental benefits."

This divergence of opinion surrounding "aqueous chemistry" and two-phase catalysis makes two things very clear: firstly the need to provide more detailed information on the extremely favorable environmental potential of this type of process using catalysts dissolved in water; and, secondly, the frequently criticized fact (cf. the Introduction (Chapter 1) of this book and [6]) that universities are far behind industrial users in their knowledge of the subject and make up for this deficiency by gleaning information from industry. Only now, when two-phase catalysis is becoming a focus of university research, does willingness to accept the arguments and experience of industry seem to be on the increase – almost 15 years after industry set out on this environmentally beneficial, important, and advantageous path.

5.2.2 The Ruhrchemie/Rhône-Poulenc (RCH/RP) Process

The Ruhrchemie/Rhône-Poulenc oxo process has been reviewed repeatedly (cf. Section 6.1 and [2, 3, 6]). This two-phase hydroformylation process using catalysts dissolved in water, and also the scientific work on this subject untertaken by universities, have resulted in hundreds of publications and patents [7] and extensive experience with the first large-scale implementation.

It is probably no coincidence that the interest of the scientific world in biphasic hydroformylation, as the only process so far to be used industrially, lags far behind the industry's acticivities, although technical implementation normally follows the pure (not the applied) scientific results.

The advances in two-phase homogeneous catalysis can be demonstrated very easily, taking the hydroformylation reaction as a large-scale, economic example. The advantages of homogeneous over heterogeneous catalysis are numerous, notably gentle reaction conditions, defined species of catalyst, the possibility of modifying the coordination sphere by varying the central atoms or ligands, and – as a consequence – high activity and selectivity. The disadvantage of homogeneous oxo catalysts, namely the difficulty of separating the catalyst from the reaction products after reaction, has so far been overcome only by using complicated recycling processes [8 a] or, taking UCC's low-pressure oxo process (LPO [9]) as an example, by thermal-stressing process steps. An oxo-active catalyst, dissolved in water but nevertheless highly active, extends the scope of hydroformylation and considerably simplifies process engineering, especially with regard to the environment [8 b]. As the catalyst is soluble in the most readily available "solvent," namely water, and the reactants and products formed are soluble in the organic phase, this allows the catalyst to be separated very easily as a "mobile phase" by decantation and simple separation of the two phases ("two-phase" or "biphase process"). Although acting homogeneously, the oxo catalyst is located in a heterogeneous phase and is thus "heterogenized" or "immobilized". A suitable idea for a process presented by Kuntz [10] has been adopted and developed since 1982 by Ruhrchemie AG ([2 d] and references therein) into an industrial-scale operation. Taking propene hydroformylation as an example (Eq. 1), the RCH/RP oxo process has been producing over 330 000 tons y^{-1} *n*-butanyraldehyde (as well as less than 4% isobutyraldehyde) since 1984.

$$H_2C{=}CH{-}CH_3 + CO/H_2 \xrightarrow{\text{cat.}} H_3C{-}CH_2{-}CH_2{-}CHO + CH_3{-}\underset{\displaystyle CH_3}{\underset{|}{CH}}{-}CHO \quad (1)$$

propylene syngas *n*-butyraldehyde *iso*-butyraldehyde

Production of more than 3 million tonnes of *n*-butanal demonstrates the strength of the aqueous-phase oxo concept, as do some other applications of two-phase homogeneous catalyst systems such as Shell's SHOP process (two-phase but not aqueous; cf. Section 7.1) or variations by Rhône-Poulenc [11], Montedison [12], Kuraray ([13] and Section 6.7), or Hoechst [14] manufacturing higher olefins, vitamin precursors, telomers, or fine chemicals (cf. Section 6.10).

The triumph of water-soluble catalysts in homogeneous catalysis follows the laborious work involved in the development of water-soluble ligands (cf. Section 3.2 and [7]). Logically, the above-mentioned foreword to the NATO workshop [5] continues:

> "The design and synthesis of aqueous transition-metal catalysis requires broad knowledge of organometallic chemistry in water as well as of the physical and chemical properties of water itself. Water is now regarded as a unique solvent for certain stoichiometric and catalytic reactions involving organometallic compounds. Because of its highly polar, protic nature, water strongly influences the acid–base behaviour of solutes, the formation and disruption of ion-pairs and hydrogen bonds, and the extent of hydrophobic interactions. These effects often lead to unexpected reaction rates and selectivities, but the phenomena are still not fully understood."

The details and backgrounds of RCH/RP's developments have been described elsewhere, especially with respect to the highly sophisticated catalyst system [2d, 3, 15]. Kuntz's work on trisulfonated triphenylphosphine (TPPTS [2d, 10]) and the industrial improvements developed by Ruhrchemie eventually laid the foundation for the subsequent successful commercialization. TPPTS is an ideal ligand modifier for the oxo-active $HRh(CO)_4$. Without any expensive preformation steps, three of the four CO ligands can be substituted by the readily soluble (1100 g L^{-1}) nontoxic (LD_{50}, oral >5000 mg kg^{-1}) TPPTS, which yields the hydrophilic oxo catalyst $HRh(CO)[P(3\text{-sulfophenyl-Na})_3]_3$ (cf. Section 6.1). TPP*MS* and TPP*DS* (the appropriate TPP mono- and disulfonates) maintain their importance, except for industrial applications, alongside TPPTS because the different degrees of sulfonation permit fine adjustment of the hydrophilic/hydrophobic ratio of the catalysts during biphase operation. Recently other authors have described an oxo process employing TPPMS [16]. The flow diagram of the RCH/RP process is shown in Section 6.1.3.1.

Because of the solubility of the Rh(I) complex in water and its insolubility in the oxo products, the oxo unit is essentially reduced to a continuous batch reactor followed by a phase separator (decanter) and a stripping column. Propene and syngas are added to the stirred, noncorrosive catalyst solution in the reactor (for an explanation see Section 6.1.3). After reaction the crude aldehyde passes to the decanter and, while being degassed, is thus separated into the aqueous catalyst solution and the organic aldehyde phase. The catalyst solution exchanges heat and produces process steam in the heat exchanger and is replaced by the same amount of water (not catalyst solution!) dissolved in the crude aldehyde and is returned to the oxo reactor. The crude aldehyde passes through a stripping column in which it is treated with countercurrent fresh syngas and, if necessary, is freed from unreacted olefin. No side-reactions occur which decrease the selectivity or yield of the crude aldehyde, since this stripping of the aldehyde is carried out in the absence of the oxo catalyst, a distinctive feature of RCH/RP's process. The crude aldehyde is fractionally distilled into *n*- and *iso*butanal in a conventional aldehyde distillation unit. The reboiler of this "*n*/*iso* column" is designed as a heat-absorbing falling film evaporator incorporated in the oxo reactor, thus providing a neat, efficient method of recovering heat by transferring the heat of reaction in the reactor to cold *n*-butanal, which subsequently heats the *n*/*iso* column. The preferred hydroformylation temperature is 110–130 °C and is therefore used for the production of process steam. Whereas other oxo processes are steam "importers", the RCH/RP process including the distillation of *n*-/*iso*butanol exports steam.

The catalyst is not sensitive to sulfur or other oxo poisons, which is another environmental advantage. Together with simple but effective decanting, which allows the withdrawal of organic and other by-products at the very moment of separation, accumulation of activity-decreasing poisons in the catalyst solution is prevented. Therefore no special pretreatment or even purification steps are necessary. This reduces the environmental burden still further. For a considerable time the oxo units at Ruhrchemie were supplied with syngas derived from coal produced by the TCGP (Texaco coal-gasification process, in Ruhrchemie/Ruhrkohle's version). In some cases this can be an important factor as far as local resources are concerned. Since syngas from widely differing sources created a highly suitable link to hydroformylations, the biphase process is best for environmentally undemanding techniques.

The oxo catalyst $[HRh(CO)(TPPTS)_3]$, its formation from suitable precursors, and its operation are described in Section 6.1.3 (see [2]). The reaction system is self-adjusting – an important consideration for safety reasons – and thus control analyses are needed only at prolonged intervals. Owing to the high degree of automation, only two employees per shift supervise two oxo units with a total capacity of currently over 330 000 tons y^{-1}. The design obviates the need for certain equipment (e.g., feed and cooling pumps); the on-stream factor of the whole system exceeds $>98\,\%$. Typical reaction conditions, crude product compositions of the RCH/RP process averaged out over a 14-year period, and a discussion of selectivities and activities are given in Table 3 of Section 6.1.3.1.

The high selectivity toward the sum of C_4 products is a special feature of two-phase operation and results from the availability of water during hydroformylation [17].

The high selectivity toward C_4 products (a maximum of 1% of higher-boiling components, "heavy ends", are formed relative to butanals) makes fractional distillation after aldehyde distillation unnecessary, reduces expenditure, and thus also minimizes the environmental load. The manufacture of the by-products becomes part of the 2-EH (2-ethylhexanal) process since the heavy ends consist mainly of 2-ethyl-3-hydroxyhexanal, which, during downstream processing, is also converted to 2-EH. This (and the avoidance of butyl formates) is the reason for the considerable simplification of the process flow diagram compared with other oxo process variants (see Figure 1).

As described earlier, as process steam permits economic heat management, the utilization of that part of the heat of reaction not used for *n*/*iso* separation is neat, saves energy, and is thus environmentally more benign. The consequences of the adjustable propene conversions and rhodium management (including losses) in respect of this heterogenized catalyst and for process factors are given elsewhere [2] together with comparative manufacturing costs. These show a crucial reduction of 10% in costs compared with other modern low-pressure oxo processes with their relatively complex operation. This demonstrates the favorable nature of the two-phase reaction system. Evidence of these advantages is apparent from the licences taken out for the process for commercial hydroformylation of butenes, and for additional new plants [15, 18], which will increase the total production figures to approx. 600000 tons $\cdot$ y^{-1}.

5.2.3 Crucial environmental improvements

The fundamental advance represented by the RCH/RP process (as a prototype of a biphasic technique) in terms of the environment, conservation of resources, and minimization of environmental pollution can be demonstrated by various criteria and proved very convincingly by means of the environmental factor, *E* [19] (which is far more suitable and constructive than Trost's "atom efficiency" [20]). Sheldon defined the *E* factor as the ratio of the amount of waste produced per kilogram of product and specified the *E* factor for every segment of industry (Table 1).

Table 1. Environmental acceptability: the *E* factor [19].

	Product tonnage	By-products/product ratio, w/w
Oil refining	10^6-10^8	0.1
Bulk chemicals	10^4-10^6	<1–5
Fine chemicals	10^2-10^4	5–50
Pharmaceuticals	10^1-10^3	25–100+

Table 2. *E* factors for oxo processes.

	By-products/product ratio, w/w	
	Isobutyraldehyde as product	Isobutyraldehyde as by-product
Co: high-pressure process	>0.6	>0.9
Rh: Ruhrchemie/Rhône-Poulenc	<0.04	<0.1

As expected and shown in Table 2, this environmental quotient for conventional oxo processes (cobalt catalyst) and the production of the bulk chemical *n*-butyraldehyde is actually about 0.6–0.9, depending on the definition of the term "target product". This range indicates that the by-product isobutyraldehyde (see Eq. 1) occurring with conventional oxo processes is further processed by a number of producers (e.g., to neopentyl glycol (2,2-dimethylpropane-1,3-diol) or isobutyric acid) so that the isobutyraldehyde thus becomes the target product and the *E* factor then falls from 0.9 to 0.6. Strictly speaking, this observation is included in Sheldon's wider assessment, according to which the *E* factor is refined and becomes the "environmental quotient" *EQ*, depending on the nature of the waste. Since such quotients "are debatable and will vary from one company to another and even from one production to another" [19] they will not be discussed here. The crucial point is that, on the same basis (taking into account all by-products, including those produced in ligand manufacture, etc.) that gives conventional oxo processes an *E* factor of 0.6–0.9, this factor falls to below 0.1 in the RCH/RP process: an important pointer to the environmental friendliness of the new process (Table 2).

The low *E* factor (<0.04) indicates that the utilization of material resources is improved more than tenfold: according to Sheldon's assessment [19] production of the bulk chemical *n*-butyraldehyde is classified alongside the highly efficient mineral-oil refining processes.

Whereas this important quotient is calculated solely from the product spectrum, process simplifications are a consequence of combining the rhodium catalyst with the special two-phase process. Compared with the conventional oxo processes and with other variants which include thermal separation of the oxo reaction products from the catalyst, the procedure is considerably simplified (see Figure 1).

Figure 1 demonstrates impressively that in the RCH/RP process most of the equipment commonly used for conventional oxo processes is not required. Items 10–15 in particular are superfluous because of the higher selectivity of conversion with Rh catalysts (also expressed in the *E* factor), whereas the efficient catalyst feed with the two-phase procedure obviates the need for items 3–8. It is therefore not surprising that a comparison of capital expenditure for the old Co high-pressure process with that for the RCH/RP process, showing that it is at least 1.9 times higher for the conventional oxo process, favors the new process.

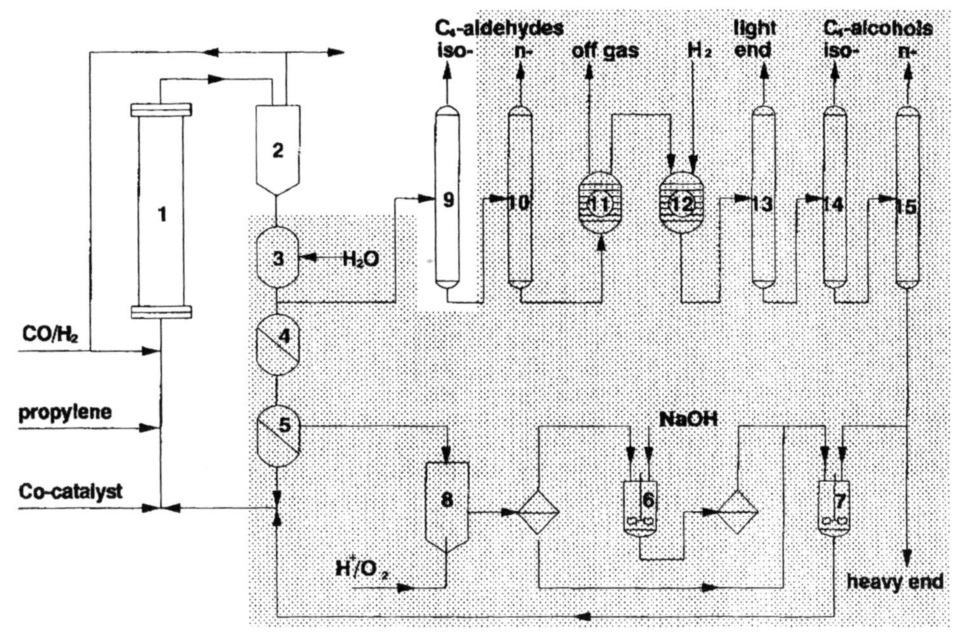

Figure 1. Schematic flowsheet of a conventional oxo process [2a]. The process steps super fluous RCH/RP's process are marked. 1, oxo reactor; 3, decobalter; 2, 4, 5, separators; 6–8, Co catalyst make-up; 9, 10, butyraldehyde distillation; 11, by-product cracking; 12, hydrogenation; 13–15, butanol distillation.

The conservation of energy resources with the RCH/RP process is dramatic. Note should be taken of the much milder reaction conditions and of the fact that the RCH/RP process is an energy exporter because of an intelligent, integrated, heat network (an unusual occurrence for conventional oxo processes – including Rh processes with triphenylphosphine as ligand). Furthermore, the steam consumption figures for the old Co process are very much higher than those for the Rh process and power consumption was twice as high as that of the RCH/RP process: both of these factors represent an environmental burden. The compression costs alone for the required syngas are 1.7:1 (old Co process compared with RCH/RP process). The volume of waste water from the new process is 70 times lower than that from the Co high-pressure process – convincing evidence of an environmentally benign process (Table 3). Other consumption figures for the new process are correspondingly lower than those for the older Co high-pressure process.

No consideration has yet been given to the special technology of the RCH/RP process. The solubility in water of the oxo catalyst, and the fact that the resultant separation of the catalyst from the reaction product proceeds very quickly, mean that to all intents and purposes the catalyst does not leave the oxo reactor and its immediate surroundings. This is an important reason why the

Table 3. Environmentally important values for the RCH/RP process.

	Old Co high-pressure process	RCH/RP Rh low-pressure process
Selectivities		
– toward C_4 products [%]	93	>99
– toward C_4 aldehydes [%]	86	99
Products other than *n*-butyraldehyde [%]	31	<5
n/*i* ratio	80:20	93–97:7–3
Manufacturing costs	140	100
Capital expenditure costs	>1.9	1
Waste water volume	70	1
Energy consumption figures		
– Steam	82	–6.5 (steam exporter)
– Power	>2	1
– Syngas compression	1.7	1
Reaction conditions		
– Pressure [bar]	300	<50
– Temperature [°C]	150	120
E factor	0.6–0.9	0.04– <0.1

low Rh losses are in the ppb range. The reduced probability of leaks also increases effectiveness and safety.

The "solvent" water reliably averts the risk of fire inherent in the old Co process as a result of leaking, highly flammable, metallic carbonyl. The reaction system with its "built-in extinguishing system" reliably prevents such fires, and the pain-staking measuring and monitoring procedure necessitated by the valuable catalyst metal rhodium, accompanied by constant simultaneous balancing of the RCH/RP process, permits any leaks from the aqueous system to be detected much earlier than was ever possible with the expensive mass and liquid balance of the old Co process. This also applies to the cooling system, in which any leak from the falling film evaporator would be noticed after a loss of only a few ppm of rhodium.

The solvent, water, is available instantly everywhere, and there is no need here to go into the special advantages of this polar solvent in respect of the process characteristics and its chemical effects [17, 21, 22]. It is not odor-free: because of its residual solubility for *n*-butyraldehyde it is as certain to be detected by the sensory/olfactory organs as are the product streams of other oxo processes. This is also the reason for the weakness of the criticisms leveled against water as a catalyst carrier (and quoted at the beginning) for being undetectable owing to the lack of odor.

5.2.4 Conclusions

Taking all the criteria into consideration, the RCH/RP biphasic oxo process is probably the soundest variant in terms of the environment; it is a "green" process which, in addition to environmental compatibility, has the advantage of being extremely cost-effective. The process is also "good-natured" from the handling aspect and thus inherently safe, as already emerged during development with a scale-up factor of $\gg 1:10000$ [2a]. The process has great development potential, whether in terms of varying the ligand TPPTS (and thus the activity and selectivity of the conversion [2] as well as the possibility of achieving asymmetrical hydroformylations), or in terms of other starting olefins. The adoption of the aqueous two-phase reaction [2d, 3] for hydroformylation in particular, and for other homogeneously catalyzed processes in general, shows the striking and unequaled advantages of this special neat type of "immobile" catalyst and its "heterogenization" by the catalyst carrier, water, as a result of which it is handled in similar fashion to a heterogeneous catalyst. This affects costs as well as environmental compatibility.

The advantages of avoiding by-products, and of successful waste management, are further good points of the process which ultimately lead to higher cost-effectiveness as a result of better environmental compatibility and less downtime. This is further proof that in the long term the most effective processes are all likely to be the environmentally sound ones. Among the homogeneously catalyzed processes the aqueous biphase variants will be in the forefront.

References

[1] G. Papadogianakis, R. A. Sheldon, *New J. Chem.* **1996**, *20*, 175.

[2] (a) E. Wiebus, B. Cornils, *Chem.Ing.Tech.* **1994**, *66*, 916; (b) B. Cornils, E. Wiebus, *CHEMTECH* **1995**, *25*, 33; (c) B. Cornils, E. Wiebus, *Hydrocarbon Proc.* **1996**, *75(3)*, 63; (d) B. Cornils, E. G. Kuntz, *J. Organomet. Chem.* **1995**, *502*, 177; (e) B. Cornils, W. A. Herrmann, R. Eckl, *J. Mol. Catal.* **1997**, *116*, 27.

[3] B. Cornils, W. A. Herrmann (Eds.), *Applied Homogeneous Catalysis with Organometallic Complexes*, VCH, Weinheim, 1996, Chapter 3.1.1.

[4] Z. Jiang, A. Sen, *Macromolecules* **1994**, *27*, 7215.

[5] (a) NATO Advanced Research Workshop *Aqueous Organometallic Chemistry and Catalysis*, Debrecen, Hungary, Aug. 29–Sept. 1, 1994, Preprints; (b) J. Haggin, *Chem. Eng. News* **1994** (October 10), 28; (c) Proceedings of the NATO Symposium (Eds. I. T. Horváth, F. Joó), Summary of the round table discussion of the state of the art and future directions of aqueous organometallic chemistry and catalysis, in *Aqueous Organometallic Chemistry and Catalysis*, Kluwer, Dordrecht, **1995**, p. 1.

[6] B. Cornils, *Nachr. Chem. Techn. Lab. (Weinheim)* **1994**, *42*, 1736.

[7] (a) A. W. Herrmann, C. W. Kohlpaintner, *Angew. Chem.* **1993**, *105*, 1588; *Angew. Chem., Int. Ed. Engl.* **1993**, *32*, 1524; (b) P. Kalck, F. Monteil, *Adv. Organomet. Chem.* **1992**, *34*, 219.

[8] (a) B. Cornils, Hydroformylation in *New Syntheses with Carbon Monoxide* (Ed.: J. Falbe), Springer, Berlin, 1980; (b) B. Cornils, *Spektr. Wissensch.*, in press.
[9] R. Fowler, H. Connor, R. A. Baehl, *Hydrocarbon Proc.* **1976**, *9*, 247.
[10] E. G. Kuntz, *CHEMTECH* **1987**, *17*, 570.
[11] C. Mercier, P. Chabardes, *Pure Appl. Chem.* **1994**, *66(7)*, 1509.
[12] (a) L. Cassar, *Chim. Ind. (Milan)* **1985**, *57*, 256; (b) Montecatini Edison SpA (M. Foa, L. Cassar, G. P. Chiusoli), DE 2.035.902 (1979).
[13] (a) Kuraray Corp. (Y. Tokitoh, N. Yoshimura), US 4.808.756 (1989); (b) *Nippon Kagaku Kaishi* **1993**, (2), 119; *Chem. Abstr.* 118, 126927; (c) N. Yoshimura in [3], p. 351.
[14] Hoechst AG (S. Haber, H.-J. Kleiner), DE-OS 19527118 A1 (1997).
[15] Cf. the contributions of the special issue of *J. Mol. Catal.* **1997**, *116*.
[16] J. Haggin, *Chem. Eng. News* **1995** (April 17), 25.
[17] A. Lubineau, J. Augé, Y. Queneau, *Synthesis* **1994**, 741.
[18] Anon., *Chemical Market Reporter* **1995** (Sept. 18).
[19] R. A. Sheldon, *CHEMTECH* **1994**, *24*, 38.
[20] B. M. Trost, *Science* **1991**, *254*, 1471; *Angew. Chem.* **1995**, *107*, 285.
[21] T. G. Southern, *Polyhedron* **1989**, *8*, 407.
[22] B. Cornils, E. Wiebus, C. W. Kohlpaintner, in *Encyclopedia of Chemical Processing and Design* (Ed.: J. J. McKetta), Marcel Dekker, New York, **1998**.

6

Typical Reactions

6.1 Hydroformylation

6.1.1 Development of Commercial Biphasic Oxo Synthesis

Boy Cornils, Emile G. Kuntz

6.1.1.1 History of Biphasic Catalysis

The history of biphasic homogeneous catalysis starts with Manassen's statement [1]:

> "... the use of two immiscible liquid phases, one containing the catalyst and the other containing the substrate, must be considered. The two phases can be separated by conventional means and high degrees of dispersion can be obtained through emulsification."

Roughly at the same time (in contradiction to a misleading publication by Papadogianakis [2]) and parallel to work done by Joó [3] and others [4], one of us (EK at Rhône-Poulenc) devoted time and effort to starting practical work on biphasic catalysis with organometallic catalysts (especially hydroformylation), developing the biphasic principle and the current well-known standard ligand triphenylphosphine trisulfonate (TPPTS, cf. Section 3.2.1).

These efforts have to be considered against an "official" background which Parshall [5] described as follows: "There was some use of transition-metal carbonyls in catalytic reactions of CO, but soluble catalysts played only a minor role in industry" (which even at that time was not correct [6b]), and "... the discovery that had the greatest impact on technology was recognition of the catalytic merit of rhodium complexes," thus indicating that only a few homogeneously catalyzed processes were commercialized before the early 1970s (among them hydroformylation with Co catalysts) and only some of them used precious metals as the catalyst base [7]. Examples were

1. the Pd/Cu catalytic system used for the oxidation of ethylene to acetaldehyde (Wacker–Hoechst process) [8];
2. the production of 1,4-hexadiene by codimerization of ethylene and butadiene catalyzed by rhodium trichloride (the scale of the production was not mentioned and the catalyst cycle was not published [9]);
3. Monsanto's processes to yield acetic acid via methanol carbonylation [10] or a chiral L-DOPA precursor via hydrogenation of substituted cinnamic acids [11].

In the field of homogeneous catalysis, the late 1960s and early 1970s saw the discovery and utilization of catalytic systems involving metals associated with phosphorus compounds and their use in oxo syntheses. After the discovery of zerovalent metal complexes in 1957 by Quin and in 1958 by Malatesta and Cariello [12], the Wilkinson school of thought opened up a new field of rhodium chemistry by applying the remarkable catalytic properties of triphenylphosphine (TPP)-substituted Rh carbonyls [$HRh(CO)(TPP)_3$] in the hydrogenation and hydroformylation of olefins [13]. The important Wilkinson complex stimulated a rapid expansion in the chemistry of the lower-oxidation-state complexes. Thus, the progress made in the area of homogeneous catalysis includes processes such as the low-pressure, highly selective hydroformylation of lower olefins on a laboratory scale since 1976 using TPP-modified Rh-carbonyls with excess TPP (commercialized by Union Carbide, Celanese, and BASF; the name "LPO" – low-pressure oxo – stems from BP [14]); the oligomerization and functionalization of dienes; the telomerization of butadiene with water to octadienols; the hydrocyanation of olefins and dienes to adiponitrile; the hydrogenation, isomerization and dimerization of olefins; and miscellaneous reactions in fine chemistry [15].

In 1970, the discovery of these new organo-soluble catalysts based on Rh, Pd, or Pt was generally considered unfeasible for industrial processes because of the prohibitive price of the metals involved. However, the high activity and productivity of these catalysts made possible production levels of 100 000 t y^{-1} with only a few dozen kilograms of precious metals needed as inventory by each single plant. Thus, the amount of precious metal involved represents only a minor part of the investment and the manufacturing costs, i.e., the price of the metal was not an important factor in the production unit cost, provided that its usage occurred without any loss. In 1972 this hypothesis was confirmed by Monsanto and its commercialization of the important process to generate acetic acid by methanol carbonylation [16].

Uncertainty still attended the handling of "commercial" quantities of precious-metal catalysts. Another point in question was the technique to be used for separating the products from the expensive catalysts, i.e., for solving the imminent problem of homogeneous catalysis, and for developing new catalyst recovery systems. One possibility for solving the separation problem mentioned was distillation, which however causes thermal stress of the products to be separated and – more seriously – of the residual catalyst. The other way was decantation as a consequence of a biphasic operation. In the oxo technique in particular, a biphase process for the homogeneously catalyzed work-up of formic acid esters by cracking with aqueous sodium formate solutions as catalyst has been employed since 1967 [17] by one of us (BC at Ruhrchemie). According to Manassen's proposal, decantation of catalyst and product meant that homogeneous catalysts dissolved in a liquid other than the organic phase to be separated. Water in particular would allow acceptable partition coefficients [18].

The principle of technically feasible two-phase catalysis with water, alongside Manassen's [1] theoretical considerations, was conceived in 1973, taking the hydroformylation of propene as an example. The metal-complex catalyst solubilized in water by a hydrophilic ligand converts the starting material into the reaction products, which can be separated from the catalyst by simple phase decantation. During the hydroformylation of propene it is important that the butyraldehyde (in which water is slightly soluble [3 % at 20 °C]) does not extract the hydrophilic complex from the water-soluble catalytic system. It was obvious that on the one hand the best hydrophilic substitute for the proven TPP as a ligand for a water-soluble complex catalyst would be the sodium salt of an appropriate TPP sulfonate, which is thermally very stable. According to Wilkinson and Joó, monosulfonated triphenylphosphine (TPP*M*S, with a carbon/sulfonate ratio of 18:1) would be slightly foamy, extractable, and thus too organophilic (or hydrophobic) [3, 4d]. It therefore proved essential to use the nondetergent, highly water-soluble (and thus hydrophilic), trisulfonated triphenylphosphine (TPP*T*S, C/sulfonate ratio 6:1) which is not extractable by an organic medium, notably not by water-saturated butyraldehyde [19].

6.1.1.2 Basic Work and Investigations by Rhône-Poulenc

Following initial studies in 1974 the preparation of TPPTS was carried out via the sulfonation of TPP with "oleum" (i.e., concentrated sulfuric acid containing 20 % by weight of SO_3) at 40 °C in one day. After hydrolysis and neutralization by NaOH, an aqueous solution of sodium sulfate and a mixture of different P compounds – consisting mainly of TPPTS and the corresponding P-oxide ("TPPOTS") as key chemical species – were obtained (Scheme 1) [20].

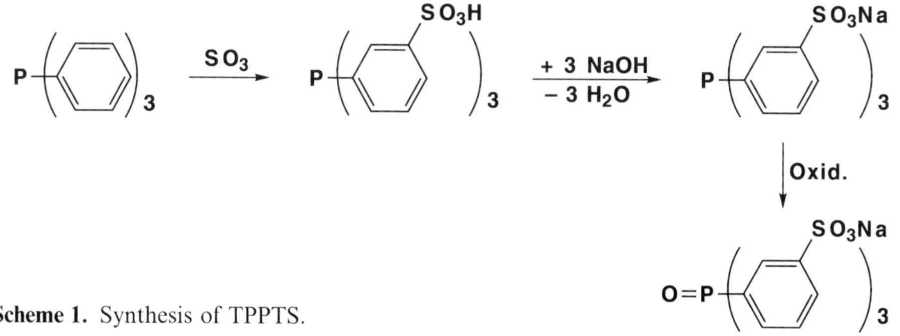

SO_3H SO_3Na P 3 SO_3 P 3 + 3 NaOH − 3 H_2O P 3 Oxid. SO_3Na O=P 3

Scheme 1. Synthesis of TPPTS.

Crude TPPTS fractions contained TPPTS and TPPOTS in ratios of roughly 55:45. Today's much improved commercial procedures [21, 59] yield TPPTS/TPPOTS ratios of 94:6 under appropriate conditions, even in the crude materi-

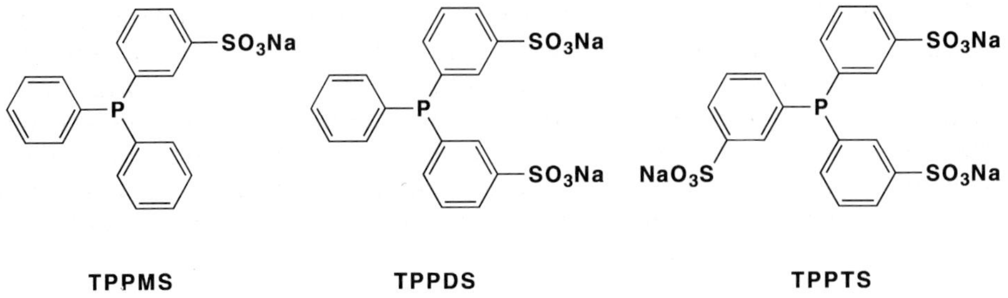

TPPMS TPPDS TPPTS

Figure 1. TPPMS, TPPDS, and TPPTS as ligands.

al. Under suitable conditions even the less preferable mono- and the disulfonated species (TPP*M*S and TPP*D*S) can be formed in addition to TPP*T*S (Figure 1).

From 1974 onwards the scope of different reactions using biphasic catalyst systems, preferably with precious metals, was tested in laboratory-scale experiments. Among these were butadiene hydrodimerization, hydrogenation of acrylonitrile or cyclohexene, hydroformylation of propene, and some other conversions to fine chemicals. Even during this initial stage of experimental work it was shown that only a small fraction of the precious metals (much less than 0.1 ppm Rh in the case of hydrogenation or hydroformylation) is leached by the organic phase. To be protected against Joó's and Beck's results on the hydrogenation of pyruvic acid to lactic acid with aqueous solutions of Ru or Ph catalysts and TPPMS [3], Rhône-Poulenc (RP) filed different patent applications on the results achieved so far, taking into consideration the first laboratory results, RP's interest in various product lines, and the economic importance of the invention [19]. Thus a series of patent applications protected three different main fields of interest [20], which included hydroformylation, hydrocyanation, and diene conversions.

Although RP was not an oxo producer, the excellent laboratory results with the biphasic system compared with conventional Co-based processes were regarded as providing an important opportunity. Taking the propene hydroformylation to *n*-butyraldehyde with a TPPTS ligand-modified rhodium catalyst as an example (Eq. 1), savings of up to 20% of propene and of syngas could be envisaged [20]. Although the reaction conditions in terms of activity, productivity, and selectivity were roughly optimized by using TPPTS instead of TPPMS or TPPDS for hydroformylations, hydrocyanations, or diene conversions [20b, c, 21] the stability of the aqueous catalyst system, the commercial design of a biphasic oxo unit, and the influences of the different reaction variables – all under conditions of a continuous process – remained unknown.

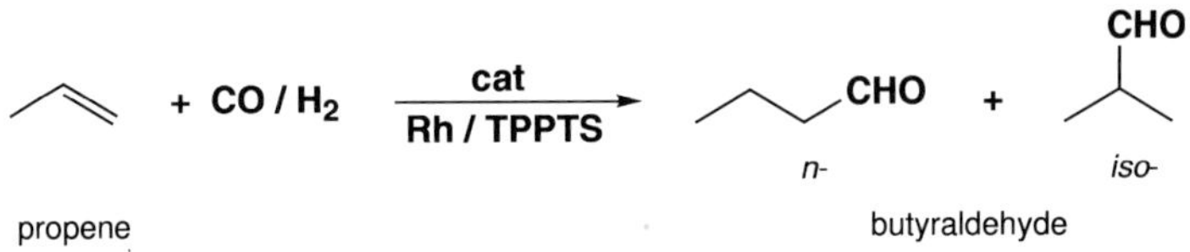

$$\text{propene} + CO/H_2 \xrightarrow[\text{Rh / TPPTS}]{\text{cat}} n\text{-butyraldehyde} + iso\text{-butyraldehyde} \quad (1)$$

Continuing the research on the reactivity of dienes with biphasic catalytic systems, Morel and Mignani discovered that the Rh/TPPTS functionalization of dienes in the 4-position has beneficial efforts [22]. With other asymmetrical dienes such as myrcene and the addition of, for example, ethyl acetoacetate, interesting regioselectivities of 99% for the desired isoprenic compounds were achieved (cf. Section 6.10, [23, 24]). A couple of subsequent reaction steps, based on TPPTS from Ruhrchemie, thus convert geranylacetone to vitamin E [25].

6.1.1.3 Investigations by Ruhrchemie AG

Following earlier contacts Ruhrchemie AG (RCH), now a part of Hoechst AG, and Rhône-Poulenc joined forces in 1982 to develop a *continuous* biphase hydroformylation process for the production of *n*-butyraldehyde from propene.

On the basis of the ideas documented in RP's applications, RCH used its own expertise with the biphasic catalytic cracking process [17] and its long experience in converting laboratory-scale syntheses to commercial processes – as Ost [26] observed as long ago as 1907: "It is one thing to discover a process that is right in principle but a very different thing to introduce it on the industrial scale." In transferring processes to the economic scale, Ruhrchemie was successful in many cases, e.g., with the Fischer–Tropsch synthesis [27], the Co-based oxo reaction (invented at RCH by Otto Roelen [6]), the UHMW variant of Ziegler-type HDPE [28], the Texaco coal gasification process [29], etc. In the case of biphasic oxo synthesis, appropriate agreements defined the tasks for RP (support) and RCH (R & D, engineering, production, licencing, and marketing and distribution agreements). Between 1982 and 1984, in a period of less than 24 months, RCH developed and tested a completely new process for which no prototype was available. Using a scale-up factor of 1:24000 the first production unit employing the "Ruhrchemie/Rhône-Poulenc oxo process" went on stream in July 1984 with an initial capacity of 100000 tons per year (see Section 6.1.3 [30]).

Because of Ruhrchemie's commitment and status as an important oxo producer, development was thus primarily driven by product and commercial considerations. It was not until the 1990s that further development became science-driven, including especially all the scientific research work currently being conducted at universities as a result of the successful implementation of the RCH/RP process.

All the development tests included the full work-up of the reactants, side-streams and products, reflecting the characteristics of the bisphasic system, the testing of reaction parameters, the dynamic responses of both the reactor and the catalytic system, product qualities, etc. [31].

The heart of the new process is the new catalyst $[HRh(CO)(TPPTS)_3]$ (Figure 2).

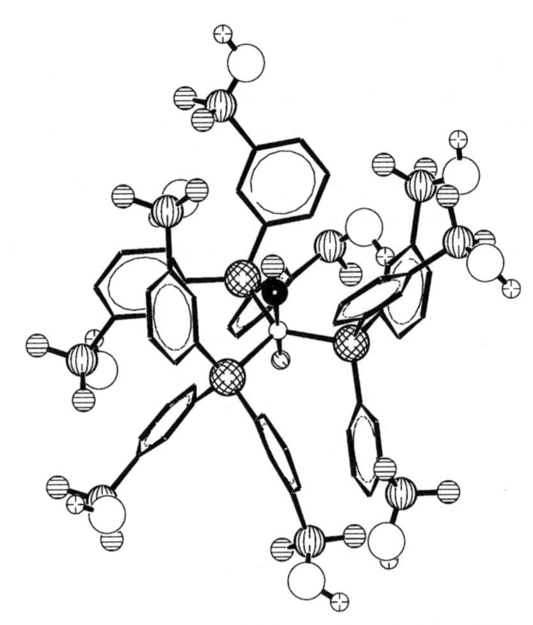

Figure 2. $[Rh(CO)(TPPTS)_3]$, the water-soluble catalyst of the RCH/RP process.

Apart from the handling of the new catalyst system with a water-soluble species, a virtually new design of a hydroformylation process had to be prepared, involving much detailed work. Most of the solutions and numerous important variants have been patented. This detailed work focused on the special aspects of the process flow chart as indicated in Figure 3.

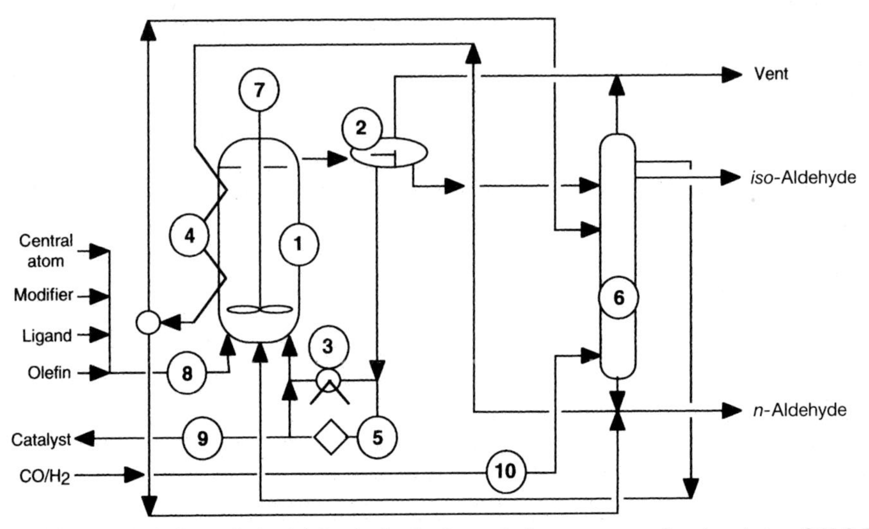

Figure 3. General design of the biphasic hydroformylation process: focal points of R & D work as indicated. (1) Reactor; (2) decantation; (3, 4) heat recovery and steam generation; (5) Rh recycle; (6) distillation; (7) control units; (8, 10) feed; (9) catalyst.

The sensitive areas of the process include the reactor and its environment (reactor, separation, control units, cooling devices, off-gases), olefin and syngas feed, preparation of the catalyst (including central atom, ligands, accompanying salts, and modifiers) and catalyst recycling (including catalyst separation from different sources and catalyst recycling at different levels). For all of the sensitive areas marked, various solutions were developed by RCH.

Some of the conditions of the new process were based on RP's patents but all had to be adjusted to the rougher operation conditions of a commercial plant, to different qualities and purities of commercially available reactants (olefins, syngas, catalyst precursors), to a normal operation of 8760 h y^{-1}, and to the fine-tuned relationships between, e.g., temperatures, pressures/partial pressures, concentrations of various organic and inorganic components in different phases, mass and heat transfer, and flow conditions of a continuous process [32]. The economics of cooling and heat recovery and the utilization of unreacted substrates in off-gases or vents are of special importance [33]. In contrast to the literature [34], a combination of a low-pressure oxo stage and a subsequent high-pressure stage proved to be more convenient than vice versa [33c].

Detailed work undertaken for the first time on this technique, which is mainly described in patents and therefore little known, has focused on special reaction conditions and special measures (even within extreme limits) which are based on the biphasic character of the conversion such as pH values, addition of CO_2, salt effects and solution ionic strengths, catalyst modifiers, spectator effects, or ultrasonic devices, etc. [35]. The measures mentioned allowed a considerably simplified process to be used compared with other oxo processes (basically consisting of a stirred tank reactor and a decanter), this being a consequence of the biphasic concept of RCH/RP. These relationships ensure a smooth, stable operation yielding high selectivities to *n*-butyraldehyde (cf. Section 6.1.3.1). The specific load of the system may be altered very unequivocally by varying the temperature, pressure, partial pressures, and concentrations (catalyst, ligands, and salts).

In contrast to other low-pressure processes, the purification of syngas is not critical [36]. Surprisingly, the presence of carbon dioxide in the syngas acts as a selectivity improver [35c], a similar effect to the one described later by Rathke and Klinger [37]. The oxo units at Oberhausen have also been supplied with syngas from a TCGP (Texaco coal gasification plant) without any problems [29]. Although the RCH/RP process is best suited to the conversion of lower olefins to the appropriate aldehydes (ethylene through pentenes) higher olefins may be converted using special precautions (cf. [38, 39c, 60]) or special ligand developments [40].

There was no prototype for the behavior and the lifetime of the biphase catalytic system and catalyst recycling. Intensive research work has been conducted to identify the optimal sets of operating conditions to ensure maximum efficiency and lifetime of the catalytic system. The advantages over other low-pressure processes are that the catalyst is on a "short circuit" around the reactor (ensuring fewer losses) and that because of mechanical separation of catalyst and product no thermal stress occurs. The catalyst may be regenerated within the reaction system using chemical or mechanical means [41]. Recycling by working-up the precious metal content as an ultimate solution is exceptional [42]. The losses of catalyst, a factor of major importance figuring in catalytic systems containing precious metals, are $< 10^{-9}$ g Rh/kg *n*-butyraldehyde. Other methods of recovering Rh from oxo crudes [43, 44 a] or distillation residues [45] are also being investigated.

The catalyst $[HRh(CO)(TPPTS)_3]$ (see Figure 2) is prepared from suitable precursors, which contain the central atom and the ligand. The catalyst can be preformed or may be prepared in situ [46]. The preparation of TPPTS is not as simple as it seems: many misinterpretations concerning catalytic strength, behavior, extractabilities, cluster or colloid formation, etc., can be explained by unsuitable methods of ligand syntheses and purification [39, 59]. For special purposes TPPTS may be modified [47]. These modifications of the original sulfonated phosphines result in, for example, different solubilities, re-immobilization [48], etc. Very different levels of phase behavior and thus various degrees of immobilization may be adjusted.

The search for water-soluble ligands other than TPPTS is a focal point of present academic and industrial work. In cooperation with Professor Herrmann of the Technical University of Munich and others, the R & D department of Ruhrchemie has focused on special water-soluble phosphines [49], phosphites [50], and other catalysts [48, 51, 52], and the most suitable way to manufacture them. The application of external or internal solubilizers may be recommended [44, 53]. Appropriate complexes with other central atoms can be used for other syntheses [54].

Although commercial emphasis has been on propene hydroformylation, other starting olefins include higher-molecular-mass substrates; see, e.g., [33 d, 38 a–d, 40, 44 b, 52, 55].

6.1.1.4 The RCH/RP Process as the Final Point of Development

As mentioned earlier, the first commercial oxo plant using the Ruhrchemie/Rhône-Poulenc biphasic process went to stream in 1984 [30] and has been reviewed recently [56]. Water-soluble catalysts offer significant advances in homogeneous catalysis: the "heterogenization" of the catalyst in a second, immiscible liquid phase ("liquid support") immobilizes the catalyst phase, thus combining the advantages of heterogeneous catalysis (e.g., long lifetime, easy separation of product from catalyst) with those of the homogeneous mode, e.g., defined species of catalyst, gentle reaction conditions, high activity, and high selectivity.

The Ruhrchemie plant reacts propene and syngas and yields *n*-butyraldehyde as the desired product besides less than 4% isobutyraldehyde. The basic flow diagram of the 330 000 t y^{-1} units is shown in Section 6.1. A smaller unit converts butenes. Additional plants are under construction.

In comparison with other oxo processes the procedure is much simpler and many steps in the original process are superfluous (cf. Section 5 [57]). The actual oxo catalyst is simply made within the oxo unit by reacting suitable Rh salts with TPPTS of appropriate quality without any additional preformation step. The reaction system is self-adjusting; the on-stream factor of the whole system is >98%. Typical reaction conditions and performances on a 14-year average are given in ref. [56, 57].

The successful implementation of the biphasic oxo process as *the* prototype of a homogeneous aqueous catalyst system will have different consequences for hydroformylation reactions as described in Chapters 3 (development of new ligands) and 4 (solvents, co-solvents, micellar techniques) and Section 6.1 (hydroformylation of higher olefins).

The suitability of the homogeneous aqueous catalysts and thus the scope of application in general will be extended in commercial or pilot-plant operation to other central atoms and reactions such as Heck reactions and other carbonylations (with Pd), hydrogenations (Pd, Pt, Rh, Ir), formation of water-soluble polymers (Pd), vinylations, metathesis conversions (Ru), Suzuki couplings, etc. (cf. Section 6.10).

A recent NATO workshop on aqueous homogeneous catalysis [58] summarized the situation:

> "A very important practical and environmentally beneficial aspect of the use of a separate aqueous phase in catalytic reactions producing water-insoluble products is the potential for relatively easy and complete recovery of water-soluble catalysts. This could lead to the elimination of further steps for removal of traces of heavy metal from the product and to considerable savings."

Reality shows this to be true.

References

[1] J. Manassen in *Catalysis Progress in Research* (Eds.: F. Basolo, R. L. Burwell), Plenum Press, London, 1973, p. 183.

[2] G. Papadogianakis, R. A. Sheldon, *New J. Chem.* **1996**, *20*, 175.

[3] (a) F. Joó, M. T. Beck, *React. Kin. Catal. Lett.* **1975**, *2*, 257; (b) F. Joó, Z. Tóth, M. T. Beck, *Inorg. Chim. Acta* **1977**, *25*, L61.

[4] For example: (a) Y. Dror, J. Manassen, *J. Mol. Catal.* **1977**, *2*, 219; (b) R. T. Smith, M. C. Baird, *Transit. Metal. Chem.* **1981**, *6*, 197; (c) R. T. Smith, R. K. Ungar, M. C. Baird, *Transit. Metal. Chem.* **1982**, *7*, 288; (d) A. F. Borowski, D. J. Cole-Hamilton, G. Wilkinson, *Nouv. J. Chem.* **1978**, *2*, 137.

[5] G. W. Parshall, *Organometallics* **1987**, *6*, 687.

[6] (a) O. Roelen, *Chem. Exp. Didakt.* **1977**, *3*, 119; (b) B. Cornils, W. A. Herrmann, M. Rasch, *Angew. Chem.* **1994**, *106*, 2219; *Angew. Chem., Int. Ed. Engl.* **1994**, *33*, 2144.

[7] N. S. Imyanitov, *Rhodium Express* **1995** (*10/11*), 3.
[8] (a) E. W. Stern, *Catal. Rev.* **1968**, *1*, 74; (b) R. Jira in *Ethylene and its Industrial Derivatives* (Ed.: S. A. Miller), Benn, London, **1969**; (c) R. Jira, Wacker-Process, in *Applied Homogeneous Catalysis with Organometallic Complexes* (Eds.: B. Cornils, W. A. Herrmann), VCH, Weinheim, 1996, Vol. 1, p. 374.
[9] (a) Dupont (T. Alderson), US 3.013.066 (1961); (b) R. Cramer, *J. Am. Chem. Soc.* **1967**, *89*, 1633; (c) A. C. L. Su, *Adv. Organomet. Chem.* **1978**, *17*, 269.
[10] J. F. Roth, J. H. Craddock, A. Hershman, F. E. Paulik, *CHEMTECH* **1971**, *1*, 600.
[11] (a) W. S. Knowles, *Acc. Chem. Res.* **1983**, *16*, 106; (b) W. S. Knowles, M. J. Sabacky, B. D. Vineyard, *Ann. N. Y. Acad. Sci.* **1977**, *295*, 274.
[12] (a) L. D. Quin, *J. Am. Chem. Soc.* **1957**, *79*, 3681; (b) L. Malatesta, C. Cariello, *J. Chem. Soc.* **1958**, 2323.
[13] (a) J. A. Osborn, G. Wilkinson, J. F. Young, *Chem. Commun.* **1965**, 17 and **1965**, 131; (b) J. A. Osborne, F. H. Jardine, J. F. Young, G. Wilkinson, *J. Chem. Soc. (A)* **1966**, 1711; (c) T. A. Stephenson, G. Wilkinson, *J. Inorg. Nucl. Chem.* **1966**, *28*, 945.
[14] BP Co. Ltd. (M. J. Lawrenson), GB 1.197.902 (1967).
[15] References are given in B. Cornils, E. G. Kuntz, *J. Organomet. Chem.* **1995**, *502*, 177.
[16] (a) Monsanto (D. Forster et al.) FR 1.573.130 (1967); (b) D. Forster, *J. Am. Chem. Soc.* **1975**, *97*, 951; (c) D. Forster, *Adv. Organomet. Chem.* **1979**, *17*, 255; (d) cf. M. Gauss et al. in [53], Vol. 1, p. 104.
[17] Ruhrchemie AG (H. Tummes, J. Meis), US 3.462.500 (1965/1969).
[18] (a) B. Cornils, C. W. Kohlpaintner in *Encyclopedia of Chemical Processing and Design* (Ed.: J. J. McKetta), Marcel Dekker, New York, in press; (b) I. T. Horváth, *J. Mol. Catal. (A)* **1997**, *106*, 1.
[19] E. Kuntz, *CHEMTECH* **1987**, 570.
[20] (a) Rhône-Poulenc Recherche (E. Kuntz), FR 2.314.910 (1975); (b) Rhône-Poulenc Industrie (E. Kuntz), FR 2.349.562 (1976); (c) Rhône-Poulenc Industrie (E. Kuntz), FR 2.338.253 (1976); (d) Rhône-Poulenc Industrie (E. Kuntz), FR 2.366.237 (1976).
[21] Rhône-Poulenc (J. Jenck), FR 2.478.078 (1980).
[22] (a) Rhône-Poulenc (D. Morel), EP 0.044.771 (1980); (b) Rhône-Poulenc (D. Morel, G. Mignani), FR 2.561.641 and 2.569.403 (1984).
[23] Rhône-Poulenc (J. L. Sabot), FR 2.532.318 (1982).
[24] (a) C. Mercier, P. Chabardes, *Pure Appl. Chem.* **1994**, *66*, 1509; (b) D. Morel, G. Mignani, Y. Colleuille, *Tetrahedron Lett.* **1985**, *26*, 6337; *ibid.* **1986**, *27*, 2591; (c) Rhône-Poulenc Ind. (D. Morel), FR 2.486.525 and 2.505.322 (1980); (d) E. Kuntz, M. Thiers, Y. Colleuille, J. Jenck, D. Morel, G. Mignani, *Actual. Chim.* **1990** (Mar./Apr.), 50.
[25] (a) J. M. Grosselin, C. Mercier, G. Allmang, F. Grass, *Organometallics* **1991**, *10*, 2166; (b) J. M. Grosselin, H. Kempf, J. P. Lecouve, EP 90/9887 (1990); (c) Rhône-Poulenc, FR 2.230.654 (1983); (d) Rhône-Poulenc (D. Morel, J. Jenck), FR 2.550.202 (1983); (e) Rhône-Poulenc (J. Jenck), FR 2.473.504 (1979); (d) Rhône-Poulenc (C. Varre, M. Desbois, J. Nouvel), FR 2.561.650 (1984).
[26] H. Ost, *Z. Angew. Chem.* **1907**, *20*, 212.
[27] B. Cornils, *Technikgeschichte* **1997**, *64*, 205.
[28] H. Käding, Brennst.-Chem. **1968**, *49(11)*, 337.
[29] B. Brunke, R. Dürrfeld, J. Langhoff, W. Gick, H. D. Hahn, N. Leder; *Erdöl, Erdgas, Kohle* **1987**, *103*, 289.
[30] (a) B. Cornils, J. Falbe, *Proc. 4th Int. Symp. on Homogeneous Catalysis,* Leningrad, Sept. 1984, p. 487; (b) H. Bach, W. Gick, E. Wiebus, B. Cornils, *Prepr. Int. Symp. High-Pressure Chem. Eng.,* Erlangen/Germany, Sept. 1984, p. 129; (c) H. Bach, W. Gick, E. Wiebus, B. Cornils, *Prepr. 8th ICC,* Berlin 1984, Vol. V, p. 417; *Chem. Abstr.* **1987**, *106*, 198.051;

cited also in A. Behr, M. Röper, *Erdgas und Kohle* **1984**, *37(11)*, 485; (d) H. Bach, W. Gick, E. Wiebus, B. Cornils, *Abstr. 1st IUPAC Symp. Org. Chemistry*, Jerusalem 1986, p. 295.

[31] Hoechst AG (G. Kessen, B. Cornils, J. Hibbel, H. Bach, W. Gick), EP 0.216.151 (1986).

[32] (a) Ruhrchemie AG (B. Cornils, J. Hibbel, W. Konkol, B. Lieder, J. Much, V. Schmidt, E. Wiebus), EP 0.103.810 (1982).

[33] (a) Ruhrchemie AG (H. Kalbfell, B. Lieder, H. Mercamp), EP 0.144.745 (1984); (b) Ruhrchemie AG (B. Lieder, V. Schmidt, S. Sedelies, H. Kalbfell), EP 0.158.196 (1985); (c) Ruhrchemie AG (B. Cornils, J. Hibbel, G. Kessen, W. Konkol, B. Lieder, E. Wiebus, H. Kalbfell, H. Bach), DE 3.245.883 (1982); (d) Hoechst AG (H. Bahrmann, W. Greb, P. Heymanns, P. Lappe, T. Müller, J. Szameitat, E. Wiebus), DE 4.333.324 (1994).

[34] BASF AG (H. Hohenschutz, M. Strohmeyer, H. Elliehausen, K. Fischer, R. Kummer, M. Herr, W. Reutemann), DE 3.102.281 (1981).

[35] (a) Ruhrchemie AG (B. Cornils, H. Bahrmann, W. Lipps, W. Konkol), DE 3.511.428 (1985); (b) Ruhrchemie AG (B. Cornils, W. Konkol, H. Bach, G. Dämbkes, W. Gick, W. Greb, E. Wiebus, H. Bahrmann), EP 0.158.246 (1985); (c) Ruhrchemie AG (B. Cornils, W. Konkol, H. Bach, G. Dämbkes, W. Gick, E. Wiebus, H. Bahrmann), DE 3.415.968 (1984); (d) Ruhrchemie AG (B. Cornils, W. Konkol, H. Bach, W. Gick, E. Wiebus, H. Bahrmann, H. D. Hahn), DE 3.546.123 (1985); (e) Ruhrchemie AG (H. Bach, B. Cornils, W. Gick, H. D. Hahn, W. Konkol, E. Wiebus), DE 3.640.614 (1986); (f) Ruhrchemie AG (H. Bach, B. Cornils, W. Gick, G. Diekhaus, W. Konkol, E. Wiebus), EP 0.269.011 (1987).

[36] Hoechst AG (W. Konkol, H. Bahrmann, G. Dämbkes, W. Gick, E. Wiebus, H. Bach), EP 0.216.258 (1986).

[37] (a) J. W. Rathke, R. J. Klinger, T. R. Krause, *Organometallics* **1991**, *10*, 1350; (b) Argonne Nat. Lab. (J. W. Rathke, R. J. Klinger), US 5.198.589 (1993).

[38] (a) Ruhrchemie AG (B. Cornils, W. Konkol, H. Bach, W. Gick, E. Wiebus, H. Bahrmann), DE 3.447.030 (1984); (b) Hoechst AG (H. Bahrmann, W. Greb, P. Heymanns, P. Lappe, T. Müller, J. Szameitat, E. Wiebus), EP 0.562.451 (1993) and EP 0.562.450 (1993); (c) Hoechst AG (B. Fell, P. Hermanns), DE 4.330.489 (1993); (d) Hoechst AG (H. Bahrmann, E. Wiebus et al.), DE 4.333.323 (1993); (e) Hoechst AG (H. Bahrmann, W. Greb, P. Heymanns, P. Lappe, T. Müller, J. Szameitat, E. Wiebus), EP 0.646.563 (1993).

[39] (a) Ruhrchemie AG (R. Gärtner, B. Cornils, H. Springer, P. Lappe), EP 0.107.006 (1983); (b) Ruhrchemie AG (L. Bexten, B. Cornils, D. Kupies), EP 0.175.919 (1985); (c) Hoechst AG (W. A. Herrmann, J. Kulpe, W. Konkol, H. Bach, W. Gick, E. Wiebus, T. Müller, H. Bahrmann), EP 0.352.478 (1989).

[40] (a) Hoechst AG (S. Bogdanovich, H. Bahrmann, C. D. Frohning, E. Wiebus), DE-Appl. 19700805.4 and 19700804.6 (1997); (b) Hoechst AG (H. W. Roesky, N. Winkhofer, U. Ritter), DE-OS 19521936 (1996) and WO 97-00132 (1997).

[41] (a) Ruhrchemie AG (R. Gärtner, B. Cornils, L. Bexten, D. Kupies), EP 0.103.845 (1983); (b) Ruhrchemie AG (W. Greb, J. Hibbel, J. Much, V. Schmidt), DE 3.630.587 (1986); (c) Hoechst AG (H. Bahrmann, M. Haubs, W. Kreuder, T. Müller), EP 0.374.615 (1989); (d) Hoechst AG (W. Konkol, H. Bahrmann, W. A. Herrmann, C. W. Kohlpaintner), EP 0.544.091 (1992).

[42] (a) Ruhrchemie AG (L. Bexten, D. Kupies), DE 3.626.536 (1986); (b) Ruhrchemie AG (G. Diekhaus, H. Kappeser), DE 3.744.213 (1987); (c) Hoechst AG (J. Weber, L. Bexten, D. Kupies, P. Lappe, H. Springer), EP 0.367.957 (1989).

[43] (a) Hoechst AG (G. Dämbkes, H. D. Hahn, J. Hibbel, W. Materne), EP 0.147.824 (1984); (b) Hoechst AG (B. Cornils, W. Konkol, H. Bahrmann, H. Bach, E. Wiebus), EP 0.183.200 (1985); (c) Hoechst AG (I. Förster, K. Mathieu), EP 0.538.732 (1992).

[44] (a) Hoechst AG (B. Cornils, W. Konkol, H. Bahrmann, H. Bach, E. Wiebus), DE 3.411.034 (1984); (b) Hoechst AG (H. Bahrmann, B. Cornils, W. Konkol, W. Lipps), EP 0.157.316 (1985); (c) Hoechst AG (H. Bahrmann, P. Lappe), EP 0.602.463 (1993).
[45] (a) Hoechst AG (P. Lappe, L. Bexten, D. Kupies), US 5.294.415 (1992); (b) Hoechst AG (P. Lappe, H. Springer), EP 0.475.036 (1991); (c) Hoechst AG (P. Lappe, H. Springer), EP 0.510.358 (1992); (d) Hoechst AG (P. Lappe, H. Springer), EP 0.424.736 (1990); (e) Hoechst AG (G. Diekhaus, H. Kappeser), EP 0.584.720 (1993).
[46] Ruhrchemie AG (H. Bach, H. Bahrmann, B. Cornils, W. Konkol, E. Wiebus), DE 3.616.057 (1986).
[47] (a) Hoechst AG (H. Bahrmann, W. Konkol, J. Weber, H. Bach, L. Bexten) EP 0.216.315 (1986); (b) Hoechst AG (H. Bach, H. Bahrmann, B. Cornils, W. Gick, V. Heim, W. Konkol, E. Wiebus), EP 0.302.375 (1988); (c) Ruhrchemie AG (H. Bahrmann, B. Cornils, W. Konkol, W. Lipps), DE 3.420.491 (1984).
[48] H. Bahrmann, M. Haubs, T. Müller, N. Schöpper, B. Cornils, *J. Organomet. Chem.* **1997**, *545/546*, 139.
[49] (a) Hoechst AG (W. Konkol, H. Bahrmann, W. A. Herrmann, J. Kulpe), EP 0.435.069 (1990) and EP 0.435.073 (1990); (b) Hoechst AG (G. Papadogianakis, B. Fell, H. Bahrmann), EP 0.489.330 (1991); (c) Hoechst AG (W. A. Herrmann, C. Kohlpaintner, H. Bahrmann), EP 0.491.239 and EP 0.491.240 (1991); (d) Hoechst AG (W. A. Herrmann, R. Manetsberger, C. Kohlpaintner, H. Bahrmann), EP 0.571.819 and EP 0.575.785 (1993); (e) Hoechst AG (H. Bahrmann, P. Lappe, W. A. Herrmann, R. Manetsberger, G. Albanese), DE 4.321.512 (1993); (f) Hoechst AG (H. Bahrmann, P. Lappe, W. A. Herrmann, G. Albanese, R. Manetsberger), DE 4.333.307 (1993); (g) Hoechst AG (H. Bahrmann, C. Kohlpaintner, W. A. Herrmann, R. Schmid, G. Albanese), DE 4.426.577 (1994); (h) Hoechst AG (G. Albanese, R. Manetsberger, W. A. Herrmann), DE 4.435.189 (1994); (i) Hoechst AG (G. Albanese, R. Manetsberger, W. A. Herrmann, C. Schwer), DE 4.435.190 (1994); (j) Hoechst AG (H. Bahrmann, P. Lappe, C. Kohlpaintner, W. A. Herrmann, R. Manetsberger), EP 0.571.819 (1992).
[50] For example, Hoechst AG (H. Bahrmann, B. Fell, G. Papadogianakis), EP 0.436.084 (1990) and EP 0.435.071 (1990).
[51] Hoechst AG (G. Albanese, R. Manetsberger, W. A. Herrmann, R. Schmid), DE 4.435.171 (1994).
[52] Cf. contributions in the special issue of *J. Mol. Catal.* **1997**, *116*.
[53] B. Cornils, W. A. Herrmann, *Applied Homogeneous Catalysis with Organometallic Complexes*, VCH, Weinheim, **1996**, Vol. 2, p. 575.
[54] Hoechst AG (W. A. Herrmann, J. Kulpe, J. Kellner, H. Riepl), EP 0.372.313 (1989).
[55] Hoechst AG (H. Bahrmann, W. Greb, P. Heymanns, P. Lappe, T. Müller, J. Szameitat, E. Wiebus), EP 0.576.905 (1993).
[56] (a) E. Wiebus, B. Cornils, *Chem.Ing.Tech.* **1994**, *66*, 916; (b) B. Cornils, E. Wiebus, *CHEMTECH* **1995**, *25*, 33; (c) E. Wiebus, B. Cornils, *Hydrocarb. Proc.* **1996**, *March*, 63; (d) B. Cornils, W. A. Herrmann, R. W. Eckl, *J. Mol. Catal.* **1997**, *116*, 27.
[57] B. Cornils, E. Wiebus, *Recl. Trav. Chim. Pays-Bas* **1996**, *115*, 211.
[58] NATO Advanced Research Workshop: *Aqueous Organometallic Chemistry and Catalysis*, Aug. 29–Sept. 1, 1994, Debrecen, Hungary; I. T. Horváth, F. Joó, *Aqueous Organometallic Chemistry and Catalysis*, Kluwer, Dordrecht, **1995**.
[59] Ruhrchemie AG (R. Gärtner, B. Cornils, H. Springer, P. Lappe), DE 3.235.030 (1982).
[60] B. Cornils, in *Topics in Current Chemistry* (Ed. P. Knochel), Springer, Heidelberg, 1998, in press.

6.1.2 Kinetics

Raghunath V. Chaudhari, Bhachandra M. Bhanage

6.1.2.1 Introduction

Biphasic catalysis using water-soluble metal complexes has been the most significant development in recent years to facilitate commercially viable separation of homogeneous catalysts from the reaction products. It is well known that several attempts to heterogenize the homogeneous catalysts, which include polymer anchoring, supported liquid-phase catalysis [1], and use of organometallic catalysts on mineral supports [2, 3], have not led to industrially viable alternatives. However, biphasic catalysis, in which water-soluble organometallic catalysts are contacted with the immiscible organic phase containing reactants and products with or without gaseous reactants in a multiphase system (gas–liquid–liquid) has proved to be the most attractive alternative to the conventional homogeneous catalysts. If was after the work of Kuntz [4] on the synthesis of triphenylphosphine trisulfonate (TPPTS) ligand and its application in the hydroformylation of olefins that the research on water-soluble catalysis gained momentum. This major breakthrough involves conducting a process such that the catalyst remains in the aqueous phase, while the reaction products are located in the organic phase, thus allowing a simple phase separation of the catalyst (from products) for recycle. The concept has been proven on a commercial scale in the Ruhrchemie/Rhône-Poulenc process for the hydroformylation of propene to butyraldehyd [5]. The role of different water-soluble ligands, their synthesis and stability as well as other means of intensifying these gas–liquid–liquid catalytic reactions have been extensively studied and the subject has been reviewed by Kalck and Monteil [6], Herrmann and Kohlpaintner [7], Cornils [8], and Beller et al. [9] among others. Herrmann and Kohlpaintner [7] have shown that Rh complexes with other water-soluble ligands, such as BISBIS (sulfonated (2,2′-bis(diphenylphosphinomethyl)-1,1′-biphenyl) and NORBOS, give exceptionally high activities and *n*/*i* ratios. A Rh catalyst prepared with a surface-active ligand such as a sulfobenzene derivative of tris(2-pyridyl) phosphine gives a 70 % yield of the aldehyde products in hydroformylation of tetradecene [10]. Although several other water-soluble ligands and catalysts have been studied, Rh–TPPTS is the most suitable and commercially proven catalyst system for biphasic hydroformylation. Several modifications of the water-soluble catalysts using co-solvents [11], supported aqueous-phase catalysis [12] and catalyst binding ligands (interfacial catalysis) [13] have been proposed to overcome the lower rates observed in biphasic catalysis due to poor solubilities of reactants in water. Most of the work done so far was focused on the design of new water-soluble catalysts to enhance the overall rate of reaction. However, limited information is available on the kinetics of biphasic hydroformylation.

Besides development of new catalysts and ligands, the understanding of the overall kinetics of biphasic catalytic reactions is an equally important aspect in the evolution of an economical process. In the case of hydroformylation of olefins using water-soluble catalysis, the rate of reaction will be governed by several factors, which include dissolution of CO, H_2 and olefins in organic and aqueous (catalyst) phases, the solubility of these components, their partition coefficients, and the intrinsic kinetics of the reaction occurring in the aqueous phase (cf. Section 4.1). The most important of these is the knowledge of kinetics, which is also essential to understanding of the reaction mechanism and elucidation of the rate-controlling step.

The aim of this contribution is to present a review of the current status of the kinetics of hydroformylation of olefins using water-soluble catalysis. Kinetic studies for various reaction systems and the role of ligands, pH, co-solvents, and surfactants are discussed.

6.1.2.2 Kinetics Using Water-soluble Catalysts

Hydroformylation of olefins using water-soluble catalysis is an example of a gas–liquid–liquid catalytic reaction in which reaction of two gaseous reactants (carbon monoxide and hydrogen) with gaseous or liquid-phase olefin occurs in the presence of a water-soluble catalyst in liquid–liquid dispersion. The reaction of dissolved gases and olefins occurs in the aqueous ("bulk") phase or organic–aqueous interface. The understanding of the overall rate of hydroformylation is important in this multiphase catalytic system, which depends on gas–liquid and liquid–liquid mass transfer, the solubility of gas-phase reactants in the organic and aqueous phases, the liquid–liquid equilibrium properties and intrinsic kinetics of the reaction in the aqueous phase. In addition the dispersion characteristics of the droplets, the droplet size and the bubble size can also influence the rate of reaction. Depending on the fractional hold-up of the aqueous phase, it will be either a continuous aqueous phase with dispersed organic droplets or a dispersed aqueous phase in a continuous organic medium. The coupled influence of mass transfer with chemical reaction is expected to be quite different in these two situations. Below, the current status of development of the kinetics of hydroformylation in two-phase systems is reviewed.

6.1.2.2.1 Kinetic Studies Without Any Additives

The hydroformylation of 1-octene using a water-soluble Rh–TPPTS catalyst in a biphasic medium was studied in the absence of any additive or co-solvent by Bhanage [14]. His experiments were carried out under conditions such that the aqueous phase containing Rh–TPPTS catalyst was dispersed in the continuous

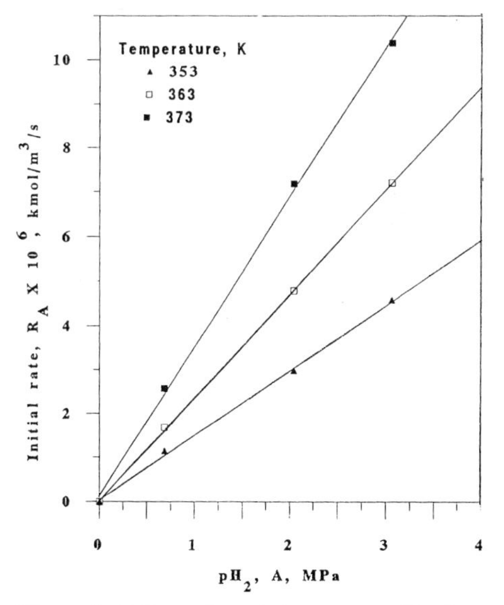

Figure 1. Effect of partial pressure of hydrogen on the rate of hydroformylation of 1-octene [14].

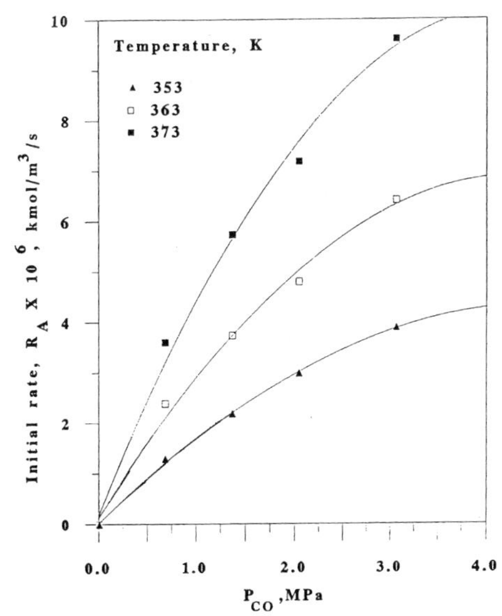

Figure 2. Effect of partial pressure of carbon monoxide on the rate of hydroformylation of 1-octene [14].

organic phase consisting of 1-octene and toluene. The results indicated absence of hydrogenation products and the selectivity to hydroformylation was greater than 98%, with the *n*/*i* ratio in the range 1–2:1. The aqueous Rh complex catalyst was found to retain its activity even after 10 recycles, indicating negligible deactivation. The effect of the concentration of catalyst precursor, TPPTS, and 1-octene, and of the partial pressure of CO and hydrogen, was studied in a temperature range of 353–373 K. The rate of reaction was found to be first order with respect to catalyst concentration. The effect of partial pressure of hydrogen on initial rate of hydroformylation is presented in Figure 1, which clearly indicates a first-order dependence. This was explained as a consequence of oxidative addition of hydrogen to acyl carbonyl rhodium species as the rate-determining step [15]. The reaction rate was found to vary with the 0.7th order with CO (Figure 2), in contrast to the CO inhibition observed for homogeneously catalyzed hydroformylation [16].

For a water-soluble catalyst, the concentration of dissolved carbon monoxide in the aqueous phase is very low compared with that in the organic phase; hence, formation of a dicarbonyl Rh species $[(RCO)Rh(CO)_2(TPPTS)_2]$, which is believed to be responsible for a negative-order dependence [15], is not very likely. Therefore, this difference in trends is not truly due to any change in the reaction mechanism. In the homogeneous catalytic reaction, the rate varies linearly with carbon monoxide pressure in the lower region and only beyond a certain pressure of CO is the inhibition in the rate observed. The 1-octene concentration dependence of the rate of hydroformylation showed an apparent

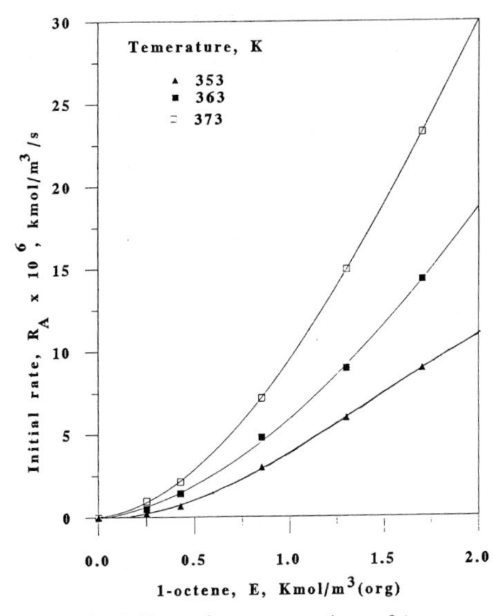

Figure 3. Effect of concentration of 1-octene on the rate of hydroformylation of 1-octene [14].

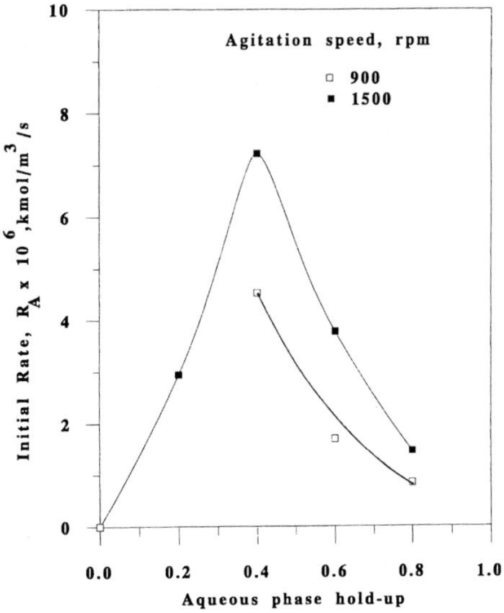

Figure 4. Effect of aqueous-phase hold-up on the rate of hydroformylation of 1-octene [14].

order of 1.7 (Figure 3), but this was due to an inappropriate account of the solubility variation with changes in 1-octene concentration. Bhanage [14] has shown that if the variation in solubility is accounted for, the rate of reaction shows first-order dependence on 1-octene concentration.

The form of rate model described by Eq. (1) was found to represent the data satisfactorily. This model is derived from the well-known mechanism of hydroformylation [15] assuming addition of hydrogen to acyl rhodium species as a rate-determining step:

$$R = \frac{k K_1 K_2 K_3 A B C D}{1 + \alpha B} \tag{1}$$

where A = partial pressure of hydrogen [Mpa]
B = partial pressure of carbon monoxide [Mpa]
C = concentration of catalyst [$\mathrm{kmol\,m^{-3}}$]
D = concentration of olefin [$\mathrm{kmol\,m^{-3}}$]
k = reaction rate constant
$K_1, K_2, K_3, K_4, \alpha$ = constants

The effect of aqueous phase hold-up on the rate of reaction for 900 and 1500 rpm is shown in Figure 4.

At 1500 rpm, the rate vs. ε_{1a} (aqueous phase hold-up) shows a maximum. For kinetic control, the rate is expected to vary linearly with catalyst loading. However, in a case where the reaction occurs essentially at the liquid–liquid interphase, it would depend on the liquid–liquid interfacial area even though liquid–liquid mass transfer is not rate-limiting. For $\varepsilon_{1a} > 0.4$, phase inversion occurs and the interfacial area would be determined by the dispersed phase, which would be the organic phase. Since, for $\varepsilon_{1a} > 0.5$, ε_1 will decrease with an increase in ε_{1a}, a reduction in liquid–liquid interfacial area is expected. Hence, the observed results of a decrease in the rates with an increase in ε_{1a} indicate a possibility of interfacial reaction rather than a bulk aqeuous-phase reaction [5]. For $d_L < 0.3$ mm, a very large interfacial area ($a_1 = 6\varepsilon_1 d/d_1$) in the range $(4-5) \times 10^4$, 1/m is likely to exist compared with the volume of the aqueous phase.

Hydroformylation of ethylene using a water-soluble Rh–TPPTS catalyst system has been investigated [17] using a toluene–water solvent system at 353 K. The effect of TPPTS concentration on rate passes through a maximum of at a P/Rh ratio of 8:1. The effect of the catalyst precursor concentration on the rate of reaction first increases and above a certain concentration it remains constant. The effect of aqueous-phase hold-up shows a maximum in the rate ($\varepsilon_{1a} = 0.4$). The apparent reaction orders for the partial pressure of hydrogen and ethylene were found to be 1 and zero respectively. A strong inhibition in the rate with an increase in P_{CO} was observed.

Herrmann et al. [18] have reported the hydroformylation of propene using a Rh–BISBIS catalyst system in a continuous reactor. The activity of this catalyst was 45.5 mol g^{-1} min^{-1}, three times higher than that of Rh–TPPTS catalyst (15 mol g^{-1} min^{-1}). The *n*/*i* ratio also improved from 94:6 (TPPTS) to 97:3 (BISBIS). They have also studied the hydroformylation of 1-hexene at 5 MPa CO/H_2 pressure and in the temperature range 395–428 K, and observed that the activity increased from 0.73 to 10.73 when the temperature was raised from 396 to 428 K whereas the *n*/*i* ratio decreased from 97:3 to 94:6. They have tested the catalyst stability after 16 h of continuous hydroformylation.

The development of supported aqueous-phase catalysis (SAPC) opened the way to hydroformylating hydrophobic alkenes such as oleyl alcohol, octene, etc. (cf. Section 4.7 [12]). SAPC involves dissolving an aqueous-phase $HRh(CO)(TPPTS)_3$ complex in a thin layer of water adhering to a silica surface. Such a catalyst shows a significantly high activity for hydroformylation. For classical liquid–liquid systems, the rate of hydroformylation decreases in the order 1-hexene > 1-octene > 1-decene; however, with SAP catalysts, these alkenes react at virtually the same rate and the solubility of the alkene in the aqueous phase is no longer the rate-determining factor [19].

6.1.2.2.2 Effect of Co-solvents on Hydroformylation of 1-Octene

It has been reported that use of a suitable co-solvent increases the concentration of the olefin in water (catalyst) while retaining the biphasic nature of the system. It has been shown that using co-solvents like ethanol, acetonitrile, methanol, ethylene glycol, and acetone, the rate can be enhanced by several times [20, 21]. However in some cases, a lower selectivity is obtained due to interaction of the co-solvent with products (e.g., formation of acetals by the reaction of ethanol and aldehyde). The hydroformylation of 1-octene with dinuclear $[Rh_2(\mu\text{-}SR)_2(CO)_2(TPPTS)_2]$ and $HRh(CO)(TPPTS)_3$ complex catalysts has been investigated by Monteil et al. [20]. The rate of reaction is controlled by the solubility of 1-octene in the aqeuous catalyst phase. To increase the reaction rate, various co-solvents were employed; ethanol was found to be the best. The selectivity to linear aldehyde was found to be proportional to the amount of ethanol in the aqeuous phase. Purwanto and Delmas [21] have reported the kinetics of hydroformylation of 1-octene using $[Rh(cod)Cl]_2$–TPPTS catalyst in the presence of ethanol as a co-solvent in the temperature range 333–353 K. First-order dependence was observed for the effect of the concentration of catalyst and of 1-octene. The effect of partial pressure of hydrogen indicates a fractional order (0.6–0.7) and substrate inhibition was observed with partial pressure of carbon monoxide. A rate equation was proposed (Eq. 2). In this case the aqueous phase was continuous and the organic phase was in the form of a dispersed phase.

$$R = \frac{kABCD}{(1+K_B B)^2(1+K_A A)} \tag{2}$$

K_A, K_B = constants

The kinetics of hydroformylation of 1-octene using $[Rh(cod)Cl]_2$ as a catalyst precursor with TPPTS as a water-soluble ligand and ethanol as a co-solvent was further studied by Deshpande et al. [11]. In this case the aqueous phase was continuous and the organic phase was in the form of dispersed droplets. The organic phase consisted of 1-octene in octane and the aqueous phase consisted of Rh/TPPTS along with the co-solvent ethanol. The effect on the initial rate of reaction of the concentration of catalyst and of 1-octene, and of the partial pressures of hydrogen and carbon monoxide, in a temperature range of 323–343 K and at a pH of 10 was studied. The rate of reaction was found to be first order with respect to catalyst concentration (Figure 5) and 1-octene concentrations (Figure 6).

The partial pressure of hydrogen dependence was shown to be 0.7th order (Figure 7). The rate vs. CO concentration passes through a maximum indicating a negative-order dependence at higher partial pressures of CO (Figure 8).

A rate equation was proposed (Eq. 3):

$$R = \frac{kABCD}{(1+k_a A)(1+k_b B)^3} \tag{3}$$

where k_a, k_b = constants

Since the catalytic cycle involves individual stoichiometric steps leading to formation of products and regeneration of catalyst, it is more meaningful to have a mechanistic rate model to predict the observed trends. In this respect a rate model (Eq. 4) derived according to the mechanism assuming addition to

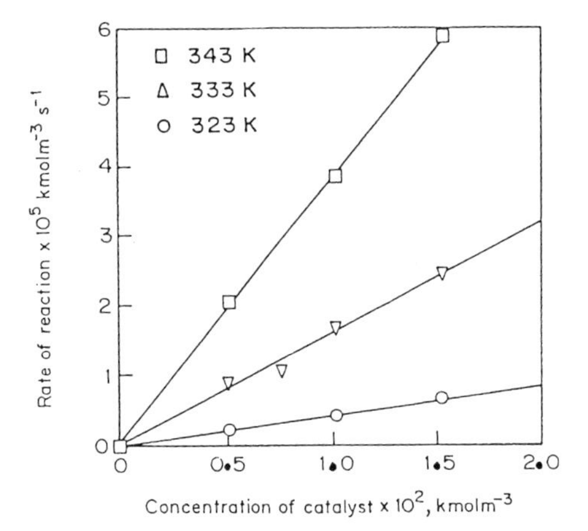

Figure 5. Effect of catalyst concentration on the rate of hydroformylation of 1-octene using ethanol as a co-solvent [11].

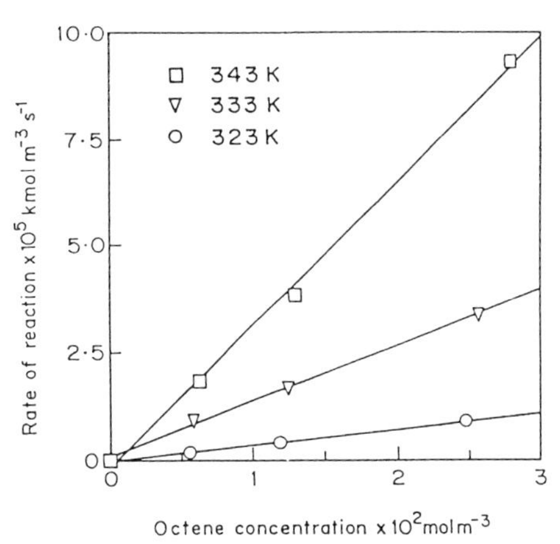

Figure 6. Effect of 1-octene concentration on the rate of hydroformylation of 1-octene using ethanol as a co-solvent [11].

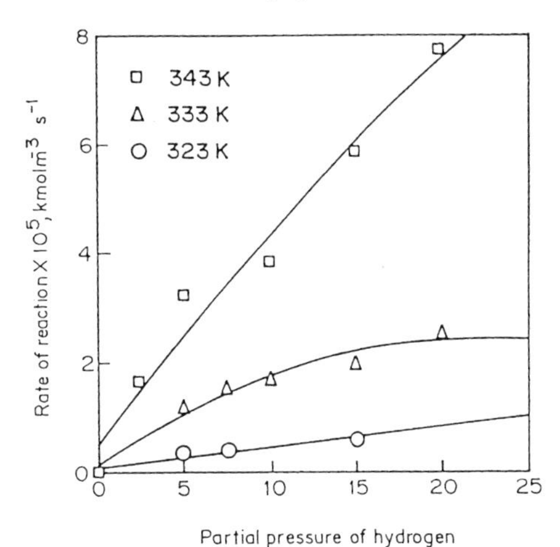

Figure 7. Effect of partial pressure of hydrogen on the rate of hydroformylation of 1-octene using ethanol as a co-solvent [11].

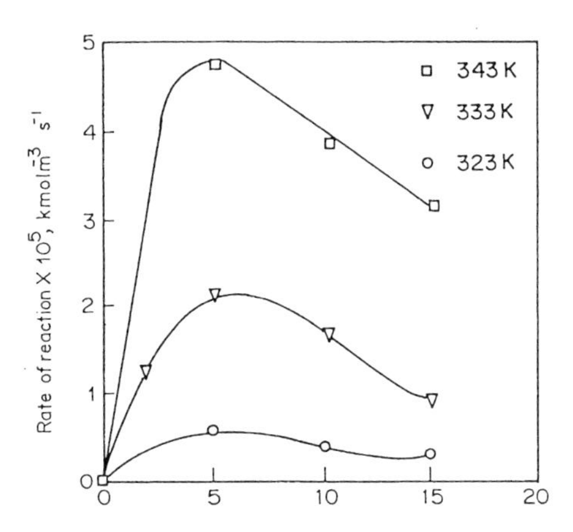

Figure 8. Effect of partial pressure of carbon monoxide on the rate of hydroformylation of 1-octene using ethanol as a co-solvent [11].

olefin to $HRh(CO)_2(TPPTS)_2$ to be the rate-controlling step was found to fit the data exceedingly well:

$$R = \frac{k_2 ABCD}{1 + k_a A + k_b AB + k_c B^2} \tag{4}$$

where k_a, k_b, k_c = constants.

The activation energy was found to be 15.7 kcal mol^{-1} using the empirical rate model (Eq. 3).

6.1.2.2.3 Effect of pH on the Rate of Biphasic Hydroformylation

Various other reaction parameters, such as pH and addition of salt, also play a significant role in the activity and selectivity performance of the catalyst. The pH of the aqueous catalyst phase shows a strong influence on the rate of reaction and the *n*/*i* ratio of the aldehyde products [25]. Smith et al. [22] have reported a drop in the activity when the pH of the reaction medium was reduced from 6.8 to 5 for an Rh/AMPHOS nitrate catalyst system (AMPHOS = 1-N,N,N trimethylamino-2-diphenyl phosphinoethane, iodide). Hydroformylation of 1-tetradecene with a water-soluble Rh–NABSDPP (NABSDPP: Na–butyl sulfonated diphenylphosphine) catalyst gave poor rates in acidic pH (2.5 to 6) medium. A seven- to eightfold increase in the rates was obtained when the pH was increased from 6 to 10 [23]. A detailed investigation into the kinetics of hydroformylation of 1-octene with an Rh/TPPTS catalyst system using ethanol as a co-solvent has been reported by Deshpande et al. [24] for various pH values. The rate increased by two- to fivefold when the pH increased from 7 to 10, while the dependence of the rate was found to be linear with olefin and hydrogen concentrations at both pH values. The rate of hydroformylation was found to be inhibited at higher catalyst concentrations at pH 7, in contrast to linear dependence at pH 10 (Figure 9). The effect of concentration of carbon monoxide was linear at pH 7, which is different from the usual observation of a negative-order dependence. At pH 10, substrate-inhibited kinetics was observed with respect to CO (Figure 10).

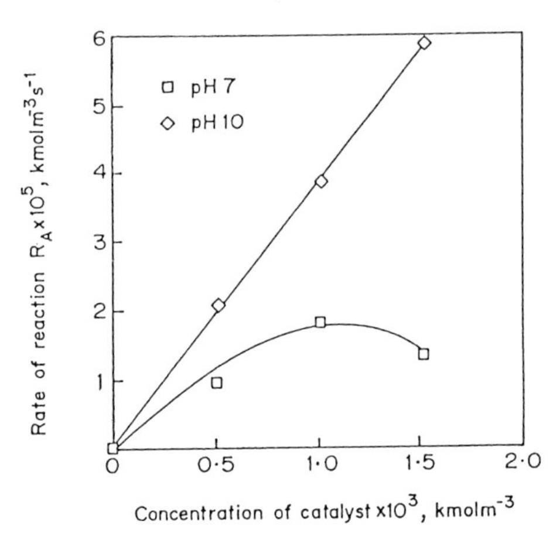

Figure 9. Effect of catalyst concentration on the rate of hydroformylation of 1-octene at different pH [24].

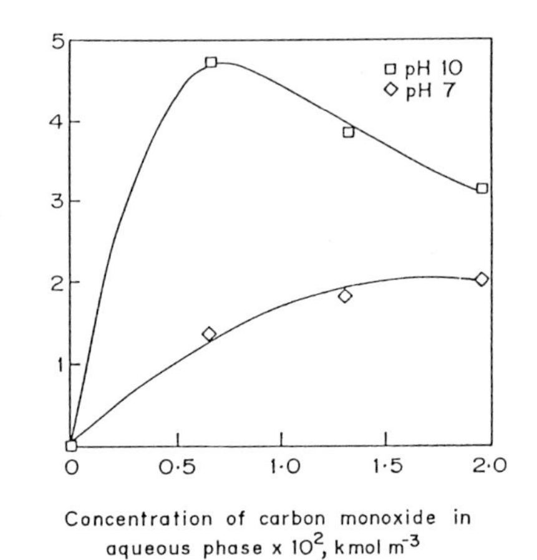

Figure 10. Effect of partial pressure of carbon monoxide on the rate of hydroformylation of 1-octene at different pH values [24].

6.1.2.2.4 Role of Catalyst-binding Ligands

In a recent report Chaudhari et al. [13] have shown that the rate of biphasic hydroformylation can be enhanced severalfold by using a catalyst binding ligand which facilitates interfacial catalytic reaction. This approach involves the use of a ligand that is insoluble in the aqueous, i.e., catalyst phase but has a strong affinity for the metal complex catalyst. The interaction of the ligand and the catalyst takes place essentially at the liquid–liquid interface (Figure 11); thus the concentration of the catalytic species will be enriched at the interface where it can access the reactants present in the organic phase, in significantly higher concentrations with respect to the aqueous phase [5]. This results in a dramatic increase in the rate of such a biphasic catalytic reaction, as indicated by experimental data on the hydroformylation of 1-octene using an Rh–TPPTS complex catalyst with triphenylphosphine as a catalyst binding ligand [13].

The rate of the reaction was enhanced by 10–50 times in the presence of catalyst binding ligands when compared with the biphasic hydroformylation reaction (Figure 12). This concept has also been demonstrated in reverse for the hydroformylation of water-soluble olefin (allyl alcohol) and organic phase con-

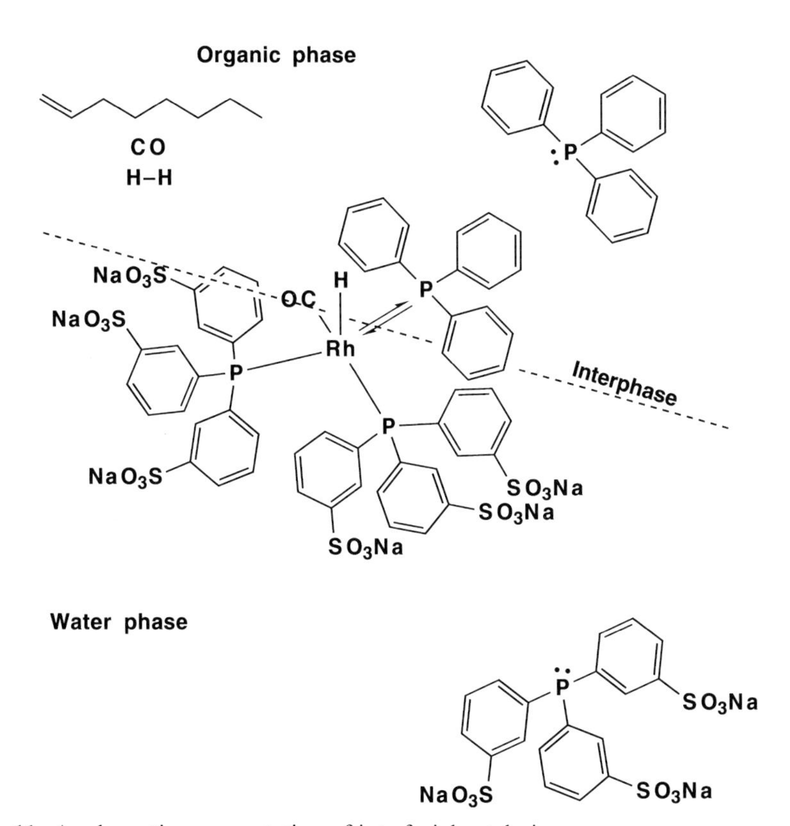

Figure 11. A schematic representation of interfacial catalysis.

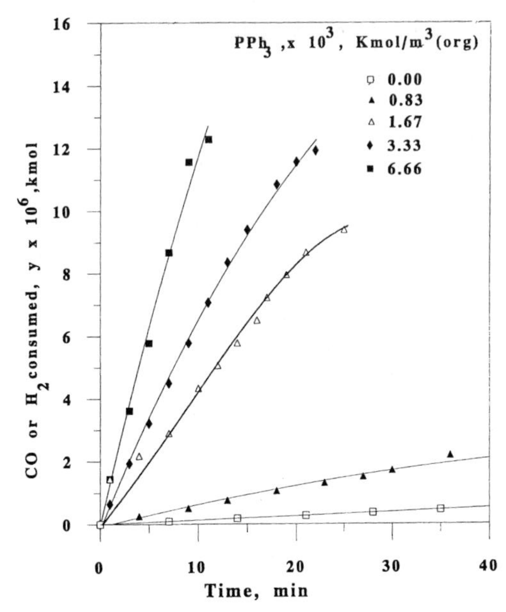

Figure 12. Effect of catalyst binding ligand on the rate of hydroformylation of 1-octene [13].

taining catalyst, $HRh(CO)(PPh_3)_3$. In this case the catalyst is present in the organic phase, whereas the catalyst binding ligand (TPPTS) is added to the aqueous phase.

The effect of reaction parameters, such as the concentrations of catalyst and olefin and the partial pressures of CO and hydrogen, on the rate of reaction has been studied at 373 K [14]. The rate varies linearly with catalyst concentration, olefin concentration, and partial pressure of hydrogen. Typical substrate-inhibited kinetics was observed with the partial pressure of carbon monoxide. Further, a rate equation to predict the obseved rate data has been proposed (Eq. 5).

$$R = \frac{kABE}{(1 + K_B B^2)(1 + K_E E)} \tag{5}$$

where E = concentration of olefin

K_B, K_E = constants

It is important to note that the kinetic trend were completely opposite for cases with and without catalyst binding ligand for carbon monoxide. Since, under conditions of interfacial catalysis, a higher CO concentration is accessible to the catalytic species, substrate inhibition is observed.

6.1.2.3 Concluding Remarks

A review of the kinetics of hydroformylation using water-soluble Rh complex catalysts demonstrates that the rate behavior varies significantly for biphasic catalytic reactions depending on the ligands, additives, and co-solvents. Particularly, the kinetics with respect to CO shows variation for different systems. A major limitation to the rate of biphasic hydroformylation is the solubility of the olefin in the aqueous catalyst phase. Using co-solvent, catalyst binding ligands and SAPC, the rates are enhanced significantly. Although sufficient information on the intrinsic kinetics is now available, further studies on understanding of the role of gas–liquid and liquid–liquid mass transfer, and the influence of dispersed-phase hold-up and drop size, and phase equilibrium properties, is necessary.

References

[1] P. R. Rony, J. F. Roth, US 3.855.307 (1974).
[2] M. E. Davis, C. Saldarria, J. A. Rossin, *J. Catal.* **1987**, *103*, 520.
[3] E. J. Rode, M. E. Davies, B. E. Hanson, *J. Catal.* **1987**, *96*, 574.
[4] (a) E. Kuntz, FR 2.314.910 (1975), US 4.248.802 (1981); (b) E. Kuntz, *CHEMTECH* **1970**, 570.
[5] (a) B. Cornils, J. Falbe, *Proc. 4th Int. Symp. on Homogeneous Catalysis*, Leningrad, Sept. **1984**, p. 487; (b) H. Bach, W. Gick, E. Wiebus, B. Cornils, Abstr. 1st IUPAC Symp. Org. Chemistry, Jerusalem **1986**, p. 295; (c) E. Wiebus, B. Cornils, *Chem.–Ing.–Tech.* **1994**, *66*, 916; (d) B. Cornils, E. Wiebus, *CHEMTECH* **1995**, *25*, 33; (e) K. Himmler, O. Wachsen et al., *J. Mol. Catal.* in press.
[6] P. Kalck, F. Monteil, *Adv. Organomet. Chem.* **1992**, *34*, 219.
[7] W. A. Herrmann, C. W. Kohlpaintner, *Angew. Chem., Int. Ed. Engl.* **1993**, *32*, 1524.
[8] B. Cornils, *Angew. Chem., Int. Ed. Engl.* **1995**, *34*, 1575.
[9] M. Beller, B. Cornils, C. Frohning, C. W. Kohlpaintner, *J. Mol. Catal.* **1995**, *104*, 17.
[10] B. Fell, G. Papadogianakis, *J. Mol. Catal.* **1991**, *66*, 143.
[11] R. M. Deshpande, Purwanto, H. Delmas, R. V. Chaudhari, *I&EC Res.* **1996**, *35*, 3927.
[12] J. P. Arhancet, M. E. Davies, J. S. Meroal, B. E. Hanson, *Nature (London)* **1988**, *339*, 454.
[13] R. V. Chaudhari, B. M. Bhanage, R. M. Deshpande, H. Delmas, *Nature (London)* **1995**, *373*, 501.
[14] B. M. Bhanage, Studies in hydroformylation of olefins using transition metal complex catalysts, Ph. D. Thesis, University of Pune, **1995**.
[15] D. Evans, J. A. Osborn, G. Wilkinson, *J. Chem. Soc. A* **1968**, 3133.
[16] R. M. Deshpande, R. V. Chaudhari, *Ind. Eng. Chem. Res.* **1988**, *27*, 1996.
[17] S. S. Divekar, Kinetic modeling of hydroformylation of olefins using homogeneous and biphasic catalysis, Ph.D. Thesis, University of Pune, **1995**.
[18] W. A. Herrmann, C. W. Kohlpaintner, H. Bahrmann, W. Konkol, *J. Mol. Catal.* **1992**, *73*, 191.

[19] I. T. Horváth, R. V. Kastrup, A. A. Oswald, E. J. Mozeleski, *Catal. Lett.* **1989**, *2*, 85.
[20] F. Monteil, R. Queau, P. Kalck, *J. Organomet. Chem.* **1994**, *480*, 177.
[21] Purwanto, H. Delmas, *Catal. Today* **1995**, *24*, 134.
[22] R. T. Smith, R. K. Ungar, L. J. Sanderson, M. C. Baird, *Organometallics* **1983**, *2*, 1138.
[23] S. Kanagasabapathy, Studies in oxidative carbonylation and hydroformylation reactions using transition metal catalysts, Ph. D. Thesis, University of Pune, **1996**.
[24] R. M. Deshpande, Purwanto, H. Delmas, R. V. Chåudhari, *J. Mol. Catal.* **1997**, (accepted).
[25] Ruhrchemie AG (B. Cornils, E. Wiebus et al.), EP 0.158.246 (1985).

6.1.3 Reaction of Olefins

6.1.3.1 Lower Olefins

Carl D. Frohning, Christian W. Kohlpaintner

6.1.3.1.1 Introduction

Although olefins with variable chain lengths have been successfully hydroformylated in aqueous two-phase reactions, a distinction between lower and higher olefins is reasonable. The solubilities of ethylene (C_2), propene (C_3) and C_4 olefins, herein referred to as lower olefins, in the aqueous catalyst phase is high enough to assure chemical reaction without phase transfer limitations. Olefins with chain lengths greater than C_4 show significantly lower solubilities, thus making special means necessary to overcome phase transfer limitations (see Section 6.1.3.2).

The following contribution will describe the basics of aqueous two-phase hydroformylation as they apply to C_3 and C_4 olefins. The focus will be on TPPTS (cf. Section 3.2.1) as a ligand and rhodium as the active metal center, e.g., the complex $HRh(CO)(TPPTS)_3$ [1]. Emphasis will be put on the commercial applications and the basic description of the processes.

6.1.3.1.2 Mechanism

The mechanism of the oxo reaction has been extensively studied in the past. A comparative study of both commercially applied oxo catalysts $HRh(CO)(TPP)_3$ (TPP = triphenylphosphine) and $HRh(CO)(TPPTS)_3$ was performed by Horváth [2]. The latter, water-soluble catalyst is considered to react according to the dissociative mechanism. However, remarkable differences exist in the catalytic activity and the selectivity of the organic- and water-soluble catalysts. The latter shows much lower specific activity but an increased selectivity to linear products in the hydroformylation of propene. From an Arrhenius plot it is concluded than the dissociation energy of TPPTS from $HRh(CO)(TPPTS)_3$ is about 30 ± 1 kcal/mol (1 kcal/mol = 4.18 kJ/mol). Compared with the dissociation energy of TPP from $HRh(CO)(TPP)_3$ (19 ± 1 kcal/mol [3]) the difference is greater than 10 kcal/mol, thus explaining the lower catalytic activity at comparable reaction conditions. Additionally it was shown that $HRh(CO)(TPPTS)_3$, in contrast to its organic-soluble derivative, does not form $HRh(CO)_2(P)_2$ (P = TPPTS; **1**) at syngas pressures up to 200 bar. By dissociation of either carbon monoxide or TPPTS the unsaturated species $HRh(CO)(TPPTS)_2$ (**2**) and $HRh(CO)_2(TPPTS)$ (**3**) are generated, which are responsible for the formation of linear or branched aldehydes (Scheme 1). As $HRh(CO)(TPPTS)_2$ is formed by dissociation of TPPTS from $HRh(CO)(TPPTS)_3$, and $HRh(CO)_2(TPPTS)$ is obtained through an equilibrium reaction from $HRh(CO)_2(TPPTS)_2$, the observed increased selectivity to linear products becomes explicable.

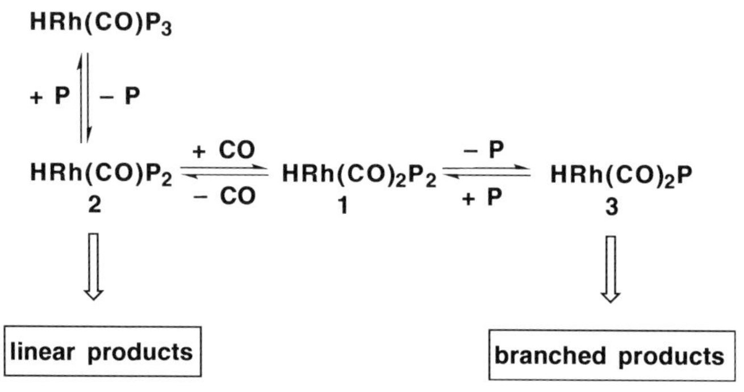

Scheme 1. Initial equilibria forming the active catalyst species; P = TPP or TPPTS.

6.1.3.1.3 Kinetics

Limited data are available for the kinetics of the oxo synthesis with $HRh(CO)(TPPTS)_3$. The hydroformylation of 1-octene was studied in a two-phase system in the presence of ethanol as a co-solvent to enhance the solubility of the olefin in the aqueous phase [4]. A rate expression was developed which was nearly identical to that of the homogeneous system, the exception being a slight correction for low hydrogen partial pressures (Eq. 1).

$$R_0 = k \frac{[\text{octene}]_0[\text{cat}][H_2][CO]}{(1 + K_{H_2}[H_2])(1 + K_{CO}[CO])^2} \qquad (1)$$

The lack of data is obvious and surprising at a time when the Ruhrchemie/Rhône-Poulenc process has been in operation for more than 13 years.

6.1.3.1.4 Recent Developments

In order to develop highly active and selective catalysts for propene hydroformylation, several ligands based on biphenyl or binaphthyl structures were synthesized and have been applied in the oxo synthesis [5]. The results are summarized in Figure 1 showing activity and selectivity of these ligands compared with TPPTS. A mixture of six-, seven- and eightfold sulfonated NAPHOS, called BINAS (**4**), together with rhodium is the most active water-soluble oxo catalyst known today. Even at very low phosphine to rhodium (P/Rh) ratios, *n*/*i* selectivities of 98:2 are achieved.

A challenge in synthesizing new water-soluble ligands is the direct functionalization of new or previously known organic phosphines. The plethora of func-

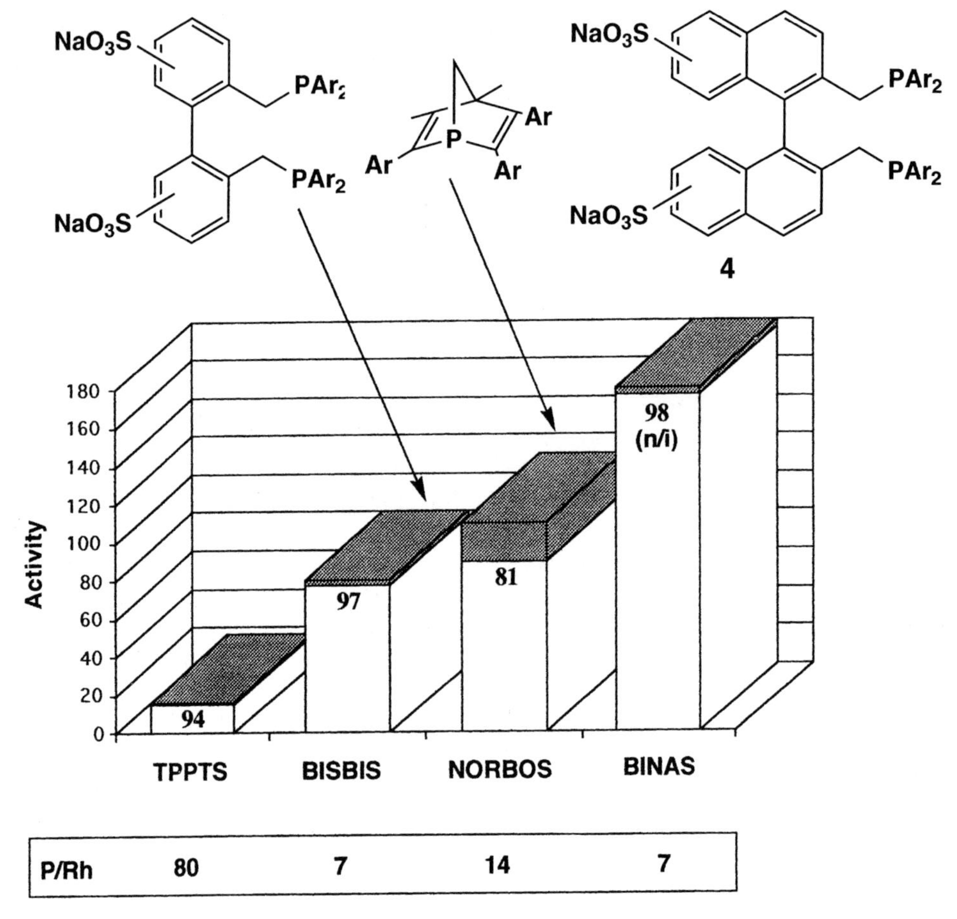

P/Rh	80	7	14	7

Figure 1. Comparison of water-soluble phosphines in the continuous hydroformylation of propene. Activity = [mol (*n* + *iso* aldehyde)]/[g atom (rhodium) × min]; Ar = *m*-$C_6H_4SO_3Na$; for NORBOS Ar = *p*-$C_6H_4SO_3Na$.

tionalized phosphines available today have been categorized and discussed in [6]. In the particular case of sulfonated phosphines the introduction of the sulfonato group is extremely difficult as the phosphines tend to oxidize during treatment with oleum (sulfur trioxide dissolved in concentrated sulfuric acid). Some information about special phosphines such as **5**–**7** and special manufacturing conditions is given in [7] (cf. Section 3.2.1).

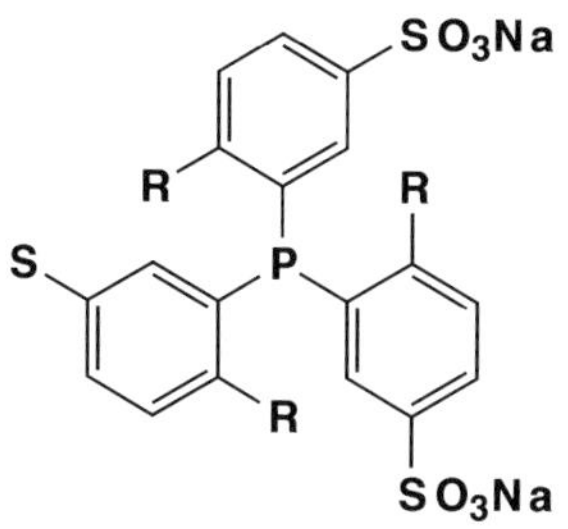

5 S = H, R = H
6 S = SO_3Na, R = CH_3
7 S = SO_3Na, R = OCH_3

6.1.3.1.5 Commercial Applications

The industrial hydroformylation of short-chained olefins such as propene and butenes is nowadays almost exclusively performed by so-called LPO (low-pressure oxo) processes, which are rhodium-based. In other words, the former high-pressure technology based on cobalt has been replaced by the low-pressure processes, which cover nearly 80% of total C_4 capacity due to their obvious advantages (cf. [8]). Nevertheless, some cobalt processes are still in operation for propene hydroformylation, for example as second stages in combination with a low-pressure process serving as the first stage [8, 9].

Two basic variants for LPO processes exist: the homogeneous processes, i.e., the catalyst and the substrate are present in the same liquid phase; and the two-phase process (Ruhrchemie/Rhône-Poulenc process, RCH/RP) using a water-soluble catalyst. The homogeneous processes dominate the field by far (Table 1), a consequence as much of their early development as of the licencing policy for the two-phase process. Both types of processes use rhodium as catalyst metal in combination with a suitable phosphine as ligand. More precisely, the ligand is triphenylphosphine (TPP) in both cases, applied as such in the homogeneous case and in its water-soluble (sulfonated) variant (TPPTS) in the RCH/RP process. No other phosphine ligands have gained commercial importance so far for LPO processes, and no metal other than rhodium has been successfully applied commercially in this technology.

Table 1. Capacities for C_4 products by various processes (excluding 2-EH).

Process	Catalyst	Capacity [1000 t]	Share [%]
UCC	Rh	2200	47
BASF	Rh	600	13
Leuna/Neftechim	Co	400	9
Mitsubishi Kasei	Rh	300	6
Ruhrchemie/RP	Rh	350	7
BASF	Co	200	4
Shell	Co	175	4
Celanese	Rh	160	3
Texas Eastman	Rh	130	3
Celanese	Co	110	2
Mitsubishi Kasei	Co	80	2
Subtotal		4705	100
Unknown		80	
Total		4785	

About 75% of the *n*-butyraldehyde generated is converted into 2-ethylhexanol, which is almost completely consumed as a phthalate ester, e.g., in plasticizers for PVC. The remainder of the butyraldehydes are either used as such for chemical synthesis, converted to acids or amines, or – more important today – hydrogenated to the butanols, which find widespread applications.

6.1.3.1.6 Economic and Ecological Aspects

The RCH/RP process has been in operation for more than 13 years now. The water-soluble catalyst $HRh(CO)(TPPTS)_3$ combines the advantages of a homogeneous catalyst (high activity, high selectivity) with those of a heterogeneous one (catalyst/product separation, cf. Section 1). These advantages, in addition to a highly efficient recovery of process heat, lead to a superior technology which also results in a cost advantage compared with the classical homogeneous processes. In Table 2 a comparison of the RCH/RP process with a process using

Table 2. Cost comparison of RCH/RP process vs. classical Rh process.

	RCH/RP	Rh-based
Raw materials	88.9	89.0
Energy costs	1.5	9.7
Credits (isobutyraldehyde, *n*-butanol)	−7.7	−11.4
Sum = material costs	82.7	87.3
Fixed costs + license fee	17.3	22.1
Manufacturing costs	100.0	109.4

a homogeneous rhodium catalyst is given [10, 11]. As can be seen, the RCH/RP process has its strengths in the efficiency of material usage (raw materials, energies, and by-product credits) along with smaller fixed costs due to the ease of operation. The overall cost advantage is estimated to be roughly 10% compared with the standard processes.

The ecological benefits of this modern process are clear and can be summarized as follows:

- usage of water as a nontoxic, nonflammable solvent;
- efficient usage of C_3 raw material (propene);
- high selectivity toward the desired products;
- energy consumption minimized, e.g., net steam exporter!
- efficient recovery of catalyst (loss factor = 1×10^{-9});
- ligand toxicity is not critical (LD_{50} (oral) > 5 g/kg);
- environmental emissions almost zero.

Along with other water-based reactions the RCH/RP process has been reviewed recently with respect to its environmental attractiveness, by Sheldon [12]. Overall the RCH/RP process is, besides a technical success, an outstanding example of the impact of modern technology on economy and ecological aspects at the same time.

6.1.3.1.7 C_3 Process Description

Ruhrchemie AG was the first to seize upon the idea of applying a water-soluble rhodium catalyst and thus commercialize a process which has been elaborated on a laboratory scale by Rhône-Poulenc earlier [13, 14]. It took only two years of intensive research to develop the technical concept and to erect the first plant, which went on-stream in 1984 [15]. By 1987 the second unit was already built and today the total capacity for *n*-butyraldehyde amounts to more than 350000 tons per year [16, 17]. An additional plant for the production of *n*-pentanal from *n*-butene was brought on-stream in 1995 (see Section 6.1.3.1.8).

Basically, the requirements for a process using an aqueous catalyst phase are the same as for the homogeneous processes. The reaction of propene, hydrogen, and carbon monoxide takes place in the aqueous catalyst solution or at the phase boundary. The second, organic, layer is formed by the reaction product, e.g., butyraldehydes. Intimate contact between the catalyst solution and the gaseous reactants has to be provided by intensive gas dispersion at the bottom of the reactor together with sufficient stirring. The two liquid phases form an intimate admixture (emulsion) upon stirring which occupies most of the reactor volume, leaving only a small headspace as an internal reservoir for the gaseous reactants. A heating/cooling device is necessary in order to enable start-up of the reactor and to control the exothermic hydroformylation reaction (about 28 kcal/mol or 118 kJ/mol). Finally, the mixture of liquid and gaseous products has to be withdrawn from the reactor, products and catalyst have to be separated, and the latter has to be recycled.

A simplified scheme of the RCH/RP unit is presented in Figure 2 [1, 10, 11]. The reactor (1) is essentially a continuous stirred tank reactor equipped with a gas inlet, a stirrer, a heat exchanger and a catalyst recycle line. Catalyst and reactants are introduced at the bottom of the reactor. Vent gas is taken from the head of the reactor and from the phase separator. Control of the liquid volume inside the reactor is simple: the liquid mixture composed of catalyst solution and aldehydes leaves via an overflow and is transferred to a phase separator (2), where it is partially degassed. The separation of the aqueous catalyst solution (density of the catalyst solution $\approx$1100 g/L) and the aldehydes occurs rapidly and completely, favored by the difference in densities (density of aldehyde layer $\approx$600 g/L due to dissolved gases). The catalyst solution passes a heat exchanger and produces process steam that is consumed in downstream operations. Some water is extracted from the catalyst solution by its physical solubility in the aldehydes (about 1.3% w/w) which may be replaced before the catalyst solution re-enters the reactor.

The most ingenious part of the RCH/RP process is the subsequent stripping column (4). From the raw organic phase coming from the phase separator and entering at the top of the stripping column, the dissolved reactants are removed by a fresh synthesis gas countercurrent stream. The pressure inside the stripping column is kept slightly higher than the pressure in the reactor, e.g., no additional mechanical compression or heating is necessary to recycle unconverted reactants. The resulting crude aldehydes are virtually free of propene as well as propane and contain only minimum amounts of dissolved synthesis gas. The head gas from the stripping column is fed back into the reactor.

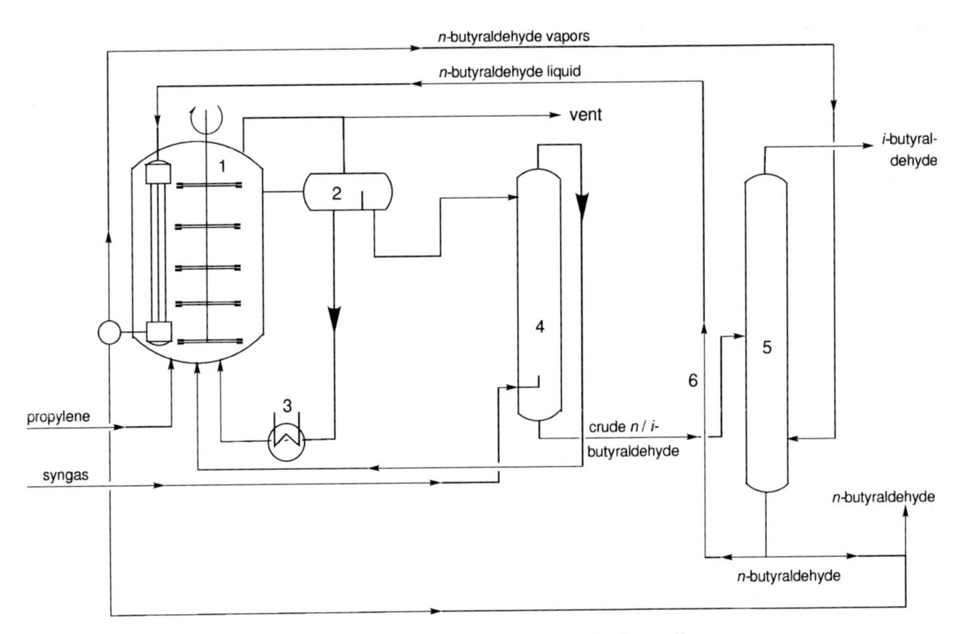

Figure 2. Ruhrchemie/Rhône-Poulenc process (RCH/RP): flow diagram.

The crude aldehydes are split into *n*- and isobutyraldehyde in the distillation column (5). The heat required is supplied by the hydroformylation itself: the re-boiler of the distillation is a falling film evaporator which is incorporated in the synthesis reactor using *n*-butyraldehyde as the heat carrier. This system is clearly advantageous over the classical hydroformylation processes, as the RCH/RP process not only efficiently uses the heat of reaction but is a net steam exporter. The favorable combination of stripping column, distillation, and heat recovery system is closely linked with the properties of the catalyst solution. The raw aldehydes are virtually free of ingredients of the catalyst solution, thus avoiding any of the well-known side reactions which take place in the presence of even traces of catalyst during thermal treatment. The fact that highly reactive *n*-butyraldehyde may be used as a heat-transfer medium underlines this statement well.

The active catalyst species $HRh(CO)(TPPTS)_3$ is generated during the start-up of the reactor. Rhodium is introduced in any suitable form, e.g., as acetate or as a salt of another organic acid. The resulting solution, with a rhodium concentration in the range 200–350 ppm, is brought to reaction temperature under synthesis gas pressure, leading to the formation of the active yellow complex. No induction period is observed when the reaction is started immediately by adding propene and synthesis gas after the reaction temperature of about 120 °C has been reached, making start-up and close-down operations extremely easy. Some typical data for the hydroformylation of propene are summarized in Table 3 [8].

Table 3. RCH/RP process: typical data.

	Range	Typical value
Reaction conditions		
Temperature [°C]	110–130	120
Pressure [bar]	40–60	50
CO/H_2 ratio	0.98–1.03:1	1.01
Propene conversion [%]	85–99	95
Propene purity [%]	85–99.9	95
Product composition [%]		
Isobutyraldehyde	4–8	4.5
n-Butyraldehyde	95–91	94.5
Isobutanol	<0.1	<0.1
n-Butanol	0.5	0.5
Butyl formates	Traces	Traces
Heavy ends	0.2–0.8	0.4
Selectivity to C_4 products [%]	>99	>99.5
Selectivity to C_4 aldehydes [%]	99	99
n/*i* ratio	93:7–97:3	95:5

6.1.3.1.8 C_4 Process Description

It is an intrinsic characteristic of the oxo reaction – not only in the presence of aqueous catalyst solutions – that the reaction rate in comparable conditions declines with increasing chain length of the olefin (see Section 6.1.3.2). This fact is attributed, *inter alia*, to the decreasing solubility of higher olefins in the aqueous catalyst solution, which correspondingly leads to low olefin concentrations and thus reduced reaction rates [18]. There are not too many options to overcome the problem: increase in temperature (with a negative impact on long-term ligand stability and *n*/*i* ratio), increase in rhodium concentration (cost factor), or addition of substances improving the solubility of the olefins (complicating the simple basic process). For the hydroformylation of *n*-butene a slight increase in rhodium concentration is sufficient to ensure appreciable space–time yields at industrially relevant conditions.

The cheapest source of *n*-1-butene is "raffinate II", a C_4 cut from which butadiene (by extraction) and isobutene (by conversion into methyl *t*-butyl ether) have been removed. The remaining mixture of C_4 hydrocarbons contains about 50–65% of *n*-1-butene, the remainder consisting of *cis*/*trans* *n*-2-butene and saturated butanes. A high concentration of *n*-1-butene in the raffinate is desirable for obvious reasons. On the other hand, the price for "raffinate II" is directly proportional to its content of *n*-1-butene. Therefore it is an unconditional requirement for the process to be compatible with different concentrations of *n*-1-butene in the feedstock.

The most valuable product of C_4 hydroformylation is *n*-pentanal, whereas the isomers 2-methylbutyraldehyde and 3-methylbutyraldehyde are less in demand and lower in value. A catalyst with high selectivity should not catalyze the hydroformylation of 2-butene and should convert 1-butene predominantly to *n*-pentanal. Both reqirements are fulfilled by the Rh/TPPTS system [19]. However, despite the high regioselectivity in hydroformylation, a side reaction occurs which diminishes the overall selectivity. Under reaction conditions, parallel to the hydroformylation reaction part of the *n*-1-butene is isomerized to *n*-2-butene, which is not hydroformylated in the presence of Rh/TPPTS under regular reaction conditions.

The choice of reaction temperature depends on several aspects. High temperatures favor the activity of the catalyst system and increase the partial pressure of *n*-1-butene (b.p. −6.1 °C) but have a negative impact on the long-term ligand stability. As a compromise a reaction temperature 5–10 °C higher than in the hydroformylation of propene is acceptable. Also, with respect to the partial pressure of *n*-1-butene the overall pressure is lower: about 40 bar has been proven suitable. The stripping column, as the central unit in the process, deserves special attention: in order to remove dissolved butenes and butane completely from the oxo crude a balance between temperature and pressure conditions has to be established. The process design corresponds to the RCH/RP process for the hydroformylation of propene, e.g., by slight adjustment of the conditions propene as well as *n*-1-butene may be processed in the same unit [19].

6.1.3.1.9 Deactivation Phenomena

All commercially applied rhodium/phosphine catalysts deactivate with time for different reasons and in several ways. Most obvious is a decline in activity, as it directly reduces the unit capacity and frequently also leads to increased materials consumption as unconverted olefins have to be vented. The reasons are found in the rhodium inventory: there are always some losses which decrease the rhodium concentration in the reaction medium, either by carry-over with the products during thermal (distillative) separation from the catalyst (homogeneous systems), or by being swept out with the products from the two-phase system. Although these losses normally are in the parts-per-billion range with respect to the rhodium concentration in the products, they may well accumulate to substantial losses if the lifetime of the catalyst charge exceeds years. In most kinetic expressions for the reaction rate the rhodium concentration is of or close to first order, thus directly influencing the reaction rate. On the other hand, the formation of inactive rhodium species may leave the rhodium inventory virtually unchanged, although the catalyst loses activity. Finally, the formation of modified phosphines has been proven to occur under reaction conditions, e.g., TPPTS can be converted to propyldi(sulfophenyl)phosphine (PDSPP), which acts as a stronger electron donor than TPPTS, thus occupying coordination sites on the rhodium [6].

The deactivation mechanism for TPPTS has been elucidated in some detail. The primary idea of *ortho* metallation of the phenyl ring has been abandoned in the meantime as it definitely plays no role. Instead the deactivation is initiated by the oxidative insertion of the rhodium metal into the P–C bond of the triphenylphosphine ligand. An analogous mechanism to that of the TPPTS degradation has been outlined for the homogeneous system with TPP as ligand [20].

In continuous operation consecutive phosphorus-containing products are formed which also influence the activity of the rhodium center and thus contribute to the catalyst deactivation. One of the main degradation products from TPPTS is the sodium salt of *m*-formylbenzenesulfonic acid, which indicates the insertion of the rhodium atom into the P–C bond. The arylrhodium species in which rhodium has replaced one phosphorus atom presumably exists as a phosphido-bridged dimer which is inactive. This compound may subsequently be converted to a series of consecutive products, e.g., alkyl-diarylphosphines, which act as catalyst poisons.

Besides the degradation reactions, all phosphines are oxidized by traces of oxygen which are always present in the olefins introduced. Synthesis gas, generated mainly by partial oxidation of hydrocarbons, may well contain small amounts of oxygen which are removed by special gas purification systems. Nevertheless, oxidation also plays a role in the losses of ligand in long-term operation. A certain concentration of "active" ligand, e.g., phosphorus in the oxidation state +III, is necessary to ensure stability of the rhodium and a sufficiently high n/i ratio; maintainance of this concentration, which is achieved

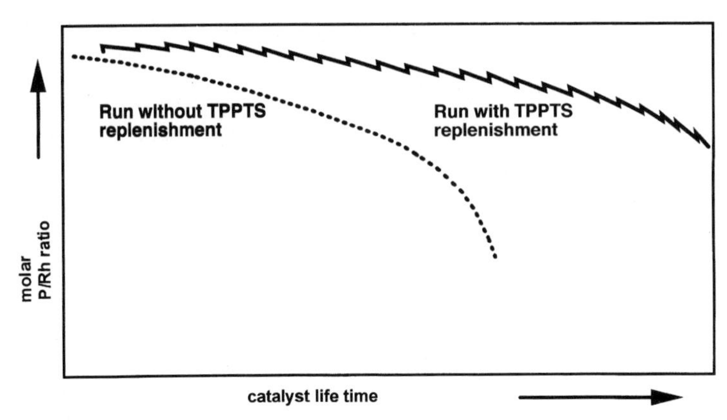

Figure 3. Characteristic course of ligand concentration vs. operating time in the RCH/RP process.

best by adding small portions of ligand over the whole run. The ligand concentration vs. operating time relationship follows a characteristic course which is depicted in Figure 3.

The TPPTS catalyst system is not sensitive toward sulfur and most of the other common poisons for hydroformylation catalysts. One reason is the continuous withdrawal of organic and other by-products with the product phase and the vent stream from the decanter (see Figure 2), avoiding the accumulation of poisons in the catalyst solution.

6.1.3.1.10 Outlook and Future Developments

The hydroformylation of propene and butene in the presence of an aqueous catalyst phase has proven successful for 13 years now. The hydroformylation of propene, especially, has acquired merit from the very beginning in 1984 and has encountered virtually no problems, even in large-scale units. Comparison with other hydroformylation processes based on the conventional homogeneous principle has shown some distinct technical and economical advantages [1, 10, 11].

- Phase separation is an elegant and efficient method to recover catalyst and oxo crude.
- Space–time yields in biphasic and homogeneous processes are on a comparable level.
- The *n*/*i* ratio is very high (95:5) and can be shifted if desired.
- Losses in rhodium and TPPTS are negligible.
- The absence of thermal strain reduces the formation of high-boiling by-products.
- The technical equipment is extremely simple and reliable.

- A process-integrated energy usage system operates (net energy exporter).
- There is high overall selectivity with respect to propene input.
- There are no environmental emissions.

However, despite the cited advantages in the hydroformylation of propene and butene, it has to be admitted that the biphasic system nears its limits when olefins with increasing chain lengths are considered. Due to the decreasing solubility of the olefins in the aqueous catalyst phase, the reaction rate slows down (cf. Section 6.1.3.2).

Several proposals have been published to solve this problem, e.g., by using polyether-substituted triphenylphosphines [21–23]. This type of phosphines show inverse-temperature dependence of their solubility in water that enables them to act as thermoregulated phase transfer ligands [24]. So far little is known about their applicability in technical operation. On the other hand, the homogeneous systems are also facing problems with long-chain olefins, but at a different process stage: whereas the hydroformylation still proceeds acceptably, the recovery of the oxo crude by distillation from the catalyst residue generates by-products and destroys the catalyst or the ligand, respectively. These facts explain the remanence of "ancient" cobalt hydroformylation processes for the hydroformylation of C_6–C_{10} olefins until a convenient solution has been found for one or the other variant. It may be noted that Celanese operates a ligand-modified homogeneous rhodium catalyst for hydroformylation of C_6 and C_8 olefins very successfully at its Bay City plant.

The extension of the biphasic principle to higher olefins may be accomplished by changing the ligand [25] from TPPTS, e.g., to bisphosphines, some of which have already been proven to be valuable tools to increase the specific activity combined with high *n/i* selectivity [5]. Increasing the specific (better: intrinsic) activity of rhodium in the aqueous two-phase system may be coupled with understanding of the relevant mass transport phenomena. In this case the role of the phase boundaries as a potential barrier for the chemical reaction will have to be carefully analyzed.

References

[1] E. Wiebus, B. Cornils, *CHEMTECH* **1995**, *25* (1), 33.
[2] I. T. Horváth, R. V. Kastrup, A. A. Oswald, E. J. Mozeleski, *Catal. Lett.* **1989**, *2*, 85.
[3] R. V. Kastrup, J. S. Merola, E. J. Mozeleski, R. J. Kastrup, R. V. Reisch, in *ACS Symp. Ser.* (Eds.: E. C. Alyea, D. W. Meek), **1982**, *196*, 34.
[4] P. Purwanto, H. Delmas, *Catal. Today* **1995**, *24*, 135.
[5] (a) W. A. Herrmann, C. W. Kohlpaintner, H. Bahrmann, W. Konkol, *J. Mol. Catal.* **1992**, *73*, 191; (b) W. A. Herrmann, C. W. Kohlpaintner, R. B. Manetsberger, H. Bahrmann, H. Kottmann, *J. Mol. Catal. A: Chem.* **1995**, *97*, 65; (c) H. Bahrmann, H. Bach, C. D. Frohning, H. J. Kleiner, P. Lappe, D. Peters, D. Regnat, W. A. Herrmann, *J. Mol. Catal. A: Chem.* **1997**, *116*, 49.
[6] W. A. Herrmann, C. W. Kohlpaintner, *Angew. Chem., Int. Ed. Engl.* **1993**, *32*, 1524.

[7] W. A. Herrmann, G. P. Albanese, R. B. Manetsberger, P. Lappe, H. Bahrmann, *Angew. Chem., Int. Ed. Engl.* **1995**, *34*, 811.
[8] C. D. Frohning, C. W. Kohlpaintner, in *Applied Homogeneous Catalysis with Organometallic Compounds* (Eds.: B. Cornils, W. A. Herrmann), VCH, Weinheim, **1996**, Vol. 1, p. 29.
[9] D. Kirchhof, Hoechst AG, private communication.
[10] B. Cornils, E. Wiebus, *Chem.–Int.–Tech.* **1994**, 916.
[11] B. Cornils, E. Wiebus, *Recl. Trav. Chim. Pays-Bas* **1996**, *115*, 211 and *Hydrocarb. Process.* **1996** (3), 63.
[12] G. Papadogianakis, R. A. Sheldon, *New J. Chem.* **1996**, *20*, 175.
[13] Rhône-Poulenc Ind. (E. G. Kuntz et al.), FR 2.230.654 (1983), FR 2.314.910 (1975), FR 2.338.253 (1976), FR 2.349.562 (1976), FR 2.366.237 (1976), FR 2.473.504 (1979), FR 2.478.078 (1980), FR 2.550.202 (1983), FR 2.561.650 (1984).
[14] E. G. Kuntz, *CHEMTECH* **1987**, *17*, 570.
[15] B. Cornils, E. G. Kuntz, *J. Organomet. Chem.* **1995**, *502*, 177.
[16] *Europ. Chem. News* **1995**, Jan. 15, 29.
[17] *Europa Chemie* **1995** (1), 10.
[18] O. Wachsen, K. Himmler, B. Cornils, *Catal. Today*, in press.
[19] H. Bahrmann, C. D. Frohning, P. Heymanns, H. Kalbfell, P. Lappe, D. Peters, E. Wiebus, *J. Mol. Catal. A: Chem.* **1997**, *116*, 35.
[20] (a) J. A. Kulpe, *Dissertation*, **1989**, Technische Universität München; (b) C. W. Kohlpaintner, Dissertation, **1992**, Technische Universität München; (c) R. A. Dubois, P. E. Garrou, K. Lavin, H. R. Allock, *Organometallics* **1984**, *3*, 649; (d) R. M. Deshpande, S. S. Divekar, R. V. Gholap, R. V. Chaudhavi, *J. Mol. Catal.* **1991**, *67*, 333.
[21] Z. Yin, Y. Yan, H. Zuo, B. Fell, *J. Prakt. Chem.* **1996**, *338*, 124.
[22] B. Fell, D. Leckel, C. Schobben, *Fat Sci. Technol.* **1995**, *97*, 219.
[23] P. Wentworth, A. M. Vandersteen, K. D. Janda, *J. Chem. Soc., Chem. Commun.* **1997**, 759.
[24] E. Monflier, G. Fremy, Y. Castanet, A. Mortreux, *Angew. Chem.* **1995**, *107*, 2450.
[25] Most recent reviews: (a) F. Jóo, À. Kathó, *J. Mol. Catal. A: Chem.* **1997**, *116*, 3; (b) B. Cornils, W. A. Herrmann, R. W. Eckl, *J. Mol. Catal. A: Chem.* **1997**, *116*, 27.

6.1.3.2 Higher Olefins

Helmut Bahrmann, Sandra Bogdanovic

6.1.3.2.1 Introduction

The Ruhrchemie/Rhône-Poulenc [1] process for the hydroformylation of short-chain olefins such as propene and butene (cf. Section 6.1.3.1) combines a facile catalyst recycling with high selectivity and sufficiently high conversion rates to provide a commercially viable large-scale manufacturing process for butyraldehyde [2] and valeraldehyde [3]. Higher olefins ($>C_8$) are not suited for the RCH/RP process as run in Oberhausen.

The hydroformylation of higher olefins with five or more carbon atoms accounts for about 25% of the world-wide capacity of oxo products. Commercial hydroformylation processes based on phosphine-modified rhodium catalysts, such as the low-pressue oxo (LPO) process of the Union Carbide Corporation, are only known for olefins with a maximum carbon atom number of 8 [4]. Processes with higher olefins encounter problems due to the thermal sensitivity of the respective aldehydes [5], which cannot be isolated from the homogeneous reaction mixture by distillation. Cobalt-based processes [6] yield primarily less thermally sensitive alcohols. Therefore about 90% of higher oxo products are manufactured by processes using unmodified or trialkylphosphine-modified cobalt catalysts. For this reason a biphasic process for the hydroformylation of higher olefins with more selective rhodium catalysts, which allows for a catalyst recycling by phase separation, would be highly desirable. So far, however, no catalytic system has been found which yields sufficiently high conversion rates and yet brings about a complete catalyst separation by decantation.

There have been many approaches to overcome the problem of low space–time yields in biphasic reaction systems with rhodium and other metals, such as the Ruhrchemie/Rhône-Poulenc process. Concerning two-phase hydroformylation of higher olefins in an aqueous–organic reaction system, the different approaches can be categorized as follows:

1. The use of water-soluble ligands with amphiphilic properties which will either improve the solubility of the higher olefins via formation of micelles or increase the reaction rate by preferential concentration of the catalyst complex close to the interface of the aqueous and the organic phase [7–10] (see Section 3.2).
2. Modification of the Ruhrchemie/Rhône-Poulenc system with co-solvents such as polar, alcoholic solvents or by use of detergent cations or modified cyclodextrins [11] to enhance the mutual solubility or mobility of the components across the phase boundary.
3. Increasing the surface area of the phase boundary by mechanical methods, such as ultrasound or cavitron.
4. The principle of thermoregulated phase-transfer catalysis (TRPTC), originally developed by Bergbreiter et al. [12], which has been applied to two-phase hydroformylation by Fell, Jin and co-workers [13], which is based on a temperature-controlled switch of the catalyst system from the aqueous phase to the organic phase (see Section 4.6.3).
5. Immobilization of unmodified rhodium catalyst (i.e., without ligands) in the aqueous phase by using polymeric or oligomeric water-soluble supports [14].
6. Supported aqueous-phase catalysis (SAPC) (see Section 4.7) [15].

Non-aqueous approaches toward two-phase hydroformylation have been demonstrated by Hórvath et al. [16] with the use of a fluorous biphasic system containing a rhodium catalyst bearing partially fluorinated "ponytail" ligands, and Olivier and Chauvin [17] with TPPMS and TPPTS dissolved in nonaqueous ionic liquids (see Sections 7.2 and 7.3).

In this section, we will report our own investigations in the two-phase hydroformylation of higher olefins with aqueous Rh–TPPTS catalyst systems. The overview on the present state of the art in two-phase hydroformylation will be confined to those investigations which are not covered by the respective original authors in this book.

6.1.3.2.2 Two-Phase Hydroformylation of Higher Olefins with Rhodium/Triphenylphosphine Trisulfonate (TPPTS) as Catalyst System

The Unmodified Ruhrchemie/Rhône-Poulenc Process

Whereas various studies have been published dealing with new water-soluble ligands and their effects on the hydroformylation of higher olefins, only little data are available in the academic literature on the Rh–TPPTS catalyst system, especially without any additives. This section will provide some information on the effect of various reaction parameters (pressure, P/Rh ratio, rhodium concentration, olefin chain length, etc.) in the two-phase hydroformylation of higher olefins with the aqueous catalyst system Rh–TPPTS. The olefin *n*-1-hexene was most thoroughly investigated. The results in the tables are avarage values from 3 to 30 single experiments (standard conditions; *T* 125 °C, *p* 25 bar, Ru concentration 300–400 ppm, P/Rh ratio 80–100:1, reaction time 3 h).

Preparation of the Basic Catalyst $HRhCO(TPPTS)_3$ [18]

There are several methods for the preparation of the active hydroformylation catalyst $HRhCO(TPPTS)_3$. In analogy to its homogeneous counterpart [19], it is possible to synthesize the catalyst directly from $RhCl_3$ and TPPTS under syngas pressure. However, purification of the resulting material has been found to be a problem since chloride ions bind strongly to the rhodium center and form less active mixed rhodium complexes. Thus, in general, water-soluble rhodium acetate was used as a starting material in the so-called "preforming" reaction. The relative consumption of P(III) and Rh (on a mol/mol basis) was investigated in a series of experiments on the preforming reaction using different amounts of rhodium and TPPTS. It was shown that despite the use of Schlenk technique the amount of P(III) consumed strongly depends on the absolute concentration of phosphorus and rhodium (see also [20]) and approaches a value of 4–7 mol P(III)/mol Rh at high catalyst concentrations.

The Relationship Between Solubility of Linear α-Olefins in Water and the Reaction Rate under Ruhrchemie/Rhône-Poulenc Conditions

If the conditions of the Ruhrchemie/Rhône-Poulenc process are applied to linear α-olefins (LAOs) with 5–12 carbon atoms, the space-time yield of the

Table 1. Variation of α-olefins in the hydroformylation of 1-hexene.

	Pressure [bar]	Yield [%]	*n*/iso Ratio	$Rh_{org. phase}$	Rate constant [min^{-1}]
1-Pentene	30	49	96:4	0.05	0.0038
1-Hexene	30	31	98:2	<0.05	0.0021
1-Octene	30	9	98:2	<0.05	0.00053
1-Decene	30	2	100:0	<0.03	0.00011
1-Dodecene	30	0	n.d.[a)]	<0.05	0
1-Pentene	50	75	96:4	<0.03	0.0077
1-Hexene	50	39	97:3	<0.03	0.0028
1-Octene	50	8	95:5	<0.03	0.00044
1-Decene	50	1	99:1	<0.03	7.83×10^{-5}
1-Dodecene	30	<1	n.d.	<0.05	$<5.58 \times 10^{-5}$
1-Pentene	80	73	94:6	0.05	0.0073
1-Hexene	80	36	96:4	<0.5	0.0025
1-Octene	80	8	95:5	<0.5	0.00046
1-Decene	80	3	94:6	<0.03	0.00015
1-Dodecene	80	<1	n.d.	<0.5	$<5.58 \times 10^{-5}$

[a)] n.d. (not determined).

hydroformylation reaction decreases with increasing chain length of the substrate. Table 1 summarizes the results of the batchwise hydroformylation of LAOs different at 30–80 bar syngas pressure.

The conversion vs. time diagram (Figure 1) illustrates the dependence of the reaction rate on the chain length of the olefin. The reaction proceeds according to first-order kinetics, i.e., the consumption rate of the substrate olefin is proportional to the concentration of the substrate.

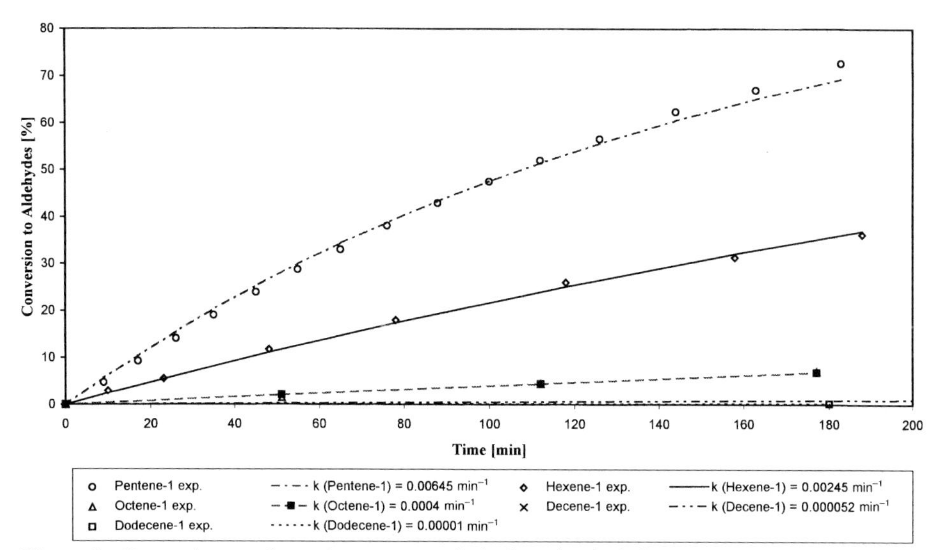

Figure 1. Dependence of reaction rate on chain length of olefins.

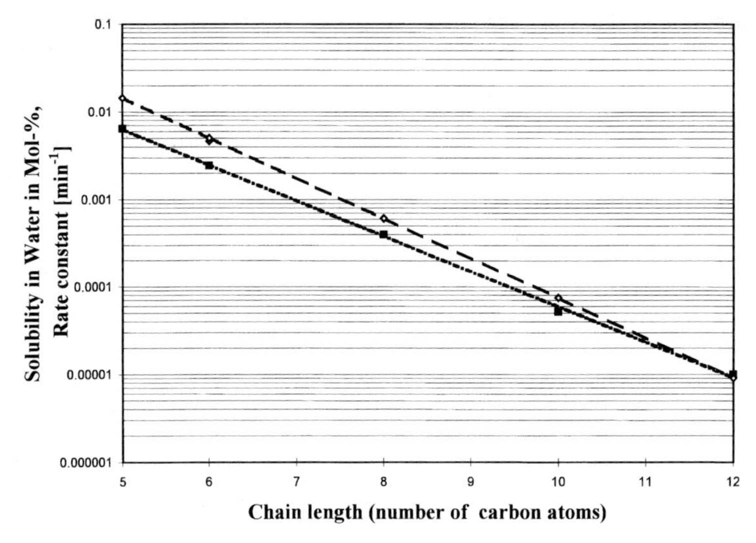

Figure 2. Solubility of olefins in water (■) and rate constants (◇) *vs*. chain length.

Figure 2 depicts the fist-order rate constants and the solubility of the respective terminal olefins in pure water at 125 °C according to a quantitative model of Brady et al. [21]. Obviously, there is a marked dependence of the reaction rate on the solubility of the olefins in hot water. It has to be kept in mind that the catalyst phase is a concentrated solution of Na–TPPTS in water whose solubility for olefins should be lower than that of pure water ("salt effects").

Obviously, the reason for the low space-time yields of the biphasic hydroformylation reaction is in some way related to the low solubility of the higher olefins in the catalyst phase. However, this finding does not necessarily imply that the catalytic reaction takes place in the bulk phase of the catalyst solution (concerning kinetics, see [22] and 6.1.2).

The Influence of the Reaction Pressure

Two-phase hydroformylation of olefins is an example of the development of a low-pressure oxo process operating at a syngas pressure of 20–50 bar. As can be seen in Table 2, the conversion rate seems to reach a maximum between 30 and 80 bar in the hydroformylation of *n*-1-hexene. By increasing the syngas pressure from 25 to 270 bar the *n*/iso-ratio decreases from 98:2 to 92:8. Whereas the increase of the conversion rate at medium pressure cannot yet be explained, the decrease of the *n*/iso ratio is probably due to the formation of a sterically less hindered rhodium dicarbonyl complex.

The Influence of the Phosphine/Rhodium Ratio

In the biphase hydroformylation of 1-hexene the conversion rate strongly depends on the phosphorus to rhodium ratio (P/Rh ratio). An increase of the P/Rh

Table 2. Variation of the pressure in the hydroformylation of 1-hexene.

	Pressure [bar]				
	25	30	50	80	270
Conversion [%]	18	31	34	35	20
n/iso ratio	98:2	98:2	97:3	96:4	92:8

Table 3. Variation of P/Rh ratio.

	P/Rh ratio				
	0	20:1	50:1	100:1	150:1
Conversion [%]	25	(27)[a]	31–35 (27)	18 (20)	9
n-iso ratio	38:62	(80:20)	97:3 (89:11)	98:2 (92:8)	98:2
Rh concn. [ppm]	100	(400)	400 (400)	400 (400)	400

[a] Values in brackets are true for 270 bar pressure.

ratio from 50:1 to 100:1 and to 150:1 results in a decrease of the conversion from 33 % to 18 % and to 9 % respectively (Table 3). This may be due to the shift of the **1**, **2** and **3** (Scheme 1). At higher phosphine concentrations the formation of the catalytically active species is suppressed. Furthermore the "salt effect" is much more pronounced in the aqueous catalyst solution than in homogeneous organic systems.

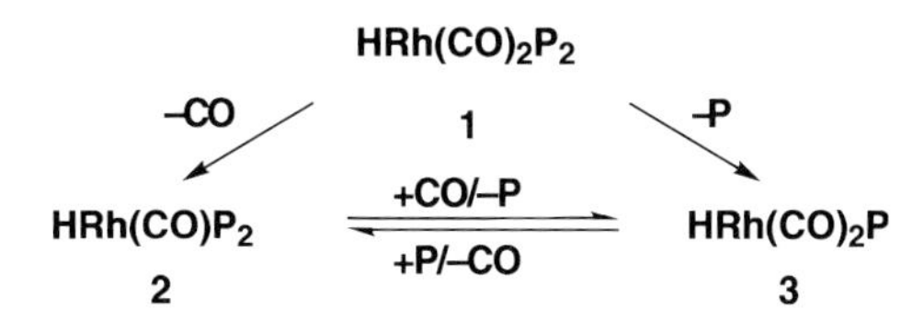

Scheme 1

Otherwise a high ligand concentration has a positive influence on the *n*/iso-ratio. This was rationalized by Hórvath et al. [23] by a shift of the equilibrium between complexes **2** and **3**.

The TPPTS ligand exhibits a higher dissociation energy compared with the TPP ligand due to hydrogen bond formation in aqueous solution. At a higher pressure level of 270 bar the influence of the P/Rh ratio on the selectivity toward linear aldehyde formation is much more pronounced because the CO as a ligand can compete better with the phosphine ligands. In the absence of phosphine ligands, the *n*/iso ratio is as low as 40:60.

Variation of the Rhodium Concentration

In the hydroformylation of *n*-1-hexene the rhodium concentration was varied at a low P/Rh ratio of 20:1 to 40:1. By increasing the rhodium concentration from 50 to 400 ppm, the conversion rate rises from 33 to 44% under standard conditions. This relatively minor effect must be due to the fact that a high rhodium concentration implies a high concentration of Na–TPPTS, which has a negative effect on the solubility of 1-hexene in the aqueous phase (salt effect). On the other hand, the salt effect shifts the equilibrium of the rhodium complexes toward the phosphine-rich complex **2**. Hence, the *n*/iso ratio is improved substantially.

Table 4. Variation of Rh concentration.

	Rh concentration [ppm]		
	50	100	400
Conversion [%]	33	36	44
n/iso ratio	86:14	90:10	96:4
P/Rh ratio	23	38	33

Variation of the Volume Ratio of Catalyst Phase to Olefin Phase

In the hydroformylation of *n*-1-hexene under standard reaction conditions the amount of catalyst solution was varied (Table 5). Clearly, by using large amounts of catalyst solution the conversion rate can be improved from 22% up to over 70%.

Table 5. Variation of the volume ratio of catalyst solution/olefin phase.

	TPPTS: *n*-1-hexene					
	1:5.1	1:3.4	1:1.7	1:0.6	1:0.15	1:0.09
Conversion [%]	4	7	10	22	66	71
n/iso ratio	97:3	97:3	98:2	98:2	98:2	98:2
Rh–TPPTS solution [mL]	130	174	261	391	478	478
n-1-Hexene [mL]	663	592	444	235	72	43
Activity [mol aldehyde/mol Rh per min]	1.1	1.2	0.9	1.0	0.74	0.50

Variation of Alkali Cations

In a series of experiments the counterion of TPPTS was varied. With K^+, Cs^+, Li^+, NH_4^+ no significant differences with respect to the catalyst performance were observed as compared with the sodium salt of TPPTS.

Comparison of TPPTS and TPPDS

By substituting TPPTS for the less water-soluble TPPDS (triphenylphosphine disulfonate, cf. Section 3.2.1) the conversion rate is increased significantly from 12% to 36%, i.e., by a factor of 3 at a low P/Rh-ratio. Whereas the leaching of rhodium into the organic phase remains unchanged, the leaching of TPPDS into the organic phase increases by a factor of 10. Therefore the use of a TPPDS-derived catalyst system cannot be considered a viable alternative for the two-phase hydroformylation of higher olefins. Higher leaching rates of both rhodium and TPPMS (triphenylphosphine monosulfonate) have been reported by Russel et al. [25].

pH Value

During the hydroformylation of higher olefins under the Ruhrchemie/Rhône-Poulenc conditions the pH value is controlled and adjusted between 5.5 and 6.2 [26]. In discontinuous operation it drops by almost one pH unit from approximately 6.5 at the beginning of the reaction to pH 5.5–6 at the end of the reaction. According to our investigations the pH shift is due to the formation of carbon dioxide, formed via the water-gas shift reaction (WGSR), which is also promoted by rhodium TPPTS complexes but to a lower extent [27].

Isomerization of Terminal Olefins to Internal Olefins

The Rh/TPPTS catalyst system is only applicable to the hydroformylation of terminal linear olefins. With branched or internal olefins as subtrates only very low conversion rates are achieved. Exceptions include strained cyclic olefins such as cyclopentene and norbornene, which are hydroformylated at moderate rates under Ruhrchemie/Rhône-Poulenc conditions.

Thus, the isomerization of linear α-olefins to internal olefins during the hydroformylation reaction is an unwanted side reaction since internal olefins accumulate in the reaction mixture. Table 6 summarizes the selectivity with respect to isomer formation in the two-phase hydroformylation of 1-pentene at different pressure levels. It is concluded that the isomerization can be suppressed most effectively by employing higher pressures.

Table 6. Isomerization in the hydroformylation of *n*-1-pentene.

	Pressure		
	30	50	80
Conversion [%]	48	75	73
Isomerization [%]	7.11	7.72	5.57
Selectivity toward isomerization [%]	14.8	10.3	7.7
n/iso ratio	98:2	96:4	95:5

Improvement of Mass Transfer by Mechanical Aids

Since the mass transfer seems to be the limiting factor in the two-phase hydroformylation of higher olefins under Ruhrchemie/Rhône-Poulenc conditions, an improvement of the mixing of the two phases can be a measure to improve the space–time yields. Two possibilities have been evaluated.

The Cavitron

The cavitron [28] is a dispersing unit or shock-wave reactor with a throughput of 8–30 $m^3 h^{-1}$ and an installed power of 25–60 kW. High shear forces are induced by an optimized rotor/stator system with passage gaps at the rotor and stator. The gaps are filled with material which is centrifugally accelerated by the rotor to the next row of gaps. Similarly to ultraound, alternating pressure fields are generated.

The cavitron device was used to study the conversion rate of *n*-1-hexene hydroformylation under biphasic conditions. Under standard reaction conditions, but at a lower pressure of 10 bar, a conversion of 17% (*n*/iso ratio 99:1) was reached in this unit compared with 12% in a regular stirred autoclave.

Ultrasound

Ultrasound denotes sound waves with a frequency of more than 20 kHz. In the field of two-phase catalysis ultrasound can be used to create stable emulsions and extremely large liquid–liquid phase boundaries. Special dipping oscillators provide a source of ultrasound in autoclaves. When the reaction mixture is exposed to ultrasonic vibration under stirring [29], the conversion rate of 1-hexene hydroformylation is increased by a factor of 2 under standard conditions.

Modification of the Rh/TPPTS System Using Additives

Alcohols

Table 7 shows that the reaction rate can be extremely enhanced by the addition of lower alcohols. The highest increase in rate was achieved with methanol. In contrast to literature data [30], the effect of ethanol is less pronounced and *n*-butanol shows no effect. If 20 wt.% of methanol are added to the olefin phase, a conversion of 90% is achieved under standard reaction conditions. The selectivity with respect to *n*-aldehyde formation drops from 98 to 92%. Formation of acetals as well as aldols can be kept below 1% if the pH is adjusted to 10.

Anionic Detergents

In order to increase the conversion rate, anionic detergents (2.5 wt.% relative to the catalyst solution) were added. A variety of anionic detergents, such as the

Table 7. Addition of alcohols.

		Ethanol	Methanol	Methanol
Amount of added alcohol [% rel. to olefin]	–	7	7	20
Conversion [%]	22	37	54	92
n/iso-ratio	98:2	97:3	95.5:4.5	92:8
Activity [mol ald./mol Rh per min]	1.0	2.9	4.6	8.3
Productivity [g ald./mL catalyst soln. per h]	0.030	0.069	0.108	0.180

sodium salts of fatty acids, alkylsulfonic acids, α-olefin sulfonates, fatty alcohol sulfates, fatty alcohol polyglycol ethers, alkylphenol ether sulfates, alkylphosphonic acids, and salicyclic acid, however, do not show any activating effects. Additionally, anionic detergents tend to produce foams and emulsions and in many chases phase separation is made impossible.

Counterion Modification [31]; TPPTS Modified with Cationic Detergents

By addition of 2.5% of a typical cationic detergent such as benzyltrimethylammonium (BTMA) to the catalyst phase under standard conditions, conversion could be increased by 45–60% [32]. Different anions such as benzenemonosulfonate, chloride, or sulfate do not have any effect (Table 8). A variation of the chain length of the alkyl group R of the detergent ammonium cation $[RN(CH_3)_3]^+$ shows the strongest effect in the typical detergent region with C_{14}. Similar results have been reported by Chen et al. [33]. At this conversion level, the *n*/iso ratio remains at 95:5. With even higher alkyl chains (C_{16}–C_{20}) the effect is less pronounced and phase separation becomes difficult due to foaming. Cations of the type $[R_2N(CH_3)_2]^+$ do not show any rate enhancement and exhibit strong foaming properties.

Table 8. Addition of cationic detergents.

Cationic detergent	Conversion [%]	*n*/iso ratio	Rate of increase[a)]	Remarks
$[BTMA][C_6H_5SO_3]$[b)]	31	98:2	1.55	
[BTMA][I]	32	98:2	1.60	
[BTMA][Cl]	30	98:2	1.50	
$[BTMA][SO_4]$	30	98:2	1.50	
[BTMA][F]	29	98:2	1.45	
$[C_{12}H_{25}N(CH_3)_3][CH_3OSO_3]$	29	95:5	1.45	
$[C_{14}H_{29}N(CH_3)_3][CH_3OSO_3]$	43	95:5	2.15	
$[C_{16}H_{33}N(CH_3)_3][CH_3OSO_3]$	32	95:5	1.60	Foaming
$[C_{20}H_{41}C_{22}H_{45}N(CH_3)_3][CH_3OSO_3]$	–	–	–	Strong foaming

Hydroformylation conditions: according to standard conditions. T1, added tenside: 2.5%, relative to catalyst solution.

a) Relative to 20% conversion.

b) BTMA, benzyltrimethylammonium.

Table 9. Use of detergent cations with TPPTS.

Detergent cation	Conversion [%]	*n*/iso ratio	Rate of increase[a)]	Remarks
$[(CH_3)_4N]$	19	98:2	–	
$[C_6H_5N(CH_3)_3]$	27	98:2	1.35	
$[HOCH_2\text{-}CH_2N(CH_3)_3]$	44	98:2	2.20	
[BTMA][b)]	82	94:6	4.10	
$[(C_4H_9)_4N]$	90	80:20	4.50	
$[(C_{12}H_{25})(C_2H_5)N(CH_3)_2]$	97	77:33	4.85	Difficult phase separation

a) Relative to a conversion of 20%.
b) BTMA, benzyltrimethylammonium.

If all sodium cations of Na–TPPTS are substituted by detergent cations [34], extremely high conversions of up to 100% can be achieved. In this case, the selectivity regarding *n*-aldehyde decreases to 77% (Table 9).

A further optimization could be achieved by only exchanging part of the sodium cations with detergent ammonium cations. An exchange of 7 to 30% of the sodium cations was found to be sufficient to reach good results in terms of conversion and selectivity [35]. Instead of ammonium cations, phosphonium cations can also be used successfully [36].

Poly(ethylene Glycol) and Poly(ethylene Glycol) Derivatives as Rate-Enhancing Additives

Poly(ethylene glycol) (PEG) of a medium molecular mass (400–1000) as an additive to the catalyst phase of the process greatly enhances the conversion rate under standard reaction conditions. Table 10 summarizes the results for the hydroformylation of the higher olefins with and without addition of the specified amount of PEG. In contrast to the addition of methanol or ethylene glycol, the leaching of PEG into the organic phase remains at a very low level (< 0.1 wt.%). Similarly there is virtually no increase in the leaching of rhodium or Na–TPPTS into the organic phase. The phase separation after the reaction is fast and as straightforward as in the classical Ruhrchemie/Rhône-Poulenc process. The modification of the catalyst phase with PEG, however, leads to a slight decrease of the *n*/iso ratio. The amount of PEG added to the catalyst phase has to be adapted to the olefinic substrate.

Interfacial Catalysis

The idea of promoter ligands which are exclusively dissolved in the organic phase of a biphasic reaction mixture was put forward by Chaudhari et al. [37]. The two-phase hydroformylation of 1-octene with $HRh(CO)((TPPTS)_3$ was

Table 10. Hydroformylation in the presence of PEG.

	Pressure [bar]	PEG-400 [wt.%]	Oxoproducts formed [%]	*n*/*iso* ratio	Leaching of rhodium [ppm]
1-Pentene	50	0	75	96:4	0.03
1-Pentene	50	12	85	95:5	0.08
1-Hexene	50	0	39	97:3	0.03
1-Hexene	50	16	71	95:5	0.05
1-Octene	50	0	8	95:5	0.05
1-Octene	50	35	63	83:17	0.11
1-Dodecene	80	0	>1	99:1	0.03
1-Dodecene	80	35	13	75:25	0.03

accelerated by a factor of 10–50 by addition of triphenylphosphine (TPP). At the phase boundary, complexes of the general formula $[HRh(CO)(TPP)_{3-x}(TPPTS)_x]$ are formed via ligand exchange reactions. According to Chaudhari these complexes preferentially concentrate close to the phase boundary of the aqueous/organic reaction mixture, giving rise to the term "interfacial catalysis".

However, it can also be argued that the TPP simply enhances the solubility in the organic phase of the mixed rhodium complexes that are formed, just as rhodium complexes with TPPDS or TPPMS instead of TPPTS do. In addition, it has to be considered, that the promoter ligand TPP will stay in the crude aldehyde mixture after phase separation and will have to be separated by a distillation step.

Modified Cyclodextrins as Phase-transfer Catalysts

A different approach toward the improvement of the mass transfer of higher olefins into the aqueous TPPTS phase has been suggested by Mortreux and co-workers [39]. By using chemically modified cyclodextrins, inverse micellar systems are obtained which show largely increased space–time yield in the two-phase hydroformylation of 1-decene. According to the authors, the effect is due to the solubility of the alkylated cyclodextrins in both the organic phase and the aqueous phase (cf. Sections 4.5 and 4.6.1).

6.1.3.2.3 Other Water-soluble Phosphines

BINAS and BISBIS

Special chelating ligands, such as BINAS (sulfonated 2,2′-bis(diphenylphosphinomethyl)1,1′-binaphthylene) [40, 41] and BISBIS (sulfonated 2,2′-bis(diphenylphosphinomethyl)-1,1′-biphenyl) [42], were found to be very useful ligands for the two-phase hydroformylation of higher olefins. Under standard

Table 11. Comparison of TPPTS and BINAS.

	No additive		$C_{14}Me_3N$ cation		
	BINAS[a]	TPPTS	BINAS[a]		TPPTS
	3 h	3 h	3 h	6 h	3 h
Additive [%]	–	–	0.24		0.86
Rate of conversion [%]	36	22	77	84	74
n/iso ratio	99:1	98:2	99:1	99:1	91:9

P/Rh ratio 15:1.
[a] Relative to the whole catalyst solution.

conditions with BINAS a relatively high conversion of 36% is achieved (Table 11). In the presence of only 0.24 wt.% of tetradecyltrimethylammonium BINAS in the catalyst phase, the conversion rate rises to 77% (3 h) or 84% (6 h). In addition it should be mentioned that no decrease of the excellent *n*/iso-ratio of 99:1 is observed with BINAS, as opposed to TPPTS.

A high selectivity toward linear aldehydes at a low P/Rh ratio could also be achieved with sulfonated fluorophosphine [Tris(*p*-fluorophenyl)phosphine] [43].

Dinuclear rhodium(I) complexes with TPPTS, containing thiolato bridging ligands in aqueous phase, were found to be transformed into the monomer $HRhCO(TPPTS)_3$ under reaction conditions [44].

6.1.3.2.4 The Combination of Triphenylphosphine Monosulfonate (TPPMS) and Polar Solvents

Several processes for the hydroformylation of higher olefins have been suggested on the basis of the water-soluble ligand TPPMS. In contrast to TPPTS, which is almost exclusively soluble in water, TPPMS can be used in both aqueous and polar organic media.

Recently, Abatjoglou et al. from the Union Carbide Corporation (UCC) presented a homogeneous process for the hydroformylation of higher olefins, combined with an aqueous two-phase catalyst recovery [45]. The key discovery is that alkali-metal salts of monosulfonated triphenylphosphine form reverse micelles in organic media in the presence of certain solubilizing agents which are stable under the reaction conditions. These systems can easily be induced to separate into a nonpolar product phase and a polar catalyst phase, thereby providing the catalyst recovery typical of two-phase reactions. The separation of the micelles can be accomplished by either raising the temperature or cooling the reaction mixture. In the case of *N*-methylpyrrolidine (NMP)-solubilized systems the addition of water brings about a sharp separation into an organic phase and a catalyst phase. The product phase, however, has to be extracted with water to eliminate traces of catalyst components completely.

Fell et al. recently published a simplified version of the above-mentioned process design of UCC in which the hydroformylation reaction of 1-tetradecene is performed homogeneously with a Rh/Li-TPPMS catalyst system in the presence of methanol. When the methanol is distilled off after almost complete conversion, the catalyst complex precipitates and can be separated by filtration or extraction with water [46].

The idea of using monosulfonated or monocarboxylated triphenylphosphines in a biphasic reaction medium in the presence of amphiphilic reagents has already been patented, in 1981, by the Johnson–Matthey Corporation. However, the recycling of the catalyst complex published in this patent was not complete so that no technical process could be established in those early days of two-phase hydroformylation [47].

6.1.3.2.5 Outlook

What are the major challenges for the two-phase hydroformylation of higher olefins?

The most important applications of higher oxo products are plasticizer alcohols in the C_8–C_{11} range and synthetic detergent alcohols in the C_{12}–C_{18} range, with a worldwide consumption of 1.5 million tons [48] and 1.2 million tons in 1995, respectively. Compared with cobalt, rhodium as catalyst metal is favorable with respect to the raw material economy and the energy balance in the hydroformylation of higher olefins. A biphasic hydroformylation process would bear the advantage that the long-chain aldehydes can be separated from the catalyst simply by phase separation. For olefins above C_{10} the crude aldehyde cannot be separated from the unreacted olefin by distillation. Therefore such a process would be required to achieve complete conversion in continuous operation.

Since the major raw materials for higher plasticizer alcohols are internal olefins from polygas units (e.g., diisobutene, tripropenes), this marked requires the development of even more efficient biphasic catalyst systems for internal and branched olefins.

Scientifically, another major challenge is the development of a biphasic hydroformylation process for internal olefins combining isomerization and hydroformylation of linear internal olefins and affording predominantly terminal hydroformylation products. Such a technology would be of primary interest for the fine chemical and the detergent alcohol markets.

References

[1] (a) W. H. Herrmann, C. W. Kohlpaintner, *Angew. Chem.* **1993**, *105*, 1588; *Angew. Chem., Int. Ed. Engl.* **1993**, *32*, 1524; (b) E. Wiebus, B. Cornils, *Chem.Ing.Tech.* **1994**, *66*, 916.

[2] B. Cornils, E. Wiebus, *CHEMTECH* **1995**, 33.

[3] H. Bahrmann, C. D. Frohning, P. Heymanns, H. Kalbfell, P. Lappe, D. Peters, E. Wiebus, *J. Mol. Catal. A: Chem.* **1997**, *116*, 35.
[4] J. D. Unruh, Ch. L. Koski, J. R. Strong, Fatty acids, in Alpha Olefins Applications Handbook (Ed.: George R. Lappin, Joe D. Sauer), Marcel Dekker, New York, **1989**, p. 311.
[5] J. Falbe, W. Payer in *Ullmanns Encyclopädie der Technischen Chemie*, 4th ed., Vol. 7, Verlag Chemie, Weinheim, **1974**, p. 118.
[6] M. Beller, B. Cornils, C. D. Frohning, C. W. Kohlpaintner, *J. Mol. Catal.* **1995**, *104*, 17.
[7] (a) T. Bartik, B. Bartik, J. Guo, B. E. Hanson, *J. Organomet. Chem.* **1994**, *480*, 15; (b) H. Ding, B. E. Hanson, T. Bartik, B. Bartik, *Organometallics* **1994**, *13*, 3761; (c) T. Bartik, B. Bartik, B. E. Hanson, *J. Mol. Catal.* **1994**, *88*, 43; (d) H. Ding, B. E. Hanson, *J. Chem. Soc., Chem. Commun.* **1994**, 2747; (e) H. Ding, B. E. Hanson, T. E. Glass, *Inorg. Chim. Acta* **1995**, *229*, 329.
[8] E. A. Karakhanov, *J. Mol. Catal.* **1996**, *107*, 235.
[9] R. T. Smith, R. K. Ungar, L. J. Sanderson, M. C. Baird, *Organometallics* **1984**, *2*, 1144.
[10] A. Buhling, P. C. J. Kamer, P. W. N. M. van Leeuwen, J. W. Elgersma, *J. Mol. Catal.* **1997**, *116*, 294.
[11] F. Monteil, R. Queau, P. Kalck, *J. Organomet. Chem.* **1994**, *480*, 177.
[12] D. E. Bergbreiter, L. Zhang, V. M. Mariagnanam, *J. Am. Chem. Soc.* **1993**, *115*, 9295.
[13] (a) Y.-Y. Yan, H.-P. Zhuo, B. Yang, Z.-L. Jin, *J. Nat. Gas Chem.* **1994**, *3*, 436; (b) Z.-L. Jin, X. Zheng, B. Fell, *J. Mol. Catal.* **1997**, *116*, 55; (c) Y. Y. Yan, H.-P. Zuo, Z.-L. Jin, *Chin. Chem. Lett.* **1996**, *7(4)*, 377.
[14] J. Chen, H. Alper, *J. Am. Chem. Soc.* **1997**, *119*.
[15] (a) J. P. Arhancet, M. E. Davis, J. S. Merola, B. E. Hanson, *J. Catal.* **1990**, *121*, 327; (b) I. Guo, B. E. Hanson, I. Toth, M. E. Davis, J. Yang, H. Zhang, Q. Cai, *Gaodeng Xuexiao Huaxue Xuebao*, **1993**, *14*, 863; *Chem. Abstr.* **1994**, *120*, 194442.
[16] (a) I. T. Horváth, J. Rabai, *Science* **1994**, *266*, 72; (b) Exxon Res. Eng. Co. (I. T. Horváth, J. Rabei) EP 0.633.062 (1994).
[17] H. Olivier, Y. Chauvin, in *Catalysis of Organic Reactions* (Ed.: R. E. Malz, Jr.) Chemical Industries Series, No. 68, Marcel Dekker, New York, **1996**, p. 256.
[18] W. A. Herrmann, J. A. Kulpe, W. Konkol, H. Bahrmann, *J. Organomet. Chem.* **1990**, *389*, 85.
[19] F. H. Jardine, Polyhedron Vol. 1, No 7–8, 569–605, 1982.
[20] C. Larpent, R. Dabard, H. Patin, *Inorg. Chem.* **1987**, *26*, 2922.
[21] C. J. Brady, J. R. Cunningham, G. M. Wilson, Water–hydrocarbon liquid–liquid–vapor equilibrium measurements to 530 °F, Research Report RR-62, Gas Processors' Association, Provo, UT, **1982**.
[22] R. M. Deshpande, P. Purwanto, H. Delmas, R. V. Chaudhari, *Ind. Eng. Chem. Res.* **1996**, *35*, 3927.
[23] I. T. Horváth, R. V. Kastrup, A. A. Oswald, E. J. Molzeleski, *Catal. Lett.* **1989**, *2*, 85.
[24] C. Larpent, R. Dabard, H. Patin, *Inorg. Chem.* **1987**, *26*, 2922.
[25] M. J. H. Russel, *Platinum Metals Rev.* **1988**, *32(4)*, 179.
[26] Ruhrchemie AG (B. Cornils, W. Konkol, H. Bach, G. Dämbkes, W. Gick, W. Greb, E. Wiebus, H. Bahrmann), EP 0.158.246 (1984).
[27] P. Kalck, F. Monteil, *Adv. Organomet. Chem. A.* **1992**, *34*, 219.
[28] Cavitron apparatus provided by v. Hagen & Funke GmbH Verfahrenstechnik, D-45549 Sprockhövel, Germany.
[29] Ruhrchemie AG (B. Cornils, H. Bahrmann, W. Lipps, W. Konkol), EP 173.219 (1985).
[30] P. Purwanto, H. Delmas, *Catal. Today* **1995**, *24*, 135.
[31] L. Lavenot, A. Roucoux, H. Patin, *J. Mol. Catal.* **1997**, *118*, 153.
[32] Hoechst AG (H. Bahrmann, B. Cornils, W. Konkol, W. Lipps), EP 157.316 (1984).

[33] H. Chen, H. Liu, *Fenzi, Cuihua,* **1995**, *9*, 145; *Chem. Abstr. 122*, 293863w.
[34] Hoechst AG (H. Bahrmann, B. Cornils, W. Konkol, W. Lipps), EP 163.234 (1984).
[35] Ruhrchemie AG (H. Bach, H. Bahrmann, B. Cornils, W. Gick, V. Heim, W. Konkol, E. Wiebus), EP 302.375, **1987**.
[36] Hoechst AG (H. Bahrmann, P. Lappe), EP 602.463 (1992).
[37] (a) R. V. Chaudhari, B. M. Bhanage, R. M. Deshpande, H. Delmas, *Nature (London)* **1995**, *373*, 501; (b) Council of Scientific and Industrial Interest (Eds.: R. V. Chaudhari, B. M. Bhanage, S. S. Divekar, R. M. Deshpande), US 5.498.801 (1996).
[38] Hoechst AG (H. Bahrmann), DE-OS 19610869 (1996).
[39] E. Monflier, G. Fremy,Y. Castanet, A. Mortreux, *Angew. Chem.* **1995**, *107*, 2450; *Angew. Chem., Int. Ed. Engl.* **1995**, *34*, 2269.
[40] (a) Hoechst AG (W. A. Herrmann, R. Manetsberger, H. Bahrmann, C. W. Kohlpaintner, P. Lappe), EP 571.819, **1992**; (b) W. A. Herrmann, C. W. Kohlpaintner, R. B. Manetsberger, H. Bahrmann, H. Kottmann, *J. Mol. Catal.* **1995**, *97*, 65; (c) H. Bahrmann, K. Bergrath, H.-J. Kleiner, P. Lappe, C. Naumann, D. Peters, D. Regnat, *J. Organomet. Chem.* **1996**, *520*, 97.
[41] (a) H. Bahrmann, H. Bach, C. D. Frohning, H.-J. Kleiner, P. Lappe, D. Peters, D. Regnat, W. A. Herrmann, *J. Mol. Catal.* **1997**, *116*, 49.
[42] W. A. Herrmann, C. W. Kohlpaintner, H. Bahrmann, W. Konkol, *J. Organomet. Chem.* **1992**, *196*.
[43] B. Fell, G. Papadogianakis, *J. Prakt. Chem.* **1994**, *336*, 591; (b) Hoechst AG (G. H. Papadogianakis, B. Fell, H. Bahrmann), EP 489.330 (1990).
[44] F. Monteil, L. Miquel, R. Queau, P. Kalck, in *Aqueous Organometallic Chemistry and Catalysis* (Eds.: I. T. Horvàth, F. Joó), Vol. 131, **1995**.
[45] (a) *Chem. Eng. News* **1995**, *73*, 25: Review on the 209th ACS National Meeting in Anaheim, **1995**; (b) Union Carbide Corporation (Eds.: A. G. Abatjoglou, D. R. Bryant), US 4.731.486, **1988**.
[46] Z. Xia, B. Fell, *J. Prakt. Chem.* **1997**, *339*, 140.
[47] Johnson Matthey Public Ltd. Co. (M. J. H. Russel, B. A. Murrer), US 4.399.312, **1983**.
[48] This figure does not include 2-ethylhexanol. *n*-Butyraldehyde-derived 2-ethylhexanol is by far the most important plasticizer alcohol, with an estimated worldwide consumption of 2.4 million t.

6.1.4 Conversion of Functionalized Olefins

Eric Monflier, André Mortreux

6.1.4.1 Introduction

Hydroformylation or oxosynthesis is a well-known homogeneous, transition metal catalyzed reaction which has known considerable and continuous development since its discovery by Otto Roelen in the laboratories of Ruhrchemie AG in 1938 [1]. This reaction, which can be considered as the addition of a formyl group and hydrogen to a double bond, has been successfully applied in the industrial context by using two basic processes: the homogeneous process where the rhodium or cobalt catalyst and the substrate are in the same phase (Shell, UCC, BASF, RCH processes) [2] and the aqueous/organic biphasic process where the water-soluble rhodium catalyst and the organic compounds are in two different phases (Ruhrchemie/Rhône-Poulenc process) [3].

Application of the homogeneous process is not limited to simple olefins like propene, butene, octene, or nonene. Indeed, a plethora of functionalized olefins can be converted to aldehydes by using this technology [4]. For instance, BASF and Hoffmann–La Roche have developed new approaches for the vitamin A synthesis which involves the hydroformylation of 1,2-diacetoxy-3-butene and 1,4-diacetoxy-2-butene, respectively [5]. Arco has commercialized a process to obtain 1,4-butanediol which requires a step in which allyl alcohol is hydroformylated to 4-hydroxybutyraldehyde by using Kuraray technology [6]. The latter company has also described some attractive routes for producing cyclic products by hydroformylation [7]. So, hydroformylation of 3-methyl-3-butenol and 3-butenol give rise to 2-hydroxy-4-methyltetrahydropyran and 2-hydroxytetrahydropyran, respectively.

Although the broad applicability of the hydroformylation in homogeneous medium has been demonstrated without ambiguity, the scope of *biphasic* catalysis for the hydroformylation of functionalized olefins remains to be investigated. Indeed, there are relatively few examples in the literature related to hydroformylation of such substrates. Furthermore, most of the work described so far has been devoted to the hydroformylation of δ-functionalized olefins (the functional group is not directly branched on the double bond but on an alkyl chain of the olefin, as in the case of 7-octen-1-al or linoleic alcohol) and little attention has been given to the hydroformylation of α-functionalized olefins (functional group directly branched on the double bond). The results described in the academic and patent literature for these two classes of olefins will be discussed separately. Although some interesting results have been reported [8], hydroformylation of water-soluble olefins in two-phase systems with water-insoluble catalysts such as $HRh(CO)(P(C_6H_5)_3)$ is far beyond the scope of this section and will not be discussed here.

6.1.4.2 Biphasic Hydroformylation of δ-Functionalized Olefins

From a regio- and chemoselectivity point of view, the behavior of this class of olefins is very similar to that observed with unfunctionalized olefins in biphasic medium. As expected for a biphasic medium, the catalytic activities are comparable with or lower than those observed under similar homogeneous conditions, and strongly dependent on the water-solubility of the olefin. Addition of solubilizing agents is often necessary for high-molecular-mass olefins. For instance, low-molecular-mass ω-alkenecarboxylic acid methyl esters such as methyl 4-pentenoate can be hydroformylated efficiently in biphasic systems whereas methyl esters of higher ω-alkenecarboxylic acids such as methyl 13-tetradecenoate require the presence of surfactants (Eq. 1) [9].

$$\text{CH}_2{=}\text{CH}(\text{CH}_2)_n\text{CH}_2\text{CH}_2\text{C(O)OCH}_3 \xrightarrow[\substack{\text{P(CO / H}_2\text{) : 100 bar; 120°C} \\ \text{Phosphine / Rh: 60; 2–10 hours}}]{\text{Rh / P(C}_6\text{H}_4\text{SO}_3\text{Na)}_3} \text{OHC}\sim(\text{CH}_2)_n\text{CH}_2\text{CH}_2\text{C(O)OCH}_3 \quad (1)$$

$n = 0, 1, 5, 6, 9$

Conversion: up to 100%
Aldehyde selectivity: up to 100%

The best results in terms of activity have been obtained with cationic surfactants such as octadecyltrimethylammonium bromide. The normal to branched (n/iso) aldehydes ratio was found to be very dependent on the nature of the surfactant. For example, methyl 9-decenoate hydroformylation gave methyl 11-formylundecanoate with an n/iso aldehydes ratio of 6.1:1, 4.0:1, 2.3:1 and with anionic, amphophilic, and cationic surfactants, respectively. Interestingly, hydroformylation of this substrate has also been achieved successfully with inverse-phase transfer catalysts such as chemically modified β-cyclodextrins. In this approach, the cyclodextrin forms an inclusion complex with methyl 9-decenoate and transfers the olefin into the aqueous phase. Under optimal conditions, the aldehydes are obtained in a 100 % yield and in an n/iso aldehydes ratio of 2.3:1 [10].

The use of cationic surfactants also makes it possible to apply the concept of biphasic catalysis in oleochemistry. For instance, the Johnson–Matthey company has reported that oleic acid methyl ester or linoleic acid methyl ester can be hydroformylated in micellar media using a water-soluble rhodium complex of monocarboxylated triphenylphosphine as catalyst [11]. Hydroformylation of linolenic acid methyl ester has also been described recently. Interestingly, with this triunsaturated fatty acid ester, the triformyl derivative selectivity can reach 55% [12] (Eq. 2).

$$\text{CH}_3(\text{H}_2\text{C}{-}\text{CH}{=}\text{CH}(\text{CH}_2)_7)_3\text{CO}_2\text{CH}_3 \xrightarrow[\substack{\text{P(CO / H}_2\text{) : 100 bar; 120°C} \\ \text{Phosphine / Rh: 20; 6 hours}}]{\text{Rh / P(C}_6\text{H}_4\text{SO}_3\text{Na)}_3} \text{CH}_3(\text{H}_2\text{C}{-}\text{CH}(\text{CHO}){-}\text{CH}_2{-}(\text{CH}_4)_7)_3\text{CO}_2\text{CH}_3 \quad (2)$$

Conversion: 100%
Triformyl selectivity: 55%

Hydroformylation of the water-insoluble oleyl alcohol into formylstearyl alcohol has also been successfully achieved with a 96.6% yield by using a rhodium/trisulfonated triphenylphosphine complex dissolved in an aqueous film supported on a high-surface-area silica gel (cf. Section 6.1) [13]. This supported catalyst has also been used to perform the hydroformylation of allyl 9-decenyl ether and 3-methyl-2-(2-pentenyl)-2-cyclopenten-1-one (*cis*-jasmone). However, with the latter substrate, the aldehyde yields did not exceed 38 % [14].

In the course of a continuous search for new strategies to produce Nylon monomers, the DSM company has described an attractive approach for the synthesis of adipic acid or 6-aminocaproic acid precursors. Indeed, in a recent patent, this company has claimed the hydroformylation of 3-pentenoic acid into 5-formylvaleric acid in biphasic medium (Eq. 3). With a water-soluble platinum complex of tetrasulfonated *trans*-1,2-bis(diphenylphosphinomethylene)cyclobutane as catalyst, the selectivity for 5-formylvaleric acids reached 62% [15]. The same catalytic system allows also the hydroformylation of *trans*-3-pentenenitrile with 91.4% slectivity.

$$H_3C-CH=CH-CH_2CO_2H \xrightarrow[\text{Pt / (cyclobutane-}(CH_2PAr_2)_2)\text{; Ar: } C_6H_4SO_3Na]{P(CO/H_2)\text{: 50 bar; 100°C; 4 hours}} H_3C{-}\underset{H}{C}{-}\underset{H}{C}{-}CH_2CO_2H \text{ (with CHO)} \quad (3)$$

Conversion: 79 %
Formylvaleric acid selectivity: 80.3 %

Recently, rhodium/poly(enolate-*co*-vinyl alcohol-*co*-vinyl acetate) catalysts have been developed for the biphasic hydroformylation of aliphatic olefins and applied to the selective hydroformylation of functionalized olefins [16]. Although the conversions were low (< 25%), excellent selectivities for the hydroformylation of *n*-butyl vinyl ether and methyl 3,3-dimethylpenten-4-onate can be achieved with such water-soluble polymer-anchored rhodium catalysts. For instance, the hydroformylation of methyl 3,3-dimethylpenten-4-onate gives only the linear aldehyde.

As far as is known, the only industrial application of the water-soluble catalyst for the hydroformylation of δ-functionalized olefins has been developed by Kuraray [17]. In this process, 7-octen-1-al is hydroformylated into nonane-1,9-dial, a precursor of nonene-1,9-diol, by using a rhodium catalyst and the monosulfonated triphenylphosphine as water-soluble ligand in a 1:1 sulfolane/water system. At the completion of reaction, the aldehydes are extracted from the reaction mixture with a primary alcohol or a mixture of primary alcohol and saturated aliphatic hydrocarbon (cf. Section 6.7).

6.1.4.3 Biphasic Hydroformylation of α-Functionalized Olefins

As reported above, a literature survey shows that the hydroformylation of this class of olefin has been scarcely investigated. Indeed, these studies have been devoted exclusively to the hydroformylation of arylic esters such as methyl acrylate, ethyl acrylate, butyl acrylate, 2-ethoxyethyl acrylate, and 2-ethylhexyl acrylate (Eq. 4) [18–21]. Most attention has been focused on the hydroformylation of methyl acrylate to 2-formylpropanoate ester since the latter is used extensively for the synthesis of pharmaceuticals and may also be considered as a potential source of methyl methacrylate [18].

$$\text{CH}_2{=}\text{CH–C(O)OR} \xrightarrow[\text{P(CO / H}_2\text{) : 50 bar; 50°C; 2 hours; Phosphine / Rh: 10}]{\text{Rh / P(C}_6\text{H}_4\text{SO}_3\text{Na)}_3} \text{aldehyde esters} \quad (4)$$

R: methyl, ethyl, 2-ethoxyethyl butyl, 2-ethylhexyl

Conversion: up to 100%
Aldehydes selectivity: up to 99%
iso / n = aldehyde ratio: 130

In contrast with the first class of functionalized olefins, immobilization of the catalyst in aqueous phase results in an enhancement of the catalytic activity [19]. Indeed, it has been observed that the hydroformylation rates of arylic esters having high solubility in water were much higher in biphasic systems than those observed under comparable homogeneous conditions. Except for 2-ethylhexyl acrylate, the initial rate was increased by a factor of 2.4, 12, 2.8, and 14 for methyl, ethyl, butyl, and 2-ethoxyethyl acrylate, respectively (see Figure 1) [20]. One of the most intriguing features is that the hydroformylation rates for ethyl and butyl acrylates in biphasic medium were respectively higher than and comparable with those observed with methyl acrylate. Actually, the water-solubilities of ethyl and butyl acrylates (18.3 and 2.0 g L^{-1} at 20 °C, respectively) are lower than that of methyl acrylate (59.3 g L^{-1} at 20 °C).

The peculiar enhancement of the catalytic activity in biphasic medium for the hydroformylation of the water-soluble acrylates seems to be due to the physical

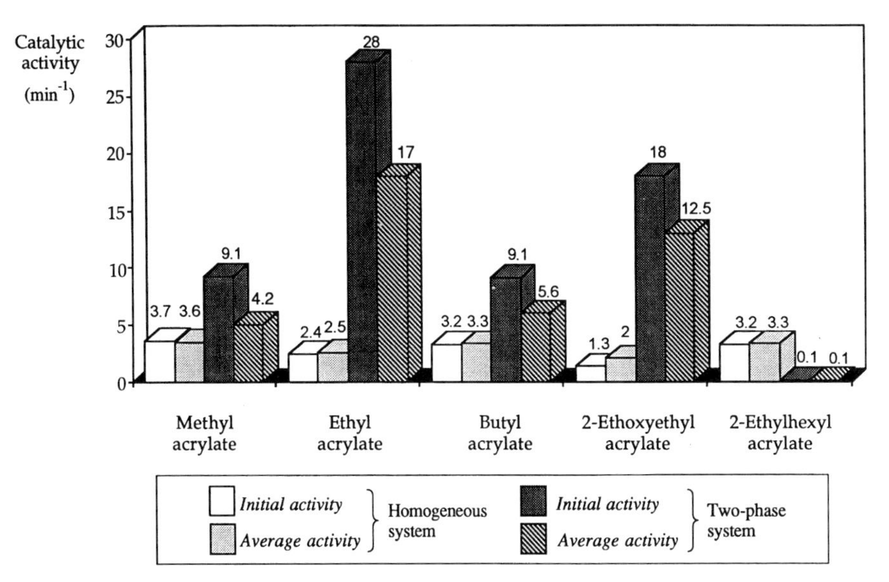

Figure 1. Initial and average catalytic activities for the hydroformylation of various acrylates in homogeneous and biphasic systems. $Rh(acac)(CO)_2$, 0.2 mmol; phosphine, 2 mmol; substrate, 100 mmol; toluene, 40 mL; water for biphasic medium, 30 mL; T: 50 °C, $P(CO/H_2)$: 50 bar.

and chemical properties of water [21]. Indeed, it is assumed that water stabilizes the catalytic species by hydrogen bonding. For instance, it has been proposed that the equilibrium between chelated and nonchelated rhodium complexes is shifted toward the reactive nonchelated species by the formation of hydrogen bonding between water and the carboxyl group of acrylate, which makes oxidative addition of hydrogen easier (see Scheme 1).

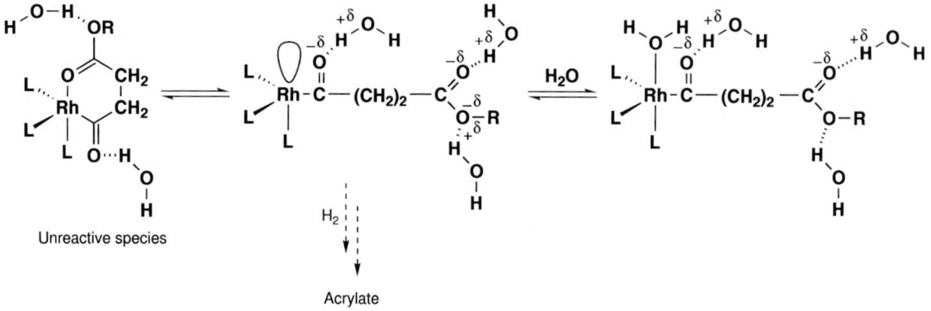

Scheme 1. Equilibrium between chelated and nonchelated rhodium complexes in water.

The decrease in the activity with 2-ethylhexyl acrylate is more familiar and can be attributed to low mass transfer between aqueous and organic phases due to the very poor solubility of this acrylate in water. As a matter of fact, it must be noticed that the hydroformylation of this substrate can be achieved by using an aqueous-phase supported rhodium catalyst [20] or inverse-phase transfer catalysts such as chemically modified cyclodextrins [22]. Owing to the formation of inclusion complexes between an appropriately modified cyclodextrin and the 2-ethylhexyl acrylate, the catalytic activity can be up to 30 times higher than those observed with an aqueous-phase supported rhodium catalyst and 50 times higher than those observed without cyclodextrin. Thus, in the presence of per(2,6-di-O-methyl)-β-cyclodextrin and a Rh/TPPTS catalyst, 2-ethylhexyl acrylate conversion and selectivity to aldehydes reached 100% and 99%, respectively.

More surprising is the unprecedented observation that immobilization of the rhodium/water-soluble phosphine catalyst on a wet silica gel yields an extremely active supported aqueous-phase catalyst for the hydroformylation of a series of acrylic esters [20]. Thus, under optimized conditions, an initial turnover frequency of 4300 h^{-1} was observed for methyl acrylate hydroformylation, compared with 545 h^{-1} in a water–toluene biphasic system and 225 h^{-1} in a homogeneous system. It has been found that the water content, the surface area, and the chemical nature of the solid support play a major role in the activity of the catalyst, whereas the pore diameter has almost no influence. The possibilities of separating the catalyst easily from the reaction medium and recycling it in further experiments were also studied. Unfortunately, a decrease of ca. 10–20% in the catalytic activity was observed between each recycling experiment. This decrease was attributed mainly to the leaching of the rhodium into the organic phase and the gradual dehydration of the silica.

6.1.4.4 Conclusion

The behavior of functionalized olefins depends strongly on the proximity of the functional group relative to the double bond to be hydroformylated. The few examples described so far reveal that most of the principles used for the hydroformylation of unfunctionalized olefins can be applied. However, unusual results can be observed with α-functionalized olefins, i.e., with acrylates. The formation of reactive nonchelated species has been suggested as an explanation of this behavior. Such a phenomenon should probably be generalized to other olefins bearing functional groups in a suitable position, but experiments still remain to be done under biphasic conditions to confirm this hypothesis.

References

[1] C. D. Frohning, C. W. Kohlpaintner, in *Applied Homogeneous Catalysis with Organometallic Compounds* (Eds.: B. Cornils, W. A. Herrmann) VCH, Weinheim, **1996**, Chapter 2.

[2] (a) B. Cornils in *New Synthesis with Carbon Monoxide* (Ed.: J. Falbe), Springer Verlag, Berlin, **1980**, pp. 1–225; (b) M. Beller, B. Cornils, C. D. Frohning, C. W. Kohlpaintner, *J. Mol. Catal. A* **1995**, *104*, 17.

[3] (a) B. Cornils, E. Wiebus, *CHEMTECH* **1995**, *25*, 33; (b) B. Cornils, E. G. Kuntz, *J. Organomet. Chem.* **1995**, *502*, 177; (c) E. Wiebus, B. Cornils, *Chem.–Ing.–Tech.* **1994**, *66*, 916.

[4] C. Botteghi, R. Ganzerla, M. Lenarda, G. Moretti, *J. Mol. Catal.* **1987**, 401, 129.

[5] (a) BASF AG (W. Himmele, W. Aquila) US 3.840.589 (1974); (b) Hoffmann La Roche (P. Fitton, H. Moffet) US 4.124.619 (1978).

[6] N. Nagato, *Encycl. Chem. Technol. (Kirk-Othmer)*, 4th ed., **1991**, Vol. 2, p. 144.

[7] Kuraray Co. (Y. Tokitoh, N. Yoshimura), EP 155.002 (1985).

[8] See for example: (a) R. M. Deshpande, S. S. Divekar, B. M. Bhanage, R. V. Chaudhari, *J. Mol. Catal.* **1992**, *75*, L19; (b) Kuraray Co. (T. Kitamura, M. Matsumoto, M. Tamura), DE 3.210.617 (1984); (c) Kuraray Co. (M. Matsumoto, T. Mashiko), US 4.215.077 (1980); (d) Ruhrchemie AG (H. Bahrmann, B. Cornils, W. Konkol, J. Weber, L. Bexten, H. Bach), EP 216.314 (1986).

[9] B. Fell, G. Papadogianakis, C. Schobben, *J. Mol. Catal.* **1995**, *101*, 179.

[10] Centre National de la Recherche Scientifique (E. Monflier, Y. Castanet, A. Mortreux), PCT Int. Appl. WO 96/22267 (1996).

[11] Johnson-Matthey Co. (M. J. H. Russel, B. A. Murrer), FR 2.489.308 (1982).

[12] B. Fell, D. Leckel, C. Schobben, *Fat. Sci. Technol.* **1985**, *6*, 219.

[13] J. P. Arhancet, M. E. Davies, J. S. Merola, B. E. Hanson, *J. Catal.* **1990**, *121*, 327.

[14] Virginia Tech. Intellectual Properties (J. P. Arhancet, M. E. Davies, B. E. Hanson), US 4.947.003 (1990).

[15] DSM Co. (O. J. Gelling, I. Toth), PCT Int. Appl. WO 95/18783 (1995).

[16] J. Chen, H. Alper, *J. Am. Chem. Soc.* **1997**, *119*, 893.

[17] Kuraray Co. (M. Matsumoto, N. Yoshimura, M. Tamura), U.S. 4.510.332 (1985).

[18] G. Frémy, E. Monflier, R. Grzybek, J. J. Ziolkowski, A. M. Trzeciak, Y. Castanet, A. Mortreux, *J. Organomet. Chem.* **1995**, *505*, 11.

[19] G. Frémy, E. Monflier, J. F. Carpentier, Y. Castanet, A. Mortreux, *Angew. Chem., Int. Ed. Engl.* **1995**, *34*, 1474.
[20] G. Frémy, E. Monflier, J. F. Carpentier, Y. Castanet, A. Morteux, *J. Catal.* **1996**, *162*, 339.
[21] G. Frémy, E. Monflier, J. F. Carpentier, Y. Castanet, A. Morteux, *J. Mol. Catal.* **1998**, in press.
[22] (a) E. Monflier, G. Frémy, Y. Castanet, A. Mortreux, *Angew. Chem., Int. Ed. Engl.* **1995**, *34*, 2269; (b) E. Monflier, G. Frémy, S. Tilloy, Y. Castanet, A. Mortreux, *Tetrahedron Lett.* **1995**, *36*, 9481.

6.1.5 Re-immobilization Techniques

Helmut Bahrmann

6.1.5.1 Introduction

Recent developments in hydroformylation [1–5] were mainly influenced and accelerated by the introduction of functionalized ligands [6–10] which enable the formation of aqueous two-phase or other biphasic catalysts [14]. Normally, the functionalization is affected by the introduction of sulfonic, carboxylic, or ammonium groups which are connected with alkali or alkaline-earth cations as counterions. If ammonium cations are used, the substituents of the nitrogen have an extreme influence on the properties of the functionalized ligands. Using trisulfonated triphenylphosphine (TPPTS) [11, 12] anions the whole system shifts from a water-soluble ("biphasic") to an insoluble ("monophasic") system by minor variations of the alkyl groups and substitution by hydrogen in the counterions (Figure 1) [13].

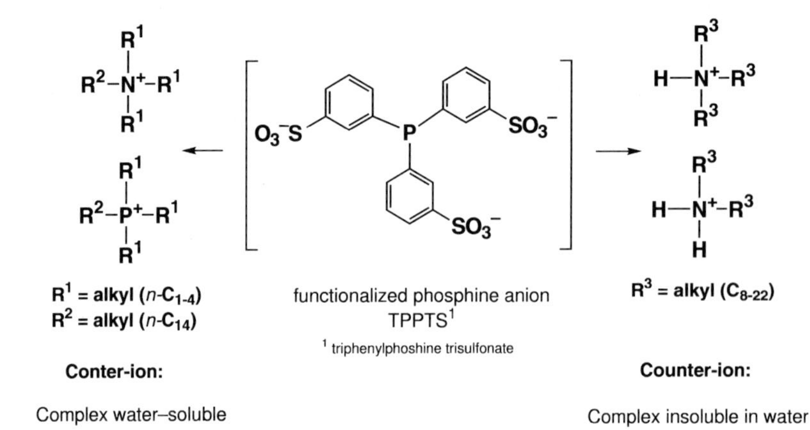

Figure 1. Influence of cations on the properties of TPPTS salts.

The chemistry and separation techniques of the insoluble ("monophasic") system are outlined in more detail below mainly for the hydroformylation, mostly on the basis of TPPTS.

6.1.5.2 Water-Insoluble, Re-immobilized Lipophilic Ligands and their Separation by Membrane Technique

The water-insoluble ammonium ligands are prepared from available water-soluble sodium salts according to Eq. (1). By introducing different amines in solvents, the sodium cations of the sulfonates can be exchanged in the presence of sulfuric acid. Sodium hydrogensulfate as a by-product can be removed smoothly by phase separation. This re-immobilization technique may also be extremely useful in the separation of salts from raw sulfonation mixtures within the preparation of sulfonated phosphines [12]. According to Eq. (1) many variations are possible, for example in the amines, functionalized phosphorus ligands, acids, solvents, and preparation procedures. The variability of the monoamines is demonstrated in Table 1.

$$P(R^1\text{-}SO_3Na)_3 \xrightarrow[-\ NaHSO_4]{NR_3,\ H_2SO_4,\ \text{toluene}} P(R^1\text{-}SO_3HNR_3)_3 \qquad (1)$$

Immobilized ligand for biphasic aqueous catalysis — Re-immobilized ligand for monophasic organic catalysis

In all cases the P(III) yield (transition into the organic phase) is sufficient. For amines with higher molecular masses the pH value is generally lower. Perhaps because of the large distance between the amine groups, the multifunctional amines do not tend to polymerize as do the lower amines (see below).

Normally, as outlined above, the "triacidic acid" TPPTS was combined with "monobasic amines" as bases. Dibasic amines such as 1,3-diaminopropane yield highly crosslinked polymeric materials, whereas other diamines such as Thancat CD® $[(CH_3)_2NCH_2CH_2]_2O$ or 3,4-8,9-Bis(dimethylaminomethylene)tricyclodecane result in the formation of water-soluble ligand salts.

Replacing the "triacidic acid" TPPTS by the disulfonated TPPDS in combination with TCD-diamine (tricyclodecane diamine), a salt partly soluble in toluene and soluble in THF was formed. The same is true with *N,N'*-dimethyl-TCD-diamine instead of TCD-diamine. These salts may be useful in water-free (although biphasic) operation. Under special circumstances even the olefin for the further catalysis may serve as a solvent itself. This concept was successfully realized with dicyclopentadiene (DCP).

Table 1. Ligand preparation with variation of amines.

Entry	Amine	Mol. mass	Temp. [°C]	pH Start	pH End	Addition of $i\text{-}C_3H_7OH$	P(III) content of phases [%] Lower aqueous	P(III) content of phases [%] Upper organic
1	Triisooctylamine	353.7	20		3.5	–	0.0	100
2	Methylditallowamine	513.6	60	7.3	3.5	+	0.8	100
3	Distearylamine	522.0	65	7.7	2.6	+	0.0	100
4	Methyldistearylamine	536.0	60	9.0	3.6	+	–	93.9
5	Jeffamine M 600	600.0	20	10.0	1.8	–	1.6	88.6
6	Tricetylamine	690.3	20	7.5	3.5	–	6.6	98.7
7	Tristearylamine	740.8	60	5.7	1.0	–	0.6	87.0
8	Tri-*n*-octadecylamine	774.5	75	6.8	1.0	–	1.8	90.6
9	Triicosylamine	858.6	75	6.8	0.0	–	1.1	95.5
10	Tridoicosylamine	942.8	80	3.5	0.1	–	–	94.3
11	Jeffamine D 2000	2000	20	9.7	3.6	+	8.9	88.0
12	Jeffamine T 3000	3000	20	9.0	3.6	+	–	93.2
13	Jeffamine D 4000	4000	20	9.5	3.4	+	16.0	73.0

Solvent, 2.7 g toluene/g amine; source of phosphorus, TPPTS, 0.33 mol/mol amine; acid, sulfuric acid.

In some cases, it is desirable and necessary to use phosphines with low basicity such as phosphites. In order to prepare ionic phosphites of the same structure as mentioned above, the preparation procedure [15] can be modified as outlined in Eqs. (2) and (3).

$$(OH)_n\text{–}R^1\text{–}(SO_3Na)_m \xrightarrow[-\ NaHSO_4]{NR_3,\ H_2SO_4,\ toluene} (OH)_n\text{–}R^1\text{–}(SO_3HNR_3)_m \quad (2)$$

$$3\ (OH)_n\text{–}R^1\text{–}(SO_3HNR_3)_m \xrightarrow[-\ 3\ HZ]{PZ_3,\ T} P[O\text{–}R^1(OH)_{n-1}(SO_3HNR^3)_m]_3 \quad (3)$$

$Z = Hal, OR$

$R^1 = C_6H_6,\ x = 6\text{-}(n+m)$

The technique is described in more detail in [38].

6.1.5.3 Separation and Use of Water-insoluble Ammonium Ligands in Hydroformylation

Whereas, on the one hand, in hydroformylation with water-soluble ligands the catalyst can be separated easily after the reaction by simple phase separation, the conversion of higher olefins is normally poor and suffers from the reduced

Table 2. Ligand testing by hydroformylation: ionic phosphites versus triphenylphosphine (TPP).

	Triphenylphosphine (TPP)		TPPpS–TIOA salt [a]	
P/Rh ratio [mol/mol]	40:1	80:1	40:1	80:1
Conversion [%]	81	83	71	70
n/iso ratio	72:28	72:28	83:17	87:13

Reaction conditions: olefinic feedstock, *n*-1-tetradecene; pressure, 50 bar; Rh concentration, 20 ppm; temp. 125 °C; reaction time, 3 h; solvent, acetone.
[a] Triisooctylamine salts of *p*-sulfonated phosphorous acid triphenyl ester.

miscibility of both phases by restricted mass transfer. This can be overcome by different supplementary methods (cf. Sections 4.6.3 and 6.1.3.2, and Chapter 7). In addition, the very active and useful phosphite ligands must be excluded from the aqueous phase due to hydrolysis. On the other hand, high conversion rates can be reached even with branched unreactive olefins using conventional homogeneous ligands in "homophasic" operation, but separation of the catalysts may be a problem, especially with high-boiling substrates, which cannot be distilled.

By using water-insoluble ammonium ligands, the advantages of both catalyst systems (easy separation under mild reaction conditions and high conversion rate) can be combined while avoiding their disadvantages. Thus, hydroformylation with ionic phosphites (e.g., the triisooctylamine salt of sulfonated phosphorous acid triphenyl ester, TPPpS–TIOA salt, Table 2) was introduced by Fell and Papadogianakis [16]. As compared with TPP, the TPPpS–TIOA salt offers a significantly better ratio of linear/branched compounds (l/b, or *n*/iso, ratio). For the separation and recycling of the water-insoluble ammonium salt ligands and Rh complexes, two different methods have been proved their worth, i.e., the phase separation by pH change after reaction and the separation of the catalyst system by membrane filtration.

6.1.5.3.1 Phase Separation by pH Change After Reaction

In this case the potential functionality of the ligands is applied to induce a phase transition of ligands and Rh complexes and subsequent phase separation by a pH change after the reaction at room temperature under mild conditions. The principle is outlined in Eqs. (4) and (5):

$$P(R'SO_3HNR_3)_3 \xrightarrow{\text{NaOH (aq.)}} P(R'SO_3Na)_3 + \text{amine} \quad (4)$$

and thus

$$\underset{\text{organic phase}}{[Rh]_{org}} \xrightarrow{\text{NaOH (aq.)}} \underset{\text{aqueous phase}}{[Rh]_{aq}} + \underset{\text{organic phase}}{\text{amine}} \quad (5)$$

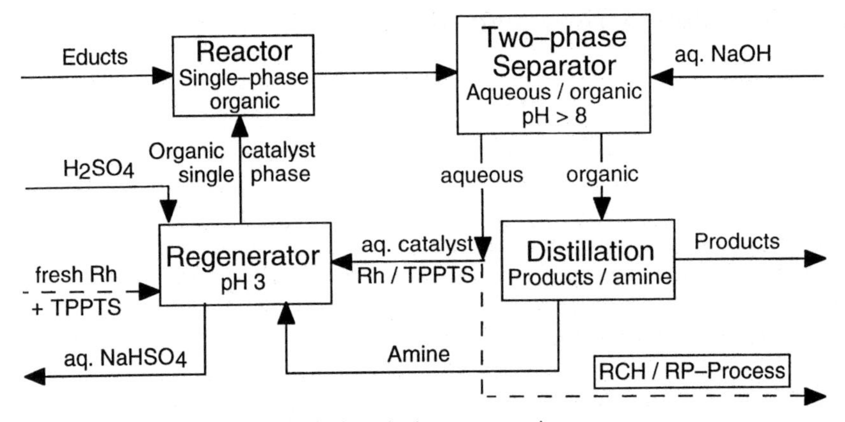

Figure 2. Catalyst recycling by pH-induced phase separation.

By changing the pH value the water-insoluble ammonium salts are reconverted into water-soluble sodium sulfonates. The phosphine and the Rh catalyst are thus transferred to the aqueous phase. They can easily be separated from the organic product. A process scheme is outlined in Figure 2.

As can be seen from Figure 2 the resulting aqueous catalyst phase can be further used in the biphase catalyst system (such as the RCH/RP process [5 g]) or may be re-immobilized in a regenerator with fresh H_2SO_4/amine for the next catalyst cycle. The amine content of the organic phase can be separated by distillation and recycled again by treatment with recycled sulfonated phosphine and sulfuric acid.

Using this procedure a variety of different olefinic compounds and structures (cycloaliphatic, internal, functionalized) have been tested with the triisooctylamine/TPPTS salt [17, 38].

Since fairly good results have been obtained by hydroformylation of oleyl alcohol, the catalyst separation and recovery were investigated successfully in more detail on a pilot-plant scale [18, 19].

6.1.5.3.2 Separation of the Catalyst by Membrane Filtration

The formation of salts according to Figure 2 is environmentally disadvantageous. Thus, a separation of the catalyst system (Rh complexes as well as an additional excess of stabilizing ligands) by membrane filtration ought to be the best way to overcome this and other disadvantages. The state-of-the-art is characterized by the transition from reverse osmosis (RO) conditions (high pressures (50–100 bar), low flow rates (1–4 L/m^2 h), and a solute rejection of about 80–95% [20–27] to ultrafiltration (UF) at lower pressures (2 bar), higher flow rates (20 L/m^2 h), and a satisfying metal retention (>95%) [28–30].

6.1.5.3.3 Recent Developments

Basics

The challenge to enable a good retention for both metal complexes and free ligands requires

- the development of new membranes resistant to organic liquids with a narrow cut-off in a molecular-mass (M) range between 1000 and 10000;
- an oxo raw material with a high process development and market potential;
- the choice and development of appropriate ligands (M 1000–10000).

At the beginning the first requirement was not fulfilled, because experiments with commercially available membranes showed insufficient resistance to organic products, especially against oxo products. Later, a new polyaramide membrane (UF-PA-5 from Hoechst AG) with outstanding properties, which meet all the requirements, was available [31]. The second requirement was met with the decision to use tricyclodecane dialdehyde ("TCD-dial") from the hydroformylation of dicyclopentadiene (DCP). TCD-dial is the starting product for TCD-diamine, which is a specialty chemical used as a curing agent for epoxy resins (Eq. 6) [37].

DCP $\xrightarrow[Rh]{CO/H_2}$ OHC–TCD–CHO (TCD dial) (6)

DCP is normally hydroformylated using unmodified Rh catalysts yielding high TCD-dial contents under disastrous Rh recycling conditions. It was found that on hydroformylating DCP an ammonium salt modified phosphine Rh catalyst can also be used.

The third requirement was achieved by the use of the "re-immobilized catalysts" which enable a fine tuning of the size of the ligands by variation of amines due to their modular ionic structure. Since the development of effective and efficient ligands is costly and exceeds the value of the transition metals, it becomes necessary to separate and recycle the ligands, too. Moreover, the structure of the re-immobilizing ligands makes it possible to remove phosphine oxides and other poisons during economical operation by a simple washing procedure. In the final treatment of the ligand, normally all of the ligand is lost, whereas in the case of ionic ligands the ammonium backbone can be re-used.

First results in membrane filtration with a polyaramide membrane and conventional Rh/TPP catalysts were unsatisfactory, even with the Rh complex itself ($M < 1000$). Only the use of the re-immobilized catalyst with a higher molecular mass (>3000) showed an encouraging Rh retention of 96% as compared with 50% with TPP [32].

Mutual Optimization of Hydroformylation and Membrane Filtration Step

This mutual optimization could be achieved via variation of amines and the P/Rh ratio as well as by introduction of chelating ligands. A variety of amines was tested with TCD-dial with respect to conversion and selectivity to dialdehyde, as well as to the separation data in the subsequent membrane separation step. Figure 3 shows the unit used and in Table 3 some results are outlined. The results are discussed intensively in [38].

In Table 3 the retention of amines is also stated. This is because in the course of the investigations traces of amine were always found in the permeate. This is a result of a very-low-temperature-dependent dissociation of the ammonium salts into amine and free acid during the hydroformylation reaction according to Eq. (7).

$$P(C_6H_4SO_3\text{–}HNR_3^+)_3 \dashrightarrow P(C_6H_4SO_3H)_3 + 3\,NR_3 \qquad (7)$$

In order to stabilize the ammonium salts it was necessary to allow the presence of some free amines. With respect to this additional requirement, an ideal system must have the same good retention for amines as well as for the rhodium and the ligands. In this light the choice of distearylamine could only be regarded as a good compromise between hydroformylation and membrane separation data. Thus, further optimization work was done by hydroformylation of DCP or propene respectively with the Rh–distearyl amine/TPPTS catalyst system.

Test runs with low P/Rh ratios and Rh concentrations while hydroformylating DCP showed excellent membrane separation results but decreasing activity data. This failure in the optimization approach of the hydroformylation and membrane separation step without regard to long-term stability again underlines a basic problem in catalyst development, the coincidental consideration of different contradictory circumstances. A second series with a high P/Rh ratio of 100 was performed with the same catalyst system and butyraldehyde from the hydroformylation of propene as feed.

This replication series (for details cf. [38]) showed overall excellent results in the hydroformylation as well as in the membrane separation step. With a high P/Rh ratio no deactivation was observed. The activity of the catalyst remained sufficient and high amounts of permeate with stable flow rates of 10 to 12 $L\,m^{-2}\,h$ in the critical first stage could be achieved. Obviously, traces of amine in the system catalyze the formation of some high boiling components.

Whereas, on the one hand, a sufficiently high excess of ligands is required for the stabilization of the catalyst system, on the other hand, this high excess of ligands leads to limitations, which depend on the physical properties of such an excess ("salt effect" of the ligands). This, in turn reduces the permeate flow rate and the permeate amount of the membrane filtration.

These contradictory effects with monodentate phosphines may be overcome by the use of bidentate ligands. Chelating ligands as strong complexing agents generally do not need a high excess of free uncomplexed ligands for the stabilization of the active catalyst complex. They make it possible to perform the hydroformylation reaction at a lower P/Rh ratio. Table 4 shows first results with

Table 3. Hydroformylation of dicyclopentadiene and membrane separation of the catalyst system (with variation of amines).

Amine	Mol. mass (M)	Conversion [%]	Selectivity, dialdehyde/monenal	P(III)/Rh ratio[a]	pH value after reaction	Permeate amount [% of input]	Permeate flux [$L\ m^{-2}\ h^{-1}$]	Retention [%] of [Rh]	Ligand [P]	Amine [N]
Triisooctylamine	353.7	99.9	99:1	74	–	5–15	65–64	89.3	69.8	16.5
Methylditallowamine	513.6	99.6	92:8	41	4.0	11	71–	91.9	95.0	76.4
Distearylamine	522.0	98.4	97:3	72	4.0	66	67–61	97.5	96.1	78.3
Methyldistearylamine	536.0	99.4	95:5	63	4.3	5–10	82–77	97.1	94.3	
Jeffamine M 600	600.0	98.8	40:60	51	3.3	23	9	99.7	98.7	63.9
Tricetylamine	690.3	98.7	96:4	78	–	22	44–	95.0	90.0	73.3
Tri-*n*-octadecylamine	774.5	98.5	91:9	84	–	53	68–49	93.0	87.0	88.7
Triicosylamine[b]	858.6	99.0	95:5	103	–	22	57–59	98.9	95.9	88.3
Tridoicosylamine[b]	942.8	99.7	90:10	–	–	48.8	44–29	96.5	94.7	81.9
Jeffamine D 2000	2000	98.1	69:31	7.6	4.0	29	23–22	99.5	89.4	93.9
Jeffamine T 3000	3000	98.4	63:37	3.9	6.2	56	45–31	99.7	97.7	91.4

Hydroformylation conditions: solvent, about 50% toluene; phosphorus source, TPPTS, 0.33 mol/mol amine; pressure 270 bar; Rh concn., 60 ppm; P/Rh ratio, 100:1; reaction temp. 130 °C; reaction time, 4 h.

Membrane separation conditions: feed, reaction product of hydroformylation of dicyclopentadiene; solvent, about 50% toluene; membrane, UF-PA-5/PET 100 from Hoechst AG; overflow, ≈200 L h^{-1}; separation temp., 40 °C; pretreatment of membranes in water at 80 °C for 10 min; transmembrane pressure, 10 bar.

[a] After reaction.

[b] Precipitation of the ligand below 40 °C.

Table 4. Replication series with Rh-distearylamine-1,3-bis(disulfonatophenyl)phosphinopropane catalyst.

Hydroformylation			Membrane filtration				
Conversion [%]	*n*/iso ratio	No of recycles	Retention [%]			Flow rate [$L\ m^{-2}\ h^{-1}$]	
			Rh	P [total]	N	1st stage	2nd stage
93.9	55:45	0	96.4	80.9	60.3	92–59	119–75
85.4	56:44	1	97.7	88.8	58.7	82–54[a)]	112–72
88.6	55:45	2					

Hydroformylation conditions: feed, propene, 270 bar; temp. 125 °C; P/Rh ratio, 2:1; reaction time 2 h; Rh concn., 20 ppm; Membrane separation conditions: feed, butyraldehyde; membrane type, UF-PA-5/PET 100 from Hoechst AG; pressure, 15 bar; temp, 40 °C; amount of permeate, 1st stage, 91–84%; 2nd stage, 95–92%.

[a)] Addition of 0.4 mmol P(III) of ligand to the permeate of the first stage.

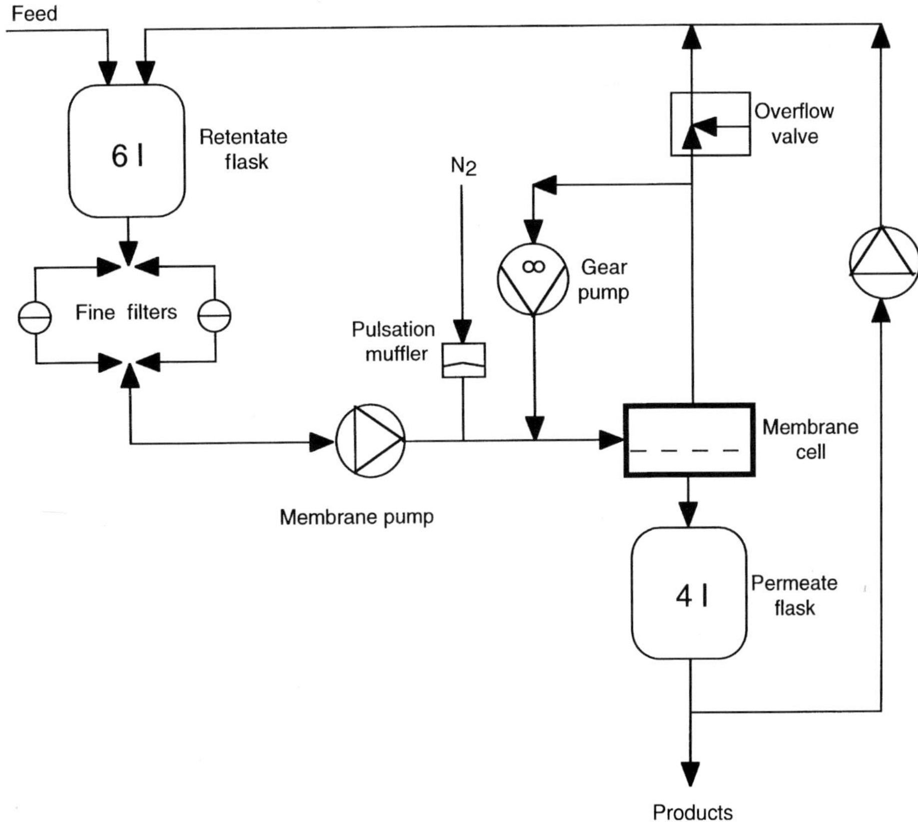

Figure 3. Laboratory-scale membrane unit.

an ammonium salt of simple functionalized chelating ligands, the distearylamine/1,3-bis(disulfonatophenyl)phosphinopropane salt and butyraldehyde from the hydroformylation of propene as feed.

Generally, retention data were somewhat poorer, but very high flow rates could be achieved. On the basis of these results, an increase of the flow rate from 10–12 up to 30–50 L m^{-2} h^{-1} by a factor of 3–4 could be reached. Additionally, the amount of ligand could be reduced by a factor of 50 and the Rh concentration by a factor of 2–3.

Membrane Techniques on a Pilot-plant Scale [33]

Finally, the separation of the Rh–distearylamine/TPPTS catalyst system by membranes was successfully tested on pilot-plant scale with crude aldehyde from the hydroformylation of DCP. The unit was continuously operated over a period of 12 weeks. No decrease in permeate flux and catalyst activity was observed. In contrast to laboratory-scale results, the Rh concentration must be increased to 100 ppm in order to obtain a selectivity of >90 % of dialdehyde. The recirculation rate of Rh was established at 99.2 % and of the ligand up to 98 %. The permeate flux was roughly 10 L m^{-2} h^{-1}. Most of the loss of ligand was due to traces of oxygen.

6.1.5.4 Separation of Phosphine Oxides and Other Degradation Products

The special functionality of the ionic phosphines enables a continuous catalyst make-up. Phosphine oxides and other degradation products can be separated from phosphine oxide by simple extraction with a dilute aqueous sodium hydroxide solution. The concentration of the aqueous sodium hydroxide and the extraction temperature have a tremendous influence on the extraction data. At low concentrations (0.01 %) only the water-soluble sodium phosphine oxides and some benzenesulfonic acid sodium salt (a degradation product of TPPTS) are extracted into the aqueous phase selectively, whereas at higher concentrations above 0.05 % some of the phosphine is discharged, too. This simple selective separation of phosphine oxides and degradation products in continuous operation can enhance the lifetime of the catalyst system. The amount of aqueous NaOH can be adjusted to the degradation rate, which is much lower than the extraction rate. An adequate amount of NaTPPTS may be added together with the aqueous NaOH for generation of new ammonium salts from the amine, which will be formed by this procedure (see Eqs. 4 and 5).

If higher concentrations of aqueous sodium hydroxide (15 %) are used, all of the re-immobilized ligands and the rhodium complex can be extracted into the aqueous phase. This can thus be submitted to the oxidative treatment for Rh recovery according to [34–36]. In this way, 92–95% of the Rh content may be recovered. The amine content of the organic phase was used again by treatment with fresh sulfonated phosphine and sulfuric acid. In the subsequent hydroformylation the same results were actually observed.

6.1.5.5 Further Developments

Because of their salt-like structure, the quaternary ammonium salts enable a selective continuous separation of phosphine oxides and degradation products during the hydroformylation reaction and thus a prolongation of catalyst lifetime [38]. The amine backbone can be re-used. The finally recovery of valuable metal, anions, or cations is possible by simple neutralization reactions.

Additionally, other types of ligands could be used which have the right molecular mass to make possible sufficient retention in the subsequent membrane separation step.

References

[1] Chemische Verwertungsgesellschaft Oberhausen mbH (O. Roelen), DE 849.548 (1938/1952) and US 2.327.066 (1943).
[2] E. Drent, *Pure Appl. Chem.* **1990**, *62(4)*, 661.
[3] Eastman Kodak (a) (T. J. Devon, G. W. Phillips, T. A. Puckette, J. L. Stavinoha, J. J. Vanderbilt), PCT Int. Appl. WO 87/07.600 (1987); (b) (T. J. Devon, G. W. Phillips, T. A. Puckette, J. L. Stavinoha, J. J. Vanderbilt, US 4.694.109 (1987); (c) (T. A. Puckette, T. J. Devon, G. W. Phillips, J. L. Stavinoha), US 4.879.416 (1989); (d) (J. L. Stavinoha, G. W. Phillips, T. A. Puckette, T. J. Devon), EP 0.326.286 (1989); (e) (T. J. Devon, G. W. Phillips, T. A. Puckette, J. L. Stavinoha, J. J. Vanderbilt), US 5.332.846 (1994).
[4] Union Carbide Corp. (a) (E. Billig, A. G. Abatjoglou, D. R. Bryant), EP Appl. 0.213.639 (1987); (b) (E. Billig, A. G. Abatjoglou, D. R. Bryant), EP App. 0.214.622 (1987); (c) (E. Billig, A. G. Abatjoglou, D. R. Bryant), US 4.769.498 (1988); (d) (J. M. Maher, J. E. Babin, E. Billig, D. R. Bryant, T. W. Leung), US 5.288.918 (1994).
[5] (a) Ruhrchemie AG (B. Cornils, J. Hibbel, W. Konkol, B. Lieder, J. Much, V. Schmidt, E. Wiebus), EP 0.103.810 (1982); (b) Ruhrchemie AG (B. Cornils, J. Hibbel, G. Kessen, W. Konkol, B. Lieder, E. Wiebus, H. Kalbfell, H. Bach) DE 3.245.883 (1982); (c) Ruhrchemie AG (H. Kalbfell, B. Lieder, H. Mercamp) DE 3.341.035 (1983) (d) B. Cornils, J. Falbe, *Proc. 4th Int. Symp. on Homogeneous Catalysis,* Leningrad, Sept. **1984**, p. 487; (e) H. Bach, W. Gick, E. Wiebus, B. Cornils, *Prepr. Int. Symp. High-Pressure Chem. Eng.,* Erlangen/Germany, Sept. **1984**, p. 129; (f) H. Bach, W. Gick, E. Wiebus, B. Cornils, *Prepr. 8th ICC*, Berlin **1984**, Vol. V, p. 417; (g) E. Wiebus, B. Cornils, *Chem.-Ing.-Tech.* **1994**, *66*, 916; B. Cornils, E. Wiebus, *CHEMTECH* **1995**, *25*, 33.
[6] E. G. Kuntz, *CHEMTECH* **1987**, *17*, 570; W. A. Herrmann, J. A. Kulpe, J. Kellner, H. Riepl, H. Bahrmann, W. Konkol, *Angew. Chem.* **1990**, *102*, 408; W. A. Herrmann,

C. W. Kohlpaintner, H. Bahrmann, W. Konkol, *J. Mol. Catal.* **1992**, *73*, 191; B. Cornils, W. A. Herrmann, C. W. Kohlpaintner, *Nachr. Chem. Tech. Lab.* **1993**, *41*, 544; W. A. Herrmann, C. W. Kohlpaintner, *Angew. Chem.* **1993**, *105*, 1588 and *Angew. Chem., Int. Ed. Engl.* **1993**, *32*, 1524.
[7] W. A. Herrmann, C. W. Kohlpaintner, R. B. Manetsberger, H. Bahrmann, H. Kottmann, *J. Mol. Catal. A* **1995**, *97*, 65.
[8] B. Cornils, W. A. Herrmann (Eds.), *Applied Homogeneous Catalysis with Organometallic Compounds*, VCH, Weinheim, **1996**.
[9] M. Beller, B. Cornils, C. D. Frohning, C. W. Kohlpaintner, *J. Mol. Catal. A* **1995**, *104*, 17.
[10] B. Cornils, *Angew. Chem.* **1995**, *107*, 1709.
[11] Rhône-Poulenc Industries (E. G. Kuntz), DE. 2.627.354 (1975).
[12] Ruhrchemie AG (R. Gärtner, B. Cornils, H. Springer, P. Lappe), DE 3.235.030 (1982).
[13] H. Bahrmann, in: *Applied Homogeneous Catalysis with Organometallic Compounds*, Vol. 2 (Eds.: B. Cornils, W. A. Herrmann), VCH, Weinheim, **1996**, p. 644.
[14] (a) NATO Advanced Research Workshop, *Aqueous Organometallic Chemistry and Catalysis*, Debrecen/Hungary, Aug./Sept. **1994**, Preprints; (b) L. T. Horváth, J. Rabai, *Science* **1994**, *266*, 72; (c) Exxon Res. Eng. Comp. (I. T. Horváth, J. Rabai), EP 0.633.062 (1994); (d) special issue of *J. Mol. Catal.* **1997**, *116*.
[15] Hoechst AG (H. Bahrmann, B. Fell, G. Papadogianakis), EP 0.435.071 (1989).
[16] Hoechst AG (H. Bahrmann, B. Fell, G. Papadogianakis), EP 0.435.084 (1989).
[17] Hoechst AG (H. Bahrmann, W. Konkol, J. Weber, H. Bach, L. Bexten), EP 0.216.315 (1985).
[18] Hoechst AG (H. Bahrmann, B. Cornils, W. Konkol, J. Weber, L. Bexten, H. W. Bach), DE-OS 3.534.317 (1985).
[19] A. Behr, *Henkel-Referate 31/1995* 31 (Henkel KGaA, Düsseldorf).
[20] BP (M. Th. Westaway, G. Walker), DE-OS 1.912.380 (1968).
[21] BP (A. Goldup, M. Th. Westaway), DE-OS 2.029.625 (1969).
[22] BP (A. Goldup, M. Th. Westaway), GB 1.266.180 (1969).
[23] BP (J. E. Ellis), GB 1.312.076 (1970).
[24] Monsanto (E. Perry), DE-OS 2.414.306 (1973).
[25] Du Pont (L. W. Gosser), US 3.853.754 (1974).
[26] Du Pont (L. W. Gosser), US 3.966.595 (1974).
[27] L. W. Gosser, W. H. Knoth, G. W. Parshall, *J. Mol. Catal.* **1977**, *2*, 253.
[28] E. Bayer, V. Schurig, *Angew. Chem., Int. Ed. Engl.* **1975**, *14*, 493; E. Bayer, W. Schumann, *J. Chem. Soc., Chem. Commun.* **1986**, 949.
[29] S. B. Halligudi et al. *Chem. Tech. Biotechnol.* **1992**, *55*, 313.
[30] Exxon Chemical Patents Inc. (H. W. Deckman, E. Kantner, J. R. Livingston, M. G. Matturro, E. J. Mozeleski), PCT WO 94/19104 (1993).
[31] Hoechst AG (M. Haubs, F. Herold, C. P. Krieg, D. Skaletz), EP 0.325.962 (1988).
[32] Hoechst AG (H. Bahrmann, M. Haubs, W. Kreuder, Th. Müller), DE-OS 3.842.819 (1988).
[33] Th. Müller, H. Bahrmann, *J. Mol. Catal.* **1997**, *116*, 39.
[34] Hoechst AG (L. Bexten, D. Kupies), EP 0.255.673 (1986).
[35] Hoechst AG (G. Diekhaus, H. Kappesser), EP 0.322.661 (1987).
[36] Hoechst AG (J. Weber, L. Bexten, D. Kupies, P. Lappe, H. Springer), EP 0.367.957 (1988).
[37] B. Cornils, R. Payer, *Chem.-Ztg.* **1974**, *98*, 70.
[38] H. Bahrmann, M. Haubs, Th. Müller, N. Schöpper, B. Cornils, *J. Organomet. Chem.* **1997**, *545/546*, 139.

6.2 Hydrogenation

Ferenc Joó, Ágnes Kathó

6.2.1 Introduction

Hydrogenation constitutes a most important class of homogeneously catalyzed reactions for several reasons. Therefore, ever since the discovery of the catalytic properties of $[Co(CN)_5]^{3-}$ and of $[RhCl(PPh_3)_3]$, hydrogenation has been *the* prototype reaction of homogeneous catalysis by transition metal complexes, leading to such important discoveries as, for example, the processes for enantioselective production of L-DOPA and naproxen [1]. The number of papers, patents, and reviews on homogeneous hydrogenation is enormous and this reaction is treated in detail in chapters of important series of books and in independent monographs.

Having learned that much about the kinetics and mechanisms of homogeneous hydrogenations in the past 60 or so years, is there anything special in hydrogenations in *aqueous systems*? The answer is "yes," and it lies in the distinct properties of water as contrasted with those of non-aqueous (mostly, but not exclusively, organic) solvents (cf. Sections 2.2 and 2.3).

Being a highly polar solvent, water is *not* a good medium in which to dissolve molecular hydrogen and the usual substrates of catalytic hydrogenations, mostly apolar organics. Conversely, its immiscibility with many of the common organic solvents makes possible the realization of hydrogenation processes in biphasic solvent systems, allowing easy and efficient isolation of products and recovery of catalysts. It is important to note that, although many biphasic reactions are known with water as one of the solvents, in the following we restrict our discussion to cases where the reaction takes place in the aqueous phase. Relation of aqueous-phase organometallic catalysis to phase-transfer catalysis is discussed in Section 4.6.1.

A specific value of water as solvent lies in the fact that there are catalysts (such as $K_3[Co(CN)_5]$) and substrates (e.g., carbohydrates) which do not dissolve in common nonpolar organic solvents. In the special cases of hydrogenations in living systems, although the constituent lipids are soluble in organic solvents, use of an aqueous medium is essential to preserve the integrity of the cell membranes (cf. Section 6.8).

Limited solubility of molecular hydrogen, organic substrates, and products has an important effect on the mechanism of hydrogenations [2]. However, water itself may also affect the formation of catalytically important intermediates by influencing acid/base equilibria of transition metal hydrides, protonation reactions, and formation of hydrogen bonds. Inorganic salts, too, may have a significant effect on the kinetics of hydrogenation ("salt effects;" cf. Section 4.3) – such phenomena are seldom observed in purely organic solutions.

Solubility of the catalysts in water can be due either to their *overall* charge, such as with $[Rh(cod)(L-L)]^+$ (cod = 1,5-cyclooctadiene, L–L = a chelating diphosphine) or to their water-soluble ligands. Frequently, derivatives of well-known tertiary phosphines (modified by sulfonation, carboxylation, etc.) serve as such ligands; however, some of the phosphines used in water-soluble hydrogenation catalysts (e.g., 1,3,5-triaza-7-phosphaadamantane) have no analogs able to dissolve in apolar organic solvents (see Chapter 3 for details).

Catalytic hydrogenation in aqueous solutions is discussed in several general overviews [3–12]. In this section, a cross-section of the field is given, with just some examples of catalysts and reactions.

6.2.2 Mechanisms and Catalysts of Hydrogenations in Aqueous Solution

6.2.2.1 Basic Mechanisms of Dihydrogen Activation

Obviously, there is a great deal of analogy between the mechanisms of hydrogen activation and hydrogenations in aqueous and non-aqueous systems. In principle, H_2 may react with a suitable transition metal complex, $[ML_n]$, in several ways (Eqs. 1–4).

$$H_2 + [ML_n] = [(H_2)ML_n] \quad (1)$$

$$H_2 + [ML_n] = [H_2ML_n] \quad (2)$$

$$H_2 + [ML_n] = [HML_n]^- + H^+ \quad (3)$$

$$H_2 + 2[ML_n] = 2[HML_n] \quad (4)$$

Reaction (1) may precede both (2) and (3), and (3) may follow either (1) or (2), or their combination.

The chemistry of transition metal dihydrogen complexes, such as formed in Eq. (1) is in the focus of very intensive studies [13, 14] and such complexes were shown to be involved in the mechanism of several hydrogenation processes [15]. Despite this fact, their role in aqueous-phase hydrogenations is largely unexplored, although examples of water-soluble (or water-stable) dihydrogen complexes are known [16, 17]. Reaction (1) does not involve the change of the oxidation state of the central metal ion and may offer a more probable pathway for hydrogen activation (via subsequent deprotonation) in the case of catalysts where oxidative addition (Eq. 2) would seem unfavorable, especially in H_2O. A suspected example is the enzyme hydrogenase containing a Ni(II) metal center [18, 19].

Oxidative addition of H_2 to low-valent metal centers, such as Rh(I) or Ir(I), results in formation of transition metal dihydrides (Eq. 2); well-characterized dihydrides are known [16, 17] and play important role in hydrogenation cycles [15].

Equation (3) depicts direct formation of a monohydrido complex from H_2 and ML_n, although the reaction is likely to proceed via deprotonation of a dihydrogen complex or that of a transition metal dihydride. Such deprotonations are obviously facilitated by the presence of sufficiently strong bases (either H_2O itself, or the bases dissolved in it).

The hydrides of radical nature, formed for example in reaction (4), are less frequently encountered in aqueous-phase catalyzed hydrogenations, despite the fact that water is an ideal solvent for free radicals in terms of its unreactivity [24]. The best-known organometallic hydrogenation catalyst acting via this route in aqueous solutions is $[Co(CN)_5]^{3-}$.

Simple thermodynamic calculations [12] show that, while in the gas phase, homolytic bond dissociation of H_2 requires much less energy (436 kJ mol^{-1}) than the heterolytic split to H^+ and H^- (1674 kJ mol^{-1}), the reverse is true for aqueous solutions (423 kJ mol^{-1} vs. 156 kJ mol^{-1}). The high hydration energies of H^+ (−1090 kJ mol^{-1}) and H^- (−435 kJ mol^{-1}) compared with the very low value for atomic H (−4 kJ mol^{-1}) explain the difference. Although these enthalpy values cannot be directly related to the activation of H_2 by metal complexes (Eqs. 1–4), they clearly show the potentially very large contribution of solvation to the driving force of the overall reaction; the heterolytic split of H_2 is clearly more preferred in the case of water than in nonpolar organic solvents.

6.2.2.2 Water-soluble Hydrogenation Catalysts with Tertiary Phosphine Ligands

A wide variety of water-soluble tertiary phosphines have been prepared and used in aqueous hydrogenations in combination with the metal ions from the platinum metals group; examples are shown as Structures **1**–**19**. Many of these compounds may have some advantages over the others in a particular application; however, most of the fundamental, general studies on catalytic hydrogenation used *sulfonated* arylphosphines because of their availability, stability and good solubility in water over a wide pH range.

In general, coordination chemistry of water-soluble phosphine complexes and their reaction with dihydrogen show many similarities to reactions of their

Ph_nPAr_{3-n} Ar = (3-sulfonatophenyl, SO_3Na)

1 n = 2; TPPMS
2 n = 0; TPPTS

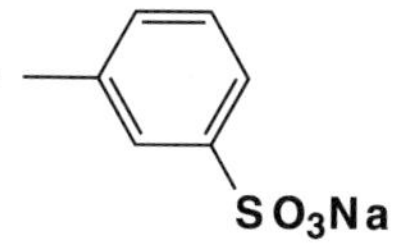

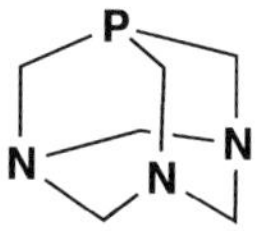

4

$[Ph_2PCH_2CH_2NMe_3]^+$
3 AMPHOS

$P[CH_2CH_2O(CH_2CH_2O)_nCH_2CH_2OCH_3]_3$
5 n = 33

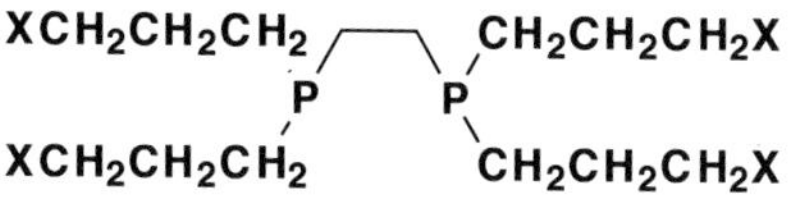

$(XCH_2CH_2CH_2)_2P\text{–}CH_2CH_2\text{–}P(CH_2CH_2CH_2X)_2$
6 X = OH
7 X = SO_3Na

$Ph_2P\text{–}(CH_2)_n\text{–}COONa$
8 n = 5 HEXNa
9 n = 7 OCTNa

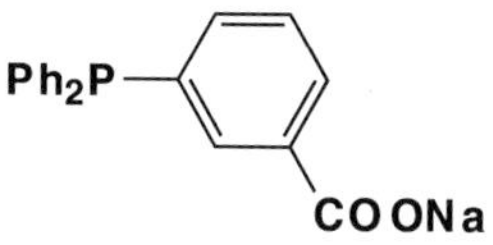

Ph_2P–(3-COONa-phenyl)
10

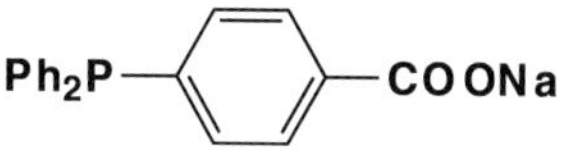

Ph_2P–(4-COONa-phenyl)
11

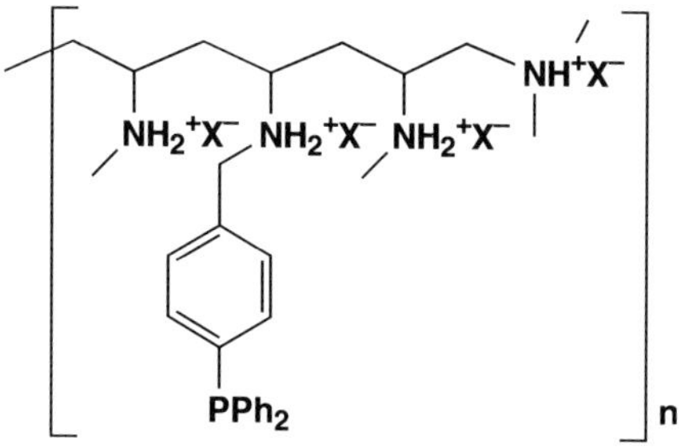

$NH_2^+X^-$ $NH_2^+X^-$ $NH_2^+X^-$ NH^+X^- PPh_2 n
12 PEI–PNH

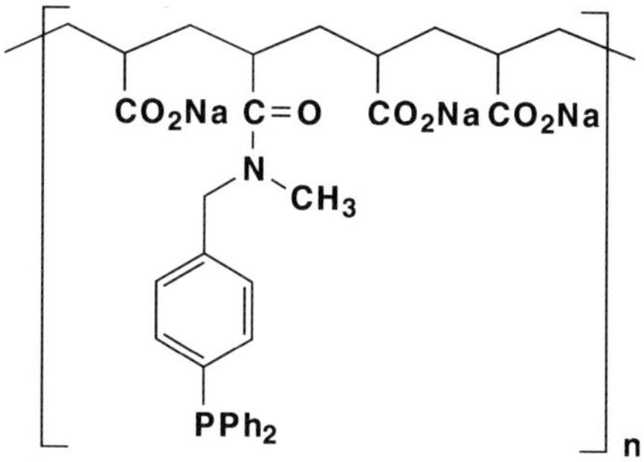

CO_2Na C=O CO_2Na CO_2Na N CH_3 PPh_2 n
13 PAA–PNH

$Ph_2P(CH_2)_2$ N $Ph_2P(CH_2)_2$ O ROH CO_2H CO_2H CO_2H n

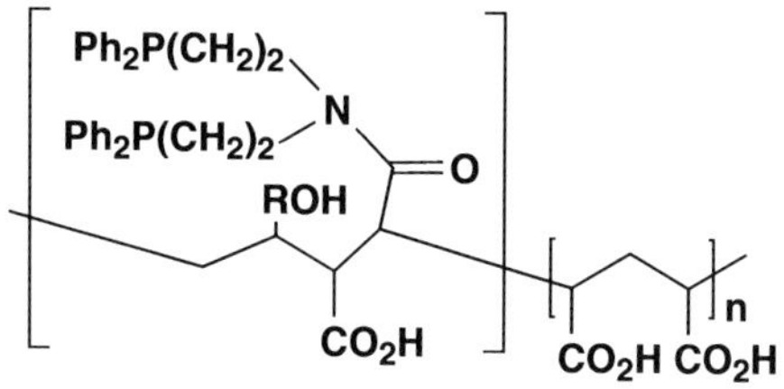

MA – MVE
n = 10, **14**
n = 12, **15**
n = 15, **16**

$Ph_2P(CH_2CH_2O)CH_2CH_2PPh_2$
17, n ~ 2000

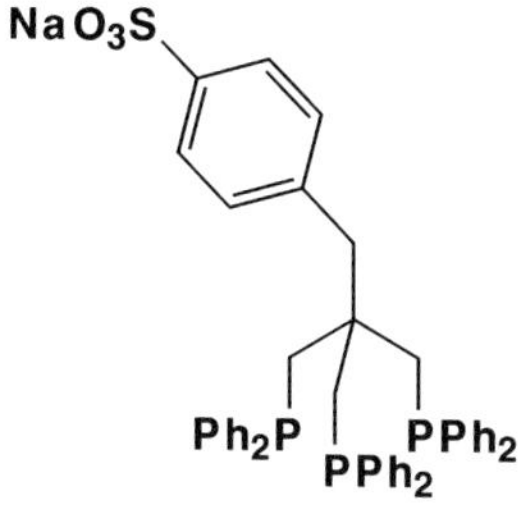

NaO_3S Ph_2P PPh_2 PPh_2
SULPHOS **18**

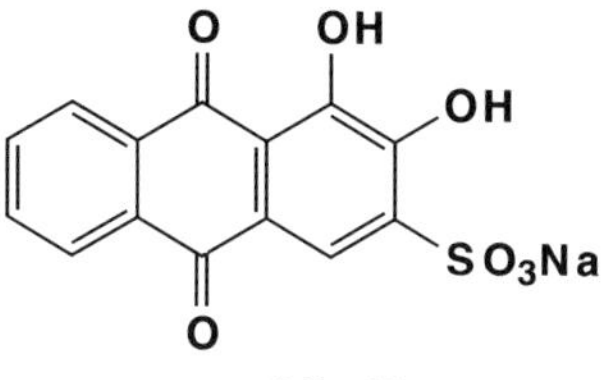

O OH OH SO_3Na O
QS **19**

water-insoluble analogs. Active hydrogenation catalysts or catalyst precursors, such as [RhCl(TPPMS)$_3$] [25], [HRu(OAc)(TPPMS)$_3$] [26], or [Rh(COD)-(BDPP$_{TS}$]$^+$ (BDPP$_{TS}$, Structure **28**) [27], can be prepared by familiar procedures, either from reduction of hydrated salts by an excess of the ligand (as to TPPMS and TPPTS cf. Section 3.2.1) (Eq. 5), or by substituting labile ligands on the metal center (Eq. 6). In several cases exchange of a PPh$_3$ ligand for TPPMS or TPPTS in tetrahydrofuran is driven to completion by the insolubility of the product in THF [28] (e.g., Eq. 7). With tertiary phosphine complexes the most common way of activation of dihydrogen is its homolytic splitting by oxidation addition, as in Eq. (8). The further fate of the resulting dihydrides is pH-dependent and formation of monohydrido complexes by dehydrochlorination or proton dissociation can be observed [29].

$$\mathrm{RhCl_{3aq} + 4\,TPPMS + H_2O = [RhCl(TPPMS)_3]_{aq} + 2\,H^+ + 2\,Cl^- + TPPMSO} \quad (5)$$

$$\mathrm{[Rh(cod)(MeOH)_2][ClO_4] + BDPP_{TS} = [Rh(cod)(BDPP_{TS})][ClO_4] + 2\,MeOH} \quad (6)$$

$$\mathit{trans}\text{-}\mathrm{[IrCl(CO)(PPh_3)_2] + 2\,TPPMS = }\mathit{trans}\text{-}\mathrm{[IrCl(CO)(TPPMS)_2] + 2\,PPh_3} \quad (7)$$

$$\mathrm{[RhCl(TPPMS)_3] + H_2 = [H_2RhCl(TPPMS)_3]} \quad (8)$$

Aqueous organometallic catalysis, however, is not a mere duplication of what had already been observed in organic solvents, and indeed, special effects of the aqueous solvent can be encountered (see examples in Section 6.2.3).

Activation of dihydrogen by oxidative addition is facilitated by complexes with low-valent metal centers; water itself may add oxidatively to the same metal ions (Eq. 9). Reductive elimination of HCl results in the formation of hydroxorhodium(I) derivatives (Eq. 10). Indeed, prolonged treatment of [RhCl(PPh$_3$)$_3$] or [RhCl(cod)]$_2$ with an excess of TPPTS in aqueous solvents gives high proportions of [Rh(OH)(TPPTS)$_3$] [23]; this compound may also be formed by direct Cl^-/OH^- exchange. An unwanted consequence of such redox and hydrolytic reactions is the degradation of the catalysts. Even in the most carefully deoxygenated water as solvent, a considerable proportion of phosphine oxide was formed upon dissolution of [RhCl(TPPTS)$_3$] under an inert atmosphere [30]. Phosphine oxidation is more pronounced with the more basic PTA [31]. Obviously, formation of hydroxo complexes is facilitated by increasing the pH of the solution. Under hydrogen, some of the complexes undergo phosphine degradation. [Pd(OH)$_2$(TPPMS)$_2$], which catalyzes the hydrogenation of alkynes and dienes to monoenes in aqueous solutions at ambient conditions, gave phosphide-bridged clusters and finally [Pd(OH)$_2$] [32] or colloidal metals [33–34].

$$\mathrm{[RhCl(TPPTS)_3] + H_2O = [HRhCl(OH)(TPPTS)_3]} \quad (9)$$

$$\mathrm{[HRhCl(OH)(TPPTS)_3] = [Rh(OH)(TPPTS)_3] + H^+ + Cl^-} \quad (10)$$

Many transition metal hydrides are sufficiently acidic to undergo proton dissociation in the presence of bases or in solvents of suitable solvation power; a thorough analysis of the phenomenon is given in [36]. Water itself can act as a base and, as discussed in Section 6.2.2.1, solvation (hydration) of H^+ is accompanied by an extraordinarily high negative enthalpy change. Consequently, formation of monohydrido complexes can often be observed without the need for an external base. Whereas in benzene solutions formation of $[HRuCl(PPh_3)_3]$ takes place only in the presence of an added base (B) such as 1,8-diaminonaphthalene ("proton sponge") or triethylamine (Eq. 11), reactions of $[RuCl_2(TPPMS)_2]_2$ in water are spontaneous (Eqs. 12–14).

$$[RuCl_2(PPh_3)_3] + H_2 + B \rightarrow [HRuCl(PPh_3)_3] + HB^+ + Cl^- \quad (11)$$

$$[RuCl_2(TPPMS)_2]_2 + 2H_2 \rightleftharpoons [HRuCl(TPPMS)_2]_2 + 2H^+ + 2Cl^- \quad (12)$$

$$[RuCl_2(TPPMS)_2]_2 + 2H_2 + 2TPPMS \rightleftharpoons 2[HRuCl(TPPMS)_3] + 2H^+ + 2Cl^- \quad (13)$$

$$[RuCl_2(TPPMS)_2]_2 + 4H_2 + 4TPPMS \rightleftharpoons 2[H_2Ru(TPPMS)_4] + 4H^+ + 4Cl^- \quad (14)$$

In addition to the temperature and phosphine excess, the positions of equilibria (12)–(14) depend critically on the pH of the solution. The number of protons produced during hydrogenation of $[RuCl_2(TPPMS)_2]_2$ with or without an excess of TPPMS was determined [37]. Below pH 3 the major species is $[HRuCl(TPPMS)_3]$, whereas $[H_2Ru(TPPMS)_4]$ becomes dominant in neutral and basic solutions; in the pH 3–7 range there is a mixture of hydrides present.

A very important conclusion arises, in that meaningful kinetic results can be obtained only in limited pH ranges, and this should be kept in mind when choosing reaction conditions. In addition, a change in the solution composition brings about changes in the rates and selectivities of the catalyzed reactions (see also Section 6.2.3.2).

It has been known for some time that polar solvents accelerate the activation of dihydrogen, as was found with $[HCo(CN)_5]^-$ [38] and $[RhCl(PPh_3)_3]$ [39]. A recent study revealed the same phenomenon, a very large rate increase, *in aqueous* solution [40]. In both dimethyl sulfoxide (DMSO) and water as solvents, the oxidative addition of H_2 to *trans*-$[IrCl(CO)(TPPMS)_2]$ yielding *trans*-$[H_2IrCl(CO)(TPPMS)_2]$ could be described by the rate law given by Eq. (15), which is identical with what had been found earlier for the reaction of trans-$[IrCl(CO)(PPh_3)_2]$ with H_2 in toluene or DMSO.

$$\text{rate} = k[\textit{trans}\text{-IrCl(CO)(TPPMS)}_2][H_2] \quad (15)$$

In DMSO, the second-order rate constants, k_{DMSO}, were the same (1.3 ± 0.02 and $1.2 \pm 0.02\ M^{-1}\ s^{-1}$) for both the TPPMS and the PPh_3 complex. However, in water *trans*-$[IrCl(CO)(TPPMS)_2]$ reacted much faster, with $k_{water} = 12 \pm 3\ M^{-1}\ s^{-1}$. (For comparison, the corresponding rate constant in toluene was as low as $0.26 \pm 0.07\ M^{-1}\ s^{-1}$, i.e. a change of solvent from toluene to water brought about a 50-fold increase.) Such a great increase should be a consequence of water favoring a polar transition state, and a pseudo-five-coordinate molecular hydrogen complex is suggested as a likely intermediate. The existence of water-stable dihydrogen complexes [16], the examples of preferential η^2-H_2 versus H_2O binding in $[W(CO)_3(PR_3)_2L]$ ($L = \eta^2$-H_2 or H_2O) [17], and the hydrogenation of acetone to 2-propanol catalyzed by $[Os(NH_3)_5(H_2)]^{3+}$ [41] do, indeed, support the hypotheses on the role of molecular hydrogen complexes along the reaction coordinate in homogeneously catalyzed aqueous hydrogenations.

$[RuCl_2(TPPMS)_2]_2$ is an active catalyst for the hydrogenation of water-soluble olefins, such as maleic, fumaric, and crotonic acids, in aqueous solutions [26]. A thorough kinetic study of crotonic acid (CA) hydrogenation was undertaken. The rate law derived is given by Eq. (16):

$$-\frac{dn[H_2]}{dt} = \frac{kK[Ru]_0[CA]_0[H_2]}{1 + K[CA]_0 + K^*[TPPMS]} \tag{16}$$

where K and K^* are the equilibrium constants for reactions (17) and (18).

$$[HRuCl(TPPMS)_2] + CA \underset{}{\overset{K}{\rightleftharpoons}} [(HCA)RuCl(TPPMS)_2] \tag{17}$$

$$[HRuCl(TPPMS)_2] + TPPMS \underset{}{\overset{K^*}{\rightleftharpoons}} [HRuCl(TPPMS)_3] \tag{18}$$

This rate law is identical with that found for the hydrogenation of maleic acid in DMF solutions catalyzed by $[RuCl_2(PPh_3)_3]$ [12]. Similar studies with an $[HRu(OAc)(TPPMS)_3]$ catalyst showed the same kinetic characteristics and again, this was analogous to the hydrogenation of 1-alkenes in benzene with $[HRu(OAc)(PPh_3)_3]$ as catalyst [12]. For both TPPMS complexes, a simple mechanism accounted for all the kinetic observations (Scheme 1).

In aqueous/organic biphasic medium the reaction rate for the hydrogenation of linear and cyclic olefins with several Ru(II) complexes including $[HRuCl(TPPMS)_2]_2$, $[HRuCl(TPPMS)_2(L)_2]$ and $[HRuCl(TPPTS)_2(L)_2]$ (L = aniline or tetrahydroquinoline) followed the order: linear C_2–$C_6 \geq$ linear C_7–C_{10} > cyclic olefins ⋙ branched olefins [54]. This reactivity pattern is similar to the case of olefin hydrogenations with $[HRuCl(PPh_3)_3]$, i.e., the least-substituted double bonds are hydrogenated the fastest.

Taking all these observations together, it could be concluded, therefore, that neither the sulfonation of the phosphine ligand nor the replacement of an organic solvent by water had any effect on the reaction mechanism of alkene hydrogenation by Ru(II)–phosphine catalysts.

Hydrogenation of maleic, fumaric, and crotonic acids in water, catalyzed by $[RhCl(TPPMS)_3]$, led to somewhat different conclusions [35]. The reaction was studied in the 30–60 °C temperature range and the pH was set by the substrates (approximately 2.5). With respect to the concentration of the catalyst and

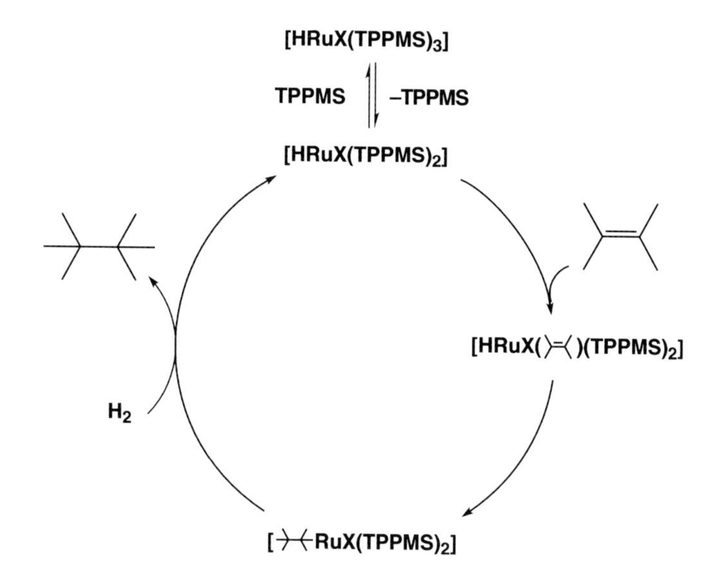

Scheme 1

substrate and to the partial pressure of H_2, the reaction showed the same behavior as $[RhCl(PPh_3)_3]$ in hydrogenation of 1-alkenes [42]. Surprisingly, however, an excess of the phosphine ligand did not affect the rate of hydrogenation of maleic and fumaric acids and the corresponding rate law was that in Eq. (19):

$$-\frac{dn[H_2]}{dt} = \frac{k^*K^*[Rh]_o[MA]_o[H_2]}{1 + K^*[MA]_o} \tag{19}$$

where K^* is defined as the equilibrium constant for reaction (20) and MA is maleic acid.

$$[RhCl(TPPMS)_3] + MA \overset{K^*}{\rightleftharpoons} [(MA)RhCl(TPPMS)_2] + TPPMS \tag{20}$$

In contrast to the case of maleic and fumaric acids, hydrogenation of crotonic acid was sharply inhibited by an excess of TPPMS – again, behavior analogous to that of $[RhCl(PPh_3)_3]$ in hydrogenation of olefins.

Taking the facile reaction of $[RhCl(TPPMS)_3]$ with H_2 (Eq. 8), the kinetic data suggest a mechanism analogous to the familiar one of olefin hydrogenation with $[RhCl(PPh_3)_3]$ [42]. Whether the reductions of maleic, fumaric and crotonic acids proceed via the "unsaturate route" or the "hydride route" was not established, but this would not effect the overall kinetics.

According to the mechanism, excess of the phosphine ligand should inhibit the reaction, and this was indeed observed in crotonic acid hydrogenation. However, as was discovered well after the cited kinetic studies, more-activated olefins, such as maleic and fumaric acids, react instantaneously with TPPMS in acidic aqueous solutions to yield alkylphosphonium salts (Scheme 2) [44]. When used as a substrate, maleic or fumaric acid is present in high excess relative to the catalyst, and therefore fast phosphonium salt formation will remove any free phosphine ligand with no appreciable change in the [substrate]/[catalyst] ratio; consequently, no inhibition by TPPMS of the hydrogenation of activated olefins is observed. In the hydrogenation of maleic acid this side reaction even helps the rhodium complex to enter the catalytic cycle: under hydrogen 18% of the TPPMS was ripped off the metal center as the corresponding phosphonium salt [43]; this results in an average composition of $[RhCl(TPPMS)_{2.5}]$. On the other hand, phosphonium salt formation by TPPMS or TPPTS and crotonic acid is sluggish [44] and will not interfere with the kinetics of hydrogenation.

$$Ph_2PAr + R_1R_2C{=}CR_3COOH \longrightarrow Ph_2ArP^+{-}CR_1R_2{-}CHR_3COO^-$$

Ar = C_6H_4-*m*-SO_3Na

Scheme 2

Another aspect of catalysis by Rh(I)–phosphine complexes is shown by proton production in the reaction of $[RhClP_3]$ complexes (P = TPPMS, or PTA, **4**) with H_2 [29]. This can be accounted for by dehydrochlorination (Eq. 21) or by deprotonation (Eq. 22) of the primary product, $[H_2RhClP_3]$, especially in the presence of excess phosphine. Although proton production could be unambigously measured, neither $[HRhP_4]$ nor the hypothetical $[HRhClP_3]^-$ could be directly detected by spectroscopic methods. In a closely related biphasic system, formation of $[HRh(PPh_3)_4]$ was assisted by stirring a benzene solution of $[RhCl(PPh_3)_3] + PPh_3$ with an aqueous solution of Et_3N; up to 85% of Cl^- was found in the aqueous phase [45].

$$[H_2RhClP_3] + P \rightleftharpoons [HRhP_4] + H^+ + Cl^- \quad (21)$$

$$[H_2RhClP_3] \rightleftharpoons [HRhClP_3]^- + H^+ \quad (22)$$

A catalytic cycle for olefin hydrogenation, involving a monohydridorhodium (I) catalyst, $[HRhL_n]$, is shown in Scheme 3. For the protonation step, H^+ originates in the solvent, and when D_2O is used in place of H_2O deuterated products do, indeed, arise [29]. Moreover, since step ② is hindered, while step ③ is facilitated by decreasing $[H^+]$, a nonlinear effect of pH may arise. In fact, with the $[RhCl(PTA)_3]$ catalyst, a sharp maximum in the hydrogenation rate of crotonic acid, as well as of allyl alcohol, was observed as a function of pH [46], furnishing kinetic evidence for the intermediacy of monohydrides in $[RhClP_3]$-catalyzed hydrogenations in aqueous solutions.

The aqueous hydrogenations catalyzed by $[RuCl_2(TPPMS)_2]$ and $[RhCl(TPPMS)_3]$ show that, while a general similarity between the hydrogenations catalyzed by analogous hydro-soluble and organo-soluble complexes may be expected, one has to be aware of the decisive influence that an aqueous environment may exert on the reaction mechanisms.

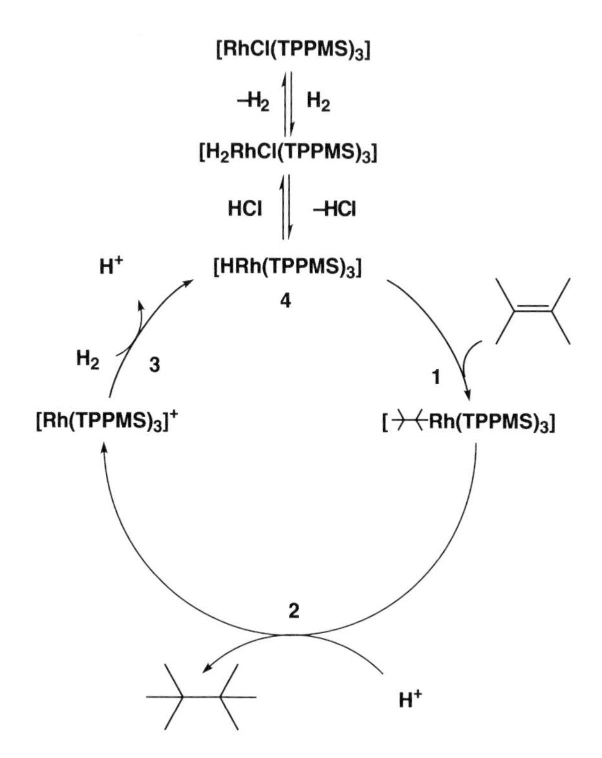

Scheme 3

Among the Rh complexes with ammonium-substituted alkylphosphines, $[Rh(nbd)(AMPHOS)_2]^{3+}$ (nbd, norbornadiene; AMPHOS, **3**) proved to be an active catalyst for hydrogenation of various alkenes in water, in methanol, or in aqueous/organic two-phase systems [47]. The kinetics of maleic acid hydrogenation was studied in detail and was found to be very similar to that of olefin hydrogenations catalyzed by $[Rh(nbd)(PPh_3)_3]^+$, although the rates were rather low due to the low solubility of H_2 in water. With 1-hexene as substrate, isomerization to internal hexenes was also observed. Similar observations were made with phosphonium phosphines, $[Ph_2P(CH)_nPMe_3]^+$, as ligands [48]. In the hydrogenation of 1-hexene the rate varied with the ligand chain length in the order of n $6 > 10 \geq 3 > 2$. Interestingly, the Rh(III) dihydride prepared with AMPHOS underwent a fast reductive elimination of H_2 in water, even under an H_2 atmosphere (Eq. 23), and no hydride species could be detected in the aqueous solutions by NMR spectroscopy.

$$[RhH_2(AMPHOS)_2(solvent)_2]^{3+} \rightarrow [Rh(AMPHOS)_2(solvent)_2]^{3+} + H_2 \qquad (23)$$

There are only a few examples of hydrogenations catalyzed by complexes with hydroxyalkylphosphines in purely aqueous solutions, monophosphines being particularly ineffective [49]. A large family of chelating hydroxyalkylphosphines, such as **6**, was prepared and $[Rh(\mathbf{6})_2]Cl$ was shown to hydrogenate 1-hexene in a biphasic system with rather low TONs ($5\ h^{-1}$) [50]. In contrast, water-soluble rhodium(I) complexes of tertiary phosphine ligands with polyether chains such as **5** showed moderate to high activity in hydrogenation of allyl alcohol [51]. Carboxyalkyl- and carboxyarylphosphines served as ligands in Rh(I)-catalyzed hydrogenation of olefins in aqueous/organic biphasic systems [52] (see also Section 6.2.3.1).

Macromolecular water-soluble phosphine ligands offer the potential to catalyst recovery via membrane ultrafiltration or solubility manipulations by adjustment of the pH or the temperature. Types of such ligands include phosphines with long-chain polyether substituents [53], phosphinated polyethylenemine (PEI–PNH) [55, 56], poly(acrylic acid) (PAA–PNH) [55–57] derivatives and phosphinated maleic anhydride/methyl vinyl ether (MA–MVE) copolymers [58]. The catalytic performance of rhodium(I) complexes prepared with these macromolecular ligands was characterized in hydrogenation of water-miscible, as well as of water-insoluble, substrates (allyl alcohol, 1-buten-4-ol, acrylic acid, 4-propenoic acid and 1-hexene, respectively) under mild conditions. PEI–PNH derivatives are soluble in acidic solutions whereas PAA–PNH derivatives dissolve in basic solutions. In case of the MA–MVE-based Rh(I) complex, a rather small drop of pH from 7.5 to 5.0 was sufficient for complete precipitation of the catalyst, which could be re-used with only a small loss of activity [58]. The solubility of the rhodium(I) complex with the polyether–phosphine ligand **17** in water shows an unusually steep inverse dependence on temperature; catalytic amounts can be easily dissolved and used for hydrogenation of allyl alcohol at 0 °C, whereas at 40–50 °C the hydration shell of the ligand is lost and the catalyst is precipitated [53].

Space constraints do not allow description of all the imaginative efforts to prepare, characterize, immobilize, and recover water-soluble transitional metal–phosphine complexes as hydrogenation catalysts. Further examples can be found in [11] and in [7]. For the mechanism of asymmetric hydrogenation of olefins and that of the hydrogenation of aldehydes, see Section 6.2.3.

6.2.2.3 Complexes of Ligands with Donor Atoms Other Than Phosphorus(III)

$RuCl_3$.aq and $RhCl_3$.aq were among the first catalysts of hydrogenation in aqueous solution [59, 60]. Spectrophotometric experiments revealed the formation of an Ru(II)–olefin complex prior to the heterolytic activation of H_2 ("unsaturated route" of hydrogenation).

Several complexes of copper, silver, ruthenium, rhodium, and cobalt, randomly exemplified by cupric acetate, silver acetate, $[RuCl_4(bipy)]^{2-}$, $[HRh(NH_3)_5]^{2+}$, and cobaloximes (bisdimethylglyoximatocobalt compounds), have been found to catalyze hydrogenations in aqueous solutions, and were studied in considerable detail [12]. Although important for the early research into homogeneous catalysis in general, and the activation of H_2 in particular, these catalysts did not gain synthetic significance. A recent study has demonstrated that ruthenium(II) carbonyl carboxylate complexes of the type $[Ru_2(CO)_4(OAc)(N-N)_2]^+$ and $[Ru(CO)_2(OAc)_2(N-N)]$ (N–N = bidentate nitrogen donor, such as for example 2,2′-bipyridine or 1,10-phenanthroline) catalyze the hydrogenation of alkenes, alkynes, and ketones in partially or fully aqueous solutions [61]. Hydrogenations were run with TOFs in the 20–200 h^{-1} range.

The hydridopentacyanocobaltate anion is readily formed under mild conditions from $Co(CN)_2$ and KCN under H_2 (Eqs. 24 and 25). The resulting complex is an active catalyst for hydrogenation of a variety of unsaturated substrates but the catalysis suffers from several drawbacks [12, 62] (rapid "aging" of the catalyst, and the necessity of using highly basic aqueous solution (see Section 6.2.3).

$$Co(CN)_2 + 3\,KCN \rightleftharpoons K_3[Co(CN)_5] \tag{24}$$

$$2\,K_3[Co(CN)_5] + H_2 = 2\,K_3[HCo(CN)_5] \tag{25}$$

The mechanism of olefin hydrogenation by $[HCo(CN)_5]^{3-}$ was worked out in very fine details [62, 63]. *Activated* olefins add across the Co–H bond with formation of an intermediate Co–alkyl species or with generation of a radical pair. Both pathways require further reaction with an additional $[HCo(CN)_5]^{3-}$ to yield the product alkane. Reaction of $[Co(CN)_5]^{3-}$ with H_2 completes the catalytic cycle.

Carbonyl complexes of transition metals or anionic carbonylmetallates, such as $[Fe(CO)_5]$, $[Co(CO)_4]^-$, or $[Rh(CO)_4]^-$, are usually not very efficient at activating molecular hydrogen in aqueous systems; however, they do even catalyze reductions with $CO + H_2O$ mixtures [11]. In most cases an aqueous base is needed as solvent, although there are a few examples of such reactions in acidic solutions, too. The explanation lies in the reaction of metal carbonyl with OH^-, e.g. reaction (26). The resulting hydrides can react with olefins similarly to the case of $[HCo(CN)_5]^{3-}$. Indeed, $[HCo(CO)_4]^-$ is a thoroughly studied hydrogenation catalyst capable of reducing alkenes and even aromatics [12]; however, its use for *hydrogenations* in aqueous solutions did not generate much interest. It should be noted here that hydrogenations using $CO + H_2O$ as reductant proceed through one-electron steps, and therefore match the preferred one-electron reduction steps of certain substrates, such as nitro compounds (see also Section 6.2.3.4).

$$[M_n(CO)_m] + OH^- \rightleftharpoons [HM_n(CO)_{m-1}]^- + CO_2 \tag{26}$$

Arene hydrido clusters of Rh and Ru are moderately active catalysts of hydrogenation of simple olefins [64, 65]. Upon hydrogenation in aqueous solution $[(\eta^6\text{-}C_6H_6)_2Ru_2Cl_4]$ gives the hexahydrido cluster, $[Ru_2(\eta^6\text{-}C_6H_6)H_6]^{2-}$, which catalyzed fumaric acid hydrogenation. No kinetic studies are known concerning the activation of H_2, in contrast to the case of $[Ru(\eta^6\text{-}C_6H_6)(CH_3CN)_3]^{2-}$, where it was established [66] that hydrogen activation takes place both on a monohydridic and a dihydridic pathway, depending – *inter alia* – on the proton solvation power of the solvent [66].

Pd(II), Pt(II) and Rh(III) complexes of certain ligands with extended conjugated π-electron systems act as very efficient catalysts for hydrogenation of olefins and nitro compounds and hydrogenolysis of organic halides. 1-Phenylazo-2-naphthol [67] and indigosulfonic acid [68], but particularly 1,2-dioxy-9,10-anthraquinone-3-sulfonic acid (Alizarin Red, QS) [69], were used in studies on such reactions. By EPR and NMR spectroscopy as well as by kinetic measurements it was established that these complexes delocalize the extra electron originating from reaction with H_2 on the ligand, from which it is eventually transferred to the substrate (peripheral mechanism of electron transfer) [70]. $[Pd(QS)_2]$ found extensive use in biomembrane hydrogenations [71] (see Sections 6.2.3 and 6.8).

6.2.3 Typical Reactions

6.2.3.1 Hydrogenation of Compounds with C=C and C≡C Bonds

Hydrogenation of simple olefins often serves as a test reaction to characterize the catalytic performance of new hydrogenation catalysts, and only in relatively few cases was it employed for straightforward synthetic purposes. In aqueous solutions the most widely employed substrates for catalyst characterization are unsaturated carboxylic acids (such as maleic, fumaric, crotonic, and itaconic acids), alcohols (i.e., allyl alcohol) and sodium 4-styrenesulfonate, while in two-phase hydrogenations α-olefins (1-hexene, 1-octene), cyclohexene, and styrene are the conventional targets of catalytic reduction. These reactions will not be discussed here in detail.

Hydridopentacyanocobaltate(II), $[HCo(CN)_5]^{3-}$, is a catalyst of choice for selective hydrogenation of conjugated dienes and polyenes to monoenes; unactivated alkenes are totally unreactive [12, 62, 63]. In general, hydrogenation proceeds with 1,4-addition of H_2 (Eq. 27). Because of the insolubility of dienes in water, such reactions are carried out in aqueous/organic biphasic systems. The possibilities for modification of the catalyst by ligand alteration are very restricted but various additives, such as KCN, KOH, lanthanide salts, cyclodextrins, phase-transfer, and micellar agents, are known to influence the selectivity of $[HCo(CN)_5]^{3-}$-catalyzed reactions.

$$R_1\text{–CH=CH–CH=CH–}R_2 \xrightarrow[H_2]{[HCo(CN)_5]^{3-}} R_1\text{–CH}_2\text{–CH=CH–CH}_2\text{–}R_2 \quad (27)$$

$$\xrightarrow[\beta\text{-CD, KOH, } C_6H_6/H_2O]{H_2,\ [HCo(CN)_5]^{3-},\ CeCl_3 \cdot 7H_2O} \quad (28)$$

In benzene/water biphasic systems a variety of dienes were hydrogenated in the presence of β-cyclodextrin (β-CD) and La, Ce, and Yb salts [72]. 2,3-Dimethyl-1,3-butadiene gave 2,3-dimethyl-1-butene with 100% yield and 97% selectivity, representing a reduction with *1,2-addition* of hydrogen (Eq. 28). In this β-cyclodextrin acted as a *reverse*-phase transfer agent, assisting the diene to enter the aqueous phase; equally selective reactions could be achieved by using poly(ethylene glycol)s, such as PEG-400.

Reaction of $[HCo(CN)_5]^{3-}$ with activated olefins, such as α,β-unsaturated acids, produces free-radical intermediates (see also Section 6.2.2.3), and therefore hydrogenation of these substrates is often characterized by low yields and numerous side-products. In the presence of β-CD the yield of acrylic acid hydrogenation increased to 81% [73]. Both neutral (Brij 35) and ionic (SDS, CTAB) micellar agents were shown to increase the rate of hydrogenation of cinnamic acid as a result of an increased local concentration of the substrate within the micelles [74] (cf. Section 4.5).

Rhodium(I) complexes, prepared from $[Rh(cod)_2][BF_4]$ and from amphiphilic, β-CD modified diphosphines, were used as catalysts for the competing hydrogenation of 4-phenyl- and 4-cyclohexyl-1-butene in two-phase water/substrate mixtures (Eq. 29) [75].

$$R_1\text{–CH}_2\text{CH}_2\text{CH=CH}_2 + R_2\text{–CH}_2\text{CH}_2\text{CH=CH}_2 \xrightarrow{\text{cat.}} R_1\text{–CH}_2\text{CH}_2\text{CH}_2\text{CH}_3 + R_2\text{–CH}_2\text{CH}_2\text{CH}_2\text{CH}_3 \quad (29)$$

cat. e.g.

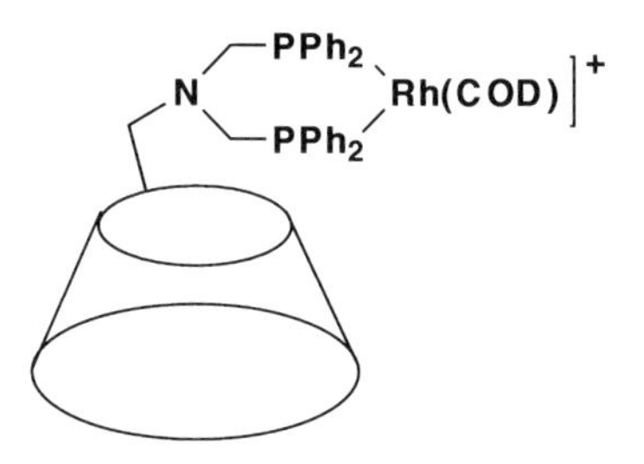

In all cases the catalyst favored the phenyl derivative; chemoselectivity was in the 80–90% range. The β-CD unit selectively incorporates the phenyl ring and this assists the transport of the substrate from the organic into the aqueous phase. Formation of such a host/guest molecular complex keeps the olefin double bond at a suitable distance to interact with the Rh–phosphine catalyst. Both effects increase the rate of hydrogenation of 4-phenyl-1-butene.

Hydrogenation of polymers results in better thermal and oxidative stability, and several attempts were made to employ homogeneous catalysis for this purpose. An ABS copolymer was reduced in the form of an aqueous emulsion with neutral and cationic Rh(I) complexes. The slightly water-soluble cationic complex was more efficiently built into the micelles of the anionic detergents used for polymer solubilization and this resulted in better conversions, e.g., 70% in contrast to 20% with $[RhCl(PPh_3)_3]$ [76].

A detailed study of hydrogenation of several olefins and polybutadiene was undertaken using the catalysts $[RhCl(HEXNa)_2]_2$ and $[RhCl(OCTNa)_2]_2$ (HEXNa and OCTNa, Structures **8** and **9**) [52] with or without an added solvent (toluene). With both catalysts the terminal olefins were hydrogenated much faster than the internal ones, and this was also reflected in the preferential hydrogenation of the pendant vinyl units (products of 1,2-addition) in polybutadiene versus the internal double bonds (from 1,4-polymerization) (Eq. 30). Internal double bonds in 2-pentene- and 3-pentenenitriles were hydrogenated unusually fast compared with simple olefins such as 1-octene, with no concomitant reduction of the nitrile group.

$$\left[CH_2-CH=CH-CH_2\right]_n \sim \left[CH_2-\underset{\substack{| \\ CH \\ \| \\ CH_2}}{CH}\right]_m \xrightarrow{H_2/cat} \left[CH_2-CH_2-CH_2-CH_2\right]_k \sim \left[CH_2-\underset{\substack{| \\ CH_2 \\ | \\ CH_3}}{CH}\right]_l \quad (30)$$

A very special application of homogeneous hydrogenation of olefins in aqueous solution is the catalytic modification of lipid membranes, either in model systems (liposomes) or in living cells [77–80]; cf. Section 6.8.

Polar lipids, such as the one depicted as Structure **20**, form aggregates (liposomes) in aqueous solutions with single or multibilayer structures. Biomembranes are bilayer arrangements of various lipids incorporating a large number of other components, such as proteins, carbohydrates, etc. Therefore the target of hydrogenation is a very complex, microheterogeneous structure, with uneven molecular and spatial distribution of the unsaturated substrate (lipids) mixed with potentially reducible or potentially inhibitory "additives" (e.g., quinones and proteins, respectively).

All kinds of selectivity of hydrogenation are important here, notably chemoselectivity (polyunsaturated versus monounsaturated lipids, unsaturated fatty acids versus quinones), regioselectivity (terminal versus internal olefinic units), and stereochemical selectivity (*cis*- versus *trans*-fatty acid units). An ideal catalyst is able to reach particular membranes in various compartments of the cell ("targeting"), does not effect transformations other than hydrogenation,

$$
\begin{array}{l}
CH_2{-}OC(=O){-}(CH_2)_7{-}CH{=}CH{-}(CH_2)_7{-}CH_3 \\
CH{-}OC(=O){-}(CH_2)_7{-}CH{=}CH{-}(CH_2)_7{-}CH_3 \\
RO{-}P(=O)(OH){-}O{-}CH_2
\end{array}
$$

$R = -CH_2CH_2N(CH_3)_3$; $-CH_2CH_2NH_2$

20

efficiently reduces the unsaturated fatty acid units at low temperatures (0–40 °C) in an aqueous environment, can be totally removed from the cell after the reaction is completed, and has no "self-effect," such as toxicity.

Although $[RhCl(TPPMS)_3]$ and $[RuCl_2(TPPMS)_2]_2$ could be used in some of the studies, the most investigated homogeneous catalyst for biomembrane hydrogenation is $[Pd(QS)_2]$ [80]. Following appropriate pretreatment [71], this catalyst will hydrogenate 10–100 mL of cell suspensions with 10^5–10^6 cells mL^{-1}, at room temperature and 1 bar H_2 in a few minutes, bringing about 10–30 % saturation of all C=C bonds. The small amounts of unsaturated lipids in such samples (usually in the micromolar range) and the similarly low catalyst loadings ensure that only a few catalyst turnovers are required to attain reasonable conversion of the substrate.

Deuteration of membrane lipids is a useful method for studying thermotropic gel-to-liquid crystalline phase transitions of the bilayer membrane structures [81]. In the reduction of dioleoylphosphatidylcholine liposomes with D_2 in H_2O as solvent and $[Pd(QS)_2]$ as catalyst, it was observed that a considerable proportion of products with nonsymmetric deuteration, i.e., having $-CD_2-CH_2-$ units instead of the expected $-CHD-CHD-$ units, was formed [82]. A very probable explanation of this finding is in the *reversible* formation/β-hydrogen elimination of a Pd–alkyl intermediate, as shown in Scheme 4.

$$
R_1HC{=}CHR_2 \underset{}{\overset{+\,PdD}{\rightleftharpoons}} R_1CHD{-}CHR_2Pd \overset{-\,PdH}{\rightleftharpoons} R_1CD{=}CHR_2 \overset{+\,PdD}{\rightleftharpoons} R_1CD_2{-}CHR_2Pd
$$

$$
R_1CHD{-}CHR_2Pd \xrightarrow{+\,H^+} R_1CHD{-}CH_2R_2 \qquad R_1CD_2{-}CHR_2Pd \xrightarrow{+\,H^+} R_1CD_2{-}CH_2R_2
$$

Scheme 4

In hydrogenations with H_2 in D_2O the product showed only $-CHD-$ stretches in the infrared. This observation excludes a fast H/D exchange on Pd, and implies a monohydridic mechanism of hydrogenation. With the same catalyst in an aqueous (D_2O) solution, itaconic acid is reduced under H_2 to yield multiply deuterated methylsuccinic acid having 1.97 deuterons at C3, 0.66 at C2 and none at C1 (Eq. 31) [83]. On the other hand, in an H_2O/ethyl acetate biphasic solvent mixture, the catalyst prepared in situ from $[Rh(cod)Cl]_2$ and TPPTS catalyzed the reduction (with D_2) of dimethyl itaconate with deuterium incorporation at C3 (2.06), C2 (0.78) *and* at C1 (0.18) [84]. Similar results were obtained in toluene/methanol (1:1) with the Rh(I)–BPPM cationic catalyst [85]. Again, these findings could be explained by a fast β-elimination from the intermediate Rh(I)–alkyl.

$$\mathrm{H_2C{=}C(COOH)CH_2COOH} \xrightarrow[\mathrm{H_2/D_2O}]{\mathrm{D_2/H_2O}} \underset{3}{\mathrm{D_mH_{3-m}C}}\text{–}\underset{2}{\mathrm{CD_nH_{1-n}(COOH)}}\text{–}\underset{1}{\mathrm{CD_kH_{2-k}COOH}} \tag{31}$$

Whereas in case of liposomes the nonsymmetric deuteration of the olefinic bond in the unsaturated fatty acids could be assisted by the liquid "cage" around the catalyst, in the homogeneous solution of itaconic acid or its esters multideuteration is facilitated by the coordination of the carboxylic group or the ester carbonyl [84].

No specific study on the homogeneous hydrogenation of acetylenes in aqueous systems has appeared [86, 87].

Although there are several important industrial syntheses based on enantioselective homogeneous hydrogenation of prochiral olefins and catalyzed by platinum metal complexes [1], at present none of them is practiced in aqueous or biphasic systems. In addition to the solubility problems, the main reason is that reaction rates and enantioselectivities in water are often much inferior than those of analogous systems using purely organic solutions. Nevertheless, due to the intensive research that has been done, the possibilities of large-scale applications are slowly emerging.

Several representatives of the widely studied tertiary phosphine ligands for enantioselective hydrogenation in aqueous solutions are shown as Structures **21**–**39**. It is seen that the most successful ligands in this field do have their water-soluble, mostly sulfonated, derivatives [7, 11]. In case of acid-sensitive compounds, solubility in water could be achieved by attaching dimethylamino substituents to the parent arylphosphines and by their further protonation or quaternization. Monophosphines, such as **21** [88], played a minor role in comparison with the chelating diphosphines.

Similarly to the purely organic solutions, the most widely studied olefinic substrates are (*Z*)-α-acetamidocinnamic and (*Z*)-α-acetamidoacrylic acids and their salts. The corresponding methyl and ethyl esters are often studied in aqueous/organic biphasic systems, one of the organic solvents most often employed being ethyl acetate.

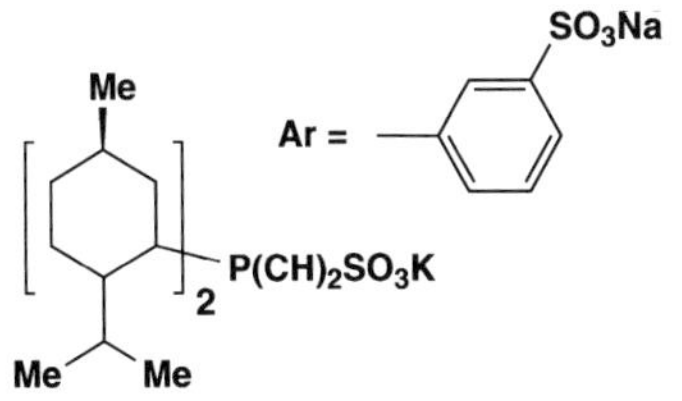

Ar = $-C_6H_4$-3-SO_3Na

$[\ \ ]_2P(CH)_2SO_3K$

21

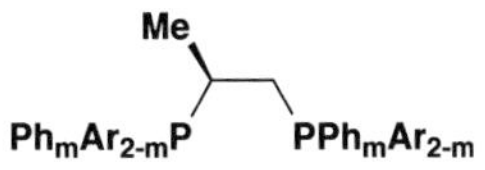

$Ph_mAr_{2-m}P$ PPh_mAr_{2-m}

m = 0, **22**

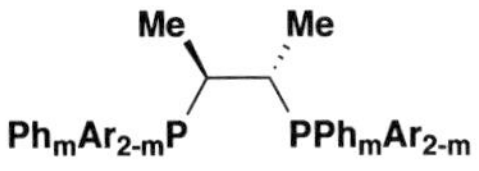

$Ph_mAr_{2-m}P$ PPh_mAr_{2-m}

m = 0, **23**

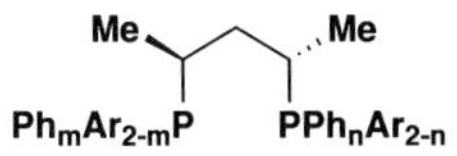

$Ph_mAr_{2-m}P$ PPh_nAr_{2-n}

m = 2	n = 2	BDPP,	**24**
m = 2	n = 1	$BDPP_{MS}$,	**25**
m = 1	n = 1	$BDPP_{DS}$,	**26**
m = 0	n = 1	$BDPP_{TRS}$,	**27**
m = 0	n = 0	$BDPP_{TS}$,	**28**

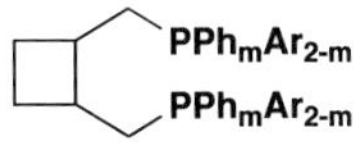

PPh_mAr_{2-m}
PPh_mAr_{2-m}

m = 0, **29**

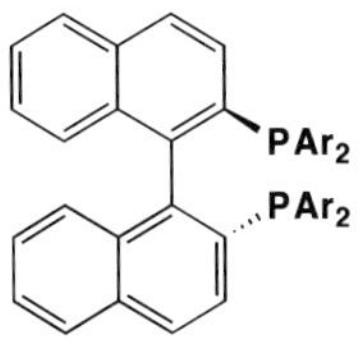

PAr_2 PAr_2

30 $BINAP_{TS}$

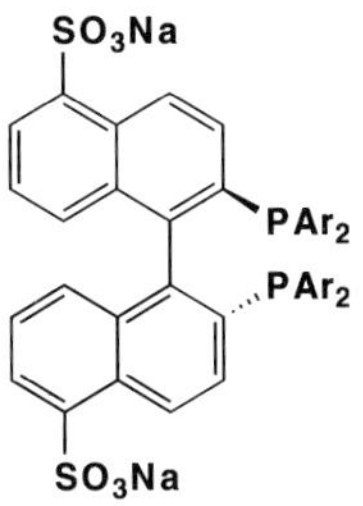

SO_3Na PAr_2 PAr_2 SO_3Na

31

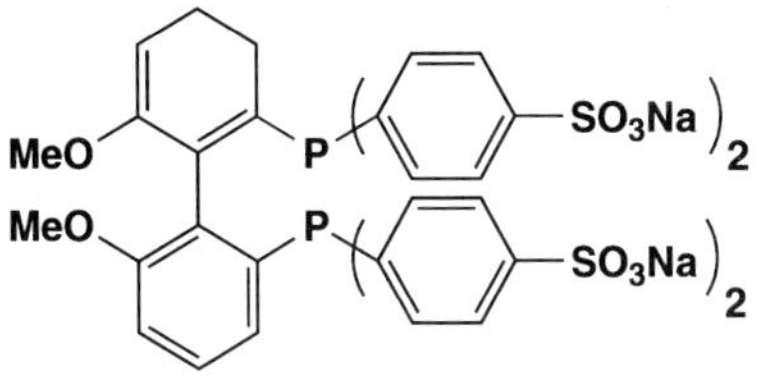

MeO MeO $P(C_6H_4$-$SO_3Na)_2$ $P(C_6H_4$-$SO_3Na)_2$

32 MeO-BIPHEP-S

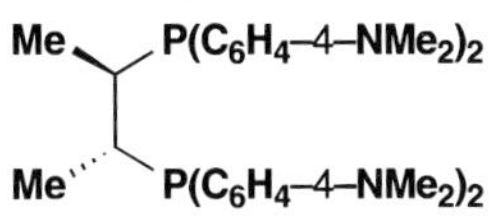

Me $P(C_6H_4$–4–$NMe_2)_2$
Me $P(C_6H_4$–4–$NMe_2)_2$

33

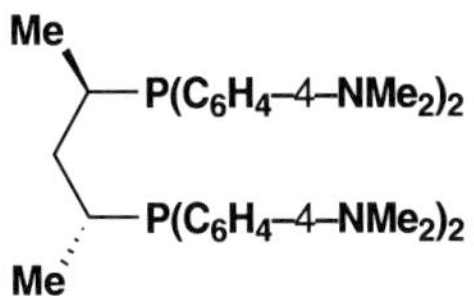

Me $P(C_6H_4$–4–$NMe_2)_2$
Me $P(C_6H_4$–4–$NMe_2)_2$

34

$[$Me $P(C_6H_4$–4–$NMe_3)_2$ Me $P(C_6H_4$–4–$NMe_3)_2]^{4+}$

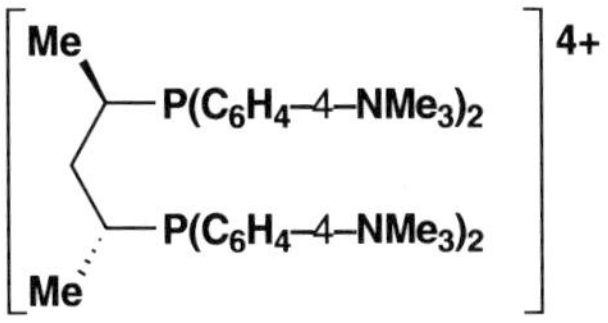

35

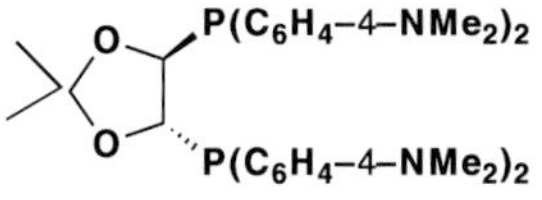

36

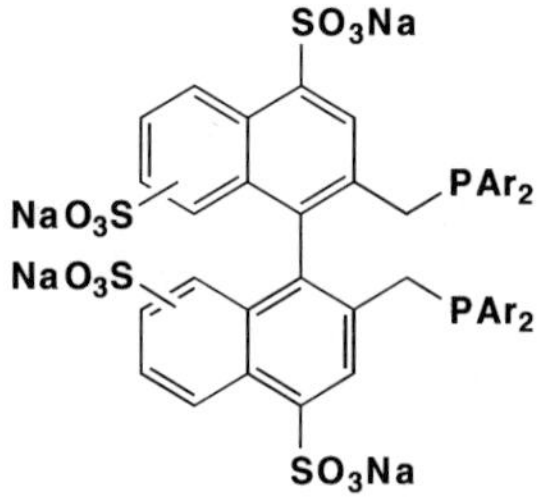

38 BINAS

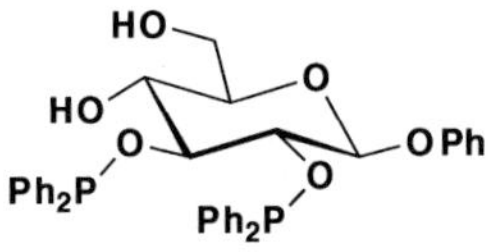

37

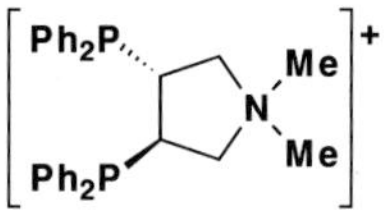

39

Prochiral enamides and dehydropeptides were biphasically hydrogenated with enantiomeric excess (*ee*) up to 88 %, with Rh(I) complexes of the chiral, sulfonated derivatives of PROPHOS, CHIRAPHOS, cyclobutaneDIOP, and BDPP, **22**–**29** (Scheme 5) [27, 89]. Using D_2O instead of H_2O, hydrogenation of methyl (*Z*)-α-acetamidocinnamate, (*Z*)-α-benzamidocinnamate, and (*Z*)-α-acetamidoacrylate with a $[Rh(cod)Cl]_2$ + phosphine catalyst (phosphine: **23**–**26**, **28**) resulted in the regiospecific deuteration at the carbon atom that is α to the amide and the ester groups (maximum incorporation of D was 76 %). The results were interpreted in terms of a fast [Rh]–H$\rightleftharpoons$[Rh]–D exchange on the σ-rhodium monohydride intermediate, although protonation (with H^+ or D^+) of the same intermediate could not be ruled out [90].

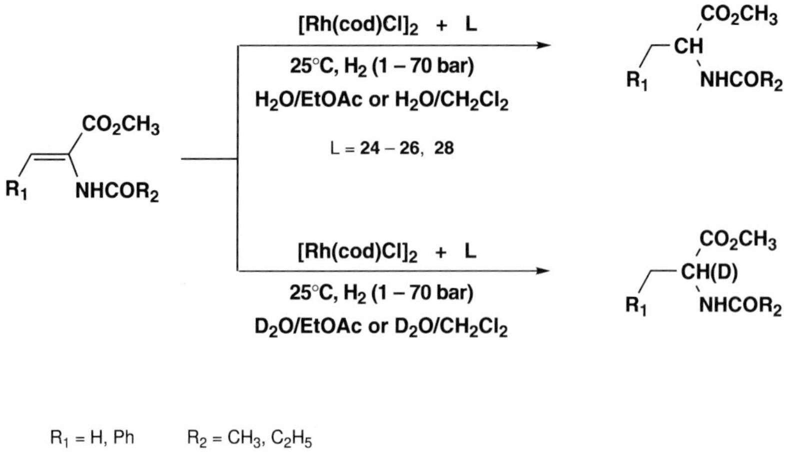

R_1 = H, Ph R_2 = CH_3, C_2H_5

Scheme 5

Quaternization of the ligand (3*R*, 4*R*)-3,4-bis(diphenylphosphino)-1-methylpyrrolidone, bound in the complex $[Rh(cod)L]^+$, (L, **39**) yielded a highly active and selective catalyst for hydrogenation (*Z*)-α-acetamidocinnamic acid; (*S*)-*N*-acetylphenylalanine was obtained in 88–96% enantiomeric excess [91]. The 4-dimethylamino derivatives of CHIRAPHOS (**33**), BDPP (**34**, **35**) and DIOP (**36**) showed similarly high activities and enantioselectivities (up to 97% *ee*) for the reduction of (*Z*)-α-acetamidoacrylic acid either in aqueous slurries or in biphasic systems [92].

Stepwise substitution in the phenyl rings of phosphines such as BDPP creates new diastereomeric pairs due to chirality on phosphorus. In a detailed investigation into the hydrogenation of (*Z*)-α-acetamidocinnamic acid, its methyl ester and dimethyl itaconate, the catalyst was prepared from $[Rh(cod)Cl]_2$ plus 2 equiv. of (2*S*,4*S*)-BDPP or its mono-, di-, tri-, or tetrasulfonated derivative (**25**–**28**) [93]. As a result, in the case of the enamides the enantioselectivity gradually decreases with the degree of sulfonation. Conversely, only in case of the monosulfonated phosphine, as ligand, was a moderate enantiomeric excess of (*R*)-methylsuccinic acid dimethyl ester obtained. (The effect is more pronounced in hydrogenation of imines; see Section 6.2.3.2.) It should be noted that the catalyst with $BDPP_{MS}$ is *not* soluble in water and moves to the organic phase during the reaction. Interestingly, the degree of quaternization of the tetrakis(4-dimethylamino) derivatives of BDPP and CHIRAPHOS did not influence the enantioselection in hydrogenation of enamides [92, 94]. Ru(II)–diphosphine catalysts are usually more robust than the related Rh(I)–diphosphine complexes, and more and wider applications of ruthenium complexes can be expected. E.g. the water-soluble variant of the highly successful Ru(II)–BINAP catalysts has been prepared using 5,5′-disulfonato-BINAP, **31** [95], and tetrasulfonated (*R*)-2,2′-bis(diphenylphosphino)-1,1′-binaphthyl ($BINAP_{TS}$, **30**) [96, 97]. In water, $[Ru(R\text{-}BINAP_{TS})Cl_2]$ catalyzed the hydrogenation of (*Z*)-α-acetamidocinnamic and (*Z*)-α-acetamidoacrylic acid with 87.7% (*R*) and 68.5% (*R*) *ee*, respectively. With respect to the hydrogen pressure the water-soluble catalyst showed the same behavior as the parent $[Ru(BINAP)Cl_2]$ (a sharp decline of *ee* with increasing p_{H_2} [97]).

Several new, water-soluble atropisomeric diphosphines in the biphenyl series, such as (*S*)-(+)- and (*R*)-(−)-MeO-BIPHEP tetrasulfonate (**32**), were prepared and used as components of Ru(II)- and Rh(I)-based hydrogenation catalysts [98]. Several C=C unsaturated substrates (enamides, unsaturated acids) could be hydrogenated biphasically with high rates and enantioselectivities (Eqs. 32 and 33). In some cases high substrate/catalyst ratios could be used (up to 10000:1), a strong requirement for practical applications [99].

One of the attractive features of the Ru(II)-based catalysts described above is that the enantioselectivities obtained in aqueous solution are usually close to those which can be observed using organic solvents; in some cases even an increase in the *ee* was observed when methanol or ethanol was replaced by water [97]. However, this is not the general case with the various Rh(I)–chiral diphosphine catalysts, where more often than not a significant drop of *ee* is seen when running the reactions in water instead of an organic solvent [105].

1. $[Ru(32)]^{2+}/H_2$(40 bar) H_2O/60 °C/21 h 2. H^+

(32)

M			
M = Et_3NH	substr/cat	10 000	98% yield, > 99% ee
M = Na	substr/cat	2000	99% yield, > 99% ee

1. $[Ru(32)]^{2+}/H_2$(60 bar) H_2O/20 °C

(33)

M			
M = Et_3NH	substr/cat	1000	84% ee
M = Na	substr/cat	1000	80% ee

So far no complete equilibrium and kinetic investigation has been published which could allow the analysis of the reaction mechanism of an aqueous enantioselective hydrogenation with the precision reached by Halpern et al. in their seminal studies [100, 101]. Nevertheless, some general facts and observations are worth discussing.

The enantioselection in certain catalytic hydrogenations is very dependent on the pressure of H_2 which, in turn, determines the equilibrium concentration, $[H_2]_{sat}$, of H_2 in the solution. A thorough recent analysis [2] of solubility and mass-transfer processes has shown, that it is the availability of dissolved H_2 in the solution, rather than $[H_2]_{sat}$, which influences the enantioselectivity, and the former is determined by the relation between the chemical and the mass-transfer rates. The solubility of hydrogen in water is 8×10^{-4} M, which is about 20 % of the solubility in MeOH, 3.8×10^{-3} M (both at 20 °C, 1 bar total pressure [102]). It follows that working in an aqueous solution is equivalent to using methanolic reaction mixtures under reduced pressure of H_2, and that reactions in which the enantioselectivity is known to be pressure-sensitive (such as those described in [92] and in [97]) cannot be strictly compared at the same partial pressure of dihydrogen. The limited solubility of H_2 may influence the chemoselectivity, too (reduction vs. isomerization of olefins) [103].

In a systematic study of solvent effects (cf. Section 4.3) on the catalytic hydrogenation of dehydro-amino acids, a good linear correlation was found between the enantioselectivity parameter, log ($\% S/\% R$), and the solvofobicity parameter, S_p [104], of the solvents; water fits this relationship well [105]. In the case of a given substrate and a given catalyst, this correlation may also reflect the solubility of H_2 in the solvents under study.

Many of the substrates are only slightly soluble or "insoluble" in water, i.e., their equilibrium concentration in saturated solutions is very low [106]. In an aqueous/organic two-phase mixture, the concentration of the substrate in the aqueous phase is lowered even further by its distribution between the phases. Addition of amphiphiles not only increases the chemical rate by the well-known solubilization phenomenon but may dramatically influence the enantioselectivity, too. A striking example of such an effect was observed with the [Rh(Me-α-glup-OH)(cod)][BF_4] catalyst (**37**), containing a phosphinated methyl-α-D-glycopyranoside ligand. On addition of sodium dodecylsulfate (SDS) the half-time ($t_{1/2}$) of the hydrogenation of (*Z*)-methyl-α-acetamidocinnamate was shortened from 390 min to 6 min together with an increase of enantioselectivity from 83.4% to 97.1% *ee* [106]. Similar improvements in the reaction rate and enantioselectivity were found with using carbohydrate-derived amphiphiles [107], but the chirality of these latter compounds did not contribute significantly to the overall increase of the *ee* [108]. The use of surface-active, chiral, tertiary phosphine ligands, such as $P[C_6H_4(CH_2)_3C_6H_4SO_3Na]_3$ eliminated the need for an additional amphiphile. With the catalyst prepared in situ from this phosphine and $[Rh(cod)Cl]_2$, (*Z*)-methyl-α-acetamidocinnamate was hydrogenated in water/ethyl acetate at a higher rate and better selectivity than with the unmodified, analogous phosphine BDPP [109]. Micellar effects are discussed in detail in Section 4.5.

6.2.3.2 Hydrogenation of Compounds with C=O and C=N Bonds

One of the most desirable chemical transformations is the aqueous selective hydrogenation of α,β-unsaturated aldehydes to the corresponding allylic alcohols. There are only a few Rh, Ru, and Ir complexes capable of catalyzing this reaction [110, 111].

Several water-soluble ruthenium complexes, with P = TPPMS, TPPTS or PTA ligands, catalyze the selective reduction of crotonaldehyde, 3-methyl-2-butenal (prenal), and *trans*-cinnamaldehyde to the corresponding unsaturated alcohols (Scheme 6) under relatively mild conditions [109–118, 127]. Chemical yields are often close to quantitative in reasonable times (from minutes to a few hours) and the selectivity toward the allylic alcohol is very high ($> 95\%$), except for the carbonyl complexes [$HRuCl(CO)P_3$] where the highest proportion of cinnamyl alcohol in the product mixture was around 30% [118]. With all three phosphine ligands TPPMS, TPPTS and PTA, the reactions are genuine two-phase processes and the catalyst-containing aqueous phase can be cleanly separated from the organic phase of the product mixture.

Both aliphatic and aromatic unsaturated aldehydes were reduced exclusively to unsaturated alcohols by hydrogen transfer from aqueous sodium formate in a two-phase system with or without an organic solvent. The reactions proceeded smoothly with either [$RuCl_2(TPPMS)_2$] [119, 120] or [$RuCl_2(PTA)_4$] catalyst

[116, 117]. Although the procedure is particularly suited for smaller-scale laboratory preparations, the inorganic by-product ($NaHCO_3$) arising from the process makes it inacceptable on the industrial scene.

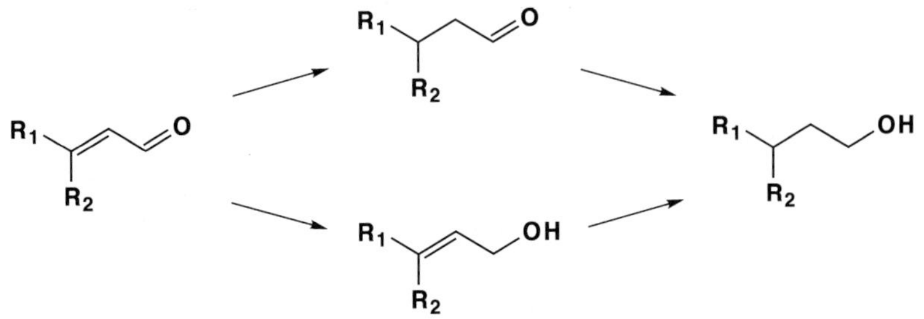

Scheme 6 e.g. $R_1 = R_2 = Me$; $R_1 = Ph$, $R_2 = H$

Much effort was devoted to exploring the mechanism of this selective reduction by varying the catalyst precursor [115] and reaction parameters [113], such as catalyst concentration, hydrogen pressure, temperature, and concentration of excess phosphine and of other additives. Strikingly, no phosphine inhibition was observed, either in the hydrogenation or in the hydrogen-transfer reduction of aldehydes [113, 120], and in some cases an excess of TPPTS or TPPMS ligand even increased the rate of reduction [120]. While this latter effect may be related to the surface activity of TPPMS [118], the explanation will not hold for TPPTS, which has no surfactant properties [118].

The most likely catalytic species for RCHO reduction is $[H_2RuP_4]$ (P = TPPTS, TPPMS, PTA), either prepared separately [22, 121] or formed in situ from the various precursors (precatalysts). However, it has not been determined unambiguously whether this complex loses a phosphine ligand before entering the catalytic cycle and coordinating the substrate, or reacts directly with the aldehydes by nucleophilic attack of one of the hydride ligands on the carbonyl oxygen. The lack of phosphine inhibition seems to support this latter suggestion (Scheme 7).

Unfortunately, most of the aldehyde reduction experiments were run in unbuffered solutions. Exceptions are the hydrogen-transfer reductions [117, 120], where the $HCOONa/NaHCO_3$ mixture has a fairly constant pH of 8; and the hydrogenation of prenal, where a phosphate buffer of pH 7 was used to prevent acid-catalyzed formation of *tert*-amyl alcohol (Scheme 8) [113].

As discussed in Section 6.2.2.2, the equilibrium between the various ruthenium hydride species, first of all between $[HRuClP_3]$ and $[H_2RuP_4]$, is critically influenced by the pH of the solutions, which – in the absence of a suitable buffer – are always slightly acidic, especially when the catalyst is prepared in situ by reducing hydrated $RuCl_3$ with an excess of the phosphine ligand. This effect may not be obvious in reduction of a saturated aldehyde, but will determine the selectivity in hydrogenation of unsaturated aldehydes. Differences in the pH of the reaction mixtures may be the cause of some apparent contradictions in the literature, such as the case of prenal hydrogenation [113, 115].

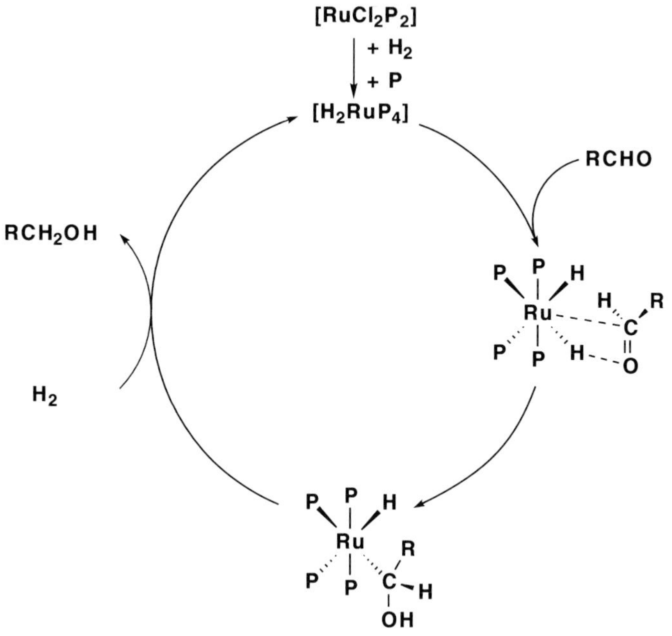

$[RuCl_2P_2]$ → $+ H_2$, $+ P$ → $[H_2RuP_4]$; RCHO; RCH_2OH; H_2

Scheme 7

P = TPPMS, TPPTS

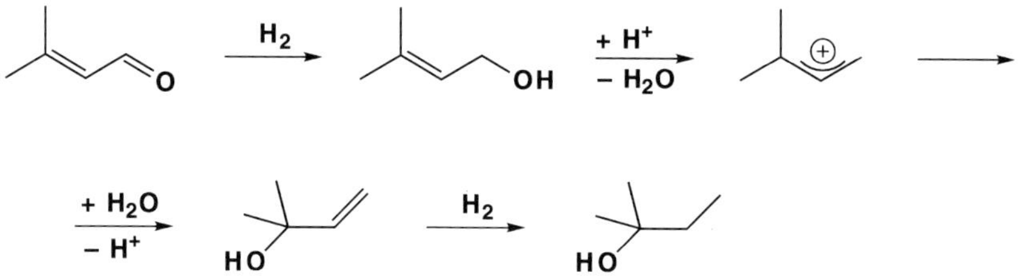

H_2; $+ H^+$, $- H_2O$; OH; $+ H_2O$, $- H^+$; HO; H_2; HO

Scheme 8

Hydrogenation of propionaldehyde, catalyzed by various ruthenium–TPPTS complexes, was dramatically influenced by the addition of certain salts [122, 123]. Whereas in the absence of salts there was no reaction at 35 °C and 50 bar H_2, in the presence of NaI TOFs of more than 2000 h^{-1} were determined. This was lowered to 300 h^{-1} when the sodium cation was selectively sequestered by a cryptand (4,7,13,16,21-pentoxa-1,10-diazabicyclo[5.8.8]tricosane). Obviously, the larger part of the "salt effect" belonged to the cation. It was concluded that electrophilic assistance by Na^+ facilitated C-coordination of the aldehyde and formation of a hydroxyalkyl intermediate.

It is noteworthy that in a homogeneous organic solution, i.e., where the catalyst and the substrate aldehyde are contained in the same phase, pronounced substrate inhibition is observed [124]. When running the reaction in a two-phase aqueous/organic mixture the solubility of aldehyde is not high enough to bring about such an inhibitory effect, and large quantities of the substrate residing in the organic phase as a reservoir can be hydrogenated. Other reactions leading to catalyst deactivation are the formation of 1-hydroxy-alkylphosphonium salts [125] and of areneruthenium(II) complexes [121].

Water-solube (rhodium(I) complexes with TPPTS, TPPMS, and PTA ligands, such as $[RhCl(TPPTS)_3]$, are capable of hydrogenating aldehydes, although their catalytic activity is inferior to the ruthenium complexes discussed above [116]. In sharp contrast to the ruthenium(II)-based catalysts, in reactions of *unsaturated* aldehydes rhodium(I) complexes preferentially promote the reduction of C=C double bonds, although the reactions are not completely selective [31, 113, 116].

In line with the expected low reactivity of osmium(II) complexes [126], $[OsCl_2(TPPMS)_2]$, $[OsH_4(TPPMS)_3]$, and $[OsHCl(CO)(TPPMS)_2]$, hydrogenated cinnamaldehyde at lower rates (and selectivities from 60 to 90%) than the analogous ruthenium(II) catalysts [127].

In general, hydrogenation of ketones is less facile and less selective than the reduction of aldehydes. Of the complexes $[RuCl_2(TPPTS)_2]_2$, $[HRuCl(TPPTS)_3]$, and $[H_2Ru(TPPTS)_4]$, the last one proved to be the most active catalyst in hydrogenation of 2-butanone, cyclohexanone and benzylacetone [115]. Under the same conditions, all three catalysts were rather selective towards the formation of the saturated ketone in hydrogenation of *trans*-4-hexen-3-one (Scheme 9), yielding only 2% to 7% of 3-hexanol. However, addition of strong alkali (LiOH or KOH) promoted reduction of the C=O function as well, probably by facilitating enolization, and as much as 76% of the saturated alcohol was formed.

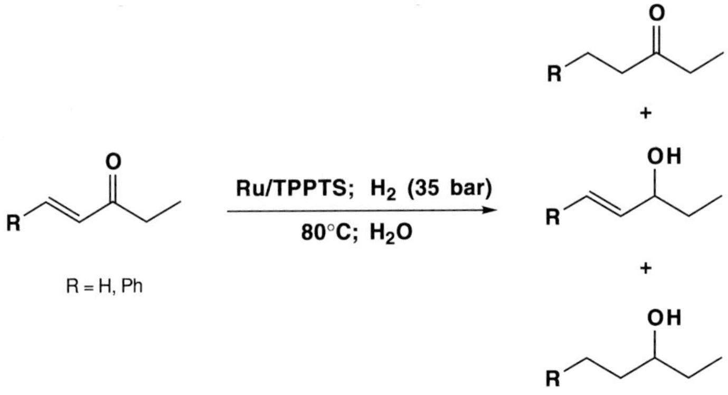

R = H, Ph

Ru/TPPTS; H_2 (35 bar)

80°C; H_2O

Scheme 9

Benzylideneacetone was hydrogenated in biphasic systems using toluene, dichloromethane or tetrahydrofuran as organic solvents, and $[RuCl_2(TPPTS)_2]_2$ and $[H_2Ru(TPPTS)_4]$ as catalysts. Very similar results were obtained with $[HRu(\eta^6\text{-}C_6H_6)(CH_3CN)][BF_4]$ as catalyst for hydrogenation of benzylideneacetone and cyclohex-1-en-2-one [66].

Ketones with functional groups facilitating enolization and/or coordination to the catalyst can be hydrogenated more smoothly. 2-Ketoacids, such as pyruvic and 2-ketoglutaric acids, were reduced by [HRuCl(TPPMS)$_3$], [RhCl(TPPMS)$_3$], and [RhCl(PTA)$_3$] [26, 35, 46], while 4- or 5-ketoacids proved unreactive. More importantly, ethyl and methyl acetoacetate were hydrogenated with Ru(II)/5,5′-disulfonato-BINAP (**31**) and Ru(II)/MeO-BIPHEP-S (**32**) with 91% *ee* and 93% *ee*, respectively [95, 98].

A rhodium complex prepared in situ from [Rh(cod)Cl]$_2$ and (1*R*,2*R*)-N,N′-dimethyldiphenylethylenediamine catalyzed the hydrogenation of methyl phenylglyoxylate with low enantioselectivity in water and in H_2O/MeOH mixtures. However, in water/methanol (30:70) solvent, in the presence of cyclodextrin, 50% *ee* was obtained, which is the same as in methanol itself [128].

Synthetic transformations of carbohydrates have enormous practical value because of their availability in large quantities from renewable sources. The first example of hydrogenation of a carbohydrate (fructose) in aqueous solution, catalyzed by [HRuCl(TPPMS)$_3$] under mild conditions (60 °C, 1 bar total pressure) originates from 1977 [26]. Under the same conditions, the simplest ketose, 1,3-dihydroxyacetone, was rapidly reduced. Efficient hydrogenation of the epimeric aldoses D-glucose and D-mannose was achieved with [HRuCl(TPPTS)$_3$] catalyst at 100 °C and 50 bar H_2 yielding D-sorbitol and D-mannitol (Eq. 34) – D-mannose being more reactive, as expected [129]. Addition of NaI accelerated the reduction of both aldoses by a factor of 6, most probably with the same mechanism as discussed before in connection with hydrogenation of propionaldehyde [123].

CHO / HO–H / HO–H / H–OH / H–OH / CH_2OH $\xrightarrow[100^\circ C,\ H_2\ (50\ bar)]{Ru/TPPTS}$ CH_2OH / HO–H / HO–H / H–OH / H–OH / CH_2OH (34)

Enantioselective hydrogenation of imines in aqueous systems generated much research interest, partly because of the practical value of the product amines, partly due to the unusual kinetic observations. Imines, such as *N*-benzylacetophenoneimine, are relatively stable to hydrolysis, and could be reduced either in a water/ethyl acetate two-phase solvent mixture [93, 130, 131], or in a benzene–AOT–water reverse micellar solution (AOT = bis(2-ethylhexyl) sulfosuccinate). With catalysts, prepared from [Rh(cod)Cl]$_2$ and the products of the stepwise sulfonation of BDPP (**25**–**28**), the highest rate and enantioselectivity were obtained with the monosulfonated ligand (**25**) [93, 130, 131]; the effect is very pronounced. It should be noted that the Rh/**25** catalyst is insoluble in water and during catalysis it moves to the organic phase, where it is supposed to have the structure shown in Eq. (35).

In the same reaction, the [Rh(NBD)(BDPP)]$^+$ catalyst showed an enantioselectivity (*ee*) of 68% both in neat methanol and in benzene. Formation of reverse micelles by AOT (0.05 M) and varying amounts of water in the benzene cat.

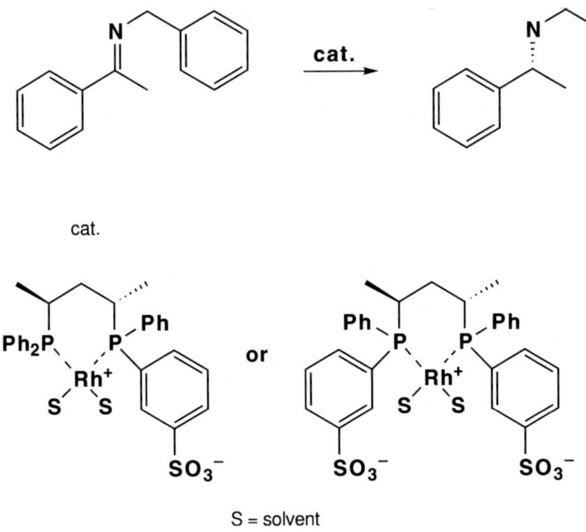

S = solvent

(35)

solution increased the selectivity to 82% *ee* (at $w = [H_2O]/[AOT] = 5$); addition of further increments of H_2O resulted in a decrease of the enantiomeric excess [132]. Note that the 14% increase in the *ee* was induced by an *achiral* additive; this implies changes in the reaction mechanism rather than in the structure of the catalytically active species (about the background cf. [132]). The high enantioselectivity, achieved with the Rh(I)/monosulfonated BDPP catalyst (with no external RSO_3) in the hydrogenation of itaconic acid and imines [93, 130, 131], may be the result of a similar coordination of the sulfonate group of the phosphine itself – a warning to bear in mind the possible effects of $-SO_3$ coordination when using sulfonated ligands.

6.2.3.3 Hydrogenolysis of C–O, C–N, C–S, and C–Halogen Bonds

Various water-soluble palladium complexes are active in the hydrogenolysis of C–O and C–N bonds, involving allylic substrates either with molecular H_2 or with other hydrogen sources such as formates or amines. Hydrogenolysis of allyl acetate with $[PdCl_2L_2]$ (L = **1**, **5**, **10**) catalyst proceeded smoothly in a heptane/aqueous sodium formate two-phase mixture at 80 °C yielding propene [133]. The catalyst, formed in situ from $[Pd(OAc)_2]$ and TPPTS, could be used for the selective removal of allylic protecting groups. With careful choice of the amount of the catalysts and of the reaction conditions, distinction between such closely related protecting groups as dimethylallylcarbamates and allyloxycarbonates was achieved. Conversely, allyloxycarbamates could be selectively deprotected with the same catalyst system in the presence of a dimethylallyl carboxylate group [134, 135].

Asymmetric hydrogenolysis of sodium *cis*-epoxysuccinate leads to malic acid derivatives which are useful building blocks in natural product synthesis. This reaction was catalyzed by $[Rh(cod)Cl]_2$ + chiral sulfonated BDPP ligands from the stepwise sulfonation of BDPP (Eq. 36) [136]. At 20 °C and 70 bar H_2 pressure the chemical yields exceeded 90% and the *ee* varied between 34% ($BDPP_{MS}$, **25**) and 26% ($BDPP_{TS}$, **28**). In contrast to the hydrogenation of prochiral imines the enantioselectivity of the hydrogenolysis of *cis*-epoxysuccinate decreased monotonically with the increasing number of sulfonate groups in the ligands. This may be the consequence of the good solubility of the ligand in the aqueous phase where the catalysis takes place. In a closely related reaction, racemic sodium *trans*-phenylglycidate was hydrogenolyzed with kinetic resolution. Using (*S,S*-$BDPP_{TS}$ as ligand, the hydrogenolysis of the (2*R*,3*S*)-epoxide enantiomer was preferred, yielding a product mixture rich in (2*R*)-2-hydroxy-3-phenylpropionate.

$$NaO_2C-\overset{O}{\triangle}-CO_2Me \xrightarrow[\substack{20°C,\ H_2,\ 70\ bar,\ 6\ h \\ H_2O\ or\ H_2O/EtOAc \\ L = 24 - 28}]{[Rh(cod)Cl]_2\ +\ L} NaO_2C-CH(H)-CH(OH)-CO_2Me \quad (36)$$

yield > 90%
ee 27 – 41%

A rare example of C–O bond scission in aqueous solution is the deoxygenation of allylic alcohols catalyzed by $[HCo(CN)_5]^{3-}$ [137]. When the two-phase reaction was carried out in the presence of *β*-cylodextrin it yielded *trans*-alkenes selectively whereas, with no cyclodextrin added, a mixture of cis- and *trans*-alkenes was formed. In a typical example, reaction of 2-hexen-1-ol yielded 91% *trans*-2-hexene and 4% 1-hexene.

The removal of sulfur from petroleum is commonly achieved by hydrogenation on heterogeneous catalysts (hydrodesulfurization, or HDS process). For biphasic processes cf. Section 6.14 [138, 139]).

Hydrogenolysis of the C–halogen bond is an important reaction both from preparative and from environmental points of view. $[HCo(CN)_5]^{3-}$ was studied in detail as a catalyst for reductive dehalogenations of organic halides, which proceed according to Eqs. (37) and (38). The results of the early experiments are summarized in [12].

$$[Co(CN)_5]^{3-} + RX \rightarrow [Co(CN)_5X]^{3-} + R^{\cdot} \quad (37)$$

$$[HCo(CN)_5]^{3-} + R^{\cdot} \rightarrow [Co(CN)_5]^{3-} + RH \quad (38)$$

Allylic chlorides can be dehalogenated with aqueous sodium formate in a heptane/water solvent mixture with $[PdCl_2L_2]$ catalysts (L = various sulfonated phosphines) [133]. $[RuCl_2(TPPMS)_2]_2$ and $[Ru(H_2O)_3(PTA)_3]^{2+}$ proved effective catalysts for hydrogenolysis of a variety of organic halides [140], including

CCl_4, $CHCl_3$, 1-hexyl and cyclohexyl halides and benzyl chloride by hydrogen transfer from aqueous sodium formate at 80 °C (Eq. 39). Initial turnover frequencies as high as 1000 h^{-1} were determined, the catalysts showing activity far superior to the analogous system [141] employing $[RuCl_2(PPh_3)_3]$. Under the same conditions, chlorobenzene was completely unreactive (cf. Section 6.16).

$$CCl_4 + HCOO^- + OH^- \rightarrow CHCl_3 + Cl^- + HCO_3^- \tag{39}$$

6.2.3.4 Miscellaneous Hydrogenations

Hydrogenation of nitro compounds can be achieved with $[HCo(CN)_5]^{3-}$ as catalyst; however, in many cases (especially with nitroarenes), products of reductive dimerization, i.e., azo and hydrazo compounds, are formed instead of the expected amino derivatives [12]. Ketoximes and oximes of 2-oxo acids are hydrogenated to amines [142]. 2-Amino acids can be prepared in high yields by reductive amination of 2-oxo acids in an aqueous NH_3 solution (Eq. 40) [143].

$$C_6H_5-CH_2C(=O)COOH \xrightarrow[40^\circ C,\ 6\%\ aq.\ NH_3]{CoCl_2,\ KCN,\ 70\ bar\ H_2} C_6H_5-CH_2CH(NH_2)COOH \tag{40}$$

Anilines can be obtained from nitroarenes under water-gas shift (WGS) conditions (Eq. 41).

$$PhNO_2 + 3CO + H_2O \rightarrow PhNH_2 + 3\,CO_2 \tag{41}$$

In most cases such reactions are conducted in strongly alkaline aqueous solutions [144] or in the presence of amines [145, 146]. $[Rh_6(CO)_{16}]$, $[Rh_{12}(CO)_{30}]^{2-}$ and $[Rh_5(CO)_{15}]^-$, formed in situ in the reaction mixture, are among the most active catalysts [147]. However, addition of a strong base to solutions of $[Rh(CO)_4]^-$ (with K^+, Cs^+, or $(PPh_3)_2N^+$ cation) diminishes the catalytic activity in the hydrogenation of nitrobenzene to aniline with $CO + H_2O$ [148]. Recently it was disclosed that the catalysts prepared from $PdCl_2$ and TPPTS or multiply-sulfonated BINAP (BINAS, **38**) catalyzed the selective reduction of nitroarenes to anilines at 100 °C under 120 bar CO [149] (cf. Section 6.3).

A similarly selective reduction of nitroarenes was achieved under WGS conditions by using $[Ru_3(CO)_{12}]$ in the presence of certain amines, such as diisopropylamine, piperidine, dibutylamine, and triethylamine [146]. Importantly, the latter reaction yielded no unwanted H_2 as by-product from a concomitant WGS reaction.

Hydrogenation of carbon dioxide can be achieved in aqueous solutions, too [150] (cf. Section 6.15).

References

[1] R. Noyori, S. Hashiguchi, in *Applied Homogeneous Catalysis with Organometallic Compounds* (Eds.: B. Cornils, W. A. Herrmann), VCH, Weinheim, **1996**, Chapter 2.9.
[2] Y. Sun, R. N. Landau, J. Wang, C. LeBlond, D. G. Blackmond, *J. Am. Chem. Soc.* **1996**, *118*, 1348.
[3] F. Joó, Z. Tóth, *J. Mol. Catal.* **1980**, *8*, 369.
[4] W. A. Herrmann, C. W. Kohlpaintner, *Angew. Chem.* **1993**, *105*, 1588; *Angew. Chem., Int. Ed. Engl.* **1993**, *32*, 1524.
[5] D. Sinou, *Bull. Soc. Chim. Fr.* **1987**, 480.
[6] P. Kalck, F. Monteil, *Adv. Organomet. Chem.* **1992**, *34*, 219.
[7] F. Joó, Á. Kathó, *J. Mol. Catal. A: Chemical* **1997**, *116*, 3.
[8] G. Papadogianakis, R. A. Sheldon, *New J. Chem.* **1996**, *20*, 175.
[9] *Aqueous Organometallic Chemistry and Catalysis* (Eds.: I. T. Horváth, F. Joó), NATO ASI Ser. 3/5, Kluwer, Dordrecht, **1995**.
[10] F. Joó, L. Vígh, in *Handbook of Nonmedical Application of Liposomes* (Eds.: Y. Barenholz, D. Lasic), CRC Press, Orlando, Fl, **1995**, 257.
[11] P. A. Chaloner, M. A. Esteruelas, F. Joó, L. A. Oro, *Homogeneous Hydrogenation*, Kluwer, Dordrecht, **1994**.
[12] B. R. James, *Homogeneous Hydrogenation*, Wiley, New York, **1973**.
[13] D. M. Heinekey, W. J. Oldham, Jr., *Chem. Rev.* **1993**, *93*, 913.
[14] A. F. Borowski, S. Sabo-Etienne, M. L. Christ, B. Donnadieu, B. Chaudret, *Organometallics* **1996**, *15*, 1427.
[15] C. Bianchini, C. Mealli, A. Meli, M. Peruzzini, F. Zanobini, *J. Am. Chem. Soc.* **1988**, *110*, 8725.
[16] Z. W. Li, H. Taube, *J. Am. Chem. Soc.* **1991**, *113*, 8946.
[17] G. J. Kubas, C. J. Burns, G. R. K. Khalsa, L. S. van der Sluys, G. Kiss, C. D. Hoff, *Organometallics* **1992**, *11*, 3390.
[18] R. T. Thauer, A. R. Klein, G. C. Hartmann, *Chem. Rev.* **1996**, *96*, 3031.
[19] M. Zimmer, G. Schulte, X. L. Luo, R. H. Crabtree, *Angew. Chem.* **1991**, *103*, 205; *Angew. Chem., Int. Ed. Engl.* **1991**, *30*, 193.
[20] C. Larpent, H. Patin, *J. Organomet. Chem.* **1987**, *335*, C13.
[21] P. J. Roman, Jr., D. P. Paterniti, R. F. See, M. R. Churchill, J. D. Atwood, *Organometallics* **1997**, *16*, 1484.
[22] E. Fache, C. Santini, F. Senocq, J. M. Basset, *J. Mol. Catal.* **1992**, *72*, 331.
[23] W. A. Herrmann, J. A. Kulpe, *J. Organometal. Chem.* **1990**, *389*, 85.
[24] D. R. Tyler, in *Aqueous Organometallic Chemistry and Catalysis* (Eds.: I. T. Horváth, F. Joó), NATO ASI Ser. 3/5, Kluwer, Dordrecht, **1995**, 47.
[25] A. F. Borowski, D. J. Cole-Hamilton, G. Wilkinson, *Nouv. J. Chim.* **1978**, *2*, 137.
[26] Z. Tóth, F. Joó, M. T. Beck, *Inorg. Chim. A* **1980**, *42*, 153.
[27] Y. Amrani, L. Lecomte, D. Sinou, J. Bakos, I. Tóth, B. Heil, *Organometallics* **1989**, *8*, 542.
[28] H. Sertchook, D. Avnir, J. Blum, F. Joó, Á. Kathó, H. Schumann, R. Weimann, S. Wernik, *J. Mol. Catal. A: Chem.* **1996**, *108*, 153.
[29] F. Joó, P. Csiba, A. Bényei, *J. Chem. Soc., Chem. Commun.* **1993**, 1602.
[30] C. Larpent, R. Dabard, H. Patin, *Inorg. Chem.* **1987**, *26*, 2922.
[31] D. J. Darensbourg, N. W. Stafford, F. Joó, J. H. Reibenspies, *J. Organomet. Chem.* **1995**, *488*, 99.
[32] A. S. Berenblyum, T. V. Turkova, I. I. Moiseev, *Izv. AN SSSR. Ser. Khim.* **1980**, 2153.
[33] C. Larpent, R. Dabard, H. Patin, *Tetrahedron Lett.* **1987**, *28*, 2507.

[34] C. Larpent, E. Bernard, F. Brisse-le Menn, H. Patin, *J. Mol. Catal. A: Chem.* **1997**, *116*, 277.
[35] F. Joó, L. Somsák, M. T. Beck, *J. Mol. Catal.* **1984**, *24*, 71.
[36] S. S. Kristjánsdóttir, J. R. Norton, in *Transition Metal Hydrides* (Ed.: A. Dedieu) VCH, New York, **1992**, Chapter 9.
[37] F. Joó, J. Kovács, A. C. Bényei, Á. Kathó, *Angew. Chem.* **1998**, *110* (7) (in print); *Angew. Chem. Int. Ed. Engl.* **1998**, *37* (7) (in print).
[38] J. Halpern, L.-H. Wong, *J. Am. Chem. Soc.* **1968**, *90*, 6665.
[39] A. S. Hussey, Y. Takeuchi, *J. Org. Chem.* **1970**, *35*, 643.
[40] D. P. Paterniti, P. J. Roman, Jr., J. D. Atwood, *J. Chem. Soc., Chem. Commun.* **1996**, 2659.
[41] W. D. Harman, H. Taube, *J. Am. Chem. Soc.* **1990**, *112*, 2261.
[42] J. A. Osborn, T. H. Jardine, J. F. Young, G. Wilkinson, *J. Chem. Soc. A* **1966**, 1711.
[43] A. Bényei, J. N. W. Stafford, Á. Kathó, D. J. Darensbourg, F. Joó, *J. Mol. Catal.* **1993**, *84*, 157.
[44] C. Larpent, H. Patin, *Tetrahedron* **1988**, *44*, 6107.
[45] F. Joó, E. Trócsányi, *J. Organomet. Chem.* **1982**, *231*, 63.
[46] F. Joó, L. Nádasdi, A. Cs. Bényei, D. J. Darensbourg, *J. Organomet. Chem.* **1996**, *512*, 45.
[47] R. T. Smith, M. C. Baird, *Transition Metal. Chem.* **1981**, *6*, 197.
[48] E. Renaud, M. C. Baird, *J. Chem. Soc., Dalton Trans.* **1992**, 2905.
[49] P. A. T. Hoye, P. G. Pringle, M. B. Smith, K. Worboys, *J. Chem. Soc., Dalton Trans.* **1993**, 269.
[50] G. T. Baxley, T. J. R. Weakley, W. K. Miller, D. K. Lyon, D. R. Tyler, *J. Mol. Catal. A: Chem.* **1997**, *116*, 191.
[51] R. G. Nuzzo, D. Feitler, G. M. Whitesides, *J. Am. Chem. Soc.* **1979**, *101*, 3683.
[52] D. C. Mudalige, G. L. Rempel, *J. Mol. Catal. A: Chem.* **1997**, *116*, 309.
[53] D. E. Bergbreiter, L. Zhang, V. M. Mariagananam, *J. Am. Chem. Soc.* **1993**, *115*, 9295.
[54] A. Andriollo, A. Bolivar, F. A. López, D. E. Páez, *Inorg. Chim. A* **1995**, *238*, 187.
[55] T. Malström, H. Weigl, C. Andersson, *Organometallics* **1995**, *14*, 2593.
[56] T. Malström, C. Andersson, *J. Mol. Catal. A: Chem.* **1997**, *116*, 237.
[57] T. Malström, C. Andersson, *J. Chem. Soc., Chem. Commun.* **1996**, 1135.
[58] D. E. Bergbreiter, Y.-S. Liu, *Tetrahedron Lett.* **1997**, *38*, 3703.
[59] J. F. Harrod, S. Ciccone, J. Halpern, *Can. J. Chem.* **1961**, *39*, 1372.
[60] B. R. James, G. L. Rempel, *Can. J. Chem.* **1966**, *44*, 233.
[61] P. Frediani, M. Bianchi, A. Salvini, R. Guarducci, L. C. Carluccio, F. Piacenti, *J. Organomet. Chem.* **1995**, *498*, 187.
[62] C. Masters, *Homogeneous Transition-metal Catalysis – A Gentle Art,* Chapman and Hall, London, **1981**.
[63] J. Kwiatek, *Catal. Rev.* **1967**, *1*, 37.
[64] G. Süss-Fink, A. Meister, G. Meister, *Coord. Chem. Rev.* **1995**, *143*, 97.
[65] G. Meister, G. Rheinwald, H. Stoeckli-Evans, G. Süss-Fink, *J. Chem. Soc., Dalton Trans.* **1994**, 3215.
[66] W.-C. Chan, C.-P. Lau, L. Cheng, Y.-S. Leung, *J. Organomet. Chem.* **1994**, *464*, 103.
[67] E. G. Chepaikin, M. L. Khidekel', *J. Mol. Catal.* **1978**, *4*, 103.
[68] Yu. A. Sakharovsky, M. B. Rosenkevich, A. S. Lobach, E. G. Chepaikin, M. L. Khidekel', *React. Kinet. Catal. Lett.* **1978**, *2*, 249.
[69] A. V. Bulatov, E. N. Izakovich, L. N. Karklin', M. L. Khidekel', *Izv. AN SSSR, Ser. Khim.* **1981**, 2032.
[70] A. V. Bulatov, G. P. Voskerchyan, S. N. Dobryakov, A. T. Nikitaev, *Bull. Acad. Sci. USSR, Chem. Sci.* **1986**, *35*, 747.
[71] F. Joó, N. Balogh, L. I. Horváth, G. Filep, I. Horváth, L. Vígh, *Anal. Biochem.* **1991**, *194*, 34.

[72] J. T. Lee, H. Alper, *J. Org. Chem.* **1990**, *55*, 1854.
[73] J. T. Lee, H. Alper, *Tetrahedron Lett.* **1990**, *31*, 1941.
[74] K. Ohkubo, T. Kawabe, K. Yamashita, S. Sakaki, *J. Mol. Catal.* **1984**, *24*, 83.
[75] M. T. Reetz, S. R. Waldvogel, *Angew. Chem.* **1997**, *109*, 870; *Angew. Chem., Int. Ed. Engl.* **1997**, *36*, 865.
[76] Johnson–Matthey Inc. (B. A. Murrer, J. W. Jenkins,) GB 2.070.023 (1981).
[77] L. Vígh, D. Los, I. Horváth, N. Murata, *Proc. Natl. Acad. Sci. USA* **1993**, *90*, 9090.
[78] L. Vígh, F. Joó, in *Applied Homogeneous Catalysis with Organometallic Compounds* (Eds.: B. Cornils, W. A. Herrmann), VCH, Weinheim, **1996**, Chapter 3.3.10.2.
[79] F. Joó, F. Chevy, O. Colard, C. Wolf, *Biochim. Biophys. A* **1993**, *1149*, 231.
[80] P. J. Quinn, F. Joó, L. Vígh, *Prog. Biophys. Molec. Biol.* **1989**, *53*, 71.
[81] Z. Török, B. Szalontai, F. Joó, C. A. Wistrom, L. Vígh, *Biochem. Biophys. Res. Commun.* **1993**, *192*, 518.
[82] B. Szalontai, F. Joó, É. Papp, L. Vígh, *J. Chem. Soc., Chem. Commun.* **1995**, 2299.
[83] F. Joó, É. Papp, Á. Kathó, *Topics in Catalysis* **1998**, *5* (in press).
[84] Á. Bucsai, J. Bakos, M. Laghmari, D. Sinou, *J. Mol. Catal. A: Chem.* **1997**, *116*, 335.
[85] D. J. Hardick, I. S. Blagbrough, B. L. V. Potter, *J. Am. Chem. Soc.* **1996**, *118*, 5897.
[86] T. X. Le, J. S. Merola, *Organometallics* **1993**, *12*, 3798.
[87] J. S. Merola, M. A. Franks, P. Chirik, P. Rafford, *209th Meet. Am. Chem. Soc., Anaheim, CA* **1995**, INOR 416.
[88] E. Paetzold, A. Kinting, G. Oehme, *J. Prakt. Chem.* **1987**, *329*, 725.
[89] M. Laghmari, D. Sinou, A. Masdeu, C. Claver, *J. Organomet. Chem.* **1992**, *438*, 213.
[90] J. Bakos, R. Karaivanov, M. Laghmari, D. Sinou, *Organometallics* **1994**, *13*, 2951.
[91] U. Nagel, E. Kinzel, *Chem. Ber.* **1986**, *119*, 1731.
[92] I. Tóth, B. E. Hanson, M. E. Davis, *Tetrahedron: Asymm.* **1990**, *1*, 913.
[93] C. Lensink, E. Rijnberg, J. G. de Vries, *J. Mol. Catal. A: Chem.* **1997**, *116*, 199.
[94] I. Tóth, B. E. Hanson, *Tetrahedron: Asymm.* **1990**, *1*, 895.
[95] Takasago Inc. (T. Ishizaki, H. Kumobayashi), EP Appl. 544.455 (1993); *Chem. Abstr. 119*, 181016d.
[96] K. Wan, M. E. Davis, *J. Chem. Soc., Chem. Commun.* **1993**, 1662.
[97] K. Wan, M. E. Davis, *Tetrahedron: Asymm.* **1993**, *4*, 2461.
[98] R. Schmid, E. A. Broger, M. Cereghetti, Y. Crameri, J. Foricher, M. Lalonde, R. K. Müller, M. Scalone, G. Schoettel, U. Zutter, *Pure Appl. Chem.* **1996**, *68*, 131.
[99] Hoffmann-La Roche AG (M. Lalonde, R. Schmid), EP Appl. 667.530, *Chem. Abstr. 124*, 8995c.
[100] J. Halpern, D. P. Riley, A. S. C. Chan, J. J. Pluth, *J. Am. Chem. Soc.* **1977**, *99*, 8055.
[101] J. Halpern, in *Asymmetric Synthesis* (Ed. J. D. Morrison), Academic Press, London, **1985**, *5*, 41.
[102] W. F. Linke, A. Seidell, *Solubilities of Inorganic and Metal-organic Compounds*, American Chemical Society, Washington, DC, **1958**.
[103] Y. Sun, C. LeBlond, J. Wang, D. G. Blackmond, *J. Am. Chem. Soc.* **1995**, *117*, 12647.
[104] M. H. Abraham, P. L. Grellier, R. A. McGill, *J. Chem. Soc., Perkin Trans. 2* **1988**, 339.
[105] L. Lecomte, D. Sinou, J. Bakos, I. Tóth, B. Heil, *J. Organomet. Chem.* **1989**, *370*, 277.
[106] A. Kumar, G. Oehme, J. P. Roque, M. Schwarze, R. Selke, *Angew. Chem.* **1997**, *106*, 106; *Angew. Chem., Int. Ed. Engl.* **1994**, *33*, 2197.
[107] R. Selke, M. Ohff, A. Riepe, *Tetrahedron* **1996**, *48*, 15079.
[108] I. Grassert, V. Vill, G. Oehme, *J. Mol. Catal. A: Chem.* **1997**, *116*, 231.
[109] H. Ding, B. E. Hanson, J. Bakos, *Angew. Chem.* **1995**, *107*, 1728; *Angew. Chem., Int. Ed. Engl.* **1995**, *34*, 1645.
[110] B. R. James, R. H. Morris, *J. Chem. Soc., Chem. Commun.* **1978**, 929.

[111] E. Farnetti, M. Pesce, J. Kaspar, R. Spogliarich, M. Graziani, *J. Mol. Catal.* **1987**, *43*, 35.
[112] J. M. Grosselin, C. Mercier, *J. Mol. Catal.* **1990**, *63*, L25.
[113] J. M. Grosselin, C. Mercier, G. Allmang, F. Grass, *Organometallics* **1991**, *10*, 2126.
[114] C. Mercier, P. Chabardes, *Pure Appl. Chem.* **1994**, *66*, 1509.
[115] M. Hernandez, P. Kalck, *J. Mol. Catal. A: Chem.* **1997**, *116*, 131.
[116] D. J. Darensbourg, F. Joó, M. Kannisto, Á. Kathó, *Organometallics* **1992**, *11*, 1990.
[117] D. J. Darensbourg, F. Joó, M. Kannisto, Á. Kathó, J. H. Reibenspies, D. J. Daigle, *Inorg. Chem.* **1994**, *33*, 200.
[118] A. Andriollo, J. Carrasquel, J. Mariño, F. A. López, D. E. Páez, I. Rojas, N. Valencia, *J. Mol. Catal. A: Chem.* **1997**, *116*, 157.
[119] F. Joó, A. Bényei, *J. Organomet. Chem.* **1989**, *363*, C19.
[120] A. Bényei, F. Joó, *J. Mol. Catal.* **1990**, *58*, 151.
[121] M. Hernandez, P. Kalck, *J. Mol. Catal. A: Chem.* **1997**, *116*, 117.
[122] E. Fache, F. Senocq, C. Santini, J. M. Basset, *J. Chem. Soc., Chem. Commun.* **1990**, 1776.
[123] E. Fache, C. Santini, F. Senocq, J. M. Basset, *J. Mol. Catal.* **1992**, *72*, 337.
[124] R. Bar, L. K. Bar, Y. Sasson, J. Blum, *J. Mol. Catal.* **1985**, *33*, 161.
[125] D. J. Darensbourg, F. Joó, Á. Kathó, J. N. W. Stafford, A. Bényei, J. H. Reibenspies, D. J. Daigle, *Inorg. Chem.* **1994**, *33*, 175.
[126] R. A. Sánchez-Delgado, M. Rosales, M. A. Esteruelas, L. A. Oro, *J. Mol. Catal A: Chem.* **1995**, *96*, 231.
[127] R. A. Sánchez-Delgado, M. Medina, F. López-Linares, A. Fuentes, *J. Mol. Catal. A: Chem.* **1997**, *116*, 167.
[128] C. Pinel, N. Gendreau-Diaz, M. Lemaire, *Chem. Ind.* **1996**, *68*, 385.
[129] S. Kolarić, V. Šunjić, *J. Mol. Catal. A: Chem.* **1996**, *110*, 189.
[130] J. Bakos, Á. Orosz, B. Heil, M. Laghmari, P. Lhoste, D. Sinou, *J. Chem. Soc., Chem. Commun.* **1991**, 1684.
[131] C. Lensink, J. G. de Vries, *Tetrahedron: Asymm.* **1992**, *3*, 235.
[132] J. M. Buriak, J. A. Osborn, *Organometallics* **1996**, *15*, 3161.
[133] T. Okano, I. Moriyama, H. Konishi, J. Kiji, *Chem. Lett.* **1986**, 1463.
[134] J.-P. Genêt, A. Linquist, E. Blart, V. Mouries, M. Savignac, M. Vaultier, *Tetrahedron Lett.* **1995**, *36*, 1443.
[135] S. Lemaire-Audoire, M. Savignac, G. Pourcelot, J.-P. Genêt, J.-M. Bernard, *J. Mol. Catal. A: Chem.* **1997**, *116*, 247.
[136] J. Bakos, Á. Orosz, S. Cserépi, I. Tóth, D. Sinou, *J. Mol. Catal. A: Chem.* **1997**, *116*, 85.
[137] J. T. Lee, H. Alper, *Tetrahedron Lett.* **1990**, *31*, 4101.
[138] C. Bianchini, P. Frediani, V. Sernau, *Organometallics* **1995**, *14*, 5458.
[139] C. Bianchini, A. Meli, V. Patinec, V. Sernau, F. Vizza, *J. Am. Chem. Soc.* **1997**, *119*, 4945.
[140] A. Cs. Bényei, Sz. Lehel, F. Joó, *J. Mol. Catal. A: Chem.* **1997**, *116*, 349.
[141] S. Xie, E. M. Georgiev, D. M. Roundhill, K. Troev, *J. Organomet. Chem.* **1994**, *482*, 39.
[142] J. Kwiatek, I. L. Mador, J. K. Seyler, *Adv. Chem. Ser.* **1963**, *37*, 201.
[143] A. J. Birch, D. H. Williamson, *Org. React.* **1977**, *24*, 1.
[144] R. M. Laine, E. J. Crawford, *J. Mol. Catal.* **1988**, *44*, 357.
[145] K. Kaneda, M. Yasumura, T. Imanaka, S. Teranishi, *J. Chem. Soc., Chem. Commun.* **1982**, 935.
[146] K. Nomura, *J. Mol. Catal. A: Chem.* **1995**, *95*, 203.
[147] F. Joó, H. Alper, *Can. J. Chem.* **1985**, *63*, 1157.
[148] F. Ragaini, S. Cenini, *J. Mol. Catal. A: Chem.* **1996**, *105*, 145.
[149] A. M. Tafesh, M. Beller, *Tetrahedron Lett.* **1995**, *36*, 9305.
[150] W. Leitner, *Coord. Chem. Rev.* **1996**, *153*, 257.

6.3 Carbonylation Reactions

Matthias Beller, Jürgen G. E. Krauter

6.3.1 Introduction

Apart from hydroformylation, the potential advantages of two-phase catalysis for other carbonylation reactions has not been thoroughly evaluated. Despite considerable industrial interest in synthesis using carbon monoxide as low-cost feedstock, so far only a few examples of carbonylation reaction under biphasic conditions have been described. Examples include reductive carbonylations; carbonylation of aryl-, benzyl- and allyl-X (X = Br, Cl); and very recently hydrocarboxylations. The main part of this section focuses on synthetic developments of new two-phase processes using water-soluble catalyst systems. Other biphasic reactions using a hydrophilic solvent, e.g., water, only as reactand or using phase-transfer conditions [1] are not recognized as biphase catalysis, and therefore are only treated in special cases.

6.3.2 Reductive Carbonylations

Metal-catalyzed reductive carbonylation of nitro aromatics using carbon monoxide has been the subject of intensive investigation in recent years because of the commercial importance of amines, urethanes, and isocyanates [2]. However, catalyst efficiency is still an unsolved problem for industrial applications. In this respect a biphasic reaction medium could offer interesting possibilities regarding the ease of catalyst recycling. Thus, palladium catalysts have been applied in the presence of water-soluble ligands such as TPPTS (sodium *m*-trisulfonated triphenylphosphine) and BINAS (**1**) for the carbonylation of substituted nitro aromatics (Eq. 1). Interestingly, the nitro group can be selectively reduced to an amino group, even in the presence of halide substituents or a vinyl group [3]. Thus, *m*-nitrostyrene yielded 3-aminostyrene in 50 % isolated yield.

The reduction of nitrobenzene to aniline using water as solvent without the addition of any base or ligand is also reported with $[Rh(CO)_4]^-$ as catalyst [4] (Eq. 2). Best turnover numbers (TON up to 1000) were achieved using $PPN^+[Rh(CO)_4]^-$ ($PPN^+ = (Ph_3P)_3N^+$). In general, the conversion is increased by the addition of tetraalkylammonium salts, such as

BINAS, $PdCl_2$; H_2O, Xylol; 120 bar, 100°C

R = $CH=CH_2$, $COCH_3$, CN, Cl

50–85% 5–8%

BINAS (1)

$PAr_{2-n}Ph_n$

NaO_3S

Ar = m–C_6H_4–SO_3Na

(1)

$[Et_3NCH_2Ph]^+Cl^-$, which act as phase-transfer agent, whereas no addition of ammonium salt is necessary if the PPN^+ catalyst is used. From this observation it is deduced that the catalysis takes place in the organic phase. Recycling of the catalyst by simple phase separation is *not* possible because the catalyst is dissolved after the reaction partly in the water and partly in the organic phase.

NO_2 → NH_2; $[Rh(CO)_4]^-$ / CO, H_2O; 200°C, 1.5 h, 40–80 bar (2)

Another reductive carbonylation in biphasic media was developed by Sheldon and co-workers [5]. While investigating the carbonylation of 5-hydroxymethylfurfural it was found that the benzylic hydroxy group can be reduced to give the corresponding methyl group (Eq. 3). The homogeneously catalyzed deoxygenation reaction takes place in acidic aqueous medium in the presence of a water-soluble palladium/TPPTS catalyst and strong coordinating anions (see also Table 1, below).

HOH_2C–(furan)–CHO + CO → (Pd^{II} / TPPTS; HI / H_2O) H_3C–(furan)–CHO + CO_2 (3)

6.3.3 Carboxylation of C–X Derivatives

The hydroxycarbonylations (carboxylations) of alkyl, aryl, benzyl and allyl halides are from a retrosynthetic and mechanistic standpoint closely related. This type of reaction is widely used in organic synthesis [6], although a stoichiometric amount of salt by-product makes these methods less attractive on a large scale. The use of water-soluble catalysts for carbonylation of organic halides was scarcely studied in the past. Up to now palladium, cobalt, and nickel compounds in combination with water-soluble ligands have been used as catalysts for various carboxylations.

Beletskaya and co-workers have shown that the reaction is possible in neat water as solvent. Thus, aryl iodides have been carbonylated with various palladium salts lacking phosphine ligands as depicted in Eq. (4) [7]. Although this reaction is not a truely biphasic process the results are remarkable regarding catalyst efficiency. Thus, a maximum turnover number (TON) of 100 000 was described (R = *p*-COOH, quantitative yield after 6 days). Quite different is the performance of a water-soluble palladium phosphine catalyst described by Kalck et al. [8]. The hydrocarboxylation of the less activated bromobenzene with either $Pd(TPPTS)_3$ or a mixture of $Pd(OAc)_2$ and TPPTS proceeds only sluggishly (turnover frequency TOF $<10\ h^{-1}$). In order to prevent decomposition of palladium an excess of phosphine has to be used. At least 15 equiv. of ligand is necessary to prevent formation of metallic palladium. Because of rapid oxidation of the ligand the re-use of the water phase is not possible.

$$R\text{-}C_6H_4\text{-}I + CO \xrightarrow[\text{1 bar, 80°C}]{Pd(OAc)_2,\ H_2O,\ \text{base}} R\text{-}C_6H_4\text{-}COOH \qquad (4)$$

R e.g.: H, *p*-NO_2, *p*-MeO, *p*-NH_2, *p*-OH

Apart from aryl–X derivatives, and more interestingly from an industrial point of view, metal-catalzed carbonylations of substituted benzyl halides to give the corresponding phenylacetic acids were investigated [9]. Two-phase systems are applied with the catalyst and substrate being dissolved in the organic phase and the product formed is dissolved in an excess of alkaline aqueous solution. Despite significant disadvantages such as indispensable addition of phase-transfer agents and additional salt as by-product, the carbonylation of benzyl chloride to give phenylacetic acid for use in perfume constituents and pesticides has been reported to be practiced on a commercial scale by Mortedison [10]. The conversion takes place in the presence of 5–10 mol% of $Co_2(CO)_8$ and a benzyltrialkylammonium surfactant in a biphasic medium employing diphenyl ether and aqueous 40% NaOH as solvents (Scheme 1).

Other metal catalysts which have been utilized for biphasic carbonylation of benzylic halides to carboxylic acids under phase-transfer conditions, besides

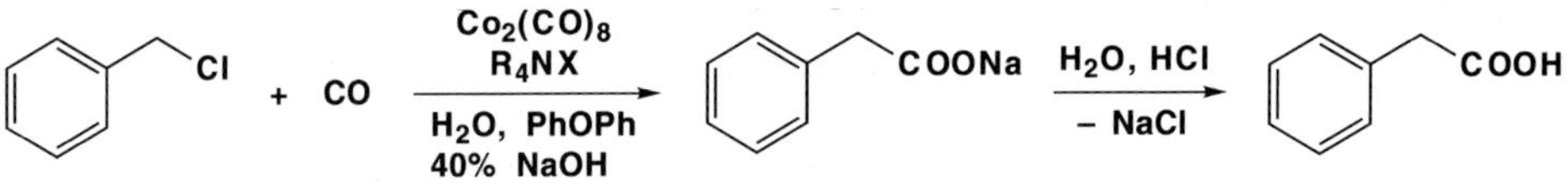

$C_6H_5CH_2Cl + CO \xrightarrow[\text{H}_2\text{O, PhOPh, 40\% NaOH}]{\text{Co}_2\text{(CO)}_8,\ \text{R}_4\text{NX}} C_6H_5CH_2COONa \xrightarrow[-\text{NaCl}]{\text{H}_2\text{O, HCl}} C_6H_5CH_2COOH$

Scheme 1. Carboxylation of benzyl chloride.

cobalt carbonyl [11], include palladium(0) complexes [12] and water-soluble nickel cyanide complexes [13]. Although not investigated in detail, it must be assumed that catalysis takes place in all these reactions in the organic phase.

However, by the use of the water-soluble ligand $Ph_2P(m\text{-}C_6H_4SO_3Na)$ (TPPMS), a palladium catalyst which is active in the water phase is formed [14]. Nevertheless, the addition of surfactants such as $n\text{-}C_7H_{15}SO_3Na$ is effective in accelerating this reaction (TON = 95; TOF ca. 10 h^{-1} at 30 °C) [15]. This effect is attributed not to a simple surface activation by the sulfonate but to counter-phase transfer catalysis [16]. Unfortunately, under the reaction conditions described the product is also soluble to a large extent in the water phase. Thus, the product/catalyst separation is difficult again. The same applies to a catalyst system using a ruthenium(III)–EDTA complex [17]. Recently, we described a new carbonylation process for substituted benzyl chlorides on the basis of water-soluble palladium catalysts which are preferable to other known systems with respect to catalytic efficiency [18]. Here, the palladium-catalyzed, atmospheric-pressure carboxylation of substituted benzylic chlorides occurs readily in an aqueous sodium hydroxide/organic solvent two-phase system, giving phenylacetic acids (Scheme 2). The catalyst, consisting of a palladium salt and a water-soluble sulfonated phosphine, is dissolved in an alkaline aqueous solution. During the reaction the corresponding phenylacetic acid is formed directly by carboxylation and in-situ neutralization in 80–94% yield. TONs up to 1500 and TOFs up to 135 h^{-1} are reached. The catalyst system could be re-used three times without significant loss of activity.

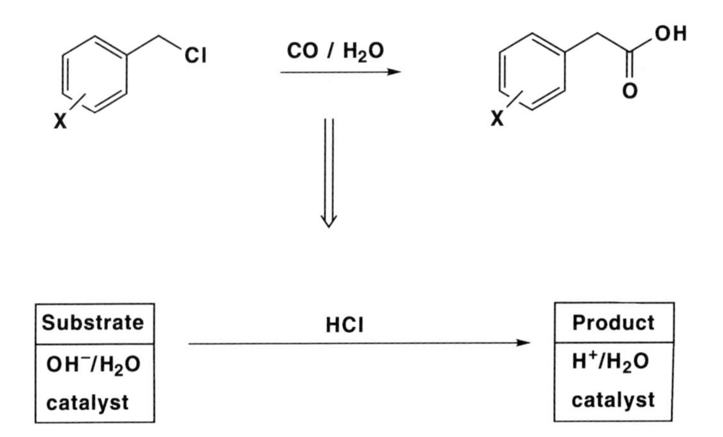

Scheme 2. Carboxylation of substituted benzyl chlorides.

Based on the palladium-catalyzed carbonylation of benzylic halides Sheldon and co-workers investigated the functionalization of 5-hydroxymethylfurfural (HMF) to 5-formylfuran-2-acetic acid (FFA) in aqueous medium in the presence of a water-soluble palladium/TPPTS catalyst (Scheme 3) [5]. Here, the hydroxy group displays similar reactivity under acidic conditions compared with benzylic halides.

Scheme 3. Carboxylation of 5-hydroxymethylfurfural.

Table 1 shows typical results obtained in the carbonylation of HMF in water at 70 °C and 5 bar CO pressure in the presence of TPPTS and various Brønsted acids. 5-Methylfurfural (MF) is obtained as the only by-product. Both selectivity and activity of HMF carbonylation are influenced by the TPPTS/Pd molar ratio, maximum efficiency being observed for a TPPTS/Pd ratio of 6:1. Interestingly, acids with weakly or noncoordinating anions afford mainly carbonylation, those with strongly coordinating anions reduction.

The key intermediate in the catalytic cycle proposed by Sheldon (Scheme 4) is an $[L_2PdR]^+X^-$ (**2**) complex, which controls the chemoselectivity of the reaction. Either coordination of CO, insertion into the Pd–C σ-bond and

Table 1. Carboxylation and reductive carbonylation of HMF by Pd/TPPTS complexes in aqueous solution.[a)]

Run	P/Pd [molar ratio]	Acid		Conversion[b)] [mol%]	Selectivity [mol%]	
		Type	[mmol]		FFA	MF
1	0	H_2SO_4	1.25	0	0	0
2	2	H_2SO_4	1.25	1	22.4	77.3
3	4	H_2SO_4	1.25	53	64.9	34.4
4	6	H_2SO_4	1.25	90	71.6	27.9
5	12[c)]	H_2SO_4	1.25	40	72.9	26.3
6	6	CF_3COOH	0.75	75	76.0	23.4
7	6	HCl	0.75	76	70.8	28.6
8	6	HBr	0.75	77	47.8	51.5
9	6	HI	0.75	54	0	99.8

a) For reaction conditions cf. [5].
b) Determined by HPLC on the basis of HMF.
c) No metallic palladium formation.

hydrolysis forms FFA or – if X^- is a strongly coordinating anion and the coordination of CO is less favorable – protonation gives MF and L_2PdX_2, which is reduced subsequently to PdL_3 by CO.

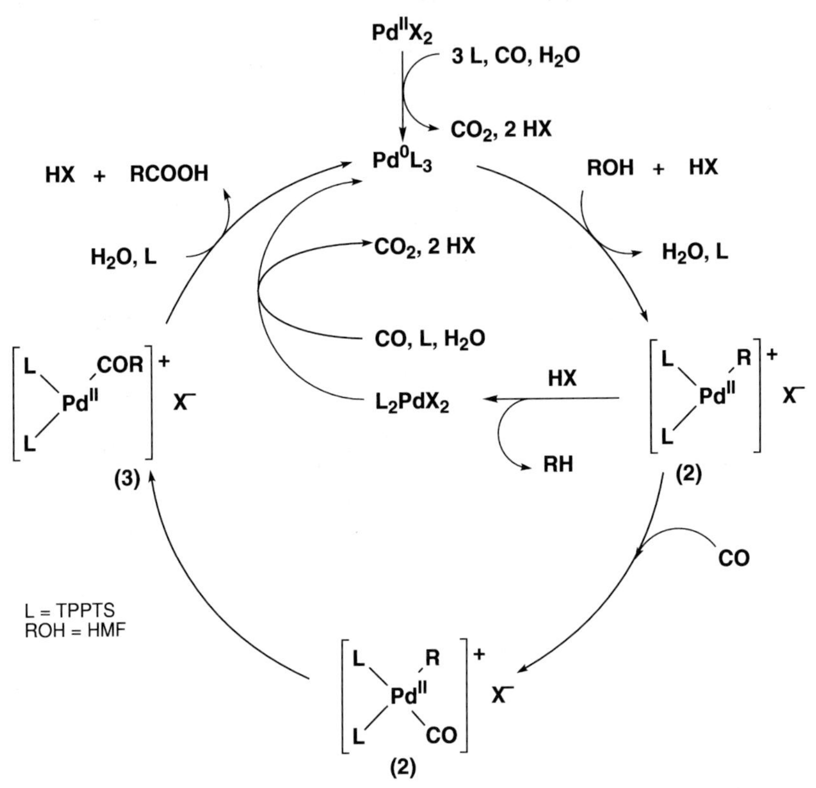

Scheme 4. Proposed catalytic cycle for the carboxylation of 5-hydroxymethylfurfural.

The transition metal-catalyzed carbonylation of allylic compounds has been developed as a simple extension of carbonylations of benzylic derivatives. In this respect nickel cyanide has been used as a catalyst precursor for carbonylation of allyl halides [19]. The key catalytic species is believed to be a water-soluble cyanotricarbonylnickel anion. Without additional phase-transfer catalyst the carbonylation of allyl bromide leads in a biphasic process to (*E*)-2-butenoic acid in 82% yield (TON = 10).

In general for carbonylations, palladium as catalyst metal is preferable to nickel with respect of catalyst efficiency. Thus, Okano, Kiji, and co-workers described some other efficient palladium-catalyzed carbonylations of allyl chloride and substituted allyl halides (Eqs. 5–10). In greater detail, the water-soluble palladium complex $PdCl_2[Ph_2P(m\text{-}C_6H_4SO_3Na)]_2$ has been used in a two-phase system (e.g., aqueous NaOH/benzene medium) at atmospheric carbon monoxide pressure, giving 3-butenoic acids [20]. In the carbonylation of allyl chloride a mixture of 2-butenoic acid, which was formed by base-catalyzed

$$\text{CH}_3\text{CH=CHCH}_2\text{Cl} + \text{CO} + \text{ROH} \xrightarrow[\text{aq. NaOH/benzene, rt, 1 bar CO}]{(\text{TPPTS})_2\text{PdCl}_2} \text{CH}_3\text{CH=CHCH}_2\text{COOR} \quad (5)$$

R = H, CH_3, C_2H_5 etc.

$$\text{CH}_2\text{=CHCH}_2\text{Br} \xrightarrow[\text{aq. NaOH/heptane, 30–50°C, 1 bar CO}]{(\text{TPPTS})_2\text{PdCl}_2} \text{CH}_2\text{=CHCH}_2\text{COOH}\ (63\%) \quad (6)$$

$$\text{CH}_2\text{=C(CH}_3\text{)CH}_2\text{Cl} \xrightarrow[\text{aq. NaOH/heptane, 30–50°C, 1 bar CO}]{(\text{TPPTS})_2\text{PdCl}_2} \text{CH}_2\text{=C(CH}_3\text{)CH}_2\text{COOH}\ (73\%) \quad (7)$$

$$\text{CH}_3\text{CH=CHCH}_2\text{Cl} \xrightarrow[\text{aq. NaOH/heptane, 30–50°C, 1 bar CO}]{(\text{TPPTS})_2\text{PdCl}_2} \text{CH}_3\text{CH=CHCH}_2\text{COOH}\ (73\%) \quad (8)$$

$$(\text{CH}_3)_2\text{C=CHCH}_2\text{Br} \xrightarrow[\text{aq. NaOH/heptane, 30–50°C, 1 bar CO}]{(\text{TPPTS})_2\text{PdCl}_2} (\text{CH}_3)_2\text{C=CHCH}_2\text{COOH}\ (4\%) \quad (9)$$

$$\text{geranyl chloride} \xrightarrow[\text{aq. NaOH/heptane, 30–50°C, 1 bar CO}]{(\text{TPPTS})_2\text{PdCl}_2} \text{geranyl-COOH}\ (64\%) \quad (10)$$

isomerization, and 3-butenoic acid was obtained in up to 90% yield (TON = 135), albeit at moderate selectivity (24:76). Clearly, the isomerization depends on the concentration of the base and was therefore suppressed by a method of continuous addition to the aqueous medium.

So far, easily available alkyl halides have not been applied to biphasic carbonylations. However, the carbonylation of phenethyl bromide with hydrophilic cobalt catalysts to give benzylpyruvic acid (BPA) and benzylacetic acid (BAA) was investigated (Eq. 11) [21].

$$\text{C}_6\text{H}_5\text{–(CH}_2)_2\text{–Br} + \text{CO} \xrightarrow[\text{H}_2\text{O}/t\text{BuOH, Ca(OH)}_2]{\text{Co}_2(\text{CO})_6\text{L}_2} \underset{\text{BPA}}{\text{C}_6\text{H}_5\text{–(CH}_2)_2\text{–C(=O)–C(=O)–OH}} + \underset{\text{BAA}}{\text{C}_6\text{H}_5\text{–(CH}_2)_2\text{–C(=O)–OH}} \quad (11)$$

Using two-phase catalysis it is found that weakly basic phosphines, such as TPPTS and TPPMS, show rather good results compared with those obtained with the water-insoluble phosphines as ligands. Interestingly, the double/mono-carbonylation selectivity strongly depends on the nature of the phosphine.

6.3.4 Hydrocarboxylation of Olefins

Olefins had not been used in biphasic hydrocarboxylation reactions until recent reports by Monflier and co-workers [22]. As shown in Eq. (12), the reaction of olefins, such as styrene, ethylene, propene, 1-hexene, and 1-decene, proceeds in the presence of a palladium/TPPTS catalyst and a Brønsted acid as a co-catalyst. Best results are given by styrene and ethylene, with more than 90% yield of acid. Detailed studies revealed that the catalytic activity is increased by lowering the pH value, with an optimum at pH 1.8. The nature of anion only slightly affects the rate of the hydrocarboxylation reaction. In case of the halogenides as coordinating anions, no precipitation of metallic palladium is observed after the reaction, so catalyst recycling is possible in principle. So far, *n*/iso-selectivities are at least moderate, thus precluding specific synthetic use of the method. Typical values for branched to linear (*n*/iso) ratios range from 1.0 to 1.4.

$$R{-}CH{=}CH_4 + CO + H_2O \xrightarrow[100^\circ C,\ 50\ bar]{Pd/TPPTS,\ HX} \underset{(n)}{R{-}CH_2{-}CH_2{-}COOH} + \underset{(i)}{R{-}\underset{COOH}{\underset{|}{CH_2}}{-}CH_3} \qquad (12)$$

$X = Cl^-, Br^-, I^-, CF_3CO_2^-, PF_6^-$

Long-chain aliphatic olefins give only insufficient conversion to the acids due to low solubility and isomerization side reactions. In order to overcome these problems the effect of co-solvents and chemically modified *β*-cyclodextrins as additives was investigated for the hydrocarboxylation of 1-decene [23]. Without such a promoter, conversion and acid selectivity are low, 10% and 20% respectively. Addition of co-solvents significantly increases conversion, but does not reduce the isomerization. In contrast, the addition of dimethyl-*β*-cyclodextrin increased conversion and induced 90% selectivity toward the acids. This effect is rationalized by a host/guest complex of the cyclic carbohydrate and the olefin which prevents isomerization of the double bond. This pronounced chemoselectivity effect of cyclodextrins is also observed in the hydroformylation and the Wacker oxidation of water-insoluble olefins [24, 25].

6.3.5 Conclusions

Apart from hydroformylations, other biphasic carbonylation reactions of organic substrates are still a relatively unexplored research area. So far, nearly all reported reactions in a two-phase system suffer similarly to their homogeneous counterparts from low catalyst efficiency. Thus, the principle advantages of biphasic catalysis have not been realized due to insufficient product/catalyst separation as well as to metal losses with product stream and rapid catalyst deactivation. However, it may be predicted for the future that careful design of water-soluble catalyst systems will make it possible to overcome this problems. Recent progress in the area of carbonylation of benzylic halides indeed indicates this possibility [18]. Interestingly, nearly all two-phase carbonylation reactions described up to now suffer from the problem that salts are a reaction by-product that is enriched in the aqueous phase if the catalyst is re-used. Clearly, carbonylation reactions such as the hydrocarboxylation of olefins (100 % atom-economics) are more suitable for catalyst recycling.

In view of the vast number of homogeneous carbonylation reactions known it is easy to say that biphasic carbonylations have not yet reached their culmination point. Moreover, the actual importance of other two-phase media, e.g., fluorous biphasic systems [26] (cf. Section 7.2), will lead to further exploitation for carbonylations.

References

[1] E. V. Dehmlow, S. S. Dehmlow, *Phase Transfer Catalysis*, 3rd ed., VCH, Weinheim, **1993**.
[2] A. M. Tafesh, J. Weiguny, *Chem. Rev.* **1996**, *96*, 2035 and literature cited therein.
[3] A. M. Tafesh, M. Beller, *Tetrahedron Lett.* **1995**, *36*, 9305.
[4] F. Ragaini, S. Cenini, *J. Mol. Catal.* **1996**, *105*, 145.
[5] G. Papadogianakis, L. Maat, R. A. Sheldon, *J. Chem. Soc., Chem. Commun.* **1994**, 2659.
[6] M. Beller, in *Applied Homogeneous Catalysis with Organometallic Compounds* (Eds.: B. Cornils, W. A. Herrmann), VCH, Weinheim **1996**, *Vol. 1*, p. 148.
[7] I. P. Beletskaya, *10th International Symposium on Homogeneous Catalysis (ISHC)* Princeton, **1996**.
[8] F. Monteil, P. Kalck, *J. Organomet. Chem.* **1994**, *482*, 45.
[9] Review: M. Beller, B. Cornils, C. D. Frohning, C. W. Kohlpaintner, *J. Mol. Catal.* **1995**, *104*, 17.
[10] L. Cassar, *Chim. Ind.* **1985**, *67*, 256.
[11] (a) L. Cassar, M. Foa, *J. Organomet. Chem.* **1977**, *134*, C15; (b) H. Alper, H. des Abbayes, *J. Organomet. Chem.* **1977**, *134*, C11.
[12] H. Alper, K. Hashem, J. Heveling, *Organometallics* **1982**, *1*, 775.
[13] I. Amer, H. Alper, *J. Am. Chem. Soc.* **1989**, *111*, 927.
[14] T. Okano, I. Uchida, T. Nakagaki, H. Konishi, J. Kiji, *J. Mol. Catal.* **1989**, *54*, 2589.
[15] T. Okano, T. Hayashi, J. Kiji, *Bull. Chem. Soc. Jpn.* **1994**, *67*, 2339.
[16] T. Okano, Y. Moriyama, H. Konishi, J. Kiji, *Chem. Lett.* **1986**, 1463.

[17] M. M. Taqui Khan, S. B. Halligudi, S. H. R. Abdi, *J. Mol. Catal.* **1988**, *44*, 179.
[18] C. W. Kohlpaintner, M. Beller, *J. Mol. Catal.* **1997**, *116*, 259.
[19] F. Joo, H. Alper, *Organometallics* **1985**, *4*, 1775.
[20] (a) J. Kiji, T. Okano, W. Nishiumi, H. Konishi, *Chem. Lett.* **1988**, 957; (b) T. Okano, N. Okabe, J. Kiji, *Bull. Chem. Soc. Jpn.* **1992**, *65*, 2589.
[21] E. Monflier, A. Mortreux, *J. Mol. Catal.* **1994**, 295.
[22] S. Tilloy, E. Monflier, F. Bertoux, Y. Castanet, A. Mortreux, *New J. Chem.* **1997**, *21*, 529.
[23] E. Monflier, S. Tilloy, F. Bertoux, Y. Castanet, A. Mortreux, *New J. Chem.* **1997**, in press.
[24] E. Monflier, G. Fremy, Y. Castanet, A. Mortreux, *Angew. Chem., Int. Ed. Engl.* **1995**, *34*, 2269.
[25] (a) E. Monflier, E. Blouet, Y. Barbaux, A. Mortreux, *Angew. Chem., Int. Ed. Engl.* **1994**, *33*, 2100; (b) E. Monflier, S. Tilloy, G. Fremy, Y. Barbaux, A. Mortreux, *Tetrahedron Lett.* **1995**, 387.
[26] I. T. Horvath, J. Rabai, *Science* **1994**, *266*, 72.

6.4 C–C Coupling by Heck-type Reactions

Wolfgang A. Herrmann, Claus-Peter Reisinger

6.4.1 Introduction

Here, the term "Heck-type reaction" summarizes palladium-catalyzed C–C coupling processes where vinyl or aryl derivatives are functionalized with olefins, alkynes, or organometallic reagents (see Eqs. 1 and 2) [1]. Aryl and vinyl chlorides are most reluctant to undergo Pd-catalyzed activation, as expected from C–X bond dissociation energies [2].

$$\text{R-C}_6\text{H}_4\text{-X} + \text{CH}_2\text{=CH-R'} + \text{B} \xrightarrow{[\text{Pd}]} \text{R-C}_6\text{H}_4\text{-CH=CH-R'} + [\text{HB}]\text{X} \tag{1}$$

$$\text{R-CH=CH-X} + \text{CH}_2\text{=CH-R'} + \text{B} \xrightarrow{[\text{Pd}]} \text{R-CH=CH-CH=CH-R'} + [\text{HB}]\text{X} \tag{2}$$

X = I, Br, N_2, BF_4 B = base: NR_3, K_2CO_3, NaOAc

Although most applications of Heck-type reactions are carried out in polar aprotic media, there are several successful approaches using partially or completely aqueous solution and aqueous–organic biphasic systems. Furthermore, the methodology was expanded to N–C and P–C bond forming reactions; in addition, cross-coupling reactions will also be discussed here. In most cases, the recovery and re-use of the water-soluble catalyst is of minor importance, because the procedures are developed for the laboratory scale with respect to fine chemical synthesis. The main advantage of this approach may be the significant change in thermodynamics, resulting in milder reaction conditions and improvements in chemo- and regioselectivity.

6.4.2 Catalysts and Reaction Conditions

Palladium is one of the most versatile and efficient catalyst metals in organic synthesis. Solubility in water is achieved by utilization of simple palladium(II) salts or water-soluble ligands, such as TPPTS and TPPMS. The active catalysts for Heck-type reactions are zerovalent palladium(0) species [3], which are often generated in-situ by thermal decomposition of a Pd(II) precursor or by the application of a reducing agent, e.g., 1–6 equiv. of a phosphine in the presence of base generates Pd(0) and the phosphine oxide (Eq. 3) [4].

$$Pd(OAc)_2 + H_2O + 2\,R'_3N + n\,PR_3 \longrightarrow Pd^0(PR_3)_{n-1} + [R'_3NH]OAc + O{=}PR_3 \tag{3}$$

The isolation of water-soluble palladium(0) complexes was achieved by Herrmann and co-workers by gel-permeation chromatography for $Pd(TPPTS)_3$ [5]. An X-ray determination was reported by Casalnuovo and Calabrese for $Pd(TPPMS)_3$ [6]; this is the first published structure of a transition metal complex containing a sulfonated phosphine.

Furthermore, the combination of palladium(II) salts with tetrabutylammonium halide additives, called "Jeffery conditions," is an efficient system for Heck-type reactions [7a], but the mechanistic implications are unknown. Also, nonionic phosphine ligands, such as triphenylphosphine which yields $Pd(PPh_3)_4$, are applied in water-miscible organic solvents, like DMF and acetonitrile. In these cases, the application of water is of crucial importance, but the role is often not well investigated.

6.4.3 Olefination

Since the pioneering work by Beletskaya and co-workers [8] the intra- and (more commonly) intermolecular arylation of olefins has been shown to proceed very smoothly in aqueous medium in the presence of palladium acetate. At the beginning, the methodology seemed to be limited to aryl iodides under a strong influence of the base: it was shown that the presence of potassium acetate instead of carbonate yielded lower reaction temperatures and higher rates (Eq. 4).

(4) 3-Iodobenzoic acid (CO_2H, I) + acrylic acid (CO_2H), H_2O:
- 2 h, 80 °C, K_2CO_3 → 97%
- 1 h, 50 °C, K_2CO_3, KOAc → 98%

Product: 3-(CO_2H-vinyl)benzoic acid (CO_2H, CO_2H)

Several years later, a similar approach succeeded even in the application of deactivated bromoanisole (because it was donor-substituted) with acrylic acid in water at 100 °C, using 1 mol% palladium chloride with 3 equiv. of sodium carbonate, which demonstrates the efficient application of water as the reaction medium [9].

Further investigations by Jeffery indicated the rate- and selectivity-enhancing ability of tetraalkylammonium salts in Heck-type reactions [7]. In particular, tetrabutylammonium chloride, bromide and hydrogensulfate are extensively applied in aqueous DMF and acetonitrile, resulting in the fast and clean conversion of phenyl iodide with acrylic acid in 96% yield.

This approach was adopted by Daves for the coupling of iodo derivatives of nitrogen heterocycles with cyclic enol ethers and furanoid glycals in a water/ethanol mixture, using tetrabutylammonium chloride as a promoter (Eq. 5) [10]. Surprisingly, the use of absolute ethanol as reaction solvent was ineffective.

10 mol% $Pd(OAc)_2$, *n*-Bu_4NCl; $NaHCO_3$, NEt_3, H_2O/EtOH; 80% (5)

Furthermore, comparative studies with arylphosphine ligands in aqueous organic media demonstrated the superior activity of palladium tri(*o*-tolyl)-phosphine complexes [11] with an unusual combination of 10 mol% tributylamine with 1.5 equiv. of potassium carbonate in water [9]. The catalyst system was successfully applied to bromobenzene and even bromoanisole with water-insoluble styrene, yielding 86% product in 6 h with 1 mol% Pd (Eq. 6).

100 °C, 6h, $PdCl_2$ + 4 P(*o*-Tol)$_3$; 0.1 eq. NBu_3, K_2CO_3, H_2O; 86% (6)

If water-soluble phosphine ligands are applied, extremely mild reaction conditions can be achieved. Especially, $Pd(TPPMS)_3$, which converts 4-iodotoluene in 2 h at 30 °C (10 mol% Pd), is tolerant of a broad range of functional groups, including those present in unprotected nucleotides and amino acids [6].

Interestingly, even the coupling of donor-substituted iodoarenes and cyclic olefins can be conducted by palladium acetate with TPPTS at only 25 °C in aqueous acetonitrile. However, the low rates observed require a reaction time of up to 48 h for high conversions [12].

The application of ethylene in Heck reactions often shows different activities from other olefins, because of Wacker-type side reactions. It was found, however, that iodo- and acceptor-substituted bromoarenes are cleanly converted in aqueous media to the corresponding styrenes utilizing a palladium-TPPMS complex [13]. Furthermore, high purity *o*- and *p*-vinyltoluenes were prepared on a large scale (in up to a 10-gallon (38-L) reactor) in a dimethylformamide/water mixture with palladium tri(*o*-tolyl)phosphine complexes [14]. Here, the role of water may be the dissolution of the inorganic base (potassium carbonate) in the organic media.

Even superheated (to 260 °C) or supercritical (to 400 °C) water was employed in the Heck reaction with several catalyst precursors and aryl halides with styrene. However, all conversions show large amounts of side products and the yields were in the 5–30% range, indicating radical intermediates and by-products from decomposition of the arene starting material [15].

The progress of tandem Heck reactions in organic synthesis [16] led to their first application in the aqueous phase. Hence, a double Heck reaction on a substrate for which β-hydride elimination is possible results in three tricyclic products (Scheme 1) [17].

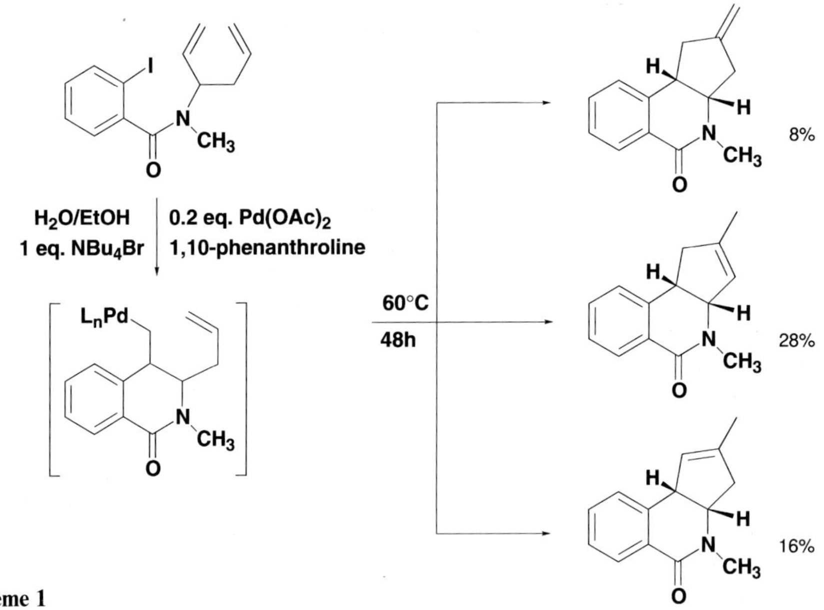
I
N
CH3
O
H2O/EtOH
1 eq. NBu4Br
0.2 eq. Pd(OAc)2
1,10-phenanthroline
LnPd
N
CH3
O
60°C
48h
H
H
N
CH3
O
8%
H
H
N
CH3
O
28%
H
H
N
CH3
O
16%

Scheme 1

Surprisingly, the application of 1,10-phenanthroline as a ligand suppresses β-hydride elimination completely and raises the total yield of double cyclization products to 52%. This may arise from the hydrophobic effect of the heterocycle and the alkene in water/ethanol, forming an aggregate in the transition state favored in water [18]. In addition, an efficient one-pot procedure for Heck reactions starting with aniline derivatives, forming arenediazonium salts with sodium nitrite in 42% aqueous HBF_4, was reported (Eq. 7) [19]. The process has several advantages: short reaction times, high catalytic turnover frequency, superior reactivity of the diazonium nucleofuge, and, most significantly, the use of aqueous reaction conditions. Therefore, this route toward ring-modified phenylalanine and tyrosine was used via a ring nitration and reduction sequence, expanding the field of artificial amino acids [20].

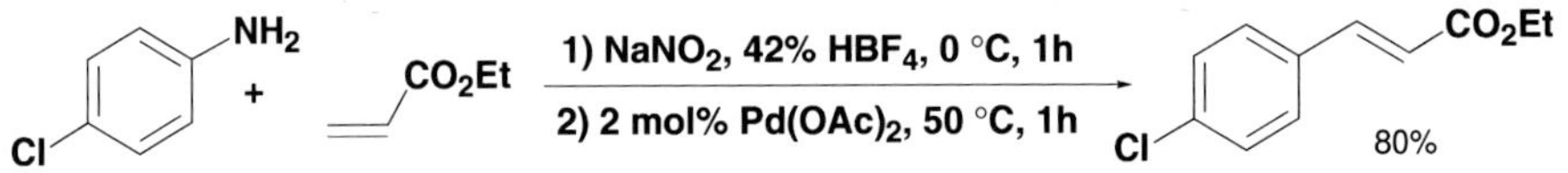
NH2
Cl
+
CO2Et
1) NaNO2, 42% HBF4, 0 °C, 1h
2) 2 mol% Pd(OAc)2, 50 °C, 1h
CO2Et
Cl
80%

(7)

6.4.4 Alkyne Coupling

The palladium-catalyzed coupling of terminal acetylenes with aryl and vinyl halides is a widely used reaction in organic synthesis [21]. Hence, the application of water-soluble palladium complexes was first reported in aqueous acetonitrile with $Pd(TPPMS)_3$ and CuI as promoter, but limited to aryl iodides [6]. The advantages of this catalyst already mentioned are low reaction temperatures and short reaction times with high yields (Eq. 8). Further ligand variations with TPPTS [12] and guanidino-functionalized phosphines [22] revealed that this methodology works also without any Cu(I) promoter, when higher amounts of palladium (10 mol%) are used.

$Pd(TPPMS)_3$, H_2O/CH_3CN; 95% (8)

Furthermore, Bumagin and Beletskaya reported the first coupling in neat water in the presence of a small amount tributylamine (10 mol%) and potassium carbonate as base [23]. Surprisingly, the catalyst system consists of water-insoluble triphenylphosphine with $PdCl_2$ and CuI at room temperature, resulting in high yields with aryl iodides and phenylacetylene. The role of cuprous iodide was noted to be important to facilitate the reaction, which may be rationalized by two connected catalytic cycles (Scheme 2).

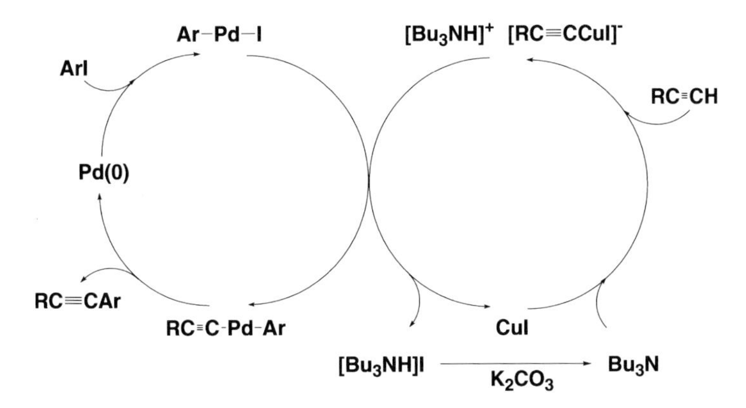

Scheme 2

In addition, the application of "Jeffery's conditions" by Sinou and co-workers, with extra triphenylphosphine and tetrabutylammonium hydrogensulfate, confirmed that CuI is not essential to success in alkyne coupling reactions [24]. Moreover, they reported the most efficient coupling of bromoanisole with propargyl alcohol in 81% yield. These results prompted the authors to apply these conditions in a cascade reaction, consisting of an intermolecular Heck reaction followed by cyclization of the intermediate σ-complex (Eq. 9). The product, which is a mixture of (*E*) and (*Z*) stereomers (approximately 1:1), was sometimes contaminated with a by-product resulting from aromatization, especially for longer reaction times.

C_6H_5 ... O ... C≡CH + ArX $\xrightarrow[\text{CH}_3\text{CN/H}_2\text{O, 80 °C; Bu}_4\text{NHSO}_4\text{/NEt}_3]{\text{Pd(OAc)}_2\text{, PPh}_3}$ (C$_6$H$_5$, Ar, Ar dihydrofuran) + (C$_6$H$_5$, Ar, Ar furan) (9)

6.4.5 Cross-coupling Reactions

The field of Heck-type reactions has been extended by a variety of cross-coupling reactions, each of which has its own name because of its uniqueness and importance in organic synthesis.

6.4.5.1 Suzuki Coupling

The Suzuki coupling is defined by the presence of boron-containing coupling reactions. Thus, the palladium-catalyzed cross-coupling reaction of aryl or alkenyl halides with alkenylboronates or arylboronic acids is a regio- and stereoselective bond formation affording, in particular, unsymmetric substituted biaryls [25]. Once again, the first application of this approach in aqueous phase was reported by Casalnuovo and Calabrese demonstrating the high efficiency of their Pd(TPPTS)$_3$-based catalyst system. Thus, 4-bromopyridine was coupled with *p*-tolylboronic acid in a water/methanol/benzene solvent mixture in 98% yield (Eq. 10) [6].

H_3C–C$_6$H$_4$–B(OH)$_2$ + Br–C$_5$H$_4$N $\xrightarrow[\text{H}_2\text{O/CH}_3\text{OH/C}_6\text{H}_6\text{; Na}_2\text{CO}_3]{\text{Pd(TPPMS)}_3\text{, 80 °C, 7 h}}$ H_3C–C$_6$H$_4$–C$_5$H$_4$N 98% (10)

Later, the same methodology was applied by Wallow and Novak for the synthesis of water-soluble poly(*p*-phenylene) derivatives via the "poly-Suzuki" reaction of 4,4′-biphenylylene bis(boronic acid) with 4,4′-dibromodiphenic acid in aqueous dimethylformamide [26]. These aromatic, rigid-chain polymers exhibit outstanding thermal stability (decomposition above 500 °C) and play an important role in high-performance engineering materials [27] conducting polymers [28] and nonlinear optical materials [29] (see also Section 6.11).

Furthermore, this regio- and stereoselective bond formation between unsaturated carbon atoms was applied to the synthesis of functionalized dienes under extremely mild conditions. Thus, even vinylic boronic esters containing an allylic acetal moiety and alkenylboronate having a chiral protected allylic alcohol were successfully accomplished with vinylic iodides under aqueous conditions in 60–90% yield [30]. In addition, an exceptionally simple and efficient synthesis of a prostaglandin (PGE_1) precursor was reported by Johnson, applying a DMF/THF/water solvent mixture with a bis(diphenylphosphino)-ferrocene palladium catalyst [31]. It is curious that the presence of water is an absolute necessity in order to succeed in this approach (Scheme 3).

TBDMSO + B—$(CH_2)_6CO_2CH_3$ → $PdCl_2(dppf)$, $AsPh_3$; DMF/THF/H_2O, 25 °C → TBDMSO, $(CH_2)_6CO_2CH_3$ → → HO, $(CH_2)_6CO_2CH_3$, C_5H_{11}, OH PGE_1

Scheme 3 TBDMSO = t - butyldimethylsilyl
dppf = 1,1 - bis (diphenylphosphinoferrocene)

It is noteworthy that 9-alkyl-9-BBN (9-BBN = 9-boracyclo[3.3.1]nonyl) reagents are easily prepared by hydroboration of the corresponding olefin, demonstrating the high variability of this approach in organic synthesis.

6.4.5.2 Stille Coupling

The Stille coupling depends on tin-containing reagents. Although the cross coupling of organotin reagents with organic halides proceeds under extremely mild conditions, it seems to be the most unexplored field of palladium-catalyzed reactions [32], because of the high toxicity of the volatile tetraorganotin compounds. Thus, the first application in aqueous medium was reported by Daves in 1993, describing the synthesis of a pyrimidine derivative formed by in-situ hydrolysis of the intermediate enol ether (Eq. 11) [33].

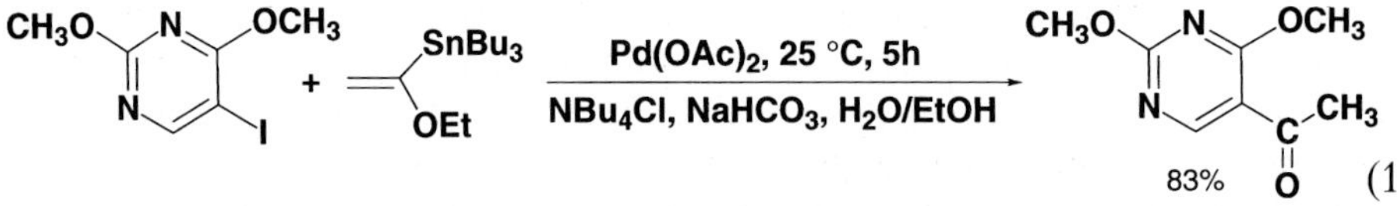

(11)

In 1995, Beletskaya [34] and Collum [35] reported independently the application of alkyltrichlorostannanes instead of tetraorganotin compounds, overcoming the disadvantage of three inert anchoring groups ("atom economy") and technologically more important, because of their lower toxicity and availability via economic direct synthesis from tin(II) compounds [36]. Furthermore, the hydrolysis of the tin–halide bond in water results in higher water-solubility, activation of the C–Sn bond toward electrophiles (e.g., in transmetallation) and less toxic by-products. The reaction may be accomplished via intermediate anionic hydroxo complexes [37], produced in situ in aqueous alkaline solution, and proceeds in most cases in 3 h at 90–100 °C (Eq. 12).

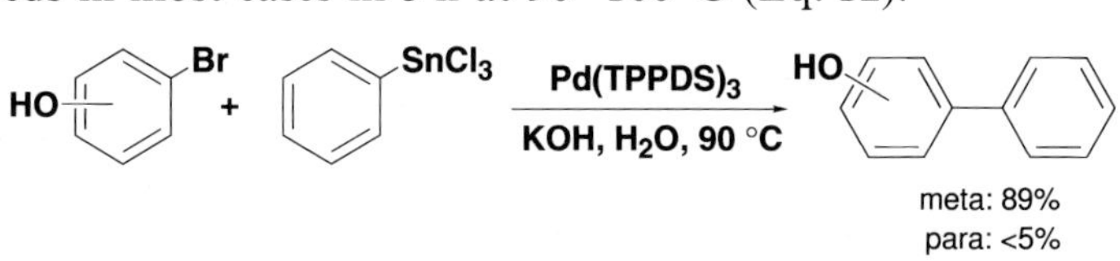

(12)

For insoluble halides to react smoothly, water-soluble phosphine ligands such as TPPMS and TPPDS have to be employed, otherwise the catalyst decomposes rapidly and palladium black is formed. The most reactive organotin reagent studied has proved to be phenyltrichlorostannane, although methyl transfer was also successful with soluble substrates.

6.4.5.3 Miscellaneous

A recent development in Heck-type reactions is P–C and N–C bond formation, which results by coupling of aryl halides with phosphorous compounds [38] and amines [39]. The first application in aqueous medium was achieved by coupling of a dialkyl phosphite with an aromatic iodide to give an arylphosphonate in 99% yield. In 1996, Stelzer and co-workers presented a P–C cross-coupling reaction between primary and secondary phosphines and functional aryl iodides to water-soluble phosphines (Eq. 13), which are potentially applicable as ligands in aqueous-phase catalysis [40]. Surprisingly, the large excess of phosphine ligands (starting material, product, and added triphenylphosphine) present in the reaction mixture obviously does not inhibit the palladium catalyst.

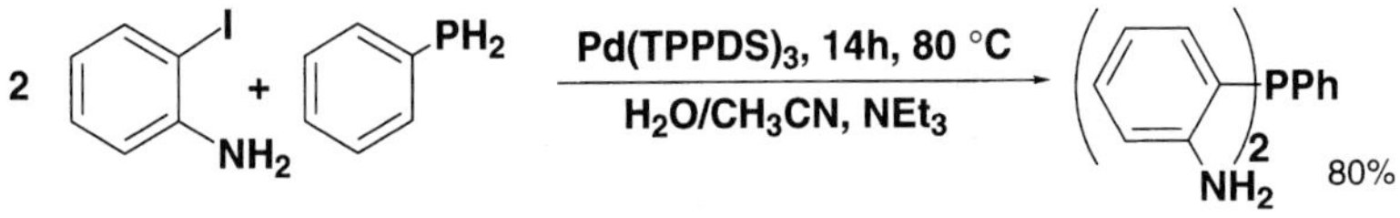

(13)

A rather unusual procedure has been published for the palladium- and copper-catalyzed synthesis of triarylamines, using an alkaline water–ethanol emulsion stabilized by cetyltrimethylammonium bromide [41]. Anyway, this method overcomes the problem in the synthesis of *N*-aryl carbazoles (Eq. 14), which are not accessible by the method developed by Hartwig and Buchwald [42].

$$\text{carbazole (N-H)} + PhI \xrightarrow[H_2O/BuOH,\ CTMAB,\ 100\ ^\circ C]{Pd(OAc)_2,\ K_2CO_3,\ CuI} \text{N-phenylcarbazole (86\%)} \quad (14)$$

6.4.6 Conclusion

The advantages of Heck-type reactions in aqueous phase are demonstrated by the large number of successful approaches presented here. The change in the thermodynamics caused by using water as reaction medium results in milder reaction conditions, higher yields, and improvements in chemo- and regioselectivity. Further progress in this field is very likely, because most of the results have only been obtained in the last five years.

References

[1] (a) H. A. Dieck, R. F. Heck, *J. Am. Chem. Soc.* **1974**, *96*, 1133; (b) R. F. Heck, *Org. React.* **1982**, *27*, 385.
[2] (a) W. A. Herrmann, *Applied Homogeneous Catalysis with Organometallic Compounds* (Eds.: B. Cornils, W. A. Herrmann), VCH, Weinheim, **1996**, pp. 712–732; (b) C.-P. Reisinger, Ph.D. Thesis, Technische Universität München, Germany, **1997**.
[3] C. Amatore, E. Blart, J. P. Genêt, A. Jutand, S. Lemaire-Audoire, M. Savignac, *J. Org. Chem.* **1995**, *60*, 6829.
[4] F. Ozawa, A. Kubo, T. Hayashi, *Chem. Lett.* **1992**, 2177.
[5] W. A. Herrmann, J. Kellner, H. Riepl, *J. Organomet. Chem.* **1990**, *389*, 103.
[6] A. L. Casalnuovo, J. C. Calabrese, *J. Am. Chem. Soc.* **1990**, *112*, 4324.
[7] (a) T. Jeffery, *Tetrahedron Lett.* **1994**, *35*, 3051; (b) T. Jeffery, *Tetrahedron* **1996**, *52*, 10113.
[8] N. A. Bumagin, P. G. More, I. P. Beletskaya, *J. Organomet. Chem.* **1989**, *371*, 397.
[9] N. A. Bumagin, V. V. Bykov, L. I. Sukhomlinova, T. P. Tolstaya, I. P. Beletskaya, *J. Organomet. Chem.* **1995**, *486*, 259.
[10] H.-C. Zhang, G. D. Daves, Jr., *Organometallics* **1993**, *12*, 1499.
[11] W. A. Herrmann, C. Broßmer, K. Öfele, C.-P. Reisinger, T. Priermeier, M. Beller, H. Fischer, *Angew. Chem., Int. Ed. Engl.* **1995**, *34*, 1844; *Angew. Chem.* **1995**, *107*, 1989.
[12] J. P. Genêt, E. Blart, M. Savignac, *Synlett* **1992**, 715.
[13] J. Kiji, T. Okano, T. Hasegawa, *J. Mol. Catal.* **1995**, *97*, 73.
[14] R. A. DeVries, A. Mendoza, *Organometallics* **1994**, *13*, 2405.
[15] (a) P. Reardon, S. Metts, C. Crittendon, P. Daugherity, E. J. Parsons, *Organometallics* **1995**, *14*, 3810; (b) J. Diminnie, S. Metts, E. J. Parsons, *Organometallics* **1995**, *14*, 4023.

[16] A. de Meijere, F. E. Meyer, *Angew. Chem., Int. Ed. Engl.* **1994**, *33*, 2379; *Angew. Chem.* **1994**, *106*, 2473.
[17] D. B. Grotjahn, X. Zhang, *J. Mol. Catal.* **1997**, *116*, 99.
[18] R. Breslow, *Acc. Chem. Res.* **1991**, *24*, 159.
[19] (a) S. Sengupta, S. Bhattacharya, *J. Chem. Soc., Perkin Trans.* **1993**, 1943; (b) K. Kikukawa, K. Nagira, N. Terao, F. Wada, T. Matsuda, *Bull. Chem. Soc. Jpn.* **1979**, *52*, 2609.
[20] S. Sengupta, S. Bhattacharya, *Tetrahedron Lett.* **1995**, *36*, 4475.
[21] (a) J. Tsuji, *Palladium Reagents and Catalysis: Innovations in Organic Synthesis,* John Wiley, Chichester, **1995**; (b) W. A. Herrmann, C.-P. Reisinger, C. Broßmer, M. Beller, H. Fischer, *J. Mol. Catal. A* **1996**, *198*, 51.
[22] H. Dibowski, F. P. Schmidtchen, *Tetrahedron* **1995**, *51*, 2325.
[23] N. A. Bumagin, L. I. Sukhomlinova, E. V. Luzikova, T. P. Tolstaya, I. P. Beletskaya, *Tetrahedron Lett.* **1996**, *37*, 897.
[24] J.-F. Nguefack, V. Bolitt, D. Sinou, *Tetrahedron Lett.* **1996**, *37*, 5527.
[25] A. Suzuki, *Pure Appl. Chem.* **1991**, *63*, 419.
[26] T. I. Wallow, B. M. Novak, *J. Am. Chem. Soc.* **1991**, *113*, 7411.
[27] *The Strength and Stiffness of Polymers* (Eds.: A. E. Zachariades, R. S. Porter), Marcel Dekker, New York, **1983**.
[28] R. L. Elsenbaumer, L. W. Shacklette, *Handbook of Conducting Polymers* (Ed.: T. A. Skotheim), Marcel Dekker, New York, **1986**.
[29] D. J. Williams, *Angew. Chem., Int. Ed. Engl.* **1984**, *23*, 640; *Angew. Chem.* **1984**, *96*, 637.
[30] J. P. Genêt, A. Linquist, E. Blart, V. Mouriès, M. Savignac, M. Vaultier, *Tetrahedron Lett.* **1995**, *36*, 1443.
[31] C. R. Johnson, M. P. Braun, *J. Am. Chem. Soc.* **1993**, *115*, 11014.
[32] (a) T. N. Mitchell, *Synthesis* **1992**, 803; (b) J. K. Stille, *Angew. Chem., Int. Ed. Engl.* **1986**, *25*, 508; *Angew. Chem.* **1986**, *98*, 504.
[33] H.-C. Zhang, G. D. Daves, Jr., *Organometallics* **1993**, *12*, 1499.
[34] A. I. Roshchin, N. A. Bumagin, I. P.Beletskaya, *Tetrahedron Lett.* **1995**, *36*, 125.
[35] R. Rai, K. B. Aubrecht, D. B. Collum, *Tetrahedron Lett.* **1995**, *36*, 3111.
[36] A. G. Davies, P. J. Smith, *Comprehensive Organometallic Chemistry* (Eds.: G. Wilkinson, F. G. A. Stone, E. W. Abel), Pergamon Press, Oxford, **1982**, *2*, 519.
[37] M. Devaud, *Rev. Chim. Miner.* **1967**, *4*, 921.
[38] (a) O. Herd, A. Heßler, M. Hingst, M. Tepper, O. Stelzer, *J. Organomet. Chem.* **1996**, *522*, 69; (b) A. L. Casalnuovo, J. C. Calabrese, *J. Am. Chem. Soc.* **1990**, *112*, 4324.
[39] (a) A. S. Guram, R. A. Rennels, S. L. Buchwald, *Angew. Chem., Int. Ed. Engl.* **1995**, *34*, 1348; *Angew. Chem.* **1995**, *107*, 1456; (b) J. Louie, J. F. Hartwig, *Tetrahedron Lett.* **1995**, *36*, 3609; (c) M. Beller, T. H. Riermeier, C.-P. Reisinger, W. A. Herrmann, *Tetrahedron Lett.*, in press.
[40] O. Herd, A. Heßler, M. Hingst, M. Tepper, O. Stelzer, *J. Organomet. Chem.* **1996**, *522*, 69.
[41] D. V. Davydov, I. P. Beletskaya, *Russ. Chem. Bull.* **1995**, *44*, 1141 (Engl. Transl.)
[42] W. A. Herrmann, C.-P. Reisinger, unpublished results, **1996**, cf. Ref. [2b].

6.5 Hydrocyanation

Henry E. Bryndza, John A. Harrelson, Jr.

6.5.1 Introduction

Hydrogen cyanide, HCN or hydrocyanic acid, is a remarkably versatile C_1 building block for the modern synthetic chemist today. Its use, however, has been limited by its relatively difficult synthesis/purification as well as by its high flammability, its tendency toward base-catalyzed explosive polymerization, and its toxicity. Most of the recent hydrocyanation literature comes from industry rather than academic laboratories.

If toxicity and handling difficulties are the drawbacks of HCN as a feedstock, the versatility of the nitrile functional group as a synthon is a significant advantage. Moreover, the full miscibility of HCN with water at 25 °C offers a potential advantage to pursuing hydrocyanation catalysis in aqueous media. Therefore, hydrocyanation reactions offer the potential for generating new nitrogenous products which are useful intermediates. These reactions include the addition of HCN to C=C, C=O, and C=N double bonds to generate new alkyl nitriles, cyanohydrins, and aminonitriles, respectively. All of these reactions have commercial impact today [1], although aqueous hydrocyanation catalysis remains an emerging area of technology.

6.5.2 HCN as a Synthon

6.5.2.1 Michael Additions of HCN to Activated Olefins

The addition of HCN to activated olefins, as in Michael additions (Eq. 1), has been known and commercially practiced for many years. However, due to the potential for competitive addition of water to Michael substrates under conditions of general base catalysis, few examples of these additions are known in aqueous media. However, the use of aqueous media for the tetraalkylammonium cyanide-catalyzed addition of HCN to isophorone (Eq. 1) was recently reported to give the corresponding nitrile in 98% yield under mild conditions [2]. Main Group organometallic cyanides as well as organic cyanide salts have been reported as catalysts for these reactions.

$$\text{isophorone} + HCN \xrightarrow[H_2O,\ 80°C,\ 2\ \text{hours}]{NR_4^+CN^-} \text{3-cyano-3,5,5-trimethylcyclohexanone} \quad (1)$$

6.5.2.2 Synthesis of Cyanohydrins from Ketones and Aldehydes

More common is the general acid- or base-catalyzed addition of HCN to ketones and aldehydes to give cyanohydrins (Eq. 2) [1]. Because of the propensity of HCN to spontaneously and exothermically polymerize under basic conditions, general acid catalysis is sometimes favored over basic media, as was the case in the recent Sumitomo [3] and Upjohn work [4, 5]. Several recent applications of aqueous media have recently been reported to lead to asymmetric hydrocyanation catalysis. In one example, Arena at Allied Signal has reported the addition of HCN to monosaccharides in which the L-sugar products are preferentially complexed by aluminates and driven out of solution to generate new sweetener intermediates [6]. Loos and co-workers have recently reported the use of hydroxynitrile lyase as a catalyst for the general asymmetric hydrocyanation of aldehydes in aqueous media (Eq. 3) [7]. These reactions proceed under mild conditions in water or water-containing solvents to produce single-enantiomer products with a wide variety of substrates.

$$\text{EtO-steroid-17-one} + HCN \xrightarrow[H_2O,\ 35°C]{H^+} \text{EtO-steroid-17-(OH)(CN)} \quad (2)$$

$$R^1C(O)R^2 + HCN \xrightarrow[H_2O\ (MTBE\ opt)]{\text{Hydroxynitrile Lyase}} R^1R^2C(OH)CN \quad (3)$$

86–99% yield
93–99% ee

6.5.2.3 Strecker Synthesis of Aminonitriles

Another variant of this HCN addition to polar C=X bonds is the Strecker synthesis of aminonitriles from ketones or aldehydes, HCN, and ammonia (Eq. 4) [8]. Given the basic environment inherent in this synthesis, these processes are normally run under conditions of general base catalysis and both organometallic (e.g., NaCN) and organic cyanide salts have been reported as catalysts. Because water is a product in Strecker syntheses and can slow the overall rate of reaction, these commercially important processes are not generally run in aqueous media. However, a number of recent patents to Distler and co-workers at BASF do teach the use of aqueous media to facilitate isolation of pure products [9], and researchers at Grace [10], Stauffer [11], Mitsui [12], and Hoechst [13] have reportedly used aqueous media to control the rates of formaldehyde aminohydrocyanation to isolate intermediate addition products in good yields. Moreover, the use of aqueous media has also proven advantageous when amides, rather than nitriles, are the desired products.

$$R^1C(=O)R^2 + R^3R^4NH + HCN \xrightarrow{H_2O,\ 0\text{–}80^\circ C} R^3R^4N\text{–}C(R^1)(R^2)\text{–}CN \quad (4)$$

99% yield

6.5.2.4 HCN Addition to Unactivated C=C Double Bonds

The addition of HCN to C=C double bonds can be effected in low yields to produce Markovnikov addition products. However, through the use of transition metal catalysts, the selective anti-Markovnikov addition of HCN to olefins can take place. The most prominent example of the use of aqueous media for transition metal-catalyzed olefin hydrocyanation chemistry is the three-step synthesis of adiponitrile from butadiene and HCN (Eqs. 5–7). First discovered by Drinkard at DuPont [14], this nickel-catalyzed chemistry can use a wide variety of phosphorus ligands [15] and is practiced commercially in nonaqueous media by both DuPont and Butachimie, a DuPont/Rhône-Poulenc joint venture. Since the initial reports of Drinkard, first Kuntz [16] and, more recently, Huser and Perron [17, 18] from Rhône-Poulenc have explored the use of water-soluble ligands for this process to facilitate catalyst recovery and recycle from these high-boiling organic products.

The initial report by Kuntz [16] teaches the use of sulfonated triphenylphosphine (usually the trisodium salt of tri(*m*-sulfonatophenyl)phosphine (TPPTS),

+ HCN —NiL$_4$, L→ CN (3-PN) + CN (2M3BN) (5)

CN (2M3BN) —NiL$_4$, L→ CN (3-PN) (6)

CN + HCN —NiL$_4$, L / Lewis acid→ NC–CN + isomers (7)

cf. Section 3.2.1) with a source of nickel(0) as the catalyst for adding HCN to butadiene in 81 % yield to give 3-pentenenitrile (3-PN), and HCN to 3-PN to produce adiponitrile, each in yields of about 60 %. The balance of the products comprises branched dinitriles such as methylglutaronitrile and ethylsuccinonitrile (Eq. 8). Kuntz has recovered and recycled these catalysts multiple times, indicating the simplicity of separations induced by the aqueous media. He also reports modest yields of hydrocyanation products when these catalysts were applied to the nonoptimized hydrocyanation of styrene and 1-hexene, indicating some likelihood of the generality of this aqueous chemistry. Finally, Kuntz provides examples in which other salts (e.g., barium) and other cyanide sources (e.g., acetone cyanohydrin) are used for these reactions with good success. The use of other metal systems (and other sulfonated ligands) is mentioned but not exemplified in this patent.

CN + HCN —Ni(TPPTS)$_4$ / exess TPPTS + Lewis acid→ NC–CN + NC–CN + NC–CN (8)

Huser and Perron have extended this work to the isomerization of 2-methyl-3-butenenitrile (2M3 BN) to 3-PN (isomerization step; Eq. (6) 92 % yield) [17]. This patent mentions the use of iron and palladium catalysts but does not provide examples beyond nickel. In other work these same inventors discuss the use of other water-soluble ligands such as those containing carboxylate, phosphate, and alkylsulfonate substituents [18], while also exploring a wide range of

Lewis acid co-catalysts for the addition of HCN to 3-pentenenitrile (Eq. 7) [19]. In general, the addition of strong Lewis acids facilitates this hydrocyanation and the promoted catalysts reported by Huser and Perron are much more active than those originally reported by Kuntz in which adventitious boron Lewis acids (generated during the reduction of Ni(II) with borohydride) were probably facilitating the HCN addition. In this light, it appears that zinc Lewis acids are more effective co-catalysts but less selective for adiponitrile than weaker species such as tin or europium halides. In general these water-soluble catalysts can be prepared by treatment of nickel(0) precursors with phosphines followed by extraction into water, by dissolution of nickel(II) salts such as chlorides or cyanides into an aqueous solution of ligand followed by chemical reduction with zinc or borohydride, or by electrochemical reduction of aqueous nickel(II) salts [20]. There is no mention in these reports about the potentially advantageous use of phase-transfer reagents to facilitate dissolution of very nonpolar olefins in the aqueous media.

Asymmetric water-soluble ligands are known for the metal-catalyzed hydrocyanation of achiral olefins. However, neither Jenck [21] or Davis [22] actually provides any examples of hydrocyanation catalysis in these patents so the performance of these mono- and bisphosphines in aqueous and supported aqueous media cannot be assessed although this may be a promising route for the synthesis of biologically active nitrile intermediates and products.

A final example of aqueous media used in the hydrocyanation of butadiene is provided by Waddan at ICI [23]. In this chemistry, copper nitrate salts in aqueous media (among many others) are used for the oxidative dihydrocyanation of butadiene to dicyanobutenes (Eq. 9). Good conversions of butadiene are reported in nonaqueous media but no examples are actually provided in which water is added as a solvent. Moreover, because of problems with olefin and HCN dimerization and the risk of explosion hazards, these reactions appear to work best when conducted stepwise (i.e., HCN addition to catalyst followed by oxidation followed, in turn, by butadiene addition), leading one to wonder about the productivity of these systems.

$$\text{butadiene} + HCN + 0.5\ O_2 \xrightarrow[\text{solvent or biphasic media}]{\text{Cu(II)salts}} NC\text{-}CH_2CH{=}CHCH_2\text{-}CN \quad (95\%\ \text{yield}) \qquad (9)$$

6.5.2.5 Cyanide Coupling Reactions

A final type of aqueous cyanide chemistry is the oxidative coupling of cyanide to produce oxamide (Eq. 10). Both batch and continuous reactions have been demonstrated at Hoechst by Riemenschneider and Wegener [24], who report advantages of aqueous media not only in concurrently hydrolyzing the coupled products but also in facilitating product isolation from the reaction medium. This clever combination of reactive solvent is reportedly the basis for the commercial production of oxamide [2].

$$2\,HCN + 0.5\,O_2 + H_2O \xrightarrow[HOAc,\ H_2O]{Cu(NO_3)_2} H_2NC(O)C(O)NH_2 \quad (95\%\ yield) \tag{10}$$

6.5.3 Summary

Even from the admittedly limited examples of hydrocyanation reactions, it is clear that HCN is a versatile reagent and that its chemistry is generally compatible with aqueous media. Some traditional advantages which recommend the use of aqueous or biphasic media include the high solubility of HCN in water, the facility in removing products from catalysts, the ability to control rates of reactions more finely and the concomitant hydrolysis of nitrile functional groups to amides along with C–C bond formation. Commercial applications of aqueous hydrocyanations are only starting to emerge, but they could well develop quickly and improved catalysts and separation technologies come into play in the chemical industry.

References

[1] K. Weissermel, H.-J. Arpe, *Industrial Organic Chemistry*, 2nd edn., VCH, Weinheim, **1993**, 43–37, 280.
[2] Nippon Chemical Co., Ltd. (H. Takahoso, N. Takahashi, K. Midorikawa, T. Sato), US 5.179.221 (1993); Nippon Chemical Co., Ltd. (H. Takahoso, N. Takahashi, K. Midorikawa, T. Sato), JP 04.279.559 (1992); Nippon Chemical Co., Ltd. (H. Takahoso, N. Takahashi, K. Midorikawa, T. Sato), EP 502.707B (1995); Nippon Chemical Co., Ltd. (H. Takahoso, N. Takahashi, K. Midorikawa, T. Sato), DE 69.205.601E (1995).
[3] Sumitomo, JP 04.198.173 (1992).
[4] Upjohn (J. G. Reid, T. Debiak-Krook), WO 9530684 (1995); Upjohn (J. G. Reid, T. Debiak-Krook), WO 9530.684 (1995).
[5] Upjohn (J. G. Reid, T. Debiak-Krook), AU 9.523.628 (1995); Upjohn (J. G. Reid, T. Debiak-Krook), EP 759.928 (1977).
[6] Allied Signal (B. J. Arena), US 4.959.467 (1990); Allied Signal (B. J. Arena), WO 8907.591 (1989); Allied Signal (B. J. Arena), ES 2.010.137 (1989); Allied Signal (B. J. Arena), EP 402.399B (1993); Allied Signal (B. J. Arena), JP 03.502.692W (1991); Allied Signal (B. J. Arena), CA 1.321.785C (1993); Allied Signal (B. J. Arena), DE 68.910.812 (1993).

[7] Duphar International Research (H. W. Geluk, W. T. Loos), US 5.350.871 (1994); Duphar International Research (H. W. Geluk, W. T. Loos), EP 547.655 (1993); Duphar International Research (H. W. Geluk, W. T. Loos), CA 2.084.855 (1993); Duphar International Research (H. W. Geluk, W. T. Loos), HUT 064.105 (1993); Duphar International Research (H. W. Geluk, W. T. Loos), JP 053.176.065 (1993); Duphar International Research (H. W. Geluk, W. T. Loos), HU 209.736B (1994); Degussa (F. Effenberger, J. Eichhorn, J. Roos), DE 19.506.728 (1976).

[8] Anon., *Cyanides (hydrogen cyanides)*, In: *Kirk–Othmer Encyclopedia of Chemical Technology*, 2nd ed., Vol. 6, pp. 5832–5383 (**1965**).

[9] BASF AG (H. Distler, K. L. Hock), US 4.478.759 (1984); BASF AG (H. Distler, K. L. Hock), EP 45.386B (1984); BASF AG (H. Distler, K. L. Hock), DE 3.162.511G (1984); BASF AG (H. Distler, E. Hartert, H. Schlecht), US 4.113.764 (1978); BASF AG (H. Distler, E. Hartert, H. Schlecht), US 4.134.889 (1979); BASF AG (H. Distler, E. Hartert, H. Schlecht), BE 854.313 (1977); BASF AG (H. Distler, E. Hartert, H. Schlecht), DE 2.620.445B (1979); BASF AG (H. Distler, E. Hartert, H. Schlecht), DE 2.621.450B (1979); BASF AG (H. Distler, E. Hartert, H. Schlecht), DE 2.621.728B (1979); BASF AG (H. Distler, E. Hartert, H. Schlecht), J 52.136.124 (1977); BASF AG (H. Distler, E. Hartert, H. Schlecht), DE 2.625.935B (1979); BASF AG (H. Distler, E. Hartert, H. Schlecht), FR 2.350.332 (1978); BASF AG (H. Distler, E. Hartert, H. Schlecht), DE 2.660.191 (1978); BASF AG (H. Distler, E. Hartert, H. Schlecht), GB 1.577.662 (1980); BASF AG (H. Distler, E. Hartert, H. Schlecht), US 4.164.511 (1979); BASF AG (H. Distler, E. Hartert, H. Schlecht), US 4.022.815 (1977); BASF AG (H. Distler, E. Hartert, H. Schlecht), BE 837.970 (1976); BASF AG (H. Distler, E. Hartert, H. Schlecht), FR 2.299.315 (1976); BASF AG (H. Distler, E. Hartert, H. Schlecht), GB 1.526.481 (1978); BASF AG (H. Distler, E. Hartert, H. Schlecht), DE 2.503.582B (1979); BASF AG (H. Distler, E. Hartert, H. Schlecht), CA 1.056.402 (1979); BASF AG (H. Distler, E. Hartert, H. Schlecht), CH 619.926 (1980); BASF AG (H. Distler), EP 36.161B (1983); BASF AG (H. Distler), DE 3.010.511 (1981); BASF AG (H. Distler), JP 56.145.251 (1981); BASF AG (H. Distler), DE 3.160.741G (1983); BASF AG (H. Distler) JP 89.036.461B (1989).

[10] W. R. Grace, J. L. Su, M. B. Sherwin, US 4.895.971 (1990); W. R. Grace, J. L. Su, M. B. Sherwin, EP 367.364B (1994); W. R. Grace, J. L. Su, M. B. Sherwin, DE 68.912.878E (1994); W. R. Grace, J. L. Su, M. B. Sherwin, ES 2.061.961T3 (1994).

[11] Stauffer (C. C. Greco, W. Stamm), US 3.862.203; Stauffer (C. C. Greco, W. Stamm), GB 1.244.329 (1971).

[12] Mitsui Toatsu Chem. Inc., JP 02.009.874 (1990); Mitsui Toatsu Chem. Inc., JP 2.575.823B (1997).

[13] Hoechst (K. Warning, M. Mitzlaff, H. Jensen), DE 2.634.048 (1978); Hoechst (K. Warning, M. Mitzlaff, H. Jensen), DE 2.634.047 (1978).

[14] Leading examples from an extensive patent literature include: DuPont (W. D. Drinkard Jr., R. V. Lindsey Jr.), US 3.536.748 (1970); DuPont (W. D. Drinkard Jr., B. W. Taylor), US 3.579.560 (1971); DuPont (Y. T. Chia), US 3.676.481 (1972); (W. D. Drinkard Jr.), US 3.739.011 (1973).

[15] C. A. Tolman, R. J. McKinney, J. D. Druliner, W. R. Stevens, *Adv. Catal. 33*, 1 (**1985**).

[16] Rhône-Poulenc (E. Kuntz), US 4.087.452 (1978); Rhône-Poulenc (E. Kuntz), DE 2.700.904C (1983); Rhône-Poulenc (E. Kuntz), BE 850.301 (1977); Rhône-Poulenc (E. Kuntz), NL 77.00262 (1977); Rhône-Poulenc (E. Kuntz), BR 77.00171 (1977); Rhône-Poulenc (E. Kuntz), FR 2.338.253 (1977); Rhône-Poulenc (E. Kuntz), JP 52.116.418 (1977); Rhône-Poulenc (E. Kuntz), GB 1.542.824 (1979); Rhône-Poulenc (E. Kuntz), SU 677.650 (1979); Rhône-Poulenc (E. Kuntz), CA 1.082.736 (1980); Rhône-Poulenc (E. Kuntz), JP 82.061.270B (1982); Rhône-Poulenc (E. Kuntz), IT 1.083.151B (1985); Rhône-Poulenc (E. Kuntz), NL 188.158B (1991).

[17] Rhône-Poulenc (M. Huser, R. Perron), US 5.486.643 (1996); Rhône-Poulenc (M. Huser, R. Perron), EP 647.619 (1995); Rhône-Poulenc (M. Huser, R. Perron), FR 2.710.909 (1995); Rhône-Poulenc (M. Huser, R. Perron), NO 9.403.748 (1995); Rhône-Poulenc (M. Huser, R. Perron), CA 2.117.725 (1995); Rhône-Poulenc (M. Huser, R. Perron), JP 07.188.143 (1995); Rhône-Poulenc (M. Huser, R. Perron), JP 2.533.073 (1996); Rhône-Poulenc (M. Huser, R. Perron), EP 0.647.619A (1994).

[18] Rhône-Poulenc (M. Huser, R. Perron), US 5.488.129 (1996); Rhône-Poulenc (M. Huser, R. Perron), WO 96.33.969 (1996); Rhône-Poulenc (M. Huser, R. Perron), EP 650.959 (1995); Rhône-Poulenc (M. Huser, R. Perron), FR 2.711.987 (1995); Rhône-Poulenc (M. Huser, R. Perron), NO 9.404.153 (1995); Rhône-Poulenc (M. Huser, R. Perron), CA 2.134.940 (1995); Rhône-Poulenc (M. Huser, R. Perron), JP 07.188.144 (1995); Rhône-Poulenc (M. Huser, R. Perron), WO 97.12.857 (1997); Rhône-Poulenc (M. Huser, R. Perron), FR 2.739.378 (1997).

[19] Only a few representative examples are given here from the many Lewis acids listed in [14].

[20] Rhône-Poulenc (D. Horbez, M. Huser, R. Perron), WO 97.24.184 (1997); Rhône-Poulenc (D. Horbez, M. Huser, R. Perron), FR 2.743.011 (1997); Rhône-Poulenc (M. Huser, R. Perron), WO 97.24.183 (1997); Rhône-Poulenc (M. Huser, R. Perron), FR 2.743.010 (1997) and examples contained in [13–15].

[21] Rhône-Poulenc (T. P. Dang, J. Jenck, D. Morel), US 4.654.176 (1987); Rhône-Poulenc (T. P. Dang, J. Jenck, D. Morel), EP 133.127B (1987); Rhône-Poulenc (T. P. Dang, J. Jenck, D. Morel), FR 2.549.840 (1985); Rhône-Poulenc (T. P. Dang, J. Jenck, D. Morel), JP 60.056.992 (1985); Rhône-Poulenc (T. P. Dang, J. Jenck, D. Morel), CA 1.213.896 (1986); Rhône-Poulenc (T. P. Dang, J. Jenck, D. Morel), DE 3.462.691G (1987); Rhône-Poulenc (T. P. Dang, J. Jenck, D. Morel), JP 92.081.595B (1992).

[22] California Institute of Technology (M. E. Davis, K. T. Wan), WO 95/22.405 (1995); California Institute of Technology (M. E. Davis, K. T. Wan), AU 95.18.467 (1995); California Institute of Technology (M. E. Davis, K. T. Wan), EP 746.410 (1996).

[23] ICI (D. Y. Waddan), EP 032.299B (1984); ICI (D. Y. Waddan), JP 56.100.755 (1981); ICI (D. Y. Waddan), DE 3.066.441 (1984).

[24] Hoechst (W. Riemenschneider, P. Wegener), US 3.989.753 (1976); Hoechst (W. Riemenschneider, P. Wegener), NL 7.402.197 (1974); Hoechst (W. Riemenschneider, P. Wegener), SE 7.402.352 (1974); Hoechst (W. Riemenschneider, P. Wegener), FR 2.219.153 (1974); Hoechst (W. Riemenschneider, P. Wegener), DD 110.854 (1975); Hoechst (W. Riemenschneider, P. Wegener), JP 50.029.516 (1975); Hoechst (W. Riemenschneider, P. Wegener), DE 2.403.120C (1982); Hoechst (W. Riemenschneider, P. Wegener), AT 7.401.402 (1976); Hoechst (W. Riemenschneider, P. Wegener), GB 1.458.871 (1976); Hoechst (W. Riemenschneider, P. Wegener), HUT 021.762 (1977); Hoechst (W. Riemenschneider, P. Wegener), IL 44.209 (1977); Hoechst (W. Riemenschneider, P. Wegener), CH 590.213 (1977); Hoechst (W. Riemenschneider, P. Wegener), CS 7.401.122 (1977); Hoechst (W. Riemenschneider, P. Wegener), CA 1.026.377 (1978); Hoechst (W. Riemenschneider, P. Wegener), RO 70.980 (1981); Hoechst (W. Riemenschneider, P. Wegener), JP 83.033.220B (1983); Hoechst (W. Riemenschneider, P. Wegener), NL 177.400B (1985); Hoechst (W. Riemenschneider, P. Wegener), SZ 7.401.011 (1974); Hoechst (W. Riemenschneider, P. Wegener), BE 811.533 (1974); Hoechst (W. Riemenschneider, P. Wegener), DE 2.308.941B (1975); Hoechst (W. Riemenschneider, P. Wegener), DE 2.402.354C (1982); Hoechst (W. Riemenschneider, P. Wegener), DE 2.402.352 (1975); Hoechst (W. Riemenschneider, P. Wegener), SU 631.069 (1978).

6.6 Allylic Substitution

Denis Sinou

6.6.1 Introduction

Since the discovery of the reaction of π-allylpalladium complexes with carbonucleophiles by Tsuji et al. [1], the organic chemistry of such complexes has attracted considerable attention and is now a usual tool in synthetic organic chemistry leading to the formation of carbon–carbon bonds as well as carbon–heteroatom bonds [2–6]. Important characteristics of this reaction are its very high chemo-, regio- and stereoselectivity and the very mild experimental conditions. In the late 1980s palladium(0)-catalyzed reaction of sodium azide with various allyl esters [7] or 1,3-diene monoepoxides [8] was reported to occur in aqueous tetrahydrofuran, leading to allylic azides in quite good yields, but it was only in 1991 that it was shown that palladium(0)-catalyzed alkylation could be performed in a two-phase system with water/organic solvent using water-soluble complexes [9].

6.6.2 Scope of the Reaction

The water-soluble palladium(0) complex was obtained from $Pd(OAc)_2$ or $Pd_2(dba)_3$ (dba = dibenzylideneacetone) in association with TPPTS in water/nitrile mixture [9]. The use of a nitrile as the co-solvent seemed very important for the recycling of the catalyst, probably because of stabilization of the palladium(0) catalyst. Effectively, if allylic alkylation using tetrahydrofuran or diethyl ether as the cosolvent of water gave the expected products, decomposition of the catalyst occurred readily; this was not the case using a nitrile (acetonitrile, butyronitrile, benzonitrile) as the cosolvent.

It was demonstrated by ^{31}P NMR spectroscopy and by cyclic voltammetry that the mixture of $Pd(OAc)_2$ and TPPTS in water and acetonitrile spontaneously afforded a palladium(0) complex [10, 11].

A detailed investigation using (*E*)-2-hexenyl methyl carbonate as the π-allyl precursor and ethyl acetoacetate as the nucleophile (Eq. 1) showed that the regioselectivity of the reaction was not affected by the $Pd(OAc)_2$/TPPTS ratio, the temperature of the reaction, the water/nitrile ratio, or the nature of the nitrile [12]; the linear and branched products were always obtained in a 90:10

ratio, quite similar to those observed in a usual organic medium. However, the catalytic activity was deeply affected by these parameters. The highest activity was obtained for a [Pd(OAc)$_2$]/TPPTS ratio of 9:1 and in acetonitrile as the solvent; this could be related to the solubility of the nitrile in water, the transfer of the reactants from the organic phase to the aqueous phase probably being the limiting step.

$CH_3COCH_2CO_2Et$; Pd(OAc)$_2$/TPPTS; RCN/H_2O

n/i = 90/10 (1)

The allylation reaction was extended to various carbonucleophiles [9, 12–14]. It was observed that the selectivity in the formation of mono- and diallylated compounds was very sensitive to the nature of carbonucleophile and its pK_a. The acyclic carbonucleophiles such as ethyl acetoacetate, acetylacetone, dimethyl malonate, dicyanomethane, and bis(phenylsulfone)methane, gave predominantly the monoallylated product (Eq. 2), although the cyclic carbonucleophiles such as tetronic acid, dimedone, and barbituric acid gave predominantly the diallylated product (Eq. 3).

+ $NO_2CH_2CO_2Et$; Pd(OAc)$_2$/TPPTS; $CH_3CN/H_2O/Et_3N$

88% (2)

Pd(OAc)$_2$/TPPTS; C_3H_7CN/H_2O

69% (3)

Alkylation of allylic acetates also occurred under these conditions, NEt_3 – or better 1,8-diazabicyclo[5.4.0]undec-7-ene (dbu) – being used as the base, as well as that of vinyl epoxides.

The alkylation reaction in a two-phase system was extended to various heteronucleophiles [14]. Secondary amines (morpholine, benzylmethylamine, etc.) as well as primary amines (*n*-butylamine, 2,2-diethylpropargylamine, cyclohexylamine, α-methylbenzylamine, etc.) reacted for example with (*E*)-cinnamyl acetate to give only the monoalkylation product in quite good yields (Eq. 4). Even the bis(*N*,*O*-boc)-protected hydroxylamine HN(boc)(Oboc) (boc = *t*-butoxycarbonyl) or hydroxylamine hydrochloride gave the *N*-allyl-protected hydroxylamine or the bis(cinnamyl)hydroxylamine, respectively. Sodium azide and sodium *p*-toluene sulfinate, which are soluble in water, reacted also under these conditions giving the expected allyl azide and allyl *p*-toluene sulfone in 92 and 95% yield, respectively; in this case this is a typical reverse-phase transfer catalysis.

$$\text{Ph-CH=CH-CH}_2\text{OAc} + \text{R-NH}_2 \xrightarrow[\text{CH}_3\text{CN/H}_2\text{O}]{\text{Pd(OAc)}_2\text{/TPPTS}} \text{Ph-CH=CH-CH}_2\text{NHR} \quad (4)$$

70-97%

Such a phase-transfer catalysis was used by Okano et al. [15] in the reduction of allyl chlorides and acetates to the corresponding alkenes (Eq. 5). Water-soluble palladium complexes containing ligands such as polyether phosphines, TPPMS, or carboxylic phosphines allowed the reduction of these substrates. The most active catalyst was $PdCl_2L_2$ with L being a polyether phosphine in a heptane/water solution at reflux, giving a mixture of nonenes in 82% yield, although $PdCl_2(P\text{-}n\text{-}Bu_3)_2$ gave lower yields under these conditions. From the mechanistic point of view, the palladium catalyst transferred the organic substrate into the aqueous phase as a π-allylpalladium species, which then reacted with sodium formate to give the corresponding olefins.

$$\text{Ph-CH=CH-CH}_2\text{Cl} \xrightarrow[\text{PdCl}_2\text{L}_2,\ \text{heptane/H}_2\text{O}]{\text{HCOONa}} \text{Ph-C}_3\text{H}_5 \quad (5)$$

Yield (%) = $P(n\text{-}Bu)_3$/22; $P[(C_2H_4O)_3CH_3]_3$/95; PPh_2mNaO_3SPh/84

All types of nucleophiles, including carbo- as well as heteronucleophiles, allowed an easy recycling of the catalyst using benzonitrile or butyronitrile as the organic solvent, without any decrease in the yields [14].

This allylic alkylation reaction was also performed using supported aqueous-phase catalysis (SAPC) [16]. Alkylation of (*E*)-cinnamyl ethyl carbonate by ethyl acetoacetate occurred in acetonitrile using $Pd(OAc)_2$-TPPTS supported on silica; no leaching of the catalyst was observed, allowing proper recycling of the catalyst.

6.6.3 Applications

A very interesting application of this palladium-catalyzed alkylation in a biphase system is the removal of the allyloxycarbonyl group from allylic esters, carbamates, and carbonates. In a homogeneous organic medium, a variety of nucleophilic species have been used for intercepting the intermediate π-allyl complexes, including carbonucleophiles [17], amines [18–22], thiols [21], carboxylates [23] and hydride donors [24–30]. Genêt et al. used the aqueous palladium catalyst obtained from $Pd(OAc)_2$ and TPPTS for the catalytic allyl transfer, diethylamine being the allyl scavenger [31–34]. Either nitrile/water or diethyl ether/water were equally suitable for removal of the aloc moiety from nitrogen or oxygen.

Deprotection of aloc-protected primary alcohols, such as (*R*)-citronellol, occurred in a few minutes upon exposure to $Pd(OAc)_2$-TPPTS in CH_3CN/H_2O, diethylamine being used in a 2–2.5-fold excess (Eq. 6) [31, 33]. Under these conditions, *t*-butyldiphenyl ether or ester functions are stable. The deprotection of secondary alcohols such as menthol proceeded smoothly.

$Pd(OAc)_2$/TPPTS; CH_3CN/H_2O; 5 eq $HNEt_2$ → OH; 94% (6)

The *N*-aloc protecting group of primary amines such as benzylamine was cleaved rapidly under these standard conditions in quantitative yields (Eq. 7) [31, 33]. It is to be noticed that the *N*-protected amino acid derived from phenylalanine was deprotected without racemization.

$PhCH_2NH$–C(O)–O–allyl → ($Pd(OAc)_2$/TPPTS; RCN/H_2O; 2–5 eq $HNEt_2$) → $PhCH_2NH_2$ 100% (7)

The deprotection of aloc derivatives of secondary amines such as *N,N*-benzylmethylamine under the above conditions gave a substantial amount of the undesired allylamine [33]. However, the use of a 40-fold excess of diethylamine as the π-allyl scavenger led to the desired benzylmethylamine in quite good yield (Eq. 8). The formation of the undesired allylamine was also suppressed using a fivefold excess of diethylamine in a butyronitrile/water system. *N*-Allyloxycarbamates derived from secondary amines, such as morpholine, piperidine, proline, and ephedrine, reacted under the above-mentioned conditions at room temperature within 15 min to give the parent amines in quantitative yields without formation of the undesired allylamine.

PhCH$_2$N(CH$_3$)C(=O)O-allyl $\xrightarrow[\text{RCN/H}_2\text{O, HNEt}_2]{\text{Pd(OAc)}_2\text{/TPPTS}}$ PhCH$_2$NHCH$_3$ + PhCH$_2$N(CH$_3$)CH$_2$CH=CH$_2$ (8)

Et$_2$NH (eq.)/solvent/ratio = **2**/CH$_3$CN/30:70; **40**/CH$_3$CN/97:3; **5**/C$_3$H$_7$CN/100:0

Deprotection of allyl groups from carboxylic allyl esters is also possible using these conditions [32, 34]. In a homogeneous CH_3CN/H_2O medium, the facility of cleavage of the allyl group follows the order allyl > cinnamyl > dimethylallyl. However, under biphasic conditions (C_3H_7CN/H_2O) the allyl group of phenylacetic acid allyl ester was still cleaved at room temperature giving phenylacetic acid in quite good yields, whereas the cinnamyl and the dimethylallyl esters remained intact even after three days at 25 °C. This procedure was used for the selective cleavage of allyloxycarbamate in the presence of substituted allyl carboxylate. For example, the allyloxycarbamate of isonipecotic acid (Eq. 9) was selectively and quantitatively cleaved under homogeneous conditions, in the presence of 1% of palladium complex, without affecting the dimethylallyl carboxylate; the resulting monodeprotected product was then treated with a higher amount of catalyst (5 mol%) to give the free amino acid. In reverse, selective cleavage of an allyloxycarbonate could be performed in the presence of an allylcarbamate, using successively C_3H_7CN and CH_3CN as the nitrile. Selective removal of the allyloxycarbonyl group of doubly protected (1*R*,2*S*)-ephedrine (Eq. 10) occurred in a biphasic butyronitrile/water medium with 5% of palladium complex, the amine being deprotected using a homogeneous acetonitrile/water medium in the presence of 5% palladium catalyst.

(9)

a) 1% mol Pd(OAc)$_2$/TPPTS, CH$_3$CN/H$_2$O, 5 eq Et$_2$NH, 20 min, 96% yield
b) 5% mol Pd(OAc)$_2$/TPPTS, CH$_3$CN/H$_2$O, 5 eq Et$_2$NH, 10 min, 100% yield

(10)

a) 5% mol Pd(OAc)$_2$/TPPTS, C$_3$H$_7$CN/H$_2$O, 5 eq Et$_2$NH, 20 min, 100% yield
b) 5% mol Pd(OAc)$_2$/TPPTS, CH$_3$CN/H$_2$O, 5 eq Et$_2$NH, 15 min, 100% yield

More recently the use of sodium azide, which is soluble in water, as the allyl scavenger allowed the cleavage of allyloxycarbonyl-protected alcohols to occur under essentially neutral conditions [35].

The use of water or a two-phase water/organic solvent system as the reaction medium could also change the selectivity of a given reaction. This was effectively observed in the allylation of uracils and thiouracils [36, 37]. Although the reaction of uracil with (*E*)-cinnamyl acetate in the presence of $Pd(PPh_3)_4$ in tetrahydrofuran gave a complex mixture of mono- and diallylated products, performing the reaction in CH_3CN/H_2O with $Pd(OAc)_2$–TPPTS as the catalyst led to allylation only at N1 in quite good yield (Eq. 11). The reaction was extended to various 5-substituted uracils.

(*E*) $PhCH{=}CHCH_2OAc$
$Pd(OAc)_2$/TPPTS/*dbu*
CH_3CN/H_2O

(11)

R	H	CH_3	Br	Cl	F
yield %	80	53	66	32	14

The same behavior was observed in the allylation of thiouracils. Performing the reaction in CH_3CN/H_2O with $Pd(OAc)_2$–TPPTS as the catalyst gave a unique product of monoallylation at sulfur, whereas the use of dioxane as the solvent in the presence of $Pd(PPh_3)_4$ again gave a complex mixture of products of allylation at N1, N3, and sulfur.

6.6.4 Conclusion

Although the first aim of the use of a water-soluble palladium catalyst in allylic alkylation in a two-phase system was the recycling of the catalyst, this methodology finds quite interesting applications in the deprotection of peptides as well as in the selective alkylation of uracils and thiouracils. It is to be expected that new applications and new developments will appear in the near future, such as the allylation of water-soluble compounds or the extensive use of supported aqueous-phase catalysis.

References

[1] J. Tsuji, H. Takahashi, M. Morikawa, *Tetrahedron Lett.* **1965**, 4387.

[2] B. M. Trost, T. R. Verhoeven, in *Comprehensive Organometallic Chemistry* (Ed.: G. Wilkinson), Pergamon Press, Oxford, **1982**, Vol. 8, p. 799.

[3] J. Tsuji, *Palladium Reagents and Catalysts: Innovations in Organic Synthesis*, Wiley, Chichester, **1995**.
[4] S. A. Godleski, in *Comprehensive Organic Synthesis* (Eds.: B. M. Trost, I. Fleming, G. Pattenden), Pergamon Press, Oxford, **1991**, Vol. 4, p. 585.
[5] P. J. Harrington, in *Comprehensive Organometallic Chemistry* (Eds.: E. W. Abel, F. G. A. Stone, G. Wilkinson), Pergamon Press, Oxford, **1995**, Vol. 12, p. 959.
[6] C. G. Frost, J. Howarth, J. M. Williams, *Tetrahedron: Asymm.* **1992**, *3*, 1089.
[7] S. I. Murahashi, Y. Taniguchi, Y. Imada, Y. Tanigawa, *J. Org. Chem.* **1989**, *54*, 3292.
[8] A. Tenaglia, B. Waegell, *Tetrahedron Lett.* **1988**, *29*, 4852–4855.
[9] M. Safi, D. Sinou, *Tetrahedron Lett.* **1991**, *32*, 2025.
[10] C. Amatore, E. Blart, J. P. Genêt, A. Jutand, S. Lemaire-Audoire, M. Savignac, *J. Org. Chem.* **1995**, *60*, 6829.
[11] F. Monteil, P. Kalck, *J. Organomet. Chem.* **1994**, *482*, 45.
[12] S. Sigismondi, D. Sinou, *J. Mol. Catal.* **1997**, *116*, 289.
[13] J. P. Genêt, E. Blart, M. Savignac, *Synlett* **1992**, 715.
[14] E. Blart, J. P. Genêt, M. Safi, M. Savignac, D. Sinou, *Tetrahedron* **1994**, *50*, 505.
[15] T. Okano, Y. Moriyama, H. Konishi, J. Kiji, *Chem. Lett.* **1986**, 1463.
[16] P. Schneider, F. Quignard, A. Choplin, D. Sinou, *New J. Chem.* **1996**, *20*, 545.
[17] H. Kunz, C. Unverzagt, *Angew. Chem., Int. Ed. Engl.* **1984**, *23*, 436.
[18] H. Kunz, H. Waldmann, *Angew. Chem., Int. Ed. Engl.* **1984**, *23*, 71.
[19] H. Kunz, H. Waldmann, U. Klinkhammer, *Helv. Chim. Acta* **1988**, *71*, 1868.
[20] A. Merzouk, F. Guibé, A. Loffet, *Tetrahedron Lett.* **1992**, *33*, 477.
[21] J. P. Genêt, E. Blart, M. Savignac, S. Lemeune, S. Lemaire-Audoire, J. M. Bernard, *Synlett* **1993**, 680.
[22] P. Lloyd-Williams, A. Merzouk, F. Guibé, F. Albericio, E. Giralt, *Tetrahedron Lett.* **1994**, *35*, 4437.
[23] P. D. Jeffrey, S. W. McCombie, *J. Org. Chem.* **1982**, *47*, 587.
[24] I. Minami, Y. Onashi, I. Shimizu, J. Tsuji, *Tetrahedron Lett.* **1985**, *26*, 2449.
[25] Y. Hayakawa, S. Wakabayashi, H. Kato, R. Noyori, *J. Am. Chem. Soc.* **1990**, *112*, 1691.
[26] F. Guibé, O. Dangles, G. Balavoine, A. Loffet, *Tetrahedron Lett.* **1989**, *30*, 2641.
[27] O. Dangles, F. Guibé, G. Balavoine, S. Lavielle, A. Marquet, *J. Org. Chem.* **1987**, *52*, 4984.
[28] F. P. Rutjes, M. M. Paz, H. Hiemstra, W. N. Speekamp, *Tetrahedron Lett.* **1991**, *32*, 6629.
[29] R. Beugelmans, S. Bourdet, A. Bigot, J. Zhu, *Tetrahedron Lett.* **1994**, *35*, 4349.
[30] M. Dessolin, M. G. Guillerg, N. Thieriet, F. Guibé, A. Loffet, *Tetrahedron Lett.* **1995**, *36*, 5741.
[31] J. P. Genêt, E. Blart, M. Savignac, S. Lemeune, J. M. Paris, *Tetrahedron Lett.* **1993**, *34*, 4189.
[32] S. Lemaire-Audoire, M. Savignac, E. Blart, G. Pourcelot, J. P. Genêt, J. M. Bernard, *Tetrahedron Lett.* **1994**, *35*, 8783.
[33] J. P. Genêt, E. Blart, M. Savignac, S. Lemeune, S. Lemaire-Audoire, J. M. Paris, J. M. Bernard, *Tetrahedron* **1994**, *50*, 497.
[34] S. Lemaire-Audoire, M. Savignac, G. Pourcelot, J. P. Genêt, J. M. Bernard, *J. Mol. Cat.* **1997**, *116*, 247.
[35] S. Sigismondi, D. Sinou, *J. Chem. Res. (S)* **1996**, 46.
[36] S. Sigismondi, D. Sinou, M. Moreno-Mañas, R. Pleixats, M. Villaroya, *Tetrahedron Lett.* **1994**, *35*, 7085.
[37] C. Goux, S. Sigismondi, D. Sinou, M. Moreno-Mañas, R. Pleixats, M. Villaroya, *Tetrahedron* **1996**, *52*, 9521.

6.7 Hydrodimerization

Noriaki Yoshimura

6.7.1 Introduction

The linear telomerization reaction [1] of dienes (the taxogen) with nucleophiles (the telogen) such as alcohols, amines, carboxylic acids, active methylene compounds, phenols, or water provides an elegant method for the synthesis of various useful compounds (Eq. 1).

2 butadiene + HX → X-substituted octadienes (linear and branched) (1)

HX = ROH, H_2O, HOAc, HNR_2, etc.

This reaction is catalyzed by nickel or palladium complexes. However, the selectivity and activity of nickel catalysts are lower than those of palladium catalysts. Palladium-catalyzed reactions give linear dimers selectively and no cyclization takes place. Not only the zerovalent palladium complexes but also certain bivalent ones can be used as active catalysts in combination with excess PPh_3. However, the zerovalent palladium complexes are somewhat tedious to prepare and unstable in oxygen, so easily available and stable bivalent palladium compounds such as $Pd(OAc)_2$ are generally used with PPh_3. A proposed mechanism is given in Figure 1 [1f].

If water serves as telogen together with 2 mol of taxogen, the telomerization becomes a special hydrodimerization (Eq. 2). The consequent product in the case of butadiene as taxogen can be hydrogenated to produce 1-octanol, which has a considerable market as a raw material for plasticizers for poly(vinyl chloride).

2 butadiene + H_2O —(Pd, TPPMS)→ octadienol (OH) —(H_2, Ni)→ 1-octanol (OH) (2)

The hydrodimerization has so far been carried out in polar organic solvents such as *t*-BuOH, THF, acetone, and acetonitrile by using triphenylphosphine-modified palladium complexes. Conventional attempts to commercialize the palladium-complex-catalyzed telomerization have failed, in spite of great ef-

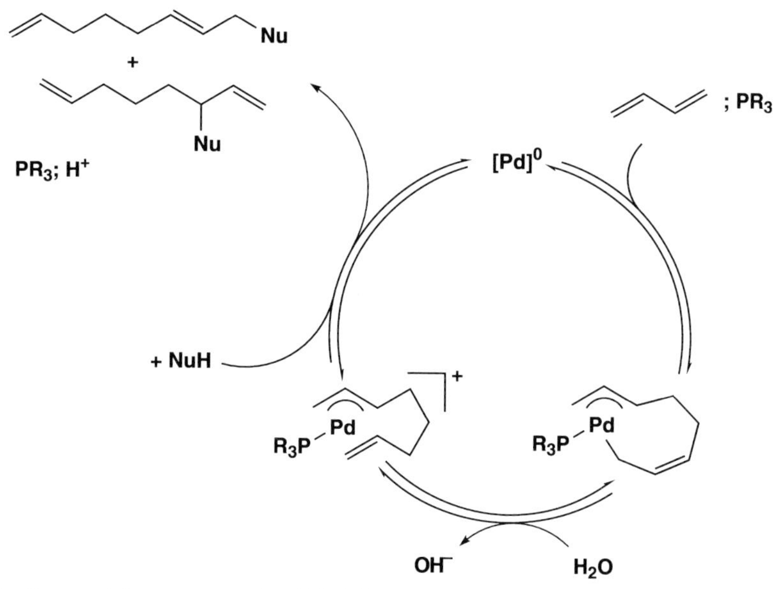

Figure 1. Proposal for a mechanism [22, 23].

forts, for the following reasons: (1) palladium complex catalysts are thermally unstable and the catalytic activity decreases markedly when, as a means of increasing the thermal stability, the ligand concentration is increased; (2) a sufficiently high reaction rate to satisfy industrial needs cannot be obtained; (3) low selectivity; and (4) distillative separation of reaction product and unreacted butadiene from the reaction mixture causes polymeric products to form and the palladium complex to metallize. The palladium complex is so expensive that its separation from the reaction medium and re-use become important. This point is especially crucial for continuous processes.

Concerning the catalyst separation from the telomerization reaction mixture, apart from direct distillation the following methods have been reported:

1. The telomerization reaction is carried out using the catalyst modified with amine-containing phosphines, then hydrochloric acid is introduced into the above reaction mixture. As a result phosphine is changed to the hydrochloric acid salt of the amine, so the modified catalyst is dissolved in water. The catalyst can be re-used after treatment with an alkaline solution [2].
2. The two-phase reaction method using an aqueous homogeneous catalyst enables products to separate by decantation or solvent extraction [3a].

In 1991, Kuraray succeeded in commercializing the production of 1-octanol by developing an aqueous homogenous catalyst [3 b].

6.7.2 Development of Technologies

Conventional dimerization of butadiene uses a trivalent phosphine as a ligand. It has been reported [4] that the catalytic activity in the telomerization is highest when the molar ratio P/Pd is kept at 1–2:1 and rapidly decreases when the ratio becomes about 6:1. Also, on telomerization with water (hydrodimerization), the catalytic activity decreases markedly with increasing P/Pd molar ratio (Figure 2).

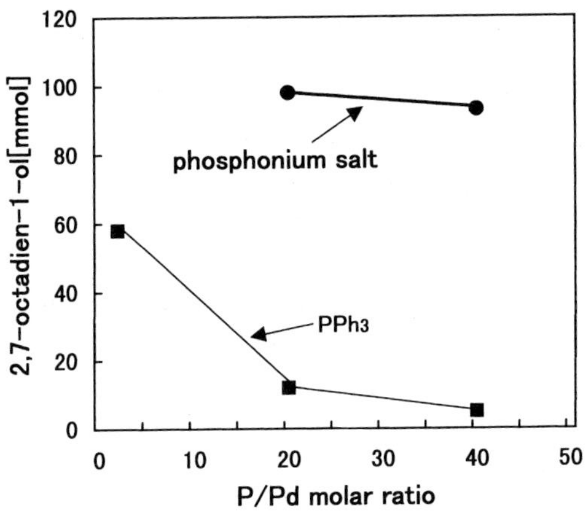

Figure 2. Effect of the ligand concentration on the telomerization of butadiene with water in sulfolane/water (50:50, wt./wt.) solution containing 2 mM $Pd(OAc)_2$ and 8 wt% triethylamine, at 75 °C for 1 h under 5 bar CO_2.

On the other hand, in order to increase the thermal stability of a palladium complex catalyst, it is necessary for a large excess of the phosphine ligand to be present together with the catalyst. This contradiction was solved by using a tetravalent phosphonium salt (**1**) as a ligand [5]. With the use of the new phosphonium salt ligand (a salt of triphenylphosphine monosulfonate, TPPMS; cf. Section 3.2.1), increasing the P/Pd molar ratio not only maintains the reaction rate at a high level (Figure 3) but shows no appreciable time-dependent deterioration of the catalytic activity upon repeated reactions, as shown in Figure 3.

Another problem associated with the use of a phosphine on a commercial scale is its conversion, due to the presence of a small amount of oxygen in the reaction zone, to the corresponding phosphine oxide, which will not act effectively as a ligand. The use of a phosphonium salt can minimize this type of conversion. Although the mechanism of the action of the phosphonium salt has not been made clear, it is considered, from the fact that aryl groups should be present, that a very rapid equilibrium with the corresponding phosphine may

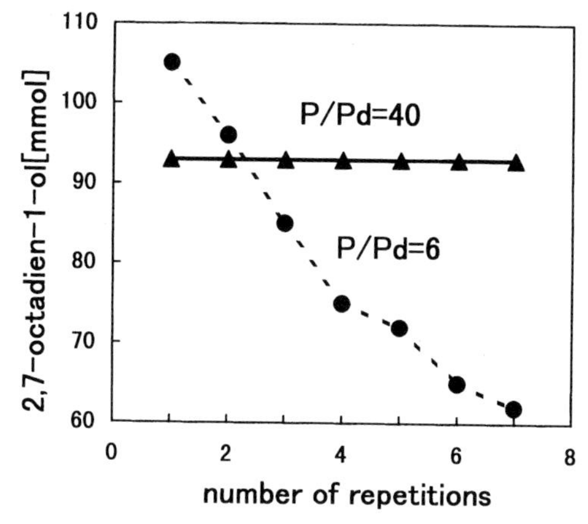

Figure 3. The catalytic activity of repeated hydrodimerization in sulfolane/water (50 : 50) solution containing 2 mM $Pd(OAc)_2$ and 8 wt% triethylamine, at 75 °C for 1 h under 5 bar CO_2. Products are extracted with hexane.

partially occur. Upon analysis of the actual reaction mixture, however, no trivalent phosphine is detected either in the reaction solution or in the palladium complex.

$[(C_6H_5)_2P(\text{allyl-R})(C_6H_4SO_3Li)]^+$ HCO_3^-

R = H, CH_3,

1

For the hydrodimerization of butadiene with water, attempts have been made to increase the reactivity by adding acidic solids [6], salts such as sodium phosphate [7], emulsifiers [8], carbon dioxide [9], or the like, with no satisfactory result. In particular, the reaction rate increases under carbon dioxide pressure, but carbonate ions, not carbon dioxide itself, are considered to play an important role in this effect. It is known that the carbonate ion concentration in water is very low even under carbon dioxide pressure. If the carbonate ion is the true reactant, the reaction rate should increase in proportion to the carbonate ion concentration. But inorganic carbonates showed little effect on the reaction rate. Therefore various tertiary amines without active hydrogen were added to the reaction mixture under carbon dioxide pressure [10]. Diamines and bifunctional amines inhibited the reaction. The reaction rate increased only in the presence of a monoamine having a pK_a of at least 7, almost linearly with its concentration (Figure 4).

Telomerization requires a solvent that can dissolve in both water and butadiene. To select a suitable solvent for this purpose the separability of the reaction

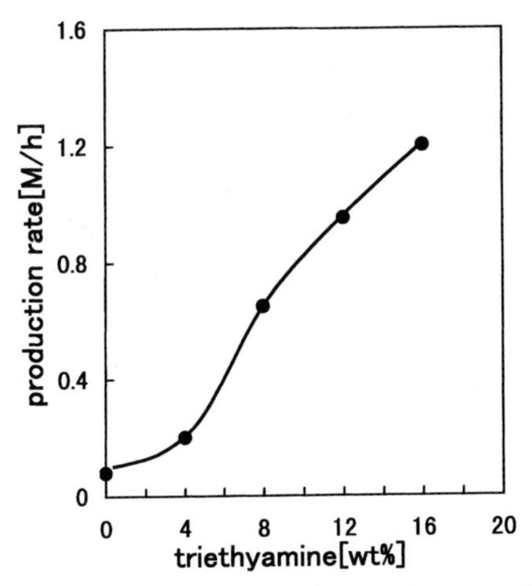

Figure 4. The effect of concentration of triethylamine on the rate of telomerization of butadiene with water in sulfolane/water (80:20) solution containing 2.7 mM $Pd(OAc)_2$ and 54 mM ligand, at 75 °C for 3 h under 15 bar CO_2.

products from the catalyst used should be considered. According to past reports, the selectivity to 2,7-octadien-1-ol, which is of high industrial value, has been 70% at most. Use of sulfolane, which had not been studied as a reaction solvent, realized a high 2,7-octadien-1-ol selectivity of at least 83%, and a high reaction rate (Table 1) [10]. Sulfolane, having both high stability in its aqueous solution and high solubility for butadiene, is considered to be the most suitable solvent for industrial long-term use. Recently, it has been reported [11] that hydrodimerization into 2,7-octadiem-1-ol can be performed in the absence of co-solvent with moderately high yields when trialkylamine with a long alkyl

Table 1. Effect of solvent on the telomerization of butadiene with water as telogen.[a]

Solvent	Octadienol		
	Yield [mmol]	Selectivity [%]	Ratio [b]
Sulfolane	91	92	92:8
CH_3CN	30	60	81:19
t-BuOH	18	67	71:29
Acetone	12	49	53:47
Dimethylsulfoxide	8	65	72:28
Dimethylformamide	9	67	82:18
Water only	1	71	81:19

[a] In solvent/water (55:45) solution (60 mL) containing 230 mmol butadiene, 0.1 mmol $Pd(OAc)_2$, 12 wt.% triethylamine, 2 mmol ligand, at 75 °C for 3 h under CO_2.

[b] 2,7-octadien-1-ol/1,7-octadien-3-ol.

chain is used in a two-phase aqueous catalyst system (Pd/TPPTS). The effect of amines having long alkyl chains is explained by an increase of mass transfer between the organic and aqueous phases, due to the formation of micelles.

For the telomerization of butadiene, distillative separation methods cannot be employed to separate the product from the reaction mixture containing catalyst, because the palladium complex catalyst has a lower thermal stability and high-boiling compounds would accumulate in the catalyst-containing solution that has been recycled. Therefore an extraction separation method has been chosen (Figure 5). In the present reaction, water acts as a nucleophile and the product hardly dissolves in water. Therefore the process is capable of retaining the catalyst component in the aqueous solution being used, and of extracting the product selectively. Thus, in order to solubilize the catalyst component in the aqueous sulfolane solution used, a hydrophilic group (e.g., sulfonic acid salt) was introduced into the phosphonium salt ligand and the product was extracted with an aliphatic saturated hydrocarbon such as hexane [5, 10, 12, 13]. This extraction separation method has the advantages that: (1) the catalyst and product can be separated without heating them, so that thermal deactivation is avoided; (2) extraction equilibrium is achieved for all compounds, so that the accumulation of catalyst poisons and high-boiling by-products is minimal. This method is commercially applicable only when the resultant catalytic lifetimes and the elution losses of catalytic components into the extractant layer containing the product are within commercially acceptable ranges.

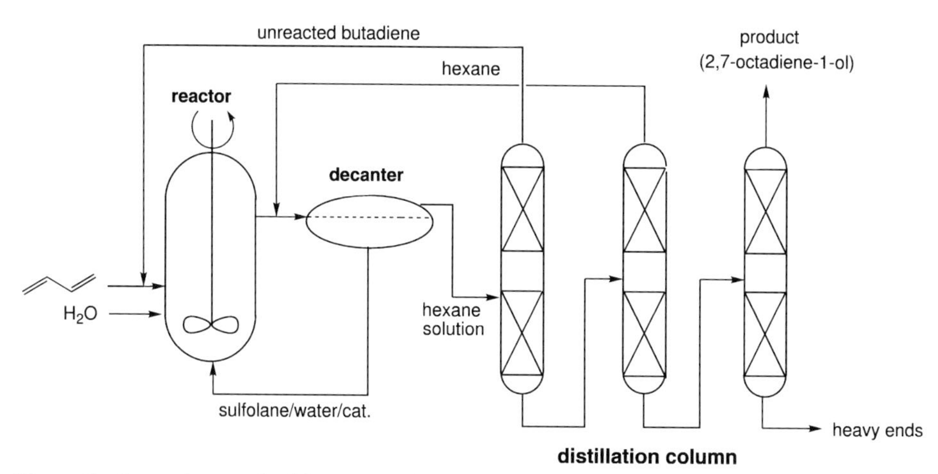

Figure 5. Extraction method for separating butadiene hydrodimerization products from catalyst-containing reaction mixture.

6.7.3 Process of the Manufacture of 1-Octanol and Other Derivatives

This process consists of four steps: hydrodimerization, extraction, hydrogenation, and distillation [12–14]. The telomerization step comprises feeding butadiene and water continuously to the reaction zone in the presence of a separately prepared palladium catalyst (1–5 mmol L^{-1}), a phosphonium salt ligand (40–50 mol/mol of palladium), and a solution of triethylammonium hydrogencarbonate in aqueous sulfolane, and reacting them at a temperature of 60–80 °C under a total pressure of carbon dioxide of 10–20 bar, to achieve a 2,7-octadien-1-ol selectivity of 90–93 % and a 2,7-octadien-3-ol selectivity of 4–5 %. In the following extraction step, 50–70 % of the reaction products are extracted with hexane, and the aqueous sulfolane containing the catalyst, part of the products, and the triethylammonium hydrogencarbonate is again circulated to the reaction step. The elution loss of the catalyst is only several parts per million. The sulfolane concentration is set at about 40 wt % in view of the extraction ratio of the products, the solubility of butadiene, and the elution loss of the catalyst. After unreacted butadiene and hexane have been recovered from the extraction mixture, 2,7-octadien-1-ol is purified by distillation. The 2,7-octadien-1-ol obtained is hydrogenated on a fixed bed in the presence of a nickel catalyst and at 30–80 bar H_2 and 130–180 °C to yield 1-octanol nearly quantitatively. A plant with a capacity of approx. 5000 t/y is on stream.

2,7-octadien-1-ol is a highly reactive compound having double bonds and a hydroxyl group. Reacting this compound in the presence of a copper chromite catalyst at a temperature of 220 °C causes intramolecular dehydrogenation/hydrogenation, to yield isomerized 7-octenal in a yield of at least 80 % [15]. This

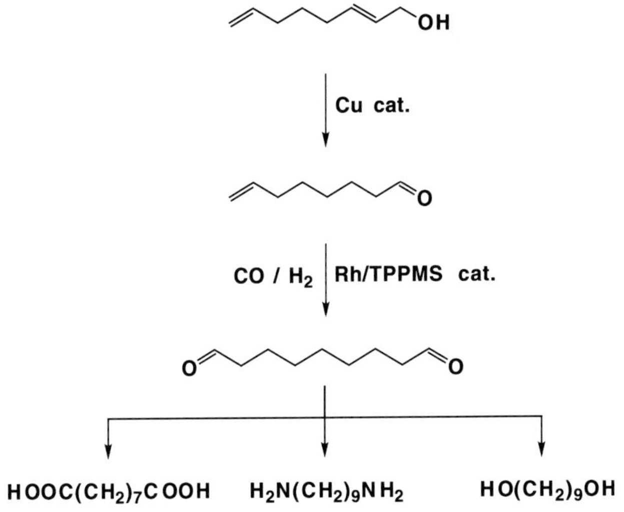

OH

Cu cat.

O

CO / H_2 Rh/TPPMS cat.

O O

Scheme 1 $HOOC(CH_2)_7COOH$ $H_2N(CH_2)_9NH_2$ $HO(CH_2)_9OH$

aldehyde is hydroformylated to the dialdehyde, which is then hydrogenated to give 1,9-nonanediol [16]. The extraction separation method is also applicable to this hydroformylation. The dialdehyde can also give, on air oxidation in an acetic solvent with a copper catalyst, azelaic acid and, on reductive amination in ammonia in the presence of a nickel catalyst, 1,9-nonanediamine (Scheme 1).

6.7.4 Application

Homogenous telomerization reaction of butadiene with ammonia produces trioctadienylamine (**5**) as a major product. It is different from the primary and secondary amines obtained (**4**) because the nucleophilicity of the alkylamine is stronger than that of ammonia (Scheme 2). When this reaction is carried out in two phases using an aqueous homogeneous palladium catalyst modified by TPPTS, primary and secondary octadienylamines are obtained as major products, as shown in Table 2. The primary amine is immediately extracted into the organic phase from the aqueous catalyst phase by using solvents such as toluene. In this way, the consecutive reactions of primary products that are slightly soluble in water are avoided. The primary octadienylamines are formed with a selectivity of 88% [18].

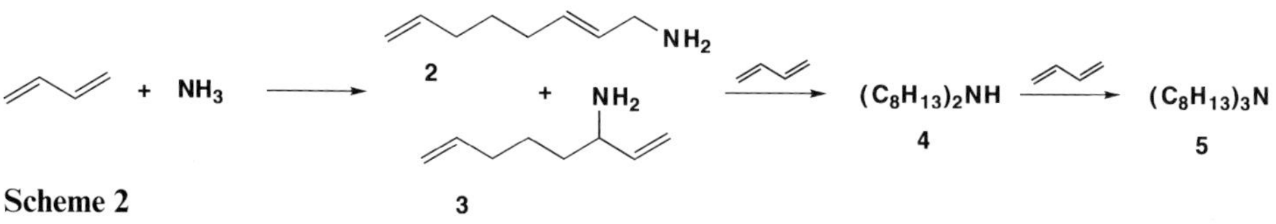

Scheme 2

Table 2. Comparison of single-phase versus two-phase catalytic reaction process in the telomerization of butadiene and ammonia.[a)]

Process	Selectivity [%]			Ratio	Total yield [%][b)]
	2 + 3	**4**	**5**	**(2 + 3):4:5**	
Single-phase	2	4	61	1:2:31	21
Two-phase	31.5	25.5	1.5	21:17:1	24

[a)] Reaction conditions: single-phase: $T = 100\,°C$, $t = 1$ h, butadiene/ammonia = 10:1, Pd/butadiene = 1:1000, 20 mL *t*-butyl alcohol, Pd/PPh_3 = 1:3.4; two-phase: $T = 100\,°C$, $t = 1$ h, butadiene/ammonia = 10:1, Pd/butadiene = 1:1000, H_2O/CH_2Cl_2 = 2:1, Pd/TPPTS = 1:4

[b)] Based on ammonia.

Telomerization of butadiene with formic acid or its salts can produce 1,7-octadiene (Eq. 3), which is useful as a modifier for polyolefins. This reaction proceeds with the same system that is used for the production of octadienol, at a temperature of 50–70 °C, to yield 1,7-octadiene/1,6-octadiene in a ratio of 88:12. Since these compounds phase-separate together, the product layer can be readily separated and the sulfolane layer containing the catalyst can be circulated for re-use [19].

+ HCOONa —Pd/TPPMS, H_2O→ + (3)

When reductive dimerization of isoprene in the presence of an alkali-metal salt of formic acid is carried out, the regioselectivity to dimethyloctadiene (Eq. 4) varies with the species of alkali metal, as shown in Table 3 [20].

+ HCOOX —Pd/TPPMS, H_2O/DMF→ **6** + **7** (4)

X = Na, K, H

Table 3. Hydrodimerization of isoprene with formic acid alkali-metal salt.[a)]

Alkali metal	Temperature [°C]	Conversion [%]	Selectivity [%] **6**	Selectivity [%] **7**
Na	60	65	14	73
K	85	91	88	11

a) 0.027 mmol $Pd(acac)_2$, 0.054 mmol TPPMS, 18.5 mmol formic acid alkali-metal salt, 5 mL DMF, 5 mL H_2O and 2.5 g isoprene for 2 h.

Etherification of carbohydrate is an important reaction. The two-phase reaction of butadiene with saccharose by an aqueous palladium complex catalyst increases the reaction yield of the desired ether products (Eq. 5) [21].

saccharose + n butadiene —Pd/TPPTS, H_2O→ (5)

R = H, ,

References

[1] (a) J. Tsuji, *Acc. Chem. Res.* **1973**, *6*, 8; (b) J. Tsuji, *Adv. Organomet. Chem.* **1979**, *17*, 141; (c) J. Beger, H. J. Reichel, *J. Prakt. Chem.* **1973**, *315*, 1067; (d) R. F. Heck, *Palladium Reagents in Organic Syntheses,* Academic Press, London, **1985**, pp. 327; (e) R. Mynott, B. Raspel, K. P. Schick, *Organometallics* **1986**, *6*, 473; (f) cf. Ref. [22, 23].

[2] Kureha Chem. Ind. (S. Enomoto, H. Wada, S. Nishida, Y. Mukaida, M. Yanaka, H. Takita), JP 51/149206 A2 (1976).

[3] (a) Rhône-Poulenc Ind. (E. Kuntz, L. France), US 4.142.060 (1979); (b) N. Yoshimura, in *Applied Homogeneous Catalysis by Organometallic Compounds* (Eds.: B. Cornils, W. A. Herrmann), VCH Weinheim, **1996**, Vol. 1, p. 351.

[4] D. Rose, H. Lepper, *J. Organomet. Chem.* **1973**, *49*, 473.

[5] Kuraray Ind. (T. Maeda, Y. Tokitoh, N. Yoshimura), US 4.927.960 (1990), 4.992.609 (1991), 5.100.854 (1991).

[6] Mitsubishi Chem. Int. (T. Onoda, H. Wada, T. Sato, Y. Kasori), JP 53/147.013 A2 (1978).

[7] Toray Ind. (J. Tsuji, T. Mitsuyasu), JP 49/35.603 B4 (1974).

[8] Kureha Chem. Ind. (S. Enomoto, H. Takita, S. Wada, Y. Mukaida), JP 51/23.210 A2 (1976).

[9] K. E. Atkins, W. E. Walker, R. M. Manyik, *Chem. Commun.* **1971**, 330.

[10] Kuraray Ind. (N. Yoshimura, M. Tamura), US 4.356.333 (1982).

[11] E. Monflier, P. Bourdauducq, J.-L. Couturier, J. Kervennal, *J. Mol. Cat.* **1995**, A97, 29.

[12] Kuraray Ind. (Y. Tokitoh, N. Yoshimura), US 5.057.631 (1991).

[13] Kuraray Ind. (Y. Tokitoh, T. Higashi, K. Hino, M. Murasawa, N. Yoshimura), US 5.118.885 (1992).

[14] Kuraray Ind. (N. Yoshimura, M. Tamura), US 4.417.079 (1983).

[15] Kuraray Ind. (N. Yoshimura, M. Tamura), US 4.510.331 (1985).

[16] Kuraray Ind. (M. Matsumoto, N. Yoshimura, M. Tamura), US 4.510.332 (1985).

[17] Kuraray Ind. (N. Yoshimura, M. Tamura), US 4.334.117 (1982).

[18] T. Prinz, W. Keim, B. Driessen-Hölscher, *Angew. Chem., Int. Ed. Engl.* **1996**, *35*, 1708.

[19] (a) Kuraray Ind. (T. Tsuda, N. Yoshimura), EU Appl. 704.417 A2 (1996); (b) Shell Internationale Res. Maatschappij BV (K. Nozaki), US 4.180.694 (1979).

[20] Shell Internationale Res. Maatschappij BV (K. Nozaki), US 42.383 (1979) and N. Yoshimura, experimental results.

[21] Eridania Beghin Say (I. Pennequin, A. Mortreux, F. Petit, J. Mentech, B. Thiriet), Demande FR 2.693.188 Al 940.107 (1992).

[22] P. W. Jolly et al., *Organometallics* **1985**, *4*, 1945.

[23] P. W. Jolly et al., *Organometallics* **1986**, *5*, 473.

6.8 Biological Conversions

Peter J. Quinn

6.8.1 Introduction

Water is the universal solvent in biology. In the order of 80% of the gross weight of living organisms consists of water. The catalysts responsible for mediating the biochemical reactions that create and sustain life depend on an aqueous environment to preserve their stability and catalytic functions. Moreover, their activity is limited to reaction conditions of temperature, pressure, pH, etc., which are compatible with survival of the living organism.

Many biological catalysts, "enzymes," often acquire their catalytic functions under these stringent reaction conditions by incorporating transition metals into their catalytic site. The metals are coordinated to ligands, which are constituents of the polypeptide chain, and participate in the formation of transition-state complexes with substrates in the performance of biochemical reactions. Whereas Nature can provide a rich diversity of organometallic catalysts that require an aqueous solvent, it is a challenge – as seen below – to the chemist to duplicate these enzymic reactions so that they can be exploited and adapted to industrial-scale processes.

6.8.2 Biological Substrates

All living cells are bounded by a membrane that divides living from nonliving matter. Higher organisms are further compartmentalized by a complex system of subcellular membranes. An important group of constituents of these membranes are lipids. The particular feature of these lipids is that the are amphipathic and are arranged in a bimolecular layer which serves to orient and support the various functional proteins. The hydrocarbon components of these lipids are usually unsaturated, straight-chain fatty acids in which *cis* double bonds are located at specific positions along the hydrocarbon chain. Such fatty acids also dominate the lipid metabolic storage depots and represent the major source of fat in our diet.

Chemical modifications of these unsaturated fatty acids in living organisms have greatly increased our knowledge of the role of these constituents in biological membranes. Their presence in food, however, presents problems with storage and catalytic hydrogenation in processing of fats and oils to prevent spoilage, and development of off-flavors is a common practice. Such processing, nevertheless, results in the creation of *trans* isomers of fatty acids which are known to present a health risk.

6.8.3 Hydrogenation of Unsaturated Lipids in Aqueous Dispersions

The idea that chemical modification of the lipids of biological membranes could be achieved *in situ* was first demonstrated in 1976 [1]. The rationale underlying the work was that if the unsaturated double bonds were largely responsible for the fluid character of the membrane lipid matrix, their saturation would result in a reduction in fluidity. Although simple in concept, the practice required application of an entirely novel approach to the catalytic hydrogenation of lipids.

It was found that conventional hydrogenation catalysts, such as Adam's catalyst, were unable to bring about hydrogenation of unsaturated lipids dispersed in aqueous systems. The reason was apparent; the catalyst was located in the aqueous phase and the substrate double bonds were sequestered in a hydrophobic domain created within dispersed aggregates of the lipid. The solution to this problem was to employ homogeneous catalysts in which complexes of transition metal atoms with suitable ligands were able to gain access to the lipid substrate arranged in a bilayer configuration. The initial work was performed using rhodium complexes with triphenylphosphines (TPP) designed for hydrogenation and hydroformylation reactions in organic solvents [2] but subsequently water-soluble homogeneous catalysts [3] were found to be active against lipid substrates in aqueous dispersions.

One of the objectives in performing hydrogenation reactions *in situ* was to modulate the fluidity of biomembranes and to examine the role of membrane lipid fluidity in biochemical and physiological functions. Secondly, because membrane lipids with six or more unsaturated double bonds were found to be significant components of some membranes, such as the retinal rod membranes of the eye, the hydrogenation of these lipids was thought to be a useful tool to identify their role in these membranes.

Many transition metal complexes capable of activating molecular hydrogen are known [4]. Most of these complexes have been shown to catalyze the efficient reduction of unsaturated bonds, e.g., olefinic ones. When such catalysts are used in biological systems, however, there are a number of factors that need to be taken into account. In the case of living organisms, for example, it is essential that the catalyst is nontoxic or at least the level of toxicity is low at concentrations required to sustain a reasonable level of hydrogenation. Toxicity can arise by breakdown of the catalyst complex and liberation of the transition metal element and/or ligands of the complex, either of which may be toxic. Furthermore, side reactions may lead to the formation of unwanted, although not necessarily toxic, by-products. Such reactions include ligand exchange with biomolecules resulting in complexes with altered catalytic properties. Side reactions are potentially damaging in the case of sulfonated derivatives of Wilkinson's catalyst, for example, where the catalytically active species, $RhH(SP\phi_2)_3$ and $[Rh(SP\phi_2)_3]^+$, are known to hydrogenate $=C=O$ functions in addition to *cis* unsaturated bonds of olefins. Reaction of biochemical compounds of a susceptible chemical configuration could have repercussions on cell viability.

Chemical catalysis is often performed under conditions of temperature, etc., that are well outside the physiological range. In biological applications the catalyst complex must be stable under the conditions required to preserve stability of biomembranes or viability of living organisms. At the same time reasonable reaction rates must be sustained under these physiological conditions. Ideally the presence of the catalyst in the system should not affect the properties of the membrane other than in the response of the membrane to the altered level of saturation of the constituent lipids. This can be achieved by removal of the catalyst complex at the completion of the hydrogenation reaction.

Because of these relatively stringent requirements there are comparatively few complexes that are suitable for biological applications. The group of complexes such as $[Co(CN)_5]_3^-$, for example, although very active under conditions appropriate for hydrogenation of biological membranes, are stable only in the presence of excess cyanide [5]. Another common ligand, 2-aminopyridine, although producing highly active catalysts under relatively mild conditions, is highly toxic to living cells. Another group of the type $RuCl_n(H_2O)_{6-n}$ require high temperatures and concentrated chloride solutions to produce even modest rates of hydrogenation of unsaturated fatty acids [6]. Finally, the classic group of organometallic compounds containing low-valent transition metal ions are largely unsuitable because of their unstable character in aqueous media.

6.8.3.1 Water-insoluble Homogeneous Catalysts

The first catalyst used in homogeneous catalytic hydrogenation of membrane lipids was Wilkinson's catalyst [2, 7, 8]. One of the features of Wilkinson's catalyst that limits its use in biomembrane systems is its low solubility in water. Nevertheless, the catalyst is active against unsaturated lipid substrates, but at a somewhat reduced rate. This can be seen in Figure 1, which shows the relationship between the initial rate of hydrogenation of soya phosphatidylcholine in aqueous mixtures of tetrahydrofuran in the presence of Wilkinson's catalyst. It can be seen that the initial rate of hydrogen uptake decreases as the proportion

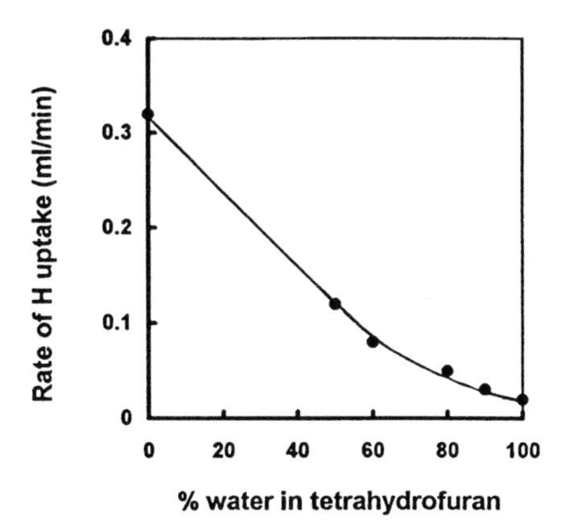

Figure 1. Initial rate of uptake of hydrogen by a dispersion of soya phosphatidylcholine in aqueous tetrahydrofuran in the presence of Wilkinson's catalyst [1].

of water in the system increases, and reaches a limiting rate when the phospholipid assumes a bilayer form. Despite differences in initial reaction rate, virtually complete hydrogenation occurs in all combinations of solvent.

In adapting water-insoluble catalysts for use in biological systems it is necessary to introduce the catalyst into the membrane using a solvent vector. Solvents such as tetrahydrofuran and dimethyl sulfoxide have been found to be useful. The introduction of catalyst in a minimum amount of solvent, which is miscible with water, causes the insoluble complex to partition into the hydrophobic domain created by the lipid substrate. The catalyst obviously cannot be removed subsequently from the substrate without destroying the integrity of the membrane. It is also important to verify that the solvent used to introduce the catalyst does not perturb the stability of the membrane.

The original studies of hydrogenation of phospholipids dispersed in aqueous systems were performed using Wilkinson's catalyst introduced in a solvent vector of tetrahydrofuran [1, 9, 10]. It was shown that complete hydrogenation of the dispersed lipid could be achieved under relatively mild conditions of temperature, hydrogen pressure, and catalyst concentration. Biophysical studies employing differential scanning calorimetry and X-ray diffraction confirmed that the solvent used to deliver the catalyst and the presence of catalyst in the lipid bilayer did not drastically alter the structural properties of the membrane. Wilkinson's catalyst has also been used to hydrogenate model membranes prepared from lipid extracts of rat liver mitochondria and microsomes and human erythrocytes [11].

It was found that the presence of cholesterol markedly influences the hydrogenation of mixed phospholipid dispersions [12] because it restricts partition of the catalyst into the lipid bilayer structure. It was also found that no dihydrocholesterol forms during hydrogenation of the phospholipid, showing that cholesterol is not a substrate for reaction under the conditions employed.

6.8.3.2 Water-soluble Homogeneous Catalysts

A major advance in the application of homogeneous catalytic hydrogenation methods to modulate lipid phase behavior was the use of water-soluble catalysts. The need to employ solvent vectors to introduce the catalyst into the membrane can be avoided and there is more scope for removal of the catalyst at the end of the reaction. Water-soluble complexes can be removed simply by washing, gel filtration, density gradient centrifugation, and, in the case of charged complexes, by adsorption to ion-exchange resins.

Synthesis of the first water-soluble catalyst complexes was reported in 1973 [3] and was based on the use of sulfonated triphenylphosphine to replace TPP. The sulfonated derivative was found to stabilize the lower oxidation states of a number of transition metals such as Rh, Ru, Ir, Pt, Ni, and Cu in aqueous systems and these water-soluble catalysts facilitated hydrogenation of soluble substrates like pyruvic acid.

The water-soluble complexes appear to have very similar chemical properties to their nonsulfonated TPP counterparts. The presence of the charged sulfonyl group renders the catalyst complex very soluble in neutral aqueous solutions [13] and solubility can be modulated by salt concentration or pH. The sulfonate group is not generally coordinated to the metal and infrared spectra indicate only minor differences in the electronic state of the central metal ion compared with the TPP complexes. The synthesis and reactivity of a range of monosulfonated triphenylphosphine (TPPMS) complexes have been reported [14, 15]. The solubility of metal complexes of phosphines in water can be increased by using multisulfonated TPP [6, 17]. These types of catalysts do not penetrate readily into lipid substrates but partition can be influenced by the use of complexes with phosphine-like ligands [18] or attachment of amphiphilic, long-chain aliphatic ternary phosphines to the metal [19].

Homogeneous catalytic complexes composed of triphenylphosphine ligands are generally unstable in the presence of oxygen and this places a major limitation on their use with living organisms under aerobic conditions. This problem has been largely overcome by synthesis of catalytic complexes based on sulfonated alizarin derivatives of Ru and Pd [20]. The Pd(II)–alizarin complex is not only resistant to inactivation by oxygen, which renders it more stable over relatively long reaction times, but it is also readily soluble in water [21]. Since it retains high activity under physiological conditions, it need only be added to biological systems in trace amounts, thereby avoiding toxicity problems. Toxicity not only arises from the metal ions and ligands but also from the detergent action of these surface-active complexes.

Hydrogenation of unsaturated phospholipids dispersed in aqueous systems using a water-soluble homogeneous catalyst was first reported in 1978 [22]. The catalyst was a sulfonated derivative of Wilkinson's catalyst which did not appear to affect the structure of bilayers with respect to their permeability barrier properties [23]. The catalyst was found to hydrogenate oil-in-water emulsions and two-phase oil/water systems without the need for organic co-solvents [24].

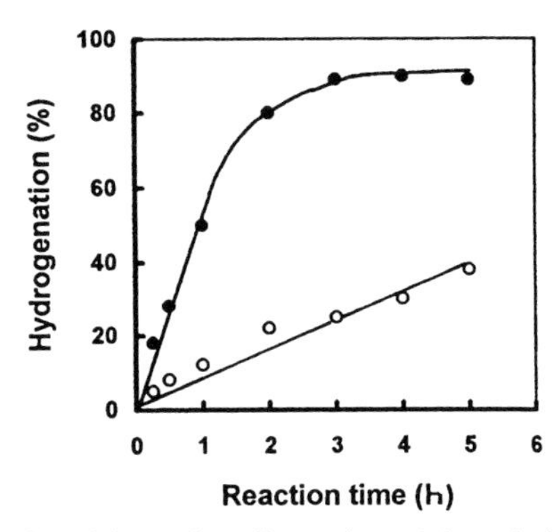

Figure 2. Hydrogenation of multilamellar dispersions (○) and single bilayer vesicles (●) of soya lecithin in the presence of the sulfonated derivative of Wilkinson's catalyst [22].

The reaction rate could be increased significantly by screening the electrostatic charge on the sulfonate groups with inorganic cations added to the aqueous phase. This allowed the catalyst to penetrate into the substrate at the interface; partition of the catalyst from the aqueous into the lipid phase could not be detected. Further evidence for exclusion of catalyst from the lipid phase can be seen in comparison of hydrogenation rates of multibilayer dispersions of unsaturated phospholipids and highly dispersed vesicular suspensions, illustrated in Figure 2. The reaction rate in multilamellar dispersions could be accelerated by dispersing the phospholipid in the presence of the catalyst rather than as shown in Figure 2, where catalyst was added to dispersed substrate.

The reactivity of water-soluble palladium catalyst, $Pd(QS)_2$ (palladium di(sodium) alizarin monosulfonate), has been examined in multilamellar dispersions of unsaturated phospholipids [25]. With substrates of dioleoylphosphatidylcholine there was a transient appearance of *trans* ω9 but no *cis* double bonds were observed when the *trans* ω9 derivative of phosphatidylcholine was used as substrate. It was suggested that hydrogenation may proceed by a *cis–trans* isomerization followed by reduction of the *trans* double bond. Hydrogenation of di-18:2 and di-18:3 derivatives of phosphatidylcholines show highly complex patterns of partially saturated molecular species including combinations of *cis* and *trans* positional isomers with little evidence of bond migration. Comparisons of the rate of hydrogenation of unsaturated molecular species of phosphatidylcholines with dioleoylphosphatidylethanolamine showed a slight preference of the reduced $Pd(QS)_2$ catalyst for phosphatidylcholines. There was a preference for polyunsaturated molecular species compared with the monounsaturated molecular species of phosphatidylcholine. Differences in accessibility of catalyst to substrates presented in bilayer form compared with those in hexagonal-II configuration may explain the different susceptibility of phosphatidylcholines and phosphatidylethanolamines. These differences persisted in mixed dispersions hydrogenated at temperatures at which phase separations of bilayer and hexagonal-II structure would be expected to occur in the substrate.

6.8.3.3 Sources of Hydrogen

The rate of homogeneous hydrogenation can be increased at physiological temperatures by increasing the pressure of hydrogen. Although many biological systems can be preserved under these conditions, hydrogen gas is not the most convenient form of hydrogen and alternative strategies have been explored. Several classes of compounds, including amines, alcohols, sugars, and silicon or tin hydrides, can serve as H-donors in catalyzed hydrogen transfer but, in general, conditions required for a meaningful conversion are not biocompatible.

Photochemical reaction of the ruthenium bipyridyl complex [$RuCl_2(bipy)_3$] together with ascorbic acid as a sacrificial electron donor [25, 26] has been shown to catalyze the reduction of water and generate molecular hydrogen in the presence of water-soluble Wilkinson's catalyst [27]. This system has been exploited to catalyze the light-dependent hydrogenation of phospholipid multilayer dispersions, emulsified triglycerides, and membranes of the living protozoan, *Tetrahymena pyriformis* [28, 29].

6.8.4 Hydrogenation of Biological Membranes

The level of unsaturation of lipids in living cells is controlled within relatively precise limits. The molecular mechanisms that are responsible for this control are not understood. Unsaturated lipids are believed to be required for preserving the fluidity of the lipid bilayer matrix and to maintain the activity of intrinsic membrane proteins with certain limits. When organisms are subjected to environmental stress or shifted to different growth conditions they often adapt by altering the level of unsaturation of their constituent lipids. All these factors have been experimentally tested by use of homogeneous catalytic hydrogenation methods.

6.8.4.1 Topology of Unsaturated Lipids in Membranes

The topology of lipids in the membrane of complex organisms or in subcellular membrane preparations can be probed by determining access to hydrogenation catalysts. Water-soluble catalyst complexes, for example, are not readily permeable to membranes and when added to suspensions of cells or closed vesicular structures their action has been shown to be largely restricted to the outer monolayer, at least at short time intervals after commencement of the reaction

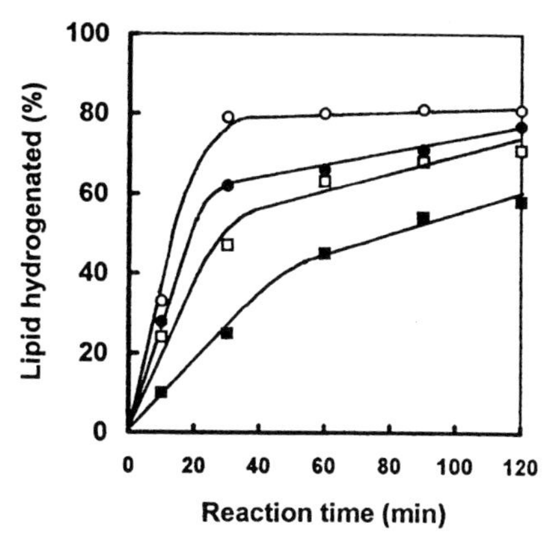

Figure 3. Time-course of hydrogenation of membrane lipids of pea chloroplasts in situ in the presence of palladium–alazarin catalyst: ○ monogalactosyldiacylglycerol; ● digalactosyl diacylglycerol; □ sulfoquinovosyldiacylglycerol; ■ phosphatidylglycerol [31].

[30] (see also Figure 2). Selective hydrogenation of lipid classes has also been observed. This is exemplified by the pattern of hydrogenation observed during incubation of suspensions of pea chloroplasts in the presence of $Pd(QS)_2$ catalyst [31]. Figure 3 shows the extent of hydrogenation of the three major galactolipid classes at intervals during the hydrogenation reaction and there is a marked difference in susceptibility of membrane lipids to hydrogenation. Galactolipids are more readily hydrogenated than the acidic lipids of the membrane. This effect could be due to a charge repulsion between the negatively charged functional groups of sulfoquinovosyldiacylglycerols and phosphatidylglycerols and the sulfonated alizarin groups on the palladium catalyst complex.

6.8.4.2 Function of Unsaturated Lipids in Membranes

Extensive hydrogenation of chloroplast suspensions has shown that inhibition of whole-chain electron transport occurs before inhibition of either photosystem-II or photosystem-I activity is observed, suggesting that removal of unsaturated double bonds may cause a decrease in fluidity of the membrane which is required to facilitate communication between the photosystems situated in different lateral domains in the membrane [32, 33]. To test the notion that plastoquinol diffusion between photosystem-II and cytochrome b_6/f complex was a rate-limiting step in photosynthetic electron transport when the fluidity of the thylakoid membrane was reduced, measurements were performed on the reduction rate of flash-oxidized cytochrome *f* in pea chloroplasts subjected to lipid hydrogenation [34]. The results of reduction of *cis*-unsaturated fatty acyl residues achieved in the presence of palladium–alizarin catalyst is shown in

Table 1. Changes in fatty acyl composition [mol %] of pea thylakoid membranes and motion of 16-deoxyl stearate resulting from hydrogenation in the presence of palladium–alizarin catalyst.

Fatty acid	Control	Hydrogenated
16:0	11	12
16:1	2	2
16:3	1	Trace
18:0	2	15
18:1	5	19
18:2	6	13
18:3	73	39
Rotational correlation time [s] of 16-deoxyl stearate at 20 °C	1.65	2.07

Table 1 together with the effect of rotational correlation time of the spin probe, 16-oxyl stearate. It can be seen that a 30% reduction in unsaturated bonds results from the treatment with catalyst and there is a significant reduction in membrane fluidity as judged by the motion of the spin probe. Nevertheless, despite a reduction in the full-chain electron transport, no change in the rate of reduction of flash-oxidized cytochrome b_6/f was observed, leading to the conclusion that the rate of diffusion of plastoquinol is unaffected by the reduced fluidity of the thylakoid membrane due to the saturation of membrane lipids. The function of the unusual fatty acid, *trans*-Δ^3-hexadecenoic acid, of phosphatidylglycerol in chloroplast membranes has also been investigated by homogeneous catalytic methods [35].

The effect of hydrogenation on membrane protein function has been examined using the Ca^{2+} pump of sarcoplasmic reticulum of rabbit hind-leg muscle [36, 37]. Up to 35% of the unsaturated bonds of the membrane lipids could be saturated in the presence of Wilkinson's catalyst. ATPase activity was completely inhibited on adding catalyst but this could be prevented by preserving the catalyst in its hydride form. When the effect of hydrogenation of sarcoplasmic reticulum on calcium pump activity was assayed in buffers saturated with H_2, it was found that removal of 25% of *cis* double bonds did not affect the activity of Ca^{2+}-ATPase.

6.8.4.3 Acclimation of Membranes to Low Temperature

Catalytic hydrogenation has been used to examine the mechanism of retailoring membrane lipids in the process of cold adaptation in *Tetrahymena pyriformis* [35]. Isolated cilial membranes, when hydrogenated in the presence of Pd(II)–sulfonated alizarin complex, showed a marked increase in order parameter and rotational correlation time of electron spin resonance probes as the constituent lipids became saturated. This was associated with a dramatic decrease in endogeneous phospholipase A activity of the membrane, even when only a small proportion of the unsaturated bonds had been hydrogenated. The way that endogenous phospholipases respond to the change in physical state of the substrate is believed to be the mechanism whereby the biochemical changes responsible for thermal adaptation are bought about.

Some organisms or differing strains of the same organism suffer loss in viability resulting from a sudden exposure to cold. These effects appear to be related to the extent of unsaturation in lipids of constituent membranes. Catalytic hydrogenation has proven to be a useful method to investigate the molecular basis of chilling sensitivity. The phase behavior of membrane lipids of the blue–green alga *Anacystis nidulans* is believed to underlie the chilling sensitivity of this organism [37] and catalytic hydrogenation studies have been undertaken to examine this hypothesis [38, 39].

With regard to the expression of desaturase genes in blue–green algae, see [40].

6.8.4.4 Membrane Unsaturation and Stability at High Temperatures

Adaptation of organisms to elevated temperature is often associated with a shift in the molecular species of membrane lipids to more saturated fatty acyl substituents. It is often argued that this change renders the membrane more stable at elevated temperatures. This hypothesis has been examined in considerable detail in chloroplast photosynthetic membranes, which are ideal for hydrogenation studies because of the highly unsaturated lipids present and the dependence on these lipids to maintain structural stability and organization. The original studies were performed using Wilkinson's catalyst [41]; it was found that loss of up to 40% of unsaturated bonds did not alter the ultrastructural features of the membrane or photosynthetic electron transport processes. Later studies using water-soluble catalysts [42, 43] showed that saturation of the lipids results in a decrease in electron transfer between the "primary" electron acceptor QA and "secondary" acceptor QB. Fluorescence induction kinetics indicated that there is an optimal level of lipid unsaturation to maintain an efficient electron transfer from QA^- to the plastoquinone pool. Furthermore, the proportion of photosystem-IIb, which has a reduced complement of light-harvesting chlorophyll-II [44–46], compared with photosystem-IIa, the form of photosystem-II with complete peripheral chlorophyll *a*/*b* light-harvesting chlorophyll-II, increases with increased hydrogenation of the membrane lipid.

As far as the hydrogenation of lipids of the photosynthetic membrane of higher plant chloroplasts is concerned, cf. [47–50].

6.8.4.5 Biochemical Homeostasis of Unsaturated Lipids

The process of regulation of the level of unsaturated lipids in biomembranes has been examined in potato tubers [51]. When membrane lipids of microsomal suspensions were hydrogenated in the presence of $Pd(QS)_2$ catalyst there was a marked rigidification of the hydrocarbon domain of the membrane as judged by electron spin resonance probe measurements and a stimulation of NADH reductase using ferricyanide as the electron acceptor. This suggests that hydrogenation results in stimulation of the electron transport pathway responsible for desaturating fatty acids. Similar findings have been reported in yeast microsomes hydrogenated with Pd complex [52]. Nevertheless, there is a risk of creating unusual molecular species of lipid which may block metabolic reactions such as desaturation by processes of competitive inhibition [53, 54].

Information on the pathways of unsaturated membrane lipid biosynthesis and processes of redistribution from the site of synthesis to the different subcellular membranes has been obtained from studies of the unicellular green alga, *Daniella salina* [55]. Membrane lipids of the alga can be extensively hydrogenated in the presence of water-soluble palladium–alizarin catalyst under conditions that permit full recovery of the cells within 24 h. When cells are incubated with the catalyst under 1 bar of hydrogen for less than 2 min, only the unsaturated lipids of the surface (plasma) membrane are reduced. Cells treated in this way cease growth for about 12 h, during which time the hydrogenated acyl chains are partially reconverted to their original level of unsaturation. Restoration of lipid unsaturation permits a resumption of growth as membrane functions are presumably restored. Subfractionation of hydrogenated cells showed that the plasma membrane component of the microsomal fraction was hydrogenated to the greatest extent and endoplasmic reticulum to a considerably lesser extent.

6.8.4.6 Hydrogenation of Living Cells

In addition to work on *Tetrahymena* [29] and *Daniella* [55], described above, hydrogenation studies have been performed using a variety of other cell types. Such studies have shown that different cells have been found to vary considerably in their ability to survive hydrogenation of the plasma membrane. The protozoan *Tetrahymena nimbres*, for example, appears to be particularly sensitive [30]. Hydrogenation of lymphocytes with Wilkinson's catalyst has been reported [56] but this catalyst was found to be highly toxic to the cells. More success has been obtained with the use of the water-soluble palladium–alizarin catalyst, which has been used to hydrogenate plasma membranes of a living murine leukemia cell line [57]. Survival of more than 80% of the cells was achieved under optimum treatment conditions which resulted in 40% reduction of total cell fatty acid *cis* double bonds, mainly in polyunsaturated 18:2, 20:4, and 22:6 compounds.

A novel catalyst consisting of colloidal palladium adsorbed onto the surface of a water-insoluble polymer, polyvinylpyrrolidone, has been shown to hydrogenate only the outer leaflet of rat platelets [58]. The effect of hydrogenation was to influence the asymmetric distribution of the phospholipids of the membrane. Catalytic hydrogenation of a phytopathogenic fungus [59] has also been reported.

6.8.5 Conclusions

The chemical modification of membranes containing unsaturated hydrocarbon substituents is a useful tool in the study of the role of these lipids in membrane structure and stability. Homogeneous catalytic hydrogenation of biological membranes in isolated organelles or living cells has developed rapidly over the past few years with the introduction of more active catalytic complexes, especially under conditions of hydrogenation more compatible with living organisms. Advances in targeting catalysts to specific membranes and localizing action to specific membrane sites are likely to be important in future developments.

References

[1] D. Chapman, P. J. Quinn, *Proc. Nat. Acad. Sci. USA* **1976**, *73*, 3971.
[2] J. A. Osborn, F. H. Jardine, G. F. Young, G. Wilkinson, *J. Chem., Soc. A* **1966**, *71*, 1711.
[3] F. Joó, M. T. Beck, *Magyar Kemiai Folyoirat* **1973**, *79*, 189.
[4] B. R. James, *Comprehensive Organometallic Chemistry* (Eds.: G. Wilkinson, F. G. A. Stone, E. W. Abel) Pergamon, Oxford, **1982**, Vol. 8, 285.
[5] B. R. James, *Homogeneous Hydrogenation*, Wiley, New York, **1973**.
[6] J. Halper, J. F. Harrod, B. R. James, *J. Am. Chem. Soc.* **1966**, *88*, 5150.
[7] J. F. Young, J. A. Osborn, F. H. Jardine, G. Wilkinson, *Chem. Commun.* **1965**, 131.
[8] P. J. Quinn, F. Joo, L. Vigh, *Prog. Biophys. Mol. Biol.* **1989**, *53*, 71.
[9] D. Chapman, P. J. Quinn, *Chem. Phys. Lipids* **1976**, *17*, 363.
[10] P. J. Quinn, D. Chapman, C. Vigo, B. R. Boar, *Biochem. Soc. Trans.* **1977**, *5*, 1132.
[11] C. Vigo, F. M. Goni, P. J. Quinn, D. Chapman, *Biochim. Biophys. Acta* **1978**, *508*, 1.
[12] L. Vigh, F. Joó, P. R. van Hasselt, P. J. C. Kuiper, *Mol. Catal.* **1983**, *22*, 15.
[13] B. Salvesen, J. Bjerrum, *Acta Chem. Scand.* **1961**, *16*, 735.
[14] A. F. Borowski, D. J. Cole-Hamilton, G. Wilkinson, *Nouv. J. Chim.* **1978**, *2*, 137.
[15] F. Joó, Z. Toth, *J. Mol. Catal.* **1980**, *8*, 369.
[16] B. Fontal, J. Orlewski, C. G. Santini, J. M. Basset, *Inorg. Chem.* **1986**, *25*, 4320.
[17] E. G. Kuntz, *CHEMTECH* **1987**, *17*, 570.

[18] M. E. Wilson, R. G. Nuzzo, G. M. Whitesides, *J. Am. Chem. Soc.* **1978**, *100*, 2269.
[19] F. Farin, H. L. M. van Gaal, S. L. Bonting, F. J. M. Daemen, *Biochim. Biophys. Acta* **1982**, *711*, 336.
[20] A. V. Bulatov, A. T. Nikitaev, M. L. Khidekel, *Izv. Akad. Nauk SSSR Ser. Khim.* **1981**, *9*, 924.
[21] (a) F. Joó, N. Balogh, L. I. Horváth, G. Filep, I. Horváth, L. Vigh, *Anal. Biochem.* **1991**, *194*, 34; (b) L. Vigh, F. Joó, in *Applied Homogeneous Catalysis with Organometallic Compounds* (Eds.: B. Cornils, W. A. Herrmann), VCH, Weinheim, Germany, **1996**, Vol. 2, p. 1142.
[22] T. D. Madden, P. J. Quinn, *Biochem. Soc. Trans.* **1978**, *6*, 1345.
[23] T. D. Madden, W. E. Peel, P. J. Quinn, D. Chapman, *J. Biochem. Biophys. Meth.* **1980**, *2*, 1.
[24] Y. Dror, J. Manassen, *J. Mol. Catal.* **1977**, *2*, 219.
[25] G. M. Brown, B. S. Brunschwig, C. Creutz, J. F. Endicott, N. Sutin, *J. Am. Chem. Soc.* **1979**, *101*, 1298.
[26] C. V. Krishnan, B. S. Brunschwig, C. Crentz, N. Sutin, *J. Am. Chem. Soc.* **1985**, *107*, 2005.
[27] S. Oishi, *J. Mol. Catal.* **1987**, *39*, 225.
[28] F. Joó, E. Csuhai, P. J. Quinn, L. Vigh, *J. Mol. Catal.* **1988**, *49*, 1.
[29] D. Chapman, W. E. Peel, P. J. Quinn, *Ann. N.Y. Acad. Sci.* **1980**, *308*, 67.
[30] Y. Pak, F. Joó, L. Vigh, A. Kathó, G. A. Thompson, *Biochim. Biophys. Acta* **1990**, *1023*, 230.
[31] I. Horváth, A. R. Mansourian, L. Vigh, P. G. Thomas, F. Joó, P. J. Quinn, *Chem. Phys. Lipids* **1986**, *39*, 251.
[32] L. Vigh, F. Joó, M. Droppa, L. A. Horváth, G. Horvath, *Eur. J. Biochem.* **1985**, *147*, 477.
[33] G. Horváth, M. Droppa, T. Szito, L. A. Mustardy, L. I. Horvath, L. Vigh, *Biochim. Biophys. Acta* **1986**, *849*, 325.
[34] Z. Gombos, K. Barabas, F. Joó, L. Vigh, *Plant Physiol.* **1988**, *86*, 335.
[35] I. Horváth, L. Vigh, T. Pali, G. A. Thompson, *Biochim. Biophys. Acta* **1989**, *1002*, 409.
[36] P. J. Quinn, R. Gomez, T. D. Madden, *Biochem. Soc. Trans.* **1980**, *8*, 38.
[37] B. Szalonati, L. Vigh, F. Joó, L. Senak, R. Mendelsohn, *Biochem. Biophys. Res. Commun.* **1994**, *200*, 246.
[38] P. V. Minorsky, *Plant Cell Environ.* **1985**, *8*, 75.
[39] L. Vigh, F. Joó, *FEBS Lett.* **1983**, *162*, 423.
[40] L. Vigh, D. A. Los, I. Horváth, N. Murata, *Proc. Nat. Acad. Sci. USA* **1993**, *90*, 9090.
[41] C. J. Restall, W. P. Williams, M. P. Percival, P. J. Quinn, D. Chapman, *Biochim. Biophys. Acta* **1979**, *555*, 119.
[42] E. Hileg, Z. Rozsa, I. Vass, L. Vigh, G. Horváth, *Photochem. Photobiophys.* **1986**, *12*, 221.
[43] G. Horváth, A. Melis, E. Hideg, M. Droppa, L. Vigh, *Biochim. Biophys. Acta* **1987**, *891*, 68.
[44] A. Melis, J. M. Anderson, *Biochim. Biophys. Acta* **1983**, *724*, 473.
[45] E. Lam, B. Baltimore, W. Ortiz, A. Melis, R. Malkin, *Biochim. Biophys. Acta* **1983**, *724*, 201.
[46] A. Melis, *Biochim. Biophys. Acta* **1985**, *808*, 334.
[47] P. G. Thomas, P. J. Dominy, L. Vigh, A. R. Mansourian, P. J. Quinn, W. P. Williams, *Biochim. Biophys. Acta* **1986**, *849*, 131.
[48] K. Gounaris, D. A. Mannock, A. Sen, A. P. R. Brain, W. P. Williams, P. J. Quinn, *Biochim. Biophys. Acta* **1983**, *732*, 229.
[49] K. Gounaris, A. P. R. Brain, P. J. Quinn, W. P. Williams, *Biochim. Biophys. Acta* **1984**, *766*, 198.
[50] L. Vigh, Z. Gombos, I. Horváth, F. Joó, *Biochim. Biophys. Acta* **1989**, *979*, 361.

[51] C. Demandre, L. Vigh, A. M. Justin, A. Jolliot, C. Wolf, P. Mazliak, *Plant Sci.* **1986**, *44*, 13.
[52] I. Horváth, Z. Torok, L. Vigh, M. Kates, *Biochim. Biophys. Acta* **1991**, *1085*, 126.
[53] M. Schlame, L. Horváth, L. Vigh, *Biochem. J.* **1990**, *265*, 79.
[54] M. Schlame, J. Horváth, Z. Torok, L. Horváth, L. Vigh, *Biochim. Biophys. Acta* **1990**, *1045*, 1.
[55] L. Vigh, I. Horváth, G. A. Thompson, *Biochim. Biophys. Acta* **1988**, *937*, 42.
[56] W. E. Peel, A. E. R. Thompson, *Leuk. Res.* **1983**, *7*, 193.
[57] S. Benko, H. Hilkmann, L.Vigh, W. J. van Blitterswijk, *Biochim. Biophys. Acta* **1987**, *896*, 1229.
[58] F. Joó, F. Chevy, O. Collard, C. Wolf, *Biochim. Biophys. Acta* **1993**, *1149*, 231.
[59] M. N. Merzylak, P. A. Kashulin, *Gen. Physiol. Biophys.* **1991**, *10*, 561.

6.9 Asymmetric Synthesis

Denis Sinou

During the last 20 years there have been very important advances in asymmetric synthesis via the use of a soluble chiral organometallic catalyst [1]. Now enantioselectivities higher than 95% have been obtained currently in reactions such as hydrogenation, isomerization, epoxidation, hydroxylation, and allylic substitution. Although water-soluble ligands were known since the 1960s and were used as catalysts in association with rhodium or ruthenium complexes in a two-phase system, it is only recently that this methodology was extended to the reduction of prochiral substrates using chiral water-soluble phosphines.

The water-solubilization of the typical chiral phosphines used (Structures **1**–**12**) are mainly due to the presence of a sodium sulfonate or a quaternary ammonium function in the molecule. The enantioselective hydrogenation of some α-amino acid precursors **13** (Eq. 1) has been thoroughly investigated using rhodium complexes associated with these chiral ligands, in water or in an aqueous/organic two-phase system (Table 1). Under these conditions the reaction rates are generally lower than in a homogeneous organic phase.

$$R_1CH{=}C(CO_2R_2)(NHCOR_3)\ (\mathbf{13}) \xrightarrow[\text{diphosphine–Rh}]{H_2} R_1CH_2CH(CO_2R_2)(NHCOR_3)\ (\mathbf{14}) \quad (1)$$

e.g.
a: $R_1 = R_2 = H$, $R_3 = CH_3$; **b**: $R_1 = H$, $R_2 = R_3 = CH_3$; **c**: $R_1 = C_6H_5$, $R_2 = H$, $R_3 = CH_3$
d: $R_1 = C_6H_5$, $R_2 = R_3 = CH_3$; **e**: $R_1 = R_3 = C_6H_5$, $R_2 = H$; **f**: $R_1 = R_3 = C_6H_5$, $R_2 = CH_3$;
g: R_1 = 3-MeO-4-AcO-C_6H_3, $R_2 = H$, $R_3 = CH_3$

Among the first chiral water-soluble ligands used in hydrogenation were PGE-17-DIOP (**1**) and ligands **2** and **3** [2–4]; however, the rhodium catalysts containing these ligands gave low enantioselectivities and could not be recycled. The most investigated chiral ligands were the sulfonated phosphines **5**–**8** [5, 6] and those possessing a quaternary ammonium function (**4** and **10**–**12**) [7–11]. It is to be noticed that rhodium complexes of water-soluble 1,2-diphosphines such as **4**, CHIRAPHOS (**7**), and **12** generally retained their high enantioselectivity in water or in a two-phase system (*ee* = 65–96%). On the other hand, rhodium complexes of 1,4-diphosphines such as CBD (**5**) or DIOP (**10**) and 1,3-diphosphines such as BDPP (**6**) and **11** gave lower enantioselectivities in aqueous phase than the insoluble analogs (8–34% for **5** and **10**, and 40–71% for **6** and **11**, respectively). More recently, Hanson and co-workers [12] prepared a surface-active tetrasulfonated chiral diphosphine derived from BDPP which in

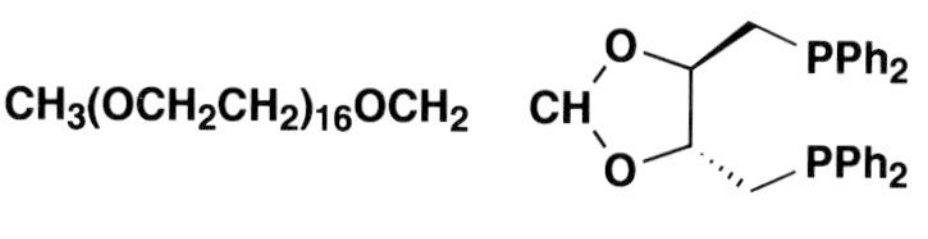

$CH_3(OCH_2CH_2)_{16}OCH_2$ CH O O PPh_2 PPh_2

(*R*,*R*)–PGE–17–DIOP **1**

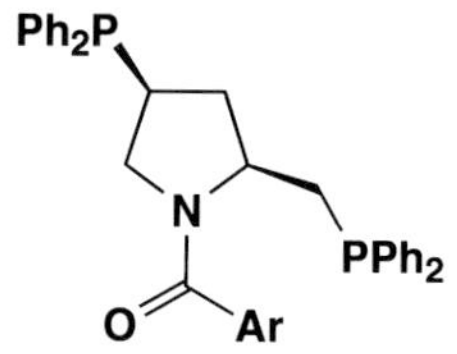

Ph_2P N PPh_2 O Ar

(*S*,*S*)–**2** Ar = C_6H_3(CO$_2$H)(CO$_2$H)

(*S*,*S*)–**3** Ar = C_6H_3(CO)(CO) $NHC_2H_4SO_3Na$ OH

BF_4^- H_3C H_3C N^+ PPh_2 PPh_2

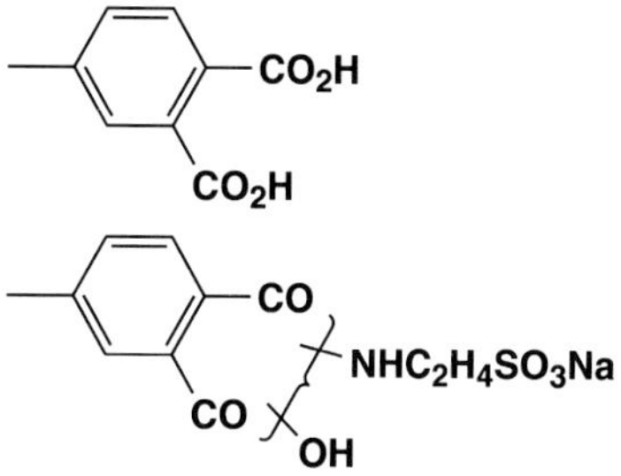

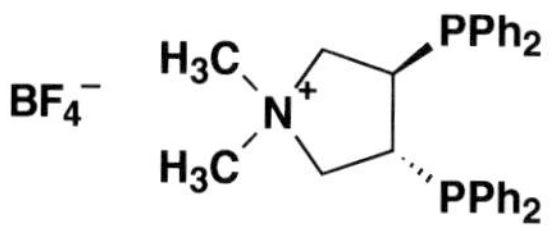

(*R*,*R*)–**4**

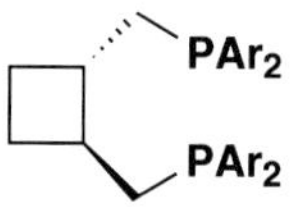

PAr_2 PAr_2 (*S*,*S*)–CBD **5**

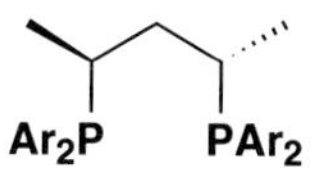

Ar_2P PAr_2 (*S*,*S*)–BDPP **6**

Ar = 3–$NaO_3SC_6H_4$

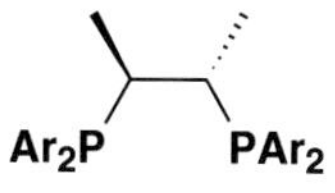

Ar_2P PAr_2 (*S*,*S*)–CHIRAPHOS **7**

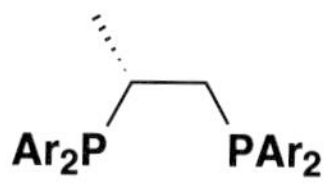

Ar_2P PAr_2 (*R*)–PROPHOS **8**

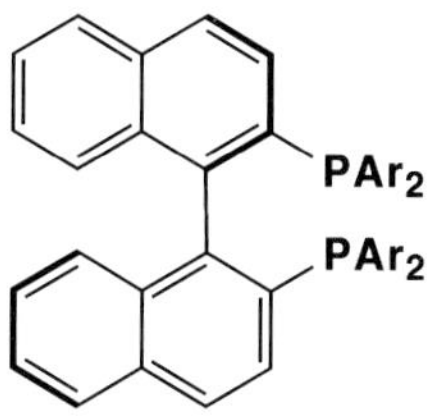

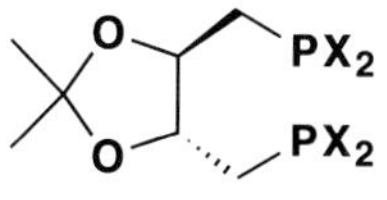

PAr_2 PAr_2 (*R*)–BINAP **9**

O O PX_2 PX_2 (*R*,*R*)–**10**

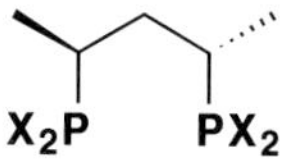

X_2P PX_2 (*S*,*S*)–**11**

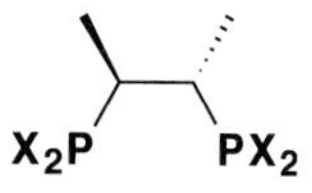

X_2P PX_2 (*S*,*S*)–**12**

a X = *p*–Me_3N-$C_6H_4{}^+BF_4^-$; **b** X = *p*–Me_2HN-$C_6H_4{}^+BF_4^-$

the reduction of **13d** in a two-phase system showed improved reactivity and similar selectivity to the unmodified BDPP (with *ee* up to 69 %).

Increasing pressure or – more importantly – a gradual change of the solvent from alcohol to water decreases the enantioselectivity, although increasing the water content gives a system more independent to increasing pressure. However, this was not true for the catalyst obtained by association of a rhodium complex with the tetrasulfonated BINAP **9** [13]; the enantioselectivities obtained (up to 70 %) are quite similar in pure water and in methanol.

The drop in enantioselectivity going from the organic system to the aqueous phase was attributed to solvent effects and the difference in enantioselectivity is due to the reaction kinetics in the two solvents [14]. When the reaction was carried out in alcohol/water solvent mixtures, an increase in water content

Table 1. Asymmetric hydrogenation of α-amino acid precursors at 25 °C using a rhodium catalyst.

Substrate	Ligand	Solvent	P_{H_2} [bar]	Product **14**		
				ee [%]	Config.	Ref.
13a	(*R*,*R*)-**1**	H_2O	1	11	*R*	[3]
13a	(*S*,*S*)-**2**	H_2O/Na_2HPO_4	1	31	*R*	[4]
13a	(*S*,*S*)-**3**	H_2O	1	34	*R*	[4]
13a	(*R*)-**9**	H_2O	1	70.4	*S*	[13]
13a	(*R*)-**9**	H_2O/CH_3OH (7:1)	1	67.0	*S*	[13]
13a	(*R*)-**9**	H_2O/CH_3OH (1:1)	1	56	*S*	[13]
13b	(*R*)-**9**	H_2O	1	69.0	*S*	[13]
13b	(*R*)-**9**	H_2O/CH_3OH (4:1)	1	61.2	*S*	[13]
13c	(*R*,*R*)-**1**	H_2O	1	30	*R*	[3]
13c	(*S*,*S*)-**2**	H_2O	5	60	*R*	[4]
13c	(*S*,*S*)-**5**	H_2O/AcOEt (1:1)	1	34	*S*	[6]
13c	(*S*,*S*)-**6**	H_2O/AcOEt (1:1)	15	65	*R*	[6]
13c	(*S*,*S*)-**7**	H_2O/AcOEt (1:2)	10	87	*R*	[6]
13c	(*R*)-**8** [a]	H_2O/AcOEt (1:2)	10	70	*S*	[6]
13c [b]	(*R*,*R*)-**4**	H_2O	1	90	*S*	[7]
13c	(*R*,*R*)-**10a**	H_2O	14	25	*S*	[9]
13c	(*R*,*R*)-**10b**	H_2O	14	34	*S*	[9]
13c	(*S*,*S*)-**11a**	H_2O	14	67	*R*	[9]
13c	(*S*,*S*)-**11a**	H_2O	91	65	*R*	[9]
13c	(*S*,*S*)-**11b**	H_2O	14	71	*R*	[9]
13c	(*S*,*S*)-**11b**	H_2O/CH_3OH (2:1)	14	70	*R*	[9]
13c	(*S*,*S*)-**11b**	H_2O/CH_3OH (1:1)	14	64	*R*	[9]
13c	(*S*,*S*)-**11b**	$H_2O/EtOAc/C_6H_6$ (2:1:1)	14	73	*R*	[9]
13c	(*S*,*S*)-**12a**	H_2O	14	94	*R*	[9]
13c	(*S*,*S*)-**12b**	H_2O	14	90	*R*	[9]
13d	(*S*,*S*)-**5**	H_2O/AcOEt (1:1)	1	20	*S*	[6]
13d	(*S*,*S*)-**6**	H_2O/AcOEt (1:1)	15	45	*R*	[6]
13d	(*S*,*S*)-**7**	H_2O/AcOEt (1:1)	10	81	*R*	[6]
13d	(*R*)-**8** [a]	H_2O/AcOEt (1:1)	10	67	*S*	[6]
13d	(*R*,*R*)-**10a**	H_2O	14	8	*S*	[6]
13d	(*R*,*R*)-**10b**	$H_2O/AcOEt/C_6H_6$ (2:1:1)	14	25	*S*	[9]
13d	(*S*,*S*)-**11a**	H_2O	14	40	*R*	[9]
13d	(*S*,*S*)-**11a**	$H_2O/AcOEt/C_6H_6$ (2:1:1)	14	45	*R*	[9]
13d	(*S*,*S*)-**11b**	$H_2O/AcOEt/C_6H_6$ (2:1:1)	14	50	*R*	[9]
13d	(*S*,*S*)-**12a**	H_2O	14	68	*R*	[9]
13d	(*S*,*S*)-**12a**	$H_2O/AcOEt/C_6H_6$ (2:1:1)	14	77	*R*	[9]
13d	(*S*,*S*)-**12b**	$H_2O/AcOEt/C_6H_6$ (2:1:1)	14	74	*R*	[9]
13e	(*S*,*S*)-**5**	H_2O/AcOEt (1:2)	1	13	*S*	[6]
13e	(*S*,*S*)-**6**	H_2O/AcOEt (1:1)	15	44	*R*	[6]
13e	(*S*,*S*)-**7** [c]	H_2O/AcOEt (1:2)	10	86	*R*	[6]
13e	(*R*)-**8** [a]	H_2O/AcOEt (1:2)	10	70	*S*	[6]
13f	(*R*,*R*)-**10a**	$H_2O/AcOEt/C_6H_6$ (2:1:1)	14	9	*S*	[9]
13f	(*R*,*R*)-**10a**	$H_2O/AcOEt/C_6H_6$ (2:1:1)	91	9	*S*	[9]
13f	(*R*,*R*)-**10b**	$H_2O/AcOEt/C_6H_6$ (2:1:1)	14	11	*S*	[9]

Table 1. (cont.)

Substrate	Ligand	Solvent	P_{H_2} [bar]	Product **14**		
				ee [%]	Config.	Ref.
13f	(*S,S*)-**11a**	$H_2O/AcOEt/C_6H_6$ (2:1:1)	14	54	*R*	[9]
13f	(*S,S*)-**11b**	$H_2O/AcOEt/C_6H_6$ (2:1:1)	14	67	*R*	[9]
13f	(*S,S*)-**12a**	$H_2O/AcOEt/C_6H_6$ (2:1:1)	14	65	*R*	[9]
13f	(*S,S*)-**12b**	$H_2O/AcOEt/C_6H_6$ (2:1:1)	14	58	*R*	[9]
13g	(*S,S*)-**5**	H_2O/AcOEt (1:2)	1	37	*S*	[6]
13g	(*S,S*)-**6**	H_2O/AcOEt (1:1)	10	58	*R*	[6]
13g	(*S,S*)-**7** [c]	H_2O/AcOEt (1:2)	10	88	*R*	[6]
13g	(*R*)-**8** [a]	H_2O/AcOEt (1:2)	10	80	*S*	[6]
13g	(*R,R*)-**10a**	H_2O	14	42	*S*	[9]
13g	(*R,R*)-**10b**	H_2O	14	67	*S*	[9]
13g	(*R,R*)-**10b**	H_2O	14	67	*S*	[9]
13g	(*S,S*)-**11a**	H_2O	14	76	*R*	[9]
13g	(*S,S*)-**11b**	H_2O	14	79	*R*	[9]
13g	(*S,S*)-**12a**	H_2O	14	93	*R*	[9]
13g	(*S,S*)-**12b**	H_2O	14	88	*R*	[9]

[a] As a mixture of 55% tetrasulfonated **8** and 35% trisulfonated **8**, together with 10% phosphine oxide.
[b] Reduced as the sodium salt.
[c] As a mixture of 60% tetrasulfonated **7** and 40% trisulfonated **7**.

induced a decrease in enantioselectivity. In a systematic study of the influence of various solvents on the enantioselectivity in the reduction of dehydroamino acids, a linear relationship was found between log (% *S*/ % *R*) and the solvophobicity parameter S_p of various solvents, log (%*S*/ %*R*) decreasing with increasing S_p.

The position of the hydrophilic substitution (*m*-SO_3Na or *p*-NMe_4) has little influence on the enantioselectivity, the tetrasulfonated ligands and the quaternary ammonium phosphines giving almost the same enantioselectivity in the reduction of the same substrate [6, 9].

One of the important points is that these catalytic solutions can be readily recycled, without loss of enantioselectivity, as shown in Table 2. This recycling was performed with little rhodium loss in a two-phase system ($< 0.1\%$), although larger quantities of rhodium ($\approx 4\%$) were entrained with the product when water alone was used as the solvent.

The biphasic hydrogenation of α-amino acid precursors was shown to be a truly homogeneous process, the reduction occurring in the aqueous phase [15]. However, water was not only a solvent, but also had a chemical effect on the reduction [16–18]. Hydrogenation of α-acetamidocinnamic acid methyl ester in $AcOEt/D_2O$ in the presence of a rhodium complex associated with a sulfonated ligand such as TPPTS occurred with a 75% regiospecific monodeuteration at the position α to the acetamido and the ester functions, the amount of deuterium

Table 2. Catalyst recycling in the asymmetric hydrogenation of α-amino acid precursors at 25 °C using a rhodium catalyst.

Substrate	Ligand	Solvent	P_{H_2} [bar]	Cycle	Product **14**		
					ee [%]	Config.	Ref.
13c	(*S,S*)-**5**	H_2O/AcOEt (1:1)	1	1	34	*S*	[6]
				2	37	*S*	
13d	(*S,S*)-**5**	H_2O/AcOEt (1:1)	1	1	20	*S*	[6]
				2	23	*S*	
13d	(*S,S*)-**7** [a]	H_2O/AcOEt (1:1)	10	1	82	*R*	[6]
				2	88	*R*	
				3	87	*R*	
13d	(*S,S*)-**12a**	H_2O/AcOEt/C_6H_6 (2:1:1)	14	1	75	*R*	[9]
				2	77	*R*	
				3	77	*R*	
13e	(*S,S*)-**12a**	H_2O	14	1	93	*R*	[9]
				2	95	*R*	
				3	90	*R*	

[a] As a mixture of 60% tetrasulfonated **7** and 40% trisulfonated **7**.

incorporation depending on the ligand used (Eq. 2). When the reduction was performed under a deuterium atmosphere in the presence of water, hydrogen incorporation occurred at the same position, the overall reaction being a *cis* addition of HD.

CO_2CH_3
Ph NHCOCH$_3$
13d
D_2/Rh/L
H_2O/AcOEt
H_2/Rh/L
D_2O/AcOEt
H D Ph CO_2CH_3 H NHCOCH$_3$
Ph CO_2CH_3 D NHCOCH$_3$

(2)

More recently the ruthenium catalyst obtained from $[Ru(C_6H_6)Cl_2]_2$ and 2 equiv. of tetrasulfonated (*R*)-BINAP (**9**) was used in the reduction of some dehydroamino acids in water at room temperature and under 1 bar of hydrogen [19]. Enantiomeric excesses as high as 85–88% have been achieved, the direction of enantioselection being the opposite of that obtained from the water-soluble catalyst rhodium-BINAP in the same solvent. The enantioselectivity in this case is lower in water than in methanol, and declines with increasing hydrogen pressure. Reduction of unsaturated acid **15** in a two-phase AcOEt/H_2O system using this Ru–BINAP (**9**) catalyst (Eq. 3) allowed the preparation of naproxen (**16**) in 78.4% *ee*, *ee* values over the range 78.0–82.7% being obtained over several recycles of the catalytic solution [20]. The supported aqueous phase catalysis (SAPC) was extended to this chiral system. If enantioselectivity up to 70% was obtained in the synthesis of naproxen in ethyl acetate as the solvent saturated with water, the use of ethylene glycol in place of water as the hydrophilic phase gave *ee* values up to 96%. In all cases the catalyst could be recycled with the same *ee* without any leaching of ruthenium.

MeO CO$_2$H —H$_2$ / Ru–L*→ MeO CO$_2$H (3)

15 16

Among the other reduced prochiral substrates, methylene succinic acid was hydrogenated with *ee* values up to 59 and 50% using respectively the catalysts Rh–**3** [3] and Ru–BINAP (**9**) [19]. Unfortunately phenyl methyl ketone was reduced in the presence of [Rh(cod)**6**]ClO$_4$ with only 22% *ee* [6].

Sinou and co-workers have reported the influence of the degree of sulfonation of chiral BDPP on the enantioselectivity in the reduction of some dehydroamino acids [6]. This drastic effect was more pronounced in the reduction of prochiral imines [21, 22]. The rhodium complex of monosulfonated BDPP gives *ee* up to 96% in the reduction of imines **17** (Eq. 4), although the tetrasulfonated or disulfonated BDPP gave only 34% and 2% *ee* respectively for the reduction of the benzylimine of acetophenone. An explanation for this surprising effect was recently proposed [23].

N C$_6$H$_5$ / Ar CH$_3$ —H$_2$, 70 bar, 25°C / Rh/sulfonated BDPP→ HN C$_6$H$_5$ / Ar CH$_3$ (4)

17 **18**

Ar/%ee: C$_6$H$_5$/96; 4–MeOC$_6$H$_4$/95; 4–ClC$_6$H$_4$/92

Dehydropeptides **19** (Eq. 5) were reduced in a two-phase system using [Rh(cod)Cl]$_2$ associated with ligands **5** and **6** [24]; the diastereomeric excess (*de*) of the dipeptide **20** obtained was strongly dependent on the absolute configuration of the substrate. For example, reduction of Ac-Δ-Ph-(*S*)-Ala-OCH$_3$ using Rh-**6** gave a *de* as high as 72% in favor of the (*R*,*S*) diastereomer, although a *de* of only 6% was obtained for Ac-Δ-Ph-(*R*)-*Ala*-OCH$_3$ in favor of the (*R*,*R*) diastereomer.

C$_6$H$_5$ H N CO$_2$Me / MeCONH O CH$_3$ —H$_2$/Rh–7 / H$_2$O/AcOEt→ C$_6$H$_5$ H N CO$_2$Me / MeCONH O CH$_3$ (5)

19 20

The binding of the chiral ligand PPM to a water-soluble polymer such as polyacrylic acid gave a macroligand **21** [25], which was used in the reduction of α-acetamidocinnamic acid (**13c**); enantioselectivities up to 56% and 74% were obtained using water and EtOAc/H_2O (1:1) respectively as the solvents.

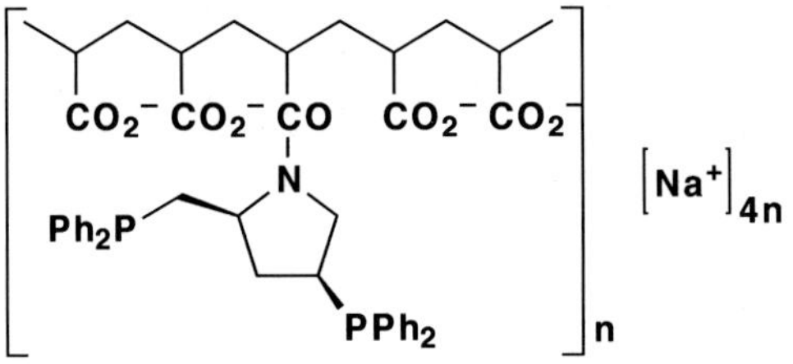

(*S*,*S*)-**21**

Formates of sodium, potassium, and ammonium can also be used as the reducing agent in the reduction of compounds **13c** and **13d** in the presence of the catalyst Rh–**5** [26] with enantioselectivity up to 43% at 50 °C.

Asymmetric hydrogenolysis of sodium *cis*-epoxysuccinate in water as the solvent in the presence of the catalyst Rh–**6** gave the corresponding sodium hydroxysuccinate in 27% *ee* [27]; kinetic resolution was also observed in the case of sodium *trans*-phenylglycidate using the same catalyst.

In conclusion, it was shown that asymmetric catalysis, and particularly asymmetric hydrogenation, occurs in a two-phase system, allowing the very easy recycling of the catalyst without loss of enantioselectivity. If some decrease in enantioselectivity was observed on going from a usual organic phase to a water organic two-phase system, the use of ethylene glycol as the hydrophilic phase using supported aqueous-phase catalysis gave enantiomeric excesses up to 96%.

References

[1] (a) *Catalytic Asymmetric Synthesis* (Ed.: I. Ojima), VCH, Weinheim, **1993**; (b) H. B. Kagan, in *Comprehensive Organometallic Chemistry* (Ed.: G. Wilkinson), Pergamon Press, London, **1993**, Vol. 8, p. 463; (c) H. Brunner, W. Zettlmeier, *Handbook of Enantioselective Catalysis with Transition Metal Compounds*, VCH, Weinheim, **1993**; (d) R. Noyori, *Asymmetric Catalysis in Organic Synthesis*, John Wiley, New York, **1994**.
[2] Y. Amrani, D. Sinou, *J. Mol. Catal.* **1984**, *24*, 231.
[3] D. Sinou, Y. Amrani, *J. Mol. Catal.* **1986**, *36*, 319.
[4] R. Benhamza, Y. Amrani, D. Sinou, *J. Organomet. Chem.* **1985**, *288*, C37.
[5] F. Alario, Y. Amrani, Y. Colleuille, T. P. Dang, J. Jenck, D. Morel, D. Sinou, *J. Chem. Soc., Chem. Commun.* **1986**, 202.
[6] Y. Amrani, L. Lecomte, D. Sinou, J. Bakos, I. Toth, B. Heil, *Organometallics* **1989**, *8*, 542.
[7] U. Nagel, E. Kingel, *Chem. Ber.* **1986**, *119*, 1731.
[8] I. Toth, B. E. Hanson, *Tetrahedron: Asymm.* **1990**, *1*, 895.
[9] I. Toth, B. E. Hanson, *Tetrahedron: Asymm.* **1990**, *1*, 913.
[10] I. Toth, B. E. Hanson, M. E. Davis, *Catal. Lett.* **1990**, *5*, 183.

[11] I. Toth, B. E. Hanson, M. E. Davies, *J. Organomet. Chem.* **1990**, *396*, 363.
[12] H. Ding, B. E. Hanson, J. Bakos, *Angew. Chem., Int. Ed. Engl.* **1995**, *34*, 1645.
[13] K. Wan, M. E. Davis, *J. Chem. Soc., Chem. Commun.* **1993**, 1262.
[14] L. Lecomte, D. Sinou, J. Bakos, I. Toth, B. Heil, *J. Organomet. Chem.* **1989**, *370*, 277.
[15] L. Lecomte, D. Sinou, *J. Mol. Catal.* **1989**, *52*, L21.
[16] M. Laghmari, D. Sinou, *J. Mol. Catal.* **1991**, *66*, L15.
[17] J. Bakos, R. Karaivanov, M. Laghmari, D. Sinou, *Organometallics* **1994**, *13*, 2951.
[18] J. Joó, P. Csiba, A. Bényei, *J. Chem. Soc., Chem. Commun.* **1993**, 1602.
[19] K. Wan, M. E. Davis, *Tetrahedron: Asymm.* **1993**, *4*, 2461.
[20] K. T. Wan, M. E. Davies, *Nature (London)* **1994**, *370*, 449; *idem, J. Catal.* **1994**, *148*, 1; *idem. J. Catal.* **1995**, *152*, 25.
[21] J. Bakos, A. Orosz, B. Heil, M. Laghmari, P. Lhoste, D. Sinou, *J. Chem. Soc., Chem. Commun.* **1991**, 1684.
[22] C. Lensink, J. G. De Vries, *Tetrahedron: Asymm.* **1992**, *3*, 235.
[23] J. M. Buriak, J. A. Osborn, *Organometallics* **1996**, *15*, 3161.
[24] M. Laghmari, D. Sinou, A. Masdeu, C. Claver, *J. Organomet. Chem.* **1992**, *438*, 213.
[25] T. Malmström, C. Anderson, *J. Chem. Soc., Chem. Commun.* **1996**, 1135.
[26] D. Sinou, M. Safi, C. Claver, A. Masdeu, *J. Mol. Catal.* **1991**, *68*, L9.
[27] J. Bakos, A. Orosz, S. Cserépi, I. Toth, D. Sinou, *J. Mol. Catal.* **1997**, *116*, 85.

6.10 Fine Chemicals Syntheses

Steffen Haber

6.10.1 Carbonylation Reactions

6.10.1.1 Fluorophenylacetic Acids

Phenylacetic acid and its substituted derivatives are valuable intermediates in the manufacture of pharmaceuticals, cosmetics, and fragrances. The current industrial production of substituted phenylacetic acids starts with the corresponding benzyl chloride, which is converted by nucleophilic substitution into the benzyl cyanide. Subsequent hydrolysis with sulfuric acid yields the phenylacetic acids. This two-step process has considerable drawbacks such as formation of a stoichiometric amount of salt, e.g., sodium chloride (exchange reaction) and ammonium sulfate (hydrolysis). Additionally, the introduction of a carbon atom by sodium cyanide is pretty expensive compared with the much cheaper carbon monoxide (US $ 1.4/kg vs. US $ 0.2/kg, respectively). Based on the principles of aqueous biphasic catalysis, Hoechst AG developed the palladium-catalyzed carbonylation of substituted benzyl chlorides [1] (Eq. 1).

$$\text{F-C}_6\text{H}_4\text{-CH}_2\text{Cl} \xrightarrow[\text{Pd / TPPTS; NaOH / H}_2\text{O / Xylene}]{\text{CO}} \text{F-C}_6\text{H}_4\text{-CH}_2\text{COOH} \qquad (1)$$

The palladium catalyst is dissolved in the alkaline aqueous solution by using water-soluble phosphine ligands. The organic phase, typically toluene or xylene, contains the substrate. During the reaction the fluorophenylacetic acid is formed with concomitant consumption of carbon monoxide and HCl production. The HCl formed during the reaction neutralizes and acidifies the aqueous solution and the product is dissolved in the organic phase. TONs are higher than 1500 and TOFs of 135 h^{-1} are reached. The reaction rate is strongly dependent on the temperature and the initial substrate concentration. Continuous addition of the benzylic chloride increases the lifetime of the catalyst.

Conversions are generally greater than 96 % with high selectivities to the acid. The high activity of the palladium/TPPTS catalyst relative to a $PdCl_2$/triphenylphosphine catalyst is founded on its presence in the water phase. The Cl^- and OH^- present in the aqueous solution stabilize the active intermediate. The sulfonated diphosphine BINAS (cf. Section 3.2.2) showed a carbonylation activity even in acidic conditions. After complete conversion of the substrate the catalyst is still active at pH = 1.

6.10.1.2 Ibuprofen

Various substituted arylpropionic acids (profenes) are important nonsteroidal anti-inflammatory drugs such as ibuprofen, flurbiprofen, ketoprofen, perprofen, fenoprofen, and naproxen. In the preparation of various types of arylacetic acid derivatives such as ibuprofen, one of the problems associated with the current processes is the use of a catalyst directly in an organic solvent to convert the 1-(4′-isobutylphenyl)ethanol (IBPE) via carbonylation to ibuprofen, i.e., 2-(4′-isobutylphenyl)propionic acid [2]. This requires, after reaction, separation of the desired end product (i.e., ibuprofen) from the catalyst in the overall reaction mass and recycling of the catalyst. There are substantial losses in catalyst as well as decreased efficiencies such as difficulties in the separation of the catalyst and the end product. Another problem associated with such a process is the requirement that a halide ion, such as chloride, be used to facilitate the overall reaction. In this case, use of HCl, for example, handling creates metallurgical problems in the equipment used. To overcome these disadvantages Sheldon and co-workers developed the carbonylation of arylcarbinols in a two-phase system wherein one phase is an aqueous medium which contains a water-soluble catalyst, and an acid, e.g., sulfuric acid (Eq. 2) [3].

OH → (CO; Pd / TPPTS; H_2O / H_2SO_4) → O, OH (2)

Ibuprofen

The procedure developed makes use of the water-soluble $Pd(TPPTS)_3$ catalyst (cf. Section 6.3) and performs under mild conditions. The highest activity is observed with $PdCl_2$ as precursor and at a TPPTS/Pd molar ratio of 10:1. At this or higher ratios, with low H^+ concentrations and a palladium concentration of 150 ppm, the catalyst remains intact without any degradation to metallic palladium. Using ligands containing fewer SO_3Na groups than in TPPTS, the catalytic activity drops; with TPPMS only traces of the carbonylation are observed. Ibuprofen was synthesized in 71.5% yield. Starting from 4-fluorobenzylic alcohol Sheldon was able to produce 4-fluorophenylacetic acid in 81% yield, compared with 84% yield starting from 4-fluorobenzylic chloride [4].

6.10.2 Carbon–Carbon Coupling

Palladium/ and nickel/water soluble complexes are effective catalytic systems for various C–C coupling reactions. The use of sulfonated arylphosphines, guanidinium-substituted phosphines, and carbohydrate-substituted arylphosphines has been reported for various palladium-catalyzed cross-coupling reactions in aqueous media [5]. Water-soluble Ni/phosphine complexes have been used in the synthesis of adiponitrile via hydrocyanation of butadiene [6].

6.10.2.1 Vitamin E Precursor Geranyl Acetone

In 1981 Rhône-Poulenc developed a new process for the addition of active methylene compounds to conjugated dienes and has been using this in a new industrial synthesis of geranyl acetone, a precursor of vitamin E, since 1988 [7]. The world market for Vitamin E is around 10000 tons/y at a price of US $ 25–30/kg for tocopheryl acetate.

The coupling of methyl acetoacetate with myrcene relies on an efficient catalyst made of rhodium sulfate and an excess of TPPTS in the presence of methanol as cosolvent (Scheme 1). The addition of alcohol facilitates the necessary contact between the hydrophilic catalyst and the lipophilic substrate. The regioselectivity of the reaction is over 99% and the isomer ratio α/β is 45:55. In the presence of triphenylphosphine instead of TPPTS, only 1:2 addition compounds were obtained in the reaction; bidentate ligands give mainly 1:1 addition products but with regioselectivity less than 80%. The reaction proceeds in a biphasic liquid/liquid system where conversion is regulated by stirring time. The organic products are easily separated by simple decantation and the aqueous phase containing the catalyst can be recycled. Rhône-Poulenc performs the reaction in an alcohol/water mixture with a TPPTS/Rh ratio of 21:1. This sigificant excess of phosphine is necessary because the reduction of rhodium(III) partially consumes the hydrosoluble phosphine, and to ensure efficient retention of rhodium in the aqueous phase. According to Bortoletto et al., the best rates are obtained with rhodium(I) complexes, e.g., $[RhCl(COD)]_2$/TPPTS, and a

P/Rh ratio of 21:1 without addition of sodium chloride [8]. The industrial process relies on the exceptional water-solubility of TPPTS and on the use of an alcoholic co-solvent. To avoid the addition of the co-solvent less water-soluble triphenylphosphine monosulfonate sodium salt (TPPMS) can be used advantageously. The conversion is maintained during recycling. With TPPTS, in contrast, methanol is necessary to ensure adequate myrcene solubilization in water, otherwise very low conversions are obtained. With 3 equiv. of phosphine (P/Rh ratio = 3:1), TPPMS is more efficient than TPPTS during the first run. However, in both cases the recycling of the aqueous phase gave poor results. TPPTS is oxidized and the more nucleophilic phosphine TPPMS is added to myrcene and is transformed into a phosphonium salt. Rhône-Poulenc produces approx. 1000 tons/y geranyl acetone.

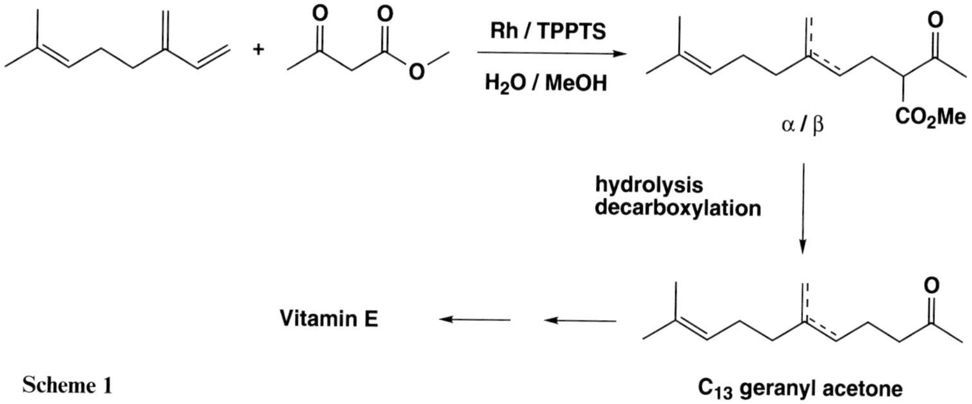

Rh / TPPTS

H_2O / MeOH

α / β CO_2Me

hydrolysis
decarboxylation

Vitamin E

Scheme 1

C_{13} geranyl acetone

6.10.2.2 Carbohydrate Octadienyl Ethers

Henkel KGaA and Eridania Beghin-Say developed the telomerization reaction of butadiene with carbohydrates as nucleophiles (Eq. 3) [9]. Beghin-Say developed the butadiene telomerization in water in the presence of a palladium salt and TPPTS. Mono- and dioctadienyl sucrose ether compounds are selectively obtained using a 1 M NaOH/isopropanol solvent mixture [10].

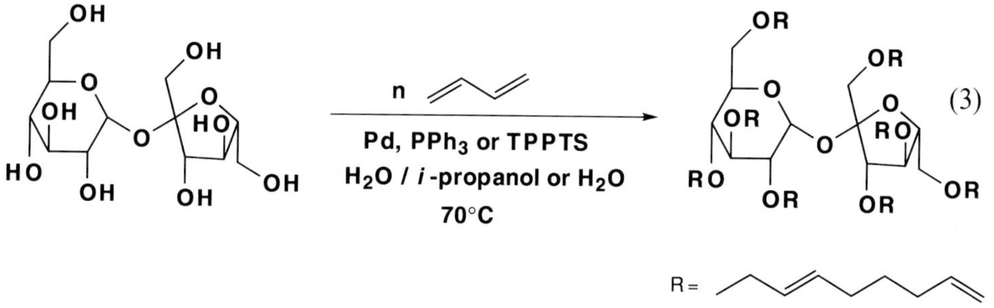

n

Pd, PPh_3 or TPPTS

H_2O / *i*-propanol or H_2O

70°C

(3)

R =

Glucose, methylglucoside, and sucrose were reacted with butadiene in 2-propanol/water (8:2) in the presence of palladium(II) 2,4-pentanedionate and triphenylphosphine as the catalyst system (Henkel KGaA). A significant increase of catalyst activity and product yield with TON up to 40000 was observed when the butadiene was added continuously to the reaction mixture at constant pressure. After 12 h at 68 °C sucrose octadienyl ethers were obtained with an average degree of substitution of 4.7–5.3. Hill found that low reaction temperatures decrease the formation of side products, e.g., 2-propyl 2,7-octadienyl ether and 2,7-octadienol, high catalyst concentrations increase the average degree of substitution, and high carbohydrate turnovers are achieved by vigorous stirring of the heterogeneous reaction mixture. A P/Pd ratio of 10:1 gave high carbohydrate conversions and only a small amount of 2-propyl 2,7-octadienyl ether was observed. Carbohydrate octyl ethers are formed via hydrogenation of the crude product prior to evaporation of solvents. Carbohydrate octyl ethers are potential emulsifiers and defoaming agents.

6.10.2.3 Substituted Biphenyls via Suzuki Cross-coupling Reactions

Aromatic, rigid-rod polymers play an important role in a number of diverse technologies, including those of high-performance engineering materials, conducting polymers, and nonlinear optical materials. Wallow and Novak prepared poly(*p*-quaterphenylene-2,2′-dicarboxylic acid) by $Pd(TPPTMS)_3$ catalyzed cross coupling of the ethylene glycol diester of 4,4′-biphenylenebis(boronic acid) and 4,4′-dibromodiphenic acid in water/DMF (7:3) (Eq. 4; cf. Section 6.11.3)

Br OH O O HO Br O O B B O O + $Pd(TPPTMS)_3$ Na_2CO_3 / H_2O / DMF HCl O OH HO O

(4)

[11]. The resulting polymer has an approximate molecular weight (M_w) of 50 000 g/mol relative to single-stranded DNA, and is isolated by addition of dilute hydrochloric acid to yield the free acid.

Unsymmetrically substituted biphenyls are also important intermediates in the syntheses of new nonpeptide angiotensin-II receptor antagonists [12]. In 1994 Hoechst developed the synthesis of 2-cyano-4′-methylbiphenyl via cross-coupling reaction of *p*-tolylboronic acid and 2-chlorobenzonitrile utilizing water-soluble phosphine/palladium complexes in the presence of a polyhydric alcohol, a sulfoxide, or a sulfone and a base (Eq. 5) [13]. The reaction is performed at 120 °C, and at the end the two-phase system is separated, whereas the organic phase contains the product, which is formed in high yields (>90 %, approx. 100 t/y).

$B(OH)_2$ + Cl (NC) —Pd / TPPTS, base→ (NC) (5)

6.10.2.4 Cinnamic Acids via Heck Reactions

The synthesis of substituted cinnamic acids and corresponding esters via Heck reaction is well known and documented [14]. Normally, donor-substituted aryl halides cause difficulties in the cross-coupling reaction. Bromine Compounds Ltd. have published a new process for the synthesis of *p*-methoxycinnamic acid 2-ethylhexyl ester [15]. Starting from *p*-bromoanisole and acrylic acid, *p*-methoxycinnamic acid was synthesized in 90 % yield by Heck reaction in water at 150 °C with KOH as base, and $PdCl_2$ as catalyst (Eq. 6). Phosphine ligands are not required in this reaction. TONs of 20 000 and TOFs up to 6020 are observed. According to this procedure *p*-methoxycinnamic acid can be isolated with high purity (98 %) by 2-ethylhexanol extraction.

MeO–Br + (O, OH) —$PdCl_2$, KOH / K_2CO_3 / H_2O, 150°C→ MeO– (O, OH) 90% (6)

References

[1] C. W. Kohlpaintner, M. Beller, *J. Mol. Catal. A: Chem.* **1997**, *116*, 259; Hoechst AG (C. W. Kohlpaintner, M. Beller), DE 4.415.681 (1994); Hoechst AG (C. W. Kohlpaintner, M. Beller), DE 4.415.682 (1994).

[2] (a) Hoechst Celanese Corporation (V. Elango, M. A. Murphy, G. L. Moss, B. L. Smith, K. G. Davenport, G. N. Mott), EP 00.284.310 (1988); *Chem. Abstr.* **1989**, *110*, 153916t; (b) Hoechst AG (S. Rittner, A. Schmidt, L. O. Wheeler, G. L. Moss, E. G. Zey), EP 00.326.027 (1989); *Chem. Abstr.* **1990**, *112*, 35448k.

[3] (a) Hoechst Celanese Corporation (R. A. Sheldon, L. Maat, G. Papadogianakis), US 5.536.874 (1996); (b) G. Papadogianakis, L. Maat, R. A. Sheldon, *J. Mol. Catal. A: Chem.* **1997**, *116*, 179.

[4] C. W. Kohlpaintner, M. Beller, *J. Mol. Catal. A: Chem.* **1997**, *116*, 259.

[5] For example, (a) A. L. Casulnuovo, J. C. Calabrese, *J. Am. Chem. Soc.* **1990**, *112*, 4324; (b) H. Dibrowski, F. P. Schmidtchen, *Tetrahedron* **1995**, *51*, 2325; (c) J. P. Genet, E. Blart, M. Savignac, *Synlett* **1992**, *715*; (d) J. P. Genet, A. Linquist, E. Blart, V. Mouries, M. Savignac, *Tetrahedron Lett.* **1995**, *36*, 1443; (e) M. Beller, J. G. E. Krauter, A. Zapf, *Angew. Chem.* **1997**, *109*, 793; (f) Bayer AG (A. Bader, D. Arlt, F. Seng), EP 0.459.258 (1990); (g) Hoechst AG (S. Haber, H. J. Kleiner), DE 19.527.118 (1995); (h) Hoechst AG (S. Haber), DE 19.535.528 (1995); (i) Hoechst AG (S. Haber, N. Egger), DE 19.620.023 (1996).

[6] (a) Rhône-Poulenc Fiber and Resin Intermediates (M. Huser, R. Perron), WO 97/12.857 (1997); (b) Rhône-Poulenc Industry, FR 2.338.253 (1977).

[7] (a) C. Mercier, P. Chabardes, *Pure Appl. Chem.* **1994**, *66*, 1509; (b) Rhône-Poulenc Ind. (D. Morel), FR 2.486.525 (1980), FR 2.505.322 (1980); (c) Anon., *Actual. Chim.* **1990** (Mar./Apr.); (d) D. Morel, G. Mignani, Y. Colleuille, *Tetrahedron Lett.* **1985**, *26*, 6337; (e) *idem, ibid.* **1986**, *27*, 2591; L. Lavenot, A. Roucoux, H. Patin *J. Mol. Catal. A: Chem.* **1997**, *118*, 153.

[8] M. H. Bortoletto, L. Lavenot, C. Larpent, A. Roucoux, H. Patin, *Appl. Catal. A* **1997**, *156*, 347.

[9] (a) K. Hill, *Henkel Referate,* **1993**, *29*, 30; (b) K. Hill, B. Gruber, K. J. Weese, *Tetrahedron Lett.* **1994**, *35*, 4541; (c) Eridania Beghin-Say (I. Pennequin, A. Mortreux, F. Petit, J. Mentech, B. Thiriet), FR 2.693.188 (1992).

[10] I. Pennequin, J. Meyer, I. Suisse, A. Mortreux, *J. Mol. Catal. A: Chem.* **1997**, *120*, 139.

[11] T. I. Wallow, B. M. Novak, *J. Am. Chem. Soc.* **1991**, *113*, 7411.

[12] R. R. Wexler, W. J. Greenlee, J. D. Irvin, M. R. Goldberg, K. Prendergast, R. D. Smith, P. B. M. W. M. Timmermans, *J. Med. Chem.* **1996**, *39*, 625, and literature cited therein.

[13] (a) Hoechst AG (S. Haber, H. J. Kleiner), DE 19.527.118 (1995); (b) Hoechst AG (S. Haber), DE 19.535.528 (1995); (c) Hoechst AG (S. Haber, N. Egger), DE 19.620.023 (1996).

[14] J. Tsuji (Ed.), *Palladium Reagents and Catalysts, Innovations in Organic Synthesis*, Wiley, New York, **1995**.

[15] Bromine Compounds Ltd. (A. O. Ewenson, B. Croitoru, A. Shushan), DE 19.633.017 (1997).

6.11 Polymers

Wolfgang A. Herrmann, Wolfgang C. Schattenmann

6.11.1 Introduction

The application of water as a reaction medium is of great current and future interest in homogeneous catalysis [1, 2a]. Ecological and economic, but also chemical reasons are to be named here. Water is cheap, nontoxic, and nonflammable. These are three major advantages, especially for large-scale industrial processes. In addition, water can be beneficial to the overall performance of a chemical reaction, with a convincing example being the Ruhrchemie/Rhône-Poulenc hydroformylation process employing the water-soluble ligand tris(*m*-sulfonatophenyl)phosphine (as the sodium salt TPPTS) at rhodium [2b].

Polymerization reactions in aqueous media are rare events indeed in organometallic catalysis. Problems associated with high viscosities and low solubilities of polymers in aqueous systems come to mind here, but the notorious instability of many organometallic compounds and catalysts toward hydrolysis is an equally severe impediment.

As a matter of principle, polymerizations in aqueous media can be conducted as homogeneous or heterogeneous reactions, or as emulsion-type processes [3]. For example, an aqueous emulsion-type polymerization would contain a water-insoluble monomer, a water-soluble initiator, and an emulsifier. The emulsion polymerization has a number of advantages; the physical state of an emulsion system (colloidal distribution) simplifies technical processing like heat transfer and viscosities. Independence of the molecular mass and the polymerization rate is the key advantage of a radical-type emulsion polymerization. The concentration of the initiator can be used to adjust the molecular mass of the product: presence of only a little initiator entails slow polymerization which, in turn, yields high-molecular-mass products. By way of contrast, bimolecular termination steps are favored by high radical concentrations. Emulsion polymerization offers the advantage of simultaneously attaining both high molecular masses and rapid reaction rates [3].

6.11.2 Butadiene Polymerization

The first emulsion polymerization catalyzed by a transition metal complex was reported as early as in 1961 [4]. Trivalent rhodium (Rh^{3+}) with an emulsifier such as sodium laurylsulfate or sodium dodecylbenzenesulfonate was used to polymerize 1,3-butadiene, yielding a highly *trans*-1,4-isomer. Other transition metals, e.g., palladium(II), iridium(III), ruthenium(III), and cobalt(I), were also suited to polymerizing 1,3-butadiene in emulsions [5], with the advantage of varying the microstructure from all-*trans*-1,4 to highly *cis*-1,4. Also, molecular masses in the range from 10^3 up to 3×10^5 g mol^{-1} were achieved: divalent palladium gives low-molecular-mass polymers ($M_n = 10^3 - 1.5 \times 10^3$ g mol^{-1}), whereas cobalt catalysts afford molecular masses of up to 3×10^5 g mol^{-1}.

Addition of 1,3-cyclohexadiene increases the reaction rates by a factor of 20. It is resonable to assume that the 1,3-cyclohexadiene stabilizes the low-oxidation-state rhodium(I) catalysts [6]. As a matter of fact, isolated rhodium(I) diene complexes are active catalysts for the emulsion polymerization of 1,3-butadiene [7]. It is well known that aqueous rhodium(III) chloride readily forms olefin complexes upon treatment with 1,5-hexadiene, 2-methyl-1,5-hexadiene, and 2,5-dimethyl-1,5-hexadiene. It can be concluded that the trivalent rhodium is reduced by the butadiene to form the active monovalent rhodium catalysts in aqueous solution [7].

It was originally tempting to assume that free-radical polymerizations occur in these cases [6]. However, both free-radical initiators and inhibitors have only little influence upon the reaction rate [7]. Also, rhodium salts are efficient inhibitors for free-radical catalysis of 1,3-butadiene polymerization, but they initiate polymerization by a coordination mechanism, so a free-radical mechanism is unlikely [8]. Only a few emulsifiers of the sulfate and sulfonate classes were found to be efficient dispersing agents in the emulsion polymerization of 1,3-butadiene by rhodium(III) chloride [9].

6.11.3 Suzuki Cross Coupling

While the above-mentioned polymerization of 1,3-butadiene belongs to the so-called chain-growth polymerizations, Suzuki cross coupling is classified as a "step-growth polymerization" (or polycondensation) [3]. The polymer is formed stepwise by intermolecular "condensation" (not necessarily by extrusion of water) of bifunctional monomers with complementary functional groups. Two different classes, depending on the type of monomer, are possible: different monomers A–A and B–B engage in an alternating connectivity mode, or A–B monomers "self-condense", provided that a selective head-to-tail connection occurs.

The stepwise formation of polymers has immediate consequences for the quantity and distribution of molecular masses. For statistical reasons, the most abundant species tend to co-condense; in fact small chains most probably react among themselves. High-molecular-mass polymers are thus only obtained at high conversions (usually higher than 99% yield); this feature differs from chain-growth polymerizations. Particularly interesting is poly(*p*-phenylene) (PPP) and its derivatives [10]. Suzuki cross coupling can be employed with A–B monomers 1 according to Eq. (1). Soluble derivatives of PPP are formed when palladium(0) phosphine catalysts are used in a biphasic water/benzene medium [10, 11].

$$n\ \text{Br–C}_6\text{H}_2\text{R}_2\text{–B(OH)}_2\ (\mathbf{1}) \xrightarrow[\text{C}_6\text{H}_6/\text{H}_2\text{O}\ (\text{Na}_2\text{CO}_3)]{\text{Pd(PPh}_3)_4} [\text{–C}_6\text{H}_2\text{R}_2\text{–}]_n\ \ \mathbf{PPP} \qquad (1)$$

poly(p-phenylene) derivatives

Analogous nickel catalysts yield lower-molecular-mass products. However, long alkyl-chain substituents R must be used in order to keep the rigid backbone soluble. Copolymers of different PPP derivatives can be obtained under biphasic conditions, too [12].

Palladium(0)-catalyzed cross coupling can also be conducted in the monophasic mode [13–15]. Thus, the monosulfonated triphenylphosphine TPPMS attached to the metal polymerizes the bifunctional biphenyl derivatives according to the scheme A–A + B–B (Eq. 2). The water-soluble catalyst is prepared from commercial palladium(II) chloride and the phosphine. The polymer is also water-soluble and exhibits the outstanding thermal stability typical of rigid-chain polymers of this type.

$$n\ \text{Br–biphenyl(CO}_2\text{H)}_2\text{–Br} + n\ \text{(dioxaborolane)B–biphenyl–B(dioxaborolane)} \xrightarrow{\text{PdLn},\ \text{H}_2\text{O}} [\text{–quaterphenyl(CO}_2\text{H)}_2\text{–}]_n \qquad (2)$$

L = TPPMS

Until now, no polymer synthesis in a biphasic system via Suzuki cross coupling has become industrially relevant. For other Suzuki couplings cf. [16–18] and Section 6.10.

6.11.4 Ring-opening Metathesis Polymerization (ROMP)

Metathesis polymerization of olefins has elegantly opened an entry to a number of otherwise inaccessible polymers. Titanium, molybdenum, tungsten, and rhenium are widely used in coordination compounds as catalysts. They polymerize olefins in a "living" fashion. However, the conventional metathesis catalysts are unstable toward functional groups and protic media. Catalysts based on "late" transition metals such as ruthenium, iridium, and osmium tolerate functional groups much better [19].

It has been known since 1965 that norbornene and several derivatives containing functional groups (alcohols, esters) undergo emulsion polymerization in the presence of $(NH_4)_3[IrCl_6]$ and $RuCl_3 \cdot 3\,H_2O$ as catalysts [20]. It was not unexpected, then, that cyclobutene and 3-methylcyclobutene undergo ring-opening polymerization in the presence of ruthenium(III) chloride under aqueous conditions to form polybutadienes [21].

It is now possible to polymerize norbornene and 7-oxanorbornene derivatives exhibiting polar functional groups (anhydrides, carboxylic acids, ethers, alcohols, esters) according to Eq. (3), again employing salts of ruthenium, of iridium, and of osmium (normally chlorides) in catalytic amounts [22–24].

O OH OH $\xrightarrow[H_2O]{RuCl_3}$ O n HO OH (3)

In the meantime, ruthenium(II) complexes such as $[Ru(H_2O)_6](Tos)_2$ (Tos = tosyl) were found to be active catalysts in the ROMP reaction of functionalized monomers under aqueous conditions [25, 26]. It is interesting to note that water appears to be a co-catalyst. In this context, carbene–ruthenium complexes were synthesized which catalyze ROMP reactions in a living fashion in organic solvents (Structures **2** and **3** [27]). Although the metal carbene catalyst **2** is insoluble in water, polymers are obtained under aqueous conditions [28]. Products of sufficiently high monodispersity are only obtained, however, if organic co-solvents such as dichloromethane are present. It is interesting to note that the polydispersity indices (PDIs) of polymers resulting from a hydrophilic monomer (in the presence of water) are narrower than those from solution polymerization.

A simple structural modification of the catalyst type **2** yields water-soluble catalysts **3**.

Neoglycopolymers are formed in biphasic media from sugar-based norbornenes in the presence of the lipophilic metal–carbene catalyst **2** [30]. This reaction also works with ruthenium(III) chloride in water. It was shown that the water-soluble polymer products inhibit cell-agglutinating properties of concanavalin A [31, 32].

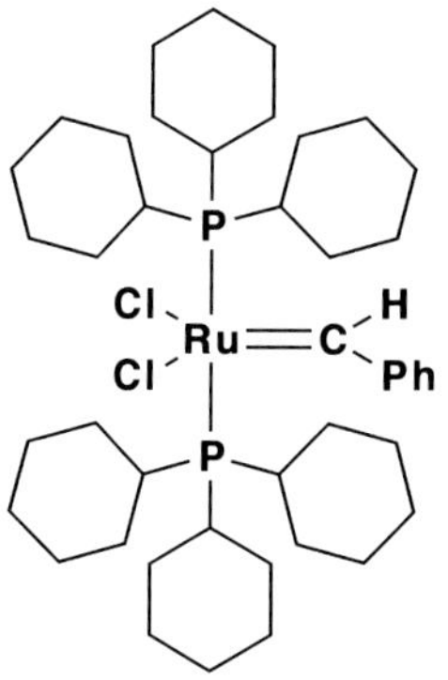

2

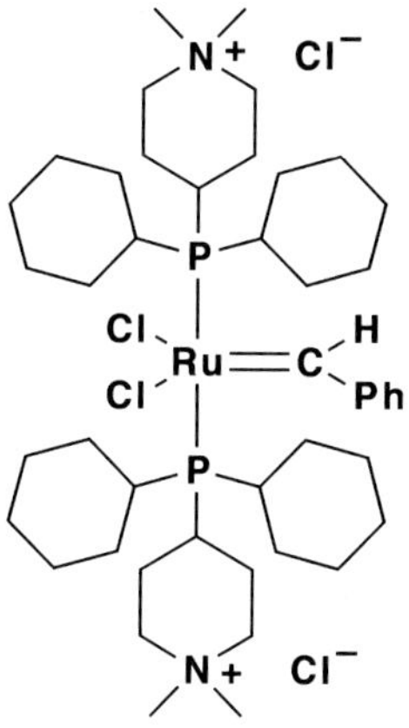

3

6.11.5 Other Polymerizations

6.11.5.1 Cyclobutene

The polymerization of cyclobutene can proceed either at the double bond with formation of enchained cyclobutane rings or by ring opening (metathesis) with formation of polymers exhibiting a 1,4-polybutadiene structure (Scheme 1); the type of catalyst (transition metal) will govern the mechanism.

Trivalent ruthenium produces 1,4-polybutadiene in a ROMP-like manner in aqueous emulsion [33]. The microstructure of the polymer is 50% 1,4-*cis*- and 50% 1,4-trans, which result contrasts with the 100% 1,4-*trans* product obtained in an ethanolic solution of ruthenium(III) chloride.

The former double bond of cyclobutene is not subject to cleavage if rhodium(III) chloride is present under emulsion conditions. Instead, a polymerization proceeds with high stereoregularity [21]. In this case, ring opening is only a minor side reaction.

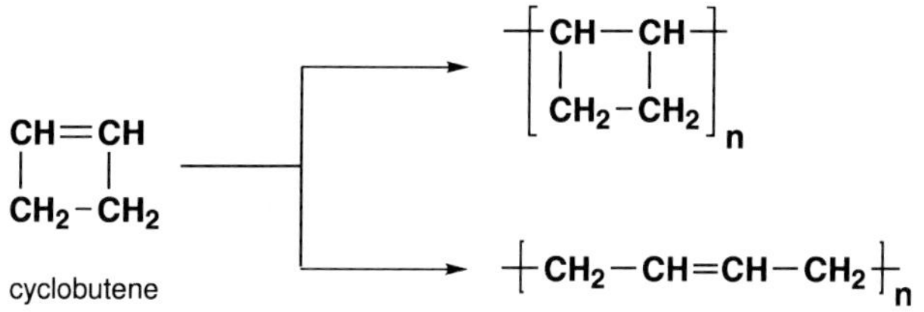

Scheme 1 cyclobutene

6.11.5.2 Norbornadiene

Palladium(II) salts catalyze the polymerization of norbornadiene in a 1,2-addition to give polymers containing 1 mol of unsaturation in the repeating unit according to Structure **4**. By way of contrast, rhodium leads to a saturated polymer with a repeating unit according to Structure **5** of nortricyclene as a result of 1,5-polyaddition [7].

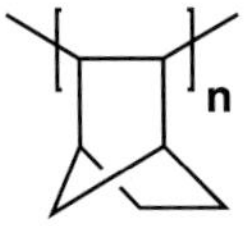

4

1,2-polyaddition

5

1,5-polyaddition

Aqueous emulsions work very well for the 1,2-insertion polymerization of 7-oxanorbornadiene derivatives in the presence of palladium(II) catalysts [34]. The resulting polymer can be converted consecutively into polyacetylene when the retro-Diels–Alder reaction is conducted at 100 °C (Scheme 2) [34].

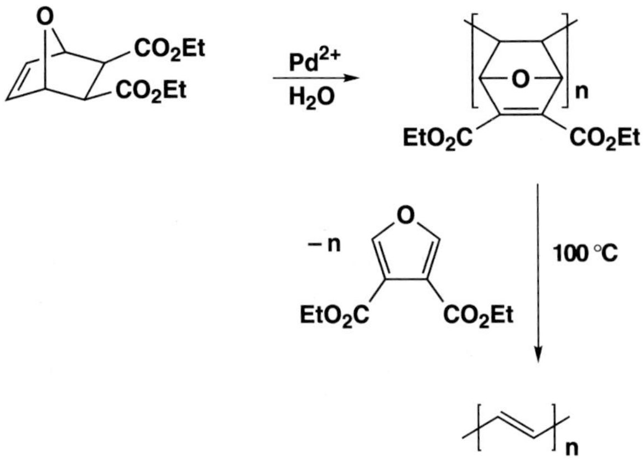

O, CO_2Et, CO_2Et, Pd^{2+}, H_2O, O, n, EtO_2C, CO_2Et, −n, O, EtO_2C, CO_2Et, 100 °C, n

Scheme 2

polyacetylene

6.11.5.3 Ethylene

There is no more than one report on the (very slow) coordination polymerization of ethylene under aqueous conditions in the presence of a rhodium catalyst [35]. However, the results are not convincing: a low-molecular-mass polyethylene (M_w = 5100) is obtained after 90 days under constant gas pressure of 60 bar.

6.11.5.4 Olefin/Carbon Monoxide Copolymerization

The alternating copolymerization of olefins with carbon monoxide [1, 36, 37] can also be performed in aqueous media [38]. The catalyst $[Pd(dppp\text{-}SO_3K)(H_2O)_2](BF_4)_2$ (dppp = 1,3-bis(diphenylphosphino)propane), generated *in situ* from standard palladium(II) salts, works very well for ethylene or propene (Eq. 4). Beyond that, the terpolymerization of ethylene, propene and carbon monoxide is possible with this catalytic system.

$$\text{R-CH=CH}_2 + \text{CO} \xrightarrow[\text{H}_2\text{O}]{[\text{Pd(L)(H}_2\text{O)}_2](\text{BF}_4)_2} -[\text{CH}_2\text{-CH(R)-C(=O)}]_n- \quad (4)$$

L = dppp–SO_3K = $Ar_{2-n}Ph_nP(CH_2)_3PPh_nAr_{2-n}$; n = 0, 1; Ar = C_6H_4–*m*–SO_3K

It appeared that the initial palladium hydride was formed from water via the water-gas shift reaction, because the terminal group of the C_2H_4/CO polymer becomes deuterated when the reaction is carried out in fully deuterated water. Polyketones are industrially desired high-melting materials, with multiple applications particularly in automobiles [1, 36, 37].

6.11.5.5 Acetylenes

The polymerization of phenylacetylene is effected by a rhodium(I) complex under biphasic conditions [39, 40]. The catalyst (cod)RhCl(TPPTS) · H_2O (cod = 1,5-cyclooctadiene; TPPTS = $Ph_2P(C_6H_4\text{-}m\text{-}SO_3Na)$ polymerizes phenylacetylene within 10 h at 40 °C to a *cis*-polymer with >96% yield in toluene/water. Upon elevation of the temperature considerable amounts of 1,3,5-triphenylbenzene were formed, and part of the polymer lost its original structure.

References

[1] W. A. Herrmann, B. Cornils, *Angew. Chem.* **1997**, *109*, 1075; *Angew. Chem. Int. Ed. Engl.* **1997**, *36*, 1049.

[2] (a) W. A. Herrmann, B. Cornils in *Applied Homogeneous Catalysis with Organometallic Compounds* (B. Cornils, W. A. Herrmann, eds.), p. 1167, VCH, Weinheim, **1996**; (b) C. D. Frohning, C. W. Kohlpaintner, in [2a], p. 29.

[3] G. Odian, *Principles of Polymerization*, 3rd ed., Wiley–Interscience, New York, **1991**.

[4] R. E. Rinehart, H. P. Smith, H. S. Witt, H. Romeyn, *J. Am. Chem. Soc.* **1961**, *83*, 4864.

[5] A. J. Canale, W. A. Hewett, T. M. Shryne, E. A. Youngman, *Chem. Int. (London)* **1962**, 1054.

[6] P. Teyssie, R. Dauby, *J. Polym. Sci. B* **1964**, *2*, 413.
[7] R. E. Rinehart, *J. Polym. Sci. C* **1969**, *27*, 7.
[8] R. E. Rinehart, H. P. Smith, H. S. Witt, H. Romeyn, *J. Am. Chem. Soc.* **1962**, *84*, 4145.
[9] M. Morton, B. Das, *J. Polym. Sci. C* **1969**, 27, 1.
[10] M. Rehahn, A.-D. Schlüter, G. Wegner, W. J. Feast, *Polymer* **1989**, *30*, 1060.
[11] A.-D. Schlüter, G. Wegner, *Acta Polym.* **1993**, *44*, 59.
[12] M. Rehahn, A.-D. Schlüter, G. Wegner, *Makromol. Chem.* **1990**, *191*, 1991.
[13] T. I. Wallow, B. M. Novak, *J. Am. Chem. Soc.* **1991**, *113*, 7411.
[14] T. A. P. Seery, T. I. Wallow, B. M. Novak, *ACS Polym. Prepr.* **1993**, *34*, 727.
[15] T. I. Wallow, T. A. P. Seery, F. E. Goodson, B. M. Novak, *ACS Polym. Prepr.* **1994**, *35*, 710.
[16] Hoechst AG (S. Haber, H.-J. Kleiner), DE 19.527.118-A1 (1997).
[17] Pd/TPPTS complexes: W. A. Herrmann, J. A. Kulpe, W. Konkol, H. Bahrmann, *J. Organomet. Chem.* **1990**, *389*, 85.
[18] Pd/TPPTS complexes: W. A. Herrmann, J. Kellner, H. Riepl, *J. Organomet. Chem.* **1990**, *389*, 103.
[19] R. H. Grubbs, *Pure Appl. Chem.* **1994**, *A3(11)*, 1829.
[20] R. E. Rinehart, H. P. Smith, *J. Polym. Sci. B* **1965**, *3*, 1049.
[21] G. Natta, G. Dall'Asta, G. Motroni, *J. Polym. Sci. B* **1964**, *2*, 349.
[22] W. J. Feast, D. B. Harrison, *J. Mol. Catal.* **1991**, *65*, 63.
[23] S.-Y. Lu, P. Quayle, F. Heatley, C. Booth, S. G. Yeates, J. C. Padget, *Eur. Polym. J.* **1993**, *29*, 269.
[24] S.-Y. Lu, J. M. Amass, N. Majid, D. Glennon, A. Byerley, F. Heatley, P. Quayle, C. Booth, S. G. Yeates, J. C. Padget, *Makromol. Chem. Phys.* **1994**, *195*, 1273.
[25] B. M. Novak, R. H. Grubbs, *J. Am. Chem. Soc.* **1988**, *110*, 7542.
[26] M. B. France, R. H. Grubbs, D. V. McGrath, R. A. Paciello, *Macromolecules* **1993**, *26*, 4742.
[27] P. Schwab, R. H. Grubbs, J. W. Ziller, *J. Am. Chem. Soc.* **1996**, *118*, 100.
[28] D. M. Lynn, S. Kanaoka, R. H. Grubbs, *J. Am. Chem. Soc.* **1996**, *118*, 784.
[29] B. Mohr, D. M. Lynn, R. H. Grubbs, *Organometallics* **1996**, *15*, 4317.
[30] C. Fraser, R. H. Grubbs, *Macromolecules* **1995**, *28*, 7248.
[31] K. H. Mortell, M. Gingras, L. L. Kiessling, *J. Am. Chem. Soc.* **1994**, *116*, 12053.
[32] M. C. Schuster, K. H. Mortell, A. D. Hegeman, L. L. Kiessling, *J. Mol. Catal.* **1997**, *116*, 209.
[33] G. Natta, G. Dall'Asta, L. Porri, *Makromol. Chem.* **1965**, *81*, 253.
[34] A. L. Safir, B. M. Novak, *Macromolecules* **1993**, *26*, 4072.
[35] L. Wang, R. S. Lu, R. Bau, T. C. Flood, *J. Am. Chem. Soc.* **1993**, *115*, 6999.
[36] E. Drent, J. A. M. v. Broekhoven, P. H. M. Budzelaar, *The Alternating Copolymerization of Alkenes and Carbon Monoxide*, 1st ed., VCH, Weinheim, **1996**, Vol. 1.
[37] W. A. Herrmann, B. Cornils, in [2], Vol. 2, p. 1167.
[38] Z. Jiang, A. Sen, *Macromolecules* **1994**, *27*, 7215.
[39] I. Amer, H. Schumann, V. Ravindar, W. Baidossi, N. Goren, J. Blum, *J. Mol. Catal.* **1993**, *85*, 163.
[40] W. Baidossi, N. Goren, J. Blum, H. Schumann, H. Hemling, *J. Mol. Catal.* **1993**, *85*, 153.

6.12 Oleochemistry

Arno Behr

6.12.1 Introduction

In fat chemistry ("olechemistry") no considerable application of homogeneous catalysis and of industrial importance is known so far [1]. One reason for this is the modest reactivity of the starting chemicals. Oleochemicals are molecules with a long carbon chain. The double bonds of unsaturated fatty compounds are always in internal positions. Hence, the steric hindrance of oleochemicals is often very high, and the coordination to metal complexes is made difficult. In addition all fatty compounds contain a substituent with a heteroatom such as carboxyl, ester, aldehyde, alcohol, or amine groups. These substituents often react with organometallic compounds and can inactivate the catalyst.

In spite of these relatively unfavourable characteristics of unsaturated oleochemicals, organometallic catalysis has – nevertheless – gained some access to fatty chemistry [2, 3], and aqueous-phase organometallic catalysis also is well known. Because the fatty compounds are organic nonpolar molecules, the addition of water always generates a liquid–liquid two-phase system (LLTP). If the organometallic catalyst is soluble in water, these LLTP systems are ideal for achieving a nearly complete recycle of the homogeneous metal catalyst (see Section 4.2).

The most typical examples in oleochemistry are the hydrogenation, the carbon monoxide reactions hydroformylation and hydrocarboxylation, and the oxidation reaction.

6.12.2 Hydrogenation

One important aim in oleochemistry hydrogenations is the selective hydrogenation of multiply unsaturated fatty compounds to singly unsaturated products. A typical example is the selective hydrogenation of linoleic acid ($C_{18:2}$) to oleic acid ($C_{18:1}$) without significant formation of stearic acid ($C_{18:0}$) as shown in Eq. (1). This reaction was studied intensively under homogeneous conditions in organic solvents. As organometallic catalysts carbonyl complexes of cobalt [4–6], iron [7, 8] and chromium [9–20] were used, and also catalyst systems based on platinum/tin [21–27], palladium/aluminum [28], iridium [29], rhodium [30–32], and ruthenium [33]. However, in all these investigations recycling of the homogeneous catalyst was not possible.

linoleic acid —COOH $\xrightarrow[+ H_2]{[cat]}$ oleic acid —COOH

[cat] $\not\downarrow$ + H_2 (1)

stearic acid —COOH

Selective hydrogenation of oleochemicals in combination with catalyst recycle was first studied by Bayer and Schumann [34, 35]. They used the LLTP system with water-soluble polymers as catalyst ligands. Polymers such as polyvinylpyrrolidone (PVP), poly(vinyl alcohol), polyoxyethylene or polyethyleneimine are the ligands of water-soluble palladium hydrogenation catalysts (see for example Structure **1**). After the oleochemical substrate is added to this aqueous catalyst solution, the mixture is stirred vigorously at room temperature. After the hydrogenation reaction, the two phases are allowed to separate, and the aqueous catalyst solution can be re-used directly without any further activation. Bayer compared the activity (reacted substrate molecules per metal atom/min) of the catalysts with slow enzymes. The same author described the selective hydrogenation of triglycerides in water using a Ni/PVP catalyst formed by the reduction of nickel sulfate with sodium boranate in the presence of PVP [36, 37].

CH—CH_2, N, O, PdL_x, n

1

A water-soluble rhodium hydrogenation catalyst has been found by Patin and co-workers [38]. Hydrogenations are carried out in very mild conditions, with Rh/TPPTS in H_2O/EtOH [1:1] and the catalyst is not air-sensitive and is easily prepared. However, the hydrogenation of linoleic acid is not selective, but yields only stearic acid.

A new development is biphasic hydrogenation using solvent-stabilized colloid (SSCs) catalysts [39–41]. Palladium colloid systems, especially, were proven to give high reactivity and selectivity. Best solvents are dimethylformamide and particularly the two cyclic carbonic acid esters, ethylene carbonate and 1,2-propene carbonate. In these solvents sodium tetrachloropalladate – stabilized by a sodium carbonate buffer – is reduced with hydrogen to yield the solvent-stabilized palladium colloid. Transmission electron microscopy of the palladium colloid demonstrates that the colloid particles are spherical with an average diameter of 4 nm.

The hydrogenation of the oleochemicals can proceed in the same apparatus in which the catalyst is formed. The fatty compound is added to the catalyst solution, thus yielding a biphasic system of two immiscible liquids. After the necessary quantity of hydrogen has been introduced, an intensive stirrer mixes the two liquid phases and the gas phase. On the laboratory scale, typical hydrogenations were carried out at room temperature and at ambient pressure allowing a reaction time of about 30 min. After the reaction the two phases are separated and the solvent/water phase is recycled with the colloid catalyst to the reactor.

Broad screening of the possible donor solvents proved that only a few solvents are suitable for use in phase separation and colloid stabilization. By far the best results are obtained with propene carbonate (Table 1), which is favored by the high selectivity to $C_{18:1}$ and the short reaction time.

Table 1. Selective hydrogenation with palladium SSCs in different carbonate solvents.

	$C_{18:0}$ [%]	$C_{18:1}$ [%]	$C_{18:2}$ [%]	T [°C]	*t* [min]
Starting mixture:					
– Sunflower-oil fatty acid methyl esters	2	22	70	–	–
Hydrogenation products in:					
– Propene carbonate	2	92	0	25	23
– Ethylene carbonate	4	91	0	60	110
– Glyceryl carbonate	5	88	0	25	180

It can be assumed that the solvent propene carbonate acts simultaneously as a solvent and as a complex ligand to the palladium. A structural proposal is given in Eq. (2): propene carbonate may coordinate to the metal via two oxygen atoms. If the hydrogen molecule adds to this complex, the carbonate chelate ligand gives rise to a heterolytic splitting of the hydrogen into a palladium hydride species and a proton, which is picked up by the ligand. The palladium hydride than acts as the starting complex for the catalytic hydrogenation cycle.

The propene carbonate-stabilized palladium colloid is an excellent catalyst for the hydrogenation of a great number of different fatty acids, fatty esters, and triglycerides. Table 2 gives a survey of results with sunflower, palm-kernel and

CH2–CH–CH3
|O| |O|
Pd
C
O
+ H2
⊖ Pd
H
O
CH2
⊕ C
CH–CH3
HO
O

(2)

Table 2. Selective hydrogenation of different oleochemicals with palladium SSCs ($T = 25\,^\circ C$, $p = 1$ bar H_2; solvent propene carbonate).

Oleochemical	$C_{18:0}$ [%]	$C_{18:1}$ [%]	$C_{18:2}$ [%]	$C_{18:3}$ [%]
Sunflower oil				
– Untreated	2	22	70	0
– Hydrogenated	4	86	4	0
– Fatty acids, hydrogenated	2	93	0	0
– Fatty esters, hydrogenated	2	86	1	0
Palm-kernel fatty acids				
– Untreated	1	81	11	0
– Hydrogenated	2	90	0	0
Rape-seed oil (new)				
– Untreated	2	60	20	8
– Hydrogenated	5	87	1	0

rapeseed oils, acids, and esters. The yield of $C_{18:1}$ products after hydrogenation is in the range of 86–93%. In all examples the reaction time is very short.

The kinetics of the selective hydrogenation with palladium SSCs is shown in Figure 1. At room temperature the linoleic acid ($C_{18:2}$) is completely hydrogenated after 45 min. During this time no additional stearic ester is formed. This result is characteristic of the high selectivity of the palladium SSC. Obviously, the double unsaturated fatty compound coordinates essentially better than the monounsaturated compounds to the palladium.

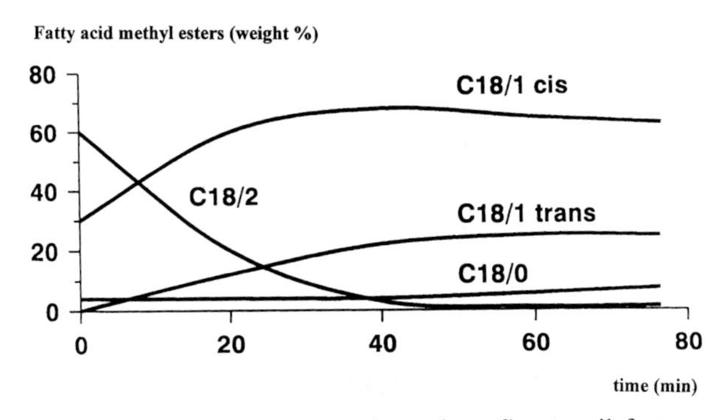

Figure 1. Kinetics of the selective hydrogenation of sunflower oil fatty acid methyl esters ($T = 25\,^\circ C$, solvent propene carbonate; Pd/ester ratio 1:5000).

These observations are summarized in a mechanistic proposal shown in Figure 2. The palladium hydrido species (compare Eq. 2) is able to add equally to the double bonds in positions 9 and 12. The olefin inserts into the Pd–H bond, yielding the corresponding alkyl complexes. After reaction with hydrogen the monounsaturated fatty acids ($C_{18:1}$) are set free, and the palladium hydride species is formed again, thus starting a new catalytic cycle.

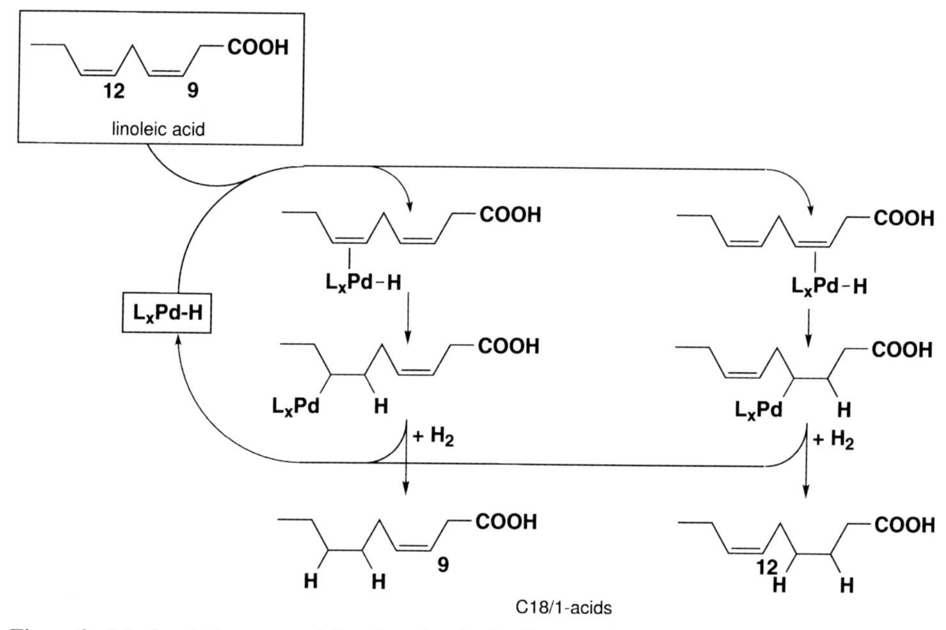

Figure 2. Mechanistic proposal for the selective hydrogenation of linoleic acid with Pd SSCs (L_xPd-H = stabilized palladium hydrido species).

The reaction time can be influenced by different parameters. By an enhanced hydrogen exposure at 5 bar the reaction time can be reduced substantially. For technical realization, it is of high economic importance that the SSC be active even in low concentrations. If the weight ratio of fatty acids to palladium is changed from 10 000:1 to 100 000:1, the reaction time increases steadily. However, at very low catalyst concentrations the reaction time is still very short. Another point of great importance is the number of possible catalyst recycles. An eight-fold recycle of the palladium SSC had no essential influence on activity and selectivity. The reaction could easily be scaled-up to the pilot scale.

6.12.3 Hydroformylation

Instead of olefins, unsaturated fat chemicals also can be hydroformylated. First studies were done in the 1960s by Lai and Ucciani [42, 43], who investigated cobalt bislaurate and dicobaltoctacarbonyl as catalyst precursors. Frankel et al. [44] also used $Co_2(CO)_8$ and obtained fatty aldehydes at reaction temperatures of about 100 °C and fatty alcohols when working at 180 °C. Starting from oleic acid methyl ester, yields of up to 84 % were obtained. Conversions of oleochemical feedstocks with Rh catalysts were possible, too [45].

To solve the problem of metal recycle, one idea was to fix the rhodium on an inactive support. This rhodium was then extracted from the support by a triphenylphosphine-containing solution, thus producing in situ the homogeneous homophase catalyst solution. Friedrich et al. [46] described such a process of rhodium-catalyzed hydroformylation including filtration of the supported catalyst, cooking of the support in a rotary furnace, and thermal stressing distillation of the filtrate, thus separating Rh from the products.

As early as 1978 Andreetta et al. [47] applied this technique to the hydroformylation of olefins and oleochemicals. They used the alternative of working in two phases, however, not during the reaction but after hydroformylation. The catalyst was composed of a rhodium component, e.g., $Rh_4(CO)_{12}$, and an aminophosphine ligand, e.g., $P(CH_2CH_2CH_2NEt_2)_3$. After the reaction the catalyst was recovered by one of the following methods. One possibility was the extraction of the catalyst and of the excess phosphines by means of dilute sulfuric acid solutions at ambient pressure. Addition of alkali to the aqueous solution causes the separation of an oily catalyst which is recovered preferentially by extraction with the starting olefin or oleochemical. The second method was the extraction of the catalyst with water under a carbon dioxide atmosphere of 3–5 bar. Carbon dioxide is then removed from the aqueous solution (which contains the catalyst and the free phosphine as phosphinoammonium bicarbonates) by boiling at atmospheric pressure in a stirred vessel, once again in presence of the starting olefin. Andreetta et al. described the recycle of rhodium catalysts by these methods and without substantial losses.

A very similar procedure has been described by Bahrmann and Cornils [48] for the hydroformylation of oleylalcohol. They carried out the reaction in a homogeneous organic phase at 130 °C and a syngas pressure of 270 bar. The catalyst was formed by rhodiumtris(2-ethylhexanoate) and aromatic phosphines and was soluble in an organic medium. The special feature was the choice of the ligand: they used the TPPTS ligand, modified as an ammonium salt with high-molecular-weight hydrocarbon chains. This ligand $TPPTS^{-}NHR_3^{+}$ with butyl to dodecyl substituents dissolved the rhodium in the starting oleyl alcohol. After the reaction (Figure 3) an excess of aqueous sodium hydroxide solution was added to the product mixture, thus forming the water-soluble sodium sulfonates. Hence, the organic rhodium-free phase could be separated and was hydrogenated in a second reaction step to the target nonadecadiol product. The aqueous phase contained the ligand and more than 99 % of the starting rhodium metal. By changing the pH value from 8.5 to 1.0 by addition of aqueous hydrochloride solution and by addition of new tertiary amine, the initial catalyst is re-formed and can be recycled to the reactor (cf. Section 6.1.5).

In the meantime, many other catalytic systems have been described for the hydroformylation of fatty compounds, especially by Fell's group [49–51]. They used water-soluble catalysts, for instance consisting of $Rh_4(CO)_{12}$ and surface-active sulfobetaine derivatives of tris(2-pyridyl)phosphine [49]. Other ligand systems are the sodium salt of TPPTS in combination with detergents [50] and the lithium salt of triphenylphosphinemonosulfonic acid TPPMS [51].

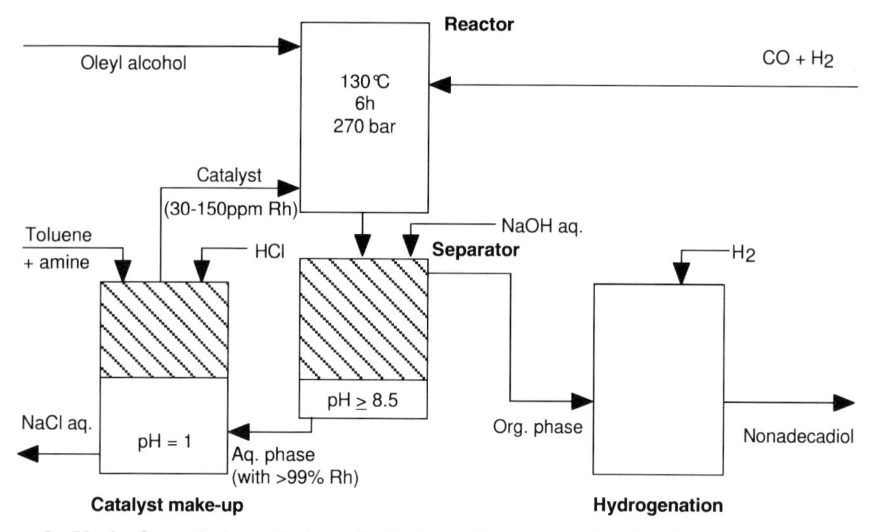

Figure 3. Hydroformylation of oleyl alcohol: catalyst recycle by LLTP technique.

Another possible way to separate the catalyst from the fatty products was found by Davis [52–54] and further investigated by Fell [55]. This new method is supported aqueous-phase catalysis (SAPC; cf. Section 4.7). On a hydrophilic support, e.g., silicon oxide with a high surface area, a thin aqueous film is applied which contains the water-soluble rhodium catalyst, for instance $HRh(CO)L_3$ with sodium TPPTS ligands. Oleyl alcohol and syngas react at the organic/aqueous interface and form the formylstearyl alcohol in a yield of 97%. The catalyst can be separated from the product by simple filtration without loss of activity.

Hydroformylation of oleochemicals, carried out thus by means of aqueous-phase catalysis, can provide a great number of different fatty derivatives which may be useful in many industrial applications [56–66].

6.12.4 Hydrocarboxylation

An interesting reaction of oleochemicals with carbon monoxide is hydrocarboxylation, the reaction of unsaturated compounds with CO and water yielding carboxylic acids. If oleic acid is used as the starting compound, the product is a branched C_{19}-dicarboxylic acid. Reppe and Kröper [67, 68] have studied this reaction using the catalysts nickel tetracarbonyl and nickel iodide and obtained products in about 70% yield. Levering [69] used dicobaltoctacarbonyl and Frankel and Thomas [70, 71] palladium phosphine catalysts. The recycle of this palladium catalyst via the LLTP technique is shown in Figure 4.

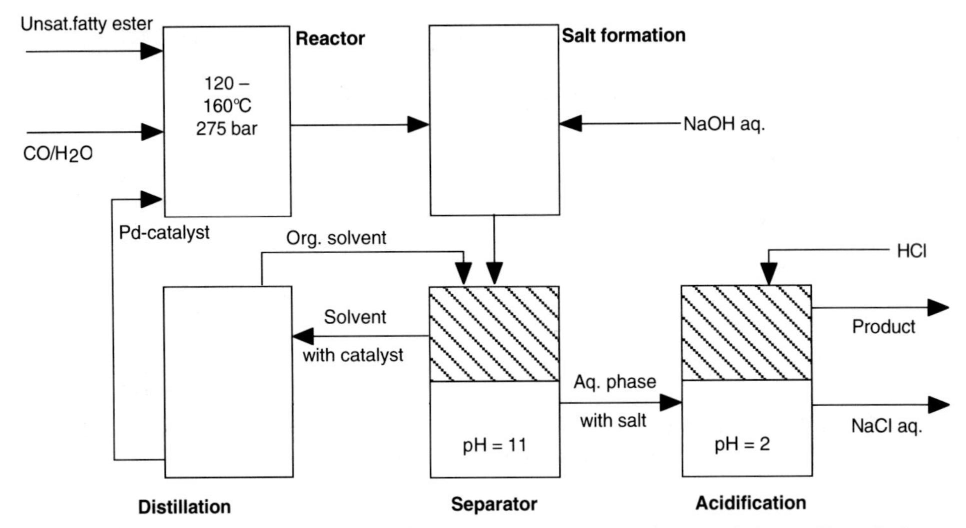

Figure 4. Hydrocarboxylation of unsaturated fatty esters: catalyst recycle by LLTP technique.

The hydrocarboxylation of an unsaturated fatty acid ester is carried out in a single organic phase. As product, the monoester of a dicarboxylic acid is formed. When this product is treated with aqueous sodium hydroxide solution this acid is converted into a water-soluble sodium salt. In the following separator the water-insoluble palladium catalyst is extracted from the aqueous phase by addition of an organic solvent. When the aqueous product phase is acidified with aqueous hydrochloride solution the released carboxylic acid forms a second organic phase and can easily be separated. The catalyst-containing solvent phase passes a distillation step, thus recycling the solvent to the separator and the palladium catalyst to the reactor.

6.12.5 Oxidation

Oxidation of unsaturated oleochemicals can proceed in different ways, and yields numerous products. Typical oxidations of fatty acids are, for instance, ketonizations yielding keto acids [72, 73], hydroxylations to bishydroxy acids [74], epoxidations to epoxy acids [75–78] and oxidative splitting reactions [72, 74] yielding mixtures of mono- and dicarboxylic acids. However, not only the double bond but also the functional group of the fatty compound, can be oxidized. One example is the ruthenium-catalyzed oxidation of fatty alcohols to fatty aldehydes or fatty acids [79]. As catalyst a ruthenium precursor is used, for instance $RuCl_2(PPh_3)_2$. A very smooth oxidation agent is *N*-methylmorpholine oxide (MMO), which can be prepared in situ starting from *N*-methylmorpholine

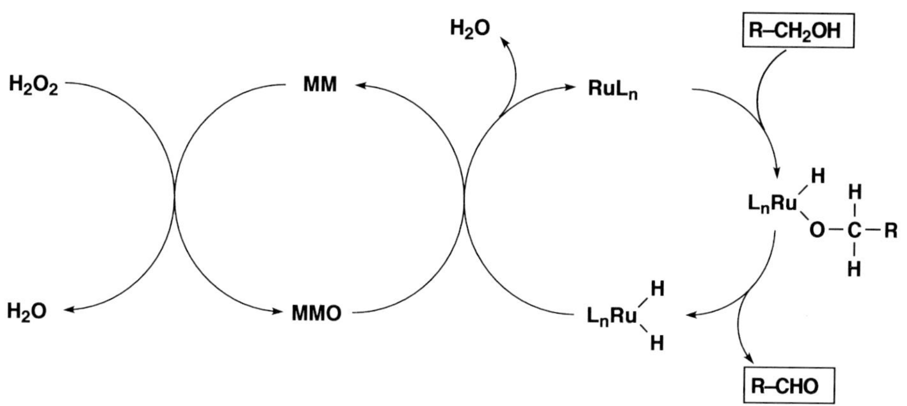

Figure 5. Mechanism of the ruthenium-catalyzed oxidation of fatty alcohols to aldehydes (MM = *N*-methylmorpholine; MMO = *N*-methylmorpholine oxide).

(MM) and aqueous hydrogen peroxide solution. The mechanistic steps are shown in Figure 5: the left-hand of the two integrated cycles demonstrates the repeated formation of MMO from MM and H_2O_2; the right-hand cycle is the catalytic cycle with the ruthenium complex catalyst. First the fatty alcohol adds to the RuL_n species yielding a ruthenium dihydride complex and the fatty aldehyde, which reacts further to the carboxylic acid. The dihydride species reacts with MMO yielding back the MM, the by-product water and the catalytic complex RuL_m, thus closing the catalytic cycle.

The technical realization of this process can once again proceed via the LLTP technique (Figure 6): the reactor contains the aqueous MM/catalyst phase; then the reaction unit is charged with fatty alcohol and an aqueous hydrogen peroxide solution. After the oxidation is finished the whole mixture is passed into a separator, where the aqueous MM/catalyst phase is detached and recycled to the reactor. The organic phase enters a multistage distillation unit, where both products, fatty aldehydes and acids, are isolated. Solvent as well as unchanged fatty alcohol is recycled to the reactor. In this way a very efficient overall oxidation with MM/H_2O_2 can be achieved.

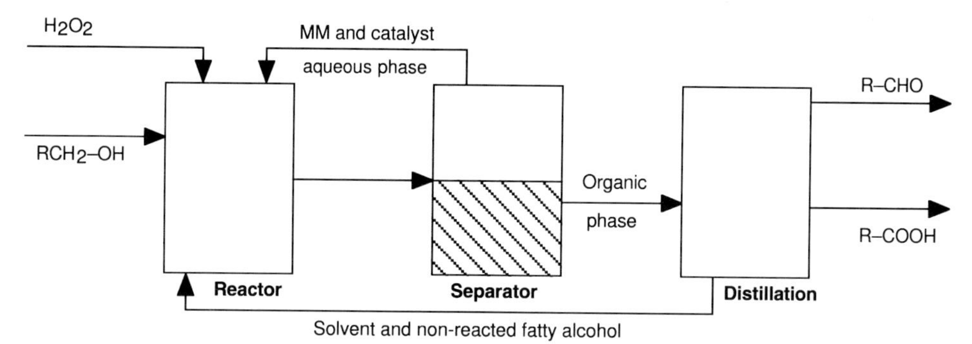

Figure 6. Oxidation of fatty alcohols: catalyst recycle by LLTP technique.

References

[1] A. Behr, *Fat Sci. Technol.* **1990**, *92*, 375.
[2] H. Eierdanz (Ed.), *Perspektiven nachwachsender Rohstoffe in der Chemie*, Wiley-VCH, Weinheim, **1996**.
[3] H. Baumann, M. Bühler, H. Fochem, F. Hirsinger, H. Zoebelein, J. Falbe, *Angew. Chem.* **1988**, *100*, 41.
[4] E. N. Frankel, E. P. Jones, V. L. Davison, E. Emken, H. J. Dutton, *J. Am. Oil Chem. Soc.* **1965**, *42*, 130.
[5] G. Cecchi, E. Ucciani, *Rev. Fr. Corps Gras* **1972**, *19*, 289.
[6] E. Ucciani, G. Cecchi, R. Phan Tan Luu, *Rev. Fr. Corps Gras* **1972**, *19*, 383.
[7] E. N. Frankel, T. L. Mounts, R. O. Butterfield, H. J. Dutton, *Adv. Chem. Ser.* **1968**, *70*, 177.
[8] G. Cecchi, E. Ucciani, *Rev. Fr. Corps Gras* **1974**, *21*, 423.
[9] E. N. Frankel, R. O. Butterfield, *J. Org. Chem.* **1969**, *34*, 3930.
[10] E. N. Frankel, F. L. Little, *J. Am. Oil Chem. Soc.* **1969**, *46*, 256.
[11] E. N. Frankel, F. L. Thomas, J. C. Cowan, *J. Am. Oil Chem. Soc.* **1970**, *47*, 497.
[12] E. N. Frankel, *J. Am. Oil Chem. Soc.* **1970**, *47*, 11.
[13] G. Ben-et, A. Dolev, M. Schimmel, R. Stern, *J. Am. Oil Chem. Soc.* **1972**, *49*, 205.
[14] E. N. Frankel, F. L. Thomas, *J. Am. Oil Chem. Soc.* **1972**, *49*, 70.
[15] E. N. Frankel, R. A. Awl, J. P. Friedrich, *J. Am. Oil Chem. Soc.* **1979**, *56*, 965.
[16] S. Koritala, E. N. Frankel, *J. Am. Oil Chem. Soc.* **1981**, *58*, 553.
[17] J. R. Tucker, D. P. Riley, *J. Organomet. Chem.* **1985**, *279*, 49.
[18] J. A. Heldal, E. N. Frankel, *J. Am. Oil Chem. Soc.* **1985**, *62*, 1044.
[19] L. J. Rubin, S. S. Koseoglu, L. L. Diosady, W. F. Graydon, *J. Am. Oil Chem. Soc.* **1986**, *63*, 1551.
[20] P. A. Bernstein, W. F. Graydon, L. L. Diosady, *J. Am. Oil Chem. Soc.* **1989**, *66*, 680.
[21] J. C. Bailar, H. Itatani, *J. Am. Oil Chem. Soc.* **1966**, *43*, 377.
[22] Unilever N.V., NE 6.604.800, NE 6.604.801, NE 6.604.802 (1966).
[23] J. C. Bailar, H. Itatani, *J. Am. Chem. Soc.* **1967**, *89*, 1592.
[24] E. N. Frankel, E. A. Emken, H. Itatani, J. C. Bailar, *J. Org. Chem.* **1967**, *32*, 1447.
[25] H. Itatani, J. C. Bailar, *J. Am. Oil Chem. Soc.* **1967**, *44*, 147.
[26] J. C. Bailar, *J. Am. Oil Chem. Soc.* **1970**, *47*, 475.
[27] E. N. Frankel, H. Itatani, J. C. Bailar, *J. Am. Oil Chem. Soc.* **1972**, *49*, 132.
[28] B. Fell, W. Schäfer, *Fat Sci. Technol.* **1990**, *92*, 264.
[29] C. Fragale, M. Gargano, T. Gomes, M. Rossi, *J. Am. Oil Chem. Soc.* **1979**, *56*, 498.
[30] W. J. Bland, T. C. Dine, R. N. Jobanputra, G. G. Shone, *J. Am. Oil Chem. Soc.* **1984**, *61*, 924.
[31] J. A. Heldal, E. N. Frankel, *J. Am. Oil Chem. Soc.* **1985**, *62*, 1117.
[32] E. A. Emken, *J. Am. Oil Chem. Soc.* **1988**, *65*, 373.
[33] C. Bello, L. L. Diosady, W. F. Graydon, L. J. Rubin, *J. Am. Oil Chem. Soc.* **1985**, *62*, 1587.
[34] Heyl & Co. (E. Bayer, W. Schumann), DE 2.835.943 (1980).
[35] E. Bayer, W. Schumann, *J. Chem. Soc., Chem. Commun.* **1986**, 949.
[36] E. Bayer, L. Tang, D. Waidelich, M. Kutubuddin, *Fat Sci. Technol.* **1992**, *94*, 79.
[37] Heyl & Co. (W. Schumann, E. Bayer, T. Ito, Y. Nawata), DE 3.727.704.A1 (1989).
[38] C. Larpent, R. Dabard, H. Patin, *Tetrahedron Lett.* **1987**, *28*, 2507.
[39] Henkel KGaA (A. Behr, N. Döring, C. Kozik, H. Schmidke, S. Durowicz), DE 4.012.873.3 (1990).
[40] Henkel KGaA (A. Behr, C. Lohr, B. Ellenberg), DE 4.109.246.5 (1991), EP 0.576.477.B1 (1992), US 5.354.877 (1994).

[41] A. Behr, N. Döring, S. Durowicz-Heil, B. Ellenberg, C. Kozik, C. Lohr, H. Schmidke, *Fat Sci. Technol.* **1993**, *95*, 2.
[42] R. Lai, M. Naudet, E. Ucciani, *Rev. Fr. Corps Gras* **1966**, *13*, 737.
[43] R. Lai, M. Naudet, E. Ucciani, *Rev. Fr. Corps Gras* **1968**, *15*, 15.
[44] E. N. Frankel, S. Metlin, W. K. Rohwedder, I. Wender, *J. Am. Oil Chem. Soc.* **1969**, *46*, 133.
[45] E. N. Frankel, *J. Am. Oil Chem. Soc.* **1971**, *48*, 248.
[46] J. P. Friedrich, G. R. List, V. E. Sohns, *J. Am. Oil Chem. Soc.* **1973**, *50*, 455.
[47] A. Andreetta, G. Barberis, G. Gregorio, *Chim. Ind.* **1978**, *60*, 887.
[48] (a) Ruhrchemie AG (H. Bahrmann, B. Cornils, W. Konkol, J. Weber, L. Bexten, H. Bach), EP 0.216.314 (1986); (b) T. Müller, H. Bahrmann, *J. Mol. Catal. A* **1997**, *116*, 39; (c) H. Bahrmann, M. Haubs, T. Müller, N. Schöpper, B. Cornils, *J. Organomet. Chem.* **1997**, *545/546*, 139.
[49] B. Fell, G. Papadogianakis, *J. Mol. Catal.* **1991**, *66*, 143.
[50] B. Fell, D. Leckel, C. Schobben, *Fat Sci. Technol.* **1995**, *97*, 219.
[51] Z. Xia, C. Schobben, B. Fell, *Fett/Lipid* **1996**, *98*, 393.
[52] J. P. Arhancet, M. E. Davis, J. S. Merola, B. E. Hanson, *Nature* (*London*) **1989**, *339*, 454.
[53] J. P. Arhancet, M. E. Davis, J. S. Merola, B. E. Hanson, *J. Catal.* **1990**, *121*, 327.
[54] M. E. Davis, *CHEMTECH* **1992**, *22*, 498.
[55] Z. Xia, K. Klöckner, B. Fell, *Fett/Lipid* **1996**, *98*, 313.
[56] E. H. Pryde, E. N. Frankel, J. C. Cowan, *J. Am. Oil Chem. Soc.* **1972**, *49*, 451.
[57] A. W. Schwab, E. N. Frankel, E. J. Dufek, J. C. Cowan, *J. Am. Oil Chem. Soc.* **1972**, *49*, 75.
[58] R. A. Awl, E. N. Frankel, E. H. Pryde, G. R. Riser, *J. Am. Oil Chem. Soc.* **1974**, *51*, 224.
[59] E. N. Frankel, W. E. Neff, F. L. Thomas, T. H. Khoe, E. H. Pryde, G. R. Riser, *J. Am. Oil Chem. Soc.* **1975**, *52*, 498.
[60] J. P. Friedrich, *J. Am. Oil Chem. Soc.* **1976**, *53*, 125.
[61] R. A. Awl, E. N. Frankel, E. H. Pryde, *J. Am. Oil Chem. Soc.* **1976**, *53*, 190.
[62] E. N. Frankel, E. H. Pryde, *J. Am. Oil Chem. Soc.* **1977**, *54*, 873A.
[63] E. J. Dufek, *J. Am. Oil Chem. Soc.* **1978**, *55*, 337.
[64] E. H. Pryde, *J. Am. Oil Chem. Soc.* **1984**, *61*, 419.
[65] E. N. Frankel, F. L. Thomas, *J. Am. Oil Chem. Soc.* **1972**, *49*, 10.
[66] E. N. Frankel, F. L. Thomas, W. K. Rohwedder, *Ind. Eng. Chem. Prod. Res. Dev.* **1973**, *12*, 47.
[67] W. Reppe, H. Kröper, DE 861.243 (1952).
[68] W. Reppe, H. Kröper, *Liebigs Ann. Chem.* **1953**, *582*, 38.
[69] D. R. Levering, *J. Org. Chem.* **1959**, *24*, 1833.
[70] E. N. Frankel, F. L. Thomas, *J. Am. Oil Chem. Soc.* **1973**, *50*, 39.
[71] E. N. Frankel, F. L. Thomas, W. F. Kwolek, *J. Am. Oil Chem. Soc.* **1974**, *51*, 393.
[72] S. Warwel, W. Pompetzki, E. A. Deckwirth, BMFT-Forschungsverbundvorhaben "Fettchemie" (1990).
[73] Y. Nakano, T. A. Foglia, *J. Am. Oil Chem. Soc.* **1982**, *59*, 163.
[74] T. A. Foglia, P. A. Barr, A. J. Malloy, M. J. Costanzo, *J. Am. Oil Chem. Soc.* **1977**, *54*, 870A.
[75] K. A. Jorgensen, *Chem. Rev.* **1989**, *89*, 431.
[76] T. Katsuki, K. B. Sharpless, *J. Am. Chem. Soc.* **1980,** *102*, 5974.
[77] J. T. Groves, R. Neumann, *J. Am. Chem. Soc.* **1987**, *109*, 5045.
[78] M.-C. Kuo, T. C. Chou, *Ind. Eng. Chem. Res.* **1987**, *26*, 277.
[79] A. Behr, K. Eusterwiemann, *J. Organomet. Chem.* **1991**, *403*, 215.

6.13 Olefin Metathesis

Robert H. Grubbs, David M. Lynn

6.13.1 Introduction

Olefin metathesis is a transition-metal-catalyzed reaction in which olefin bonds are cleaved and redistributed to form new olefins [1–3]. The reaction proceeds through the formal [2 + 2] cycloaddition of an olefin and a metal alkylidene to yield a metallocyclobutane intermediate (Scheme 1). The productive retrocycloaddition of this intermediate generates a new metal alkylidene and a new olefin product. These processes are generally reversible, and the reaction is under thermodynamic control.

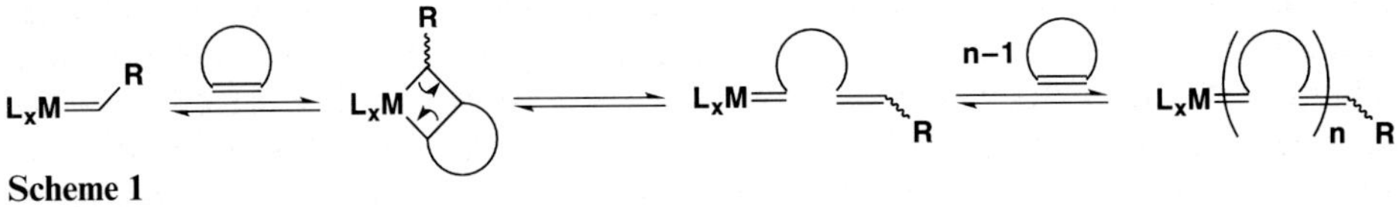

L_xM R L_xM R L_xM R n−1 L_xM n R

Scheme 1

The versatility of the metathesis reaction has attracted the attention of chemists with interests ranging from polymer chemistry to organic synthesis. For example, cyclic olefins will undergo ring-opening metathesis polymerization (ROMP) to yield polymers having backbone unsaturation (Scheme 1) [1, 4, 5]. ROMP is particularly well suited to the polymerization of highly strained monomers as the reaction is driven by relief of monomer ring strain. Recently, the polymerization of acyclic α,ω-dienes *via* acyclic diene metathesis has been demonstrated [6], and applications to the synthesis of small molecules and natural products *via* ring-closing metathesis and cross-metathesis have received much attention [7, 8].

Over the last 20 years, metathesis catalysts have evolved from poorly defined, heterogeneous mixtures to well-defined, single-component metallocycles and alkylidenes [9–18]. These complexes react in controlled, consistent ways, and their activities can be attenuated through simple ligand substitution. In contrast to heterogeneous systems, for example, several of these well-defined alkylidenes initiate living ROMP [4]. An important result of having catalysts based on a variety of transition metals was the observation of trends in functional-group tolerance [19, 20]. In particular, it was found that as the metal centers in these complexes were chosen from further right in the Periodic Table, the resulting alkylidenes reacted more selectively with olefins in the presence of other harder, Lewis-basic functional groups (Table 1) [3].

Table 1. Relationship between metal center and olefin selectivity.

Metal center: Titanium	Tungsten	Molybdenum	
Acids	Acids	Acids	↑
Alcohols, water	Alcohols, water	Alcohols, water	
Aldehydes	Aldehydes	Aldehydes	Increasing
Ketones	Ketones	*Olefins*	reactivity
Esters, amides	*Olefins*	Ketones	
Olefins	Esters, amides	Esters, amides	

Historically, olefin metathesis has been limited to the polymerization of cyclic hydrocarbons in highly purified organic solvents, due to the extreme sensitivities of early transition metal catalyst systems to oxygen, water, and polar functional groups [21]. It was quickly realized that the development of metathesis catalysts that were particularly tolerant of polar and protic functional groups would offer several advantages, the most obvious being the use of substrates and solvents without rigorous purification and drying. More tolerant catalysts would also broaden the scope of the reaction, enabling the ROMP of highly functionalized monomers and obviating the need for functional group protection/deprotection schemes. Additionally, the possibility of water-tolerant and, ultimately, water-soluble catalysts could enable the metathesis of olefinic substrates in aqueous solution. These potential benefits prompted a search for active catalysts based on late transition metals that would be stable toward functional groups and which might function homogeneously in water.

6.13.2 "Classical" Group VIII Catalysts

Soon after the discovery of the olefin metathesis reaction, reports indicated that complexes of ruthenium, osmium, and iridium could initiate ROMP. Michelotti et al. initially reported the polymerization of norbornene and its derivatives catalyzed by the hydrates of $RuCl_3$, $OsCl_3$, and $IrCl_3$ in refluxing ethanol (Eq. 1) [22, 23].

n (norbornene, X) —[Ru(III), Os(III), Ir(III); C_2H_5OH]→ polymer (cyclopentane, X)$_n$ (1)

X = H, CH_2Cl, CH_2OH, CO_2H

These complexes also functioned well in benzene, although small amounts of ethanol were necessary to initiate polymerization. The order of activity for these catalysts was Ir(III) > Os(III) > Ru(III), and they were found to polymerize monomers with *exo* substituents more readily than *endo* isomers. Rinehart and Smith later demonstrated that these complexes initiated the aqueous polymerization of a substituted norbornene derivative in the presence of anionic emulsifiers and suitable reducing agents [24]. This reaction gave particularly low yields of polymer (typically less than 9%), but the overall tolerance of these complexes to polar and protic functionalities made them ideal candidates for further study.

For the ROMP of functionalized 7-oxanorbornenes, possible only *via* these late transition metal catalysts, it was found that the best catalyst was $RuCl_3$ in a mixture of ethanol and benzene [20, 25]. However, the lack of a preformed alkylidene in this "precatalyst" limited its practical usefulness, as polymerizations were preceded by lengthy initiation periods ranging from several hours to several days [25, 26]. Rigorous exclusion of water and oxygen from these systems was found to lengthen initiation periods, while the addition of small amounts of water substantially increased initiation rates. The conclusion that water functioned as a co-catalyst in these systems eventually led to the discovery that $RuCl_3$ functioned as an excellent ROMP initiator in entirely aqueous environments [27, 28]. For example, the emulsion polymerization of *exo*-5,6-bis-methoxymethyl-7-oxanorbornene **1** (Eq. 2) proceeded quantitatively with initiation periods as short as 30 min [27].

n **1** (OMe, OMe) —[$RuCl_3/H_2O$; > 95% Yield]→ poly(**1**) (MeO, OMe)$_n$ (2)

The molecular masses of poly(**1**) synthesized in aqueous media were typically high ($M_n \approx 10^6$), and polydispersity indices (PDIs) were often lower (< 2.0) than the polydispersities of polymers produced by classical systems in organic solvents [27, 29, 30]. These low polydispersities have been attributed to the low occurrence of termination reactions during the polymerization and the relative inactivity of the propagating species toward the acyclic olefins in the polymer, which suppresses chain-transfer reactions [25]. Although initiation occurred more quickly in aqueous systems, the extent of initiation was still low. In fact, it has been estimated that fewer than 1% of the metal centers are converted to active alkylidenes. As a result, polymer properties were inconsistent from run to run and depended heavily on the purity of the notoriously impure $RuCl_3$ complex [25, 29, 30]. Molecular masses were generally independent of monomer/catalyst ratios, indicating that these aqueous polymerizations were not living.

The problems outlined above prompted an investigation of the mechanism through which initiation occurred in these ill-defined systems. During the examination of used aqueous ruthenium solutions, it became evident that these solutions could be used to initiate additional polymerizations, and that the catalytic species in these solutions became more active upon successive use [25, 27]. Initiation periods for these recycled solutions were as low as 10 s after two or three polymerizations, and catalyst solutions could be re-used up to 14 times without a decrease in activity. This effect was observed for aqueous polymerizations initiated by other Ru(III) complexes, including K_2RuCl_5 and $[Ru(NH_3)_5Cl]Cl_2$, which reached the same limiting initiation time of 10 s upon re-use [25].

Studies employing Ru(II) complexes, such as $[(C_6H_6)Ru(H_2O)_3]tos_2$ (tos = *p*-toluenesulfonate), revealed similar effects on recycling, although they were initially more active than their Ru(III) counterparts. For example, in aqueous polymerizations of **1** catalyzed by $Ru(H_2O)_6tos_2$, induction periods were initially as short as 50 s. An important step in the identification of the active species in this polymerization was made when a ruthenium–olefin complex (Structure **2**) was observed after polymerization of **1** initiated by $Ru(H_2O)_6tos_2$ [25–27].

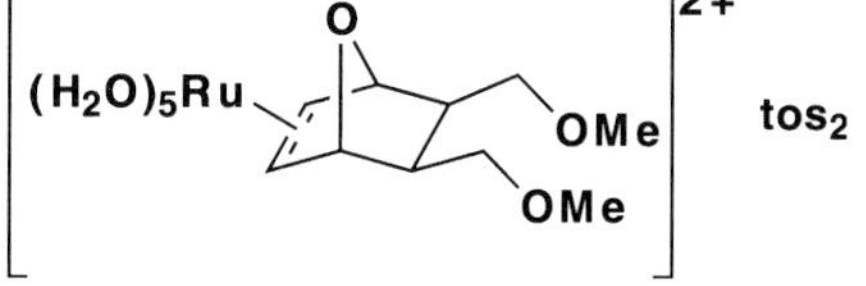

$[(H_2O)_5Ru(\text{olefin})]^{2+}\ tos_2$

2

Recycled solutions of **2** initiated ROMP as quickly as the recycled Ru(III) solutions, closer examination of which revealed NMR resonances identical to those of the olefin protons in **2** [25]. It was therefore suggested that a key step in the initiation process using Ru(III) was the in-situ formation of a Ru(II)–olefin complex [27]. Current evidence supports the disproportionation of the Ru(III) species to form Ru(II) and Ru(IV) species, followed by formation of a Ru(II)–olefin complex [25]. The equilibrium constant for disproportionation is small, accounting for the poor initiation efficiency of the Ru(III) systems [30]. An alternative, the disproportionation of an equilibrium amount of Ru(III)–olefin complex to a Ru(II)–olefin complex and a Ru(IV) species, is unlikely since Ru(III)–olefin complexes are generally unstable. Formation of a ruthenium alkylidene, the requisite active species in these polymerizations, *via* rearrangement of **2** has been proposed (Eq. 3), although the mechanism for this reaction is not known [20].

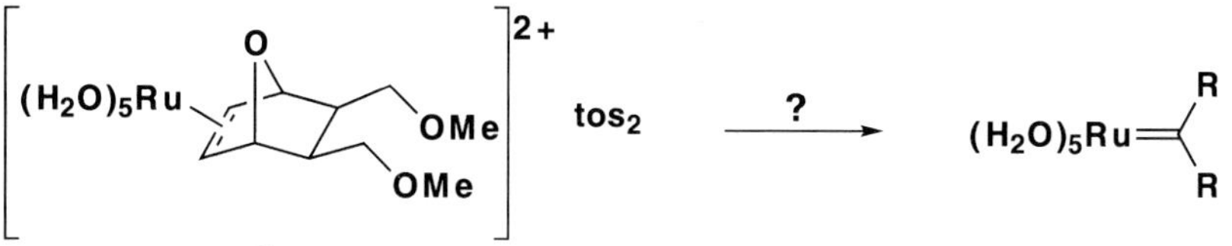

$$[\mathbf{2}]^{2+}\ tos_2 \xrightarrow{?} (H_2O)_5Ru{=}CR_2 \quad (3)$$

Water-soluble bisallyl Ru(IV) complexes **3** and **4** also initiate the emulsion polymerization of norbornene [31]. The lack of preformed alkylidenes in these complexes limits initiation efficiency, although the onset of initiation is not subject to lengthy initiation periods. Speculation on the active species in polymerizations initiated by these bisallyl complexes has not been reported.

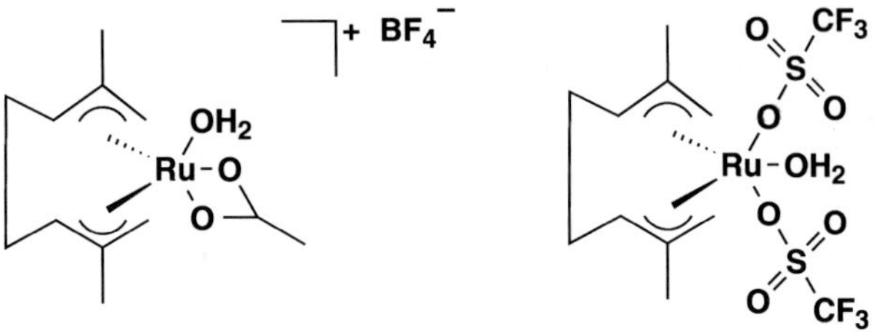

3

4

Karlen and co-workers have described a photoinitiated ROMP (PROMP) system in water/ethanol mixtures using a variety of cationic ruthenium complexes with photolabile ligands [32, 33]. For example, the irradiation of $[Ru(CH_3CN)_6](tos)_2$ or $[(C_6H_6)_2Ru](tos)_2$ leads to partially and fully solvated Ru(II) species which initiate the ROMP of highly strained olefins, presumably in the manner outlined below (Eq. 4).

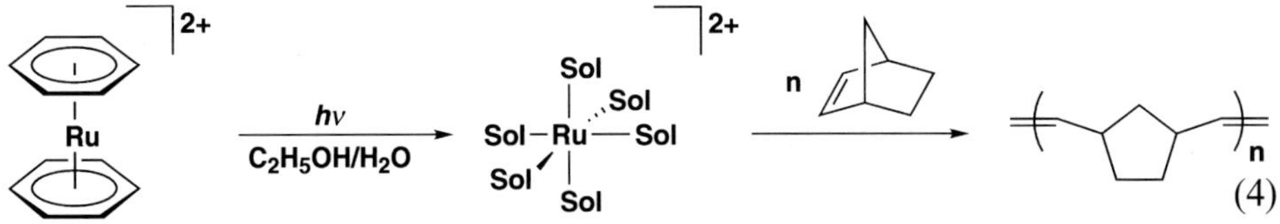

(4)

6.13.3 Polymers Prepared via Aqueous ROMP

The Group VIII complexes discussed above are generally limited to the ROMP of functionalized norbornenes and 7-oxanorbornenes. These complexes have been used to initiate the polymerization and copolymerization of monomers containing alkyl [22, 23, 34], aryl [30], ether [27, 29, 35], alcohol [25], ester [30, 34], anhydride [25, 34, 36, 37], carboximide [38, 39], and fluoromethyl [30] functionalities in aqueous environments. Aqueous polymerization of dicarboximide functionalized monomer **5** initiated by $Ru(H_2O)_6tos_2$ gave quantitative yields of a polymer having excellent thermal properties (Eq. 5) [38]. Additionally, polyacid materials were synthesized *via* the ROMP of anhydride-functionalized monomer **6**, which spontaneously opened to the diacid upon polymerization in aqueous environments (Eq. 6) [25, 26, 37]. Kiessling *et al.* have recently used $RuCl_3$ to initiate the polymerization of **7** and other saccharide-containing monomers to produce a variety of new glycopolymers having biological activities (Eq. 7) [40–42]. Many polymers based on the ROMP of functionalized 7-oxanorbornenes have been investigated for their potentials as ionophoric materials [26].

n **5** $\xrightarrow[H_2O]{Ru(H_2O)_6tos_2}$ (5)

n **6** $\xrightarrow[H_2O]{Ru(H_2O)_6tos_2}$ (6)

n **7** $\xrightarrow[H_2O]{RuCl_3}$ (7)

As previously mentioned, the molecular masses of the polymers obtained from these aqueous reactions are generally higher than desired due to the small number of active species. Although the propagating species in these polymerizations do not typically react with acyclic olefins, modest control over molecular mass is possible when certain acyclic chain-transfer agents are employed [35, 44]. For example, Feast and Harrison have used very high concentrations of *cis*-2-butene-1,4-diol or its dimethyl ether as chain-transfer agents [35]. The chain-transfer constants in these reactions were small, and inclusion of these olefins in the reaction mixture was shown to affect initiation periods and catalyst activites in complex ways.

The microstructures of polymers synthesized in aqueous media have been well studied by ^1H and ^{13}C NMR. Although most polymers prepared using these catalysts contain a high degree of *trans* olefin bonds, the ratio of *trans* to *cis* olefins has been found to vary considerably from catalyst to catalyst [35]. For example, poly**1** prepared in water using $RuCl_3$ generally contains 60% *trans* olefins, while polymer samples prepared from $OsCl_3$ and $IrCl_3$ contain 75% and

90% *trans* olefins, respectively. In all cases, poly**1** prepared by these catalysts was atactic. For the polymerization of *exo*-5,6-bis(methoxycarbonyl)-7-oxanorbornene, $RuCl_3$ gave a polymer having 88% *trans* olefin bonds, while $[Ru(\eta^6\text{-}C_6H_6)(H_2O)_3]tos_2$ and $Ru(H_2O)_6tos_2$ gave polymers containing equal amounts of *cis* and *trans* olefins [30]. In the $RuCl_3$-initiated polymerizations, the ratio of *trans* to *cis* olefins remained constant over prolonged reaction times, again demonstrating the relative absence of chain-transfer reactions that would eventually result in thermodynamic equilibration. A notable reversal in the *trans/cis* selectivities in these reactions is observed in the emulsion polymerization of norbornene initiated by bisallylruthenium complexes **3** and **4**, which yield polynorbornene having 85–90% *cis* olefin bonds [31].

6.13.4 Alkylidenes as Catalysts

6.13.4.1 Well-defined Ruthenium Alkylidenes

The insight derived from the investigation of ill-defined ruthenium ROMP initiators was successfully applied to the development of Ru(II) alkylidenes **8** and **9** [15–18]. In contrast to the classical complexes, these well-defined alkylidenes initiated ROMP quickly and quantitatively, reacted readily with acyclic olefins, and could be used to initiate living polymerizations in organic solvents.

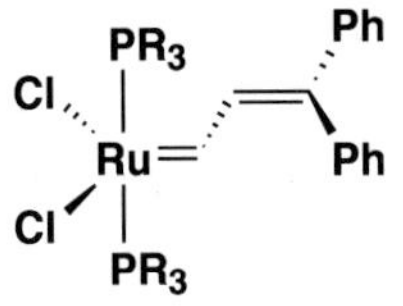

8a R = Ph
8b R = Cy

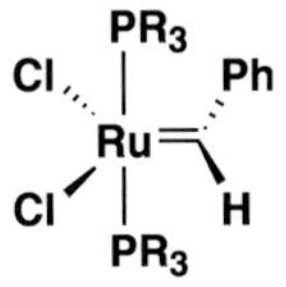

9a R = Ph
9b R = Cy

Although these well-defined complexes were insoluble in water, they were highly water-tolerant, and complexes **8b** and **9b** were used to catalyze living ROMP in aqueous emulsions (Eq. 8) [45, 46]. Introduction of a small amount of organic solvent was necessary to achieve controlled initiation, and these reactions worked best in the presence of cationic emulsifiers. Initiation was fast and complete, and polymer molecular masses were found to vary linearly with the ratios of monomer to initiator, indicating that these aqueous ROMP systems were indeed living [45]. Polymers having narrow polydispersities (PDI ≤ 1.10) were prepared from hydrophilic and hydrophobic monomers, and well-defined block copolymers were synthesized via sequential monomer addition. Kiessling and co-workers have recently applied this methodology to the synthesis of water-soluble, biologically active glycopolymers [47].

5 —(**8b** or **9b**; H_2O, CH_2Cl_2, DTAB)→ (8)

6.13.4.2 Water-soluble Alkylidenes

Water-soluble derivatives of alkylidenes **8** and **9** were prepared *via* phosphine ligand substitution reactions. Exchange of the phosphines in **8a** for $PhP(p\text{-}C_6H_4SO_3Na)_2$ afforded a water-soluble vinyl alkylidene [20]. This alkylidene was soluble in water, but the triarylphosphine ligands were too small and insufficiently electron-donating to produce an active catalyst [48]. Analogous substitution of the phosphines in **9a** for more sterically demanding, electron-rich, water-soluble phosphines yielded ruthenium alkylidenes **10** and **11** (Scheme 2), which were soluble in both water and methanol [49].

In contrast to the "classical" catalysts, alkylidenes **10** and **11** initiated ROMP quickly and quantitatively [49, 50]. A propagating alkylidene species was directly observed by ^1H NMR during the polymerization of most monomers, although the catalyst often decomposed before polymerization was complete. In the ROMP of water-soluble monomers **12** and **13** initiated by **10**, for example, conversions ranging from 45 to 80% were usually observed (Eq. 9) [50].

12 X = C; **13** X = O —(**10**; H_2O, 42 °C)→ (9)

It was found, however, that monomers could be *quantitatively* polymerized by **10** and **11** in the presence of stoichiometric amounts of a strong Brønsted acid [50]. The effect of the acid in these systems was determined to be twofold,

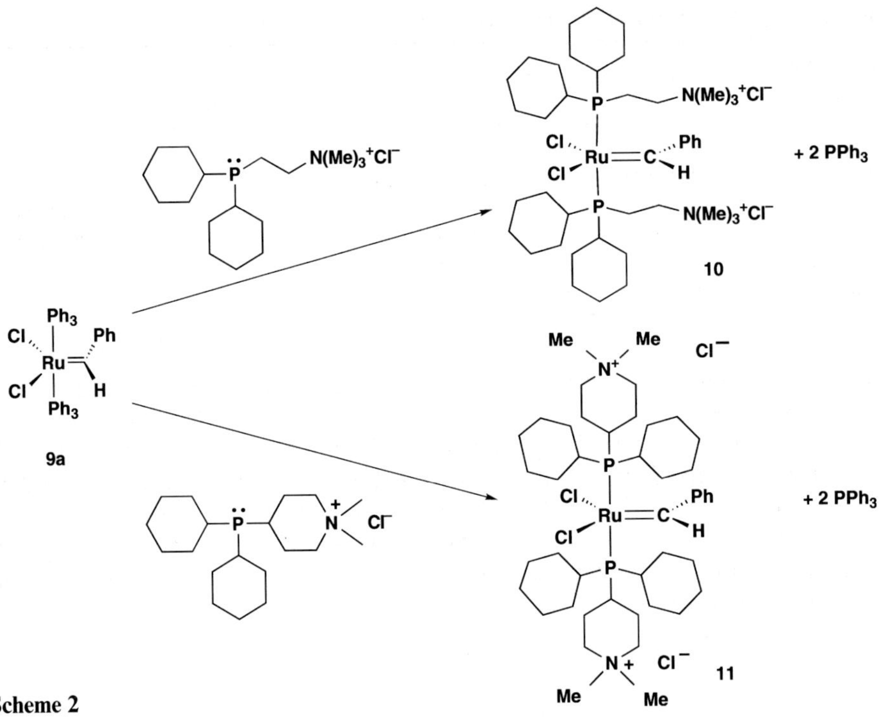

Scheme 2

eliminating small concentrations of detrimental hydroxide ions, and accelerating the rate of metathesis with respect to termination reactions. Remarkably, the acids did not react to decompose the ruthenium alkylidene, and a propagating alkylidene species was clearly observed following complete consumption of monomer. Addition of more monomer to the reaction mixture resulted in further quantitative polymerization, enabling the synthesis of block copolymers.

6.13.5 Summary

The synthesis of alkylidenes incorporating late transition metals has resulted in olefin metathesis catalysts having unprecedented functional group tolerance. In particular, the discovery that complexes of Group VIII transition metals were efficient ROMP catalysts introduced several advantages. Relative to their early transition metal counterparts, these "classical" catalysts functioned well in the presence of a variety of polar and protic functional groups (Table 2), and they functioned homogeneously in water.

The lessons learned from these complexes were eventually applied to the synthesis of well-defined ruthenium alkylidenes **8** and **9**. Although they were

Table 2. Early vs. late transition metal selectivity.

Metal center: Titanium	Tungsten	Molybdenum	Ruthenium	
Acids	Acids	Acids	*Olefins*	↑
Alcohols, water	Alcohols, water	Alcohols, water	Acids	
Aldehydes	Aldehydes	Aldehydes	Alcohols, water	Increasing
Ketones	Ketones	*Olefins*	Aldehydes	reactivity
Esters, amides	*Olefins*	Ketones	Ketones	
Olefins	Esters, amides	Esters, amides	Esters, amides	

insoluble in water, these alkylidenes could be used to initiate the living ROMP of functionalized norbornenes and 7-oxanorbornenes in aqueous emulsions. Substitution of the phosphine ligands in **9** for bulky, electron-rich, water-soluble phosphines produced water-soluble alkylidenes **10** and **11**, which served as excellent initiators for the ROMP of water-soluble monomers in aqueous solution. These new ruthenium alkylidene complexes are powerful tools in the synthesis of highly functionalized polymers and organic molecules in both organic and aqueous environments.

References

[1] K. J. Ivin, J. C. Mol, *Olefin Metathesis,* Academic Press, London, **1997**.
[2] V. Dragutan, A. T. Balaban, M. Dimonie, *Olefin Metathesis and Ring Opening Polymerization of Cyclo-Olefins,* 2nd ed., Wiley-Interscience, New York, **1985**.
[3] R. H. Grubbs, in *Comprehensive Organometallic Chemistry*, Pergamon Press, Oxford, **1982**, Chapter 54.
[4] R. H.Grubbs, W. Tumas, *Science* **1989**, *243*, 907.
[5] R. R. Schrock, *Acc. Chem. Res.* **1990**, *23*, 158.
[6] K. B. Wagener, J. M. Boncella, J. G. Nel, *Macromolecules* **1991**, *24*, 2649.
[7] R. H. Grubbs, S. J. Miller, G. C. Fu, *Acc. Chem. Res.* **1995**, *28*, 446.
[8] R. H. Grubbs, in *Organic Synthesis via Organometallics* (Eds.: K. H. Dötz, R. W. Hoffman), Vieweg, Braunschweig, **1991**, pp. 1–14.
[9] L. R. Gilliom, R. H. Grubbs, *J. Am. Chem. Soc.* **1986**, *108*, 733.
[10] R. R. Schrock, J. Feldman, L. F. Cannizzo, R. H. Grubbs, *Macromolecules* **1987**, *20*, 1169.
[11] R. R. Schrock, R. T. DePue, J. Feldman, C. J. Schaverien, J. C. Dewan, A. H. Liu, *J. Am. Chem. Soc.* **1988**, *110*, 1423.
[12] G. Bazan, R. R. Schrock, M. O'Regan, *Organometallics* **1991**, *10*, 1062.
[13] A. Aguero, J. Kress, J. A. Osborn, *J. Chem. Soc., Chem. Commun.* **1986**, 531.
[14] J.-L. Couturier, C. Paillet, M. Leconte, J. M. Basset, K. Weiss, *Angew. Chem., Ind. Ed. Engl.* **1992**, *31*, 628.
[15] S. T. Nguyen, L. K. Johnson, R. H. Grubbs, *J. Am. Chem. Soc.* **1992**, *114*, 3974.
[16] S. T. Nguyen, R. H. Grubbs, *J. Am. Chem. Soc.* **1993**, *115*, 9858.
[17] P. Schwab, R. H. Grubbs, J. W. Ziller, *J. Am. Chem. Soc.* **1996**, *118*, 100.

[18] P. Schwab, M. B. France, J. W. Ziller, R. H. Grubbs, *Angew. Chem., Int. Ed. Engl.* **1995**, *34*, 2039.
[19] R. H. Grubbs, *Pure Appl. Chem.* **1994**, *A31(11)*, 1829.
[20] R. H. Grubbs, in *Aqueous Organometallic Chemistry and Catalysis* (Eds.: I. T. Horváth, F. Joó), Kluwer Academic Publishers, Dordrecht, **1995**, pp. 15–22.
[21] R. Streck, *J. Mol. Catal.* **1988**, *46*, 305.
[22] F. W. Michelotti, W. P. Keaveney, *J. Polym. Sci. A*, **1965**, *3*, 895.
[23] F. W. Michelotti, J. H. Carter, *Polym. Prepr.* **1965**, *6* (*1*), 224.
[24] R. E. Rinehart, H. P. Smith, *Polym. Lett.* **1965**, *3*, 1049.
[25] B. M. Novak, Ph. D. Thesis, California Institute of Technology, **1989**.
[26] B. M. Novak, R. H. Grubbs, *J. Am. Chem. Soc.* **1988**, *110*, 960.
[27] B. M. Novak, R. H. Grubbs, *J. Am. Chem. Soc.* **1988**, *110*, 7542.
[28] B. M. Novak, W. Risse, R. H. Grubbs, *Adv. Polym. Sci.* **1992**, *102*, 47.
[29] S.-Y. Lu, P. Quayle, F. Heatley, C. Booth, S. G. Yeates, J. C. Padget, *Macromolecules* **1992**, *25*, 2692.
[30] A. Mühlebach, P. Bernhard, N. Bühler, T. Karlen, A. Ludi, *J. Mol. Catal.* **1994**, *90*, 143.
[31] S. Wache, *J. Organomet. Chem.* **1995**, *494*, 235.
[32] T. Karlen, A. Ludi, A. Mühlebach, P. Bernhard, C. Pharisa, *J. Polym. Sci., Part A: Polym. Chem.* **1995**, *33*, 1665.
[33] A. Hafner, P. A. van der Schaaf, A. Mühlebach, *Chimia* **1996**, *50*, 131.
[34] C. M. McArdle, J. G. Hamilton, E. E. Law, J. J. Rooney, *Macromol. Rapid Commun.* **1995**, *16*, 703.
[35] W. J. Feast, D. B. Harrison, *J. Mol. Catal.* **1991**, *65*, 63.
[36] W. J. Feast, D. B. Harrison, A. F. Gerard, D. R. Randell, UK 2.235.460, 1991.
[37] T. Viswanathan, J. Jethmalani, A. Toland, *J. Appl. Polym.* **1993**, *47*, 1477.
[38] M. A. Hillmyer, C. Lepetit, D. V. McGrath, B. M. Novak, R. H. Grubbs, *Macromolecules* **1992**, *25*, 3345.
[39] T. Viswanathan, J. Jethmalani, *J. Appl. Polym.* **1993**, *48*, 1289.
[40] K. H. Mortell, M. Gingras, L. L. Kiessling, *J. Am. Chem. Soc.* **1994**, *116*, 12053.
[41] K. H. Mortell, R. V. Weatherman, L. L. Kiessling, *J. Am. Chem. Soc.* **1996**, *118*, 2297.
[42] M. C. Schuster, K. H. Mortell, A. D. Hegeman, L. L. Kiessling, *J. Mol. Catal. A: Chem.* **1997**, *116*, 209.
[43] R. H. Grubbs, B. M. Novak, US 4.883.851, 1989.
[44] M. B. France, R. H. Grubbs, D. V. McGrath, R. A. Paciello, *Macromolecules* **1993**, *26*, 4742.
[45] D. M. Lynn, S. Kanaoka, R. H. Grubbs, *J. Am. Chem. Soc.* **1996**, *118*, 784.
[46] C. Fraser, R. H. Grubbs, *Macromolecules* **1995**, *28*, 7248.
[47] D. D. Manning, X. Hu, P. Beck, L. L. Kiessling, *J. Am. Chem. Soc.* **1997**, *119*, 3161.
[48] E. L. Dias, S. T. Nguyen, R. H. Grubbs, *J. Am. Chem. Soc.* **1997**, *119*, 3887.
[49] B. Mohr, D. M. Lynn, R. H. Grubbs, *Organometallics* **1996**, *15*, 4317.
[50] D. M. Lynn, B. Mohr, R. H. Grubbs, *J. Am. Chem. Soc.* **1998**, *120*, 1627.

6.14 Hydrogenation and Hydrogenolysis of Thiophenic Molecules

Claudio Bianchini, Andrea Meli

6.14.1 Introduction

With the advent of stricter pollution laws concerning sulfur in gasoline and diesel oil, an outburst of research on the hydrodesulfurization (HDS) process is currently pervading chemical laboratories worldwide. In the HDS process (Eq. 1), sulfur is removed from fossil materials upon treatment with a higher pressure of H_2 (35–170 bar) in the presence of heterogeneous catalysts at high temperature (300–425 °C) [1].

$$C_xH_yS + 2\,H_2 \longrightarrow C_xH_{y+2} + H_2S \qquad (1)$$

With the existing HDS technology, the sulfur contents in gasoline can only with difficulty be reduced to the marketing limit ($\leq$ 60 ppm) without drastic changes in the octane and cetane rating, especially when the cracking and coking naphthas come from fossil materials containing large amounts of thiophenic molecules. These comprise an enormous variety of substituted thiophenes, benzo[*b*]thiophenes, and dibenzo[*b,d*]thiophenes as well as other fused-ring thiophenes, all of which are generally less easily disulfurized over heterogeneous catalysts than any other sulfur compound in petroleum feedstocks (e.g., thiols, sulfides, and disulfides). In addition to the inherent refractoriness, it is now apparent that each category of thiophenes may undergo HDS through a distinct mechanism [1]. As a consequence, specific catalysts for each type of thiophene should be developed for the deep desulfurization of refined fuels as these may contain different distributions of thiophenic contaminants. A problem of this type cannot find a general solution as HDS reactors in refineries are commonly fed with petroleum feedstocks of various natures. Moreover, the design of an heterogeneous catalyst for the degradation of a specific substrate, in trace amounts and in the presence of similar substrates, is an arduous task, certainly much more demanding than tailoring a soluble homogeneous catalyst for an analogous action. It is probably for all of these reasons that the HDS of thiophenes and other organosulfur compounds catalyzed by water-soluble metal complexes is attracting increasing interest from petrochemical industries [2], especially in relation to the development of catalysts for improving gasoline quality once the major part of the sulfur has already been eliminated by hydrotreatment on conventional heterogeneous catalysts. Over the last few years, in fact, aqueous biphase catalysis has made impressive progress, particularly as regards its aplication to large-volume reactions [3].

6.14.2 Hydrogenation Reactions

Researchers at INTEVEP SA (Ven) have recently shown that the regioselective hydrogenation of benzo[*b*]thiophene (BT) to dihydrobenzo[*b*]thiophene (DHBT) (Eq. 2) can be performed in a 1:1 water/decalin mixture using an in-situ catalyst system formed by addition of an excess of either *m*-monosulfonated triphenylphosphine (TPPMS) or trisulfonated triphenylphosphine (TPPTS) to various Ru(II) precursors [2a]. It is believed that the catalytically active species is a mononuclear Ru(II) complex with chloride and hydride ligands.

$$\text{BT} \xrightarrow[\text{H}_2]{\text{TPPMS/Ru}} \text{DHBT} \qquad (2)$$

Interestingly, the rate almost quadruples if nitrogen bases such as quinoline (Q) or aniline are used as co-catalysts. This kinetic effect may be due to several beneficial actions by the nitrogen bases on either the catalyst (faster formation and/or better stabilization) or the phase system (stabilization of the emulsion, improved solubility of BT in water). The quality of the emulsion seems to be of particular importance as shown by the fact that TPPTS, which is a worse surfactant than TPPMS, also gives worse catalytic results. Other nitrogen bases such as acridine, tetrahydroquinoline (THQ), piperidine, and triethylamine give lower conversions to DHBT as compared with analogous reactions co-catalyzed by aniline or quinoline.

The tolerance of the Ru/TPPMS or TPPTS systems toward organic nitrogen bases has also allowed their use as catalysts for the hydrogenation of Q/BT mixtures in which Q kinetically favors the hydrogenation of BT [2a]. This result is quite surprising as it is generally observed that the presence of nitrogen compounds in fuel feedstocks inhibits HDS reaction over heterogeneous catalysts [1a, f].

More recent work on the hydrogenation of thiophenic molecules catalyzed by water-soluble metal complexes is pursuing the use of polyphosphine ligands (cf. Section 3.2.2). These studies follow the success obtained with the tridentate phosphine $MeC(CH_2PPh_2)_3$ (TRIPHOS), which forms rhodium and iridium catalysts for the hydrogenation, hydrogenolysis, and desulfurization of various thiophenic substrates in homogeneous phase [4–7]. The two water-soluble chelating polyphosphines $NaO_3S(C_6H_4)CH_2C(CH_2PPh_2)_3$(NaSULPHOS, Structure **1**) [8] and $(NaO_3S(C_6H_4)CH_2)_2C(CH_2PPh_2)_2$ (Na_2DPPPDS; Structure **2**) [9] have already proved to be effective ligands for the formation of transition metal complexes capable of catalyzing both the hydrogenation and hydrogenolysis of thiophenic substrates.

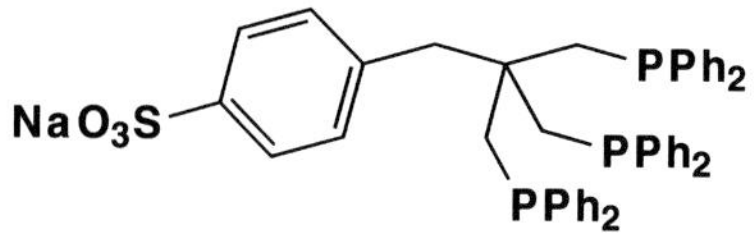

NaSULPHOS

1

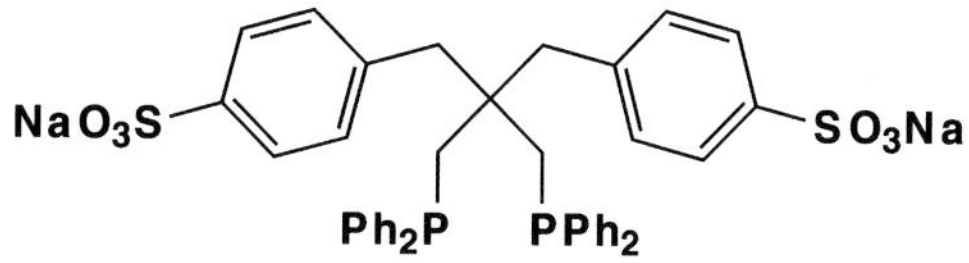

Na_2DPPPDS

2

The triphosphine **1** is the water-soluble version of TRIPHOS and was initially developed to study, in association with rhodium, the liquid-biphase hydrogenation and hydroformylation of alkenes [8]. The Rh catalyst employed in these reactions, e.g., the zwitterionic Rh(I) complex (SULPHOS)Rh(cod) (**3**), has recently been used as catalyst precursor for the hydrogenation of BT to DHBT in biphase systems comprising water and methanol as polar phase and *n*-heptane as hydrocarbon phase [10]. Although the catalyst system is very robust and can tolerate relatively drastic reaction conditions for a long time (30 bar H_2, 200 °C), the catalytic activity is quite modest (TOF $\simeq$ 2). Increased catalytic efficiency has been obtained by substituting Ru for Rh. Under comparable conditions (30 bar H_2, BT/Ru ratio 100:1), the tris-(acetonitrile)complex $[(SULPHOS)Ru(MeCN)_3]SO_3CF_3$ (**4**) catalyzes the regioselective hydrogenation of BT to DHBT in H_2O–MeOH/*n*-heptane with a TOF of ca. 40 already at 100 °C [11]. The addition of MeOH to water may be necessary to enhance the mutual solubility and mobility of the components across the phase boundary. Analogously to the TPPMS/ and TPPTS/Ru systems developed earlier, both the Rh and Ru complexes **3** and **4** have proved effective catalyst precursors for the hydrogenation of Q to THQ, thus showing that the hydrodenitrogenation (HDN) of fossil fuels may also be pursued by aqueous-biphase catalysis.

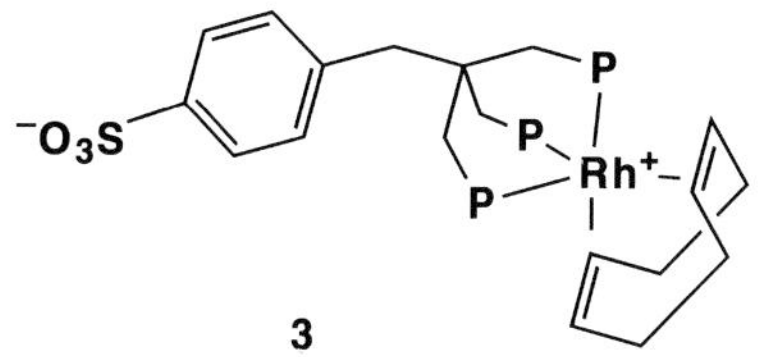

3

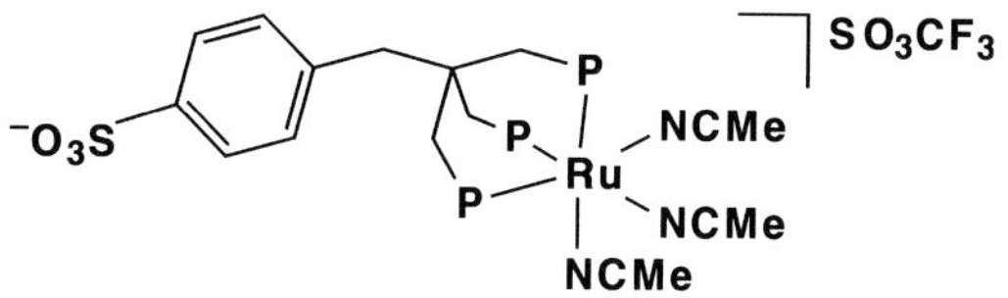

4

Unlike the tripodal ligand **1**, the chelating diphosphine **2** forms rhodium and ruthenium catalysts (by simple reaction with the corresponding hydrated trichlorides) that catalyze the hydrogenation of BT in water/hydrocarbon biphase systems with no need for alcohol co-solvents [11]. The catalytic activity is generally lower than that with the tridentate ligand **1**, however.

In summary, BT can be hydrogenated in liquid biphase systems using water-soluble catalysts with transition metals belonging to the family of the HDS promoters such as Ru and Rh [12]; the ruthenium catalysts are generally more efficient than those with rhodium, a finding that is consistent with heterogeneous HDS reactions [1]. The fact that all the reported studies involve BT is most probably attributable to the pronounced "olefinic" character of the C_2–C_3 bond that favors hydrogenation pathways [4]. Moreover, the reduction products of BT are easier to handle. This, however, does not exclude the application of aqueous biphase catalysis to thiophenes or dibenzothiophenes, for which homogeneous processes have already been developed [5, 7]. Although no mechanistic study of aqueous hydrogenation reactions of BT has ever been reported, it is very likely that the reactions comprise the usual steps of H_2 oxidative addition, hydride transfer to η_2-C_2, C_3-coordinated BT to form dihydrobenzothienyl species, and reductive elimination of DHBT to complete the cycle (Scheme 1) [13].

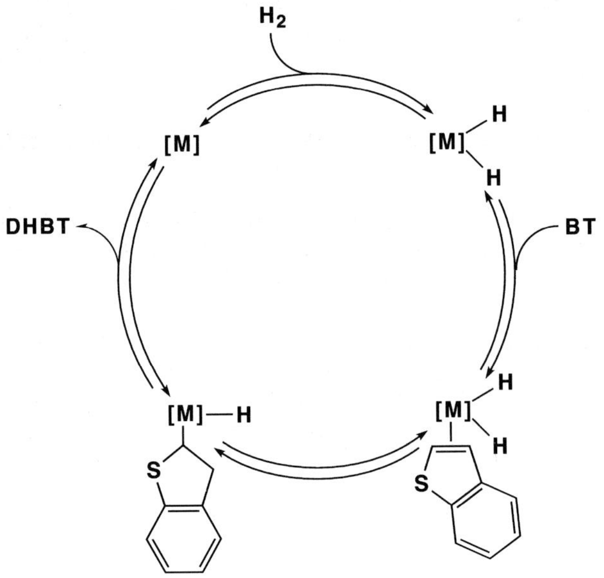

Scheme 1

6.14.3 Hydrogenolysis Reactions

The catalytic conversion of thiophenic substrates to the corresponding thiols (hydrogenolysis) (Eq. 3) is a reaction of much relevance in the HDS process as the thiols can subsequently be desulfurized over conventional HDS catalysts with greater efficiency and under milder reaction conditions than those required to accomplish the HDS of the thiophene precursors [1]. This aspect is particularly important for fused-ring thiophenes because the conventional catalysts can desulfurize the corresponding thiols without affecting the benzene rings necessary to preserve a high octane number.

$$C_xH_yS + H_2 \longrightarrow C_xH_{y+1}SH \qquad (3)$$

The hydrogenolysis of various thiophenes (thiophene, benzo[*b*]thiophene, dibenzo[*b,d*]thiophene, dinaphtho[2,1-*b*:1′,2′-*d*]thiophene) has been achieved in homogeneous phase with the $16e^-$ catalyst [(TRIPHOS)RhH] [4–7, 13d]. This can be generated from appropriate TRIPHOS/Rh precursors by either thermal elimination of H_2 [4, 6, 14] or base-assisted heterolytic splitting of H_2 [7–10]. Depending on the thiophenic substrate, it has been observed that the addition of strong Brønsted bases (NaOH, KO*t*Bu) to the catalytic mixtures may increase the reaction rate by as much as tenfold [7]. Besides generating M–H moieties in a reductive manner (e.g., $[M]^{n+} + H_2 + \text{base}^- \rightarrow [M{-}H]^{(n-1)+} + \text{base}{-}H$), the base serves to speed up the catalysis rate by aiding the elimination of the thiol product from the metal center, i.e. the rate-determining step for the hydrogenolysis reactions assisted by [(TRIPHOS)RhH] [4] (Eq. 4).

$$(TRIPHOS)Rh(H)_2SR + NaOH \longrightarrow [(TRIPHOS)RhH] + NaSR + H_2O \qquad (4)$$

The application of the "strong base" method to liquid biphase hydrogenolysis reactions has provided quite interesting results using the SULPHOS complex **3** as catalyst precursor, BT as substrate and NaOH as base [10]. In a typical reaction in H_2O–MeOH/*n*-heptane (BT/NaOH/**3** = 100:100:1, 30 bar H_2), all the substrate is selectively transformed into 2-ethylthiophenol sodium salt (ETPNa; the production of DHBT is generally less than 1%) (Scheme 2). At 200 °C, however, appreciable decomposition of the catalyst has been observed to give metal particles that are responsible for heterogeneous HDS of BT to ethylbenzene and H_2S.

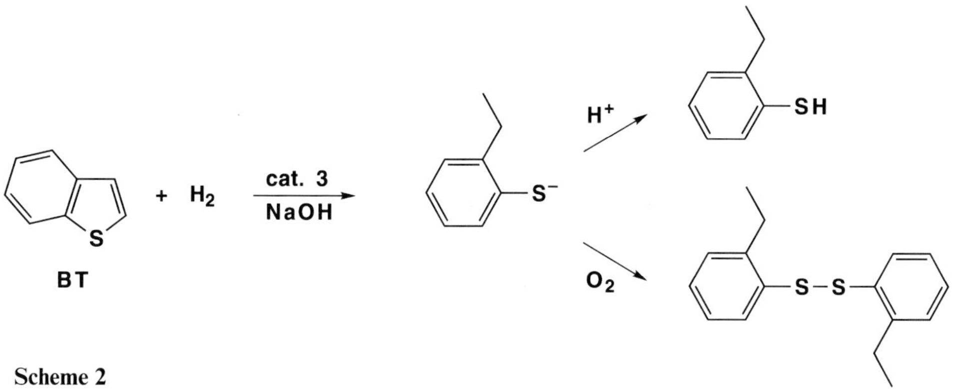

BT + H2 —cat. 3 / NaOH→ 2-ethylthiophenolate (S⁻); H⁺ → 2-ethylthiophenol (SH); O2 → bis(2-ethylphenyl) disulfide (S–S)

Scheme 2

When all the BT has been consumed, the hydrogenolysis product accumulates in the polar phase as Na[o-S(C_6H_4)C_2H_5] leaving the hydrocarbon phase completely desulfurized and also devoid of any rhodium. The hydrogenolysis product can also be separated as the disulfide [o-S(C_6H_4)C_2H_5]$_2$ if the polar phase is exposed to air oxidation. Alternatively, the reaction mixture can be acidified under an inert atmosphere and the thiol extracted from the polar phase with n-heptane so as to leave the complexed rhodium in the polar phase for use in a further catalytic run. Due to the inherent surfactant properties of SULPHOS metal complexes [8], very efficient emulsions are obtained and the hydrogenolysis rates are independent of the stirring rate in the 650–1800 rpm range as well as of the addition of surfactants. Methanol, however, is a necessary co-solvent to ensure the mutual solubility and mobility of the components across the phase boundary.

The effect of the H_2 pressure and of the substrate, catalyst, and base concentrations on the conversion rate of BT has been studied. The hydrogenolysis reactions are first-order in both the catalyst and BT concentration. They are also first-order in the base concentration as long as the NaOH/BT ratio is lower than or equal to 0.6:1. For greater concentrations of base, the rates tend to flatten because the high ionic strength of the polar phase disfavors the mixing of the phases as well as the diffusion of BT. The hydrogenolysis rates increase linearly also with the H_2 pressure in the range from 5 to 30 bar. Above 30 bar, the rates decrease due to the formation of increasing equilibrium concentrations of the trihydride complex [(SULPHOS)RhH_3]$^-$ (**5**) ([10, 15]; cf. Scheme 3).

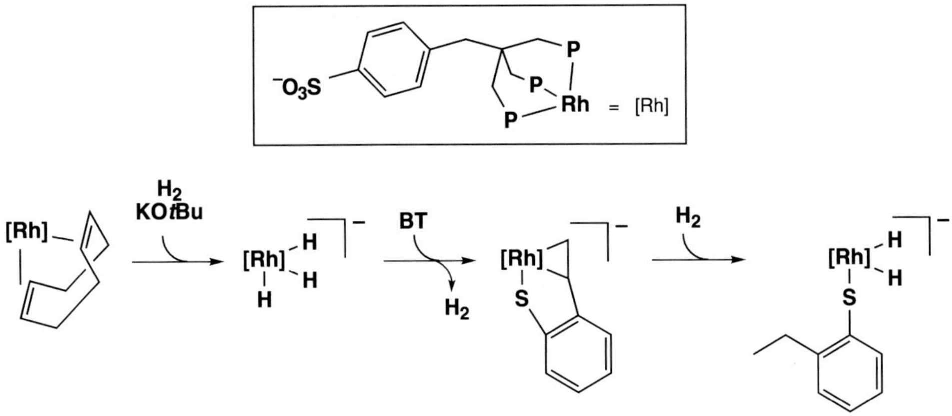

Scheme 3

The trihydride **5** is the first Rh compound to be formed upon hydrogenation of the precursor **3** in the presence of KOtBu. The Rh(I) catalytically active species [(SULPHOS)RhH]$^-$, generated by thermal elimination of H_2 from **5**, then reacts with BT to give the C–S insertion 2-vinylthiphenolate complex

$[(SULPHOS)Rh(\eta^3\text{-}S(C_6H_4)CH{=}CH_2)]^-$ (**6**), which can be isolated as the potassium salt. As in monophase systems [4], **6** would form by regioselective insertion of rhodium into the $C_2{-}S$ bond from BT, followed by intramolecular hydride migration to the α-carbon atom of a metallathiacycle intermediate [14, 16]. Reaction of **6** with H_2 finally yields the (2-ethylthiophenolate)dihydride complex $[(SULPHOS)Rh(H)_2(o\text{-}S(C_6H_4)C_2H_5)]^-$ (**7**).

The catalysis cycle proposed for the opening and hydrogenation of BT to ETPNa does not diverge much from those previously reported for the hydrogenolysis of BT to 2-ethylthiophenol [4, 13d] and of thiophene to 1-butanethiol [7] catalyzed by [(TRIPHOS)RhH]. The mechanism illustrated in Scheme 4 involves the usual steps of C–S insertion, hydrogenation of the C–S-inserted BT to 2-ethylthiophenolate, and base-assisted reductive elimination of the thiol to complete the cycle.

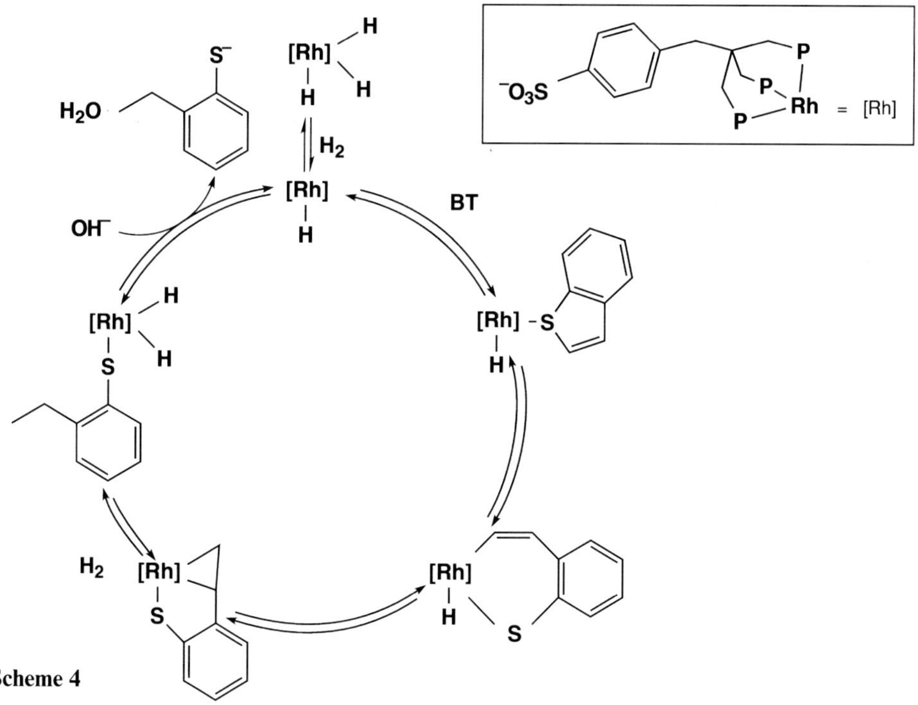

Scheme 4

At variance with the homogeneous reactions with [(TRIPHOS)RhH] in which the base simply accelerates the catalytic process, in the biphase reactions the presence of NaOH is mandatory for catalysis to occur. This is most probably because the concentration of BT in the polar phase is too low to promote effectively the elimination of the thiol product. A similar dependence of the reaction rate on Brønsted bases has been observed for the homogeneous hydrogenolysis of thiophene by TRIPHOS/Rh catalysis, and was attributed to the weak nucleophilic properties of thiophene [7].

Unlike the hydrogenation of BT to DHBT, the substitution of (SULPHOS)Ru for (SULPHOS)Rh in the base-assisted hydrogenolysis to ETPNa seems to decrease the conversion rate. Preliminary results with the tris(acetonitrile) complex **4** show that the transformation of the substrate does not exceed 10% of that obtained with comparable Rh catalysts in analogous conditions [11]. This poor activity is attributed to the formation of stable μ-OH ruthenium species rather than to the minor propensity of (polyphosphine)Ru(II) moieties to catalyze the hydrogenolysis of thiophenes. As a matter of fact, good conversion and selectivity of BT to ETPNa are being observed in water/*n*-hexane using Ru(II) catalysts with the water-soluble diphosphine Na_2DPPPDS [9, 11].

6.14.4 Future Developments

If business pull has played the dominant role in the innovation process by which HDS has led to a doubling of the catalytic activity of conventional heterogeneous catalysts over the last 20 years, environmental push will largely contribute to drive future research efforts aimed at developing more efficient technologies for sulfur removal form crude oil. Emerging technologies of great interest are aqueous biphase catalysis and its variations such as supported liquid-phase catalysis [17]. The introduction of aqueous biphase techniques to industrial HDS will require an enormous amount of research work to be particularly directed to the synthesis of water-soluble catalysts containing inexpensive metals (Fe, Co, Ni, Ru) and endowed with specific characteristics such as inherent emulsifying attributes and great thermal and chemical stability.

References

[1] (a) H. Topsøe, B. S. Clausen, F. E. Massoth, *Hydrotreating Catalysis,* Springer-Verlag, Berlin, **1996**; (b) J. Scherzer, A. J. Gruia, *Hydrocracking Science and Technology,* Marcel Dekker, **1996**; (c) M. L. Occelli, R. Chianelli, *Hydrotreating Technology for Pollution Control*, Marcel Dekker, **1996**; (d) P. T. Vasudevan, J. L. G. Fierro, *Catal. Rev. – Sci. Eng.* **1996**, *38*, 161; (e) B. C. Wiegand, C. M. Friend, *Chem. Rev.* **1992**, *92*, 491; (f) M. J. Girgis, B. C. Gates, *Ind. Eng. Chem. Res.* **1991**, *30*, 2021; (g) R. Prins, V. H. J. de Beer, G. A. Somorjai, *Catal. Rev. – Sci. Eng.* **1989**, *31*, 1; (h) C. M. Friend, J. T. Roberts, *Acc. Chem. Res.* **1988**, *21*, 394.

[2] (a) INTEVEP S. A. (D. E. Páez, A. Andriollo, R. Sánchez-Delgado, N. Valencia, F. López-Linares, R. E. Galiasso), US 08/657.960 (1996); (b) T. G. Southern, *Polyhedron* **1989**, *8*, 407.

[3] B. Cornils, W. A. Herrmann, in *Applied Homogeneous Catalysis with Organometallic Compounds* (Eds.: B. Cornils, W. A. Herrmann), VCH, Weinheim, **1996**, Vol. 2, p. 575.

[4] C. Bianchini, V. Herrera, M. V. Jiménez, A. Meli, R. A. Sánchez-Delgado, F. Vizza, *J. Am. Chem. Soc.* **1995**, *117*, 8567.

[5] C. Bianchini, M. V. Jiménez, A. Meli, S. Moneti, F. Vizza, V. Herrera, R. A. Sánchez-Delgado, *Organometallics* **1995**, *14*, 2342.

[6] C. Bianchini, D. Fabbri, S. Gladiali, A. Meli, W. Pohl, F. Vizza, *Organometallics* **1996**, *15*, 4604.

[7] C. Bianchini, J. A. Casares, A. Meli, V. Sernau, F. Vizza, R. A. Sánchez-Delgado, *Polyhedron* **1997**, *16*, 3099.

[8] C. Bianchini, P. Frediani, V. Sernau, *Organometallics* **1995**, *14*, 5458.

[9] CNR (C. Bianchini, A. Meli, F. Vizza), IT FI96A000.272 (1996).

[10] C. Bianchini, A. Meli, V. Pantinec, V. Sernau, F. Vizza, *J. Am. Chem. Soc.* **1997**, *119*, 4945.

[11] C. Bianchini, A. Meli, unpublished results.

[12] A. N. Startsev, *Catal. Rev. – Sci. Eng.* **1995**, *37*, 353.

[13] (a) E. Baralt, S. J. Smith, I. Hurwitz, I. T. Horváth, R. H. Fish, *J. Am. Chem. Soc.* **1992**, *114*, 5187; (b) R. A. Sánchez-Delgado, V. Herrera, L. Rincón, A. Andriollo, G. Martín, *Organometallics* **1994**, *13*, 553; (c) V. Herrera, A. Fuentes, M. Rosales, R. A. Sánchez-Delgado, C. Bianchini, A. Meli, F. Vizza, *Organometallics* **1997**, *16*, 2465; (d) C. Bianchini, A. Meli, in *Applied Homogeneous Catalysis with Organometallic Compounds* (Eds.: B. Cornils, W. A. Herrmann), VCH, Weinheim, **1996**, Vol. 2, p. 969.

[14] C. Bianchini, P. Frediani, V. Herrera, M. V. Jiménez, A. Meli, L. Rincón, R. A. Sánchez-Delgado, F. Vizza, *J. Am. Chem. Soc.* **1995**, *117*, 4333.

[15] CNR (C. Bianchini, A. Meli, A. Traversi), IT FI97A000.025 (1997).

[16] (a) C. Bianchini, A. Meli, M. Peruzzini, F. Vizza, S. Moneti, V. Herrera, R. A. Sánchez-Delgado, *J. Am. Chem. Soc.* **1994**, *116*, 4370; (b) C. Bianchini, M. V. Jiménez, A. Meli, S. Moneti, F. Vizza, *J. Organomet. Chem.* **1995**, *504*, 27.

[17] M. E. Davis, *CHEMTECH* **1992**, 498.

6.15 CO_2 Chemistry

Walter Leitner, Eckhard Dinjus, Franz Gaßner

6.15.1 Introduction

Carbon dioxide (CO_2) has found considerable and ever-increasing attention in recent years as an economically interesting and ecologically benign C_1 building block for chemical synthesis [1, 2]. It provides a number of important advantages over commonly used feedstocks like carbon monoxide (CO) or phosgene ($COCl_2$), especially within the frameworks of a "sustainable development". Furthermore, CO_2 is readily available in very large quantities from flue gases of chemical processes or power plants [3, 4]. Basic aqueous solutions are used in technically mature processes for scrubbing the CO_2-containing emissions [5, 6] and the direct use of such solutions as a chemical feedstock appears to be an attractive option (see below).

In sharp contrast to the still very limited use of carbon dioxide as a raw material in catalytic synthetic chemistry, all life on our planet ultimately depends on CO_2 as the primary source of carbon. Naturally, all biochemical processes involving CO_2 fixation proceed in aqueous solution and a number of important conversions rely on the catalytic activity of transition metal centers [1 a, 7]. The most striking example is probably the active site in carbon monoxide dehydrogenase, which contains two essential transition metals, nickel and iron [8]. The insertion of CO_2 in a $Ni-CH_3$ fragment, i.e., in a true organometallic species, appears to be a key step in the acetyl-CoA synthesis by enzymes of this class [9].

Carbonic anhydrase, a zinc-containing class of enzymes [10], has been developed by Nature to facilitate equilibration of the reactions shown in Scheme 1 which occur upon dissolution of CO_2 in water. It is not surprising that these enzymes are among the most effective biocatalysts ($k_{cat}/k_{uncat} \geq 10^8$) as these equilibria are of paramount importance for regulation of the CO_2 concentration in the cell. The CO_2/carbonate equilibrium is of course also important for potential homogeneously catalyzed synthetic processes. The individual concentrations are highly dependent on parameters such as pH, temperature, and pressure, thus representing an additional complication in mechanistic studies.

In spite of the overwhelming precedence of biological systems, surprisingly few efforts have been made so far to establish homogeneously transition-metal-catalyzed processes involving CO_2 as a C_1-building block in aqueous solution.

$$CO_2 + H_2O \rightleftharpoons HCO_3H \rightleftharpoons HCO_3^- + H^+ \rightleftharpoons CO_3^{2-} + 2\,H^+$$

Scheme 1

A remarkably efficient catalytic reaction in this context is the hydrogenation of CO_2 to formic acid. A description of the related large and prospering field of photo- [11 a, b] and electrocatalytic [11 c, d] reduction of CO_2 is beyond the scope of this review. Some examples for transition-metal-catalyzed reactions involving CO_2 without its incorporation into the final products are addressed in Section 6.15.3.

6.15.2 Transition-metal-catalyzed Reaction of CO_2 with Dihydrogen in Aqueous Media

6.15.2.1 General Background

In principle, the reaction of CO_2 with dihydrogen can lead to the reduction products outlined schematically in Scheme 2. Reduction of CO_2 to oxidation states beyond the formate level is still rare with homogeneous catalysts in organic [12] or aqueous solutions [13]. The formation of CO and water from CO_2 and H_2 is referred to as the reverse water-gas shift (WGS) reaction. The technically important WGS equilibrium is adjusted by soluble transition metal catalysts in both directions [14]. The reaction pathway in general proceeds via hydroxycarbonyl complexes [15] (often referred to as metallacarboxylic acids), but may also involve formate species or free HCO_2H [13]. In any case, the WGS equilibrium must be kept in mind for mechanistic considerations of reactions based on H_2/CO_2 mixtures in aqueous solution. Its utility as an alternative entry to C_1 chemistry would require the availability of H_2 from nonfossil sources, however [1 b].

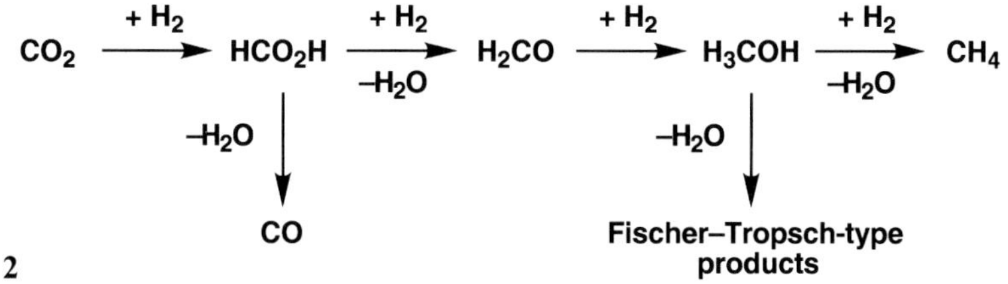

$$CO_2 \xrightarrow{+H_2} HCO_2H \xrightarrow[-H_2O]{+H_2} H_2CO \xrightarrow{+H_2} H_3COH \xrightarrow[-H_2O]{+H_2} CH_4$$

$HCO_2H \xrightarrow{-H_2O}$ CO

$H_3COH \xrightarrow{-H_2O}$ Fischer–Tropsch-type products

Scheme 2

Transition metal complexes in homogeneous solution are known to catalyze the reaction of CO_2 with 1 equiv of H_2 to give formic acid or its derivatives very efficiently in organic solvents and in supercritical CO_2 as reaction media [16]. The high yields of formic acid obtained in some of these systems are quite remarkable, considering the unfavorable entropic situation that renders the overall process endergonic ($\Delta G^0 < 0$) under standard conditions and shifts the equilibrium shown in Eq. (1) far to the side of the gaseous substrates, despite the exothermicity ($\Delta H^0 > 0$) of the addition of H_2 to CO_2 [17].

$$H_2\,(g) + CO_2\,(g) \xrightleftharpoons[]{\Delta H^0_{298} = -32\ \text{kJ mol}^{-1},\ \Delta G^0_{298} = +33\ \text{kJ mol}^{-1}} HCO_2H\,(l) \quad (1)$$

$$H_2\,(aq) + CO_2\,(aq) \xrightleftharpoons[]{\Delta G^0_{298} = -4\ \text{kJ mol}^{-1}} HCO_2H\,(aq) \quad (2)$$

The known efficient catalytic systems rely therefore on derivatization in situ of formic acid to esters or formamides or on a judicious choice of the reaction conditions in order to overcome the thermodynamic limitations [16]. Apart from the addition of amines to form either formates or formic acid/amine adducts [18], the choice of the solvent has been shown to be of special importance for defining the equilibrium constant of Eq. (1) [17]. One can readily conclude from standard thermodynamic data [19] that water should be an ideal reaction medium for the hydrogenation of CO_2 to formic acid (Eq. 2).

6.15.2.2 Transition-metal-catalyzed Hydrogenation of CO_2 to Formic Acid in Aqueous Reaction Media

6.15.2.2.1 Development of the Catalytic System

The hydrogenation of CO_2 or CO_3^{2-} in aqueous solutions using homogeneous or heterogeneous catalysts based on $PdCl_2$ [20] and Pd/C [21], respectively, has been described. The activities of these catalysts are, however, considerably lower than those observed with late-transition-metal phosphine catalysts in other reaction media [16]. Owing to the poor solubility of most ordinary phosphine complexes in aqueous solution, none of the latter systems could be operated in H_2O as a solvent. Nevertheless, a beneficial effect of small amounts of water has been observed already in very early studies on the hydrogenation of CO_2 to formic acid using $[(Ph_3P)_3RhCl]$ as the catalyst [22]. Many other systems known to date show a similar behavior and in some cases reaction media containing up to 20% (v/v) H_2O have been successfully employed [23, 24]. There is spectroscopic evidence for the active participation of H_2O in the catalytic cycle using the cationic complex $[(Me_2PPh)_3Rh(nbd)](BF_4)$ (nbd = norbornadiene) as a catalyst precursor [24]. On the other hand, addition of water has been demonstrated to result in a lower catalytic efficiency in the rhodium-catalyzed hydrogenation of CO_2 in reaction media consisting of nonprotic solvents and tertiary amines [17b].

It was not until recently that catalysts containing the well-known water-soluble phosphine ligand $(m\text{-}C_6H_4SO_3^-Na^+)_3P$ (TPPTS) [25] were discovered to combine high solubility and high catalytic efficiency for the hydrogenation of CO_2 in aqueous solution in a unique manner [26, 27]. A maximum yield of HCO_2H per mole of rhodium (turnover number, TON) of 3440 and a maximum catalytic activity (turnover frequency, TOF = TON per hour) of 1365 h^{-1} was achieved using [(TPPTS)$_3$RhCl] (Structure **1**) under the mild conditions summarized in Eq. (3) [26]. Catalytic activities of several thousand turnovers per hour can be achieved at higher temperatures (Table 1). The corresponding Ru(II) complex [(TPPTS)$_3$RuCl$_2$] is also a catalyst for CO_2 hydrogenation, albeit with a TOF not exceeding 6 h^{-1} under otherwise identical conditions, whereas the combination Pd(0)/TPPTS showed no catalytic activity at all [27].

$$H_2 + CO_2 \underset{[(TPPTS)_3RhCl]\ \mathbf{1}\ (cat.)}{\overset{H_2O/amine,\ p^0 = 40\ bar,\ T = 23°C}{\rightleftharpoons}} HCO_2H \tag{3}$$

The concentration/time profile for a typical reaction using **1** as a catalyst in $H_2O/HNMe_2$ solution is depicted in Figure 1 [26]. Most notably, there is no induction period [28] and the maximum catalytic activity (maximum TOF) is obtained immediately. The catalytic activity is reduced by a factor of about 10 at a TPPTS/Rh ratio of 30:1 and also decreases slightly at Rh concentrations above 3 mM. The maximum TOF is directly proportional to the partial pressure of H_2, while the partial pressure of CO_2 has a nonlinear influence due to complications at low pressures arising partly from Scheme 1 and from mass-transport limitations. The overall activation barrier was determined from the

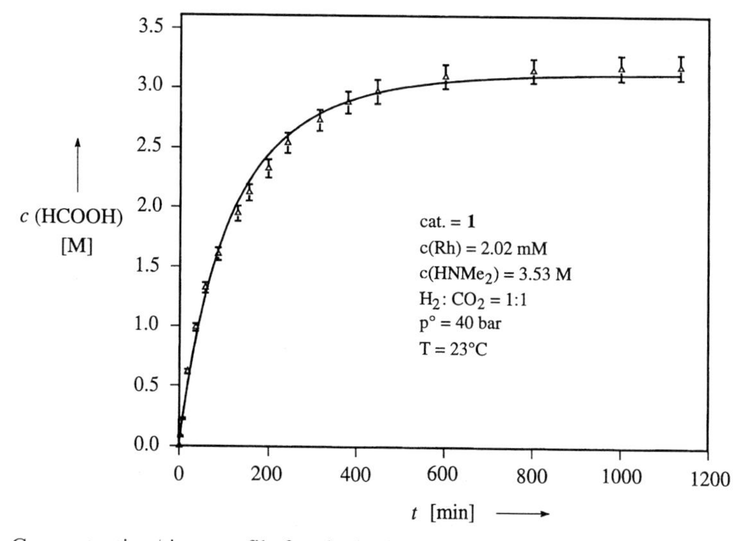

Figure 1. Concentration/time profile for the hydrogenation of CO_2 to formic acid in aqueous solution using [(TPPTS)$_3$RhCl] (**1**) as the catalyst.

Table 1. Representative results for the hydrogenation of CO_2 to formic acid using Rh(I)/TPPTS catalysts in H_2O/amine mixtures[a]

Catalyst	c°(Rh) [mM]	amine	c°(amine) [M]	T [°C]	t [h]	$c(HCO_2H)$ [M]	TON[b]	TOF[c]	Ref.
[{(cod)Rh(μ-Cl)}$_2$]/TPPTS	5.81	NEt_3	1.20	23	18	0.45	83	–	[26]
[{(cod)Rh(μ-H)}$_4$]/TPPTS	0.83	NEt_3	1.20	23	18	0.84	1068	–	[26]
[(TPPTS)$_3$RhCl] (**1**)	4.73	NEt_3	1.20	23	12	1.19	264	–	[26]
[(TPPTS)$_3$RhCl] (**1**)	5.41	–	–	23	12	n.d.[d]	–	–	[26]
[(TPPTS)$_3$ThCl] (**1**)	0.54	$HNMe_2$	3.97	23	12	1.76	3439	–	[26]
[(TPPTS)$_3$RhCl] (**1**)	2.02	$HNMe_2$	3.53	23	24	3.50[e]	(1733)	1364	[27]
[(TPPTS)$_3$RhCl] (**1**)	2.02	$HNMe_2$	3.53	42	0.5	1.89[f]	(936)	3089	[27]
[(TPPTS)$_3$RhCl] (**1**)	2.02	$HNMe_2$	3.53	61	0.5	1.42[f]	(703)	4681	[27]
[(TPPTS)$_3$RhCl] (**1**)	2.02	$HNMe_2$	3.53	81	0.5	1.09[g]	(540)	7260	[27]

[a] $H_2/CO_2 = 1:1$, $p^\circ = 40$ bar.
[b] Turnover number; the values in parentheses are not corrected for changes in volume during reaction.
[c] Turnover frequency; the values are determined from data points within the first 10 min of reaction and corrected for variation of CO_2 solubility at higher temperature.
[d] n.d., not detectable.
[e] Maximum value under these conditions.
[f] Catalyst still active.
[g] Catalyst deactivated after approx. 10 min.

temperature dependence of the reaction rates to be $E_a = 25$ kJ mol^{-1} and is thus in the same range as observed for homogeneously Ru-catalyzed CO_2 hydrogenation (31 kJ mol^{-1} [13 b]) and for heterogeneously Pd-catalyzed reduction of $NaHCO_3$ (25 kJ mol^{-1} [21 b]).

The remarkable efficiency of Rh(I)/TPPTS catalysts in H_2O/amine mixtures is demonstrated in Table 1 and Figure 2. The use of the highly water-soluble secondary amine $HNMe_2$ ensures homogeneous reaction mixtures under all conditions and allows formation of formic acid concentrations up to 3.5 M. On the other hand, no formation at all of HCO_2H is observed in pure H_2O, indicating that the amine component plays an important role within the catalytic cycle (see Section 6.15.2.2.2) [16 b, 26]. The ethanolamines $(HOCH_2CH_2)_xNH_{2-x}$ ($x = 0$, MEA; $x = 1$, DEA; $x = 2$, TEA) are also suitable as additives; as such, their efficiency decreases with increasing x [27]. The MEA solutions are of special interest as they are used on a large scale for the scrubbing of CO_2 from flue gases [5, 6]. Most notably, HCO_2H was formed with a maximum TOF of 35 h^{-1} using catalytic amounts of **1** in H_2O/MEA solution (3.53 M) saturated with CO_2 at room temperature (MEA/CO_2 = 2.4:1) under 20 bar of H_2 without an additional supply of CO_2.

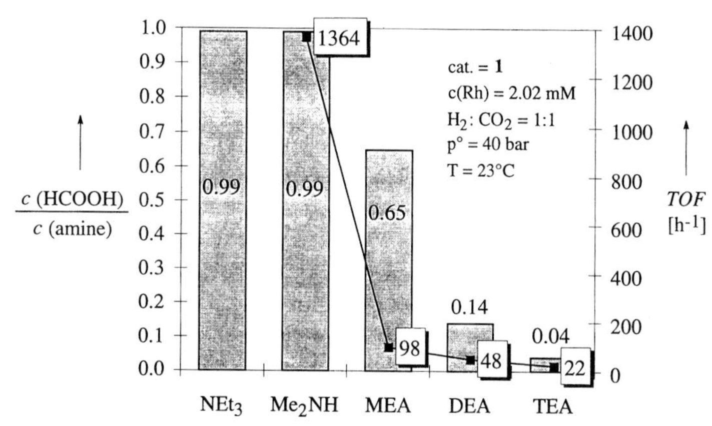

Figure 2. Influence of the amine component on the final yield and the catalytic activity of the hydrogenation of CO_2 to formic acid in aqueous solution using $[(TPPTS)_3RhCl]$ (**1**) as the catalyst. MEA, DEA, TEA is mono-, di- and triethanolamine, respectively.

Even with trialkylamines like NEt_3, the final concentration of HCO_2H never exceeds the concentration of the amine [26]. This is in contrast to the situation in dipolar nonprotic solvents [17b] or in supercritical CO_2 [23 c] where HCO_2H/NEt_3 ratios of 1.6:1 are obtained under otherwise similar conditions owing to the formation of stable HCO_2H/amine adducts. As expected, the ^{13}C-NMR signal of the reaction product in an aqueous reaction mixture showed typical values ($\delta = 170.6$, $^1J_{C,H} = 194$ Hz) for the formate ion HCO_2^- when $HNMe_2$ was used as the amine component [29]. The distinct influence of the amine on the yield and rate of the CO_2 hydrogenation cannot be directly associated with the basicity of the amines, but seems to depend also on the rate of CO_2 uptake [6].

The most significant advantage of the hydrogenation of CO_2 in aqueous dimethylamine solution is the enhanced stability of HCO_2H with respect to the back reaction to yield CO_2 and H_2. In contrast to reactions carried out in DMSO/NEt_3 [17b], no decomposition of the product formed under 40 bar initial pressure is observed after the pressure in the reaction vessel is reduced to 1 bar, although the catalyst is still active (Figure 3) and deuterium labeling studies clearly demonstrate the reversibility of the C–H bond-forming step. Even more significantly, appreciable amounts of HCO_2H are formed – albeit very slowly (TON 3 per day) – in a water/MEA mixture containing **1** at room temperature under a 1:1 atmosphere of H_2 and CO_2 at ambient pressure [27].

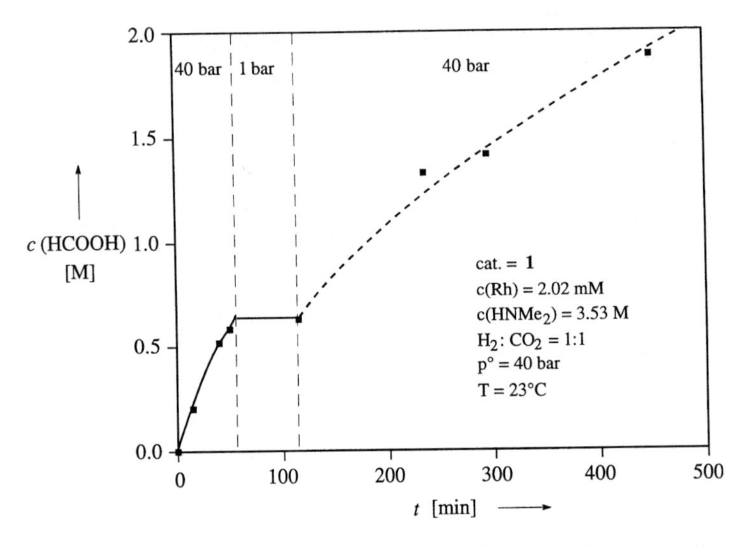

Figure 3. The reversibility of the C–H bond formation during the hydrogenation of CO_2 to formic acid in aqueous solution using [$(TPPTS)_3RhCl$] (**1**) as the catalyst does not lead to decomposition of formic acid at ambient pressure.

The stability of HCO_2H toward decarboxylation is an important aspect for potential technical application of the CO_2 hydrogenation, as it greatly facilitates the handling and the workup of the product mixture. It should be mentioned that the isolation of free formic acid from the reaction mixture in an economically and ecologically feasible manner appears to be currently the major problem for a large-scale application of this reaction, although some interesting solutions have been discussed [30].

6.15.2.2.2 Mechanistic Aspects of the Rhodium-catalyzed Hydrogenation of CO_2 in Aqueous Solution

The current understanding of the catalytic cycle of the CO_2 hydrogenation in aqueous solution is depicted in Scheme 3. Representation of the reversibility of the individual steps within the catalytic cycle has been omitted for clarity. This scheme is consistent with the data described above and based on results of high-pressure NMR investigations and labeling studies [27] as well as closely related studies on the mechanism of CO_2 hydrogenation in DMSO/NEt_3 solutions [31].

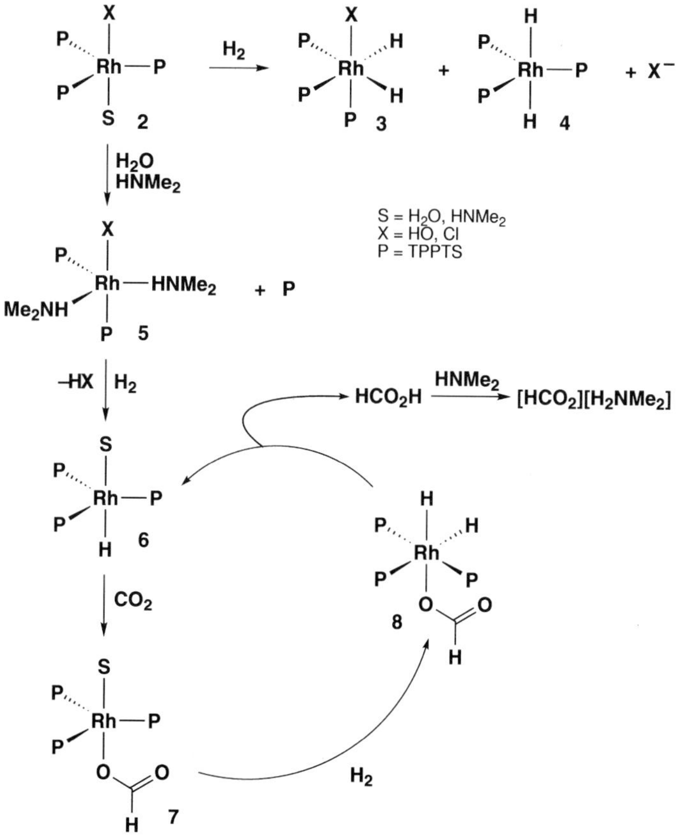

Scheme 3. Catalytic cycle for the hydrogenation of CO_2 to formic acid in aqueous solution using [$(TPPTS)_3RhCl$] (**1**) as the catalyst, explaining the role of the amine. All species except **7** and **8** have been detected by NMR spectroscopy.

The species **2** (as derived from **1**) is converted to **3** [32] and **4** in neutral aqueous solutions under 10 bar H_2. In agreement with the lack of catalytic activity in the absence of amine, both complexes were found to remain unreactive upon introduction of CO_2 up to 20 bar total pressure. Phosphine substitution from complex **2** was observed to give **5** in the presence of $HNMe_2$ under an argon atmosphere. This species reacted quantitatively with H_2 under 10 bar to give the neutral rhodium(I) monohydride complex **6** [33]. Complex **6** is also formed upon treatment of a mixture of **3** and **4** with $HNMe_2$. In presence of H_2 and CO_2, catalytic formation of HCO_2H occurs readily in solutions containing **6**, but only broad signals are observed in the NMR spectra.

Although no direct evidence for the subsequent steps of the catalytic cycle are available, it is reasonable to assume that reversible insertion of CO_2 into the Rh–H bond [31a] of **6** to form **7** is the productive step for the C–H bond formation. Indeed, it was found that HCO_2H formation occurs with appreciable rates only under conditions where free dissolved CO_2 is present in solution. An alternative mechanism involving reaction of **6** with CO_2 to give a bicarbonato complex and CO via reverse WGS [34] followed by subsequent hydration of CO seems unlikely as the formation of HCO_2H from CO and H_2O/amine mixtures requires much higher temperatures [35]. Furthermore, CO could not be detected by GC/MS during CO_2 hydrogenation using **2** in either the gas phase or solution. The liberation of formic acid from **7** may then proceed via a classical oxidative addition/reductive elimination pathway involving **8** as depicted in Scheme 3. It should be noted, however, that σ-bond metathesis must be considered as an alternative low-energy pathway [31b, c].

Overall, there are striking similarities between the mechanism of the rhodium(I)-catalyzed hydrogenation of CO_2 in water/amine and DMSO/amine mixtures. The well-established ligand effects on the catalytic activities in the latter system [36], together with the large body of known water-soluble phosphine ligands [25b, d], makes it likely that catalysts with even higher activity than **1** (or **2** respectively) can be developed in future research.

6.15.3 Transition-metal-catalyzed Reactions Without Incorporation of CO_2

The transition-metal-catalyzed hydrogenation of CO_2 to formic acid proceeds with high efficiency in homogeneous aqueous solutions. The use of water as a medium for transition-metal-catalyzed reactions using CO_2 as a C_1 building block seems highly promising from these results. At the same time, it seems worth mentioning that CO_2 has also been observed to be able to modify the course of catalytic reactions without being incorporated into the final products. Although these effects have yet to be fully explored, one can envisage several possible reasons for the observed modifications: changes in the pH of the reaction mixture (see Scheme 1), interaction of CO_2 with the catalytic active metal center as found in stable CO_2 complexes [2i, 37], or a reversible reaction of CO_2 with highly reactive intermediates (e.g., M–H or M–C units) resulting in different catalyst resting states [2i].

A technically important example of the modifying effect of CO_2 is found in two-phase hydroformylation using Rh(I)/TPPTS catalysts [25]. Small amounts of CO_2 ($< 4\%$, v/v) added to the CO/H_2 gas mixture help to suppress undesired side reactions that lead to high-boiling by-products [38]. This beneficial effect seems to rely not on pH control alone, as increased selectivities have also been described for hydroformylation in the presence of CO_2 in non-aqueous media [39].

Another striking example of the influence of CO_2 on catalytic organometallic reactions in aqueous solution is the Pd-catalysed telomerization of 1,3-dienes (Eqs. 4 and 5). In organic solvents, CO_2 can be incorporated efficiently into products from butadiene telomerization to yield cyclic or linear unsaturated esters [40]. In the presence of water, however, telomerization of butadiene occurs exclusively with incorporation of H_2O even under CO_2 pressure to yield 2,7-octadien-1-ol (**9**) as the main product [41]. Compound **9** is an attractive intermediate for special and fine chemical synthesis and is produced by Kuraray in Japan [42] (cf. Section 6.7).

2 [butadiene] —[CO_2 (10 bar); H_2O, NEt_3, surfactant 90°C; cat.: $Ph_2P(m\text{-}C_6H_4SO_3Na)/Pd(OAc)_2$]→ [2,7-octadien-1-ol] OH **9** (4)

2 [butadiene] —[CO_2 (3 bar); CH_3CN/H_2O (4:1), 90°C; cat.: $Ph_3P/Pd(OAc)_2$]→ **10** (5)

It is well established that the carbon dioxide atmosphere exhibits a remarkable activating effect on the Pd-catalyst used for the synthesis of **9** [41, 42]. This effect can be reasonably assigned to formation of HCO_3^- and its nucleophilic attack at an allylpalladium intermediate [43]. More recently, a two-phase system using TPPTS and related ligands in the presence of neutral and cationic surfactants has been developed (Eq. 4) [44]. The combination of ligand and surfactant has been shown to exhibit a subtle influence on the selectivity and activity of this telomerization process [42]. Furthermore, it was observed that 2-vinylmethylenecylopentane (**10**) can also be obtained as the main product from Pd-catalyzed cyclodimerization of butadiene in acetonitrile/H_2O solutions (Eq. 5), whereby the choice of the phosphine ligand and the CO_2 atmosphere was found to be critical for the selectivity toward **10** [45].

6.15.4 Conclusions

Aqueous solutions provide ecologically benign and economically interesting reaction media for catalytic organometallic chemistry involving CO_2. The hydrogenation of CO_2 to formic acid represents a promising approach for the use of CO_2 as a C_1 building block in water as a solvent. The specific chemical properties of CO_2 can also result in changes of selectivities or activities in catalytic reactions without incorporation of CO_2 by interaction with catalytic intermediates or with the solvent. Some of these aspects have just emerged during the last few years and future research efforts are clearly needed to explore fully their synthetic utility and to understand the underlying principles.

References

[1] (a) S. Inoue, N. Yamazaki (Eds.), *Organic and Bio-organic Chemistry of Carbon Dioxide*, Wiley, New York, **1982**; (b) A. Behr, *Carbon Dioxide Activation by Metal Complexes*, VCH, Weinheim, **1988**; (c) W. M. Ayers (Eds.) *Catalytic Activation of Carbon Dioxide*, ASC Symposium Series, No. 363, **1988**; (d) M. Aresta, J. V. Schloss (Eds.) *Enzymatic and Model Carboxylation and Reduction Reactions for Carbon Dioxide Utilization*, NATO ASI Series C, No. 314, **1990**; (e) M. M. Halmann, *Chemical Fixation of Carbon Dioxide*, CRC Press, Boca Raton, **1993**; (f) J. Paul, C.-M. Pradier (Eds.), *Carbon Dioxide Chemistry: Environmental Issues*, Royal Society of Chemistry, London, **1994**.

[2] (a) D. J. Darensbourg, R. A. Kudaroski, *Adv. Organomet. Chem.* **1983**, *22*, 129; (b) D. R. Palmer, R. van Eldik, *Chem. Rev.* **1983**, *83*, 651; (c) D. Walther, *Coord. Chem. Rev.* **1987**, *79*, 135; (d) A. Behr, *Angew. Chem.* **1988**, *100*, 681; *Angew. Chem., Int. Ed. Engl.* **1988**, *27*, 661; (e) P. Braunstein, D. Matt, D. Nobel, *Chem. Rev.* **1988**, *88*, 747; (f) I. S. Kolomnikov, T. V. Lysak, *Russ. Chem. Rev. (Engl. Transl.)* **1990**, *59*, 344; (g) M. Aresta, E. Quaranta, I. Tommasi, *New. J. Chem.* **1994**, *18*, 133; (h) E. Dinjus, R. Fornika, in *Applied Homogeneous Catalysis with Organometallic Compounds* (Eds.: B. Cornils, W. A. Herrmann), VCH, Weinheim, **1996**, Vol. 2, p. 1048; (i) W. Leitner, *Coord. Chem. Rev.* **1996**, *153*, 257.

[3] X. Xiaoding, J. A. Moulijn, *Energy Fuels* **1996**, *10*, 305.

[4] Anon., *Greenhouse Gas Emissions from Power Stations*, IEA Greenhouse Gas R & D Programme, Cheltenham, **1993**.

[5] (a) Anon., *Carbon Dioxide Capture From Power Stations*, IEA Greenhouse Gas R & D Programme, Cheltenham, **1994**; (b) G. Sartori, W. S. Ho, W. A. Thaler, G. R. Chludzinski, J. C. Wilbur, in [1f], p. 205; (c) T. Suda, M. Fujii, T. Miura, S. Shimojo, M. Iijima, S. Mitsuoka, in [1f], p. 222; (d) R. A. Reck, K. J. Hoag, *Energy (Oxford)* **1997**, *22*, 115.

[6] G. F. Versteeg, L. A. J. van Dijck, W. P. M. van Swaaij, *Chem. Eng. Commun.* **1996**, *144*, 113.

[7] M. Aresta, E. Quaranta, I. Tommasi, P. Giannoccaro, A. Ciccarese, *Gazz. Chim. Ital.* **1995**, *125*, 509.

[8] J. R. Lancaster, Jr. (Ed.), VCH, Weinheim, **1988**, Chapters 13 and 14.

[9] M. Kumar, D. Qiu, T. Spiro, S. W. Ragsdale, *Science* **1995**, *270*, 628.

[10] (a) *Zinc Enzymes* (Eds.: I. Bertini, C. Luchinat, W. Maret, M. Zeppezauer), Birkhäuser, Boston, **1986**, Vol. I; (b) X. Zhang, C. D. Hubbard, R. van Eldik, *J. Phys. Chem.* **1996**, *100*, 9161.

[11] (a) I. Willner, B. Willner, *Top. Curr. Chem.* **1991**, *159*, 153; (b) T. Ogata, S. Yanagida, B. S. Brunschwig, E. Fujita, *J. Am. Chem. Soc.* **1995**, *117*, 6708, and references therein; (c) *Electrochemical and Electrocatalytic Reactions of Carbon Dioxide* (Eds.: E. P. Sullivan, K. Krist, H. E. Guard), Elsevier, Amsterdam, **1993**; (d) J. A. R. Sende, C. R. Arana, L. Hernandez, K. T. Potts, M. Keshevarzk, H. D. Abruña, *Inorg. Chem.* **1995**, *34*, 3339, and references therein.

[12] (a) S. Schreiner, J. Y. Yu, L. Vaska, *Inorg. Chim. Acta* **1988**, *147*, 139; (b) T. C. Eisenschmid, R. Eisenberg, *Organometallics* **1989**, *8*, 1822; (c) K. Tominaga, A. Sasaki, T. Watanabe, M. Saito, *Bull. Chem. Soc. Jpn.* **1995**, *68*, 2837.
[13] (a) M. M. Taqui Khan, S. B. Halligudi, S. Shukla, *J. Mol. Catal.* **1989**, *53*, 305; (b) M. M. Taqui Khan, S. B. Halligudi, S. Shukla, *J. Mol. Catal.* **1989**, *57*, 47.
[14] (a) P. C. Ford, *Acc. Chem. Res.* **1981**, *14*, 31; (b) P. C. Ford, A. Rokicki, *Adv. Organomet. Chem.* **1988**, *28*, 139; (c) R. M. Laine, E. J. Crawford, *J. Mol. Catal.* **1988**, *44*, 357.
[15] (a) T. Yoshida, T. Okano, Y. Ueda, S. Otsuka, *J. Am. Chem. Soc.* **1981**, *103*, 3411; (b) B. S. Lima Neto, K. Howland Ford, A. J. Pardey, R. G. Rinker, P. C. Ford, *Inorg. Chem.* **1991**, *30*, 383, and references therein.
[16] (a) P. G. Jessop, T. Ikariya, R. Noyori, *Chem. Rev.* **1995**, *95*, 259; (b) W. Leitner, *Angew. Chem.* **1995**, *107*, 2391; *Angew. Chem., Int. Ed. Engl.* **1995**, *34*, 2207.
[17] (a) E. Graf, W. Leitner, *J. Chem. Soc., Chem. Commun.* **1992**, 623; (b) W. Leitner, E. Dinjus, F. Gaßner, *J. Organomet. Chem.* **1994**, *475*, 257.
[18] (a) K. Wagner, *Angew. Chem.* **1970**, *82*, 73; *Angew. Chem., Int. Ed. Engl.* **1970**, *9*, 50; (b) K. Narita, M. Sekija, *Chem. Pharm. Bull.* **1977**, *25*, 135.
[19] R. C. Weast (Ed.) *Handbook of Chemistry and Physics,* 65th ed., CRC Press, Boca Raton, **1984**.
[20] K. Kudo, N. Sugita, Y. Takezaki, *Nippon Kagaku Kaishi* **1977**, 302.
[21] (a) C. J. Stalder, S. Chao, D. P. Summers, M. S. Wrighton, *J. Am. Chem. Soc.* **1983**, *105*, 6318; (c) D. C. Engel, G. F. Versteeg, W. P. M. van Swaaij, *Chem. Eng. Res. Des.* **1975**, *73*, 701; and references therein.
[22] Y. Inoue, H. Izumida, Y. Sasaki, H. Hashimoto, *Chem. Lett.* **1976**, 863.
[23] (a) BP Ltd. (A. G. Kent) EP 151.510 (1985); *Chem. Abstr.* **1986**, *104*, 109029h; (b) M. Sakamoto, I. Shimizu, A. Yamamoto, *Organometallics* **1994**, *13*, 407; (c) P. G. Jessop, Y. Hsiao, T. Ikariya, R. Noyori, *J. Am. Chem. Soc.* **1996**, *118*, 344; (d) J. Z. Zhang, Z. Li, H. Wang, C. Y. Wang, *J. Mol. Catal. A: Chem.* **1996**, *112*, 9; (e) E. Lindner, B. Keppeler, P. Wegner, *Inorg. Chim. Acta* **1997**, *258*, 97.
[24] J.-C. Tsai, K. M. Nicholas, *J. Am. Chem. Soc.* **1992**, *114*, 5117.
[25] (a) E. Kuntz, *CHEMTECH* **1987**, 570; (b) W. A. Herrmann, C. W. Kohlpaintner, *Angew. Chem.* **1995**, *105*, 1588; *Angew. Chem., Int. Ed. Engl.* **1993**, *32*, 1524; (c) B. Cornils, E. G. Kuntz, *J. Organomet. Chem.* **1995**, *502*, 177; (d) see also Chapter 2 of this book.
[26] F. Gaßner, W. Leitner, *J. Chem. Soc., Chem. Commun.* **1993**, 1465.
[27] (a) F. Gaßner, Dissertation, Universität Jena, **1994**; (b) F. Gaßner, E. Dinjus, W. Leitner, *Proceedings of the XXVIII. Jahrestreffen deutscher Katalytiker*, Dechema, Frankfurt, **1995**, p. 52.
[28] F. Gaßner, E. Dinjus, H. Görls, W. Leitner, *Organometallics* **1996**, *15*, 2078.
[29] One might argue that the reaction should therefore be correctly referred to as hydrogenation of CO_2 "to ammonium formates" rather than "to formic acid." We prefer the latter for simplicity as the amine is not a stoichiometric reagent in most cases, and in order to avoid additional distinction between the products of the type R_3NH/HCO_2^- and R_3N/HCO_2H [18].
[30] (a) BP Ltd. (J. J. Anderson, J. E. Hamlin), EP 126.524 (1984); *Chem. Abstr.* **1985**, *102*, 95259c; (b) BP Ltd. (J. J. Anderson, D. J. Drury, J. E. Hamlin, A. G. Kent), WO 2.066 (1986); *Chem. Abstr.* **1986**, *105*, 210757q; (c) BP Ltd. (M. J. Green, A. R. Lucy, M. Kitson, S. J. Smith), EP 329.337 (1989); *Chem. Abstr.* **1990,** *112*, 58742r; (d) BP Ltd. (R. G. Beevor, D. J. Gulliver, M. Kitson, R. M. Sorrell), EP 357.243 (1989); *Chem. Abstr.* **1990**, *113*, 114638w; (e) University Twente (D. C. Engel, W. P. van Swaji, G. F. Versteeg), EP 597.151 (1994); *Chem. Abstr.* **1995**, *122*, 58814x.
[31] (a) T. Burgemeister, F. Kastner, W. Leitner, *Angew. Chem.* **1993**, *105*, 781; *Angew. Chem., Int. Ed. Engl.* **1993**, *32*, 739; (b) F. Hutschka, A. Dedieu, W. Leitner, *Angew. Chem.* **1995**,

107, 1905; *Angew. Chem., Int. Ed. Engl.* **1995**, *34*, 1742; (d) F. Hutschka, A. Dedieu, M. Eichberger, R. Fornika, W. Leitner, *J. Am. Chem. Soc.* **1997**, *119*, 4432.
[32] C. Larpent, H. Patin, *J. Organomet. Chem.* **1987**, *335*, C13.
[33] (a) For the corresponding monohydride with TPPMS see: F. Joó, C. Csiba, A. Bényei, *J. Chem. Soc., Chem. Commun.* **1993**, 1602; (b) It is interesting to note that the hydride ligands of the neutral hydrides **3**, **6** and **6**-TPPMS cannot be detected by ^{1}H-NMR in H_2O/H_2 owing to fast exchange processes, whereas the cationic species **4** shows the expected high-field signals.
[34] T. Yoshida, D. L. Thorn, T. Okano, J. A. Ibers, S. Otsuka, *J. Am. Chem. Soc.* **1979**, *101*, 2412.
[35] BASF AG (F. Lippert, A. Höhn), EP 583.695 (1994); *Chem. Abstr.* **1994**, *120*, 216715f.
[36] (a) R. Fornika, H. Görls, B. Seemann, W. Leitner, *J. Chem. Soc., Chem. Commun.* **1995**, 1479; (b) E. Graf, W. Leitner, *Chem. Ber.* **1996**, *129*, 91; (c) K. Angermund, W. Baumann, E. Dinjus, R. Fornika, H. Görls, M. Kessler, C. Krüger, W. Leitner, F. Lutz, *Chem. Eur. J.* **1997**, *3*, 755.
[37] D. H. Dibson, *Chem. Rev.* **1996**, *96*, 2063.
[38] Ruhrchemie AG (B. Cornils, W. Konkol, H. Bach, G. Dämbkes, W. Gick, E. Wiebus, H. Bahrmann), DE 3.415.968 (1984); *Chem. Abstr.* **1986**, *104*, 209147p.
[39] (a) Universal Oil Products Co. (S. N. Massie, J. A. Vesely), DE 2.415.902 (1974); *Chem. Abstr.* **1975**, *82*, 16305g; (b) Argonne Nat. Lab. (J. W. Rathke, R. J. Klingler), US 5.198.589 (1993); *Chem. Abstr.* **1993**, *119*, 141597n.
[40] (a) A. Behr, R. He, K.-D. Juszak, C. Krüger, Y.-H. Tsay, *Chem. Ber.* **1986**, *119*, 991, and references therein; (b) see also: E. Dinjus, W. Leitner, *Appl. Organomet. Chem.* **1995**, *9*, 43; (c) S. Pitter, E. Dinjus, B. Jung, H. Görls, *Z. Naturforsch. Teil B* **1996**, *51*, 934.
[41] (a) K. E. Atkins, W. E. Walker, R. M. Manvik, *J. Chem. Soc., Chem. Commun.* **1971**, 330; (b) K. Kaneda, H. Kurosaki, M. Terasaw, T. Imanaka, S. Teranishi, *J. Org. Chem.* **1981**, *46*, 2356; (c) J. P. Bianchini, B. Waegell, E. M. Gaydou, H. Rzehak, W. Keim, *J. Mol. Catal.* **1981**, *10*, 247; (d) U. M. Dzhemilev, V. V. Sidorova, R. V. Kunakova, *Izv. Akad. Nauk SSR, Ser. Khim.* **1983**, *3*, 584; (e) J. P. Bianchini, B. Waegell, E. M. Gaydou, A. Eisenbeis, W. Keim, *J. Mol. Catal.* **1985**, *30*, 197.
[42] E. Monflier, P. Bourdauducq, J. L. Couturier, J. Kervenal, A. Morteux, *Appl. Catal. A: General* **1995**, *131*, 167; and references therein.
[43] (a) P. W. Jolly, *Angew. Chem.* **1985**, *97*, 279; *Angew. Chem., Int. Ed. Engl.* **1985**, *22*, 283; (b) M. Sakamoto, I. Shimizu, A. Yamamoto, *Bull. Chem. Soc. Jpn.* **1996**, *69*, 1065.
[44] (a) E. Monflier, J. L. Couturier, P. Bourdauducq, J. Kervenal, A. Morteux, *J. Mol. Catal.* **1995**, *97*, 29; (b) E. Monflier, P. Bourdauducq, J. L. Couturier, J. Kervenal, I. Suisse, A. Morteux, *Catal. Lett.* **1995**, *34*, 201.
[45] F. Bergamini, F. Panella, R. Santi, E. Antonelli, *J. Chem. Soc., Chem. Commun.* **1995**, 931.

6.16 Halogen Chemistry

Mario Bressan, Antonino Morvillo

6.16.1 Introduction

A large number of halogenated organics have been produced commercially in the past few decades, which have been used for a variety of purposes. The quest for environmentally friendly technology in general has risen to a substantial thrust to get away from chlorocarbons and halogenated materials altogether, due to the generic deleteriousness associated to them. Halogenated compounds continue to be produced and utilized, however, as they are still both the best solvents for some high-tech processes and flexible starting materials for a variety of organic syntheses.

6.16.2 Reductive and Oxidative Dehalogenation

Dehalogenation of halo-organics before their release from current processes, and the retrospective treatment of the mess created as a result of past expediency, are important issues because of the environmental persistence of these species in groundwater: compounds involved vary from volatile aliphatics to heavy aromatics such as polychlorophenols, polychlorinated biphenyls (PCBs), and dioxins. Indeed, serious problems are associated with the combustion of chlorinated organics, since these classes of compounds, although thermodynamically unstable, are characterized by a high degree of chemical inertness and thermal stability: total conversion of chlorinated organics to innocuous materials (HCl, CO_2, and H_2O) cannot be achieved without considerable expense.

The ubiquitous presence of methane monooxygenase (MMO) and cytochrome *P*-450 (CyP450) enzymic systems indicates that they may be of principal importance on a global scale for the oxidative degradation of halo-organics in biological systems. In many instances, however, current techniques of bioremediation of contaminated groundwaters involving entire microorganisms are inadequate because of toxic pollutant concentrations, nutrient limitations, and the lack of membrane permeability for halo-organics (which, coupled with absence of suitable extracellular enzymes, is the accepted basis for a molecule being recalcitrant) [1]. MMO is capable of oxidizing a wide variety of haloalkenes at rates comparable with those of other substrates for the enzyme, by being the competitive inhibitor tetrachloroethylene the only chlorinated ethylene not turned over [2]. The rates observed are at least two orders of magnitude faster than the rates reported for whole-cell oxidation reactions by nonmethanotrophs and 2–10 times faster than comparable oxidations catalyzed by CyP450 containing mixed-function oxidase systems [3].

CyP450 and other reduced iron porphyrins are also reported to mediate reductive dehalogenation of a variety of haloalkanes and -alkenes [4], and of the aliphatic portion of DDT [5]. Reductive dehalogenation refers specifically to the reaction in which two electrons and a proton act as substrates along with the halogenated compound to yield a reduced product and the corresponding halide (Eq. 1).

$$RX + 2e^- + H^+ \rightarrow RH + X^- \quad (1)$$

Compounds with high halogen substitution and therefore with carbon atoms in high formal oxidation states are expected to be resistant to the degradation under aerobic conditions and more susceptible to a reductive degradation. Products similar to those resulting from treatment with anaerobic bacteria or CyP450 were obtained by depositing a stable ordered film of myoglobin (Mb) and a surfactant on electrodes: a highly reduced form of Mb was produced in the films, which were used to catalyze reduction of organohalide pollutants such as trichloroacetic acid and polyhaloethylenes [6]. Lindane dechlorination is effected by iron-containing hemin and hematin, and by cobalt-containing protoporphyrin and various cobalamins [7]. Vitamin B_{12}, but not hematin, was also shown to reductively dechlorinate pentachlorophenol and trichlorophenoxyacetic acid [8]; cobalamins have also been used in the reductive dechlorination of CCl_4, polychlorinated ethanes, polychlorobenzenes and PCBs [9]. Reductive dehalogenation of chlorinated aliphatic hydrocarbons and freons occurs with participation of coenzyme F430, a nickel porphyrinoid present in anaerobic bacteria [10].

Liquid-phase catalysts are close models both to the monooxygenase and to the reductive enzymes and can be a gentle alternative method of destruction of halogenated hydrocarbons. Transition metal complexes, in particular metal porphyrins, corrins and phthalocyanines, have been studied in homogeneous abiotic aqueous systems as potential remediation catalysts, but further identification of degradative products is necessary, since innocuous products must result if synthetic catalysts are to be used effectively. Moreover, the implementation of homogeneous catalysts is still impractical because of problems with separating the catalyst; in principle these can be overcome by immobilizing the complexes on a solid support. There are numerous examples of supported catalysts, but their application in aqueous systems has rarely been investigated [11].

Cationic water-soluble iron, cobalt and nickel porphyrins with various functional groups in *meso* positions suitable for immobilization have been tested as catalysts for reductive dehalogenation of CCl_4 with dithiothreitol: $CHCl_3$, CH_2Cl_2, and CO were found to be breakdown products [12]. Nickel(I) octaethyl isobacteriochlorin has been used as model for F430 factor for the reduction and coupling of alkyl halogenide [13]. Photoreductive dehalogenation of aqueous $CHBr_3$ is mediated by the anionic cobalt macrocycle CoPcS (PcS = tetrasulfophthalocyanine anion) adsorbed on the positively charged titania surface [14]. Nonspecific biomimetic macrocycles, CoTMPyP (TMPyP = tetrakis-(*N*-methyl-4-pyridiniumyl) porphyrin cation) and CoPcS, were used as homogeneous and mineral-supported catalysts: the study demonstrates that the charged, mineral-supported macrocycles reductively dechlorinate CCl_4 in water, even at high concentrations that would inhibit microbial activity [15]. A C–X bond in aliphatic or benzylic halides can be transformed into a C–H bond in a transfer hydrodehalogenation reaction with formate as hydrogen donor and water-soluble ruthenium catalysts, $RuCl_2$(TPPMS) (TPPMS = *m*-sulfophenyl-diphenylphosphine anion) or $Ru(H_2O)_3(PTA)_3(tos)_2$(PTA = 1,3,5-triaza-7-phosphaadamantane; tos = tosyl) [16].

The oldest catalytic oxidative system for dehalogenation is Fenton's reagent; essentially the hydroxyl radical is one of the few chemical species capable of attacking refractory halo-organic compounds; the scope of the reaction in terms of effective substrate-oxidation vs. H_2O_2-dismutation is often limited by sensitivity to pH and a narrow H_2O_2/Fe^{2+} ratio [17]. Various water-soluble iron or manganese sulfophenylporphyrins catalyze with exceedingly high activity (turnover frequency, TOF, up to $20\ s^{-1}$) the oxidative dechlorination of trichlorophenol (TCP) with $KHSO_5$ in aqueous acetonitrile [18]. The more easily accessible MnPcS or FePcS catalysts equally behave in water also in the presence of the environmentally safe oxidant H_2O_2 wih TOFs in excess of $0.1\ s^{-1}$. Products of dechlorination (up to two chloride ions were released per TCP molecule), of aromatic ring cleavage, and of oxidative coupling have been detected (Scheme 1). The catalysts substantially maintain their activities with H_2O_2 when immobilized on cationic resins [19].

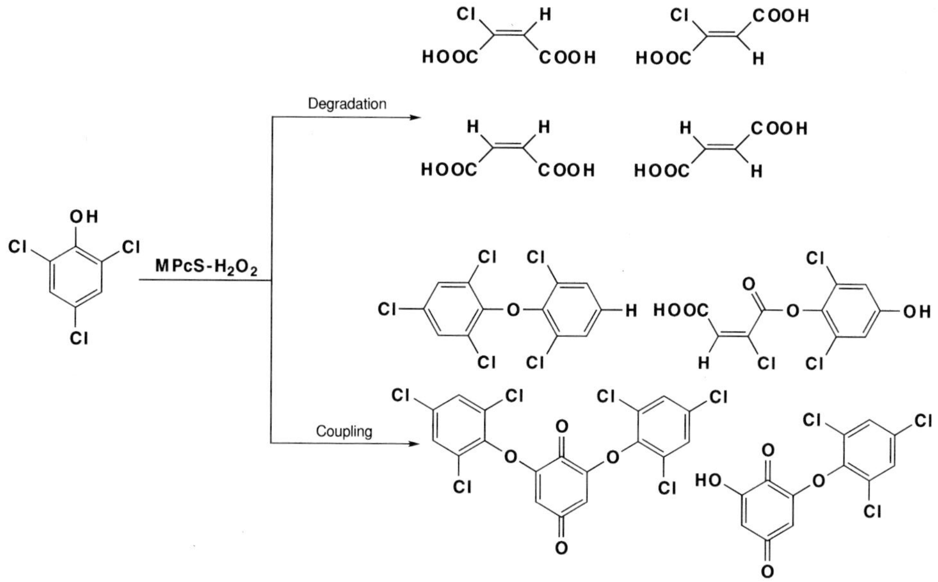

Scheme 1. Oxidation of TCP by MPcS catalysts (M = Mn, Fe) and hydrogen peroxide.

Ruthenium tetroxide was shown to oxidize PCBs in water [20]. Water-soluble ruthenium complexes, such as $[Ru(H_2O)_2(DMSO)_4]^{2+}$, are effective catalysts for the $KHSO_5$ deep oxidation of a number of chloroaliphatics, of α-chlorinated olefins, polychlorobenzenes, and polychlorophenols. When the reactions are carried out in water in the presence of surfactant agents, degradation of the substrates is definitely faster. Aromatic substrates are mainly converted into HCl and CO_2, polychlorophenols being more sensitive to oxidation than substituted benzenes [21]. α-Chlorinated olefins are degraded to HCl and the appropriate carboxylic acid and/or CO_2 with TOFs in excess of 1 s^{-1} (Scheme 2) [22].

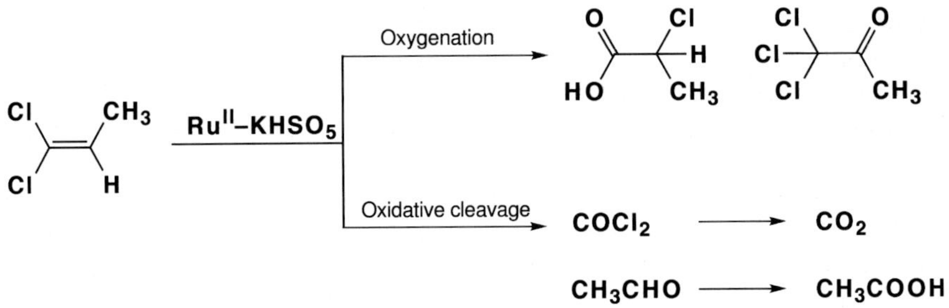

Oxygenation
Ru^{II}–$KHSO_5$
Oxidative cleavage
$COCl_2$ ⟶ CO_2
CH_3CHO ⟶ CH_3COOH

Scheme 2. Oxidation of 1,1′-dichloropropene by Ru(II) catalysts and monopersulfate.

6.16.3 Coupling and Carbonylation Reactions

Halo-organics could be simple and inexpensive starting materials for many organic syntheses, such as the important carbon–carbon bond formation, and have been considered significant as intermediates for the prebiotic synthesis of organic compounds in the primitive aqueous environment [23].

A number of important palladium-catalyzed coupling reactions starting from aryl and vinyl halides have been successfully imported into aqueous media [24]. Suzuki coupling between 1-iodocycloalkenes and vinylboronic esters to give the corresponding 1-vinylcycloalkenes derivatives has been carried out in the presence of palladium salts and water-solube tri(*m*-sulfophenyl)phosphine anion (TPPTS) and sulfonated diphosphines [25]. The synthesis of a terphenyl derivative complexed by the cationic moiety $[Cp^*Ru]^+$ by Suzuki coupling of $[Cp^*Ru(BrC_6H_4Br)]^+$ with phenylboronic acid under catalysis by $[Pd(PPh_3)_4]$ in a DME/water mixture is reported to occur in quantitative yield [26].

The palladium-catalyzed reaction of haloarenes with alkenes or alkynes (Heck reaction) is a useful synthetic method and several variations have been devised in recent years to perform the reaction in aqueous media: coupling of iodobenzene and methylacrylate to methyl cinnamate proceeds in exceedingly good yields when carried out in a mixture of the organic phase of the reactants and water [27]. Iodo- and bromobenzenes react cleanly with ethylene to yield substituted styrenes in good yields, when $[PdCl_2(TPPMS)_2]$ catalyst is used in water or in a water/acetonitrile mixture: these reactions provide a synthetic route to poly(phenylenevinylenes) [28]. Palladium-catalyzed coupling of terminal alkynes with aryl halides occurs in quite good yields in the presence of a quaternary ammonium salt and a base in an acetonitrile/water solution, without any added cuprous iodide [29]. Reactions of 4-carboxyphenylacetylenes with 4-iodobenzoic acid are carried out in water or water/acetonitrile in the presence of palladium acetate and water-soluble aryl alkyl guanidinium phosphines, yielding various amounts of cross- or homocoupling products. A comparative study of TPPTS and the aryl alkyl guanidinium phosphines in the aqueous-phase palladium-catalyzed C–C coupling between 4-iodobenzoate and trifluoroacetylpropargylamine shows the latter to be of superior activity [30].

Aqueous palladium-catalyzed Stille coupling shows much of the versatility found for the original version using the lipophilic n-Bu_3SnR in organic solvents, with some notable advantages. Unlike the organic conditions, where the n-Bu_3SnR reagent transfers vinyl and aryl groups preferentially over the n-Bu groups, the aqueous variant does not require group-selective transfer; furthermore, whereas halogen substituents on the tin dramatically retard the Stille coupling in organic media, hydrolysis of the $RSnCl_3$ reagent apparently facilitates both solubilization and C–Sn bond activation. Two relevant and closely related studies were reported recently, dealing with Stille coupling of water-soluble aryl and vinyl halides with allyl-, aryl-, and vinyltrichlorostannanes to give a variety of coupling products, among them alkyl-, aryl- and vinyl-substituted benzoic acids: the reaction is carried out in alkaline media and catalyzed by palladium salts in the presence of the water-soluble phosphines TPPMS and di(m-sulfophenyl)phenylphosphine anion (TPPDS) [31].

Palladium-catalyzed hydrocycarbonylation of aryl halides can be conducted in biphasic conditions, even if the palladium complex is a normal PPh_3 derivative, probably insoluble in the aqueous phase [32]. The selective hydroxycarbonylation of bromobenzene to benzoic acid has also been successfully explored with the water-soluble $[Pd(TPPTS)_3]$ complex, which leads to better yields, probably by improving the contact between the catalytic species and the nucleophilic agent and preventing the precipitation of metallic palladium under the reaction conditions [33]. The zwitterionic $[Rh(cod)]BPh_4$ complex is reported to catalyze the carbonylation of benzylic and allylic bromides to carboxylic esters in a biphasic system (aqueous $NaOH/CH_2Cl_2$, with phase-transfer agents) [34]. Another class of useful – even if complex – reactions is represented by the $Co_2(CO)_8$-catalyzed carbonylation of phenethyl bromide to benzylpyruvic and benzylacetic acids: this reaction has been investigated in the aqueous phase and in the presence of water-soluble phosphines TPPTS and TPPMS, and exhibits variable reactivity and selectivity [35].

References

[1] D. B. Janssen, B. Witholt, in *Metal Ions in Biological Systems* (Eds.: H. Sigel, A. Sigel), M. Dekker, New York, **1992**, Vol. 28, p. 158; R. B. Winter, H. Zimmermann, *ibid.* p. 300.
[2] B. G. Fox, J. G. Borneman, L. P. Wackett, J. D. Lipscomb, *Biochemistry* **1990**, *29*, 6419.
[3] R. E. Miller, F. P. Guengerich, *Biochemistry* **1982**, *21*, 1090.
[4] S. Li, L. P. Wackett, *Biochemistry* **1993**, *32*, 9355; C. E. Castro, R. S. Wade, N. O. Belser, *ibid.* **1985**, *24*, 204; R. S. Wade, C. E. Castro, *J. Am. Chem. Soc.* **1973**, *95*, 226.
[5] J. A. Zoro, J. M. Hunter, G. Eglinton, *Nature (London)* **1974**, *247*, 235.
[6] A.-E. F. Nassar, J. M. Bobbitt, J. D. Stuart, J. F. Rusling, *J. Am. Chem. Soc.* **1995**, *117*, 10986.
[7] T. S. Marks, J. D. Allpress, A. Maule, *Appl. Environ. Microbiol.* **1989**, *55*, 1258.
[8] N. Assaf-Anid, L. Nies, T. M. Vogel, *Appl. Environ. Microbiol.* **1992**, *58*, 1057.
[9] N. Assaf-Anid, K. F. Hayes, T. M. Vogel, *Environ. Sci. Technol.* **1994**, *28*, 246; C. J. Gantzer, L. P. Wackett, *ibid.* **1991**, *25*, 715.
[10] U. E. Krone, R. K. Thauer, H. P. C. Hogenkamp, K. Steinbach, *Biochemistry* **1991**, *30*, 2713; U. E. Krone, K. Laufer, R. K. Thauer, H. P. C. Hogenkamp, *ibid.* **1989**, *28*, 10061; U. E. Krone, R. K. Thauer, H. P. C. Hogenkamp, *ibid.* **1989**, *28*, 4908.
[11] A. Maldotti, R. Amadelli, C. Bartocci, V. Carassiti, E. Polo, G. Varani, *Coord. Chem. Rev.* **1993**, *125*, 143.
[12] T. A. Lewis, M. J. Morra, J. Habdas, L. Czuchajowski, P. D. Brown, *J. Environ. Qual.* **1995**, *24*, 56.
[13] M. C. Helveston, C. E. Castro, *J. Am. Chem. Soc.* **1992**, *112*, 8490.
[14] R. Kuhler, G. A. Santo, T. R. Caudill, E. A. Betterton, R. G. Arnold, *Environ. Sci. Technol.* **1993**, *27*, 2104.
[15] L. Ukrainczyk, M. Chibwe, T. J. Pinnavaia, S. A. Boyd, *Environ. Sci. Technol.* **1995**, *29*, 439.
[16] A. Cs. Benyei, S. Lehel, F. Joó, *J. Mol. Catal. A* **1997**, *116*, 349.
[17] M. A. Jafar Khan, R. J. Watts, *Water, Air, Soil Pollut.* **1996**, *88*, 247; S. W. Leung, R. J. Watts, G. C. Miller, *J. Environ. Qual.* **1992**, *21*, 377; E.g./D. L. Sedlak, A. W. Andren, *Environ. Sci. Technol.* **1991**, *25*, 777.
[18] G. Labat, J.-L. Series, B. Meunier, *Angew. Chem., Int. Ed. Engl.* **1990**, *29*, 1471.
[19] A. Sorokin, B. Meunier, *J. Chem. Soc., Chem. Commun.* **1994**, 1799; A. Sorokin, J.-L. Seris, B. Meunier, *Science* **1995**, *268*, 1163.
[20] C. S. Creaser, A. R. Fernandes, D. C. Ayres, *Chem. Ind. (London)* **1988**, 499.
[21] A. Morvillo, L. Forti, M. Bressan, *New J. Chem.* **1995**, *19*, 951.
[22] M. Bressan, L. Forti, A. Morvillo, *J. Chem. Soc., Chem. Commun.* **1994**, 253.
[23] C. Estevez, paper presented at the symposium *11th International Conference on the Origins of Life*, Orleans, France, July **1996**.
[24] F. Joó, A. Kathó, *J. Mol. Catal.* **1997**, *116*, 3.
[25] J. P. Genet, A. Linquist, E. Blart, V. Mouries, M. Savignac, M. Vaultier, *Tetrahedron Lett.* **1995**, *36*, 1443.
[26] K. Harre, V. Enkelmann, M. Schulze, U. H. F. Bunz, *Chem. Ber.* **1996**, *129*, 1323.
[27] T. Jefferty, *Tetrahedron Lett.* **1994**, *35*, 3051.
[28] J. Kiji, T. Okano, T. Kasegawa, *J. Mol. Catal. A* **1995**, *97*, 73; J. Kiji, *Macromolecular Symposia* **1996**, *105*, 167.
[29] J. F. Nguefack, V. Bolitt, D. Sinou, *Tetrahedron Lett.* **1996**, *37*, 5527.
[30] H. Dibowski, F. P. Schmidtchen, *Tetrahedron* **1995**, *51*, 2325; A. Hessler, O. Stelzer, H. Dibowski, K. Worm, F. P. Schmidtchen, *J. Org. Chem.* **1997**, *62*, 2362.

[31] A. I. Roshchin, N. A. Bumagin, I. P. Beletskaya, *Tetrahedron Lett.* **1995**, *36*, 125; R. Rai, K. B. Audrecht, D. B. Collum, *Tetrahedron Lett.* **1995**, *36*, 3111.
[32] V. V. Grushin, H. Alper, *Organometallics* **1993**, *12*, 1890; *Chem. Rev.* **1994**, *94*, 1407; *J. Am. Chem. Soc.* **1995**, *117*, 4305.
[33] F. Monteil, P. Kalck, *J. Organomet. Chem.* **1994**, *482*, 45.
[34] S. Amaratunga, H. Alper, *J. Organomet. Chem.* **1995**, *488*, 25.
[35] M. Monflier, A. Mortreux, *J. Mol. Catal.* **1994**, *88*, 295.

6.17 Oxidations

Roger A. Sheldon, Georgios Papadogianakis

6.17.1 Introduction

The main difficulty in discussing oxidation reactions within the context of aqueous-phase organometallic catalysis is to define the frame of reference. Metal-catalyzed hydrogenations, carbonylations, hydroformylations, etc., involve transition metals in low oxidation states coordinated to soft ligands, e.g., phosphines, as the catalytically active species and organometallic compounds as reactive intermediates. Performing such reactions in aqueous/organic biphasic media generally involves the use of water-soluble variants of these ligands, e.g., sulfonated triarylphosphines. In contrast, catalytic oxidations involve transition metals in high oxidation states as the active species, generally coordinated to relatively simple hard ligands, e.g., carboxylate. Reactive intermediates tend to be coordination complexes rather than organometallic species.

Water often has an inhibiting effect on catalytic oxidations, owing to strong coordination to the hard metal center hampering coordination of a less polar substrate, e.g., a hydrocarbon (cf. Section 2.1). Coordination of complex nitrogen- and/or oxygen-containing ligands can lead to the generation of more active oxidants by promoting the formation of high oxidation states. For example, in heme-dependent oxygenases and peroxidases the formation of active high-valent oxoiron complexes is favored by coordination to a macrocyclic porphyrin ligand. Hence, the frame of reference has been limited to systems in which the reaction takes place in the aqueous phase with two types of catalyst: transition-metal complexes of water-soluble ligands and transition-metal salts/complexes in micellar systems.

The vast body of literature is devoted to catalytic oxidations of carbohydrates, which by necessity are performed in water but generally involve the use of heterogeneous noble-metal catalysts or simple metal salts [1]. These oxidations, as well as the use of phase-transfer agents to solubilize inorganic oxidants, e.g. H_2O_2, NaOCl or metal catalysts in an organic phase [2] have been excluded from this section. Although such systems have considerable synthetic utility they fall outside the scope of the present discussion. One example of a catalytic oxidation which takes place in water and involves organometallic intermediates is the palladium-catalyzed oxidation of olefins to the corresponding ketones (Wacker process). This reaction is discussed in the following section (Section 6.18) and will not be dealt with here.

Just as with hydrogenations, hydroformylations, etc., a major reason for performing catalytic oxidations in water is to provide for facile recovery of the catalyst, by simple phase separation, from the product which is in an organic phase. However, many examples of catalytic oxidations in water involve water-soluble substrates and/or products. In this case catalyst recovery can be facilitated by using polymeric water-soluble ligands (see below) in conjunction with separation with an ultrafiltration membrane [3] or by other measures (cf. Section 4.2).

6.17.2 Water-soluble Ligands

Much of the research devoted to catalytic oxidations mediated by metal complexes of water-soluble ligands falls into the category of biomimetic oxidations. Hence, water-soluble porphyrins and the structurally related phthalocyanines have been widely used (see Figure 1 for examples).

For example, the cobalt(II) complex of phthalocyanine tetrasodium sulfonate (PcTs) catalyzes the autoxidation of thiols, such as 2-mercaptoethanol (Eq. 1) [4] and 2,6-di(*t*-butyl)phenol (Eq. 2) [5]. In the first example the substrate and product were water-soluble whereas the second reaction involved an aqueous

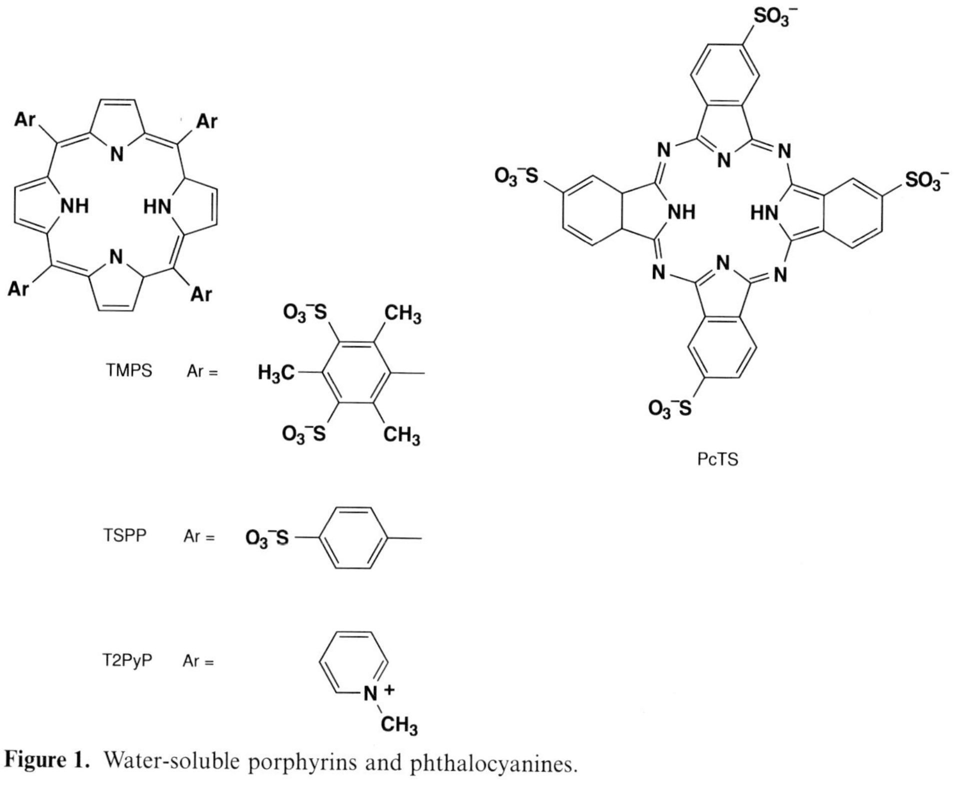

Ar NH HN N N Ar Ar Ar

TMPS Ar = O_3^-S, CH_3, H_3C, O_3^-S, CH_3

TSPP Ar = O_3^-S

T2PyP Ar = N^+ CH_3

SO_3^- O_3^-S SO_3^- N N N NH HN N N N O_3^-S

PcTS

Figure 1. Water-soluble porphyrins and phthalocyanines.

suspension. In both cases the activity of the Co(PcTs) was enhanced by binding it to an insoluble polymer, e.g., polyvinylamine [4] or a styrene–divinylbenzene copolymer substituted with quaternary ammonium ions [5]. This enhancement of activity was attributed to inhibition of aggregation of the Co(PcTs) which is known to occur in water, by the polymer network. Hence, in the polymeric form more of the Co(PcTs) will exist in an active monomeric form. In Eq. (2) the polymer-bound Co(PcTs) gave the diphenoquinone (**1**) with 100% selectivity whereas with soluble Co(PcTs) small amounts of the benzoquinone (**2**) were also formed. Both reactions involve one-electron oxidations by Co(III) followed by dimerization of the intermediate radical (RS· or ArO·).

$$2\ RSH + 1/2\ O_2 \xrightarrow{CoPcTs} RSSR + H_2O \quad (1)$$

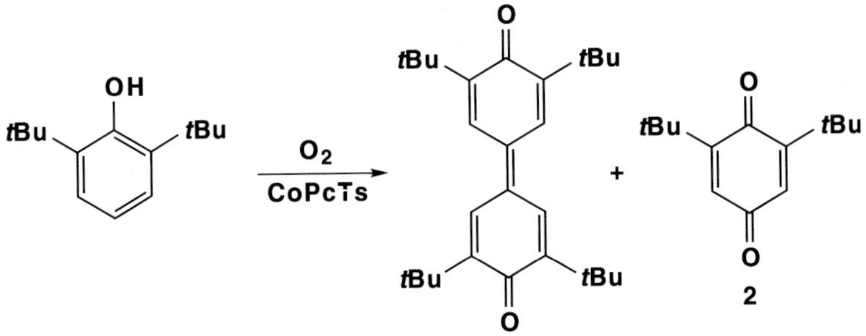

(2)

1 **2**

Much of the research on water-soluble metalloporphyrins and metallophthalocyanines stems from the interest in their use as catalysts in environmentally friendly O_2-based delignification of wood pulp in paper manufacture. Conventional processes involve the use of Cl_2 or ClO_2 as oxidants and produce effluents containing chlorinated phenols. For example, Wright and co-workers [6] oxidized lignin model compounds such as **3** with O_2 (Eq. 3) in the presence of $Na_3Fe(III)(PcTS)$, $Na_3Co(III)(TSPP)$ and $Na_3Rh(III)(TSPP)$ catalysts. The latter gave the highest rates and selectivities.

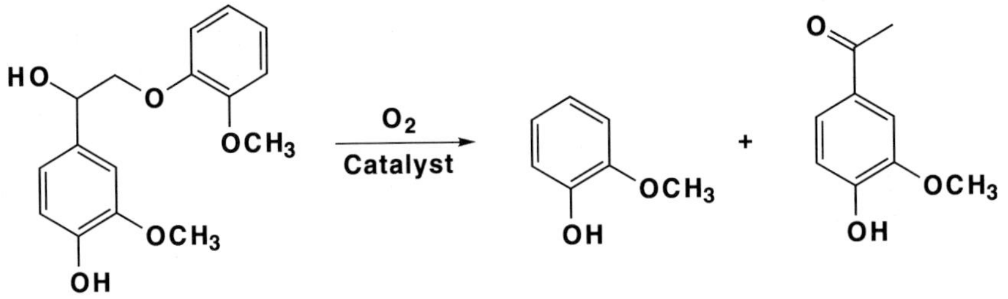

(3)

3 **4** **5**

Similarly, Hampton and Ford [7] studied the Fe(PcTS)-catalyzed autoxidation of 3,4-dimethoxybenzyl alcohol as a model for delignification. They concluded, however, that the catalyst degrades too fast to be useful for delignification. In this context it is worth mentioning that water-soluble polyoxometallates such as $PV_2Mo_{10}O_{40}^{5-}$ have also been used as catalysts for delignification with O_2 [8]. Fe(PcTS) [9] and Fe(TMPS) [10] have also been examined as catalysts for the oxidative destruction of chlorinated phenols in waste water, using H_2O_2 or $KHSO_5$ as the oxidant. For example, 2,4,6-trichlorophenol underwent facile oxidation to 2,6-dichloro-1,4-benzoquinone with Fe(PcTS)/H_2O_2 [9] or Fe(TMPS)/$KHSO_5$ [10]. Similarly, Fe(III) and Mn(III) complexes of T2MPyP catalyzed the oxidation of phenols with $KHSO_5$ [11].

Water-soluble manganese complexes of 1,4,7-trimethyl-1,4,7-triazacyclononane (**6**) and related ligands are highly effective catalysts for low-temperature bleaching of stains [12]. Polyphenolic compounds were used as appropriate models for tea stains. The same complexes were shown to catalyze the selective epoxidation of styrene and 4-vinylbenzoic acid with aqueous H_2O_2 (Eq. 4) in aqueous MeOH or water, respectively. However, large amounts of H_2O_2 (10 equiv.) were required, indicating that considerable nonproductive decomposition occurs.

Ar–CH=CH$_2$ + H_2O_2 —Catalyst→ Ar-epoxide + H_2O

Ar = C_6H_5, 4–(CO_2H)C_6H_4 98–99%

Catalyst: $[LMn^{IV}(\mu\text{-O})_3Mn^{IV}L]^{2+}$ L = **6** (1,4,7-trimethyl-1,4,7-triazacyclononane; H_3C, N, N, CH_3, N, CH_3) (4)

As noted earlier, coordination of transition-metal ions to water-soluble polymers can allow for facile catalyst recovery, by ultrafiltration, from water-soluble substrates and/or products. For example, Han and Janda [13] used an osmium complex of the water-soluble polymeric chiral ligand **7** as a catalyst for the asymmetric dihydroxylation of olefins in aqueous acetone (Eq. 5). However, they suggested that the catalyst should be recovered by precipitation with methylene chloride. Obviously the use of an ultrafiltration membrane for catalyst separation would be far more attractive.

A variant on this theme is to attach a transition-metal complex of a "smart" polymer, the solubility of which can be dramatically influenced by a change in a physical parameter, e.g., temperature [14] (cf. Sections 4.6 and 4.7). Catalyst recovery can be achieved by simply lowering or raising the temperature. For example, block copolymers of ethylene oxide and propene oxide show an inverse dependence of solubility on temperature in water [15]. Karakhanov et al. [16] prepared water-soluble polymeric ligands comprising bipyridyl (bipy) or acetylacetonate (acac) moieties covalently attached to poly(ethylene glycol)s (PEGs) or ethylene oxide/propene oxide block copolymers.

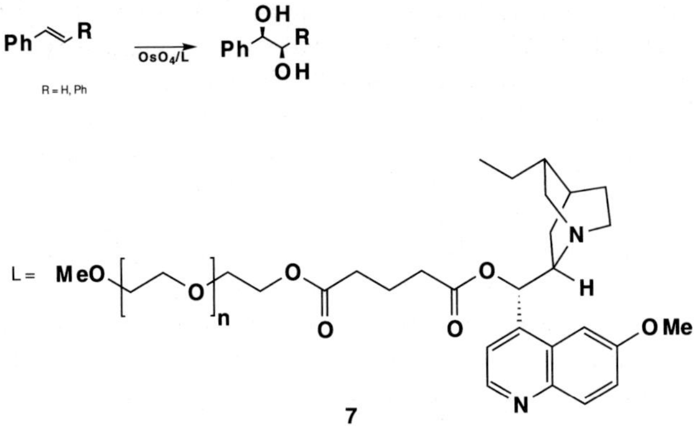

Ph⁀R (R = H, Ph) —OsO_4/L→ Ph–CH(OH)–CH(OH)–R (5)

L = MeO–[CH₂CH₂O]ₙ–CH₂CH₂O–C(O)–CH₂CH₂CH₂–C(O)–O–(quinine-derived alkaloid, OMe)

7

Iron(III) and cobalt(II) complexes of these polymeric ligands were found to be effective catalysts for the oxidation of cyclohexane and ethylbenzene with H_2O_2 or O_2 in biphasic media. The authors proposed that the oxidation takes place inside polymer micelles which can be regarded as microreactors. However, no recycle experiments were performed to ascertain the stability of these catalysts. *A priori* one would expect acetylacetonate ligands to undergo facile degradation under oxidizing conditions.

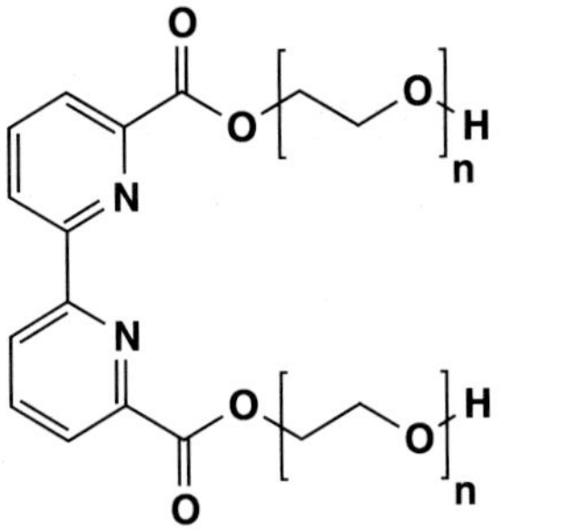

8

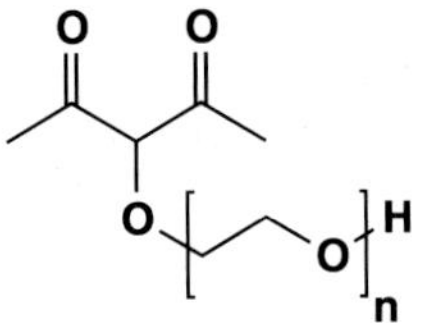

9

6.17.3 Micellar Catalysis (cf. Section 4.5)

Autoxidations of substituted toluenes to the corresponding carboxylic acids, catalyzed by cobalt in conjunction with bromide, are generally performed in acetic acid as solvent. Hronec and co-workers [17] showed that, in the presence of catalytic amounts of lipophilic quaternary ammonium or phosphonium compounds, cobalt bromide catalysts exhibited high activity in aqueous/organic biphasic media. Nonionic surfactants, produced by reaction of C_{14}–C_{18} alcohols with ethylene oxide, were also effective. The authors proposed a micellar effect to explain the results, i.e., the formation of micelles strongly influences the local concentrations of reactants and Co/Br species.

Fish and co-workers [18] studied the oxidation of cyclohexane with *t*-butyl hydroperoxide (TBHP) catalyzed by a methane monooxygenase (MMO) mimic, $[Fe_2O(H_2O)(OAc)(TPA)_2]^{3+}$ (TPA = tris(2-pyridyl)methane), in an aqueous micelle system. Catalytic amounts of cetyltrimethylammonium hydrogensulfate were used to create the micelles. Experimental observations were consistent with the reaction taking place within the aqueous micelles via intermediate *t*-butoxy radicals. Similarly, a water-soluble ruthenium complex, $[Ru(H_2O)_2(Me_2SO)_2]$-$(BF_4)_2$ in the presence of cetyltrimethylammonium hydrogensulfate was found to be a very effective catalyst for the oxidative degradation of chlorinated olefins by $KHSO_5$ [19]. TOFs were in excess of 6000 h^{-1}. The authors speculated that the anionic oxidant could be strongly bound to the cationic head groups of the surfactant while the hydrophobic substrates undergo co-micellization.

6.17.4 Concluding Remarks

So far only sporadic attention has been devoted to the use of water-soluble catalysts, in aqueous/organic biphasic media, for selective oxidations. As more emphasis is placed on this methodology as a means of facilitating catalyst recovery and recycling, we expect it to receive more attention in the future. In particular the use of water-soluble polymeric ligands with built-in surfactant properties appears to offer definite advantages.

References

[1] For a recent review, see: S. J. H. F. Arts, E. J. M. Mombarg, H. van Bekkum, R. A. Sheldon, *Synthesis* **1997**, *6*, 597.

[2] C. M. Starks, C. L. Liotta, M. Halpern, *Phase Transfer Catalysis: Fundamentals, Applications and Industrial Perspectives,* Chapman and Hall, New York, 1994, p. 500, and references cited therein.

[3] U. Kragl, C. Dreisbach, C. Wandrey, in *Aqueous Organometallic Chemistry*, Vol. 2 (Eds.: B. Cornils, W. F. Herrmann), VCH, Weinheim, **1996**, p. 832.

[4] W. M. Brouwer, P. Piet, A. L. German, *J. Mol. Catal.* **1985**, *31*, 169.

[5] H. Turk, W. T. Ford, *J. Org. Chem.* **1988**, *53*, 460.

[6] P. A. Watson, L. J. Wright, T. J. Fullerton, *J. Wood Chem. Technol.* **1993**, *13*, 371, 391, 411.

[7] K. W. Hampton, W. T. Ford, *J. Mol. Catal. A: Chem.* **1996**, *113*, 167.

[8] I. A. Weinstock, R. H. Atilla, R. S. Reiner, M. A. Moen, K. E. Hammel, C. J. Houtman, C. L. Hill, M. K. Harrup, *J. Mol. Catal. A: Chem.* **1997**, *116*, 59.

[9] A. Sorokin, B. Meunier, *J. Chem. Soc., Chem. Commun.* **1994**, 1799; A. Sorokin, J.-L. Séris, B. Meunier, *Science* **1995**, *268*, 1163.

[10] R. S. Shakla, A. Robert, B. Meunier, *J. Mol. Catal. A: Chem.* **1996**, *113*, 45.
[11] N. W. J. Kamp, J. R. Lindsay Smith, *J. Mol. Catal. A: Chem.* **1996**, *113*, 131.
[12] R. Hage, J. E. Iburg, J. Kerschner, J. H. Koek, E. L. M. Lempers, R. J. Martens, U. S. Racheria, S. W. Russell, T. Swarthoff, M. R. P. van Vliet, J. B. Warnaar, L. van der Wolf, B. Krijnen, *Nature (London)* **1994**, *369*, 637.
[13] H. Han, K. D. Janda, *J. Am. Chem. Soc.* **1996**, *118*, 7632; see also C. Bolm, A. Gerlach, *Angew. Chem., Int. Ed. Engl.* **1997**, *36*, 741.
[14] D. E. Bergbreiter, *CHEMTECH* **1987**, 686, and references cited therein; see also Z. Jin, X. Zheng, and B. Fell, *J. Mol. Catal. A: Chem.* **1997**, *116*, 55.
[15] D. E. Bergbreiter, L. Zhang, V. M. Mariagnanam, *J. Am. Chem. Soc.* **1993**, *115*, 9295.
[16] E. A. Karakhanov, Yu. S. Kardasheva, A. L. Maksimov, V. V. Predeina, E. A. Runova, A. M. Utukin, *J. Mol. Catal. A: Chem.* **1996**, *107*, 235.
[17] M. Harustiak, M. Hronec, J. Ilavsky, S. Witek, *Catal. Lett.* **1988**, *1*, 391.
[18] A. Rabion, R. M. Buchanan, J. L. Seris, R. H. Fish, *J. Mol. Catal. A: Chem.* **1997**, *116*, 43.
[19] M. Bressan, L. Forti, A. Morvillo, *J. Chem. Soc., Chem. Commun.* **1994**, 253.

6.18 Wacker-type Oxidations

Eric Monflier, André Mortreux

6.18.1 Possibilities of Wacker-type Oxidations

With oxo synthesis, Wacker-type oxidations of olefins is one of the older homogeneous transition-metal-catalyzed reactions [1]. The most prominent example of this type of reaction is the manufacture of acetaldehyde from ethylene. This well-known reaction, which has been successfully developed on an industrial scale (Wacker process), combines the stoichiometric oxidation of ethylene by palladium(II) in aqueous solution with the in-situ reoxidation of palladium(0) by molecular oxygen in the presence of a copper salt (Eqs. 1–4) [2].

$$C_2H_4 + H_2O + PdCl_2 \longrightarrow H_3C{-}CHO + Pd^0 + 2\,HCl \quad (1)$$

$$Pd^0 + 2\,CuCl_2 \longrightarrow PdCl_2 + Cu_2Cl_2 \quad (2)$$

$$Cu_2Cl_2 + 2\,HCl + \tfrac{1}{2}O_2 \longrightarrow 2\,CuCl_2 + H_2O \quad (3)$$

$$C_2H_4 + \tfrac{1}{2}O_2 \longrightarrow H_3C{-}CHO \quad (4)$$

In an industrial context, this oxidation proceeds either in one stage or in two stages [3]. The single-stage Hoechst process involves feeding an oxygen/ethylene mixture into an aqueous solution of palladium and copper. The two-stage Wacker Chemie process requires the oxygen-free stoichiometric oxidation of ethylene by a mixture of palladium and copper salts in a first stage and the reoxidation of the catalytic mixture with air in a separate reactor in a second stage. Both processes give small amounts of by-products such as 2-chloroethanol, acetic acid, oxalic acid, and chloroacetaldehyde. However, the formation of chlorinated products can be avoided by using chlorine-free oxidants such as ferric sulfate [4], heteropolyacid [5], benzoquinone [6] or two-component systems, consisting of benzoquinone and iron(II) phthalocyanine [7]. The nature of the products in the palladium-catalyzed oxidation of ethylene is also strongly dependent on the reaction conditions. For instance, 2-chloroethanol becomes the major product of ethylene oxidation at high pressure and high cupric chloride concentration [8].

When media other than water are used, different but related processes operate. Thus, the oxidation of ethylene in acetic acid can be directed to give vinyl acetate, ethylene glycol acetate, or 2-chloroethyl acetate [9]. Similarly, the synthesis of acetals or ketals can be achieved in an alcoholic medium [10]. Although the oxidation of olefins in such a medium is closely parallel to the Wacker process, the chemistry of these reactions is far beyond the scope of this section, which is limited to Wacker-type reactions in aqueous media, and will not be discussed here.

The Wacker reaction has also been applied to numerous simple olefins such α-olefins or cycloalkenes, or to functionalized olefins such nitroethylene, acrylonitrile, styrene, allyl alcohol, or maleic acid [3]. The carbonyl group is formed at the carbon atom of the double bond where the nucleophile would add in a Markovnikov addition (Eq. 5). Among these different olefins, the oxidation of propene to acetone is the only oxidation which has been developed to an industrial scale.

$$RCH=CHX + O_2 \xrightarrow{Pd\ /\ Cu} RC(=O)CH_2X \quad (5)$$

X: H, NO_2, CN, CO_2H
R: H, Alkyl

Conversion and selectivity for the oxidation of these olefins were found to be very dependent on the water-solubility of the olefin. Indeed, high-molecular-mass olefins do not react under the standard biphasic Wacker conditions due to their low solubility in water. Furthermore, the products obtained are often highly contaminated by chlorinated products and isomerized olefins [11, 12]. In order to overcome these problems, numerous studies have been undertaken. Only the most significant approaches for the oxidation of higher α-olefine (1-hexene and larger) will be developed here.

- Co-solvent or solvent is generally used to oxidize these water insoluble olefins efficiently [12]. Among the different solvents described in the literature (dimethyl sulfoxide, acetone, tetrahydrofuran, dioxane, acetonitrile), dimethylformamide appears to be the most suitable one for the oxidation. Indeed, by employing a water/dimethylformamide mixture, oxidation of 1-dodecene to 2-dodecanone can be achieved with yields greater than 80%. However, hydrolysis of dimethylformamide is also possible during the reaction, yielding dimethylamine which can lead to the formation of inactive complexes.

- Phase-transfer catalysis constitutes an alternative technique to avoid poisoning of the catalyst and for its easily recovery. Moreover, under suitable phase-transfer catalysis conditions (benzene/water system; cf. Section 4.6.1), terminal alkenes can be converted in high yields to the corresponding 2-alkanones (Eq. 6) [13]. The reaction is very sensitive to the nature of the phase-transfer agent. Although some results are contradictory in the literature [14], it seems that only quaternary ammonium salts containing at least one long-chain alkyl group, such as tetradecyltrimethylammonium bromide or dodecyltrimethyl ammonium chloride, are suitable as phase-transfer catalysts.

Although isomerization of alkenes occurs simultaneously with the oxidation, rhodium and ruthenium complexes can also be used instead of palladium for the oxidation of terminal olefin [15]. With these catalysts, symmetrical quaternary ammonium salts such as tetrabutylammonium hydrogensulfate are effective. Interestingly, the rate of palladium-catalyzed oxidation of terminals olefins can be improved by using poly(ethylene glycol) (PEG) instead of quaternary ammonium salts [16]. Thus, the rates of PEG-400-induced oxidation of 1-decene are up three times faster than those observed with cetyltrimethylammonium bromide under the same conditions. Interestingly, internal olefins can be efficiently oxidized in this polyethylene glycol/water mixture.

$$R-CH=CH_2 + O_2 \xrightarrow[R_4N^+X^-,\ C_6H_6,\ H_2O]{PdCl_2,\ CuCl_2} R-\overset{\overset{O}{\|}}{C}-CH_3 \qquad (6)$$

T: 80°C, Po_2: 1 atm

Yields: 33–73%

- Microemulsion systems have also been proposed to perform the oxidation of sparingly water-soluble olefins [17]. The microemulsion system consisted of formamide/1-hexene/2-propanol as co-surfactant and $C_9H_{19}C_6H_4$-$(OCH_2CH_2)_8OH$ as surfactant. In such a medium, the oxidation rates of 1-hexene to 2-hexanone were three times faster than those observed in the water/dimethylformamide mixture under similar conditions.

- Immobilized catalysts have also been described for the oxidation of water-insoluble olefins. Most studies have been done with polymer-anchored palladium catalysts [18]. For instance, palladium supported on a highly rigid cyanomethylated polybenzimidazole produces a highly effective catalyst for the Wacker oxidation of 1-decene, with activity higher than homogeneous systems in some cases [19]. This water-soluble polymer-anchored palladium catalyst can be recycled and used at high temperature without metallic palladium precipitation. The reoxidation agent can also be incorporated into the polymer support in the form of *p*-quinone groups [20]. In this case, reoxidation of the resulting hydroquinone groups seems to be the rate-limiting step. Recently, supported aqueous-phase catalysts have also been proposed to perform the olefin oxidation [21]. In this approach, palladium and copper salts are dissolved in an aqueous film supported on a high-surface-area silica gel (see Section 4.7). With this supported catalyst, the ketone yields are rather low (< 25%) and significant isomerization of the olefin occurs. The activity and the selectivity of the catalyst are sensitive to the water content of the catalyst and to the oxygen pressure. The high yields in isomeric olefins have been attributed to the slow reoxidation of palladium(0) to palladium(II) due to low oxygen diffusion in this heterogeneous material.

- Inverse-phase Transfer Catalysis (cf. Section 4.6.2) has been successfully applied to perform the quantitative and selective oxidation of 1-decene in an aqueous two-phase system [22]. The success of this oxidation is mainly due to the use of β-cyclodextrin – a cyclic oligosaccharide composed of seven glucopyranose units – functionalized wih hydrophilic or lipophilic groups. The best results have been obtained with a multicomponent catalytic system composed of $PdSO_4$, $H_9PV_6Mo_6O_{40}$, $CuSO_4$, and per(2,6-di-*O*-methyl)-*β*-cyclodextrin (Eq. 7) [23].

$$CH_3(CH_2)_7-CH{=}CH_2 \;+\; 1/2\,O_2 \xrightarrow{PdSO_4\,/\,H_9PV_6Mo_6O_{40}\,/\,CuSO_4} CH_3(CH_2)_7-\overset{O}{\overset{\|}{C}}{=}CH_2 \qquad (7)$$

(below arrow: per(2,6-di-O-methyl)-β-cyclodextrin, repeat unit with OCH_3, O, HO, OCH_3, subscript 7)

Yields: 99%

In this reaction, the chemically modified *β*-cyclodextrins behave mainly as inverse-phase transfer catalysts. Actually, owing to the formation of inclusion complexes, the chemically modified cyclodextrins transfer the higher olefins into the aqueous phase, and thus improve the mass transfer between aqueous and organic layers. It has been found that the mass-transfer efficiency is strongly dependent on the nature of the substituent group and on the degree of substitution the dyclodextrin (Figure 1). These results have been mainly correlated with the solubility of cyclodextrins in water and with the weak stability of the host–guest complexes which would facilitate the dissociation–association reactions between the organic compounds and cyclodextrin.

This multicomponent catalytic system also made it possible to oxidize in high yields a wide range of straight-chain higher *α*-olefins [24]. However, the oxidation rate and the ketone selectivity were strikingly dependent on the nature of

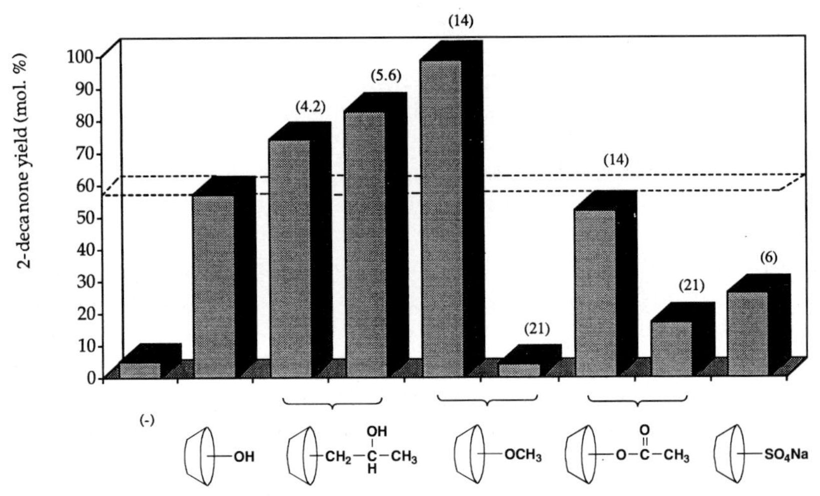

Figure 1. Effect of the nature of the substituent group and of the substitution degree of the cyclodextrin on 2-ketone yields. The value in bracket indicates the number of functionalized hydroxy group. $PdSO_4$ (0.86 mmol), $CuSO_4$ (10 mmol), $H_9PV_6Mo_6O_{40}$ (10 mmol), DMCD (1 mmol), water (30 mL), 1-alkene (40 mmol), 6 hours, T: 80 °C, PO_2: 1 bar.

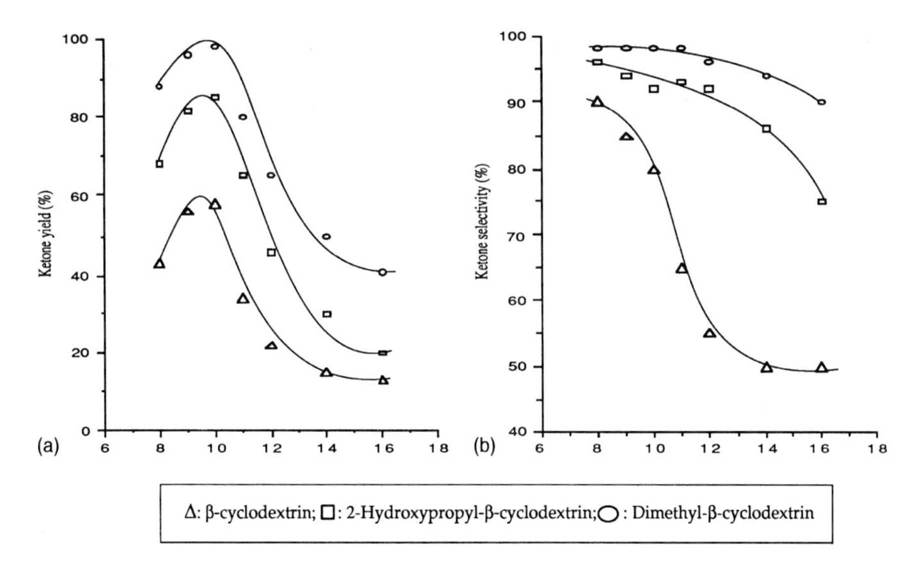

Figure 2. Effect of chain length of olefin on 2-ketone yield and selectivity after 6 hours of reaction in the presence of various cyclodextrins. $PdSO_4$ (0.86 mmol), $CuSO_4$ (10 mmol), $H_9PV_6Mo_6O_{40}$ (10 mmol), DMCD (1 mmol), water (30 mL), 1-alkene (40 mmol), 6 hours, T: 80 °C, PO_2: 1 bar.

the cyclodextrin and on the chain length of the olefin (Figure 2). The olefin optimal size and shape was reached with 1-decene. Attempts to oxidize internal olefins in biphasic medium wih cyclodextrins modified in this way failed due to the lack of accessibility of the double bond in the inclusion complex.

6.18.2 Conclusions

Although the Wacker-type oxidation of olefins has been applied for more than four decades using ethylene as substrate, processes involving higher olefins are still the subject of investigations because of their poor solubility in water. This problem can be overcome using different approaches. Particularly interesting in this context is the inverse-phase transfer system using host molecules such as cyclodextrins which, upon careful choice of the substituent, avoid the isomerization into internal olefins. This recent concept proves so far that new ideas may lead to major improvements in a well-known and intensively studied system. For industrial applications, it is probable that the activities obtained so far with higher olefins should be increased by at least one order of magnitude while keeping the same selectivity: this may be the challenge of this reaction in the near future.

References

[1] G. W. Parshall, S. D. Ittel, in *Homogeneous Catalysis. The Applications and Chemistry of Catalysis by Soluble Transition Metal Complexes*, 2nd ed., Wiley–Interscience, New York, **1992**, p. 138.

[2] (a) Consortium für Elektrochemische Industrie (J. Smidt, W. Hafner, R. Jira, J. Sedlmeier, R. Rüttinger) DE 1.049.845 (1959); (b) R. Jira, W. Blau, D. Grimm, *Hydrocarbon Proc.* **1976**, 97.

[3] R. Jira, in *Applied Homogeneous Catalysis with Organometallic Compounds* (Eds.: B. Cornils, W. A. Herrmann), VCH, Weinheim, **1996**, p. 374.

[4] Maruzen Oil Co. (H. Hasegawa, M. Triuchijima), GB 1.240.889 (1971).

[5] (a) K. I. Matveev, *Kinet. Catal. (Engl. Transl.)* **1977**, *18,* 716; (b) Catalytica Associates (J. Vasilevskis, J. C. De Decken, R. J. Saxton, J. D. Fellmann, L. S. Kipnis), US 4.723.041 (1988).

[6] W. A. Clement, C. M. Selwitz, *J. Org. Chem.* **1964**, *29,* 241.

[7] (a) J. E. Bäckvall, R. B. Hopkins, *Tetrahedron Lett.* **1988**, *29,* 2885; (b) J. E. Bäckvall, R. B. Hopkins, H. Grennberg, M. M. Mader, A. K. Awasthi, *J. Am. Chem. Soc.* **1990**, *112*, 5160; (c) J. E. Bäckvall, A. K. Awasthi, Z. D. Renko, *J. Am. Chem. Soc.* **1987**, *109*, 4750.

[8] H. Stangl, R. Jira, *Tetrahedron Lett.* **1970**, *471*, 3589.

[9] (a) P. M. Henry, *J. Org. Chem.* **1967**, *32*, 2575; (b) P. M. Henry, *J. Org. Chem.* **1974**, *39*, 3871.

[10] I. I. Moiseev, M. N. Vargaftik, *Dokl. Akad. Nauk. SSSR* **1966**, *166*, 370; (b) W. G. Lloyd, B. J. Luberoff, *J. Org. Chem.* **1969**, *34*, 3949.

[11] P. M. Henry, *J. Am. Chem. Soc.* **1966**, *88*, 1595.

[12] (a) W. H. Clement, C. M. Selwitz, *J. Org. Chem.* **1964**, *29*, 241; (b) J. Tsuji, H. Nagashima, H. Nemoto, *Org. Synth.* **1984**, *62*, 9; (c) D. R. Fahey, E. A. Zuech, *J. Org. Chem.* **1974**, *39*, 3276.

[13] (a) K. Januskiewicz, H. Alper, *Tetrahedron Lett.* **1983**, *47*, 5159; (b) Phillips Petroleum Co. (T. P. Murtha, T. K. Shioyama), US 4.434.082 (1984).

[14] (a) C. Lapinte, H. Riviere, *Tetrahedron Lett.* **1977**, *43*, 3817; (b) Phillips Petroleum Co. (P. R. Stapp), US 4.237.071 (1980).

[15] K. Januszkiewicz, H. Alper, *Tetrahedron Lett.* **1983**, *24*, 5163.

[16] H. Alper, K. Januszkiewicz, D. J. H. Smith, *Tetrahedron Lett.* **1985**, *26*, 2263.

[17] I. Rico, F. Couderc, E. Perez, J. P. Laval, A. Lattes, *J. Chem. Soc., Chem. Commun.* **1987**, 1205.

[18] F. R. Hartley, in *Supported Metal Complexes, Catalysis by Metal Complexes* (Eds.: R. Ugo, B. R. James), D. Reidel Publishing Company, **1985**, p. 293.

[19] (a) D. C. Sherrington, H. G. Tang, *J. Catal.* **1993**, *142*, 540; (b) D. C. Sherrington, H. G. Tang, *J. Mol. Catal.* **1994**, *94*, 7.

[20] H. Arai, M. Yashiro, *J. Mol. Catal.* **1978**, *3*, 427.

[21] J. P. Arhancet, M. E. Davis, B. E. Hanson, *Catal. Lett.* **1991**, *11*, 129.

[22] E. Monflier, E. Blouet, Y. Barbaux, A. Mortreux, *Angew. Chem., Int. Ed. Engl.* **1994**, *33*, 2100.

[23] E. Monflier, S. Tilloy, E. Blouet, Y. Barbaux, A. Mortreux, *J. Mol. Catal. A: Chem.* **1996**, *109*, 27.

[24] E. Monflier, S. Tilloy, G. Fremy, Y. Barbaux, A. Mortreux, *Tetrahedron Lett.* **1995**, *36*, 387.

6.19 Lanthanides in Aqueous-phase Catalysis

Shū Kobayashi

6.19.1 Introduction

Lewis acid-catalyzed reactions have been of great interest in organic synthesis because of their unique reactivities and selectivities, and for the mild conditions used [1]. Although various kinds of Lewis acid-promoted reactions have been developed and many have been applied in industry, these reactions must be carried out under strictly anhydrous conditions despite the general recognition of the utility of aqueous reactions [2]. The presence of even a small amount of water stops the reaction, because most Lewis acids immediately react with water rather than with the substrates and decompose or deactivate, and this fact has restricted the use of Lewis acids in organic synthesis.

Yet lanthanide trifluoromethanesulfonates (triflates), including scandium and yttrium triflates, were recently found to be stable Lewis acids in water, and many useful aqueous reactions using lanthanide triflates as catalysts have been reported. Lanthanides have a larger radius and specific coordination number than typical transition metals. They have been expected to act as strong Lewis acids because of their hard character and to have strong affinity toward carbonyl oxygens [3]. Among these compounds, lanthanide trifluoromethanesulfonates (lanthanide triflates) were expected to be one of the strongest Lewis acids because of the strongly electron-withdrawing trifluoromethanesulfonyl group, but their hydrolysis was postulated to be slow on the basis of their hydration energies and hydrolysis constants [4]. In fact, unlike most metal triflates, which are prepared under strictly anhydrous conditions, lanthanide triflates have been reported to be prepared in aqueous solution [5–7]. Many synthetic reactions using these triflates as catalysts have been developed [8].

6.19.2 Aldol Reactions

Formaldehyde is a versatile reagent as one of the most highly reactive C_1 electrophiles in organic synthesis [9, 10] and *dry* gaseous formaldehyde has been required for many reactions. For example, the titanium tetrachloride ($TiCl_4$)-

promoted hydroxymethylation reaction of a silyl enol ether was carried out by using trioxane as an HCHO source under strictly anhydrous conditions [11, 12]. Formaldehyde/water solution could not be used because $TiCl_4$ and the silyl enol ether reacted with water rather than with HCHO in that aqueous solution.

It was found that the hydroxymethylation reaction of silyl enol ethers with commercial formaldehyde solution proceeded smoothly when lanthanide triflates were used as Lewis acid catalysts (Eq. 1) [7, 13]. The reactions were first carried out in commercial formaldehyde solution THF media.

$$R^1CHO + R^2C(OSiMe_3)=CHR^3 \xrightarrow[\text{67–93\%}]{\text{cat. Ln(OTf)}_3,\ H_2O\text{–THF, rt}} R^2C(=O)CH(R^3)CH(OH)R^1 \quad (1)$$

R^1 = H, Ph, $Ph(CH_2)_2$, CH_3, CH_2=CH, $ClCH_2$, PhCO, 2-hydroxyphenyl, 2-pyridyl, etc.

Ln = Sc, Y, La, Ce, Pr, Nd, Sm, Eu, Gd, Tb, Dy, Ho, Er, Tm, Yb, Lu

Lanthanide triflates are effective for the activation of aldehydes other than formaldehyde [13–15]. The aldol reaction of silyl enol ethers with aldehydes proceeds smoothly to afford the aldol adducts in high yields in the presence of a catalytic amount of scandium triflate ($Sc(OTf)_3$), ytterbium triflate ($Yb(OTf)_3$), gadolinium triflate ($Gd(OTf)_3$), lutetium triflate ($Lu(OTf)_3$), etc., in aqueous media (water/THF). Diastereoselectivities are generally good to moderate. One feature in the present reaction is that water-soluble aldehydes, for instance, acetaldehyde, acrolein, and chloroacetaldehyde, can be reacted with silyl enol ethers to afford the corresponding cross-aldol adducts in high yields. Some of these aldehydes are commercially supplied as aqueous solutions and are appropriate for direct use. Phenylglyoxal monohydrate also works well. It is known that water often interferes with the aldol reactions of metal enolates with aldehydes and that in the cases where such water-soluble aldehydes are employed, some troublesome purifications including dehydration are necessary.

6.19.2.1 Recovery and Re-use of the Catalysts

Lanthanide triflates are more soluble in water than in organic solvents such as dichloromethane. Very interestingly, almost 100% of the lanthanide triflate is quite easily recovered from the aqueous layer after the reaction is completed and it can be re-used. The reactions are usually quenched with water and the products are extracted with an organic solvent (for example, dichloromethane). Lanthanide triflates are in the aqueous layer and only removal of water gives the catalyst which can be used in the next reaction (Scheme 1). It is noteworthy that lanthanide triflates are expected to solve some severe environmental problems induced by mineral acid- or Lewis acid-promoted reactions in the industrial chemistry [16].

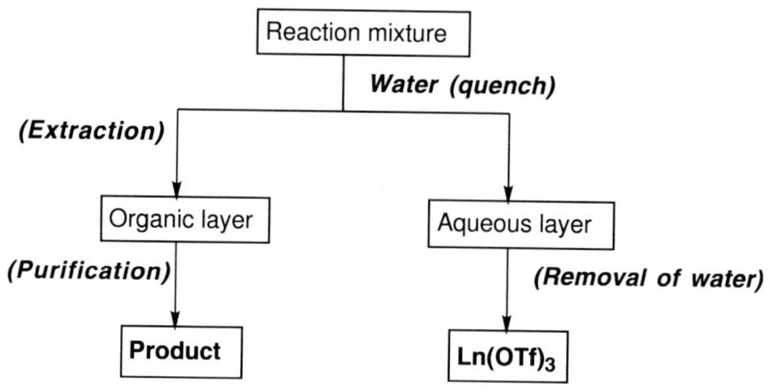

Scheme 1. Recovery of the catalyst.

The aldol reactions of silyl enol ethers with aldehydes also proceed smoothly in water/ethanol/toluene [17]. The reactions proceed much faster in this solvent than in water/THF (Eq. 2). Furthermore, the new solvent system involves continuous use of the catalyst by a very simple procedure. Although the water/ethanol/toluene (1:7:4) system is one phase, it easily becomes two phases by adding toluene after the reaction is completed. The product is isolated from the organic layer by a usual work-up. On the other hand, the catalyst remains in the aldueous layer, which is used directly in the next reaction without removing water. It is noteworthy that the yields of the second, third, and fourth runs are comparable with that of the first run.

CHO, N + $OSiMe_3$, Ph $\xrightarrow[\text{solvent, rt, 4h}]{Yb(OTf)_3\text{ (10 mol\%)}}$ OH, O, N, Ph (2)

6.19.3 Mannich-type Reactions

The Mannich and related reactions provide one of the most fundamental and useful methods for the synthesis of β-amino ketones. Although the classical protocols include some severe side reactions, new modifications using preformed iminium salts and imines have been developed [18]. These materials however, are often hygroscopic and are not stable at high temperatures. The direct preparation of β-amino ketones from aldehydes is desirable from a synthetic point of view.

Mannich-type reactions between aldehydes, amines, and vinyl ethers proceed smoothly using $Ln(OTf)_3$ in aqueous media [19] (Eq. 3). Commercially available formaldehyde and chloroacetaldehyde aqueous solutions are used directly and the corresponding *β*-amino ketones are obtained in good yields. Phenylglyoxal monohydrate, methyl glycoxylate, an aliphatic aldehyde, and an *α,β*-unsaturated ketone also work well to give the corresponding *β*-amino esters in high yields. Other lanthanide triflates can also be used; in the reaction of phenylglyoxal monohydrate, *p*-chloroaniline, and 2-methoxypropene, 90% ($Sm(OTf)_3$), 94% ($Tm(OTf)_3$), and 91% ($Sc(OTf)_3$) yields are obtained. In some Mannich reactions with preformed iminium salts and imines, it is known that yields are often low because of the instability of the imines derived from these aldehydes or because troublesome treatments are known to be required for their use [20]. The present method provides a useful route for the synthesis of *β*-amino ketones.

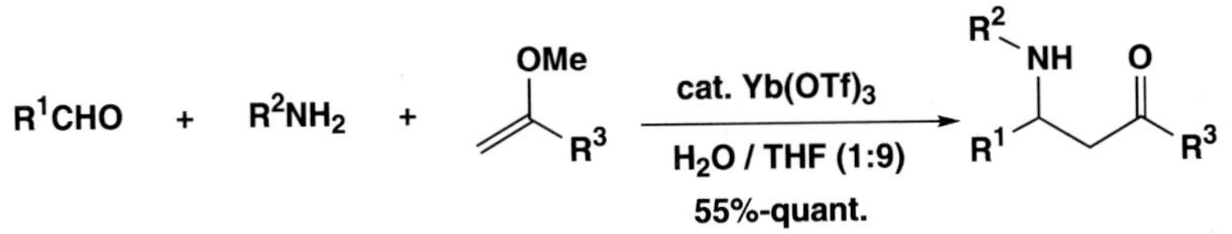
R1CHO + R2NH2 +
OMe
R3
cat. Yb(OTf)3
H2O / THF (1:9)
55%-quant.
R2
NH O
R1
R3

(3)

A possible mechanism for the present reaction is shown in Scheme 2. It should be noted that dehydration accompanied by imine formation and successive addition of a vinyl ether proceed smoothly in aqueous solution. Use of lanthanide triflate, a water-tolerant Lewis acid, is key and essential in this reaction.

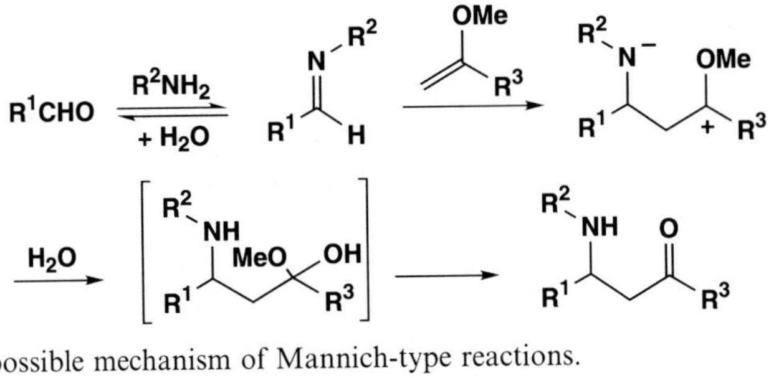
R1CHO
R2NH2
+ H2O
N R2
R1 H
OMe
R3
R2
N
OMe
R1
+ R3
H2O
R2
NH
MeO OH
R1 R3
R2
NH O
R1 R3

Scheme 2. A possible mechanism of Mannich-type reactions.

6.19.4 Diels–Alder Reactions

Although many Diels–Alder reactions have been carried out at higher reaction temperatures without catalysts, heat-sensitive compounds in complex and multistep syntheses cannot be employed. While Lewis acid catalysts allow the reactions to proceed at room temperature or below with satisfactory yields in organic solvents, they are often accompanied by diene polymerization and excess amounts of the catalyst are often needed to catalyze carbonyl-containing dienophiles [21, 22].

It was found that the Diels–Alder reaction of naphthoquinone with cyclopentadiene proceeded in the presence of a catalytic amount of a lanthanide triflate in H_2O/THF at room temperature to give the corresponding adduct in a 93% yield (Eq. 4) [23].

O O + Sc(OTf)$_3$ (10 mol%) / H_2O / THF (1:9) → O O (4)

93% yield, *endo/exo* = 100/0

6.19.5 Micellar Systems

6.19.5.1 Aldol Reactions

Quite recently, it has been found that scandium triflate (Sc(OTf)$_3$-catalyzed aldol reactions of silyl enol ethers with aldehydes can be successfully carried out in micellar systems [24]. While the reactions proceeded sluggishly in pure water (without organic solvents), remarkable enhancement of the reactivity was observed in the presence of a small amount of a surfactant (cf. Section 4.5).

Lewis acid catalysis in micellar systems was first found in the model reaction of the silyl enol ether of propiophenone with benzaldehyde. Although the reaction proceeded sluggishly in the presence of 0.2 equiv. Yb(OTf)$_3$ in water, remarkable enhancement of the reactivity was observed when the reaction was carried out in the presence of 0.2 equiv. Yb(OTf)$_3$ in an aqueous solution of sodium dodecylsulfate (SDS, 0.2 equiv., 35 mM), and the corresponding aldol adduct was obtained in a 50% yield. In the absence of the Lewis acid and in surfactant (water-promoted conditions) [11], only 20% yield of the aldol adduct was isolated after 48 h, while a 33% yield of the aldol adduct was obtained after 48 h in the absence of the Lewis acid in an aqueous solution of SDS. The amounts of the surfactant also influenced the reactivity, and the yield was improved when Sc(OTf)$_3$ was used as a Lewis acid catalyst. Judging from the critical micelle concentration, micelles would be formed in these reactions, and it is noteworthy that the Lewis acid-catalyzed reactions proceeded smoothly in micellar systems [25].

Several examples of Sc(OTf)$_3$-catalyzed aldol reactions in micellar systems are shown in Table 1. Not only aromatic, but also aliphatic and α,β-unsaturated aldehydes, react with silyl enol ethers to afford the corresponding aldol adducts

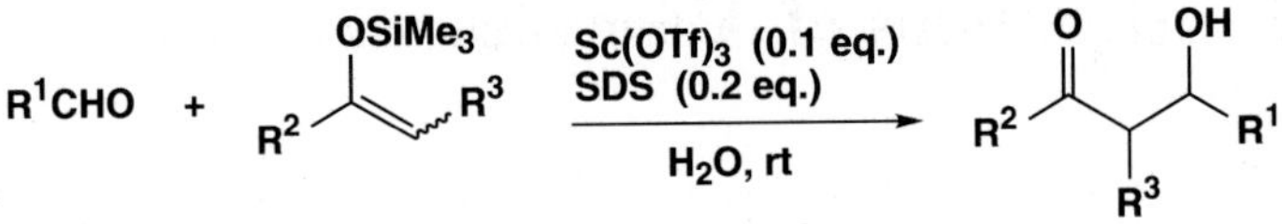

R^1CHO + $R^2C(OSiMe_3)=CHR^3$ $\xrightarrow[\text{H}_2\text{O, rt}]{\text{Sc(OTf)}_3 \text{ (0.1 eq.)}, \text{ SDS (0.2 eq.)}}$ $R^2C(=O)CH(R^3)CH(OH)R^1$

Table 1. $Sc(OTf)_3$-catalyzed aldol reactions in micellar systems.

Aldehyde	Silyl enol ether	Yield [%]
PhCHO	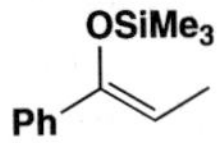 OSiMe$_3$, Ph **1**	88[a)]
 Ph(CH$_2$)$_2$CHO	**1**	86[b)]
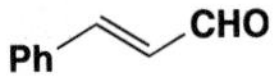 PhCH=CHCHO	**1**	88[c)]
HCHO	**1**	82[d)]
PhCHO	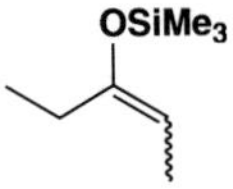 OSiMe$_3$	88[e)]
PhCHO	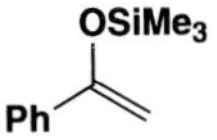 OSiMe$_3$, Ph	75[f,g)]
PhCHO	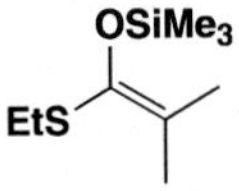 OSiMe$_3$, EtS	94
PhCHO	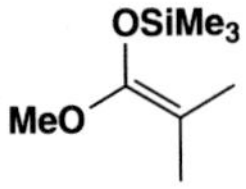 OSiMe$_3$, MeO **2**	84[g)]

a) *Syn/anti* = 50/50.
b) *Syn/anti* = 45/55.
c) *Syn/anti* = 41/59.
d) Commercially available HCHO aq. (3 ml), **1** (0.5 mmol), $Sc(OTf)_3$ (0.1 mmol), and SDS (0.1 mmol) were combined.
e) *Syn/anti* = 57/43.
f) $Sc(OTf)_3$ (0.2 equiv.) was used.
g) Additional silyl enolate (1.5 equiv.) was charged after 6 h.

in high yields. Formaldehyde/water solution also works well. It is exciting that ketene silyl acetal **2**, which is known to hydrolyze very easily even in the presence of a small amount of water, reacts with an aldehyde in the present micellar system to afford the corresponding aldol adduct in high yields.

6.19.5.2 Allylations of Aldehydes

It was also found that the allylation reactions of aldehydes (cf. Section 6.6) with tetra-allyltin proceeded smoothly in micellar systems using $Sc(OTf)_3$ as a catalyst [26]. Utilities of organometallic reagents are now well recognized in organic synthesis, and a variety of organometallics have been developed to achieve unique reactivities as well as selectivities [27]. In general, however, most organometallic reagents are hygroscopic and therefore they are deactivated or decomposed in the presence of even a small amount of water, which sometimes limits their use in organic synthesis. On the other hand, the allylation reaction of 2-deoxyribose (an unprotected sugar) was found to proceed smoothly in water under the influence of 0.1 equiv. of $Sc(OTf)_3$ and 0.2 equiv. of SDS (sodium dodecylsulfate) by using tetra-allyltin (0.5 equiv.) as an allylating reagent (Eq. 5) [28]. The reaction proceeded sluggishly without the Lewis acid (surfactant only) or without the surfactant (Lewis acid only). It is noted that the sugar reacted smoothly without protecting any hydroxyl groups in this system, and that the reaction proceeded smoothly in water without using any organic solvents. Several examples of the allylation reactions were examined, and in all cases they were successfully carried out in micellar systems to afford the corresponding homoallylic alcohols in high yields. With reagard to the stoichiometry of an aldehyde and tetra-allyltin, it was found that 0.25 equiv. of tetraallyltin was enough to achieve high yields of the homoallylic alcohols. Either SDS or TritonX-100 could be used as a surfactant. Not only aromatic but also aliphatic and α,β-unsaturated aldehydes reacted with tetra-allyltin to afford the corresponding allylated adducts in high yields. Under the present reaction conditions, salicylaldehyde and 2-pyridinecarboxaldehyde reacted with tetra-allyltin to afford the homoallylic alcohols in good yields. Unprotected sugars other than 2-deoxyribose also reacted directly to give the adducts, which are intermediates for the synthesis of higher sugars [29], in high yields.

HO, O, OH, OH — 1.0 eq. + (allyl)$_4$Sn — 0.5 eq. → 1. cat. $Ln(OTf)_3$, surfactant, H_2O, rt, 48 h; 2. Ac_2O, pyridine → AcO, OAc, OAc, OAc (5)

6.19.5.3 Three-component Reactions of Aldehydes, Amines, and Allyltributyltin

The reaction of imines with allyltributyltin provides a useful route for the synthesis of homoallylic amines [30]. The reaction is generally carried out in the presence of a Lewis acid in organic solvents under strictly anhydrous conditions

[31], because most imines, Lewis acids, and the organotin reagents used are hygroscopic and decompose easily in the presence of even a small amount of water [32]. It was found that three-component reactions of aldehydes, amines, and allyltributyltin proceed smoothly in micellar systems using $Sc(OTf)_3$ as a Lewis acid catalyst (Eq. 6) [33]. The reaction of benzaldehyde, aniline, and allyltributyltin was chosen as a model, and several reaction conditions were examined. While the reaction proceeded sluggishly in the presence of $Sc(OTf)_3$ without SDS or in the presence of SDS without $Sc(OTf)_3$, a 77% yield of the desired homoallylic amine was obtained in the coexistence of $Sc(OTf)_3$ and SDS. A satisfactory yield was obtained when 0.2 equiv. $Sc(OTf)_3$ and 35 mM (0.2 equiv.) SDS were used and the reaction was carried out at room temperature (rt) for 20 h [34]. No homoallylic alcohol (an adduct between an aldehyde and allyltributyltin) was produced under these conditions.

$$R^1CHO + R^2NH_2 + \text{CH}_2{=}\text{CHCH}_2\text{SnBu}_3 \xrightarrow[\substack{H_2O,\ rt,\ 20\ h \\ 66\text{–}90\%}]{\substack{Sc(OTf)_3\ (0.2\ eq.) \\ SDS\ (0.2\ eq.)}} R^1CH(NHR^2)CH_2CH{=}CH_2 \quad (6)$$

The present three-component reactions of aldehydes, amines, and allyltributyltin proceeded smoothly in water without using any organic solvents in the presence of a small amount of $Sc(OTf)_3$ and SDS, to afford the corresponding homoallylic amines in high yields. Not only aromatic aldehydes but also aliphatic, unsaturated, and heterocyclic aldehydes worked well. It is known that severe side reactions occur to decrease yields in the reactions of imines having α-protons with allyltributyltin [30]. It should be noted that aliphatic aldehydes, especially unbranched aliphatic aldehydes, reacted smoothly under these conditions to afford the homoallylic amines in high yields. In all cases, no aldehyde adducts (homoallylic alcohols) were obtained. It was suggested that imine formation from aldehydes and amines was very fast in the presence of $Sc(OTf)_3$ and SDS [19], and that the selective activation of imines rather than aldehydes was achieved using $Sc(OTf)_3$ as a catalyst [35].

6.19.6 Conclusions

Lanthanide triflates are stable Lewis acids in water and are successfully used in several carbon–carbon bond-forming reactions in aqueous solutions. The reactions proceed smoothly in the presence of a catalytic amount of the triflate under mild conditions. Moreover, the catalysts can be recovered after the reactions are completed and can be re-used. Lewis acid catalysis in micellar systems will lead to clean and environmentally friendly processes, and it will become a more important topic in the future.

References

[1] (a) *Selectivities in Lewis Acid Promoted Reactions* (Ed.: D. Schinzer), Kluwer, Dordrecht, **1989**; (b) M. Santelli, J.-M. Pons, *Lewis Acids and Selectivity in Organic Synthesis,* CRC, Boca Raton, **1995**.

[2] (a) C.-J. Li, *Chem. Rev.* **1993**, *93*, 2023; (b) A. Lubineau, J. Ange, Y. Queneau, *Synthesis* **1994**, 741.

[3] G. A. Molander, *Chem. Rev.* **1992**, *92*, 29.

[4] C. F. Baes, Jr., R. E. Mesmer, *The Hydrolysis of Cations*, John Wiley, New York, **1976**, p. 129.

[5] K. F. Thom, US 3.615.169 (1971); *Chem. Abstr.* **1972**, *76*, 5436a.

[6] (a) J. H. Forsberg, V. T. Spaziano, T. M. Balasubramanian, G. K. Liu, S. A. Kinsley, C. A. Duckworth, J. J. Poteruca, P. S. Brown, J. L. Miller, *J. Org. Chem.* **1987**, *52*, 1017. See also (b) S. Collins, Y. Hong, *Tetrahedron Lett.* **1987**, *28*, 4391; (c) M.-C. Almasio, F. Arnaud-Neu, M.-J. Schwing-Weill, *Helv. Chim. Acta* **1983**, *66*, 1296; Cf.: (d) J. M. Harrowfiedld, D. L. Kepert, J. M. Patrick, A. H. White, *Aust. J. Chem.* **1983**, *36*, 483.

[7] S. Kobayashi, *Chem. Lett.* **1991**, 2187.

[8] S. Kobayashi, *Synlett* **1994**, 689.

[9] (a) B. B. Snider, D. J. Rodini, T. C. Kirk, R. Cordova, *J. Am. Chem. Soc.* **1982**, *104*, 555; (b) B. B. Snider, in *Selectivities in Lewis Acid Promoted Reactions* (Ed.: D. Schinzer), Kluwer, Dordrecht, **1989**, p. 147; (c) K. Maruoka, A. B. Concepcion, N. Hirayama, H. Yamamoto, *J. Am. Chem. Soc.* **1990**, *112*, 7422; (d) K. Maruoka, A. B. Concepcion, N. Murase, M. Oishi, H. Yamamoto, *ibid.* **1993**, *115*, 3943.

[10] TMSOTf-mediated aldol-type reaction of silyl enol ethers with dialkoxymethanes was also reported: S. Murata, M. Suzuki, R. Noyori, *Tetrahedron Lett.* **1980**, *21*, 2527.

[11] Lubineau reported the water-promoted aldol reactions of silyl enol ethers with aldehydes, but the yields and the substrate scope were not yet satisfactory: (a) A. Lubineau, *J. Org. Chem.* **1986**, *51*, 2142; (b) A. Lubineau, E. Meyer, *Tetrahedron* **1988**, *44*, 6065.

[12] (a) T. Mukaiyama, K. Narasaka, K. Banno, *Chem. Lett.* **1973**, 1011; (b) T. Mukaiyama, K. Banno, K. Narasaka, *J. Am. Chem. Soc.* **1974**, *96*, 7503.

[13] S. Kobayashi, I. Hachiya, *J. Org. Chem.* **1994**, *59*, 3590.

[14] S. Kobayashi, I. Hachiya, *Tetrahedron Lett.* **1992**, 1625.

[15] S. Kobayashi, I. Hachiya, H. Ishitani, M. Araki, *Synlett* **1993**, 472.

[16] J. Haggin, *Chem. Eng. News* **1994**, *Apr. 18*, 22.

[17] S. Kobayashi, I. Hachiya, Y. Yamanoi, *Bull. Chem. Soc. Jpn.* **1994**, *67*, 2342.

[18] E. F. Kleinman, *Comprehensive Organic Synthesis* (Ed.: B. M. Trost), Pergamon Press, Oxford, **1991**, Vol 2, p. 893.

[19] S. Kobayashi, H. Ishitani, *J. Chem. Soc., Chem. Commun.* **1995**, 1379.

[20] Grieco et al. reported *in situ* generation and trapping of immonium salts under Mannich-like conditions: S. D. Larsen, P. A. Grieco, *J. Am. Chem. Soc.* **1985**, *107*, 1768; P. A. Grieco, D. T. Parker, *J. Org. Chem.* **1988**, *53*, 3325, and references cited therein.

[21] Review: W. Carruthers, *Cycloaddition Reactions in Organic Synthesis,* Pergamon Press, Oxford, **1990**.

[22] (a) D. Yates, P. E. Eaton, *J. Am. Chem. Soc.* **1960**, *82*, 4436; (b) T. K. Hollis, N. P. Robinson, B. Bosnich, *J. Am. Chem. Soc.* **1992**, *114*, 5464.

[23] S. Kobayashi, I. Hachiya, M. Araki, H. Ishitani, *Tetrahedron Lett.* **1993**, *34*, 3755.

[24] S. Kobayashi, T. Wakabayashi, S. Nagayama, H. Oyamada, *Tetrahedron Lett.* **1997**, *38*, 4559.

[25] (a) J. H. Fendler, E. J. Fendler, *Catalysis in Micellar and Macromolecular Systems,* Academic Press, London, **1975**; (b) *Mixed Surfactant Systems* (Eds.: P. M. Holland, D. N.

Rubingh), American Chemical Society, Washington, DC, **1992**; (c) C. J. Cramer, D. G. Truhlar (Eds.), *Structure and Reactivity in Aqueous Solution*, American Chemical Society, Washington, DC, **1994**; (d) *Surfactant-enhanced Subsurface Remediation* (Eds.: D. A. Sabatini, R. C. Knox, J. H. Harwell), American Chemical Society, Washington, DC, **1994**.
[26] S. Kobayashi, T. Wakabayashi, H. Oyamada, *Chem. Lett.* **1997**, 831.
[27] For example, *Organometallics in Synthesis* (Ed.: M. Schlosser), John Wiley, Chichester, **1994**.
[28] As for allylation using tetra-allyltin: (a) W. G. Peet, W. Tam, *J. Chem. Soc., Chem. Commun.* **1983**, 853; (b) G. Daude, M. Pereyre, *J. Organomet. Chem.* **1980**, *190*, 43; (c) D. N. Harpp, M. Gingras, *J. Am. Chem. Soc.* **1988**, *110*, 7737; (d) S. Fukuzawa, K. Saito, T. Fujinami, S. Sakai, *J. Chem. Soc., Chem. Commun.* **1990**, 939; (e) A. Yanagisawa, H. Inoue, M. Morodome, H. Yamamoto, *J. Am. Chem. Soc.* **1993**, *115*, 10356; (f) T. M. Cokley, P. J. Harvey, R. L. Marshall, A. McCluskey, D. J. Young, *J. Org. Chem.* **1997**, *62*, 1961.
[29] (a) W. Schmid, G. M. Whitesides, *J. Am. Chem. Soc.* **1991**, *113*, 6674; (b) T. H. Chan, M. B. Isaac, *Pure Appl. Chem.* **1996**, *68*, 919.
[30] Y. Yamamoto, N. Asao, *Chem. Rev.* **1992**, *93*, 2207.
[31] (a) G. E. Keck, E. J. Enholm, *J. Org. Chem.* **1985**, *50*, 147; (b) Y. Yamamoto, T. Komatsu, K. Maruyama, *J. Org. Chem.* **1985**, *50*, 3115; (c) M. A. Ciufolini, G. O. Spencer, *J. Org. Chem.* **1989**, *54*, 4739; (d) C. Bellucci, P. G. Cozzi, A. Umani-Ronchi, *Tetrahedron Lett.* **1995**, *36*, 7289; (e) M. Yasuda, Y. Sugawa, A. Yamamoto, I. Shibata, A. Baba, *Tetrahedron Lett.* **1996**, *37*, 5951.
[32] (a) P. A. Grieco, A. Bahsas, *J. Org. Chem.* **1987**, *52*, 1378; (b) H. Nakamura, H. Iwama, Y. Yamamoto, *J. Am. Chem. Soc.* **1996**, *118*, 6641.
[33] S. Kobayashi, T. Busujima, S. Nagayama, *J. Chem. Soc., Chem. Commun.* **1998**, 19.
[34] In some other substrates, yields were improved a little with increasing amounts of SDS.
[35] S. Kobayashi, S. Nagayama, *J. Org. Chem.* **1997**, *62*, 232.

6.20 Methyltrioxorhenium(VII) in Oxidation Catalysis

Wolfgang A. Herrmann, Fritz E. Kühn, Gerhard M. Lobmaier

6.20.1 Introduction

Less than ten years ago the importance of rhenium compounds in oxidation catalysis was minimal [1]. This picture has been changing dramatically ever since then. Organorhenium(VII) oxides, especially the water-soluble (30 g/L) methyltrioxorhenium (MTO, **1**), have proven to be excellent oxidation catalysts. This section summarizes the basic knowledge on this particular topic.

6.20.2 Methyltrioxorhenium(VII)

The first preparation of MTO was reported in 1979 [2]. A breakthrough came in 1987 when an efficient synthesis, starting from dirhenium heptoxide and tetramethyltin (Eq. 1) [3] was developed.

$$Re_2O_7 + Sn(CH_3)_4 \xrightarrow{THF} \underset{\mathbf{1}}{CH_3ReO_3} + (CH_3)_3SnOReO_3 \qquad (1)$$

The method was further improved [4, 5]. Recently, it was extended to perrhenates so that the moisture-sensitive Re_2O_7 can be replaced by starting materials that are more convenient to handle [6]. This synthetic progress was accompanied by the discovery of a plethora of catalytic applications of MTO and related derivatives [7–11]. It was also possible to immobilize MTO on special surfaces for catalytic purposes [12]. Very important for the understanding of the role of MTO and other organorhenium oxides in oxidation catalysis were the isolation and characterization of the reaction product of MTO with 2 equiv. of H_2O_2 [13–16]. According to Eq. (2) a bis(peroxo) complex of stoichiometry $(CH_3)Re(O_2)_2O$ (**3**) is formed.

$$\text{MTO; } \mathbf{1} \xrightarrow[-H_2O]{+H_2O_2} \mathbf{2} \xrightarrow[-H_2O]{+H_2O_2} \mathbf{3} \qquad (2)$$

Experiments with the isolated bis(peroxo) complex **3** have shown that it is an active species in oxidation catalysis, e.g. in the oxidation of olefins [15, 16]. Experiments *in situ* indicated that the reaction of MTO with 1 equiv. of H_2O_2 leads to a monoperoxo complex (**2**) according to Eq. (2) which is also catalytically active in certain oxidation processes.

The amazing versatility of MTO as an oxidation catalyst precursor is reflected in the broad variety of oxidation reactions catalyzed by **2** and **3** (Scheme 1).

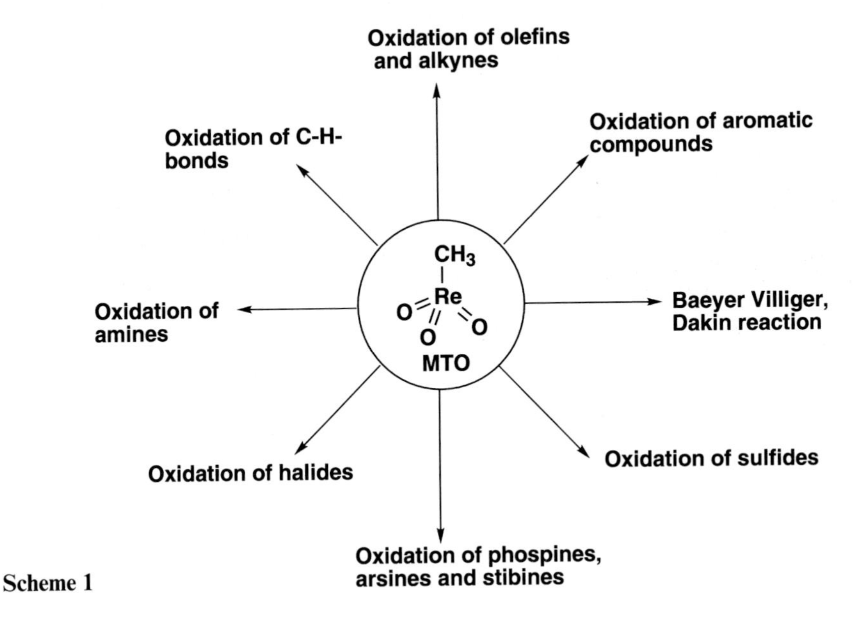

Scheme 1

6.20.3 Oxidation Reactions

6.20.3.1 Oxidation of Alkenes

One of the best examined catalytic processes using MTO as the catalyst precursor is olefin epoxidation [12–23]. Usually less than 85 wt. % of H_2O_2 is used and MTO is typically in concentrations of 0.2–0.4 mol%. Turnover numbers up to 2000 and turnover frequencies of ca. 1500 h^{-1} can be achieved.

The catalytic MTO/H_2O_2 system is already active below room temperature, e.g. at $-30\,°C$. The reactions between **3** and alkenes are approx. one order of magnitude faster in semiaqueous solvents (e.g. 85% H_2O_2) than in methanol. The rate constants for the reaction of **3** with aliphatic alkenes correlate closely with the number of alkyl groups on the olefinic carbons. The reactions become slower when electron-attracting groups, such as –OH, –CO, –Cl, and –CN are present.

Based on Eq. (2), two catalytic pathways may be described corresponding to the concentration of the hydrogen peroxide used. If this concentration is 85%, the equilibrium is toward the right of Eq. (2) and only compound **3** seems to be responsible for the epoxidation activity (Scheme 2).

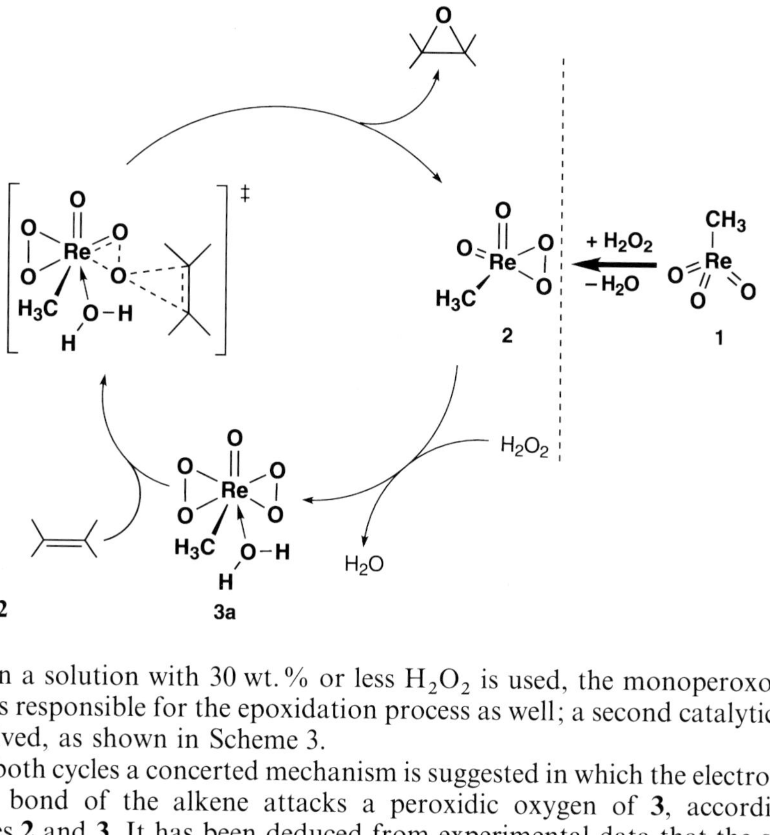

Scheme 2

When a solution with 30 wt.% or less H_2O_2 is used, the monoperoxo complex **2** is responsible for the epoxidation process as well; a second catalytic cycle is involved, as shown in Scheme 3.

For both cycles a concerted mechanism is suggested in which the electron-rich double bond of the alkene attacks a peroxidic oxygen of **3**, according to Schemes **2** and **3**. It has been deduced from experimental data that the system might show a spiro arrangement [13, 16, 17, 22]. The selectivity toward epoxides can be enhanced by the addition of Lewis bases such as quinuclidine, pyridine or 2,2′-bipyridine to the system [16, 24, 25]. Lewis acids catalyze ring-opening reactions and diol formation. These reactions are suppressed after the addition of Lewis bases. It has also been shown that the selectivity toward epoxides is dependent on the pK_b values of the Lewis bases used. The lower the pK_b values, the higher the selectivity [26]. There is evidence that an excess of N-base ligands

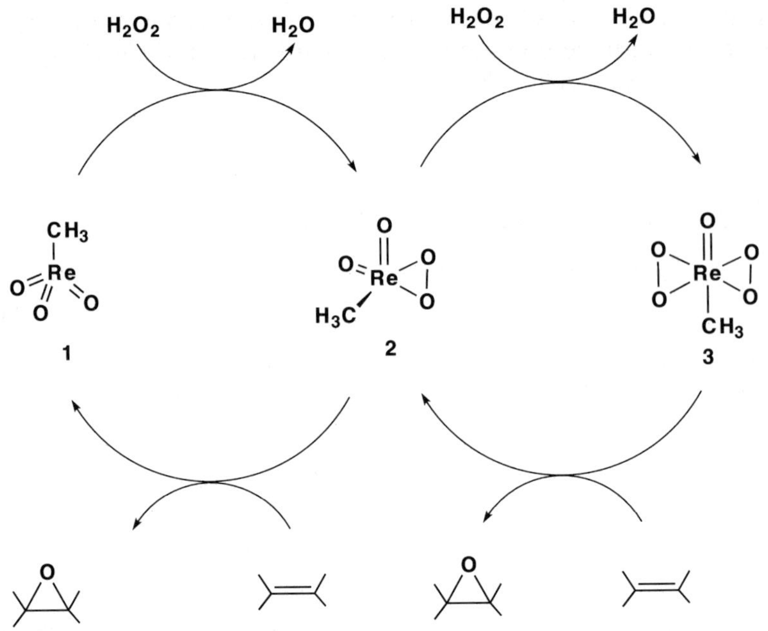

Scheme 3

may lead to accelerated reactions [24, 25]. Another possibility for enhancing the selectivity toward epoxides is the use of the urea–H_2O_2 adduct. It enables the oxidation to be carried out in water-free solutions, thus avoiding formation of any diols and other side reactions. In the case of the oxidation of chiral allyl alcohols, high diastereoselectivities have been achieved [19, 22].

6.20.3.2 Oxidation of Aromatic Compounds

The mechanistic aspects of the oxidation of aromatics have not been examined in great detail up to now but obviously complex **3** plays an important role as catalyst [27–33]. Noteworthy is the high regioselectivity, most notably in the industrially interesting synthesis of vitamin K_3. Since water is an inhibitor, concentrated (85 wt.%) H_2O_2 is preferred. Alternatively, commercially available 35% H_2O_2 in acetic anhydride can be employed; a considerable regioselectivity is obtained with this system. The conversion is higher for electron-rich arenes (nearly 100%) and selectivities of more than 85% have been reached [27]. Hydroxy-substituted arenes can be oxidized by aqueous hydrogen peroxide (85 wt.%) in acetic acid to afford the corresponding *p*-quinones in isolated yields of up to 80% [30]. It has been shown recently that using a mixture of acetic acid and acetic anhydride further improves the product yield [33]. Anisole was also found to undergo selective oxidation with the MTO/H_2O_2 system to yield *o*- and *p*-methoxyphenols [28].

6.20.3.3 Baeyer–Villiger Oxidation and Dakin Reaction

It has been shown that **3** also acts as an active species in the Baeyer–Villiger oxidation of ketones (Eq. 3) and in the Dakin reaction [34–36].

$$\text{cyclic ketone } (CH_2)_n \xrightarrow{H_2O_2 \text{ / cat. } 3} \text{lactone } (CH_2)_n \qquad (3)$$

It is somewhat surprising that the MTO/H_2O_2 system presents this activity since these oxidations involve nucleophilic attack at the carbonyl group, which contrasts with all the preceding examples where the substrates attacked the electrophilic peroxo rhenium complexes, e.g., in the olefin epoxidation. Nevertheless, **3** reacts stoichiometrically with cyclobutanone in the absence of H_2O_2 as shown in Eq. (3). This reversed behavior may be due to substrate binding to rhenium.

The reaction was found to be strongly solvent-dependent by means of the mechanistic probe for oxygen transfer reactions, thiantrene-5-oxide [34, 35]. Donor solvents such as acetonitrile seem to enhance the nucleophilicity of the peroxo groups. It has been suggested that the double bond of the enol form (the major tautomer) attacks a peroxo oxygen of the peroxo rhenium complex **2** or **3**. This reaction affords a 2-hydroxy-1,3-dicarbonyl intermediate which can be detected by ^{1}H-NMR. This hydroxy intermediate is susceptible to cleavage via Baeyer–Villiger oxidation to yield carboxylic acids as final products. Low H_2O_2 concentrations are sufficient and no H_2O_2 decomposition is observed at temperatures up to 70 °C. This is an advantage of the catalytic MTO/H_2O_2 system over the known transition metal catalysts containing V, Mo, Mn, or Os. However, the nucleophilic character of the peroxidic atoms in **3** is not as pronounced as in Pt or Ir peroxo complexes that react with CO_2 or SO_2 to give isolable cycloaddition products [34, 35]. In the case of **3**, turnover frequencies of 18000 h^{-1} are obtained for cyclobutanone, but in other cases turnover numbers of up to 100 are usual [36]. Cycloketones can be converted into lactones even below room temperature by dilute hydrogen peroxide.

6.20.3.4 Oxidation of Sulfur Compounds

Organic sulfides can be oxidized to the corresponding sulfoxides by hydrogen peroxide in the presence of MTO [37]. Both complexes **2** and **3** seem to be active in this reaction but kinetic results indicate that **2** might be more active than **3**. The kinetic results also point to a mechanism that involves the nucleophilic attack of the sulfur atom on a coordinated peroxide oxygen since electron-donating substituents have accelerating effects. The first reports on the scope and selectivity of the reaction [38] have been recently re-examined. Using ethanol as solvent, the MTO/H_2O_2 system can be used to oxidize dialkyl, diaryl, and alkyl aryl sulfides to sulfoxides ($R_2S/H_2O_2 = 1:1.1$) or sulfones ($R_2S/H_2O_2 = 1:2.2$) with excellent yields and selectivity even in the presence of oxidatively sensitive functions on the sulfide side chain [34].

This mild selectivity has also been demonstrated in the oxidation of thioether Fischer carbene complexes [39]. Oxidation of thiophene and its derivatives has been achieved with this system. However, the rate constants are two to four orders of magnitude below those reported for the "aliphatic" sulfides where the S atom is not part of an aromatic heterocyclic ring [40]. Oxidation studies of coordinated thiolates were carried out on the model Co(III) complex $[(en)_2Co(SCH_2CH_2NH_2)]^{2+}$ (en = ethylene-1,2-diamine) [41].

6.20.3.5 Oxidation of Phosphines, Arsines, and Stilbines

Tertiary phosphines, triarylarsines, and triarylstilbines are converted to their oxides, R_3EO (E = P, As, Sb) by MTO/H_2O_2. Kinetic studies lead to the assumption that **2** and **3** have similar catalytic activities in all cases. The kinetic data support a mechanism involving nucleophilic attack of the substrate at the rhenium peroxides. The proposed catalytic cycle is given in Scheme 4 [42].

6.20.3.6 Oxidation of Amines

The MTO/H_2O_2 system also catalyzes the oxidation of anilines. The major product of the oxidation of aniline is nitrosobenzene. For 4-substituted *N,N*-dimethylanilines, the N-oxide is the only oxidation product. Electron-withdrawing substituents inhibit the reaction. Kinetic results suggest that both compounds **2** and **3** are involved in the oxidation process [43]. It is suggested that the rate-controlling step is the nucleophilic attack of the nitrogen lone-pair

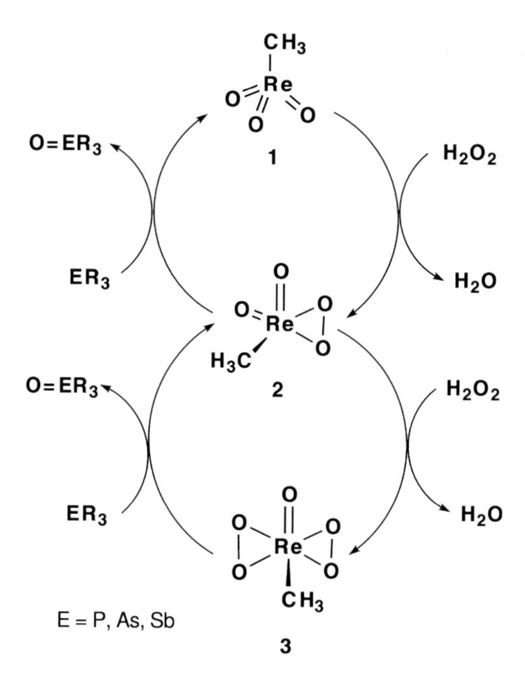

Scheme 4

electrons of the anilines on a peroxidic oxygen of the catalyst. Electron-donating groups attached to the nitrogen atom of aniline increase the rate constant. In the case of $ArNH_2$ derivatives, the oxidation proceeds ca. 50 times faster than without catalyst [43]. Furthermore, not only anilines but a broad variety of other aromatic and aliphatic amines are oxidized to the corresponding *N*-oxides [44–47]. The reactions are facile and high-yielding at or below room temperature [48].

6.20.3.7 Oxidation of Halide Ions

Another application of the MTO/H_2O_2 system is the catalytic oxidation of chloride and bromide ions in acidic aqueous solutions. The chloride oxidation steps are three to four orders of magnitude slower than the corresponding bromine oxidation steps. Both compounds **2** and **3** have been shown to be active catalysts in these processes. In both cases the catalyzed reactions were about 10^5 times faster than the uncatalyzed reactions under similar conditions. In a first step HOX is formed, then HOX reacts with X^- to form X_2. When H_2O_2 is used in excess the reaction yields O_2 [49, 50].

6.20.3.8 Oxidation of C–H Bonds

MTO/H_2O_2 also catalyzes the insertion of oxygen into a variety of activated and unactivated C–H bonds, with yields varying from good to excellent. Alcohols or ketones are formed e.g. Eq. (4). Suitable substrates proved that the reaction is stereospecific with retention of the configuration. Alcohols, e.g., ethanol and *t*-butanol, are used as solvents, the reaction temperatures range between 40 and 60 °C, and nearly quantitative yields have been obtained. However, the reaction times are generally longer than those used for most epoxidations [11].

CH3, H, CH3, H —MTO / H_2O_2, *t*–BuOH; 98%→ CH3, OH, CH3, H (4)

6.20.3.9 Oxidation of Alkynes

Internal alkynes yield carboxylic acids and α-diketones when oxidized with the MTO/H_2O_2 system. Rearrangement products were observed only for aliphatic alkynes. Terminal alkynes give carboxylic acids, derivatives thereof and α-keto acids as the major products. The yields of these products vary with the solvent used [51].

6.20.4 Perspectives

Methyltrioxorhenium(VII), a water-soluble and water-stable metal alkyl, has become the most versatile oxidation catalyst in organic chemistry [52–58]. The secret of its outstanding reactivity is the rhenium–methyl bond, which is thermally and chemically very stable. In addition, the ligands surrounding the rhenium center require only little space, so the heptavalent rhenium can exploit its pronounced Lewis acidity with regard to incoming substrates. It is striking that destruction of the methyl-to-rhenium bond effects a sudden, complete breakdown of some of the reported reactions, e.g., catalytic epoxidation. It is thus the organometallic moiety that guarantees unique reactivities MTO is now commercially available.

References

[1] K. A. Jørgensen, *Chem. Rev.* **1989**, *89*, 447.
[2] I. R. Beattie, P. J. Jones, *Inorg. Chem.* **1979**, *18*, 2318.
[3] W. A. Herrmann, J. G. Kuchler, J. K. Felixberger, E. Herdtweck, W. Wagner, *Angew. Chem.* **1988**, *100*, 420; *Angew. Chem., Int. Ed. Engl.* **1988**, *27*, 394.
[4] W. A. Herrmann, F. E. Kühn, R. W. Fischer, W. R. Thiel, C. C. Romão, *Inorg. Chem.* **1992**, *31*, 4431.
[5] W. A. Herrmann, W. R. Thiel, F. E. Kühn, R. W. Fischer, M. Kleine, E. Herdtweck, W. Scherer, J. Mink, *Inorg. Chem.* **1993**, *32*, 5188.
[6] W. A. Herrmann, R. Kratzer, R. W. Fischer, *Angew. Chem.* **1997**, *109*, 2767; *Angew. Chem., Int. Ed. Engl.* **1997**, *36*, 2652.
[7] W. A. Herrmann, *Angew. Chem.* **1988**, *100*, 1269; *Angew. Chem., Int. Ed. Engl.* **1988**, *27*, 394.
[8] W. A. Herrmann, *J. Organomet. Chem.* **1995**, *500*, 149.
[9] W. A. Herrmann, F. E. Kühn, *Acc. Chem. Res.* **1997**, *30*, 169.
[10] C. C. Romão, F. E. Kühn, W. A. Herrmann, *Chem. Rev.* **1997**, *97*, 3197.
[11] R. W. Murray, K. Iyanar, L. Chen, J. T. Wearing, *Tetrahedron Lett.* **1995**, *36*, 6415.
[12] Z. Zhu, J. H. Espenson, *J. Mol. Catal.* **1997**, *121*, 139.
[13] W. A. Herrmann, R. W. Fischer, D. W. Marz, *Angew. Chem., Int. Ed. Engl.* **1991**, *30*, 1638.
[14] W. A. Herrmann, J. D. G. Correia, G. R. J. Artus, R. W. Fischer, C. C. Romão, *J. Organomet. Chem.* **1996**, *520*, 139.
[15] W. A. Herrmann, R. W. Fischer, W. Scherer, M. U. Rauch, *Angew. Chem., Int. Ed. Engl.* **1993**, *32*, 1157.
[16] W. A. Herrmann, R. W. Fischer, M. U. Rauch, W. Scherer, *J. Mol. Catal.* **1994**, *86*, 243.
[17] A. Al-Ajlouni, J. H. Espenson, *J. Am. Chem. Soc.* **1995**, *117*, 9243.
[18] W. A. Herrmann, in *Organic Peroxygen Chemistry* (Ed.: W. A. Herrmann) Springer, Berlin, **1993**, Vol. 164, p. 130.
[19] T. R. Boehlow, C. D. Spilling, *Tetrahedron Lett.* **1996**, *37*, 2717.
[20] R. W. Murray, M. Singh, B. L. Williams, H. M. Moncrieff, *Tetrahedron Lett.* **1995**, *36*, 2437.
[21] A. M. Al-Ajlouni, J. H. Espenson, *J. Org. Chem.* **1996**, *61*, 3969.
[22] W. Adam, C. M. Mitchell, *Angew. Chem., Int. Ed. Engl.* **1996**, *35*, 533.
[23] R. W. Murray, K. Iyanar, *Tetrahedron Lett.* **1997**, *38*, 335.
[24] K. B. Sharpless, J. Rudolph, K. L. Redding, J. P. Chiang, *J. Am. Chem. Soc.* **1997**, *119*, 6189.
[25] B. Schmidt, *J. Prakt. Chem.* **1997**, *339*, 439.
[26] W. A. Herrmann, F. E. Kühn, M. R. Mattner, G. J. R. Artus, M. Geisberger, J. D. G. Correia, *J. Organomet. Chem.* **1997**, *118*, 33.
[27] W. Adam, W. A. Herrmann, J. Lin, C. R. Saha-Möller, R. W. Fischer, J. D. G. Correia, *Angew. Chem., Int. Ed. Engl.* **1994**, *33*, 2475.
[28] E. I. Karasevich, A. V. Nikitin, V. L. Rubailo, *Kinet. Katal.* **1994**, *35*, 810; *Kinet. Katal.* **1994**, *35*, 878.
[29] S. Yamazaki, *Chem. Lett.* **1995**, 127.
[30] W. Adam, W. A. Herrmann, J. Lin, C. R. Saha-Möller, *J. Org. Chem.* **1994**, *59*, 8281.
[31] W. Adam, W. A. Herrmann, C. R. Saha-Möller, M. Shimizu, *J. Mol. Catal.* **1995**, *97*, 15.
[32] U. Suchhardt, D. Mandelli, G. B. Shul'pin, *Tetrahedron Lett.* **1996**, *37*, 6487.
[33] W. A. Herrmann, J. J. Haider, R. W. Fischer, *J. Mol. Cat.* **1998**, in press.
[34] W. A. Herrmann, R. W. Fischer, J. D. G. Correia, *J. Mol Catal.* **1994**, *94*, 213.
[35] M. M. Abu-Omar, J. H. Espenson, *Organometallics* **1996**, *15*, 3543.

[36] R. W. Fischer, Ph. D. Thesis, Technische Universität München, **1994**.
[37] S. Yamazaki, *Bull. Chem. Soc. Jpn.* **1996**, *69*, 2955.
[38] W. Adam, C. M. Mitchell, C. R. Saha-Möller, *Tetrahedron* **1994**, *50*, 13121.
[39] R. L. Beddoes, J. E. Painter, P. Quayle, P. Patel, *Tetrahedron Lett.* **1996**, *37*, 9385.
[40] K. N. Brown, J. H. Espenson, *Inorg. Chem.* **1996**, *35*, 7211.
[41] P. Huston, J. H. Espenson, A. Bakac, *Inorg. Chem.* **1993**, *32*, 4517.
[42] M. M. Abu-Omar, J. H. Espenson, *J. Am. Chem. Soc.* **1995**, *117*, 272.
[43] Z. Zhu, J. H. Espenson, *J. Org. Chem.* **1995**, *60*, 1326.
[44] A. Goti, L. Nannelli, *Tetrahedron Lett.* **1996**, *37*, 6025.
[45] R. W. Murray, K. Iyanar, I. Chen, T. Wearing, *J. Org. Chem.* **1996**, *61*, 8099.
[46] S. Yamazaki, *Bull. Chem. Soc. Jpn.* **1997**, *70*, 877.
[47] R. W. Murray, K. Iyanar, I. Chen, T. Wearing, *Tetrahedron Lett.* **1996**, *37*, 809.
[48] K. A. Vassell, J. H. Espenson, *Inorg. Chem.* **1994**, *33*, 5491.
[49] J. H. Espenson, O. Pestovsky, P. Huston, S. Staudt, *J. Am. Chem. Soc.* **1994**, *116*, 2869.
[50] P. J. Hansen, J. H. Espenson, *Inorg. Chem.* **1995**, *34*, 5389.
[51] Z. Zhu, J. H. Espenson, *J. Org. Chem.* **1995**, *60*, 7728.
[52] W. A. Herrmann, F. E. Kühn, M. R. Mattner, G. R. J. Artus, M. Geisberger, J. D. G. Correia, *J. Organomet. Chem.* **1997**, *538*, 203.
[53] W. A. Herrmann, J. D. G. Correia, M. U. Rauch, G. R. J. Artus, F. E. Kühn, *J. Mol. Cat.* **1997**, *118*, 33.
[54] T. H. Zauche, J. H. Espenson, *Inorg. Chem.* **1997**, *36*, 5257.
[55] R. Neumann, T. J. Wang, *Chem. Commun.* **1997**, 1915.
[56] A. K. Yudin, K. B. Sharpless, *J. Am. Chem. Soc.* **1997**, *119*, 11536.
[57] C. Copéret, H. Adolfsson, K. B. Sharpless, *Chem. Commun.* **1997**, 1565.
[58] W. A. Herrmann, H. Ding, R. M. Kratzer, F. E. Kühn, J. J. Haider, R. W. Fischer, *J. Organomet. Chem.* **1997**, *549*, 319.

7

Other Biphasic Concepts

7.1 SHOP Process

Dieter Vogt

7.1.1 Introduction

The Shell higher-olefin process was undoubtedly the first commercial catalytic process taking benefit from two-phase (but nonaqueous) liquid/liquid technology. In this special case two immiscible organic phases are used to separate the catalyst from the products formed, with the more or less pure products forming the upper phase.

The basis for this nickel-catalyzed oligomerization of ethylene goes back to Ziegler and his school at the Max Planck Institute at Mülheim. There the so-called "nickel effect" [1, 2] was found, and Wilke and co-workers learned how to control the selectivity of nickel-catalyzed reactions by use of ligands. Keim introduced $\widehat{PO}$ chelate ligands and on this basis carried out the basic research for the oligomerization process at the Shell research company at Emeryville [3–12]. The whole process was developed in a collaboration between Shell Development, USA, and the Royal Shell Laboratories at Amsterdam in the Netherlands [13–22]. The SHOP is not only a process for ethylene oligomerization, but a very efficient and flexible combination of three reactions: oligomerization, isomerization, and metathesis. It was designed to meet the market needs for linear α-olefins for detergents [23].

Table 1. Linear α-olefin capacities via ethylene oligomerization.

Technology	Company	Location	Capacity [10^3 t/y^{-1}]		
			Initial (year)	Expansion (year)	Present total [a])
Zielger-type	Chevron	Cedar Bayou, TX	125 (1966)	125 (1990)	249
	Ethyl	Pasadena, TX	400 (1971)	55 (1989)	472
	Ethyl	Feluy, Belgium	200 (1992)		200
	Chemopetrol	Czech Republic	120 (1992)		120
	Mitsubishi Kasei Corp.	Kurashiki, Okayama Pref., Japan	50		50
SHOP	Shell	Geismar, LA	200 (1977)	390 (1989)	590
	Shell	Stanlow, UK	170 (1982)	100 (1989)	278
Zr	Idemitsu Petrochemicals	Ichihara, Chiba Pref., Japan	50 (1989)		50

[a]) In 1992.

The first commercial plant was built at Geismar, LA, USA, in 1977. The development of this plant and that at Stanlow (UK) is summarized in Table 1, together with other oligomerization capacities based on other technology [24]. The two operational SHOP sites today have a total capacity of nearly 1 million tons of α-olefins per year. This is about one-half of the total amount made by oligomerization. Today linear α-olefines are produced mainly by ethylene oligomerization because of the high product quality and the good availability of ethylene. The wide application and increasing need for short-chain α-olefins as co-monomers for polymers cause the linear olefin market to continue growing.

7.1.2 Process Description

The oligomerization is carried out in a polar solvent in which the nickel catalyst is dissolved but the nonpolar products, the α-olefins, are almost insoluble. Preferred solvents are alkanediols, especially 1,4-butanediol (1,4-BD). This use of a biphasic organic liquid/liquid system is one of the key features of the process. The nickel catalyst is prepared in situ from a nickel salt, e.g., $NiCl_2 \cdot 6\,H_2O$, by reduction with sodium borohydride in 1,4-BD in the presence of an alkali hydroxide, ethylene, and a chelating $\widehat{PO}$ ligand such as *o*-diphenylphosphinobenzoic acid (Structure **1**) [11, 20]. Suitable ligands are the general type of diorganophosphino acid derivatives (**2**).

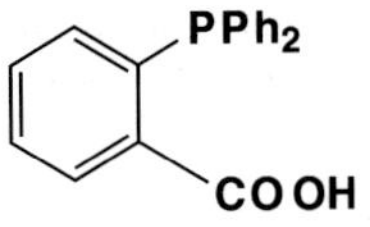

PPh₂ / COOH

1

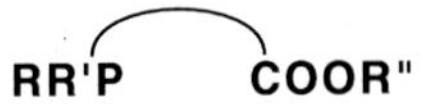

RR'P COOR"

2

The nickel concentration in the catalyst system is in the range 0.001–0.005 mol% (approx. 10–50 ppm). The oligomerization is carried out in a series of reactors at temperatures of 80–140 °C and pressures of 7–14 MPa. The rate of the reaction is controlled by the rate of catalyst addition [19]. A high partial pressure of ethylene is required to obtain good reaction rates and high product linearity [11]. The linear α-olefins produced are obtained in a Schulz–Flory-type distribution with up to 99% linearity and 96–98% terminal olefins over the whole range from C_4 to C_{30}^+ (cf. Table 2) [23].

The shape of the Schulz–Flory distribution and the chain length of the α-olefins are controlled by the geometric chain-growth factor *K*, defined as $K = n(C_{n+2})/n(C_n)$ (Figure 1).

For the economy of the whole process it is very important that the *K*-factor can easily be adjusted by varying the catalyst composition. Usually the value is between 0.75 and 0.80. The heat of the reaction is removed by water-cooled heat

Table 2. Comparison of product qualities of technical C_6-C_{18} α-olefins [25].

Product	Wax-cracking	Quality [wt.% α-olefin]		
		Chevron	Ethyl	SHOP
α-Olefins	83–89	91–97	63–98	96–98
Branched olefins	3–12	2–8	2–29	1–3
Paraffins	1–2	1.4	0.1–0.8	0.1
Dienes	3–6	–	–	–
Monoolefins	92–95	99	>99	99.9

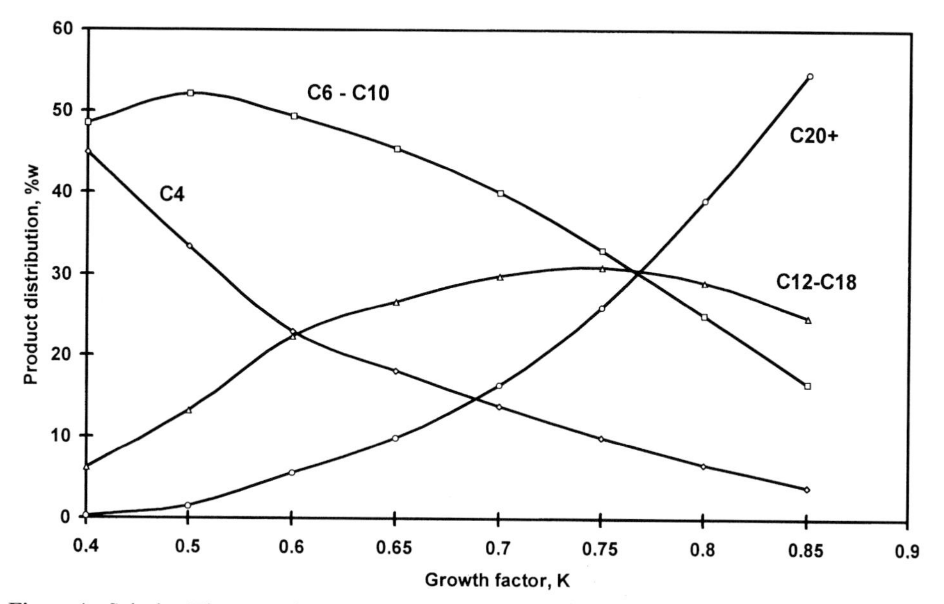

Figure 1. Schulz–Flory product distribution dependence on the chain-growth factor K.

exchangers between the reactors (Figure 2). In a high-pressure separator the insoluble products and the catalyst solution as well as unreacted ethylene are separated.

The catalyst solution is fed back into the oligomerization reactor. Washing of the oligomers by fresh solvent in a second separator removes traces of the catalyst. This improves product quality and the catalyst utilization [26]. Traces of remaining catalyst in the product can lead to the formation of insoluble polyethylene during upstream processing, resulting in fouling of process equipment [27].

The formation of insoluble polyethlene cause problems also in other parts of the process. During catalyst preparation this can be avoided by adding the preformed, stable nickel complex and the chelate ligand separately to the oligomerization reactor. By this simple change, the catalyst utilization is enhanced markedly, resulting in a significant reduction of nickelt salt and borohy-

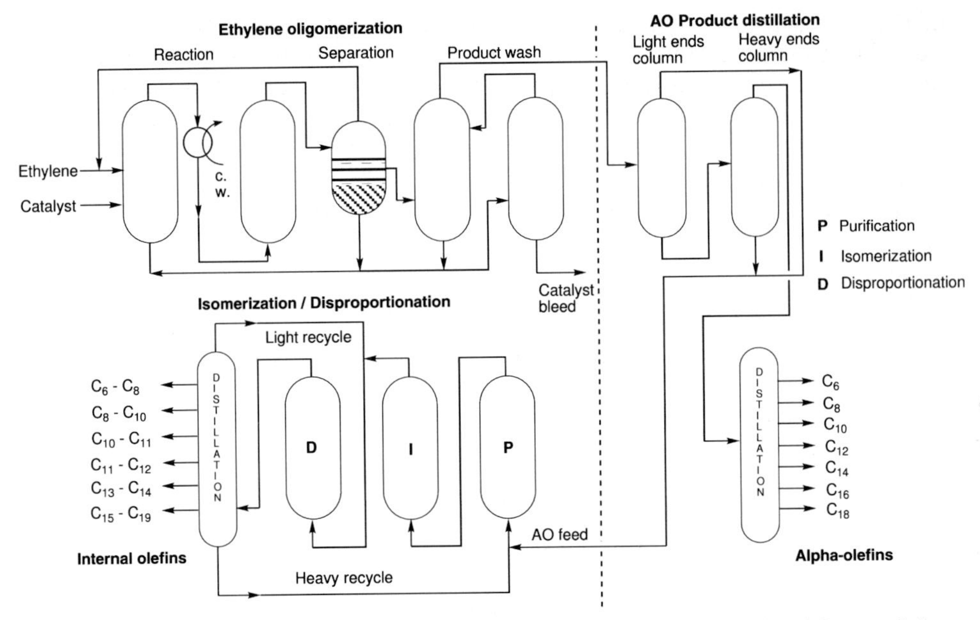

Figure 2. Flow scheme of the Shell higher-olefin process (SHOP). AO = α-olefin; C.w. = cooling water.

dride consumption [28–30]. So one major problem is the complete separation of the catalyst. Many attempts have been made to improve this [19]. One approach was using methanol/water solvent mixtures together with sulfonated ligands [31–37]. In the course of this it was shown that the catalyst is not deteriorated by water [38], which might lead to new approaches in the future.

Completely different liquid/liquid two-phase applications for oligomerization of ethylene were reported recently. Chauvin used ionic liquids as solvents for oligomerization catalysts [39] (cf. Section 7.3). Another approach is the use of perfluorinated solvents together with catalysts bearing perfluorinated ligands [40, 41] (cf. Section 7.2).

Further processing of the product α-olefins involves separation into the desired product fractions in a series of distillation columns. First the lower C_4–C_{10} α-olefins are stripped off. In a heavy-ends column the C_{20+} α-olefins are removed from the desired C_{12}–C_{20} α-olefins. Finally the middle-range products meeting the market needs are separated into the desired cuts and blends. The very high flexibility of the "SHOP" results from the following steps. The C_4–C_{10} and the C_{20+} fractions are combined to be isomerized to internal linear olefins and then subjected to a metathesis reaction. Both steps require about 80–140 °C and 0.3–2 MPa. Isomerization is accomplished by a typical isomerization catalyst such as Na/K on Al_2O_3 or a MgO catalyst in the liquid phase [42], where about 90 % of the α-olefins are converted to internal olefins. Metathesis of the lower and higher internal olefins gives a mixture of olefins with odd and even carbon chain lengths. The mixture comprises about 11–15 %

of the desired $C_{11}-C_{14}$ linear internal olefins, which are separated by distillation. The undesired fractions can be recycled, feeding the light olefins directly back to metathesis while the higher-boiling fractions are again subjected to isomerization. Because of the high proportion of short-chain olefins in the metathesis feed, the double bonds in the end product are shifted toward the chain ends. Altogether the different possibilities of shifting products to the desired chain length and double-bond position makes the "SHOP" the most elegant and flexible process operating today. It is furthermore one of the larger applications of homogeneous catalysis.

The mechanism of the nickel-chelate complex-catalyzed oligomerization has been investigated in detail by Keim and co-workers [43–52]. Based on these results, the mechanism shown in Scheme 1 was postulated.

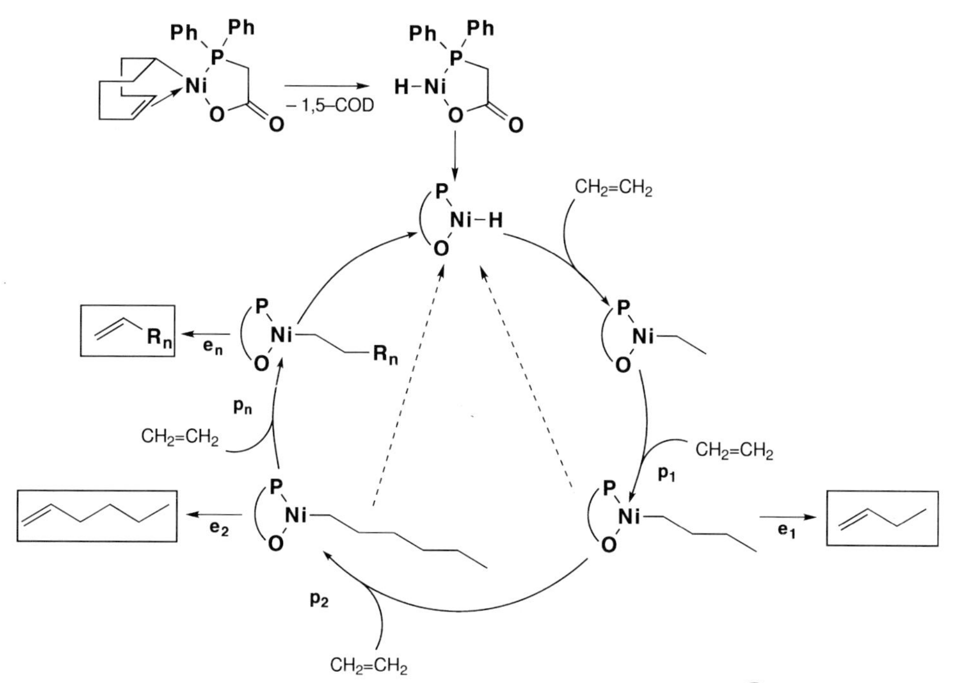

Scheme 1. Postulated meachanism for ethylene oligomerization via a PO-stabilized nickel hydride species: $\mathbf{p}_1$, $\mathbf{p}_2 \cdots \mathbf{p}_n$ = propagation steps; $\mathbf{e}_1$, $\mathbf{e}_2 \cdots \mathbf{e}_n$ = elimination steps.

References

[1] G. Wilke, *Angew. Chem.* **1988**, *100*, 189; *Angew. Chem., Int. Ed. Engl.* **1988**, *27*, 185.

[2] G. Wilke, in *Fundamental Research in Homogeneous Catalysis* (Ed.: M. Tsutsui), Plenum, New York, **1979**, Vol. 3, p. 1.

[3] Shell Dev. Co. (S. R. Baur, H. Chung, P. W. Glockner, W. Keim, H. van Zwet), US 3.635.937 (1972).

[4] Shell Dev. Co. (S. R. Baur, H. Chung, D. Camel, W. Keim, H. van Zwet), US 3.637.636 (1972).
[5] Shell Dev. Co. (S. R. Baur, P. W. Glockner, W. Keim, H. van Zwet, H. Chung), US 3.644.563 (1972).
[6] Shell Dev. Co. (H. van Zwet, S. R. Baur, W. Keim), US 3.644.564 (1972).
[7] Shell Dev. Co. (P. W. Glockner, W. Keim, R. F. Mason), US 3.647.914 (1972).
[8] Shell Dev. Co. (S. R. Baur, P. W. Glockner, W. Keim, R. F. Mason), US 3.647.915 (1972).
[9] Shell Dev. Co. (S. R. Baur, H. Chung, W. Keim, H. van Zwet), US 3.661.803 (1972).
[10] Shell Dev. Co. (S. R. Baur, H. Chung, C. Arnett, P. W. Glockner, W. Keim), US 3.686.159 (1972).
[11] E. R. Freitas, C. R. Gum, *Chem. Eng. Prog.* **1979**, *75*, 73.
[12] E. L. T. M. Spitzer, *Seifen Öle Fette Wachse* **1981**, *107*, 141.
[13] Shell Oil (R. F. Mason), US 3.676.523 (1972).
[14] Shell Oil (R. F. Mason), US 3.686.351 (1972).
[15] Shell Oil (A. J. Berger), US 3.726.938 (1973).
[16] Shell Oil (R. F. Mason), US 3.737.475 (1973).
[17] Shell (R. F. Mason, G. R. Wicker), DE 2.264.088 (1973).
[18] Shell Oil (E. F. Lutz), US 3.825.615 (1974).
[19] Shell Oil (E. F. Lutz), US 4.528.416 (1985).
[20] Shell (E. F. Lutz, P. A. Gautier), EP 177.999 (1986).
[21] W. Keim, in *Fundamental Research in Homogeneous Catalysis* (Eds.: M. Graziani, M. Giongo), Plenum, New York, **1984**, Vol. 4, p. 131.
[22] E. F. Lutz, *J. Chem. Educ.* **1986**, *63*, 202.
[23] A. H. Turner, *J. Am. Oil Chem. Soc.* **1983**, *60*, 594.
[24] C. S. Read, R. Willhalm, Y. Yoshida, in *Linear Alpha-Olefins, The Chemical Economics Handbook Marketing Research Report*, SRI International, Oct. **1993**, 681.5030 A.
[25] A. M. Al-Jarallah, J. A. Anabtawi, M. A. B. Siddique, A. M. Aitani, A. W. Al-Sa'doun, *Catal. Today* **1992**, *14*, 1.
[26] Shell Oil (A. T. Kister, E. F. Lutz), US 4.020.121 (1975).
[27] Shell Oil (C. R. Gum, A. T. Kister), US 4.229.607 (1980).
[28] Shell Oil (A. E. O'Donnell, C. R. Gum), US 4.260.844 (1981).
[29] Shell Oil (A. E. O'Donnell, C. R. Gum), US 4.377.499 (1983).
[30] Shell Oil (E. F. Lutz), US 4.284.837 (1981).
[31] Gulf Res. Dev. Co. (D. L. Beach, J. J. Harrison), US 4.301.318 (1981).
[32] Gulf Res. Dev. Co. (D. L. Beach, J. J. Harrison), US 4.310.716 (1982).
[33] Chevron/Gulf Co. (D. L. Beach, J. J. Harrison), EP 46.330 (1982).
[34] Gulf Res. Dev. Co. (D. L. Beach, J. J. Harrison), US 4.382.153 (1983).
[36] Chevron/Gulf Co. (D. L. Beach, Y. V. Kissin), US 4.686.315 (1987).
[37] Y. V. Kissin, *J. Polym. Sci., Polym. Chem. Ed.* **1989**, *27*, 147.
[38] Chevron Res. Co. (D. L. Beach, J. J. Harrison), US 4.711.969 (1987).
[39] (a) Institut Français du Petrole (Y. Chauvin, D. Commereue, I. Guibard, A. Hirschauer, H. Olivier, L. Saussine), US 5.104.840 (1992); (b) Y. Chauvin, S. Einloft, H. Olivier, *Ind. Eng. Chem. Res.* **1995**, *34*, 1149; (c) Y. Chauvin, H. Olivier in *Applied Homogeneous Catalysis with Organometallic Compounds* (Eds. B. Cornils, W. A. Herrmann), VCH, Weinheim, **1996**, Vol. 1, p. 258.
[40] M. Vogt, Ph.D. Thesis, RWTH Aachen, Germany (**1991**).
[41] I. T. Horváth, J. Rábai, *Science* **1994**, *266*, 72.
[42] Shell Oil (F. F. Farley), US 3.647.906 (1972).
[43] W. Keim, F. H. Kowaldt, R. Goddard, C. Krüger, *Angew. Chem.* **1978**, *90*, 493; *Angew. Chem., Int. Ed. Engl.* **1978**, *17*, 466.

[44] W. Keim, A. Behr, B. Limbäcker, C. Krüger, *Angew. Chem.* **1983**, *95*, 505; *Angew. Chem., Int. Ed. Engl.* **1983**, *22*, 503.
[45] W. Keim, A. Behr, B. Gruber, B. Hoffmann, F. H. Kowaldt, U. Kürschner, B. Limbäcker, F. P. Sistig, *Organometallics* **1986**, *5*, 2356.
[46] W. Keim, *New J. Chem.* **1987**, *11*, 531.
[47] W. Keim, *J. Mol. Catal.* **1989**, *52*, 19.
[48] A. Behr, W. Keim, *Arab. J. Sci. Eng.* **1985**, *10*, 377.
[49] W. Keim, *Angew. Chem.* **1990**, *102*, 251; *Angew. Chem. Int. Ed. Engl.* **1990**, *29*, 235.
[50] W. Keim, *Ann. N.Y. Acad. Sci.* **1983**, *415*, 191.
[51] U. Müller, W. Keim, C. Krüger, P. Betz, *Angew. Chem.* **1989**, *101*, 1066; *Angew. Chem., Int. Ed. Engl.* **1989**, *28*, 1011.
[52] W. Keim, R. P. Schulz, *J. Mol. Catal.* **1994**, *92*, 21.

7.2 Fluorous Phases

István T. Horváth

7.2.1 Introduction

The increasing demand for environmentally benign processes with high product selectivities at economically favorable reaction rates has renewed interest in homogeneous organometallic catalysis [1]. The development of simple and efficient separation of the generally thermally sensitive organometallic catalysts from the products under mild conditions is crucial for their industrial applications. The use of biphasic systems, in which one of the phases contains the dissolved catalyst and the other products, could allow easy separation of the products. Since the formation of a liquid-liquid biphase system is due to the sufficiently different intermolecular forces of two liquids [2], the selection of a catalyst phase depends primarily on the solvent properties of the product phase at high conversion levels. For example, if the product is apolar the catalyst phase should be polar, and vice versa: if the product is polar the catalyst phase should be apolar. The success of any biphasic system depends on whether the catalyst could be designed to dissolve preferentially in the catalyst phase. Perhaps the most important rule for such design is that the catalyst has to be like the catalyst phase, since it has been known for centuries that "similia similibus solvuntur," or "like dissolves like" [3].

7.2.2 The Fluorous Concept

Perfluorinated alkanes, dialkyl ethers, and trialkylamines are unusual because of their nonpolar nature and low intermolecular forces. Their miscibility, even with common organic solvents such as toluene, THF, acetone, and alcohols, is low at room temperature, so these materials could form fluorous biphase systems [2]. The term "fluorous" was introduced [4, 5], as the analog to the term "aqueous," to emphasize the fact that one of the phases of a biphase system is richer in fluorocarbons than the other. Fluorous biphase systems can be used in stoichiometric and catalytic chemical transformations by immobilizing reagents [4–13] and catalysts [4, 5, 14–19] in the fluorous phase. The most effective fluorous solvents are perfluorinated alkanes, perfluorinated dialkyl ethers, and perfluorinated trialkylamines. Their remarkable chemical inertness, thermal stability, and nonflammability coupled with their unusual physical properties make them particularly attractive for catalyst immobilization. Furthermore, these materials are practically nontoxic by oral ingestion, inhalation, or intraperitoneal injection [20]. Although their thermal degradation can produce toxic decomposition products, such decomposition generally begins only at very high temperatures well above the thermal stability limits of most organometallic compounds.

A fluorous organometallic catalyst system consists of a fluorous phase containing a preferentially fluorous-soluble organometallic compound and a second product phase, which may be any organic or inorganic solvent with limited solubility in the fluorous phase (Figure 1). Organometallic complexes can be made fluorous-soluble by attaching fluorocarbon moieties to ligands in appropriate size and number. The most effective fluorocarbon moieties are linear or branched perfluoroalkyl chains with high carbon numbers that may contain other heteroatoms (the "fluorous ponytails"). It should be emphasized that perfluoroaryl groups do offer dipole–dipole interactions [21], making perfluoroaryl-containing ligands soluble in common organic solvents and therefore less compatible with fluorous biphase systems. Fluorous-soluble phosphines [4, 5, 14, 22–24], phosphites [5], porphyrins [5, 6, 15, 16, 25], phthalocyanines [5, 26, 27], diketonates [14], tris(pyrazolyl)borates [28], bipyridines [29], and cyclopentadienes [30] have been prepared.

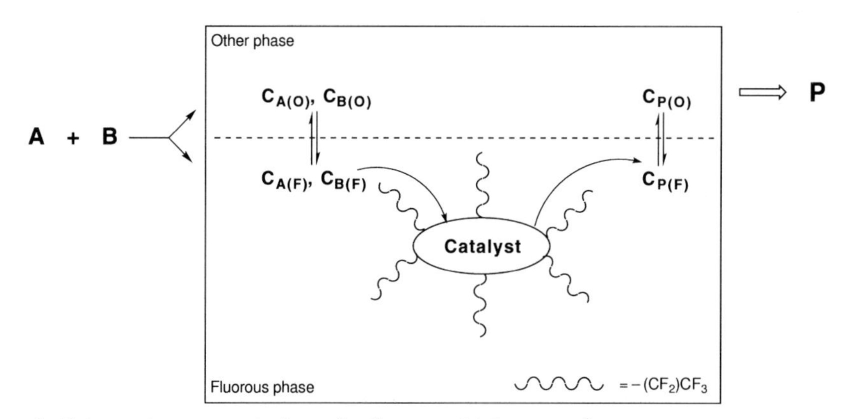

Figure 1. Schematic representation of a fluorous biphase catalyst system.

Because of the well-known electron-withdrawing properties of the fluorine atom, the attachment of fluorous ponytails to ligands could change significantly the electronic properties and consequently the coordinating power of the ligands. Insertion of insulating groups before the fluorous ponytail may be necessary to decrease the strong electron-withdrawing effects, an important consideration if catalyst reactivity is desired to approximate that observed for the unmodified species in hydrocarbon solvents. For example, theoretical calculations show that the electronic properties of $P[(CH_2)_x(CF_2)_yCF_3]_3$ ($x = 0$, $y = 2$ or 4 and $x = 0–5$, $y = 2$) can be tuned by varying the number of methylene groups [$-(CH_2)_x-$] between the phosphorus atom and the perfluoroalkyl moi-

Table 1. [a])

Phosphine	P Mulliken population [q]	P lone pair level [eV]	Protonation energy [eV]	P–H [Å]	H-P-L angle [deg]
$P[CF_2CF_3]_3$	0.83	−11.7	−6.5	1.189	85.9
$P[CF_2CF_2CF_2CF_3]_3$	0.83	−11.7	−6.4	1.192	85.4
$P[CH_2CF_2CF_3]_3$	0.62	−10.6	−7.7	1.205	86.3
$P[(CH_2)_2CF_2CF_3]_3$	0.48	−9.9	−8.3	1.218	92.3
$P[(CH_2)_3CF_2CF_3]_3$	0.40	−9.5	−8.6	1.225	91.8
$P[(CH_2)_4CF_2CF_3]_3$	0.38	−9.3	−8.8	1.226	92.0
$P[(CH_2)_5CF_2CF_3]_3$	0.36	−9.2	−8.9	1.228	91.8
$P[(CH_2CH_2CH_2CH_3]_3$	0.33	−8.7	−9.3	1.230	91.7

[a]) The calculations were performed using the UniChem version of MNDO93 and employed the PM3 parameter set. Full geometry optimizations were performed.

ety [19] (Table 1). The effect of perfluoroethyl ponytails, which are adequate models for longer perfluoroalkyl groups, is small for two ($x = 2$) and essentially negligible for three ($x = 3$) methylene units. However, the differences between the electronic properties of $P[(CH_2)_xCF_2CF_3]_3$ with more than three methylene groups ($x > 3$) and $P[(CH_2)_3CH_3]_3$ are small but finite.

The fluorous analogs of several organometallic complexes have been prepared and spectroscopically characterized, including $HRh(CO)\{P[CH_2CH_2(CF_2)_5$-

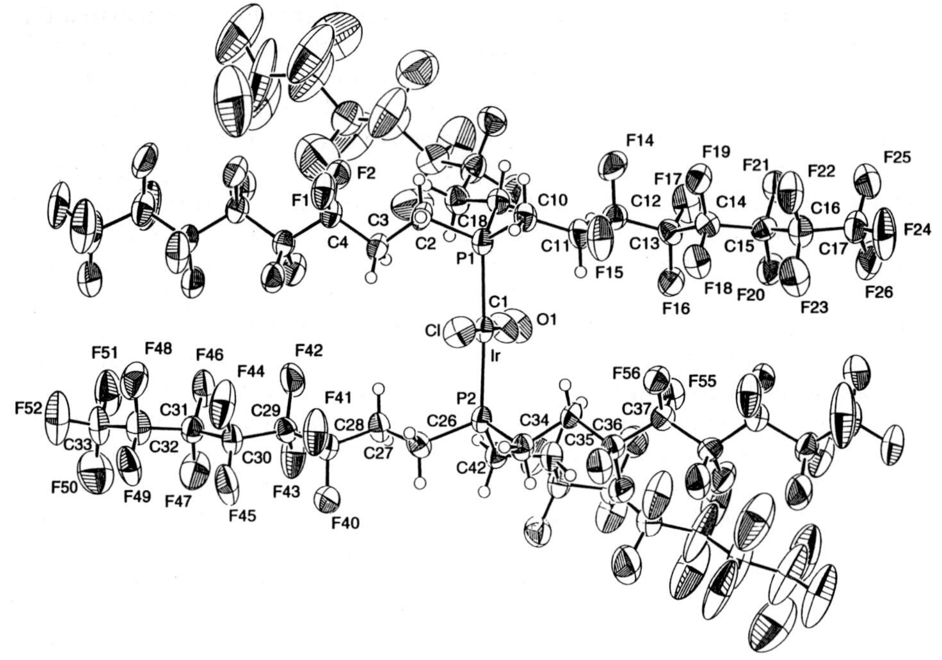

Figure 2. Molecular structure of $ClIr(CO)\{P[CH_2CH_2(CF_2)_5CF_3]_3\}_2$.

$CF_3]_3\}_3$ [4, 5], the fluorous Wilkinson's catalyst, $ClRh\{P[CH_2CH_2(CF_2)_5$-$CF_3]_3\}_3$ [17], the fluorous Vaska's complex, $ClIr(CO)\{P[CH_2CH_2(CF_2)_5CF_3]_3\}_2$ [31], and its rhodium analog [32], fluorous prophyrins with Co [16], Fe [5, 25], and Mn [15] metal centers, and fluorous cyclopentadienyl complexes with Mn, Re, Fe, and Co metal centers [30]. The molecular structure of $ClM(CO)$-$\{P[CH_2CH_2(CF_2)_5CF_3]_3\}_2$ (M = Rh [32], Ir [31]) has recently been established. Both compounds contain six perfluorohexyl groups which provide a fluorous blanket around the hydrocarbon domain of about 30% of the total volume of the molecule (Figure 2). Rhodium K-edge EXAFS studies of $ClRh(CO)$-$\{P[CH_2CH_2(CF_2)_5CF_3]_3\}_2$ in the solid state and in fluorous solution indicate that it maintains its structure in the fluorous phase [32].

7.2.3 Process and Applications

A fluorous biphase reaction could proceed either in the fluorous phase or at the interface of the two phases, depending on the solubilities of the reactants in the fluorous phase and the relationship of mass transport to the chemical reaction velocity. When the solubilities of the reactants are very low in the florous phase, the chemical reaction may still occur at the interface or appropriate phase-transfer agents may be added to facilitate the reaction. It should be emphasized that a fluorous biphase system might become a one-phase system on increasing the temperature. Thus, a fluorous catalyst could combine the advantages of one-phase catalysis with biphase product separation by running the reaction at higher temperatures and separating the products at lower temperatures.

Fluorous organometallic catalysts are best suited for converting apolar reactants to products of higher polarity, as the partition coefficients of the reactants and products will be higher and lower, respectively, in the fluorous phase. The net results are no or little solubility limitation on the reactants and easy separation of the products. Furthermore, as the conversion level increases, the proportion of polar products increases, further enhancing separation. One of the most important advantages of the fluorous biphase catalyst concept is that many well-established hydrocarbon-soluble catalysts can be converted to fluorous-soluble ones. Accordingly, several fluorous analogs of hydrocarbon-soluble catalysts have been prepared and shown to have comparable catalytic performance with the additional benefit of facile catalyst recycling [4, 5, 14–19].

The application of hydrocarbon-soluble phosphine-modified rhodium catalysts for the hydroformylation of higher olefins such as 1-decene is limited by catalyst degradation during distillation of the aldehyde from the catalysts [33]. Although the use of water-soluble catalysts could provide easy separation for heavy aldehydes, the low solubility of the higher olefins in water could limit the application of aqueous catalysts [34]. In contrast, the fluorous-soluble

$P[CH_2CH_2(CF_2)_5CF_3]_3$-modified rhodium catalyst system is an excellent catalyst for the hydroformylation of 1-decene at 100 °C under 11 bar CO/H_2 (1 : 1) in perfluoromethylcyclohexane and the aldehydes can be easily separated from the fluorous catalyst phase [4, 5]. High-pressure NMR and IR have revealed that $HRh(CO)\{P[CH_2CH_2(CF_2)_5CF_3]_3\}_3$ and $HRh(CO)_2\{P[CH_2CH_2(CF_2)_5CF_3]_3\}_2$ are the two species in solution under CO/H_2 (1 : 1) pressure [19]. Comparative kinetic studies have shown that the fluorous catalyst is about ten times slower than the triphenylphosphine-modified rhodium catalyst. The *n*/*i* selectivity is about the same for these two systems. Rhodium analysis of the product phases obtained by simply removing the product phase from the reactor through a deep-leg at 15 °C during nine subsequent (semi-continuous) hydroformylation of 1-decene showed the loss of about 1 ppm rhodium/mol aldehyde product [19]. While the use of heavier fluorous solvents and longer fluorous ponytails on the phosphine will further decrease the rhodium loss, the recovery of such low levels of rhodium can be achieved during the final purification of the aldehydes from the unreacted and isomerized olefins and other heavier side products.

The fluorous analog of the Wilkinson's catalyst, $ClRh\{P[CH_2CH_2(CF_2)_5CF_3]_3\}_2$, was used a a catalyst, in fluorous biphasic media, in hydrogenation, hydroboration, and hydrosilation reactions [35]. For example, the catalytic hydroboration of a great variety of alkenes and alkynes was achieved using catecholborane [17]. Catalyst loadings of 0.01–0.25 mol% were effective (25–40 °C, 1–40 h) and gave turnover numbers as high as 8500! The alkylboranes or alkenylboranes were readily separated from the fluorous catalyst and converted to alcohols by the subsequent addition of H_2O_2/NaOH to the product phase and the fluorous catalyst solution was re-used with minimal activity loss.

Earlier, several fluorous-soluble nickel catalysts were developed for the oligomerization of ethylene [14]. For example, a very active catalyst was obtained by the reaction of $Ni(cod)_2$ with $HOOCCOCH_2COCF(CF)_3$-$OCF_2(CF_3)CF_3]_3F$. Compared with conventional nickel catalysts, increased formation of *trans*-2-butene was observed for the fluorous catalysts and the products were readily separated from the catalyst phase.

The fluorous medium is especially suitable for oxidation reactions, as the solubility of dioxygen is very high in fluorous solvents [36] and perfluoroalkanes are extremely resistant to oxidation. In addition, most oxidation reactions lead to highly polar products, which are inherently less soluble in fluorous solvents, thus resulting in easy separation. The oxidation of cyclohexene using O_2 in the presence of fluorous-soluble cobalt phthalocyanato [5] and cobalt carboxylato $Co\{OOCCF(CF)_3$-[-$OCF_2(CF_3)CF$-$]_4F\}$ [14] complexes has been investigated. While the cobalt phthalocyanine complex remained intact during oxidation and could be separated from the products [5], the cobalt carboxylato complex decomposed under the reaction conditions [14]. It has recently been shown that the addition of 1 equiv of tris$\{N$-$[(CH_2)_2(CF_2)_7CF_3]\}$-1,4,7-triazacyclononane to $M\{OOC(CH_2)_2(CF_2)_7CF_3\}_2$ (M = Mn, Co) provides fluorous-soluble catalysts for alkane, arene, and alkene oxidation in the presence of *t*-BuOOH and O_2 [18]. The epoxidation of olefins in the presence of fluorous cobalt-porphyrins using O_2 and 2-methylpropionaldehyde at room temperature proceeds with excellent selectivities at high conversion levels [16] and the fluorous catalyst can be re-used without decreasing either catalytic activity or selectivity. Finally, diphenyl sulfide and dibenzothiophene were catalytically oxidized to the corresponding sulfones with fluorous-soluble iron phthalocyanine complexes [5]; the fluorous concept has also been extended to some other reactions [37].

Although fluorous biphase organometallic catalysis is still is its infancy, the results are already indicating that it may be a complementary approach to aqueous and ionic biphase organometallic catalysis. With the aim of finding for each catalytic chemical reaction its own perfectly designed catalyst (the chemzyme), the possibility of selecting from biphase systems ranging from fluorous to aqueous provides a powerful portfolio for catalyst designers.

References

[1] W. A. Herrmann, B. Cornils, *Angew. Chem., Int. Ed. Engl.* **1997**, *36*, 1049.

[2] J. H. Hildebrand, J. M. Prausnitz, R. L. Scott, *Regular and Related Solutions,* Van Nostrand Reinhold, New York, **1970**, Chapter 10.

[3] C. Reichardt, *Solvents and Solvent Effects in Organic Chemistry*, 2nd ed., VCH, Weinheim, **1990**.

[4] I. T. Horváth, J. Rábai, *Science* **1994**, *266*, 72.

[5] Exxon Research and Engineering Co. (I. T. Horváth, J. Rábai), EP 633.062.A1 (1995), US 5.463.082 (1995).

[6] S. G. DiMagno, P. H. Dussault, J. A. Schultz, *J. Am. Chem. Soc.* **1996**, *118*, 5312.

[7] D. P. Curran, S. Hadida, *J. Am. Chem. Soc.* **1996**, *118*, 2531.

[8] D. P. Curran, M. Hoshino, *J. Org. Chem.* **1996**, *61*, 6480.

[9] A. Studer, S. Hadida, R. Ferritto, S. Y. Kim, P. Wipf, D. P. Curran, *Science* **1997**, *275*, 823.

[10] J. H. Horner, F. N. Martinez, M. Newcomb, S. Hadida, D. P. Curran, *Tetrahedron Lett.* **1997**, *38*, 2783.

[11] A. Studer, P. Jeger, P. Wipf, D. P. Curran, *J. Org. Chem.* **1997**, *62*, 2917.

[12] A. Studer, D. P. Curran, *Tetrahedron* **1997**, *53*, 6681.

[13] D. E. Bergbreiter, J. G. Franchina, *J. Chem. Soc. Chem. Commun.*, submitted.

[14] M. Vogt, Ph.D. Thesis, Rheinisch-Westfälische Technische Hochschule, Aachen, Germany (1991).

[15] G. Pozzi, I. Colombani, M. Miglioli, F. Montanari, S. Quic, *Tetrahedron* **1996**, *52*, 11879.

[16] G. Pozzi, F. Montanari, S. Quic, *J. Chem. Soc., Chem. Commun.* **1997**, 69.

[17] J. J. J. Juliette, I. T. Horváth, J. A. Gladysz, *Angew. Chem.* **1997**, *109*, 1682; *Angew. Chem., Int. Ed. Engl.* **1997**, *36*, 1610.

[18] J.-M. Vincent, A. Rabion, V. K. Yachandra, R. H. Fish, *Angew. Chem.*, in press; *Angew. Chem., Int. Ed. Engl.*

[19] I. T. Horváth, G. Kiss, P. A. Stevens, J. E. Bond, R. A. Cook, E. J. Mozeleski, J. Rábai, I. Pelczer, unpublished results.

[20] J. W. Clayton, Jr., *Fluorine Chem. Rev.* **1967**, *1*, 197.

[21] R. Filler, in *Fluorine Containing Molecules* (Eds.: J. F. Liebman, A. Greenberg, W. R. Dolbier, Jr.), VCH, Weinheim, **1988**, Chapter 2.

[22] US Air Force (C. Tamborski, C. E. Snyder, Jr., J. B. Christian), US 4454349 (1984).

[21] R. Filler, in *Fluorine Containing Molecules* (Eds.: J. F. Liebman, A. Greenberg, W. R. Dolbier, Jr.), VCH, Weinheim, **1988**, Chapter 2.
[22] US Air Force (C. Tamborski, C. E. Snyder, Jr., J. B. Christian), US 4454349 (1984).
[23] S. Benefice-Malouet, H. Blancou, A. Commeyras, *J. Fluorine Chem.* **1985**, 30, 171.
[24] J. J. Kampa, J. W. Nail, R. J. Lagow, *Angew. Chem., Int. Ed. Engl.* **1995**, *34*, 1241.
[25] S. G. DiMagno, R. A. Williams, M. J. Therien, *J. Org. Chem.* **1994**, *59*, 6943.
[26] Minnesota Mining and Manufg. Co., GB 840.725 (1960).
[27] T. M. Keller, J. R. Griffith, *J. Fluorine Chem.* **1978**, 12, 73.
[28] H. V. R. Dias, H. J. Kim, *Organometallics* **1996**, *15*, 5374.
[29] N. Garelli, P. Vierling, *J. Org. Chem.* **1992**, *57*, 3036.
[30] R. P. Hughes, H. A. Trujillo, *Organometallics* **1996**, *15*, 286.
[31] M. A. Guillevic, A. M. Arif, I. T. Horváth, J. A. Gladysz, *Angew. Chem.* **1997**, *109*, 1685; *Angew. Chem., Int. Ed. Engl.* **1997**, *36*, 1612.
[32] J. Fawcett, E. G. Hope, R. D. W. Kemmitt, D. R. Paige, D. R. Russel, A. M. Stuart, D. J. Cole-Hamilton, M. J. Pane, *J. Chem. Soc., Chem. Commun.* **1997**, 1127.
[33] J. D. Jamerson, R. L. Pruett, E. Billig, R. A. Fiato, *J. Organomet. Chem.* **1980**, *193*, C43.
[34] I. T. Horváth, *Catal. Lett.* **1990**, *6*, 43.
[35] J. J. J. Juliette, I. T. Horváth, J. A. Gladysz, 10th Int. Symposium on Homogeneous Catalysis, Princeton, NJ, USA (**1996**), Abstract PP-A59.
[36] J. G. Riess, M. Le Blanc, *Pure Appl. Chem.* **1982**, *54*, 2383.
[37] B. Cornils, *Angew. Chem.* **1997**, *109*, 2147; *Angew. Chem., Int. Ed. Engl.* **1997**, *36*, 2057.

7.3 Nonaqueous Ionic Liquids (NAILs)

Hélène Olivier

7.3.1 Introduction

The role of the solvent in organic and catalyzed homogeneous reactions is of the utmost importance. Interactions of solutes with the solvent are often specific. The energy associated with solvation is often large, especially for ions, and is a major factor influencing the rate and selectivity of reactions [1] (cf. Section 2.3).

Many coordination chemists are concerned with enhancing the chemical reactivity and/or selectivity of metal complexes, particularly for applications in catalysis. For example, a coordinatively unsaturated cationic metal complex, which is often the intermediate implicated in activation of organic molecules, is considered. Wanted is a solvent

- able to solubilize the metal ion while maintaining its ionic character, and
- which involves a weak, labile metal–solvent bond.

The solvent has not to compete with the reactant for the coordination on the active metal center. This is why the development of biphasic catalysis should parallel that of highly polar new solvents. The use of water as a second phase may have its limitations. Water can be coordinating toward the active metal center or react with the metal–carbon bond. In addition, the water-solubility of the starting material often proves to be too low.

7.3.2 The Search for Suitable Polar Solvents

From this point of view, nonaqueous room-temperature ionic liquids (NAILs), or "molten salts," are of particular interest. They are generally composed of large organic cations, e. g., tetraalkylammonium, tetraalkylphosphonium, trialkylsulfonium, *N*-alkylpyridinium, pyrazolium or *N*,*N'*-dialkylimidazolium, associated with inorganic or organic anions. They proved to be a new class of solvents, with a spectrum of physical and chemical properties much larger than that of organic or inorganic classical solvents such as water [2]. The anions are essentially responsible for the chemical properties, the most important of which,

Table 1. Classification of some anions according to their coordination ability toward some transition metal complexes.

Relative anion coordination ability Hardness/softness	Examples of typical anions	Typical transition metal complexes
Strong/Hard	Cl^-, $SnCl_3^-$, $GaCl_3^-$	A high proportion of transition metal complexes
Weak/Hard	$AlCl_4^- > Al_2Cl_7^- > Al_3Cl_{10}^-$	Ni^{2+}, W
Weak/Borderline	OTf $\sim$ $BF_4^- > PF_6^- > SbF_6^-$, $CuCl_2^-$	Transition metal groups 8, 9, 10
Weak/Soft	BAr_4^- (the hardness depends on the Ar substituents)	Many complexes, including the hard Cp_2ZrR^+

for coordination chemists, is the coordinating ability and/or reactivity of anions toward the metal center. The coordination properties of anions have been the subject of many studies in recent reviews [3]. These properties depend in large part on the nature of the anions themselves (size, charge), but also on the hardness of the metal center, its oxidation state, and its surrounding ligands. Here the qualitative hard and soft acids and bases (HSAB) concept developed by Pearson can be applied to classify some typical anions in respect of their softness or hardness to transition metal complexes (Table 1).

NAILs are becoming increasingly important as a nonconventional, novel class of solvents for biphasic catalytic reactions [4].

Ionic transition metal complexes of the type $[L_{n-1}M]^+X^-$ are particularly suitable for use in these media because of their excellent ability to dissolve ionic metal salts. The characteristics of NAILs have to be tuned to the nature of the counterion X^- in order to avoid competition with the anions involved. Catalytic reactions using nonionic transition metal complexes can also be performed in NAILs. However, in that case, ligands have to be specially designed to render the complex soluble and nonextractable in the product second phase. Finally, some noncatalyzed organic reactions can also be successfully performed in these media.

7.3.3 Some Typical Room-temperature Ionic Liquids

NAILs were developed, many years ago, mainly by electrochemists looking for "ideal" electrolytes for technical use in batteries or in metal electrodeposition. The properties required for such applications are a wide liquid-range temperature and low vapor pressure, a large electrochemical window, and a high ionic mobility [5]. In addition, NAILs show a wide applicability for spectroscopic studies because their chemical nature can preclude the occurrence of both solvation and solvolysis [6]. Finally, due to their large spectrum of miscibilities with other liquids, NAILs have proven to be useful as solvents in separation technology, in particular, some of them behave as liquid clathrates [7].

Cations are responsible to a large extent for the melting point of the ionic liquid. *N,N'*-Dialkylimidazolium cations, e.g., 1-butyl-3-methylimidazolium or 1-ethyl-3-methylimidazolium, are very often preferred because of their low melting point and high thermal and chemical stabilities. The most frequently described anions are:

- halides, tin and germanium halides,
- BF_4^-, SbF_6^-, PF_6^-, FSO_3^-,
- $CF_3SO_3^-$, $(CF_3SO_2)_2N^-$, $CF_3CO_2^-$ [8]
- tetraalkyl borides [9]

Some other anions have been envisioned [10].

Most of these anions are not water-sensitive. Miscibilities of the corresponding ionic liquids with organic reactants can be altered by varying the alkyl chain length of the dialkylimidazolium cation.

On the other hand, with ammonium halides aluminum halides can form stoichiometric and nonstoichiometric mixtures containing the X^-, AlX_4^-, $Al_2X_7^-$ and/or $Al_3X_{10}^-$ anions [11]. Dramatic changes in the solvation characteristics and electrochemical windows of these ionic liquids can be achieved simply by varying the ratio of aluminum halide to organic salt in the mixture. Haloaluminate salts containing polynuclear anions are potentially strong Lewis acids due to the ability of polynuclear anions to dissociate into aluminum halide and mononuclear anion. Gutmann acceptor numbers have been used as a qualitative nonthermodynamic measure of their Lewis acidity. Acidic chloroaluminates are extremely poor donor and strong acceptor media (with a value around 100, between water and trifluoroacetic acid, depending on the melt composition) compared with other classical organic solvents [12]. These ionic liquids have been used in few organic reactions [13]. In a similar way, trialkylaluminum and alkylaluminum chloride derivatives form with dialkylimidazolium chloride low-melting liquids over a wide range of composition. The presence of an aluminum–carbon bond proves useful in catalysis [14]. Liquid salts containing fluorinated anions such as $AlEt_2F_2^-$ and $Al_2Et_5F_2^-$ have also been obtained [15]. Obviously, aluminum-halide-based salts are water-sensitive. Mixtures of LiCl with $AlEtCl_2$ also form room-temperature liquids for appropriate molar ratios [16].

In a similar way, mixtures of copper chloride and ammonium or phosphonium chlorides give liquid salts. Depending on the molar ratio, various chlorocuprate anions may be formed, e.g., $CuCl_2^-$, $Cu_2Cl_3^-$ [17].

Many physico-chemical properties used to quantify the properties of solvents often prove inappropriate to characterize NAILs [18]. Although their dielectric constant is generally quite small, ionic liquids behave as strongly dissociating solvents. Solvation in these systems differs from that in molecular solvents. For example, it is common to state that polar solvents solvate mostly polar or ionic solutes, while nonpolar solvents solvate weakly polar and nonpolar solutes. According to this principle, "molten salts" should solvate only strongly polar or ionic species. However, a wide variety of species, ranging from metal chlorides to benzene, are soluble in these media.

In addition to their tunable coordinating ability toward transition metal complexes, ionic liquids present other useful properties for chemical reactions and catalytic applications. Among these are a maximum working range between melting and boiling point and a high heat conductivity which permits a rapid dispersal of heat of reaction. A further advantage is their adjustable miscibility with organic liquids.

7.3.4 Applications in Organic Synthesis

7.3.4.1 Salts Containing Strongly Coordinating Anions Stabilize Anionic Complexes

The first report of catalytic reactions carried out in ionic liquids was published by Parshall [19], using $PtCl_2$ as a hydrogenation catalyst precursor dissolved in tetrabutylammonium tin trichloride. In this way, the catalyst could be re-used several times. Remarkably Parshall was able to characterize in situ the active species, i.e., $[HPt(SnCl_3)_4]^{3-}$.

Subsequently, Knifton described various carbonylation reactions catalyzed by ruthenium and palladium complexes, such as hydroformylation and carboxylation of olefins, and glycol synthesis, using phosphonium halides as the polar phase [20]. In the same way Jenck [21] has carried out the carboxylation reaction of diacetoxybutene into dialkyl adipate catalyzed by palladium in tetrabutylammonium bromide. Very recently, the Heck reaction of aryl halides and butyl acrylate, catalyzed by palladium, has been described in tetraalkylammonium or phosphonium halides as reaction medium [22].

The main advantage in using these coordinating anions is that they stabilize the active species. This is particularly obvious in the case of palladium complexes, whose tendency to decompose into black metal is well documented. However, the melting point of such salts is rather high.

7.3.4.2 Salts Containing Weakly Coordinating Anions for Cationic and Molecular Complexes

7.3.4.2.1 Hard Anions (Chloroaluminates)

Acidic chloroaluminates (hard anions) have been used in reactions which need the presence of Lewis acidity (Ziegler–Natta-type catalysts) and for which there is no interaction between the reaction products and aluminum chloride. Polymerization of ethylene catalyzed by $(C_5H_5)_2TiCl_2/Al_2Me_3Cl_3$ complexes has been carried out in acidic chloroaluminate ionic liquids [23].

The cationic nickel complex $[\eta\text{-allylNi}(PR_3)]^+$, already described by Wilke et al. [24] as an efficient catalyst precursor for olefin dimerization when dissolved in chlorinated organic solvents, proved to be very active in acidic chloroaluminate ionic liquids. In spite of the strong potential Lewis acidity of the medium, a similar phosphine effect is observed. However, there is competition for the phosphine between the soft nickel complex and the hard aluminum chloride coming from the dissociation of polynuclear chloroaluminate anions. Aromatic hydrocarbons, when added to the system, can act as competitive bases, thus preventing the decoordination of phosphine ligand from the nickel complex [25]. Performed continuously, dimerization of propene and/or butenes with this biphasic system compares favorably to the Dimersol® homogeneous process in terms of yields and selectivity into olefinic dimers.

The same acidic chloroaluminate ionic liquids have been used as a solvent of tungsten aryloxide complexes for metathesis of olefins [26] (cf. Section 6.13).

7.3.4.2.2 Borderline Anions

The cationic rhodium complex $[Rh(nbd)(PPh_3)_2]^+$ (nbd = norbornadiene), previously described by Osborn as a catalyst precursor for the hydrogenation of olefins when dissolved in acetone for example [27], has proved to be active in PF_6^-, $CuCl_2^-$ or SbF_6^- based salts. No rhodium is extracted in the hydrocarbon phase. The catalyst can be used repeatedly [28]. Ruthenium and cobalt complexes have also been used for hydrogenation of olefins and diolefins [29a].

Asymmetric hydrogenation has been successfully achieved in the same type of medium using $[RuCl_2(BINAP)]_2NEt_3$ or $[Rh(nbd)(DIOP)]^+$ (DIOP = 2,3-O-Isopropylidene-2,3-dihydroxy-1,4-bis(diphenylphosphino)butane) complexes as catalyst precursors [28, 29b].

The same weakly coordinating anion-based salts can be used to perform reactions catalyzed by molecular transition metal complexes. For example, the hydroformylation of olefins using $Rh(acac)(CO)_2$ and PPh_3 as catalyst precursors has a high reaction rate when carried out in the biphasic system. However, in contrast to the hydrogenation reaction, a small part of the active catalytic species dissolves in the organic phase. By rendering the phosphine ligand soluble in the ionic liquid by the introduction of ionic substituents, the extraction of rhodium can be avoided.

Another example is the dimerization of conjugated diene. When dissolved in tetrahydrofuran (THF), "$Fe(NO)_2$" is a well-known catalyst for the selective dimerization of butadiene into 4-vinylcyclohexene. When dissolved in PF_6^-- or SbF_6^--based salts, the solvent effect on reaction rate is very important. However, part of the catalyst dissolves in the hydrocarbon phase and slowly deactivates. Despite these drawbacks, turnover numbers are higher than in THF [30].

7.3.4.3 Salts Containing Acidic Chloroaluminate Anions as Solvents for Protons or Carbenium Ions

Pure chloroaluminate-based ionic liquids have no proton-donating ability. However, these liquids can demonstrate Brønsted acidity as a result of intentional or unintentional protonic species. These protons behave as superacids with Hammett acidities ranging from -12.6 to -18. The proton speciation in acidic salts has been the subject of numerous papers [31]. With increasing $AlCl_3$ molar fraction in the ionic liquid, the solvent interaction is weaker and the Brønsted superacidity of the proton is increased. Protonation of aromatic hydrocarbons to generate several arenium ions and carbocations has been achieved in these solvents [32]. Thus it is expected that cationic catalyzed reactions could also be achieved in these media. Chloroaluminates are in fact good solvents and acidic catalysts for paraffin/olefin alkylation, e. g., isobutane/ethylene or isobutane/butene [33]. Other acid-catalyzed reactions such as isobutene or polymerization of light olefins and aromatic hydrocarbon alkylation have been achieved in these media [34]. Oxidative electropolymerization of benzene to polyphenylene has been described using haloaluminates [35].

These media have also proved to be suitable stoichiometric solvents for Friedel-Crafts acylation reactions of aromatic hydrocarbons and ferrocene [36].

7.3.4.4 Supported Ionic Liquid Catalysis

NAILs have been used as reaction media for supported catalysis. For this application, the main advantage, compared with conventional solvents, is their low volatility which allows the operating temperatures to be extended. This technique has been applied, for the Wacker catalyst, to the oxidation ethylene to acetaldehyde (cf. Section 6.18). It also proven to be useful in developing efficient catalysts for other processes [37].

7.3.4.5 Solvents for Pericyclic Processes: the Diels–Alder Reaction

Polar solvents, particularly water, dramatically enhance rate and stereoselectivity of $[4\pi + 2\pi]$ cycloaddition reactions [38]. A similar effect could be expected in NAILs. In fact, the condensation rate of, for example, cyclopentadiene with methyl vinyl ketone is much faster in BF_4^--based salts than in water [39]. Crotonaldehyde (methacrolein) reacts with cyclopentadiene at $-25\,^{\circ}C$ affording the desired products in the presence of dialkylimidazolium salts. The *endo:exo* selectivity is always greater than 90:10 for crotonaldehyde–cyclopentadiene [40]. As for water and other polar solvents, solvophobic interactions or hydrogen bonding could be used to explain this rate acceleration.

7.3.5 Concluding Remarks

NAILs constitute a unique class of versatile solvents for organic biphasic reactions. They are the only solvents capable of dissociating the ionic pairs without solvating the cation. The nature of quaternary ammonium or phosphonium cations can be varied between wide limits, thus altering the miscibility of the corresponding salts with organic reactants and products. The use of these solvents could probably be extended to many other catalyzed and noncatalyzed reactions. From a practical point of view, their low vapor pressure prevents atmospheric pollution, thereby complying with the criteria required for industrial development. From a fundamental point of view, we still have to elucidate the types of interactions between active species or transition states and the anions and cations present in these salts, in order to draw up a rationalization of the selection of the solvent.

References

[1] C. Reichardt, *Solvents and Solvent Effects in Organic Chemistry,* VCH, Weinheim, **1990**.

[2] T. A. O'Donnell, *Superacids and Acidic Melts as Inorganic Chemical Reaction Media,* VCH, New York, **1993**.

[3] (a) S. H. Strauss, *Chem. Rev.* **1993**, *93*, 927; (b) P. K. Hurburt, D. Van Seggen, J. J. Rack, S. H. Strauss, *Inorganic Fluorous Chemistry Towards the 21st Century* (Eds.: J. S. Thrasher, S. H. Strauss), ACS Symp. Series No. 555, **1996**, 338.

[4] (a) Y. Chauvin, H. Olivier, *CHEMTECH* **1995**, *25*, 26; (b) K. R. Seddon, *Kinet. Catal.* **1996**, *37*, 693; (c) K. R. Seddon, *J. Chem. Tech. Biotechnol.* **1997**, *68*, 351; (d) H. E. Brynzda, E. B. Coughlin, G. Proulx, *Proc. 212nd ACS National Meeting*, Orlando, FA, **1996**, INORG-007.

[5] (a) C. L. Hussey, *Pure Appl. Chem.* **1988**, *60*, 1763; (b) R. A. Osteryoung, in *NATO ASI Series C*, Vol. 202 (Eds.: G. Mamantov, R. Marassi), D. Reidel, Amsterdam, **1986**, 329; (c) C. L. Hussey, in *Chemistry of Nonaqueous Solutions* (Eds.: G. Mamantov, A. I. Popov), VCH, New York, **1994**.
[6] K. R. Seddon, in *Molten Salt Chemistry* (Eds.: G. Mamantov, R. Marassi), D. Reidel, Boston, **1987**, 365.
[7] J. L. Atwood in *Coordination Chemistry of Aluminum* (Ed.: G. H. Robinson), VCH, New York, **1993**.
[8] (a) J. S. Wilkes, M. J. Zaworotko, *J. Chem. Soc., Chem. Commun.* **1992**, 965; (b) J. Fuller, R. T. Carlin, H. C. De Long, D. Haworth, *J. Chem. Soc., Chem. Commun.* **1994**, 299; (c) R. T. Carlin, H. C. De Long, J. Fuller, P. C. Trulove, *J. Electrochem. Soc.* **1994**, *141*, 7, L73; (d) P. Bonhôte, A. P. Dias, N. Papageorgiou, K. Kalyanasundaram, M. Grätzel, *Inorg. Chem.* **1996**, *35*, 1168.
[9] W. T. Ford, R. J. Hauri, D. J. Hart, *J. Org. Chem.* **1973**, *39*, 3916.
[10] T. B. Scheffer, M. S. Thomson, *Proc. 7th Int. Symposium on Molten Salts* (Eds.: C. L. Hussey, J. S. Wilkes, S. N. Flengas, Y. Ito), Physical Electrochemistry and High Temperature Materials Division. The Electrochemical Society, Pennington NJ, **1990**, Vol. 90-17, p. 281.
[11] R. A. Osteryoung, in *Molten Salt Chemistry* (Eds.: G. Mamantov, R. Marassi), D. Reidel, Boston, **1987**, 329.
[12] A. Zawodzinski, R. A. Osteryoung, *Inorg. Chem.* **1989**, *28*, 1710.
[13] R. M. Pagni, in *Advances in Molten Salts Chemistry* (Eds.: G. Mamantov, C. B. Mamantov, J. Braunstein), Elsevier, New York, **1987**, p. 211.
[14] (a) Y. Chauvin, F. Di Marco-Van Tiggelen, H. Olivier, *J. Chem. Soc., Dalton Trans.* **1993**, 1009; (b) B. Gilbert, Y. Chauvin, H. Olivier, F. Di Marco-Van Tiggelen, *ibid* **1995**, 3867.
[15] Siemens-Schuckertwerke AG, FR 1.461.819 (1966).
[16] Institut Français du Pétrole (Y. Chauvin, H. Olivier, R. F. De Souza), FR 2.736.562 (1997).
[17] D. D. Axtell, J. T. Joke, *Inorg. Chem.* **1973**, *12*, 1265.
[18] W. B. Harrod, N. J. Pienta, *J. Phys. Org. Chem.* **1990**, *3*, 534.
[19] G. W. Parshall, *J. Am. Chem. Soc.* **1972**, *94*, 8716; Du Pont de Nemours (G. W. Parshall), US 3.657.368 (1972); US 3.832.391 (1974); US 3.919.271 (1975).
[20] J. F. Knifton, *J. Am. Chem. Soc.* **1981**, *103*, 3959; *idem, J. Mol. Catal.* **1988**, *47*, 99; Texaco Inc. (J. F. Knifton), US 4.332.914 (1982), US 4.362.822 (1982).
[21] S. Duprat, H. Deweerdt, J. Jenck, P. Kalck, *J. Mol. Catal.* **1993**, *80*, L9.
[22] D. E. Kaufmann, M. Nouroozian, H. Henze, *Synlett* **1996**, 1091.
[23] (a) R. T. Carlin, J. S. Wilkes, *J. Mol. Catal.* **1990**, *63*, 125; (b) R. T. Carlin, R. A. Osteryoung, J. S. Wilkes, J. Rovang, *Inorg. Chem.* **1990**, *29*, 3003.
[24] G. Wilke, B. Bogdanovic, P. Hart, O. Heimbach, W. Kroner, W. Oberkirch, K. Tanaka, E. Steinrücke, D. Walter, H. Aimmerman, *Angew. Chem., Int. Ed. Engl.* **1966**, *5*, 151.
[25] (a) Y. Chauvin, B. Gilbert, I. Guibard, *J. Chem. Soc., Chem. Commun.* **1990**, 1716; (b) Y. Chauvin, S. Einloft, H. Olivier, *Ind. Eng. Chem.* **1995**, 1149; (c) Y. Chauvin, H. Olivier, C. Wyrvalski, L. C. Simon, R. F. De Souza, *J. Catal.* **1997**, *165*, 275; (d) S. Einloft, F. F. Dietrich, R. F. De Souza, J. Dupont, *Polyhedron* **1996**, *15*, 3257.
[26] Y. Chauvin, H. Olivier, F. Di Marco-Van Tiggelen, B. Gilbert *Proc. 9th Int. Symposium on Molten Salts* (Eds: C. L. Hussey, D. S. Newman, G. Mamantov, Y. Ito), The Electrochemical Society, Pennington, NJ, **1994**, p. 617.
[27] R. R. Schrock, J. A. Osborn, *J. Am. Chem. Soc.* **1976**, *98*, 2134.
[28] Y. Chauvin, L. Mussmann, H. Olivier, *Angew. Chem., Int. Ed. Engl.* **1995**, *34*, 23.

[29] (a) P. A. Z. Suarez, J. E. L. Dullius, S. Einloft, R. F. De Souza, J. Dupont, *Polyhedron* **1996**, *15*, 1217; *idem*, *Inorg. Chim. Acta*, **1997**, *255*, 207; (b) A. L. Monteiro, F. Zinn, R. F. De Souza, J. Dupont, *Tetrahedron: Asymm.* **1997**, *8*, 177.
[30] Institut Français du Pétrole (H. Olivier, Y. Chauvin, R. F. De Souza), FR 2.728.180 (1995).
[31] (a) J. E. Campbell, K. E. Johnson, *J. Am. Chem. Soc.* **1995**, *117*, 7791; *idem, Inorg. Chem.* **1993**, *32*, 3809; (b) J. E. Campbell, K. E. Johnson, J. R. Torkelson, *Inorg. Chem.* **1994**, *33*, 3340; (c) P. C. Trulove, D. K. Sukumaran, R. A. Osteryoung, *J. Phys. Chem.* **1994**, *98*, 141.
[32] M. Ma, K. E. Johnson, *J. Am. Chem. Soc.* **1995**, *117*, 1508.
[33] (a) H. Olivier, Y. Chauvin, A. Hirschauer, *Proc. 203rd ACS National Meeting,* Division of Petroleum Chemistry, San Francisco **1992**, *37*, 780; (b) Y. Chauvin, A. Hirschauer, H. Olivier, *J. Mol. Catal.* **1994**, *92*, 155.
[34] (a) BP Chemicals (A. K. Abdul-Sada, K. R. Seddon, N. J. Steward), PCT Int. WO 95/21.872 (1995); (b) BP Chemicals (A. K. Abdul-Sada, P. W. Ambler, K. R. Seddon, N. J. Steward), PCT Int. WO 95/21.871 (1995); BP Chemicals (A. K. Abdul-Sada, M. P. Atkins, B. Ellis), PCT Int. WO 95/21.806 (1995); (c) M. Goledzinowski, V. I. Birss, *Ind. Eng. Chem. Res.* **1993**, *32*, 1795.
[35] L. M. Goldenberg, R. A. Osteryoung, *Synth. Metals* **1994**, *64*, 63.
[36] (a) J. A. Boon, J. A. Levisky, J. L. Plug, J. S. Wilkes, *J. Org. Chem.* **1986**, *51*, 480; (b) J. K. D. Surette, L. Green, R. D. Singer, *J. Chem. Soc., Chem. Commun.* **1996**, 2753.
[37] (a) V. Rao, R. Datta, *J. Catal.* **1988**, 377; (b) R. Datta, H. Joshi, G. Deo, in *Spring Meeting of The Electrochemical Society,* Los Angeles, **1996**, Abstracts 1139, 1413.
[38] U. Pindur, G. Lutz, C. Otto, *Chem. Rev.* **1993**, *93*, 741.
[39] Instiut Français du Pétrole (H. Olivier, A. Hirschauer), FR Demande 96.1692 (1996).
[40] J. Howarth, K. Hanlon, D. Fayne, P. McCormac, *Tetrahedron Lett.* **1997**, 3097.

7.4 The Amphiphilic Approach

Paul C. J. Kamer, Piet W. N. M. van Leeuwen

7.4.1 Separation Methods

Especially when a catalyst is expensive and produces low-value bulk chemicals or when it is toxic, its separation from the reaction medium is pivotal [1 a]. In some cases, the high costs of sophisticated ligands, e. g., for asymmetric catalysis, render recycling of the ligand equally important. A challenging separation principle is based on the difference in solubility of the reagents and the catalyst in two immiscible solvents. There are basically two ways of applying this principle. Most extensively investigated is two-phase catalysis (Figure 1; cf. Section 4.2). The catalyst is located in a solvent, usually water, that is immiscible with the solvent containing the reagents. The catalytic reaction occurs in the catalytic phase or at the phase boundary. In some instances the product forms the second, immiscible solvent, making the use of a second solvent redundant [1 b, c].

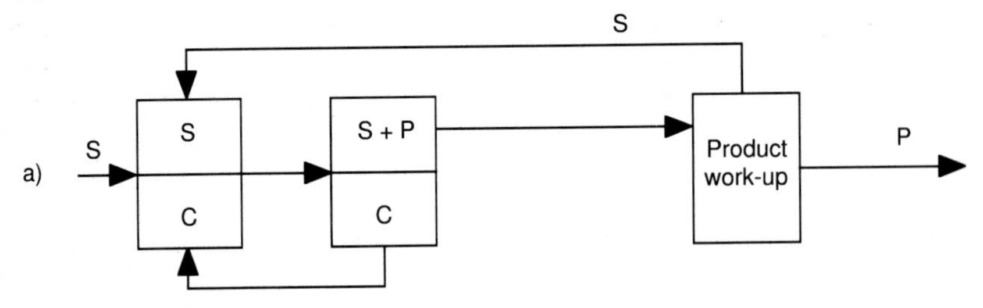

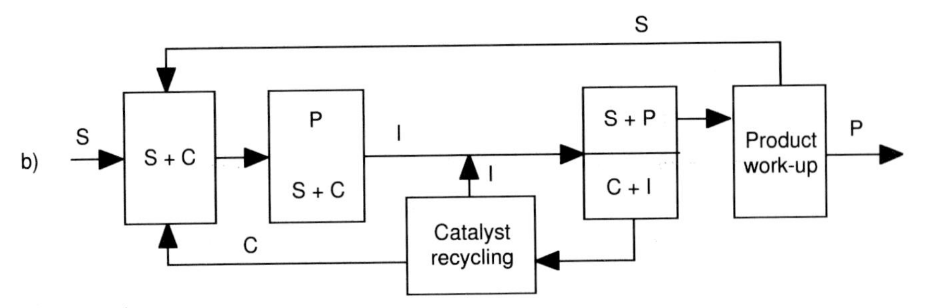

Figure 1. Simplified flow diagrams for (a) two-phase catalysis and (b) the extraction concept [1 b, c] (C = catalyst, S = substrate, P = product, I = immiscible solvent).

In the other approach, the catalytic reaction is initially performed in a homogeneous reaction medium and in a second step, i.e., not in the reactor, the catalyst is extracted (Figure 1 b). Several variations on this extraction concept can be envisaged; only the most common ones will be discussed.

7.4.1.1 Two-phase Catalysis

This concept was first realized industrially in the Shell higher-olefin process (SHOP) designed by Keim et al. [2, 3] (cf. Section 7.1). Ethylene is oligomerized in a polar phase of 1,4-butanediol that also contains the nickel complex of a phosphino carboxylate. The product α-olefins (C_{4-20}) are not soluble in the polar solvent and can be separated easily. In general, the distribution factor of catalysts and reagents between two organic phases is not very high and the use of water, which is immiscible with most organic products, as the phase containing the catalyst is an apparent remedy. Catalysis in water required the development of water-soluble ligands (cf. Section 3.2).

Aqueous biphasic catalysis allows easy separation of product and catalyst. Conducting the catalysis in water can have both detrimental and beneficial effects on the selectivity. The application of this system, however, is limited by the (in)solubility of the substrate in water. The use of co-solvents or surface-active agents may enhance phase mixing, but is likely to induce catalyst loss by increasing the catalyst concentration in the product phase or giving rise to stable emulsions. Interestingly, Horváth [4] recently introduced a novel concept for biphasic catalysis which may be especially suitable for the hydroformylation of hydrophobic higher alkenes. This system consists of a fluorocarbon-rich phase containing rhodium complexed to a fluorinated ligand $P[CH_2CH_2(CF_2)_5CF_3]_3$, and a common organic solvent. The advantage of this fluorous biphasic system (FBS) is that the product aldehydes are less soluble in the fluorous phase than the alkenes because of their higher polarity (cf. Section 7.2).

7.4.1.2 The Extraction Concept

The amount of work done in this field is rather modest compared with that in twophase catalysis, although it offers some major advantages. Catalysis is conducted in a homogeneous medium, which may be the pure substrate, with concomitantly high reaction rates, since substrate solubility is not limited. Naturally, the advantages have to offset the extra costs arising from the continuous catalyst recycle. For many years, the extraction concept has been applied industrially in the Du Pont adiponitrile process [5 a]. The nickel arylphosphite catalyst that is exploited in the hydrocyanation of butadiene is separated from the polar high-boiling adiponitrile by extraction with cyclohexane (cf. Section 6.5).

The extraction concept mostly utilizes an amphiphilic ligand system that allows transfer of the catalyst back and forth from an organic to an aqueous layer by varying the pH of the system. After the reaction the organic phase is washed with water of an appropriate pH. At this pH the catalyst and excess ligand become water-soluble, either by protonation or deprotonation, and are extracted in the aqueous phase. The organic products can thus be separated from the catalyst. For some catalytic reactions it can be advantageous that water is not present during the reaction. The cycle can be completed by neutralization of the aqueous phase and extracting the catalyst and excess ligand into a new batch of substrate. A disadvantage of this system is that it also produces salts, albeit in catalytic quantities only.

For cobalt, this method has been commercialized in the Kuhlmann process, using amphiphilic $HCo(CO)_4$ [5b]. The Kuhlmann process (now the Exxon process) involves cobalt-catalyzed hydroformylation of higher alkenes, for which the flowsheet – a liquid/liquid separation – is shown in Figure 2. In this process the hydroformylation is done in the organic phase consisting of alkene and aldehyde. A loop reactor, or a reactor with an external loop to facilitate heat transfer, is often used.

A liquid/liquid separation of product and catalyst is performed in separate vessels after the reaction has taken place. The reaction mixture is sent to a gas separator and from there to a countercurrent washing tower in which the effluent is treated with aqueous Na_2CO_3. The acidic $HCo(CO)_4$ is transformed into the water-soluble conjugate base $NaCo(CO)_4$. The product is scrubbed with water to remove the traces of base. The oxo crude goes to the distillation unit.

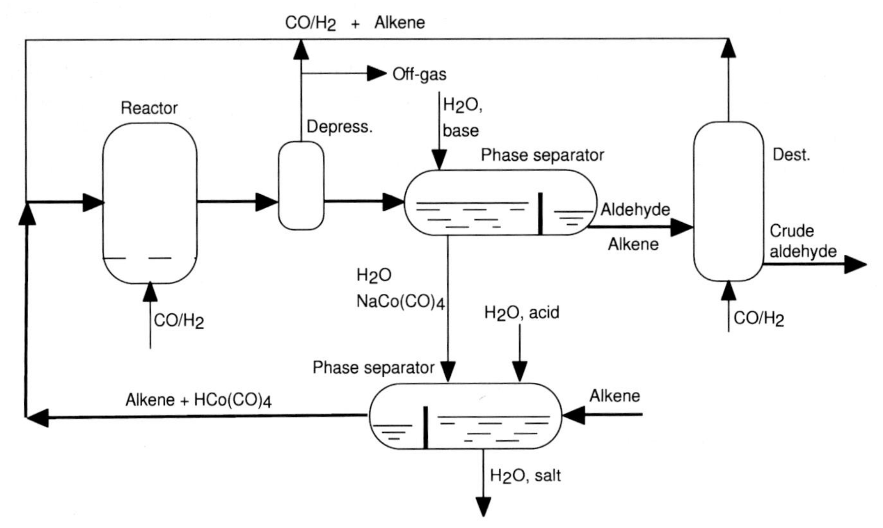

Figure 2. Simplified scheme for the Kuhlmann hydroformylation process (Dest. = distillation unit).

The basic solution in water containing $NaCo(CO)_4$ is treated with sulfuric acid in the presence of syngas, and $HCo(CO)_4$ is regenerated. This can be extracted from water into the substrate, alkene, and is returned to the reactor. Compared with other schemes (former processes of BASF, Ruhrchemie) the elegant detail of the Kuhlmann process is that the cobalt catalyst is not decomposed via (partial) oxidation but is left in the system as the tetracarbonylcobaltate.

There are several ways of applying the extraction concept *without* utilization of an amphiphilic ligand system. Three interesting alternatives will be discussed. The first concerns the hydroformylation of higher alkenes ($>C_7$) using a hydrido rhodium carbonyl catalyst [6]. Rhodium can be removed from the organic reaction medium by extraction with an aqueous solution containing sulfonated or carboxylated nitrogen ligands, such as bipyridines or phenanthrolines, at a reduced syngas pressure. Contacting the separated aqueous phase with a new organic layer under high CO/H_2 pressure leads to reversible decomplexation of the nitrogen ligand and regenerates the desired rhodium carbonyl catalyst. Evidently this method is limited to metal carbonyl catalysts. A similar recycling procedure has been patented for hydrido cobalt carbonyl catalysts [7], in which case TPPTS serves as complexing agent.

A second alternative for the separation of hydroformylation products from a rhodium [8] or cobalt [9] catalyst is to perform the catalytic reaction in a polar solvent using complexes of monosulfonated trialkyl- or triarylphosphines (e. g., TPPMS). Addition of both water and an apolar solvent such as cyclohexane gives a biphasic system. After separation of the apolar layer, the added apolar solvent must be stripped from the products. In order to form a homogeneous system with new substrate alkene, the polar catalytic phase must be freed from water, e. g., by azeotropic or extractive distillation. Clearly, these extra co-distillation steps are energy-consuming.

Thirdly, separation of catalyst and products can be conveniently achieved by extraction with water if the products are water-soluble, e. g., in the hydroformylation of allyl alcohol [10, 11]. In the ARCO process the hydroformylation products are extracted (more than 99 %) from the organic reaction medium containing the conventional $HRh(CO)(PPh_3)_3$ catalyst, excess PPh_3, and diphosphines. The losses of rhodium and phosphines due to solubility in water are negligible.

7.4.2 Use of Amphiphilic Phosphines

7.4.2.1 Catalysis Using Amphiphilic Ligands

A relatively new approach involves the functionalization of phosphines or other ligands with weakly basic or acidic functionalities. An early study reports on a modified cobalt catalyst [12]. This cobalt carbonyl complex containing the

$P(CH_2CH_2NEt_2)_3$ ligand was claimed to be extractable into dilute carbonic acid and could be re-extracted into an organic phase by simply reducing the CO_2 pressure.

For rhodium, several studies concerning the use of amphiphilic ligands have been reported. Rhodium catalysts dervied from tris(2-pyridyl)phosphine achieve selective hydroformylation of 1-hexene both in a homogeneous acetophenone system and, at a much lower rate, in a two-phase water/1-hexene system [13]. Attempts to extract the rhodium complex from the homogeneous system with water were not successful; the use of HCl or HBF_4 resulted in rapid evolution of H_2 and about half the rhodium could not be extracted from the orange, organic phase.

By using thiolate ligands containing an amino group, recoverable dirhodium μ-thiolato complexes have been obtained [14] (cf. Section 3.2.3). Addition of dilute aqueous sulfuric acid to the hydroformylation reaction mixture causes the immediate and complete precipitation of the yellow ammonium sulphate salt. This solid is insoluble in water and common organic solvents, but can be easily regenerated – by addition of an aqueous base and extraction with an organic solvent – and re-used without loss of activity.

Amino-derivatized SKEWPHOS (2,4-bis(diphenylphosphino)pentane) has been used in asymmetric hydrogenation catalysis [15–17] (cf. Sections 6.2 and 6.9). NMR analysis showed that a ten-fold excess of HBF_4 is sufficient to protonate reversibly all four amino groups in the [Rh(diene)(SKEWPHOS)]BF_4 complex. Recycling of the catalyst after enantioselective hydrogenation of dehydroamino acid derivatives in methanol is achieved by acidification with aqueous HBF_4 followed by extraction of the product with Et_2O. Immobilization of the protonated SKEWPHOS rhodium complex on a Nafion support has been studied as well [18].

An extensive study on rhodium-catalyzed hydroformylation using amphiphilic phosphines has been performed by Van Leeuwen and co-workers [19–22] (cf. Section 3.2.2). The amphiphilic ligands based on triphenylphosphine form complexes having an amphiphilic character when coordinated to rhodium (Structures **1**–**9**). In the hydroformylation of 1-octene the selectivity is unaffected by these additional functionalities, as compared with the parent PPh_3 [19]. Also, most ligands showed reaction rates comparable with that of PPh_3. The use of pyridylphosphines as ligands resulted in faster hydroformylation catalysis; this was ascribed to the higher χ-value [23], which is known to increase the reaction rate as a result of a less strong bonding of the CO molecules to the rhodium centre facilitating alkene coordination [24].

For the rhodium-catalyzed hydroformylation of higher alkenes, novel amphiphilic diphosphines have been reported (Structures **10**–**16**), based on BISBI (2,2′-bis(diphenylphosphino)methyl-1,1′-biphenyl), XANTHAM, POPpy and POPam, which can be used in the rhodium recycling system [21, 22].

The series of new diphosphines have been tested in the rhodium-catalyzed hydroformylation (Table 1) of 1-octene. The reaction rates for the catalysts derived from the pyridyl-modified ligands **10**–**12** and POPpy are higher than those of BISBI and **13**, which is consistent with earlier observations made for the pyridyl-modified triphenylphosphines [19].

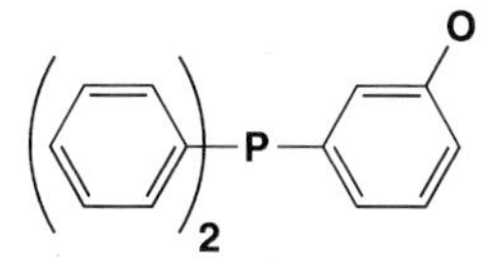

1 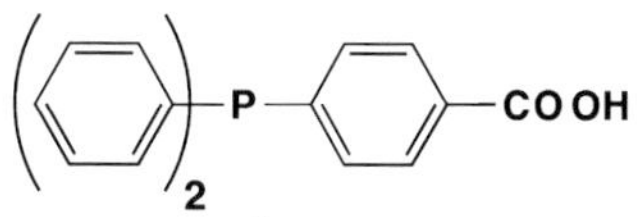 2

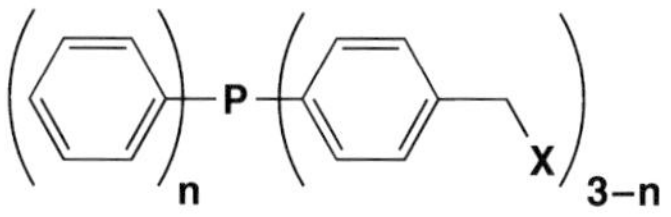

3 n = 2, X = NEt_2
4 n = 1, X = NEt_2
5 n = 2, X = NPh_2
6 n = 1, X = NMePh

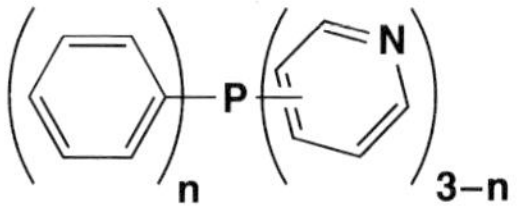

7 n = 2, *4*-pyridyl
8 n = 2, *3*-pyridyl
9 n = 1, *3*-pyridyl

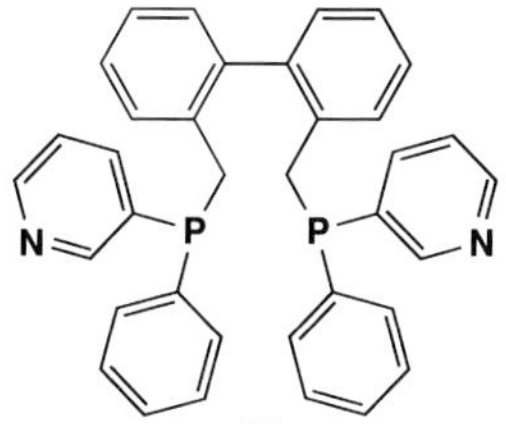

10 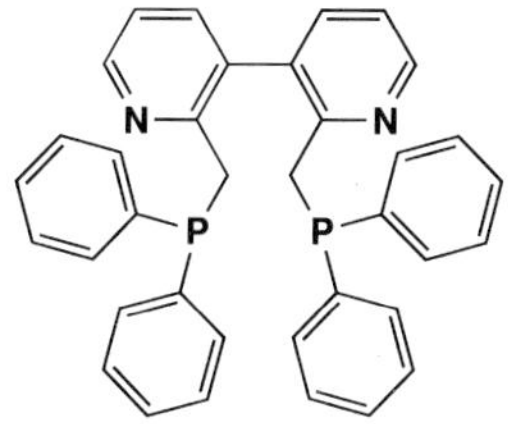 11

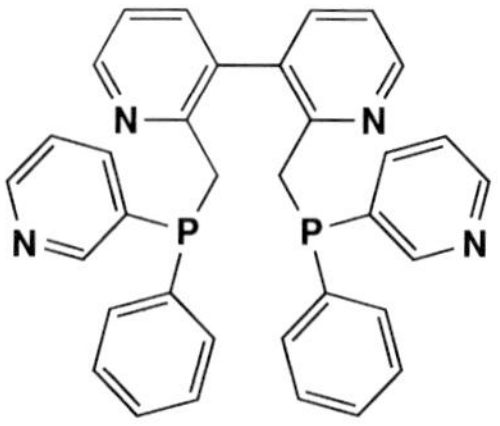

12 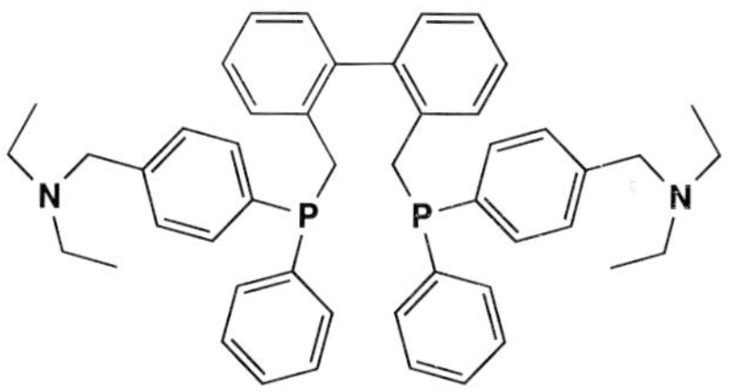 13

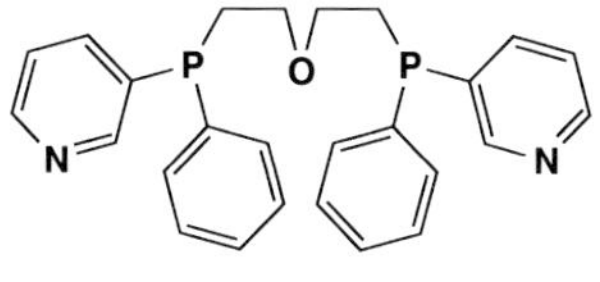

14 POPpy 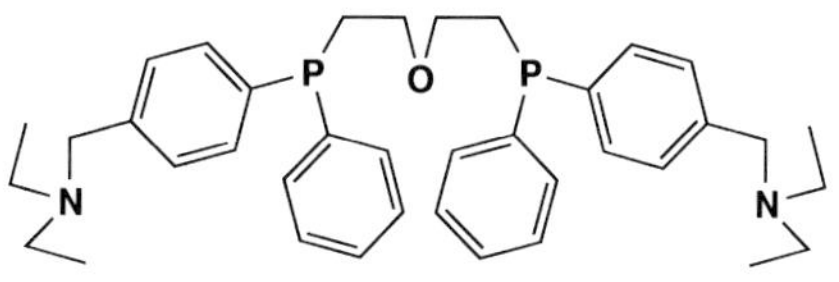 **15** POPam

All BISBI-type diphosphine-modified catalysts give selectivities around 90% to linear aldehydes. As can be seen in Table 1, POP (2,2′-bis(diphenylphosphino)diethyl ether) and its amphiphilic derivatives give rise to mainly linear aldehyde (88–89%). Although the selectivity for linear aldehydes is moderate compared with XANTHOS (9,9-dimethyl-4,5-bis(diphenylphosphino)-xanthene), virtually no isomerization is observed.

It can be seen that the rhodium catalyst derived from XANTPHOS and XANTHAM gives rise to an even higher yield of linear aldehyde owing to both

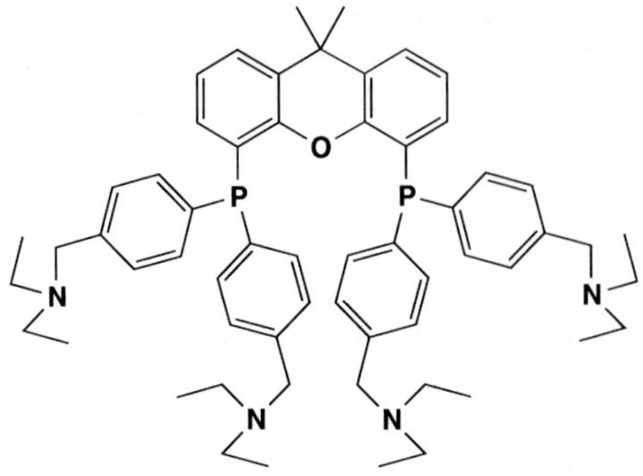

16 XANTHAM

Table 1. Hydroformylation of 1-octene under standard conditions.[a)]

Ligand	Time [h]	Conversion [%][b)]	Selectivity [%]			n/i	TOF[d)]
			Isomers[c)]	n-Ald.	i-Ald.		
PPh_3[e)]	2	81.2	1.5	72.6	25.6	2.8	2000
1[e)]	2	76.6	0.8	73.1	26.1	2.8	1900
3[e)]	2	76.2	0.3	73.3	26.0	2.8	1900
4[e)]	2	72.6	0.8	73.0	26.1	2.8	1800
8[e)]	2	86.3	2.7	71.7	25.6	2.8	2100
9[e)]	1	80.2	2.7	71.7	25.6	2.8	3900
BISBI	21	82.8	7.6	90.2	2.2	41	182
10	20	91.4	7.9	90.3	1.8	51	210
11	21	91.5	7.2	89.0	3.8	24	204
12	21	93.6	7.8	89.2	3.0	30	206
13	21	76.9	6.5	90.6	2.9	32	172
POP	20	67.0	0.0	88.2	11.8	7.5	168
POPpy	20	88.0	0.7	89.3	10.0	8.9	207
POPam	21	71.5	0.0	88.0	12.0	7.3	178
XANTPHOS	24	61.6	3.9	94.1	2.0	46	123
XANTHAM	24	67.9	4.0	94.1	1.9	49	137

a) Conditions: 20 bar H_2/CO (1 : 1), 80 °C, toluene (20 mL), $[L] = 17 \times 10^{-4}$ M, $[Rh] = 1.7 \times 10^{-4}$ M, [1-octene] = 0.84 M.
b) Percentage of 1-octene converted.
c) Percentage of 2-, 3- and 4-octenes formed.
d) Turnover frequency in mol aldehydes/mol Rh per hour, averaged over the time given.
e) $[L] = 34 \times 10^{-4}$ M.

a high linear/branched (n/i) ratio and a relatively low activity for isomerization. This is ascribed to the well-defined template structure of both rigid ligands, which preferentially occupy two equatorial sites in the catalytically active rhodium hydride owing to their relatively large natural bite angle. This geometry leads to a higher proportion of linear (n-)aldehyde formation as compared with geometries with axial–equatorial chelates [25, 26]. XANTPHOS and XAN-

THAM give similar results in the hydroformylation of 1-octene. The reaction rate of the XANTHAM-derived catalyst, however, is somewhat higher despite the electron-donating effect of the four aminomethyl groups, which has a negative effect on the reaction rate [27, 28].

7.4.2.2 Distribution Characteristics of the Free Ligands

The distribution characteristics of the free ligands have been reported as a function of the pH. These data are of interest for the following reasons. First, they can serve to decide which ligands are most suitable for the present goal. Secondly, the distribution characteristics of the free ligand can give an indication of the behavior of the corresponding rhodium complex. The complex is expected to be extracted at a milder pH than the ligand since it contains several ligands with accordingly more functional groups.

The distribution characteristics of the free ligands were determined by measuring the UV-absorption spectra. In Figure 3 the D–pH plots are depicted for the phosphines functionalized with the benzylic amines **3–6**. A clear correlation between the ease of extraction and the basicity of the amino group was observed. At pH values higher than 5.5 the ligands **10** and **11** are mostly located in the Et_2O layer, and only at pH values below 1 are both ligands extracted into the aqueous phase. Ligand **10** was extracted at a slightly higher pH than **11**.

Ligands **12** and **13** were more readily extracted: they were extracted quantitatively at the relatively high pH of 2.

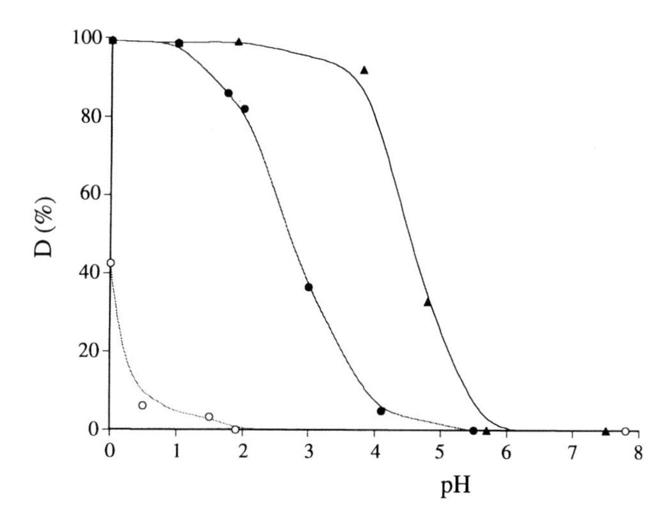

Figure 3. Extraction curves of ligands **3** (●), **4** (▲) and **6** (○). D = distribution coefficient = $C_{H_2O}/(C_{H_2O}+C_{org}) \cdot 100\%$.

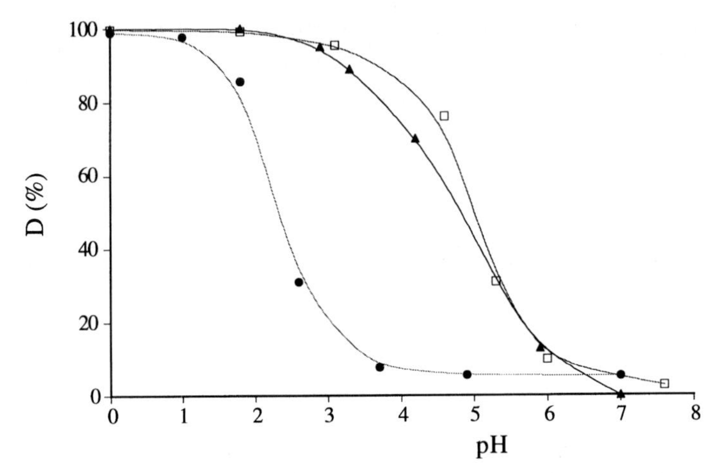

Figure 4. Extraction curves of POPpy (●, POPam (□) and XANTHAM (▲).

In Figure 4 the D–pH plots for POPpy, POPam am XANTHAM are depicted. POPpy is located mostly in the Et_2O layer at pH values of 4 to 7. Extraction of the ligand into the aqueous phase occurs at a pH of 3 and is complete at pH 1. The curve closely resembles that of $PhP(3\text{-pyridyl})_2$ [20]. The recorded extraction curves for POPam and XANTHAM are almost the same and closely resemble that of $PhP(C_6H_4CH_2NEt_2)_2$ [20]. Both ligands are completely located in the organic layer at pH 7, but are extracted more than 60% at pH 4.5. Extraction is complete at pH 2.5.

7.4.2.3 Rhodium Recycling

The efficiency of the rhodium recycling system using the new amphiphilic ligands was determined by re-use of the catalyst solution in a second hydroformylation run and by determination of the absolute amount of rhodium (metal) recovered using ICP–AES.

The results of the optimization of the recycling procedure are summarized in Table 2. The turnover frequency of the first run (TOF1) is taken as the number of moles of aldehyde formed per mole of rhodium averaged over the reaction time. Since the TOF of the second run (TOF2) was determined after a similar conversion of 1-octene, the quotient of both TOF values is a measure for the recovery of catalytically active rhodium and is referred to as the retention of activity (RA).

It can be seen that rhodium and excess ligand can be recycled and 1-octene can indeed be hydroformylated in a second run. The selectivity is generally

Table 2. Results of the recycling experiments: rhodium measurements by ICP–AES and retention of activity.[a]

Ligand	Acidic extraction	Rhodium content [μg]			Rh recovery[b]	Rh balance[c]	RA[d]
		First organic layer	Aqueous layer	New organic layer	[%]	[%]	[%]
4[e]	Titrations: 6 × extraction at pH 2.2	46	17	2258	97	96	87
9[e]	Titrations: 5 × extraction at pH 1.8	862	18	1356	57	98	60
10	Titrations: 5 × extraction at pH 1	252	719	219	18	96	14.1
11	Titrations: 5 × extraction at pH 1	51	1027	144	12	99	10.9
12	Titrations: 7 × extraction at pH 4–4.5	34	1150	47	41	100	2.5
13	Titrations: 7 × extraction at pH 5	11	8[f]	1108	91	99	72.1
POPpy	8 acidic extractions at pH 3	30	996	216	17	100	16
POPam	8 acidic extractions at pH 5–5.5	<0.7	432	791	65	99	58
XANTHAM	9 acidic extractions at pH 5–5.5	<0.7	26	1193	98	99	86

[a] The 95% confidence interval of the mean measured values is ±4.5% for contents >10 μg, otherwise ±10%.
[b] Rhodium recovered in the new organic layer as percentage of the total amount measured.
[c] Rhodium mass balance, defined as the total amount of rhodium measured as a percentage of the starting amount (1235 μg).
[d] Retention of activity, defined as the turnover frequency of the recovered rhodium as a percentage of the original turnover frequency as measured in the first run (see text).
[e] Total amount of rhodium 2414 μg.
[f] Persistent emulsion contained 97 μg.

exactly the same as in the first run. The recovery of rhodium as measured by rhodium analyses by ICP–AES can be nearly complete. Apparently, the rhodium and excess ligand are mostly recycled, but the rhodium in the eventual toluene phase is only partly active due to a certain amount of irreversible catalyst decomposition. Using amphiphilic bidentate ligands the originally highly selective and active rhodium hydrides can be regenerated up to 72% (ligand **13**). In contrast with ligand **13**, ligands **10–12** are unsuitable for application in rhodium recovery. Treatment with basic solutions leads to extensive, irreversible decomposition. The decomposition reactions of the rhodium complexes have not been elucidated but NMR analysis established that the excess ligand did not decompose and was recycled. Presumably, the presence of a pyridyl nitrogen atom, whether positioned in a bipyridine backbone or a pyridyl ring, facilitates irreversible decomposition of the rhodium hydride species to trivalent rhodium species which cannot be extracted from the aqueous layer. The small amount of rhodium which was re-extracted in toluence could be largely regenerated to the active hydride, since the same high selectivity was observed as in the first hydroformylation.

The catalytic mixture of POPpy was not recovered from the aqueous phase. The recovered toluene phase accordingly showed a low retention of activity (13%). Evidently, POPpy is also unsuitable for rhodium re-use, as observed earlier for pyridyl-modified PPh_3 and BISBI ligands [20, 21]. Also the recovery of POPam proved to be troublesome; only 59% of the total amount of rhodium charged was found in the new organic layer.

Successive extraction of the catalytic mixture of XANTHAM affected the recovery of rhodium. Although the overall rhodium recovery was almost quantitative (97%), a retention of activity of 86% was observed when the acidic titration procedure was performed at pH 5–5.5. This result equals the success of $PhP(C_6H_4CH_2NEt_2)_2$ [20].

7.4.3 Conclusions

Efficient separation of catalyst from the reaction medium can be achieved by extraction of polar rhodium complexes of TPPMS with water [8]. However, many catalytic processes for the production of fine chemicals require the use of modified tailor-made catalysts. The introduction of amphiphilic substituents on phosphines opens the way to easy separation of ligands that induce high selectivity and/or activity in catalytic reactions simply by extraction with acidic or basic water.

The novel amphiphilic diphosphines give rise to hydroformylation catalysts which are highly active and selective. Functionalization of the ligands with the diethylamino groups allows the separation of the product aldehydes from the rhodium catalyst, and the recovery of rhodium to the extent of 99.95% from the organic reagents by simple acidic extraction. XANTHAM is very promising for use as a ligand in the rhodium-catalyzed hydroformylation of higher alkenes as it is even somewhat more active than the catalyst derived from the hydrophobic XANTPHOS ligand. For commercial application in a process producing commodity aldehyde, a rhodium recovery of 99.98% [29] or even more [30] is mentioned in the literature. The 86% retention of activity is high, but far form satisfactory from an industrial point of view. Avoidance of irreversible decomposition could be envisaged by performing all recycling steps under syngas pressure. Finally, the present recovery procedure in combination with the XANTHAM ligand may already be applied to the batchwise hydroformylation of fine chemicals.

References

[1] (a) G. W. Parshall, W. A. Nugent *CHEMTECH* **1988**, *18*, 184, 314, 376; (b) W. A. Herrmann, C. W. Kohlpaintner, *Angew. Chem.* **1993**, *105*, 1588; Angew. Chem., Int. Ed. Engl. **1993**, *32*, 1524; (c) P. Kalck, F. Monteil, *Adv. Organomet. Chem.* **1992**, *34*, 219.

[2] Shell (W. Keim, T. M. Shryne, R. S. Bauer, H. Chung, P. W. Glockner, H. van Zwet), DE 2.054.009 (1969).

[3] W. Keim, *Chem. Ing. Tech.* **1984**, *56*, 850.

[4] I. T. Horváth, J. Rábai, *Science* **1994**, *266*, 72.

[5] (a) V. D. Luedeke, *in Encyclopedia of Chemical Processing and Design*, M. Dekker, New York, **1977**, Vol. 2; (b) B. Cornils in *New Syntheses with Carbon Monoxide* (Ed.: J. Falbe), Springer, Berlin, **1980**, p. 165. (c) Cf. [5b], p. 80.

[6] BASF (H. J. Kneuper, M. Roeper, R. Paciello) EP Appl. 588.225 (1994) to BASF.

[7] DSM N.V. (O. E. Sielcken, N. F. Haasen), EP WO 94 14.747 (1992).

[8] Union Carbide (A. G. Abatjoglou, D. R. Bryant, R. R. Peterson), EP 350.922 (1990).

[9] Union Carbide (A. G. Abatjoglou, D. R. Bryant), EP 350.921 (1990).

[10] R. M. Deshpande, S. S. Divekar, B. M. Bhanage, R. V. Chaudhari, *J. Mol. Catal.* **1992**, *75*, L19.

[11] M. Tamura, *Prepr. Meeting of the Chemical Society of Japan*, Okayama, **1989**, p. 11.

[12] A. Andreetta, G. Barberis, G. Gregorio, *Chim. Ind. (Milan)* **1978**, *60*, 887.

[13] K. Kurtev, D. Ribola, R. A. Jones, D. J. Cole-Hamilton, G. J. Wilkinson, *J. Chem. Soc. Dalton Trans.* **1980**, 55.

[14] J. C. Bayón, J. Real, C. Claver, A. Polo, A. Ruiz, *J. Chem. Soc., Chem. Commun.* **1989**, 1056.

[15] I. Tóth, B. E. Hanson, M. E. Davis, *J. Organomet. Chem.* **1990**, *396*, 363.

[16] I. Tóth, B. E. Hanson, M. E. Davis, *Catal. Lett.* **1990**, *5*, 183.

[17] I. Tóth, B. E. Hanson, M. E. Davis, *Tetrahedron: Asymm.* **1990**, *1*, 913.

[18] I. Tóth, B. E. Hanson, M. E. Davis, *J. Organomet. Chem.* **1990**, *397*, 109.

[19] A. Buhling, P. C. J. Kamer, P. W. N. M. van Leeuwen, *J. Mol. Catal.* **1995**, *98*, 69.

[20] A. Buhling, J. W. Elgersma, P. C. J. Kamer, P. W. N. M. van Leeuwen, *J. Mol. Catal. A: Chem.* **1997**, *116*, 297.

[21] A. Buhling, S. Nkrumah, J. W. Elgersma, P. C. J. Kamer, P. W. N. M. van Leeuwen, *J. Chem. Soc., Dalton Trans.* **1996**, 2143.

[22] A. Buhling, J. W. Elgersma, P. C. J. Kamer, P. W. N. M. van Leeuwen, K. Goubitz, J. Fraanje, *Organometallics* **1997**, *16*, 3027.

[23] C. A. Tolman, *Chem. Rev.* **1977**, *77*, 313.
[24] P. W. N. M. van Leeuwen, *Catalysis. An Integrated Approach to Homogeneous, Heterogeneous and Industrial Catalysis* (Eds.: J. A. Moulijn, P. W. N. M. van Leeuwen and R. A. van Santen), 2nd ed, Elsevier, Amsterdam, **1995**, Vol. 79, p 201.
[25] M. Kranenburg, Y. E. M. van der Burgt, P. C. J. Kamer, P. W. N. M. van Leeuwen, K. Goubitz, J. Fraanje, *Organometallics* **1995**, *14*, 3081.
[26] J. M. Brown, A. G. Kent, *J. Chem. Soc., Perkin Trans. 2* **1987**, 1597.
[27] W. R. Moser, C. J. Papite, D. A. Brannon, R. A. Duwell, *J. Mol. Catal.* **1987**, *41*, 271.
[28] J. D. Unruh, J. R. Christenson, *J. Mol. Catal.* **1982**, *14*, 19.
[29] E. G. Kuntz, *CHEMTECH* **1987**, *17*, 570.
[30] B. Cornils, E. Wilbus, Chem.–Ing.–Techn. **1994**, *66*, 916.

7.5 Water-soluble Polymer-bound Catalysts

Matthias Beller, Jürgen G. E. Krauter

7.5.1 Introduction

To overcome the difficulties of removing the reaction product and starting materials from a soluble homogeneous catayst, intensive work has focused on the attachment of ligands to essentially insoluble inorganic carriers or inert organic resins [1]. Despite considerable efforts so far, most of the resulting heterogenized or polymer-supported catalysts are significantly less active and show leaching of the precious metal. A different concept for separating a polymer-bound catalyst from the reaction mixture makes use of soluble polymers [2, 3]. Such polymeric catalysts mimic to some extent the reactivity of the conventional homogeneous catalyst because they are in solution during reaction. However, the catalyst can be recycled after the reaction by membrane filtration, because of large molecular weight differences (cf. Section 4.4). This approach was first introduced by Bayer and Schurig in 1976 [2]. Interestingly, it took some time until other groups adopted this idea. Later it was mainly Bergbreiter and co-workers who extended this concept elegantly [4]. Of particular interest is the development of so-called "smart"-polymeric catalysts, which undergo a significant change of solubility in a narrow temperature range [4b]. Thus, the polymeric catalyst can simply be separated by either cooling or heating of a biphasic mixture.

Until now only selected examples of the use of soluble polymeric catalysts in laboratory processes have been described.

7.5.2 Synthesis and Catalytic Applications of Soluble Polymer-bound Ligands

The first soluble polymer-bound catalysts were described by Bayer and Schurig [2], who demonstrated that macromolecular phosphine ligands from non-crosslinked polystyrene are active in the rhodium-catalyzed hydrogenation of 1-pentene as well as in the hydroformylation of styrene. Since then, polymer-bound catalysts have been used in cyclooligomerization [3], hydrogenation [5], reductions of alkyl halides [6], cyclopropanation [7], and inter- and intramolecular Kharasch reactions [8]. Regarding water-soluble polymers, various kinds of polymers have been introduced as supports for ligands (**1**–**5**).

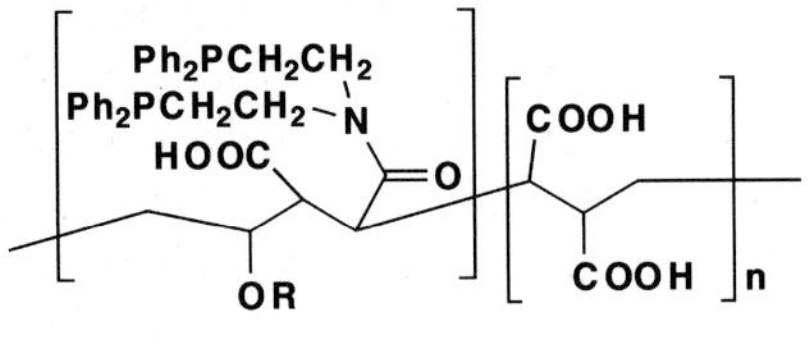

1 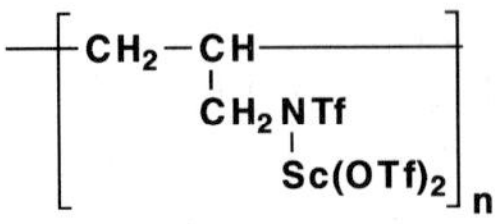 **2**

$Ph_2P{-}(CH_2CH_2O)_n{-}PPh_2$

3

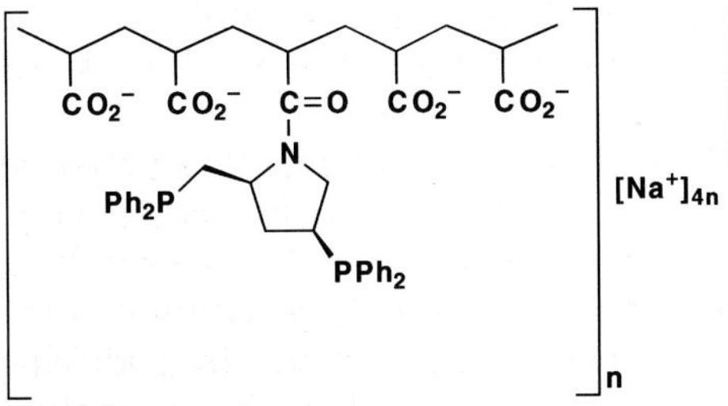

4 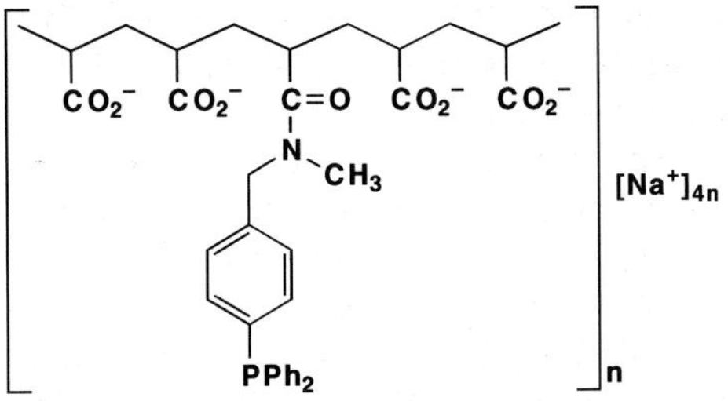 **5**

Poly(ethylene oxide) has been applied most often [4, 9]. Other water-soluble polymers used include poly(*N*-alkylacrylamide)s [10], poly(acrylic acid)s [11], derivatives of polyamines [12], and most recently a copolymer of maleic anhydride and methyl vinyl ether [13]. Apart from membrane filtration, recyling of such polymers can be effected by temperature-dependent phase separation (cf. Section 4.6.3), solvent precipitation, or a change in the pH. For example, poly(ethylene oxide) and poly(*N*-alkylacrylamide)s are known to undergo a temperature-dependent phase change wherein they separate from a water phase by heating [14]. A simple explanation for this inverse-temperature dependence of the solubility is that positive entropy effects arise from the loss of water molecules by the polymer. Thus, the polymer becomes more hydrophobic as the temperature increases. In order to use this physical property for catalysis, Bergbreiter et al. described the modification of commercially available block copolymers of ethylene oxide and propene oxide (Eqs. 1 and 2) [4].

$$HO{-}(CH_2CH_2O)_n{-}H \xrightarrow[NEt_3,\ CH_2Cl_2]{CH_3SO_2Cl} CH_3SO_2O{-}(CH_2CH_2O)_n{-}SO_2CH_3 \xrightarrow{LiPPh_2} \underset{\mathbf{3}}{Ph_2P{-}(CH_2CH_2O)_n{-}PPh_2} \quad (1)$$

$$\underset{HOCH_2{-}7{-}CH_2OH}{HO{-}(CH_2CH_2O)_n((CH_3)CHCH_2O)_m{-}(CH_2CH_2O){-}H} \xrightarrow[H_2SO_4]{CrO_3} HOOC{-}7{-}COOH \xrightarrow{SOCl_2} ClCO{-}7{-}COCl$$

$$ClCO{-}7{-}COCl \xrightarrow[NEt_3,\ CH_2Cl_2]{HN[CH_2CH_2P(C_6H_5)_2]_2} \underset{\mathbf{6}}{[(C_6H_5)_2PCH_2CH_2]_2NCO{-}7{-}CON[CH_2CH_2P(C_6H_5)_2]_2} \quad (2)$$

Poly(alkylene oxide)phosphine ligands were synthesized either by end-group oxidation using chromium trioxide in sulfuric acid, subsequent acid chloride formation, and amidation with 2-diphenylphosphinoethylamine or by mesylation and nucleophilic substitution with $LiPPh_2$. Indeed, the resulting polymeric ligands **3** and **6** have inverse-temperature-dependent solubility. A cationic rhodium complex of ligand **3** also exhibited this inverse-temperature-dependent solubility. Thus, the catalyst was soluble at 0 °C in a water–ethanol mixture (80:35, v/v) but insoluble at 50 °C leading to an anti-Arrhenius reactivity which was described as "smart" behavior. The implications of these solubility properties on catalysis have been demonstrated in the hydrogenation reaction of allyl alcohol in water. An approximately 400-fold decrease in rate is observed when the temperature is raised from 0 °C to 40–50 °C [4]. The same effect was observed in hydrogenations of α-acetamidoacrylic acid, albeit in different solvent systems.

Very recently, Janda and co-workers [9b] and Bolm and co-workers [9c] also applied poly(ethylene oxide)-modified dihydrophthalazines as chiral polymeric ligands for asymmetric Sharpless dihydroxylations. Re-use up to six times of the ligand system has been described without loss of selectivity.

Another type of water-soluble phosphine-containing polymer was prepared from commercially available poly(maleic anhydride)-*co*-poly(methyl vinyl ether) by reaction with bis(2-diphenylphosphinoethyl)amine [4]. The resulting amphoteric ligand was used for the rhodium-catalyzed hydrogenation of various olefins (Eq. 3). In water, the rhodium catalyst can be recovered by acidifying the solution to pH < 5. After filtration the recycled catalyst can be redissolved in water at pH > 7.5. As shown in Table 1, the catalyst has been reused for up to three cycles according to this principle. The observed loss of activity appears to be a result of phosphine oxidation.

$$\text{R-CH=CH}_2 \xrightarrow[\text{1/[Rh(COD)}_2]^+\text{OTf}^-]{\text{H}_2\text{ / H}_2\text{O}} \text{R-CH}_2\text{CH}_3 \qquad (3)$$

Table 1. Activity of base-soluble, acid-insoluble Rh(I) hydrogenation catalyst **1**/$[Rh(COD)]^+$-OTf^- in water.[a)]

Substrate	TOF [mol H_2/mol Rh h]	Yield [%]
Allyl alcohol		
1st cycle	8	> 95
2nd cycle	7.4	> 95[b)]
3rd cycle	6.9	> 95[b)]
Acrylic acid	19	> 95[b)]
N-Isopropyl acrylamide	84	94[c)]
α-Acetamidoacrylic acid	42	90[c)]
Sodium *p*-styrenesulfonate	8	> 90[b)]

a) Hydrogenation reactions were run at 25 °C using **1**/$[Rh(COD)]^+OTf^-$ as catalyst ($[RH]_{aqueous} = 8 \times 10^{-4}$ *M*).
b) Estimated by H_2 uptake. c) Isolated yield.

7.5.3 Conclusions

The application of ligands to control the steric and electronic environment of metal complexes is a fundamental principle in homogeneous catalysis. Also established is the utilization of low-molecular-weight ligands to control the solubility behavior of catalysts. The use of soluble polymeric ligands for this purpose is a relatively new and unexplored research area. It seems clear that this interdisciplinary area will attract increasing attention because the concept of polymeric catalyst recovery is very appealing. Of special interest are the so-called "smart" ligands which show an inverse-temperature-dependent solubility in water resulting in temperature-induced phase separation. However, one has to consider also the disadvantages of this approach, such as more complicated ligand design and thus more cost-intensive ligand synthesis, and lower activity compared with "molecular" catalysts. Moreover, it has not been demonstrated unequivocally that the catalyst systems are stable in terms of an industrial time scale.

References

[1] (a) P. Panster, S. Wieland in *Applied Homogeneous Catalysis with Organometallic Compounds*, Vol. 2 (Eds.: B. Cornils, W. A. Herrmann), VCH, Weinheim, **1996**, p. 605; (b) R. A. Sheldon, *J. Mol. Catal.* **1996**, *107*, 75; (c) C. U. Pittman Jr., Polymer-supported catalysts, in *Comprehensive Organometallic Chemistry*, Vol. 8 (Ed.: G. Wilkinson), Pergamon, Oxford **1982**, p. 553; (d) F. R. Hartley, *Supported Metal Complexes. A New Generation of Catalysts,* D. Reidel, Dordrecht, **1985**; (e) J. Manassen, *J. Plat. Met. Rev.* **1971**, *15*, 142.

[2] E. Bayer, V. Schurig, *CHEMTECH* **1976**, 212.

[3] D. E. Bergbreiter, R. Chandran, *J. Org. Chem.* **1986**, *51*, 4754.

[4] (a) D. E. Bergbreiter, V. M. Mariagnanam, L. Zhang, *Adv. Mater.* **1995**, *7*, 69; (b) D. E. Bergbreiter, L. Zhang, V. M. Mariagnanam, *J. Am. Chem. Soc.* **1993**, *115*, 9295; (c) D. E. Bergbreiter, B. Chen, D. Weatherford, *J. Mol. Catal.* **1992**, *74*, 409; (d) D. E. Bergbreiter, R. Chandran, *J. Am. Chem. Soc.* **1987**, *109*, 174.

[5] D. E. Bergbreiter, R. Chandran, *J. Am. Chem. Soc.* **1987**, *109*, 174.

[6] (a) D. E. Bergbreiter, S. A. Walker, *J. Org. Chem.* **1989**, *54*, 5138; (b) D. E. Bergbreiter, J. R. Blanton, *J. Org. Chem.* **1987**, *52*, 472.

[7] D. E. Bergbreiter, M. Morvant, B. Chen, *Tetrahedron Lett.* **1991**, 2731.

[8] J. C. Phelps, D. E. Bergbreiter, G. M. Lee, R. Villani, S. M. Weinreb, *Tetrahedron Lett.* **1989**, 3915.

[9] (a) E. A. Karakhanov, Y. S. Kardasheva, A. L. Maksimov, V. V. Predeina, E. A. Runova, A. M. Utukin, *J. Mol. Catal.* **1996**, *107*, 235; (b) H. S. Han, K. D. Janda, *J. Am. Chem. Soc.* **1996**, *118*, 7632; (c) C. Bolm, A. Gerlach, *Angew. Chem., Int. Ed. Engl.* **1997**, *36*, 741; (d) D. E. Bergbreiter, J. W. Caraway, *J. Am. Chem. Soc.* **1996**, *118*, 6092.

[10] T. Malmström, C. Andersson, *J. Chem. Soc., Chem. Commun.* **1996**, 1135.

[11] T. Malmaström, H. Weigl, C. Andersson, *Organometallics* **1995**, *14*, 2593.
[12] (a) S. Kobayashim, S. Nagayama, *J. Am. Chem. Soc.* **1996**, *118*, 8977; (b) F. Joo, L. Somasak, M. T. Beck, *J. Mol. Catal.* **1984**, *24*, 71.
[13] D. E. Bergbreiter, Y.-S. Liu, *Tetrahedron Lett.* **1997**, *38*, 3703.
[14] (a) F. E. Bailey, J. V. Koleske, *Poly(ethylene oxide)*, Academic Press, New York, **1976**; (b) F. M. Winnik, F. M. Ottaviani, S. H. Bossmann, M. Garcia-Garibay, N. J. Turro, *Macromolecules* **1992**, *25*, 6007.

8

Aqueous-phase Catalysis: What Should be Done in the Future

8.1 State of the Art

Boy Cornils, Wolfgang A. Herrmann

This book is focused on the technique of aqueous, homogeneous two-phase catalysis, with the active catalyst for the reaction being (and – if not otherwise stated – remaining) dissolved in water. Thus, the reactants can be separated from the reaction products, which are typically organic in nature and relatively nonpolar, after the reaction is completed by simply separating the second phase from the catalyst solution. The catalyst can be recirculated without any problems (cf. Chapter 1, Sections 4.1 and 4.2).

As described, aqueous biphase catalysis makes it possible to utilize fully the inherent advantages of homogenous catalysis. A major aspect of this new technology is the tailoring of organometallic complexes as catalysts, as is nowadays becoming increasingly important both for industrial applications and for new reactions, and for new products [1] (cf. Chapters 2, 3 and 5). The fact that selectivity- and yield-reducing operations (such as thermal stress caused by chemical catalyst removal or distillations) for separating product and catalyst are avoided makes it possible to use sensitive reactants and/or to obtain sensitive reaction products from homogeneous catalysis. Figure 1 of Section 5.2 demonstrates the advantages of the aqueous technique over the conventional one, taking the hydroformylation reaction as an example (cf. also Section 6.1): The aqueous variant gets rid of all technical equipment except for the column, 9 – a tremendous cost-saving [2].

Furthermore, this spontaneous separation of catalyst and product is the most effective and the only successful method of immobilizing a homogeneous catalyst and thus making it "heterogeneous" – just by anchoring it on the "liquid support," water [3].

The concept and development of aqueous, homogeneous two-phase catalysis followed unconventional routes for chemical processes. After the idea was first expressed by Manassen [4] (not Bailar [5], as erroneously stated by Papadogianakis and Sheldon [6]), it was very quickly taken over by the university-based researcher Joó [7] (Debrecen, Hungary) and in the industrial field by Kuntz [8]. However, these studies remained curiosities and the far-sighted visions of Kuntz, particularly, found no echo in the academic community. The reason may have been that the idea of organometallic complex catalysts in the presence of water seemed unnatural or even perverse, although even air-stable aquocarbonyl complexes of transition metals such as rhenium [9] were known in the meantime. In addition, metal–carbon bonds are thermodynamically unstable relative to their hydrolysis products (cf. Section 2.2), although it was well known that reaction rates could be increased by a factor of up to 10^{11} by aqueous media [10]. In simple terms, organometallic chemistry was not ready for aqueous operation before the early 1980s.

It was not until the work at Ruhrchemie AG (and thus the occupation of a skilled and experienced team in industry) that development led to the first large-scale utilization of the aqueous, homogeneous catalysis technique at the beginning of the 1980s, viz. in hydroformylation (the oxo process) [11]. The generally used embodiment of two-phase catalysis, for example as practised in Shell's SHOP method [12], was thus extended to *aqueous* two-phase catalysis. These (and the other industrial applications; see Chapter 6) have led to the literature concerning these particularly attractive aqueous variants being dominated by publications from industry, particularly patent literature, rather than from academia, for virtually a decade.

It was only in 1996 that the sleepy-headed academic literature provocatively asked "Why water?" [13]. This development sequence was quite unusual: the idea from academia, initial experimental work in the 1970s by an industrial chemist (Kuntz, then at Rhône-Poulenc), first industrial use in the 1980s (i.e. 14 years ago) by another industrial company – Ruhrchemie – and only then (for about the last eight years) more detailed scientific study. The fact that some "re-invention of the wheel" is occurring in these intensified academic studies is due not least to the widespread unwillingness of the academic community to read and take note of patents, especially non-English ones. Joó and Kathó [14] politely referred to this time delay between academic and industrial research as an "induction period", although it was more like a serious case of mass transfer inhibition.

The difference between biphasic catalysis in general (cf. processes mentioned in Chapter 7) and the aquous biphasic technique in particular is enormous. The aqueous variant is not just a special version, because the utilization of water as a solvent and a liquid support offers different advantages quite apart from the fact of simply going "biphasic" (cf. Table 1).

The various properties of water in different aspects (being important for the reactivity, reaction kinetics or mechanisms, reaction engineering, or other concerns) are discussed elsewhere. The procedures for tailoring the water-solubility of the catalysts are many-sided and may be generalized much more easily than the corresponding methods for SHOP (cf. Section 7.1), fluorous phase (Section 7.2), or NAIL utilization (Section 7.3): no wonder that all other biphasic applications remain singular or are still just proposals. Both the scientific and industrial communities expect realistic applications (other than the ones effected so far) due to the versatility of possible catalyst tailoring.

Many of the different possibilities of adjusting to the proper degree of water-solubility by balancing polar and apolar properties between the aqueous and the organic phase through suitable chemical modification of the ligand or the catalyst have been proven (cf. Chapters 2, 3, 5, and 6). The status of hydroformylation (oxo synthesis) as the first, most advanced and quantitatively important application (annual output some 600000 tons) has been described in some detail (Section 6.3.1 and [2, 14, 15]). The new process configuration, completely different from the status beforehand, confirmed the following focal points:

Table 1. Properties of water as a liquid support of aqueous two-phase catalysis.

Property	
1	Polar and easy to separate from apolar solvents or products; polarity may influence (improve) reactivity
2	Nonflammable, incombustible
3	Widely available in suitable quality
4	Odorless and colourless, making contamination easy to recognize
5	Formation of a hexagonal two-dimensional surface structure and a tetrahedral three-dimensional molecular network, which influence the mutual (in)solubility significantly, chaotropic compounds lower the order by H-bond breaking
6	High Hildebrand parameter as unit of solubility of nonelectrolytes in organic solvents
7	A density of 1 g/cm^3 provides a sufficient difference from most organic substances
8	Very high dielectric constant
9	High thermal conductivity, high specific heat capacity, and high evaporation enthalpy
10	Low refractive index
11	High solubility of many gases, especially CO_2
12	Formation of hydrates and solvates
13	Highly dispersible and high tendency toward micelle formation, stabilization by additives
14	Amphoteric behaviour in the Brønsted sense
15	Advantageously influencing chemical reactivity

- the catalyst – for the first time to be supplied in water-soluble form, in satisfactory quality, and inexpensively
- the process procedure – two-phase with phase separation during continuous operation
- the completely new, simple circulation of the aqueous catalyst solution
- the reactor with special mixing of two phases
- the possibility of a particular energy balance, as a result of which the process (in the case of the hydroformylation) becomes a net steam supplier by utilizing the exothermic heat of reaction

The reaction engineering is advanced insofar that many alternative process concepts [11e, 15d] were tested during the development phase; however, only a few were pursued further in the concept stage up to the pilot plant reactor. There remained the principle of the biphasic Ruhrchemie/Rhône-Poulenc oxo process which was developed in long-term tests and protected by a matched patent strategy, as described in Section 6.1.3.

Aqueous, two-phase catalysis is also utilized industrially in a number of other processes apart from hydroformylation. The hydrodimerization of butadiene and water, a telomerization variant yielding 1-octanol or 1,9-nonanediol (cf. Section 6.7), is carried out at a capacity of 5000 tonnes per annum by the Kuraray Corporation in Japan. Rhône-Poulenc is operating two-phase, aqueous, catalytic C–C coupling processes (using TPPTS obtained from Ruhrchemie) for small-scale production of various vitamin precursors such as geranylacetones (Section 6.10). Moreover, TPPTS-modified Ru catalysts have been proposed for the homogeneously catalyzed hydrogenation to convert unsaturated ketones into saturated ones.

Homogeneous, aqueous two-phase catalysis has also gained industrial significance in the production of the important intermediate, phenylacetic acid (PAA). The previous process (benzyl chloride to benzyl cyanide with hydrolysis of the latter) suffered from the formation of large amounts of salt (1400 kg/kg of PAA). The new carbonylation method reduces the amount of salt by 60% and makes use of the great cost difference between –CN (approx. US $ 1.4/kg) and –CO (less than US $ 0.2/kg) [16]. Finally, the Suzuki coupling of aryl halides and arylboronic acids substituting Pd/TPPMS by Pd/TPPTS catalysts deserves to be mentioned (Section 6.10).

The field is in flux. As a consequence of the increased scientific study of aqueous biphasic homogeneous catalysis an increasing number of commercial applications may be expected in the future [14, 17]. Several processes are the subject of detailed surveys and reviews [17].

8.2 Improvements to Come

Further methodological progress in aqueous-phase homogeneous catalysis will comprise both the improvement of the technology and extension of the respective reactions to other applications.

8.2.1 Other Technologies

Theoretically, aqueous, two-phase homogeneous catalyses with their underlying principle of phase separation during continuous operation can be achieved by combining various basic and auxiliary unit operations such as extraction, extraction using solvent mixtures, reactive extraction, distillation, osmosis, reverse

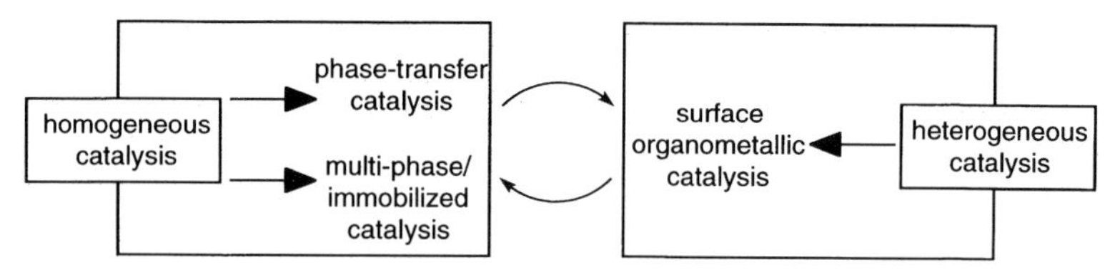

Figure 1. Technologies at the borderline between heterogeneous and homogeneous catalysis [19].

osmosis, phase transfer, absorption or adsorption, immobilization or partial immobilization, etc. (cf. Section 4.2). Depending on the properties – or more precisely: the property differences – of the reactants and the reaction products (e. g., solubility, polarity, boiling point), complex processes have been proposed [18]. However, every additional process step and every additional ingredient introduced into the process results in growing material losses at increased costs. For this reason, phase separation will remain the only viable low-cost option for industrial-scale applications.

Combination of aqueous homogeneous catalysis with some characteristics of heterogeneous catalysis ("supported aqueous-phase catalysis," SAPC; cf. Section 4.7) also lead to a fundamentally different technology. This variant, which is particularly seductive for scientists (because it seems logical), has indeed produced good initial results in hydrogenation and hydroformylation. However, in industrial use the problems inherent in the system need to be overcome, especially stability questions regarding the support similar to those, in its time, in the SLPC variant of the oxo process. The future has to show how far the gap between homogeneous and heterogeneous catalysis (Figure 1) may be bridged by aqueous variants of SOMC ("surface organometallic chemistry" [20]), combinations of catalytic active sites of zeolites and water-soluble precursors, colloids, clusters, or mesoporous organometals [19].

A serious drawback of this special variant originates from the notorious "leaching" of the catalyst, concerning the (usually expensive) transition metal, the ligand (which, as a tailor-made compound to meet special demands, may be even more expensive), or both. Leaching normally results from the dissociation of the metal from one of the "anchoring" ligands, thus liberating the (active) molecular catalyst. But leaching may also originate from structural changes with concomitant weakening of certain bonds during the catalytic cycle, during which the coordination sphere of a metal undergoes continuous changes in composition and structure (e.g., the change between tetragonal *tetra*carbonylhydridocobalt and trigonal-bipyramidal *tri*carbonylhydridocobalt during the catalytic cycle of the classical oxo process). This is immanent for hydroformylation and has so far prevented (despite some hundreds of different approaches) this immobilization by anchoring. It is – at the least – in doubt whether an aqueous technique can solve this problem.

8.2.2 Other Feedstocks and Reactions

In the case of hydroformylation, there is great interest in the use of olefins higher than hexene. The previous technology is restricted to the use of propene and butenes (cf. Section 6.1). A reason which has been postulated for this (and other conversions) is the decreasing solubility in water with increasing number of carbon atoms in the starting olefins and the products (Figure 2) and the associated mass-transfer problems in the two-phase reaction.

Quite obviously, the problem is solved if the reaction is carried out in a homogeneous phase using thermoreversible catalysts, and only the catalyst/product separation is carried out in the heterogeneous phase. This is the great advantage of the use of FBSs or ionic liquids and a tremendous stimulus for various *non*aqueous, "biphasic" alternatives (cf. Section 7). However, for various reasons, neither of the methods mentioned will be used for the hydroformylation of higher olefins (which is under great cost pressure) or for other large-tonnage chemicals.

On the other hand, the possibility of achieving thermoreversibility, i.e., reaction in a homogeneous phase at higher temperature and in two phases at lower temperature, by means of appropriately tailored catalysts (specifically by means of tailored ligands), appears to have better prospects. Corresponding developments by Jin and Fell (cf. Section 4.6.3) based on ethoxylated phosphines underline this elegant procedure. The advantage of this method is that the desired property, i.e., the thermoreversibility of the phase behavior, is introduced by the ligand itself and not by costly auxiliaries and additives. The ability to recycle the

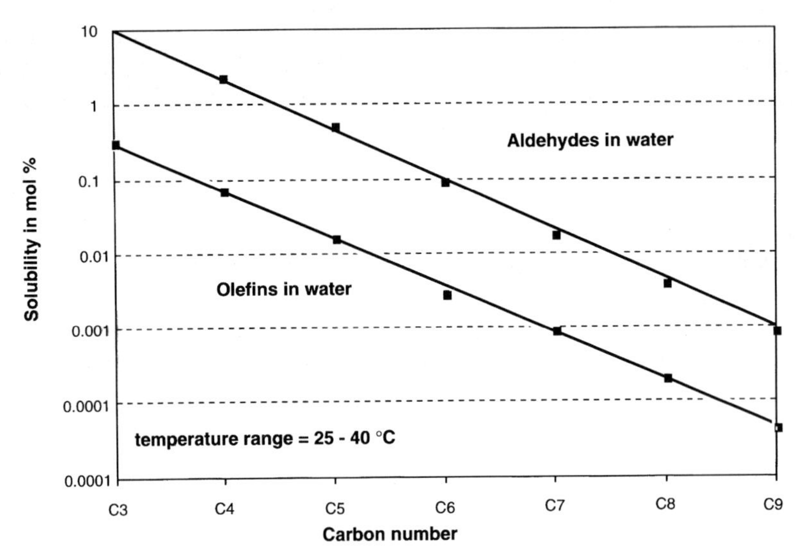

Figure 2. Solubility of olefins and aldehydes in water.

catalyst is therefore ensured in principle and the above warnings reagarding additives and additional process steps do not apply. The above-mentioned limitations are, however, valid for solubilizing/thermoregulating additives such as (poly)ethyleneglycol) [21].

Combinations of thermoregulating reaction mixtures with the use of membrane processes are discussed in Sections 4.4 and 6.1.5. and ref. [22]. Although the membrane process constitutes an additional process step, its use allows regeneration of the catalyst simultaneously and continuously and could thus be acceptable in terms of cost. The integration of membrane techniques in an aqueous-phase process is demonstrated in Figure 3 [22].

The academic world is working intensively on extending the aqueous, two-phase, homogeneously catalyzed method to reactions of other substrates. These include hydrogenations (these important reactions will have a hard time because of the dominance of *heterogeneous* hydrogenation processes; cf. Section 6.2), selective hydrogenations (in which tailor-made ligands are most likely to be able to show their full potential, which is also true for the interesting area of enantiomeric reactions; cf. Section 3.2.5 and 6.9), other carbonylations and C–C couplings (which can make direct use of the know-how developed for hydroformylation; cf. Sections 6.3, 6.4, and 6.10), hydrocyanations (Section 6.5), the production of polymers (especially by ROMP reactions, cf. Section 6.11 and 6.13), oxidations (cf. Sections 6.17, 6.18, and 6.20), hydrations and aminations (cf. Sections 6.17, 6.18, and 6.20), hydrations and aminations (preferably anti-Markovnikov, cf. Section 6.10), allylations (Section 6.6), etc. [17]. In the case of hydroformylation, work is being carried out on variants of the earlier Aldox process (simultaneous hydroformylation, aldolization, and hydrogenation, e. g. propene → 2-ethylhexanol). This work leads into the field of the less-studied multifunctional (also multicomponent) homogeneous catalysis for which there are no models, especially not in the aqueous variant.

Although of great scientific attractivity and of importance for fine chemicals, the industrial implementation of these developments will depend much on the additional expense of the tailored, water-soluble catalysts being compensated

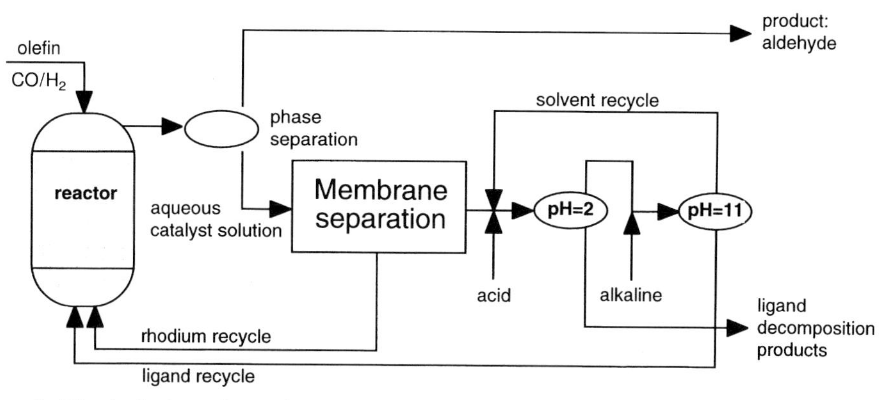

Figure 3. The inclusion of membrane techniques in aqueous-phase processes.

for by greater activity and selectivity (including enantiomeric selectivity) and by the simple process configuration of the aqueous variants which can be achieved by their use. Particularly in the case of catalyst and/or ligand mixtures, this is often questionable. New approaches to special modifications of water-soluble ligands may be helpful, as well as chemical engineering knowledge of the mechanism in the presence of water, as the key to catalyst design, activity, performance, recycle, and lifetime/reactivation.

8.3 Focal Future Developments

Within only 14 years of commercial application of aqueous, two-phase homogeneous catalysis, over four million tonnes of building blocks have been produced. Nevertheless, the state of development and the degree of maturity are such that the following further developments may be expected in due course.

The technique of aqueous catalytic reactions has had such an impact on the field of more general two-phase reactions that scientists have now also proposed and tested other solutions. "Fluorous" systems (FBS, perfluorinated solvents; cf. Section 7.2) and "nonaqueous ionic liquids" (NAILs, molten salts; cf. Section 7.3) meet the demand for rapid separation of catalyst and product phases and, owing to the thermoreversibility of their phase behavior, have advantages in the "homogeneous" reaction and the "heterogeneous" separation. However, it is safe to predict that the specially tailored ligands necessary for these technologies will be too expensive for normal applications. Compared to the cheap and ubiquitous solvent water, with its unique combination of properties (cf. Table 1), other solvents may well remain of little importance, at least for industrial applications. Other ideas are mentioned in Section 7.5.

Until now, TPPTS has proven itself the standard ligand for aqueous two-phase systems. High water-solubility (approx. 1.1 kg/L) and ready availability (sulfonation of TPP) are advantages which favor industrial use. The high level of scientific interest has led to the development of a series of other ligands which are able to increase the activity of the Rh catalysts that they are used to modify, while retaining sufficient solubility in water, However, the fact that they enable the amount used to be decreased and the *n*/*i* ratio of the aldehyde mixtures produced to be increased [23] has not been sufficient to compensate for their high costs in everyday operation. The situation may be different for enantiomeric reactions (cf. Section 3.2.5). A particularly promising candidate for aqueous Ph-catalyzed hydroformylation, BINAS-8 (**1**) [24], has four sulfo groups per phosphorus atom despite being a diphosphine (compared with three in the case of TPPTS).

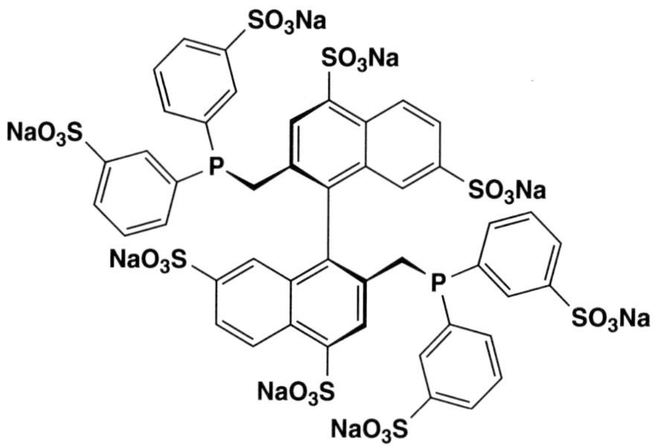

Exotic ligands, e.g., dextrins, supramolecular compounds, and templates are attractive for academic applications in aqueous homogeneous reactions and can be expected to gain much publicity. However, so far there is no experience of their practical use so that questions of long-term stability, recycling, and dependence on parameter changes and also of toxicity, environmental behavior, costs, etc., remain to be solved. The cost of such ligands need not be prohibitive (as when the "expensive" rhodium is used as the central atom in oxo catalysts) if virtually loss-free recycling of the ligands as well as the central atoms can be guaranteed. Interest is concentrated especially on the dextrins as potential inexpensive candidates.

Ligand *mixtures* with tailored activities (for example TPP as a "promoter ligand" and additive to TPPTS [25]) are attractive for scientific work and also enable particular effects to be achieved. Mixtures might be controlled in academic work. However, for industrial use it must be ascertained that *all* catalyst components can be recirculated simultaneously and in the same way with little expense, which can virtually never be ensured in the case of complicated mixtures. The same applies to the use of soluble polymers (both as phase separation agents and as ligands; cf. Section 7.5 and [26]).

Apart from rhodium, interest so far concentrates on palladium and ruthenium (cf. Section 3.1), and in special cases on rhenium (cf. Section 6.20) or the lanthanides (Section 6.19). For hydroformylations no other central atom (cobalt or central atom mixtures such as Pt/Sn) has so far been able to qualify for use in the industrial aqueous, biphasic oxo process. Should good results be obtained using central-atom mixtures in experiments on multifunctional, aqueous homogeneous catalysis, again the problem of simultaneous recycling of both components must be overcome. However, the prospects for industrial use of such mixtures are poor for the reasons stated.

The addition of additives has frequently been tested in scientific laboratories, sometimes not without success. The patent literature also lists a whole series of additives (cf. Sections 4.3 and 6.1.1). Although these additives are interesting in basic research, in industrial practice detailed engineering and cost studies have resulted in none of them being accepted for economic use. The same applies to special technical solutions, e. g., the use of ultrasonic reactors to solubilize the reactants [27].

The level of scientific work on supercritical CO_2 ($scCO_2$) has increased and overlaps with the properties of media for FBSs [28]. However, new and specific process variants with disadvantageous properties arising from perfluorinated compounds are no longer introduced by the solvents (e. g., $scCO_2$), but only by the ligands present in catalytic amounts. In these developments too, simultaneous catalyst regeneration with the aid of membrane separation is conceivable.

The environmental aspects of the aqueous biphasic processes have been intensively studied and discussed taking Ruhrchemie/Rhône-Poulenc's oxo process as an example of economic relevance. Homogeneously catalyzed aqueous processes are particularly environmentally friendly [15b]. According to Sheldon [29] the "*E*-factor" of the aqueous oxo process is between 0.04 and 0.1, depending on the way in which the by-products are considered. It thus falls into the product category of "oil chemicals", i.e., this is one of the production operations which, owing to their size and importance, are both economic and very environmentally sensitive. Other oxo processes with *E*-factors $\gg 0.1$ correspond to the category of "bulk chemicals." The environmental friendliness expressed by a favorable *E*-factor of the new biphasic process shows up in many places: the high selectivity of the chemical reaction (and thus low by-product formation), the low capital costs, the great reduction in amounts of waste gas and waste water (in the case of water, by a factor of 70) and the energy consumption (the oxo process is changed from a steam consumer to a net steam supplier, and the power consumption is more than halved), etc. [2a].

Further progress is expected from new developments and combinations of processes. Thus, it would be possible to make the disposal of the gaseous (and highly pure) waste gas streams (residual propane content of the proylene feed) cost-effective and a source of electric power by connection to novel, compact, membrane fuel cells. Potential synergisms would also occur in the operating temperature of the cells (medium-temperature cells at 120 °C using the residual exothermic heat of reaction from the oxo reaction), the membrane costs by means of combined developments (e. g., for membrane separations of the catalysts [22]), and also in the development of the "zero-emission" automobile by the automotive industry. The combination of hydroformylation with fuel cells would further reduce the *E*-factor – thus approaching a "zero-emission chemistry."

Moreover, examination of the Sheldon *E*-factors convincingly demonstrates the practicability of this concept as compared with Trost's [30] enigmatic "atom economy" which is concerned with esoteric examples of little relevance. Some people believe that the newer version of the "atom economy", justifiably emphasizing homogeneous catalysis with exotic examples of "the myriad of substances that are required to serve the needs of society" (!) [30b], are better founded.

Homogeneous catalysis has never been the prototype of a "heat-and-beat" technology in chemistry. Given the opportunity of quickly and easily separating the homogeneous catalyst from the reaction products in an aqueous process, it will be even more interesting (and more feasible) to make use of all the intrinsic features of homogeneous catalysis. This applies particularly to the possibility of tailoring the catalyst by varying the central atom, the ligand, or the phase in which it is used.

In this sense, homogeneously catalyzed aqueous syntheses will lead to an even more sophisticated field of chemistry under the most environmentally friendly conditions.

References

[1] B. Cornils, W. A. Herrmann (Eds.), *Applied Homogeneous Catalysis with Organometallic Compounds*, Vols. 1 and 2, VCH, Weinheim, **1996**.

[2] (a) E. Wiebus, B. Cornils, *Chem. Ing. Tech.* **1994**, *66*, 916; (b) *ibid.*, *Hydroc. Proc.* **1996**, *March*, 63.

[3] B. Cornils, *Modern Solvent Systems in Industrical Homogeneous Catalysis* (P. Knochel, Ed.), in *Topics in Current Chemistry*, Springer, Heidelberg, **1999**.

[4] J. Manassen, in *Catalysis Progress in Research* (F. Basolo, R. L. Burwell, Eds.), Plenum Press, London, **1973**, p. 177 ff.

[5] J. C. Bailar, *Catal. Rev.-Sci. Eng.* **1974**, *10* (1), 17.

[6] G. Papadogianakis, R. A. Sheldon, *New J. Chem.* **1996**, *20*, 175.

[7] F. Joó, T. Beck, *React. Kinet. Catal. Lett.* **1975**, *2*, 257.

[8] Rhône-Poulenc (E. G. Kuntz), FR 2.349.562, 2.366.237, 2.733.516 (1976).

[9] a) R. Alberto, A. Egli, V. Abram, K. Hegetsweiler, V. Gramlich, P. A. Schubiger, *J. Chem. Soc., Dalton Trans.* **1994**, 2815; (b) R. Alberto, R. Schibli, A. Egli, P. A. Schubiger, W. A. Herrmann, G. Artus, V. Abram, T. A. Kade, *J. Organomet. Chem.* **1995**, *492*, 217.

[10] For example, M. H. Abraham, P. L. Grellier, A. Nasehzadeh, R. A. C. Walker, *J. Chem. Soc. Perkin Trans. 2* **1988**, 1717.

[11] (a) B. Cornils, J. Falbe, *Proc. 4th Int. Symp. on Homogeneous Catalysis*, Leningrad, Sept. **1984**, p. 487; (b) H. W. Bach, W. Gick, E. Wiebus, B. Cornils, *Prepr. Int. Symp. High-Pressure Chem. Eng.*, Erlangen/Germany, Sept. **1984**, p. 129; (c) H. W. Bach, W. Gick, E. Wiebus, B. Cornils, *Prepr. 8th ICC, Berlin* **1984**, vol. V, p. 417; *Chem. Abstr.* **1987**, *106*, 198 051; cited also in A. Behr, M. Röper, *Erdöl-Kohle* **1984**, *37* (11), 485; (d) H. W. Bach, W. Gick, E. Wiebus, B. Cornils, *Absts. 1st IUPAC Symp. Org. Chemistry*, Jerusalem, **1986**, p. 295; (e) B. Cornils, E. G. Kuntz, *J. Organomet. Chem.* **1995**, *502*, 177.

[12] W. Keim F. H. Kowaldt, R. Goddard, C. Krüger, *Angew. Chem.* **1978**, *90*, 493; cf. D. Vogt, in [1], Vol. 1, p. 245.

[13] A. Lubineau, *Chem. & Ind.* **1996**, *Feb.* (4), 123 (not p. 125, as printed).

[14] F. Joó, A. Kathó, *J. Mol. Catal.* **1997**, *116*, 3.

[15] (a) B. Cornils, E. Wiebus, *CHEMTECH* **1995**, *25*, 33; (b) B. Cornils, E. Wiebus, *Recl. Trav. Chim. Pays-Bas* **1996**, *115*, 211; (c) O. Wachsen, K. Himmler, B. Cornils, paper presented at the 1997 ACS Fall Meeting (Ind. Engng. Div.), Las Vegas, cf. *Catal. Today*, in press; (d) B. Cornils, *Org. Prod. Res. Dev.*, in press.

[16] C. W. Kohlpaintner, M. Beller, *J. Mol. Catal.* **1997**, *116*, 259.

[17] (a) W. A. Herrmann, C. W. Kohlpaintner, *Angew. Chem.* **1993**, *105*, 1588; *Angew. Chem. Int. Ed. Engl.* **1993**, *32*, 1524; (b) P. Kalck, F. Monteil, *Adv. Organomet. Chem.* **1992**, *34*, 219; (c) cf. [1], Vol. 2, p. 575; (c) cf. special issue of *J. Mol. Catal.* **1997**, *116*, No. 1–2.

[18] (a) J. Haggin, *Chem. Eng. News* **1995**, 25; (b) A. G. Abajatoglou, 209 ACS National Meeting, Anaheim, USA, **1995**.

[19] W. A. Herrmann, B. Cornils, *Angew. Chem.* **1997**, *109*, 1075; *Angew. Chem. Int. Ed. Engl.* **1997**, *36*, 1049.
[20] J. M. Basset, G. P. Niccolai, in [1], Vol. 2, p. 624.
[21] Hoechst AG (A. Bahrmann, B. Cornils, W. Konkol, N. Lipps), EP 0.157.316 (1984); Johnson Matthey (M. J. H. Russel, B. A. Murrer), US 4.399.312 (1983); Hoechst AG S. Bogdanovic, C. D. Frohning, H. Bahrmann, E. Wiebus), DE-Appl. 19.700.805.4 and 19.700.804.6 (1997).
[22] (a) Th. Müller, H. Bahrmann, *J. Mol. Catal.* A **1997**, *116*, 39; (b) H. Bahrmann et al., *J. Organomet. Chem.* **1997**, *545/546*, 139.
[23] W. A. Herrmann, C. W. Kohlpaintner, R. B. Manetsberger, H. Bahrmann, *J. Mol. Catal.* **1995**, *97*, 65.
[24] W. A. Herrmann, R. W. Eckl, unpublished work.
[25] R. V. Chaudhari, B. M. Bhanage, R. M. Deshpande, H. Delmas, *Nature (London)* **1995**, *373*, 501.
[26] D. E. Bergbreiter, *CHEMTECH* **1987**, 686; and paper presented at the 1997 ACS Fall Meeting (Ind. Engng. Div.), Las Vegas, Sept. 11, 1997.
[27] Ruhrchemie AG (B. Cornils, H. Bahrmann, W. Lipps, W. Konkol), EP 0.173.219 (1984).
[28] S. Kainz, D. Koch, W. Baumann, W. Leitner, *Angew. Chem.* **1997**, *109*, 1699; *Angew. Chem. Int. Ed. Engl.* **1997**, *36*, 1628.
[29] R. A. Sheldon, *CHEMTECH* **1994**, *24*, 38.
[30] (a) B. M. Trost, *Science* **1991**, *254*, 1471; (b) *ibid., Angew. Chem.* **1995**, *107*, 285; *Angew. Chem., Int. Ed. Engl.* **1995**, *34*, 259.

Index

B

D

P

Q

R